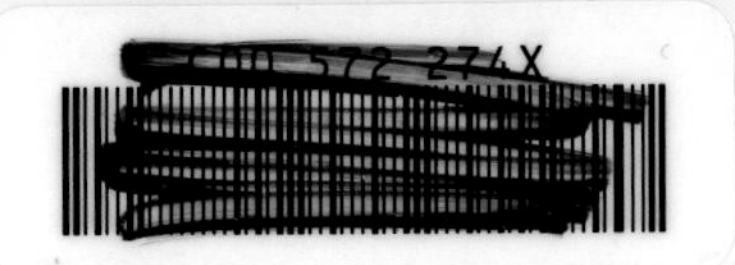

The Essential Guide to Video Processing

Second Edition

The Essential Guide to Video Processing

Second Edition

EDITOR

Al Bovik

Department of Electrical and Computer Engineering
The University of Texas at Austin
Austin, Texas

AMSTERDAM • BOSTON • HEIDELBERG • LONDON
NEW YORK • OXFORD • PARIS • SAN DIEGO
SAN FRANCISCO • SINGAPORE • SYDNEY • TOKYO

Academic Press is an imprint of Elsevier

Academic Press is an imprint of Elsevier
30 Corporate Drive, Suite 400, Burlington, MA 01803, USA
525 B Street, Suite 1900, San Diego, California 92101-4495, USA
84 Theobald's Road, London WC1X 8RR, UK

Library of Congress Cataloging-in-Publication Data
Application submitted

British Library Cataloguing-in-Publication Data
A catalogue record for this book is available from the British Library.

ISBN: 978-0-12-374456-2

Typeset by: diacriTech, India

Printed in the United States of America
09 10 11 12 9 8 7 6 5 4 3 2 1

Contents

Preface xvii

About the Author xix

CHAPTER 1 Introduction to Digital Video Processing 1

1.1 Sampled Video 3

1.2 Video Transmission 6

1.3 Objectives of this Guide 7

1.4 Organization of the Guide 8

CHAPTER 2 Video Sampling and Interpolation 11

2.1 Introduction 11

2.2 Spatiotemporal Sampling Structures 12

2.3 Sampling and Reconstruction of Continuous Time-Varying Imagery 16

2.4 Sampling Structure Conversion 20

2.4.1 Frame-rate Conversion 22

2.4.2 Spatiotemporal Sampling Structure Conversion 26

2.5 Conclusion 28

References 28

Further Information 29

CHAPTER 3 Motion Detection and Estimation 31

3.1 Introduction 31

3.2 Notation and Preliminaries 32

3.2.1 Binary Hypothesis Testing 32

3.2.2 Markov Random Fields 33

3.2.3 MAP Estimation 34

3.2.4 Variational Formulations 34

3.3 Motion Detection 35

3.3.1 Hypothesis Testing with Fixed Threshold 36

3.3.2 Hypothesis Testing with Adaptive Threshold 38

3.3.3 MAP MRF Formulation 41

3.3.4 MAP Variational Formulation 42

3.3.5 Experimental Comparison of Motion Detection Methods 43

3.4 Motion Estimation 44

3.4.1 Motion Models 45

3.4.2 Estimation Criteria 51

3.4.3 Search Strategies 54

3.5 Practical Motion Estimation Algorithms 56
3.5.1 Global Motion Estimation 56
3.5.2 Block Matching 59
3.5.3 Phase Correlation 61
3.5.4 Optical Flow via Regularization 62
3.5.5 MAP Estimation of Dense Motion 63
3.5.6 Experimental Comparison of Motion Estimation Methods 64
3.6 Perspectives 65
3.7 Acknowledgments 66
References 66

CHAPTER 4 Video Enhancement and Restoration 69
4.1 Introduction 69
4.2 Spatiotemporal Noise Filtering 72
4.2.1 Linear Filters 72
4.2.2 Order-Statistic Filters 76
4.2.3 Multiresolution Filters 79
4.3 Coding Artifact Reduction 82
4.3.1 Artifact Reduction in the Spatial Domain 83
4.3.2 Artifact Reduction in the Frequency Domain 83
4.4 Blotch Detection and Removal 84
4.4.1 Blotch Detection 85
4.4.2 Motion Vector Repair and Interpolating Corrupted Intensities 88
4.4.3 Video Inpainting 91
4.4.4 Restoration in Conditions of Difficult Object Motion 92
4.5 Vinegar Syndrome Removal 94
4.6 Intensity Flicker Correction 97
4.6.1 Flicker Parameter Estimation 98
4.6.2 Estimation on Sequences with Motion 99
4.7 Kinescope Moiré Removal 101
4.8 Scratch Removal 103
4.9 Conclusions 104
Acknowledgements 105
References 105

CHAPTER 5 Video Stabilization and Mosaicing 109
5.1 Introduction 109
5.1.1 Video Stabilization 110
5.1.2 Outline 110

5.2 Biological Motivation: Insect Navigation 111
5.2.1 Centering Behavior and Collision Avoidance........... 111
5.2.2 Control of Flight Speed and Stabilization 112
5.2.3 Measuring Distance by Integrating Optical Flow 112
5.3 Camera Model and Image Motion Model 113
5.3.1 Camera Model 113
5.3.2 Effect of Camera Motion................................ 114
5.3.3 Image Features 115
5.3.4 Structure from Motion 116
5.3.5 Feature based Algorithms 117
5.4 Flow-Based Approaches 118
5.4.1 Global Flow Models 118
5.4.2 Flow-Based Algorithm 119
5.5 Stabilization and Mosaicing...................................... 122
5.5.1 Video Mosaicing 124
5.6 Stabilization and Mosaicing with Additional Information 127
5.6.1 VIVID Metadata 127
5.6.2 Stabilization with Metadata............................. 128
5.6.3 Inertial Measurements 130
5.6.4 Stabilization with Inertial Measurements.............. 130
5.6.5 Motion Segmentation 134
5.7 Motion Super-resolution 134
5.8 Three-dimensional Stabilization.................................. 137
5.9 Summary 137
Acknowledgements 138
References 138

CHAPTER 6 Video Segmentation 141

6.1 Introduction.. 141
6.2 Scene Change Detection.. 142
6.3 Spatiotemporal Change Detection............................... 144
6.3.1 Spatial Change Detection Using Two Frames 144
6.3.2 Temporal Integration.................................... 145
6.3.3 Combination with Spatial Segmentation 146
6.4 Motion Segmentation 146
6.4.1 Dominant Motion Segmentation 147
6.4.2 Multiple Motion Segmentation.......................... 150
6.5 Simultaneous Motion Estimation and Segmentation 160
6.5.1 Motion-Field Model and MAP Framework 161
6.5.2 Two-Step Iteration Algorithm 162
6.6 Semantic Video Object Segmentation........................... 164
6.6.1 Chroma-Keying.. 164
6.6.2 Semiautomatic Segmentation 164

6.7 Examples 165
6.8 Performance Evaluation of Video Segmentation 170
Acknowledgements 170
References 170

CHAPTER 7 Motion Tracking in Video 175
7.1 Introduction 175
7.2 Rigid Object Tracking 180
7.2.1 2D Rigid Object Tracking 180
7.2.2 3D Rigid Object Tracking 205
7.3 Articulated Object Tracking 209
7.3.1 3D Articulated Object Tracking 209
7.3.2 2D Articulated Object Tracking 219
References 220

CHAPTER 8 Basic Transform Video Coding 231
8.1 Introduction to Video Compression 232
8.2 Video Compression Application Requirements 237
8.3 Digital Video Signals and Formats 241
8.3.1 Sampling of Analog Video Signals 241
8.3.2 Digital Video Formats 243
8.4 Video Compression Techniques 245
8.4.1 Entropy and Predictive Coding 246
8.4.2 Block Transform Coding—The DCT 248
8.4.3 Quantization 250
8.4.4 MC and Estimation 253
8.5 Transform Coding: Introduction to the Video Encoding Standards 256
8.5.1 Transform Coding Standard Example: The H.261 Video Encoder 258
8.6 Closing Remarks 265
References 265

CHAPTER 9 MPEG-1 and MPEG-2 Video Standards 267
9.1 MPEG-1 Video Coding Standard 267
9.1.1 Introduction 267
9.1.2 MPEG-1 Video Coding versus H.261 268
9.1.3 MPEG-1 Video Structure 270
9.1.4 Summary of the Major Differences between MPEG-1 Video and H.261 274
9.1.5 Simulation Model 275
9.1.6 MPEG-1 Video Bit-Stream Structures 276
9.1.7 Summary 277

9.2 MPEG-2 Video Coding Standard 277
- 9.2.1 Introduction 277
- 9.2.2 MPEG-2 Profiles and Levels 279
- 9.2.3 MPEG-2 Video Input Resolutions and Formats 280
- 9.2.4 MPEG-2 Video Coding Standard Compared to MPEG-1 281
- 9.2.5 Scalable Coding 286
- 9.2.6 Data Partitioning 288
- 9.2.7 Other Tools for Error Resilience 289
- 9.2.8 Test Model 289
- 9.2.9 MPEG-2 Video and System Bit-Stream Structures 290
- 9.2.10 Summary 291

References 292

CHAPTER 10 MPEG-4 Visual and H.264/AVC: Standards for Modern Digital Video 295

10.1 Introduction 295

10.2 Terminology 296

10.3 MPEG-4 Part 2 298
- 10.3.1 Object-based Representation 298
- 10.3.2 Video Object Coding 299
- 10.3.3 Mesh Object Coding 304
- 10.3.4 Model-based Coding 305
- 10.3.5 Still Texture Coding 307
- 10.3.6 Scalability 307
- 10.3.7 Error Resilience 308
- 10.3.8 MPEG-4 Part 2 Profiles 308

10.4 MPEG-4 Part 10: H.264/AVC 310
- 10.4.1 H.264/AVC Video Coding Layer: Technical Overview 310
- 10.4.2 Profiles 322

10.5 MPEG-4 Compression Performance 323
- 10.5.1 MPEG-4 Part 2 323
- 10.5.2 MPEG-4 Part 10: H.264/AVC 325

10.6 MPEG-4 Video Applications 326

10.7 Conclusions and Outlook 327

Acknowledgements 328

References 328

CHAPTER 11 Interframe Subband/Wavelet Scalable Video Coding 331

11.1 Introduction 331

11.2 Motion Estimation and Compensation for MCTF 333
- 11.2.1 Connected and Unconnected Blocks 335

11.2.2 Using Chroma for Motion Estimation 337
11.2.3 Improving the Haar MCTF 338
11.3 New Haar MCTF 342
11.3.1 Overlapped Block Motion Compensation 344
11.3.2 Scalable Motion Vector Coding 345
11.4 EZBC Coder 348
11.4.1 Coding Process 349
11.4.2 Context Modeling 351
11.4.3 Scalability 352
11.4.4 Packetization 353
11.4.5 Frequency Roll-Off 353
11.5 Extension to LeGall and Tabatabai 5/3 Filtering 355
11.6 Objective and Visual Comparisons 357
11.6.1 Some Visual Results 357
11.7 Multiple Adaptations 360
11.8 Related Coders 361
11.9 Conclusions 363
References 363

CHAPTER 12 Digital Video Transcoding **367**
12.1 Introduction 367
12.2 Video Transcoding for Bit Rate Reduction 369
12.2.1 Transcoding of Intracoded Frame 370
12.2.2 Transcoding of Intercoded Frame 372
12.2.3 Fast Video-Transcoding Architectures 372
12.2.4 DCT Domain IMC 375
12.3 Heterogeneous Video Transcoding 377
12.3.1 MV Estimation for Spatial Resolution Reduction 379
12.3.2 MV Estimation for Temporal Resolution Reduction .. 380
12.3.3 Spatial Resolution Reduction 381
12.3.4 Macro-block-Coding Type Decision 383
12.4 Bit Rate Control in Video Transcoding 383
12.5 Error-Resilient Video Transcoding 384
12.6 Concluding Remarks 385
References 386

CHAPTER 13 Embedded Video Codecs **389**
13.1 Introduction 389
13.2 Block-Based Video Coding 391
13.3 Embedded Video Codec Requirements and Constraints 393
13.4 Embedded Video Codec Design Flow 397
13.4.1 Understanding the Chip Architecture 397
13.4.2 Understanding the Codec Algorithms 400
13.4.3 Modularity and APIs Definitions 401

13.4.4 Reference Codec Software Development in Golden C 404
13.4.5 Platform-Specific Development and Porting 405
13.4.6 Kernel Optimization and Integration 406
13.4.7 Concurrent Processing 407
13.4.8 Overall Optimization 408
13.4.9 Stress and Conformance Testing 410
13.5 New Trends 411
13.6 Summary 413
References 414

CHAPTER 14 **Video Quality Assessment** **417**
14.1 Introduction 417
14.2 HVS Modeling Based Methods 419
14.3 Feature Based Methods 422
14.3.1 VQM 422
14.4 Motion Modeling Based Methods 425
14.5 Performance 431
14.6 Conclusions 433
References 434

CHAPTER 15 **A Unified Framework for Video Indexing, Summarization, Browsing, and Retrieval** **437**
15.1 Introduction 437
15.1.1 Content Categories 438
15.1.2 Storage and Compression 439
15.1.3 Terminology 440
15.2 Image and Video Features 442
15.2.1 Statistical Features 442
15.2.2 Compressed-Domain Features 450
15.2.3 Content-Based Features 452
15.3 Video Analysis 459
15.3.1 Shot Boundary Detection 459
15.3.2 Key-Frame Extraction 459
15.3.3 Play/Break Segmentation 460
15.3.4 Audio Marker Detection 460
15.3.5 Video Marker Detection 460
15.4 Video Representation 460
15.4.1 Video Representation for Scripted Content 460
15.4.2 Video Representation for Unscripted Content 462
15.5 Video Browsing 463
15.5.1 Video Browsing Using ToC-Based Summary 463
15.5.2 Video Browsing Using Highlights-Based Summary 463

15.6 Video Retrieval 463
15.6.1 Feature-Based Retrieval (Statistical and Compressed) 464
15.6.2 Content-Based Retrieval 464
15.6.3 Relevance Feedback 465
15.6.4 Query-Concept Learner 466
15.6.5 Efficient Annotation through Active Learning 466
15.6.6 Considerations in Multimedia Databases 467
15.7 A Unified Framework for Indexing, Summarization, Browsing, and Retrieval 468
15.8 Conclusions and Promising Research Directions 469
Acknowledgements 470
References 470

CHAPTER 16 Video Communication Networks 473
16.1 Introduction 474
16.2 Video Compression Standards 475
16.2.1 Introduction 475
16.2.2 Overview 476
16.2.3 MPEG-2 Video Compression Standard 478
16.2.4 MPEG-2 Systems Standard 479
16.3 Video Communication Networks 485
16.3.1 Introduction 485
16.3.2 Hybrid Fiber-Coax Networks 486
16.3.3 Digital Subscriber Loop 487
16.3.4 Wireless Networks 488
16.3.5 Fiber Optics 491
16.3.6 Integrated Services Digital Network 492
16.3.7 ATM Networks 492
16.4 Internet Protocol Networks 499
16.4.1 Introduction 499
16.4.2 Multicast Backbone 502
16.4.3 Real-Time Transport Protocol 503
16.4.4 Real-Time Transport Control Protocol 511
16.4.5 Real-Time Transport Streaming Protocol 517
16.4.6 H.323 518
16.4.7 Session Initiation Protocol 520
16.4.8 Integrated Services—Resource Reservation Protocol 521
16.4.9 Differentiated Services—DiffServ 523
16.5 Summary 525
References 525

CHAPTER 17 Video Security and Protection 527

17.1 Introduction ... 527
17.2 Video Encryption ... 527
17.2.1 Candidate Domains for Encrypting Multimedia ... 529
17.2.2 Building Blocks for Media Encryption ... 530
17.2.3 Security Evaluation of Media Encryption ... 536
17.2.4 Video Encryption System Design ... 539
17.3 Video Authentication ... 545
17.3.1 Background ... 545
17.3.2 Content Level Authentication ... 546
17.3.3 Stream Level Authentication ... 551
17.4 Video Fingerprinting for Traitor Tracing ... 553
17.4.1 The Background ... 554
17.4.2 Coded Fingerprinting ... 556
17.4.3 Experimental Results of Video Fingerprinting ... 563
17.4.4 Intravideo Collusion ... 566
References ... 566

CHAPTER 18 Wireless Video Streaming 571

18.1 Introduction ... 571
18.2 On Joint Source-Channel Coding ... 575
18.2.1 Rate-Distortion Theory ... 575
18.2.2 Operational Rate-Distortion Theory ... 576
18.2.3 Practical Constraints in Video Communications ... 577
18.2.4 Illustration ... 578
18.3 Video Compression and Transmission ... 579
18.3.1 Video Transmission System ... 579
18.3.2 Video Compression Basics ... 581
18.3.3 Channel Models ... 584
18.3.4 End-to-End Distortion ... 585
18.3.5 Error Resilient Source Coding ... 587
18.4 Channel Coding ... 589
18.4.1 Forward Error Correction ... 590
18.4.2 Retransmission ... 592
18.5 Joint Source-Channel Coding ... 594
18.5.1 Problem Formulation ... 594
18.5.2 Internet Video Transmission ... 596
18.5.3 Wireless Video Transmission ... 599
18.6 Distributed Multimedia Communications ... 604
18.6.1 Video Streaming over Multiuser Networks ... 604
18.6.2 Mobile TV Standards ... 608
18.6.3 Peer-to-Peer Internet Video Broadcasting ... 610
18.6.4 Video Streaming over Multihop Wireless Networks ... 611

18.7	Discussion	612
	References	613
CHAPTER 19	**Video Surveillance**	**619**
19.1	Introduction	619
19.2	Categorizing Applications, Target Scenes, and Video Analytics	620
	19.2.1 Video Surveillance Applications	620
	19.2.2 Video Surveillance Target Scenes	624
	19.2.3 Video Analytics for Video Surveillance	626
19.3	Review of Video Analytic Algorithms	628
	19.3.1 Motion and Change Detection	629
	19.3.2 Object Detection	635
	19.3.3 Object Tracking	637
	19.3.4 Behavioral Analysis Tools	644
	19.3.5 Gait Recognition	645
	19.3.6 Face Recognition	648
19.4	Conclusion	648
	References	649
CHAPTER 20	**Face Recognition From Video**	**653**
20.1	Introduction	653
20.2	Properties and Literature Review	655
	20.2.1 Set of Observations	655
	20.2.2 Temporal Continuity/Dynamics	661
	20.2.3 3D Model	663
20.3	A General Framework of Probabilistic Identity Characterization	665
	20.3.1 Recognition Setting and Issues	668
20.4	Instances of Probabilistic Identity Characterization	670
	20.4.1 ER From a Group of Still Images	670
	20.4.2 ER From a Video Sequence	675
20.5	A System Identification Approach	682
	20.5.1 The ARMA Model	682
	20.5.2 Framework for Recognition	683
	20.5.3 Experiments, Results, and Discussion	684
20.6	Conclusions	685
	References	685
CHAPTER 21	**Audiovisual Speech Processing**	**689**
21.1	Introduction	689
21.2	Analysis of Visual Signals	691
	21.2.1 Face Detection, Mouth, and Lip Tracking	692
	21.2.2 Visual Features	694
	21.2.3 Two Visual Feature Extraction Systems	698

21.3 Audiovisual Information Fusion 700
21.3.1 Speech Classes in Audiovisual Integration 700
21.3.2 Classifiers in Speech Applications 702
21.3.3 Feature and Classifier Fusion 704
21.4 Audiovisual Automatic Speech Recognition 707
21.4.1 Bimodal Corpora for ASR 708
21.4.2 Experimental Results 709
21.5 Audiovisual Speech Synthesis 712
21.5.1 Coarticulation Modeling 713
21.5.2 Facial Animation 714
21.5.3 Visual Text-to-Speech 717
21.5.4 Speech-to-Video Synthesis 718
21.5.5 Visual Speech Synthesis Evaluation 721
21.6 Audiovisual Speaker Recognition 723
21.7 Summary and Discussion 729
References 731

Index **739**

Preface

Welcome to *The Essential Guide to Video Processing*! You are about to embark upon learning one of the most interesting and timely topics in the high-tech universe. After all, digital video is becoming the face of television, the Internet and handheld mobile devices. Digital video is also the future of medical imaging, personal photography, industrial inspection, visual surveillance, and a wealth of other areas that affect our daily lives.

If you are looking for the latest and most complete resource for learning about digital video processing – you've come to the right place. Thanks to an assembled team of experts – each a leader in the area of their chapter contribution – this book offers the most complete treatment of digital video processing available. *The Essential Guide to Video Processing* contains 21 chapters ranging from the basics of video processing to very advanced topics, and everything in-between. Starting with the basics of video sampling and motion estimation and tracking, the book covers essential topics including video stabilization, segmentation, and enhancement. The important topic of video compression is covered next, with chapters for the beginner as well as the expert, including the standards MPEG and H.264, transcoding between standards, and embedded software implementations.

The important practical topics of video quality assessment, video indexing and retrieval, video networking, video security, and wireless are then explained by leading experts and educators in those fields.

Finally, the *Guide* concludes with four exciting chapters dealing explaining video processing applications on such diverse topics as video surveillance, face tracking and recognition from video, motion tracking in medical videos, and using video to assist speech recognition. These have been selected for their timely interest, as well as their illustrative power of how video processing can be effectively applied to problems of today's interest.

The Essential Guide to Video Processing is intended to reach a broad audience ranging from those interested in learning video processing for the first time, to professional experts needing an easy and authoritative reference, to educators that require a comprehensive treatment of the topic. However, for those that have had no exposure at all to the processing of images, including still images, then it is advisable to first read the companion volume, *The Essential Guide to Image Processing*. In that book, the broad range of still image processing are explained, including very basic tutorial chapters.

As Editor and Co-Author of *The Essential Guide to Video Processing*, I thank the wonderful co-authors who have contributed their work to this *Guide*. They have all been models of exceptional professionalism and responsiveness. This book is much greater than I think that any individual could have created.

I would also like to thank the people at Elsevier who did so much to make this book come to fruition: Tim Pitts, the Senior Commissioning Editor, Tim Pitts, for his enthusiasm and ideas; Melanie Benson for her tireless efforts and incredible organization and accuracy in making the book happen; and Eric DeCicco, the graphic artist for his imaginative and energetic cover design.

Al Bovik
Austin, Texas
April, 2009

About the Author

Al Bovik currently holds the Curry/Cullen Trust Endowed Chair Professorship in the Department of Electrical and Computer Engineering at The University of Texas at Austin, where he is the Director of the Laboratory for Image and Video Engineering (LIVE). He has published over 500 technical articles and six books in the general area of image and video processing and holds two US patents.

Dr. Bovik has received a number of major awards from the IEEE Signal Processing Society, including the Education Award (2007); the Technical Achievement Award (2005), the Distinguished Lecturer Award (2000); and the Meritorious Service Award (1998). He is also a recipient of the IEEE Third Millennium Medal (2000), and has won two journal paper awards from the Pattern Recognition Society (1988 and 1993). He is a Fellow of the IEEE, a Fellow of the Optical Society of America, and a Fellow of the Society of Photo-Optical and Instrumentation Engineers. Dr. Bovik has served Editor-in-Chief of the *IEEE Transactions on Image Processing* (1996–2002) and created and served as the first General Chairman of the *IEEE International Conference on Image Processing*, which was held in Austin, Texas, in 1994.

CHAPTER

Introduction to Digital Video Processing

1

Alan C. Bovik

Laboratory for Image and Video Engineering (LIVE),
Department of Electrical and Computer Engineering,
The University of Texas at Austin

In this age of dramatic technology shifts, one of the most significant has been the emergence of *digital video* as an important aspect of daily life. While the Internet has certainly dramatically changed the way in which we obtain information, how much more attractive is it because of the powerful medium of video? How many of us now watch the morning news on Yahoo! Video? Are there many readers who have not shared or viewed a *Utube* video? Indeed, the average person on the street has become an amateur video processor, now that handheld digital camcorders that record directly onto mini-DVD are standard, and software is available for simple film editing.

Developments in wireless and cellular telephony certainly represent the greatest recent sea change in global technology over the past five years. It is widely held that ongoing advances in handheld mobile devices will be dominated by easier wireless Internet access that will be enhanced by rich video and multimedia content. The great leaps in mobile bandwidths that are being engineered ensure that these developments will reach fruition. It is no mistake that the wildly popular Apple *iphone* has such a large display, a realization by visionary engineers that the future of wireless relies heavily on visual communication.

An increasing percentage of homes now own large-screen televisions receiving digital signals. Indeed, the age of analog TV has been declared dead. High-definition television (HDTV), heralded long ago but never realized until recently is now becoming widely available, and recent newspaper and Internet headlines herald the emergence of an agreed-upon HD DVD standard known as *Blueray.*

Beyond these notable examples, there are a wealth of other applications that involve digital video technologies. It is safe to say that digital video is very much the current frontier and the future of image processing research and development. Some of the notable digital video applications not mentioned above are the following:

- Video teleconferencing using modern software codecs that run on your PC.
- Medical video, where dynamic processes in the human body, such as the motion of the heart, can be viewed live.
- Video telephony, an application coming to your cell phone.
- Dynamic scientific visualization, such as viewable displays of galaxy formation over accelerated time.
- Multimedia video, where video is combined with graphics, speech, and other sensor modalities for an enriched experience.
- Video instruction, both in the classroom and on the Internet.
- Digital cinema, not just in the making of graphical marvels such as *Shrek* but for delivery of movies to theatres in digital format.

With all these developments, it is a particularly exciting and rewarding time to be a video processing engineer. *Digital video processing* is the study of algorithms for processing moving images that are represented in digital format. Here, we distinguish *digital video* from *digital still images,* which are images that do not change with time – basically, digital photographs. A companion volume entitled *The Essential Guide to Image Processing* covers the processing and analysis of digital still images in great detail and offers material ranging from a very basic level to much more advanced topics and applications. For the novice to the field, it is recommended that *The Essential Guide to Image Processing* be studied first, since the field of digital video processing first builds upon the principles of still image processing before advancing significantly further.

A digital video is a moving picture, or movie, that has been converted into a computer-readable binary format consisting of logical 0s and 1s. Since video is dynamic, the visual content evolves with time and generally contains moving and/or changing objects. The information contained in video is naturally much richer than is contained in still images, since our world is constantly changing owing to the movement of people, animals, vehicles, and other objects. Indeed, the richness of video results in a significant data glut, which, until recently has been difficult to overcome. However, with faster sensors and recording devices, it is it is becoming easier to acquire and analyze digital video data sets.

However, it is important to realize that digital videos are *multidimensional signals,* meaning that they are functions of more than a single variable. Indeed, digital video is ordinarily a function of three dimensions – two in space and one in time, as depicted in Fig. 1.1. Because of this, digital video processing is data intensive: significant bandwidth, computational, and storage resources are required to handle video streams in digital format.

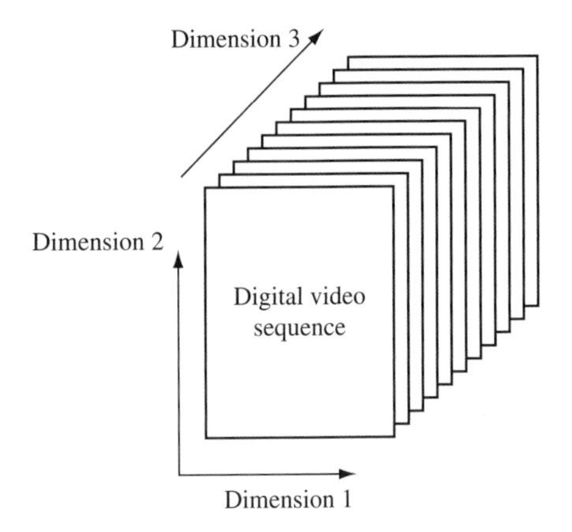

FIGURE 1.1

The dimensionality of video.

In the companion volume, *The Essential Guide to Image Processing*, a tutorial is given on the processes of digitization of images, or *analog-to-digital conversion* (A/D conversion). This conversion process consists of two distinct subprocesses: *sampling*, or conversion of a continuous-space/time video signal into a discrete-space/time video signal, and *quantization*, which is the process of converting a *continuous-valued video* that has a continuous range (set of values that it can take) of intensities and/or colors into a *discrete-valued video* that has a discrete range of intensities and/or colors. This is ordinarily done by a process of rounding, truncation, or some other irreversible, nonlinear process of information destruction. Quantization is a necessary precursor to digital processing or display, since the image intensities and/or colors must be represented with a finite precision (limited by word length) in a digital video processor or display device.

1.1 SAMPLED VIDEO

Just as with still images, digital processing of video requires that the video stream be in a digital format, meaning that it must be sampled and quantized. Video quantization is essentially the same as image quantization. However, video sampling involves taking samples along a new and different (time) dimension. As such, it involves some different concepts and techniques.

First and foremost, the time dimension has a direction associated with it, unlike the space dimensions, which are ordinarily regarded as directionless until a coordinate system is artificially imposed upon it. Time proceeds from the past toward the future, with an origin that exists only in the current moment. Video is often processed in "real time," which (loosely) means that the result of processing appears effectively "instantaneously"

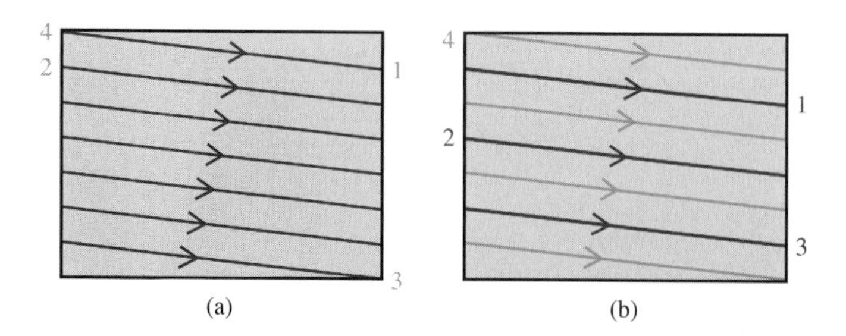

FIGURE 1.2

Video scanning. (a) Progressive video scanning. At the end of a scan (1), the electron gun spot snaps back to (2). A blank signal is sent in the interim. After reaching the end of a frame (3), the spot snaps back to (4). A synchronization pulse then signals the start of another frame. (b) Interlaced video scanning. Red and blue fields (shown in this illustration as gray and black) are alternately scanned left-to-right and top-to-bottom. At the end of scan (1), the spot snaps to (2). At the end of the blue field (3), the spot snaps to (4) (new field).

(usually in a perceptual sense) once the input becomes available. Such a processing system cannot depend more than a few future video samples. Moreover, it must process the video data quickly enough that the result appears instantaneous. Because of the vast data volume involved, the design of fast algorithms and hardware devices is a major priority.

In principle, an analog video signal $I(x,y,t)$, where (x,y) denote continuous space coordinates and t denotes continuous time, is continuous in both the space and time dimensions, since the radiation flux that is incident on a video sensor is continuous at normal scales of observation. However, the analog video that is viewed on display monitors is *not* truly analog, since it is sampled along one space dimension and along the time dimension. Practical so-called analog video systems, such as television and monitors, represent video as a one-dimensional electrical signal $V(t)$. Prior to display, a one-dimensional signal is obtained by sampling $I(x,y,t)$ along the vertical (y) space direction and along the time (t) direction. This is called scanning and the result is a series of time samples, which are complete pictures or *frames*, each of which is composed as space samples, or *scan lines.*

Two types of video scanning are commonly used: *progressive scanning* and *interlaced scanning*. A progressive scan traces a complete frame, line-by-line from top-to-bottom, at a scan rate of Δt s/frame. High-resolution computer monitors are a good example, with a scan rate of $\Delta t = 1/72$ s. Figure 1.2(a) depicts progressive scanning on a standard monitor.

A description of interlaced scanning requires that some other definitions be made. For both types of scanning, the *refresh rate* is the frame rate at which information is displayed on a monitor. It is important that the frame rate be high enough, since otherwise the displayed video will appear to "flicker." The human eye detects flicker if the refresh rate is less than about 50 frames/s. Clearly, computer monitors (72 frames/s) exceed this rate by almost 50%. However, in many other systems, notably television, such fast refresh rates are not possible unless spatial resolution is severely compromised because of bandwidth limitations. Interlaced scanning is a solution to this. In $P:1$ interlacing, every Pth line is refreshed at each frame refresh. The subframes in interlaced video are called *fields*, and hence P fields constitute a frame. The most common is $2:1$ interlacing, which is used in standard television systems, as depicted in Fig. 1.2(b). In $2:1$ interlacing, the two fields are usually referred to as the top and bottom fields. In this way, flicker is effectively eliminated provided that the field refresh rate is above the visual limit of about 50 Hertz (Hz). Broadcast television in the United States uses a frame rate of 30 Hz, hence the field rate of 60 Hz, which is well above the limit. The reader may wonder whether there is a loss of visual information, since the video is being effectively subsampled by a factor of 2 in the vertical space dimension to increase the apparent frame rate. In fact there is a loss of visual information, since image motion may change the picture between fields. However, the effect is ameliorated to a significant degree by standard monitors and TV screens, which have screen phosphors with a *persistence* (glow time) that just matches the frame rate; hence, each field persists until the matching field is sent.

Digital video is obtained either by sampling an analog video signal $V(t)$ or by directly sampling the three-dimensional space–time intensity distribution that is incident on a sensor. In either case, what results is a time sequence of two-dimensional spatial intensity arrays or equivalently a three-dimensional space–time array. If progressive analog video is sampled, then the sampling is rectangular and properly indexed in an obvious manner, as illustrated in Fig. 1.3. If interlaced analog video is sampled, then the digital video is interlaced also as shown in Fig. 1.4. Of course, if an interlaced video stream is sent to a system that processes or displays noninterlaced video, then the video data must first be converted or *deinterlaced* to obtain a standard progressive video stream before the accepting system will be able to handle it.

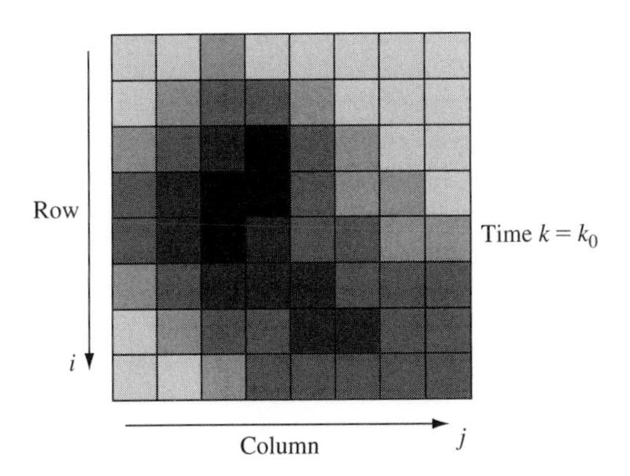

FIGURE 1.3

A single frame from a sampled progressive video sequence.

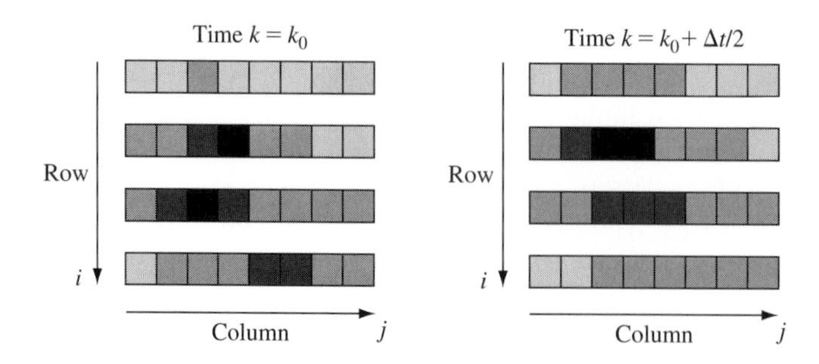

FIGURE 1.4

A single frame (two fields) from a sampled 2:1 interlaced video sequence.

1.2 VIDEO TRANSMISSION

The data volume of digital video is usually described in terms of bandwidth or bitrate. The bandwidth of digital video streams (without compression) that match the current visual resolution of current television systems exceeds 100 megabits/s (mbps). Modern television formats such as HDTV can multiply this by a factor of 4 or more. By contrast, the networks that are currently available to handle digital data are limited. Conventional telephone lines (plain old telephone service [POTS]) deliver only 56 kilobits/second (9 kbps), although digital subscriber lines (DSL) multiply this by a factor of 30 or more. Integrated Services Digital Network (ISDN) lines allow for data bandwidths equal to $64p$ kbps, where $1 \leq p \leq 30$, which falls short of the necessary data rate to handle full digital video. Dedicated T1 lines (1.5 mbps) also handle only a small fraction of the necessary bandwidth. Ethernet and cable systems, which deliver data in the gigabit/second (gbps) range, are capable of handling raw digital video but have problems delivering multiple video streams over the same network. The problem is similar to that of delivering large amounts of water to through small pipelines. Either the data rate (water pressure) must be increased or the data volume must be reduced.

Fortunately, unlike water, digital video can be compressed very effectively because of the redundancy inherent in the data and because of an increased understanding of what components in the video stream are actually visible. Because of many years of research into image and video compression, it is now possible to transmit digital video data over a broad spectrum of networks, and we may expect that digital video will arrive in a majority of homes in the near future. Based on research developments along these lines, a number of world standards have recently emerged, or are under discussion, for video compression, video syntax, and video formatting. The use of standards allows for a common protocol for video and ensures that the consumer will be able to accept the same video inputs using products from different manufacturers. The current and emerging video standards broadly extend standards for still images that have been in use for a number of years. Several chapters are devoted to describing these standards, whereas others deal with emerging techniques that may affect future standards. It is certain, in any case, that we have entered a new era where digital visual data will play an important role in education, entertainment, personal communications, broadcast, the Internet, and many other aspects of daily life.

1.3 OBJECTIVES OF THIS GUIDE

The goals of this *Guide* are ambitious, since it is intended to reach a broad audience that is interested in a wide variety of video processing applications. Moreover, it is intended to be accessible to readers that have a diverse background and that represent a wide spectrum of levels of preparation and engineering/computer education. A Guide format is ideally suited for this multiuser purpose, since it allows for a presentation that adapts to the readers needs. However, for the reader who has not yet been introduced to the concepts of digital *still* image processing, which in many ways forms the basis of digital video processing, it is strongly encouraged that the companion volume, *The Essential Guide to Image Processing,* be read first.

Because of its broad spectrum of coverage, we expect that the current book, *The Essential Guide to Video Processing,* will serve as an excellent textbook as well as an important reference for video processing engineers. The material contained here is appropriate to be used for classroom presentations ranging from the upper-division undergraduate level through the graduate level. Of course, the Guide provides an easy and complete reference for video processing professionals working in industry and research.

The specific objectives are to

- provide the practicing engineer and student with a highly accessible resource for learning and using video processing algorithms and theory,
- provide the essential understanding of the various existing and emerging video processing standards,
- provide an understanding of what videos are and how they are modeled,
- provide the necessary practical background to allow the engineer student to acquire and process his/her own digital video data.
- Provide a diverse set of example applications, as separate complete chapters, that are explained in sufficient depth to serve as extensible models to the readers own potential applications.

The Guide succeeds in achieving these goals, primarily because of the many years of broad educational and practical experience that the many contributing authors bring to bear in explaining the topics contained herein.

1.4 ORGANIZATION OF THE GUIDE

Following this introductory chapter, the Guide immediately launches into the basic topic of video sampling and interpolation in Chapter 2. Then, Chapter 3 covers the essential topic of motion estimation that is used in a large proportion of video processing algorithms. Chapter 4 covers methods for enhancing and improving digital videos that have been degraded by some kind of noise or distortion process, while Chapter 5 details methods for stabilizing and combining video streams. Since many video analysis tasks require that the video stream be segmented into separate objects, or into object versus background, Chapter 6 covers the important topic of video segmentation. Since in video analysis applications it is also important to be able to track the motion of moving objects, this topic is covered in Chapter 7.

Owing to the importance of handling the high data volume of digital video, the next several chapters cover concepts of *video compression.* Chapter 8 is a marvelous introduction to the principles of video compression that are used in every standard algorithm. Chapter 9 explains two of the most important and early video compression standards, Moving Picture Experts Group (MPEG)-1 and MPEG-2. MPEG-1 is most familiar to those who download Internet videos, whereas MPEG-2 is familiar to those who watch movies on DVD. Chapter 10 describes the exciting new video compression standards, MPEG-4 Visual and H.264 (also known as MPEG-4/Advanced Video Coding [AVC]). Chapter 11 looks beyond to video codecs that use multiresolution or wavelet processing methods to achieve compression. Since there is a diversity of video compression algorithms, even amongst the standards, and since it is often of interest to communicate between devices or programs that use different codecs, Chapter 12 covers the topic of *video transcoding,* where a video that has been compressed using one type of encoder is efficiently transformed into a format that is compatible with another type of decoder. Since practical implementations of video compression algorithms in special-purpose processors or in Digital Signal Processing (DSP) chips is necessary for practical applications, Chapter 13 describes embedded video codecs of these types.

The next chapters involve practical aspects of video processing in real-world situations. Module 14 covers *video quality assessment,* whereby the quality of videos is measured using objective algorithms using methods that agree with human visual perception of quality. This is important for benchmarking video processing algorithms, such as codecs, as well as video communication networks. With the rise in interest in digital video, the sheer volume of videos is dramatically increasing and methods must be found for archiving and efficiently recalling them, as well as for searching for specific types of video content in video libraries. This important and difficult topic is covered in Chapter 15. Chapter 16 discusses the important topic of efficiently handling and routing video streams in communication networks and Chapter 17 describes modern methods for protecting the content of digital videos from theft or copyright violation. Chapter 18 discusses the issues and problems encountered in deploying video over the wireless medium, where many of the exciting ongoing and future applications are arising.

Finally, several chapters are given that describe important and vital diverse video processing applications. These include Chapter 19 (Video Surveillance) and Chapter 20 (Face Tracking and Recognition from Video), both of which represent applications that are important for domestic security; Chapter 21, Tracking in Medical Videos, describing important new developments in medical video, and Chapter 22 that describes how video processing techniques can be used to assist in speech processing.

Video processing is certainly one of the most exciting technology areas to be involved in. It is our hope that this Guide finds you well on your way to understanding and participating in these developments.

CHAPTER

Video Sampling and Interpolation

2

Eric Dubois

School of Information Technology and Engineering (SITE),
Faculty of Engineering, University of Ottawa,
Ontario, Canada

2.1 INTRODUCTION

This chapter is concerned with the sampled representation of time-varying imagery, often referred to as video. Time-varying imagery must be sampled in at least one dimension for the purposes of transmission, storage, processing, or display. Examples are one-dimensional temporal sampling in motion-picture film, two-dimensional vertical-temporal scanning in the case of analog television, and three-dimensional horizontal-vertical-temporal sampling in digital video. In some cases, a single sampling structure is used throughout an entire video processing or communication system. This is the case in standard analog television broadcasting, where the signal is acquired, transmitted, and displayed using the same scanning standard from end to end. However, it is becoming increasingly more common to have different sampling structures used in the acquisition, processing, transmission, and display components of the system. In addition, the number of different sampling structures in use throughout the world is increasing. Thus, sampling structure conversion for video systems is an important problem.

The initial acquisition and scanning is particularly critical because it determines what information is contained in the original data. The acquisition process can be modeled as a continuous space-time prefiltering followed by ideal sampling on a given sampling structure. The sampling structure determines the amount of spatio-temporal information that the sampled signal can carry, while the prefiltering serves to limit the amount of aliasing. At the final stage of the system, the desired display characteristics are closely related to the properties of the human visual system. The goal of the display is to convert the sampled signal to a continuous time-varying image that, when presented to the viewer, approximates the original continuous scene as closely as possible. In particular, the effects caused by sampling should be sufficiently attenuated so as to lie below the threshold of perceptibility.

In this chapter, we consider only scalar-valued images, such as the luminance or one of the RGB components of a color image. All the theory and results can be applied in a straightforward way to multicomponent (or vector-valued) signals, such as color video by treating the sampling and reconstruction of each scalar component independently. However, interesting additional problems and opportunities arise when the individual components have different sampling structures, which is in fact the case for most cameras and displays employing color mosaics. These issues are beyond the scope of this chapter.

The chapter has three main sections. First, the sampling lattice, the basic tool in the analysis of spatiotemporal sampling, is introduced. The issues involved in the sampling and reconstruction of continuous time-varying imagery are then addressed. Finally, methods for the conversion of image sequences between different sampling structures are presented.

2.2 SPATIOTEMPORAL SAMPLING STRUCTURES

A continuous time-varying image $f_c(x,y,t)$ is a scalar real-valued function of two spatial dimensions x and y and time t, usually observed in a rectangular spatial window $\mathcal{W}$ over some time interval $\mathcal{T}$. The spatiotemporal region $\mathcal{W} \times \mathcal{T}$ is denoted as $\mathcal{W}_\mathcal{T}$. The spatial window is of dimension $pw \times ph$, where pw is the picture width and ph is the picture height. Since the absolute physical size of an image depends on the display device used, and the sampling density for a particular video signal may be variable, we choose to adopt the picture height ph as the basic unit of spatial distance, as is common in the broadcast video industry. The ratio pw/ph is called the aspect ratio, the most common values being 4/3 for standard TV and 16/9 for HDTV. The image f_c can be sampled in one, two, or three dimensions. It is almost always sampled in at least the temporal dimension, producing an *image sequence.* An example of an image sampled only in the temporal dimension is motion picture film. Analog video is typically sampled in the vertical and temporal dimensions using one of the scanning structures shown in Fig. 2.2 of Chapter 1, while digital video is sampled in all three dimensions. The subset of $\mathbb{R}^3$ on which the sampled image is defined is called the *sampling structure* Ψ; it is contained in $\mathcal{W}_\mathcal{T}$.

The mathematical structure most useful in describing sampling of time-varying images is the *lattice.* A discussion of lattices from the point of view of video sampling can be found in [1] and [2]. Some of the main properties are summarized here. A lattice Λ in D dimensions is a discrete set of points that can be expressed as the set of all linear combinations with integer coefficients of D linearly independent vectors in $\mathbb{R}^D$ (called basis vectors),

$$\Lambda = \{n_1 \mathbf{v}_1 + \cdots + n_D \mathbf{v}_D \mid n_i \in \mathbb{Z}\}, \tag{2.1}$$

where $\mathbb{Z}$ is the set of integers. For our purposes, D will be 1, 2, or 3 dimensions. The matrix $V = [\mathbf{v}_1 \mid \mathbf{v}_2 \mid \cdots \mid \mathbf{v}_D]$ whose columns are the basis vectors $\mathbf{v}_i$ is called a sampling matrix and we write $\Lambda = \text{LAT}(V)$. However, the basis or sampling matrix for a given

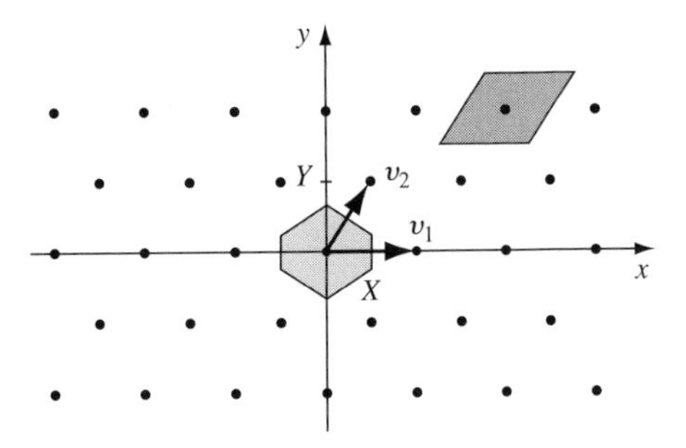

FIGURE 2.1

Example of a lattice in two dimensions with two possible unit cells.

lattice is not unique, since $\text{LAT}(V) = \text{LAT}(VE)$, where E is any unimodular ($|\det E| = 1$) integer matrix. Figure 2.1 shows an example of a lattice in two dimensions, with basis vectors $\boldsymbol{v}_1 = [2X, 0]^T$ and $\boldsymbol{v}_2 = [X, Y]^T$. The sampling matrix in this case is

$$V_\Lambda = \begin{bmatrix} 2X & X \\ 0 & Y \end{bmatrix}.$$

A *unit cell* of a lattice Λ is a set $\mathcal{P} \subset \mathbb{R}^D$ such that copies of $\mathcal{P}$ centered on each lattice point tile the whole space without overlap: $(\mathcal{P} + \boldsymbol{s}_1) \cap (\mathcal{P} + \boldsymbol{s}_2) = \emptyset$ for $\boldsymbol{s}_1, \boldsymbol{s}_2 \in \Lambda$, $\boldsymbol{s}_1 \neq \boldsymbol{s}_2$, and $\cup_{\boldsymbol{s} \in \Lambda} (\mathcal{P} + \boldsymbol{s}) = \mathbb{R}^D$. The volume of a unit cell is $d(\Lambda) = |\det V|$, which is independent of the particular choice of sampling matrix. We can imagine that there is a region congruent to $\mathcal{P}$ of volume $d(\Lambda)$ associated with each sample in Λ, so that $d(\Lambda)$ is the reciprocal of the sampling density. The unit cell of a lattice is not unique. In Fig. 2.1, the shaded hexagonal region centered at the origin is a unit cell, of area $d(\Lambda) = 2XY$. The shaded parallelogram in the upper right is also a possible unit cell.

Most sampling structures of interest for time-varying imagery can be constructed using a lattice. In the case of 3D sampling, the sampling structure can be the intersection of $\mathcal{W}_T$ with a lattice, or in a few cases, with the union of two or more shifted lattices. The latter case occurs relatively infrequently (although there are several practical situations where it is used) and so the discussion here is limited to sampling on lattices. The theory of sampling on the union of shifted lattices (cosets) can be found in [1]. In the case of one or two-dimensional (partial) sampling ($D = 1$ or 2), the sampling structure can be constructed as the Cartesian product of a D-dimensional lattice and a continuous $(3 - D)$ dimensional space. For one-dimensional temporal sampling, the 1D lattice is $\Lambda_t = \{nT \mid n \in \mathbb{Z}\}$, where T is the frame period. The sampling structure is then $\mathcal{W} \times \Lambda_t = \{(\boldsymbol{x}, t) \mid \boldsymbol{x} \in \mathcal{W}, t \in \Lambda_t\}$. For two-dimensional vertical-temporal sampling (scanning) using a 2D lattice Λ_{yt}, the sampling structure is $\mathcal{W}_T \cap (\mathcal{H} \times \Lambda_{yt})$, where $\mathcal{H}$ is a one-dimensional subspace of $\mathbb{R}^3$ parallel to the scanning lines. In video systems, the scanning spot is moving down as it scans from left to right, and of course is moving forward in time. Thus, $\mathcal{H}$ has both a vertical and temporal tilt, but this effect is minor and can usually be ignored; we assume that $\mathcal{H}$ is the line $y = 0, t = 0$. Many digital video signals are obtained by three-dimensional subsampling of signals that have initially been sampled with one or two-dimensional sampling as above. Although the sampling structure is space limited, the analysis is often simplified if the sampling structure is assumed to be of infinite spatial extent, with the image either set to zero outside of $\mathcal{W}_T$ or replicated periodically in some way.

Much insight into the effect of sampling time-varying images on a lattice can be achieved by studying the problem in the frequency domain. To do this, we introduce the Fourier transform for signals defined on different domains. For a continuous signal f_c, the Fourier transform is given by

$$F_c(u, v, w) = \iiint f_c(x, y, t) \exp[-j2\pi(ux + vy + wt)]\, dx\, dy\, dt \tag{2.2}$$

or more compactly, setting $\boldsymbol{u} = (u, v, w)$ and $\boldsymbol{s} = (x, y, t)$,

$$F_c(\boldsymbol{u}) = \int_{\mathcal{W}_T} f_c(\boldsymbol{s}) \exp(-j2\pi \boldsymbol{u} \cdot \boldsymbol{s})\, d\boldsymbol{s}, \quad \boldsymbol{u} \in \mathbb{R}^3. \tag{2.3}$$

The variables u and v are horizontal and vertical spatial frequencies in cycles/picture height (c/ph), and w is temporal frequency in Hz.

Similarly, a discrete signal $f(\boldsymbol{s})$, $\boldsymbol{s} \in \Lambda$ has a lattice Fourier transform (or discrete space-time Fourier transform)

$$F(\boldsymbol{u}) = \sum_{\boldsymbol{s} \in \Lambda} f(\boldsymbol{s}) \exp(-j2\pi \boldsymbol{u} \cdot \boldsymbol{s}), \quad \boldsymbol{u} \in \mathbb{R}^3. \tag{2.4}$$

With this nonnormalized definition, both $\boldsymbol{s}$ and $\boldsymbol{u}$ have the same units as in Eq. (2.3). As with the 1D discrete-time Fourier transform, the lattice Fourier transform is periodic. If $\boldsymbol{k}$ is an element of $\mathbb{R}^3$ such that $\boldsymbol{k} \cdot \boldsymbol{s} \in \mathbb{Z}$ for all $\boldsymbol{s} \in \Lambda$, then $F(\boldsymbol{u} + \boldsymbol{k}) = F(\boldsymbol{u})$. It can be shown that $\{\boldsymbol{k} \mid \boldsymbol{k} \cdot \boldsymbol{s} \in \mathbb{Z} \text{ for all } \boldsymbol{s} \in \Lambda\}$ is a lattice called the *reciprocal lattice* Λ^*, and that if V is a sampling matrix for Λ, then $\Lambda^* = \mathrm{LAT}((V^T)^{-1})$. Thus, $F(\boldsymbol{u})$ is completely specified by its values in a unit cell of Λ^*.

For partially sampled signals, a mixed Fourier transform is required. For the examples of temporal and vertical-temporal sampling mentioned previously, these Fourier transforms are

$$F(\boldsymbol{u}, w) = \int_{\mathcal{W}} \sum_{n} f(\boldsymbol{x}, nT) \exp[-j2\pi(\boldsymbol{u} \cdot \boldsymbol{x} + wnT)]\, d\boldsymbol{x} \tag{2.5}$$

and

$$F(u, v, w) = \int_{\mathcal{H}} \sum_{(y,t) \in \Lambda_{yt}} f(x, y, t) \exp[-j2\pi(ux + vy + wt)]\, dx. \tag{2.6}$$

These Fourier transforms are periodic in the temporal frequency domain (with periodicity $1/T$) and in the vertical-temporal frequency domain (with periodicity lattice Λ_{yt}^*), respectively.

The terminology is illustrated with two examples that will be discussed in more detail further on. Figure 2.2 shows two vertical-temporal sampling lattices: a rectangular lattice Λ_R in Fig. 2.2(a) and a hexagonal lattice Λ_H in Fig. 2.2(b). These correspond to progressive scanning and interlaced scanning, respectively, in video systems. Possible sampling matrices for the two lattices are

$$V_R = \begin{bmatrix} Y & 0 \\ 0 & T \end{bmatrix} \quad \text{and} \quad V_H = \begin{bmatrix} Y & 0 \\ T/2 & T \end{bmatrix}. \tag{2.7}$$

Both lattices have the same sampling density, with $d(\Lambda_R) = d(\Lambda_H) = YT$. Figure 2.3 shows the reciprocal lattices Λ_R^* and Λ_H^* with several possible unit cells.

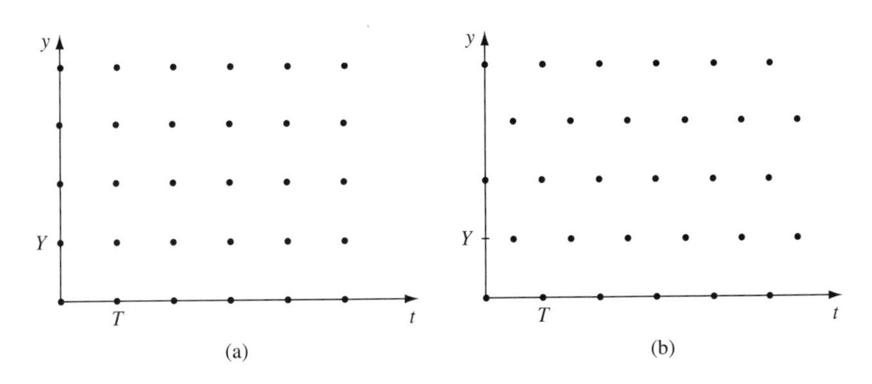

FIGURE 2.2

Two-dimensional vertical-temporal lattices. (a) Rectangular lattice Λ_R. (b) Hexagonal lattice Λ_H.

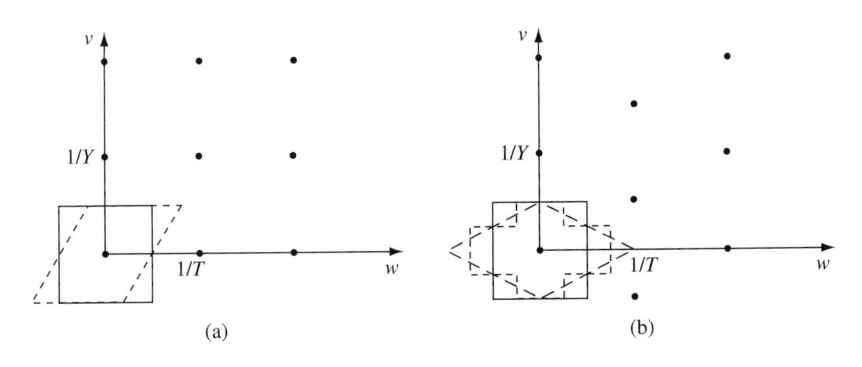

FIGURE 2.3

Reciprocal lattices of the two-dimensional vertical-temporal lattices of Fig. 2.2 with several possible unit cells. (a) Rectangular lattice Λ_R^*. (b) Hexagonal lattice Λ_H^*.

2.3 SAMPLING AND RECONSTRUCTION OF CONTINUOUS TIME-VARYING IMAGERY

The process for sampling a time-varying image can be approximated by the model shown in Fig. 2.4. In this model, the light arriving on the sensor is collected and weighted in space and time by the sensor aperture $a(\mathbf{s})$ to give the output

$$f_{ca}(\mathbf{s}) = \int_{\mathbb{R}^3} f_c(\mathbf{s}+\mathbf{s}')a(\mathbf{s}')\,d\mathbf{s}', \tag{2.8}$$

where it is assumed, here, that the sensor aperture is space and time invariant. The resulting signal $f_{ca}(\mathbf{s})$ is then sampled in an ideal fashion on the sampling structure Ψ,

$$f(\mathbf{s}) = f_{ca}(\mathbf{s}), \qquad \mathbf{s} \in \Psi. \tag{2.9}$$

By defining $h_a(\mathbf{s}) = a(-\mathbf{s})$, it is seen that the aperture weighting is a linear shift-invariant filtering operation, that is, the convolution of $f_c(\mathbf{s})$ with $h_a(\mathbf{s})$

$$f_{ca}(\mathbf{s}) = \int_{\mathbb{R}^3} f_c(\mathbf{s}-\mathbf{s}')h_a(\mathbf{s}')\,d\mathbf{s}'. \tag{2.10}$$

Thus, if $f_c(\mathbf{s})$ has a Fourier transform $F_c(\mathbf{u})$, then $F_{ca}(\mathbf{u}) = F_c(\mathbf{u})H_a(\mathbf{u})$, where $H_a(\mathbf{u})$ is the Fourier transform of the aperture impulse response. In typical acquisition systems, the sampling aperture can be modeled as a rectangular or Gaussian function.

If the sampling structure is a lattice Λ, then the effect in the frequency domain of the sampling operation is given by [1]

$$F(\mathbf{u}) = \frac{1}{d(\Lambda)} \sum_{\mathbf{k}\in\Lambda^*} F_{ca}(\mathbf{u}+\mathbf{k}), \tag{2.11}$$

in other words, the continuous signal spectrum $F_{ca}(\mathbf{u})$ is replicated on the points of the reciprocal lattice. The terms in the sum of Eq. (2.11), other than for $\mathbf{k} = \mathbf{0}$, are referred to as *spectral repeats.* There are two main consequences of the sampling process. The first is that these spectral repeats, if not removed by the display/viewer system, may be visible in the form of flicker, line structure, or dot patterns. The second is that, if the regions of support of $F_{ca}(\mathbf{u})$ and $F_{ca}(\mathbf{u}+\mathbf{k})$ have nonzero intersection for some values $\mathbf{k} \in \Lambda^*$, we

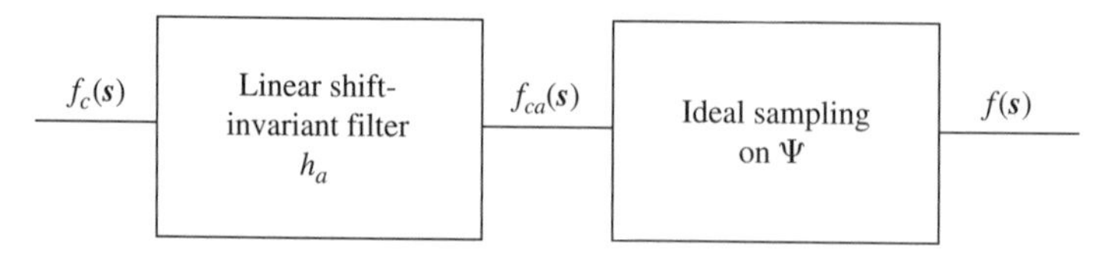

FIGURE 2.4

System for sampling a time-varying image.

have aliasing; a frequency $\boldsymbol{u}_a$ in this intersection can represent both the frequencies $\boldsymbol{u}_a$ and $\boldsymbol{u}_a - \boldsymbol{k}$ in the original signal. Thus, to avoid aliasing, the spectrum $F_{ca}(\boldsymbol{u})$ should be confined to a unit cell of Λ^*; this can be accomplished to some extent by the sampling aperture h_a. Aliasing is particularly problematic because once introduced it is difficult to remove, since there is more than one acceptable interpretation of the observed data. Aliasing is a familiar effect that tends to be localized to those regions of the image with high frequency details. It can be seen as moiré patterns in such periodic-like patterns as fishnets and venetian blinds, and as staircase-like effects on high-contrast oblique edges. The aliasing is particularly visible and annoying when these patterns are moving. Aliasing is controlled by selecting a sufficiently dense sampling structure and through the prefiltering effect of the sampling aperture.

If the support of $F_{ca}(\boldsymbol{u})$ is confined to a unit cell $\mathcal{P}^*$ of Λ^*, then it is possible to reconstruct f_{ca} exactly from the samples. In this case, we have

$$F_{ca}(\boldsymbol{u}) = \begin{cases} d(\Lambda)F(\boldsymbol{u}) & \text{if } \boldsymbol{u} \in \mathcal{P}^* \\ 0 & \text{if } \boldsymbol{u} \notin \mathcal{P}^* \end{cases} \tag{2.12}$$

and it follows that

$$f_{ca}(\boldsymbol{s}) = \sum_{\boldsymbol{s}' \in \Lambda} f(\boldsymbol{s}')t(\boldsymbol{s} - \boldsymbol{s}'), \tag{2.13}$$

where

$$t(\boldsymbol{s}) = d(\Lambda) \int_{\mathcal{P}^*} \exp(j2\pi \boldsymbol{u} \cdot \boldsymbol{s})\, d\boldsymbol{u} \tag{2.14}$$

is the impulse response of an ideal lowpass filter (with sampled input and continuous output) having passband $\mathcal{P}^*$. This is the multidimensional version of the familiar Sampling Theorem.

In practical systems, the reconstruction is achieved by

$$\widehat{f_{ca}}(\boldsymbol{s}) = \sum_{\boldsymbol{s}' \in \Lambda} f(\boldsymbol{s}')d(\boldsymbol{s} - \boldsymbol{s}') \tag{2.15}$$

where $d(\boldsymbol{s})$ is the display aperture, which generally bears little resemblance to the ideal $t(\boldsymbol{s})$ of Eq. (2.14). The display aperture is usually separable in space and time, $d(\boldsymbol{s}) = d_s(x, y)d_t(t)$, where $d_s(x, y)$ may be Gaussian or rectangular, and $d_t(t)$ may be exponential or rectangular, depending on the type of display system. In fact, a large part of the reconstruction filtering is often left to the spatiotemporal response of the human visual system. The main requirement is that the first temporal frequency repeat at zero spatial frequency (at $1/T$ for progressive scanning and $2/T$ for interlaced scanning, Fig. 2.2) be at least 50 Hz for large area flicker to be acceptably low.

If the display aperture is the ideal lowpass filter specified by Eq. (2.14), then the optimal sampling aperture is also an ideal lowpass filter with passband $\mathcal{P}^*$; neither of these are realizable in practice. If the actual aperture of a given display device operating on a lattice Λ is given, it is possible to determine the optimal sampling aperture according to a weighted-squared-error criterion [3]. This optimal sampling aperture, which will not be an ideal lowpass filter, is similarly not physically realizable, but it could at least form the design objective rather than the inappropriate ideal lowpass filter.

If sampling is performed in only one or two dimensions, the spectrum is replicated in the corresponding frequency dimensions. For the two cases of temporal and vertical-temporal sampling, respectively, we obtain

$$F(\boldsymbol{u}, w) = \frac{1}{T} \sum_{l=-\infty}^{\infty} F_{ca}(\boldsymbol{u}, w + \frac{l}{T}) \tag{2.16}$$

and

$$F(u, v, w) = \frac{1}{d(\Lambda_{yt})} \sum_{\boldsymbol{k} \in \Lambda_{yt}^*} F_{ca}(u, (v, w) + \boldsymbol{k}). \tag{2.17}$$

Consider first the case of pure temporal sampling, as in motion-picture film. The main parameters in this case are the sampling period T and the temporal aperture. As shown in Eq. (2.16), the signal spectrum is replicated in temporal frequency at multiples of $1/T$. In analogy with one-dimensional signals, one might think that the time-varying image should be bandlimited in temporal frequency to $1/2T$ before sampling. However, this is not the case. To illustrate, consider the spectrum of an image undergoing translation with constant velocity ν. This can model the local behavior in a large class of time-varying imagery. The assumption implies that $f_c(\boldsymbol{x}, t) = f_{c0}(\boldsymbol{x} - \nu t)$, where $f_{c0}(\boldsymbol{x}) = f_c(\boldsymbol{x}, 0)$. A straightforward analysis [4] shows that $F_c(\boldsymbol{u}, w) = F_{c0}(\boldsymbol{u})\delta(\boldsymbol{u} \cdot \nu + w)$, where $\delta(\cdot)$ is the Dirac delta function. Thus, the spectrum of the time-varying image is not spread throughout spatiotemporal frequency space but rather it is concentrated around the plane $\boldsymbol{u} \cdot \nu + w = 0$. When this translating image is sampled in the temporal dimension, these planes are parallel to each other and do not intersect, that is, there is no aliasing, even if the temporal bandwidth far exceeds $1/2T$. This is most easily illustrated in two dimensions. Consider the case of vertical motion only. Figure 2.5 shows the vertical-temporal projection of the spectrum of the sampled image for different velocities ν. Assume that the image is vertically bandlimited to B c/ph. It follows that, when the vertical velocity reaches $1/2TB$ picture heights per second (ph/s), the spectrum will extend out to the temporal frequency of $1/2T$ as shown in Fig. 2.5(b). At twice that velocity ($1/TB$), it would extend to a temporal frequency of $1/T$ which might suggest severe aliasing. However, as seen in Fig. 2.5(c), there is no spectral overlap. To reconstruct the continuous signal correctly, however, a vertical-temporal filtering adapted to the velocity is required. Bandlimiting the signal to a temporal frequency of $1/2T$ before sampling would effectively cut the vertical resolution in half for this velocity. Note that the velocities mentioned above are not really very high. To consider some typical numbers, if $T = 1/24$ s, as in film, and $B = 500$ c/ph (corresponding to 1000 scanning lines), the velocity $1/2TB$ is about $1/42$ ph/s. It should be noted that, if the viewer is tracking the vertical movement, the spectrum of the image on the retina will be far less tilted, again arguing against sharp temporal bandlimiting. (This is in fact a kind of motion-compensated filtering by the

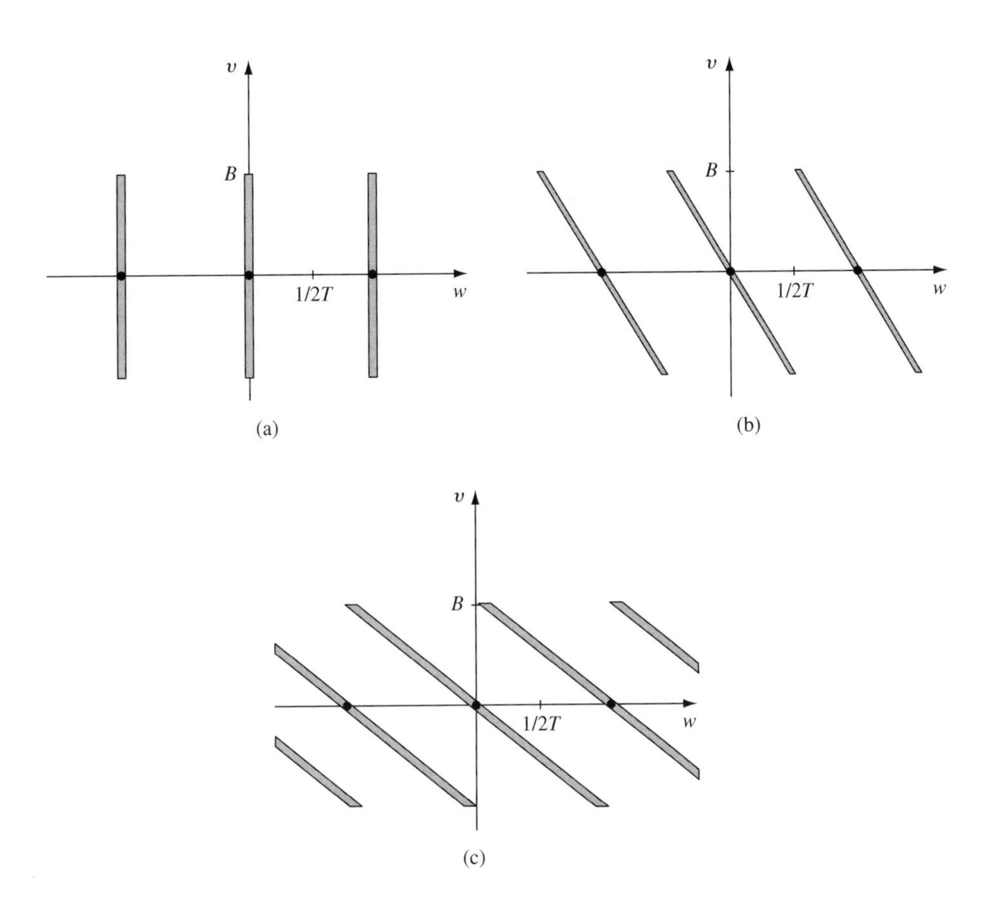

FIGURE 2.5

Vertical-temporal projection of the spectrum of temporally sampled time-varying image with vertical motion of velocity ν. (a) $\nu = 0$. (b) $\nu = 1/2TB$. (c) $\nu = 1/TB$.

visual system.) The temporal camera aperture can roughly be modeled as the integration of f_c for a period $T_a \leq T$. The choice of the value of the parameter T_a is a compromise between motion blur and signal-to-noise ratio.

Similar arguments can be made in the case of the two most popular vertical-temporal scanning structures, progressive scanning, and interlaced scanning. Referring to Fig. 2.6, the vertical-temporal spectrum of a vertically translating image at the same three velocities (assuming that $1/Y = 2B$) is shown for these two scanning structures. For progressive scanning, there continues to be no spectral overlap, while for interlaced scanning, the spectral overlap can be severe at certain velocities (e.g., 1/TB as in Fig. 2.6(f)). This is a strong advantage for progressive scanning. Another disadvantage of interlaced scanning is that each field is spatially undersampled and pure spatial processing or interpolation is very difficult. An illustration in three dimensions of some of these ideas can be found in [5].

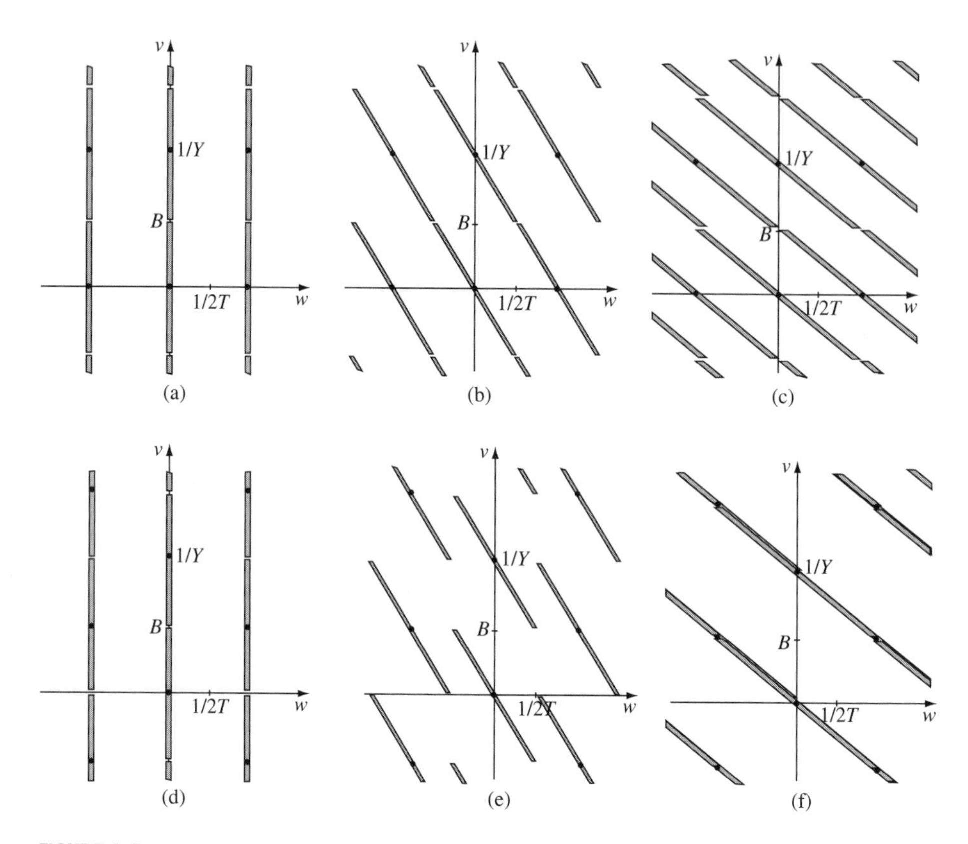

FIGURE 2.6

Vertical-temporal projection of spectrum of vertical-temporal sampled time-varying image with progressive and interlaced scanning. Progressive: (a) $\nu = 0$. (b) $\nu = 1/2TB$. (c) $\nu = 1/TB$. Interlaced: (d) $\nu = 0$. (e) $\nu = 1/2TB$. (f) $\nu = 1/TB$.

2.4 SAMPLING STRUCTURE CONVERSION

There are numerous spatiotemporal sampling structures used for the digital representation of time-varying imagery. However, the vast majority of those in use fall into one of two categories corresponding to progressive or interlaced scanning with aligned horizontal sampling. This corresponds to sampling matrices of the form

$$\begin{bmatrix} X & 0 & 0 \\ 0 & Y & 0 \\ 0 & 0 & T \end{bmatrix} \quad \text{or} \quad \begin{bmatrix} X & 0 & 0 \\ 0 & Y & 0 \\ 0 & T/2 & T \end{bmatrix},$$

respectively. A three-dimensional view of these two sampling lattices is shown in Fig. 2.7. It can be observed how the odd numbered horizontal lines in each frame from the

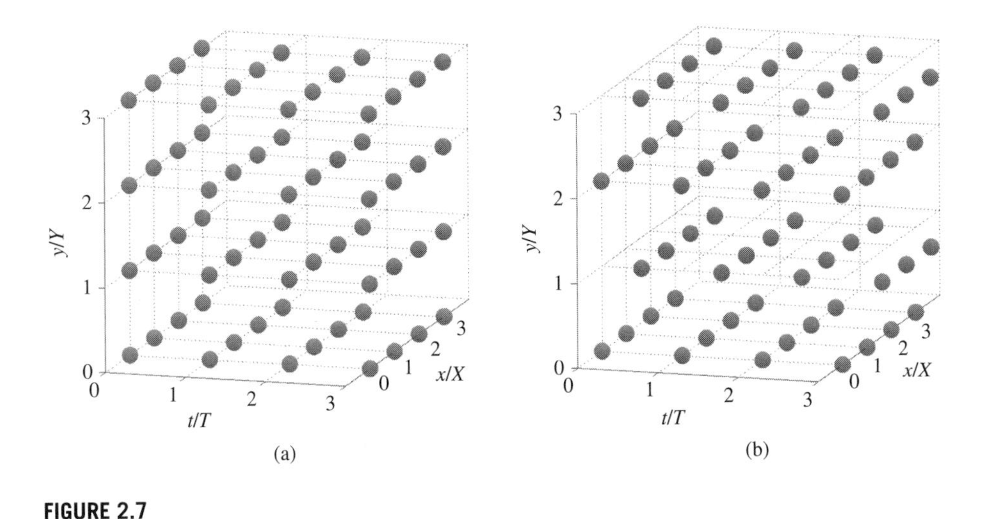

FIGURE 2.7

Three-dimensional view of spatiotemporal sampling lattices. (a) Progressive. (b) Interlaced.

progressive lattice (y/Y odd) in Fig. 2.7(a) have been delayed temporally by $T/2$ for the interlaced lattice of Fig. 2.7(b).

Table 2.1 shows the parameters for a number of commonly used sampling structures covering a broad range of applications from low-resolution QCIF used in videophone to HDTV and digitized IMAX film (the popular large-format film, about 70 mm by 52 mm, used by Imax Corporation). Considering the many additional computer and digital cinema formats, this is but a small sample of the current formats in use. Note that of the formats in the table, only HDTV and IMAX formats have $X = Y$ (that is, square pixels). It is frequently required to convert a time-varying image sampled on one such structure to another. The first applications came up in converting from film to video and converting among international analog television formats, such as PAL, NTSC, and SECAM. An input image sequence $f(\mathbf{x})$ sampled on lattice Λ_1 is to be converted to the output sequence $f_o(\mathbf{x})$ sampled on the lattice Λ_2. This situation is illustrated in Fig. 2.8. The continuous signal $f_c(\mathbf{x})$ is acquired on the lattice Λ_1 using a physical camera modeled as in Fig. 2.4 with impulse response $h_a(\mathbf{x})$ to yield $f(\mathbf{x})$. It is desired to estimate the signal $f_o(\mathbf{x})$ that would have been obtained if $f_c(\mathbf{x})$ was sampled on the lattice Λ_2 with an ideal or theoretical camera having impulse response $h_{oa}(\mathbf{x})$. Note that since this camera is *theoretical*, the impulse response $h_{oa}(\mathbf{x})$ does not have to be realizable with any particular technology. It can be optimized to give the best displayed image on Λ_2 [3]. A system $\mathcal{H}$, which can be linear or nonlinear, is then required to estimate $f_o(\mathbf{x})$ from $f(\mathbf{x})$.

Besides converting between different standards, sampling structure conversion can also be incorporated into the acquisition or display portions of an imaging system to compensate for the difficulty in performing adequate prefiltering with the camera aperture or adequate postfiltering with the display aperture. Specifically, the time-varying

TABLE 2.1 Parameters of several common spatiotemporal sampling structures.

System	X	Y	T	Structure	Aspect Ratio
QCIF	$\frac{1}{176}$ pw = $\frac{1}{132}$ ph	$\frac{1}{144}$ ph	$\frac{1}{10}$ s	P	4:3
CIF	$\frac{1}{352}$ pw = $\frac{1}{264}$ ph	$\frac{1}{288}$ ph	$\frac{1}{15}$ s	P	4:3
ITU-R-601 (30)	$\frac{1}{720}$ pw = $\frac{1}{540}$ ph	$\frac{1}{480}$ ph	$\frac{1}{29.97}$ s	I	4:3
ITU-R-601 (25)	$\frac{1}{720}$ pw = $\frac{1}{540}$ ph	$\frac{1}{576}$ ph	$\frac{1}{25}$ s	I	4:3
HDTV-P	$\frac{1}{1280}$ pw = $\frac{1}{720}$ ph	$\frac{1}{720}$ ph	$\frac{1}{60}$ s	P	16:9
HDTV-I	$\frac{1}{1920}$ pw = $\frac{1}{1080}$ ph	$\frac{1}{1080}$ ph	$\frac{1}{30}$ s	I	16:9
IMAX	$\frac{1}{4096}$ pw = $\frac{1}{3002}$ ph	$\frac{1}{3002}$ ph	$\frac{1}{24}$ s	P	1.364

P indicates progressive scanning and I indicates interlaced scanning.

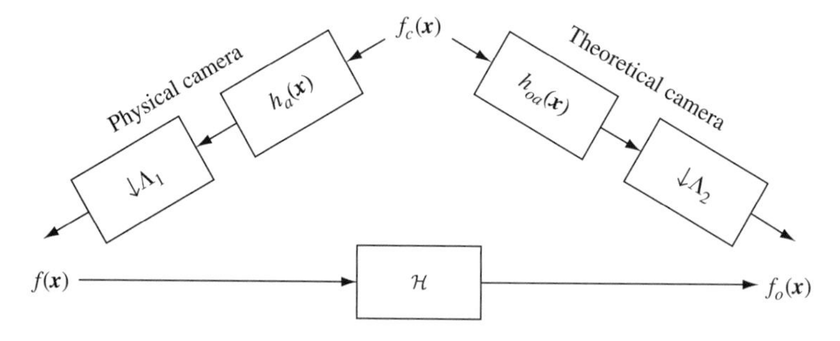

FIGURE 2.8

Acquisition models for the observed signal $f(\boldsymbol{x})$ on Λ_1 and the desired output signal $f_o(\boldsymbol{x})$ on Λ_2.

image can initially be sampled at a higher density than required, using the camera aperture as prefilter, and then downsampled to the desired structure using digital prefiltering, which offers much more flexibility. Similarly, the image can be upsampled for the display device using digital filtering, so that the subsequent display aperture has a less critical task to perform.

2.4.1 Frame-rate Conversion

Consider first the case of pure frame-rate conversion. This applies when both the input and the output sampling structures are separable in space and time with the same spatial sampling structure, and where spatial aliasing is assumed to be negligible. The temporal sampling period is to be changed from T_1 to T_2. This situation corresponds to input and output sampling lattices

$$\Lambda_1 = \begin{bmatrix} v_{11} & v_{12} & 0 \\ 0 & v_{22} & 0 \\ 0 & 0 & T_1 \end{bmatrix}, \quad \Lambda_2 = \begin{bmatrix} v_{11} & v_{12} & 0 \\ 0 & v_{22} & 0 \\ 0 & 0 & T_2 \end{bmatrix}. \tag{2.18}$$

2.4.1.1 Pure Temporal Interpolation

The most straightforward approach is pure temporal interpolation, where a temporal resampling is performed independently at each spatial location $\boldsymbol{x}$. A typical application for this is increasing the frame rate in motion picture film from 24 frames/s to 48 or 60 frames/s, giving significantly better motion rendition. Using linear filtering, the interpolated image sequence is given by

$$f_o(\boldsymbol{x}, nT_2) = \sum_m f(\boldsymbol{x}, mT_1)h(nT_2 - mT_1), \tag{2.19}$$

where $h(t)$ is a continuous-time function that we refer to as the interpolation kernel. If the temporal spectrum of the underlying continuous time-varying image satisfies the Nyquist criterion, the output points can be computed by ideal *sinc* interpolation:

$$h(t) = \frac{\sin(\pi t/T_1)}{\pi t/T_1}. \tag{2.20}$$

However, aside from the fact that this filter is unrealizable, it is unlikely, and in fact undesirable according to the discussion of Section 3, for the temporal spectrum to satisfy the Nyquist criterion. Thus, high order interpolation kernels that approximate Eq. (2.20) are not found to be useful and are rarely used. Instead, simple low-order interpolation kernels are frequently applied. Examples are zero-order and linear (straight-line) interpolation kernels given by

$$h(t) = \begin{cases} 1 & \text{if } 0 \leq t \leq T_1 \\ 0 & \text{otherwise} \end{cases} \tag{2.21}$$

and

$$h(t) = \begin{cases} 1 - |t|/T_1 & \text{if } 0 \leq |t| \leq T_1 \\ 0 & \text{otherwise,} \end{cases} \tag{2.22}$$

respectively. Note that Eq. (2.22) defines a noncausal filter and that in practice a delay of T_1 must be introduced. Zero-order hold is also called frame repeat and is the method used in film projection to go from 24 to 48 frames/s. These simple interpolators work well if there is little or no motion, but as the amount of motion increases they will not adequately remove spectral repeats causing effects, such as jerkiness, and they may also remove useful information, introducing blurring. The problems with pure temporal

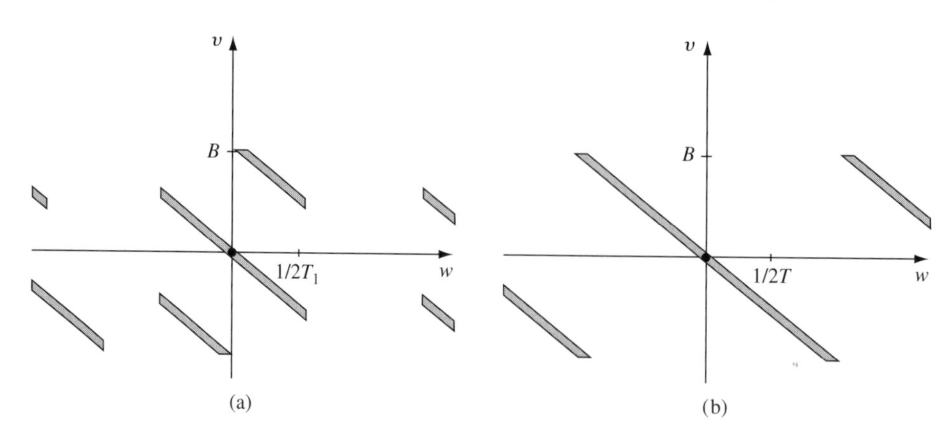

FIGURE 2.9

Frequency domain interpretation of 2:1 temporal interpolation of an image with vertical velocity $1/TB$. (a) Pure temporal interpolation. (b) Motion-compensated interpolation.

interpolation can easily be illustrated for the image corresponding to Fig. 2.5(c) for the case of doubling the frame rate, that is, $T_2 = T_1/2$. Using a one-dimensional temporal lowpass filter with cutoff at about $1/2T_1$ removes the desired high vertical frequencies in the baseband signal above $B/2$ (motion blur) and leaves undesirable aliasing at high vertical frequencies, as shown in Fig. 2.9(a).

2.4.1.2 Motion-compensated Interpolation

It is clear that to correctly deal with a situation, such as in Fig. 2.4(c), it is necessary to adapt the interpolation to the local orientation of the spectrum, and thus to the velocity, as suggested in Fig. 2.9(b). This is called motion-compensated interpolation. An auxiliary motion analysis process determines information about local motion in the image and attempts to track the trajectory of scene points over time. Specifically, suppose we wish to estimate the signal value at position $\mathbf{x}$ at time nT_2 from neighboring frames at times mT_1. We can assume that the scene point imaged at position $\mathbf{x}$ at time nT_2 was imaged at position $\mathbf{c}(mT_1;\mathbf{x},nT_2)$ at time mT_1 [6]. If we know $\mathbf{c}$ exactly, we can compute

$$f_o(\mathbf{x},nT_2) = \sum_m f(\mathbf{c}(mT_1;\mathbf{x},nT_2),mT_1)h(nT_2 - mT_1). \tag{2.23}$$

Since we assume that $f(\mathbf{x},t)$ is very slowly varying along the motion trajectory, a simple filter, such as the linear interpolator of Eq. (2.22), would probably do very well. Of course, we do not know $\mathbf{c}(mT_1;\mathbf{x},nT_2)$, so we must estimate it. Furthermore, since the position $(\mathbf{c}(mT_1;\mathbf{x},nT_2),mT_1)$ probably does not lie on the input lattice Λ_1, $f(\mathbf{c}(mT_1;\mathbf{x},nT_2),mT_1)$ must be spatially interpolated from its neighbors. If spatial aliasing is low as we have assumed, this interpolation can be done well.

If a two-point temporal interpolation is used, we only need to find the correspondence between the point at $(\mathbf{x}, nT_2)$ and points in the frames at times lT_1 and $(l+1)T_1$, where $lT_1 \leq nT_2$ and $(l+1)T_1 > nT_2$. This is specified by the backward and forward displacements

$$\mathbf{d}_b(\mathbf{x}, nT_2) = \mathbf{x} - \mathbf{c}(lT_1; \mathbf{x}, nT_2) \tag{2.24}$$

$$\mathbf{d}_f(\mathbf{x}, nT_2) = \mathbf{c}((l+1)T_1; \mathbf{x}, nT_2) - \mathbf{x}, \tag{2.25}$$

respectively. The interpolated value is then given by

$$\begin{aligned} f_o(\mathbf{x}, nT_2) &= f(\mathbf{x} - \mathbf{d}_b(\mathbf{x}, nT_2), lT_1)h(nT_2 - lT_1) \\ &+ f(\mathbf{x} + \mathbf{d}_f(\mathbf{x}, nT_2), (l+1)T_1)h(nT_2 - (l+1)T_1). \end{aligned} \tag{2.26}$$

There are a number of key design issues in this process. The main one relates to the complexity and precision of the motion estimator. Since the image at time nT_2 is not available, the trajectory must be estimated from the existing frames at times mT_1, and often just from lT_1 and $(l+1)T_1$ as defined above. In the latter case, the forward and backward displacements will be collinear. We can assume that better motion estimators will lead to better motion-compensated interpolation. However, the tradeoff between complexity and performance must be optimized for each particular application. For example, block-based motion estimation (say one motion vector per 16×16 block) with accuracy rounded to the nearest pixel location will give very good results in large moving areas with moderate detail, giving significant overall improvement for most sequences. However, areas with complex motion and higher detail may continue to show quite visible artifacts, and more accurate motion estimates would be required to get good performance in these areas. Better motion estimates could be achieved with smaller blocks, parametric motion models, or dense motion estimates, for example. Motion estimation is treated in detail in Chapter 3. Some specific considerations related to estimating motion trajectories passing through points in between frames in the input sequence can be found in [6].

If the motion estimation method used sometimes yields unreliable motion vectors, it may be advantageous to be able to fall back to pure temporal interpolation. A test can be performed to determine whether pure temporal interpolation or motion-compensated interpolation is expected to yield better results, for example, by comparing $|f(\mathbf{x}, (l+1)T_1) - f(\mathbf{x}, lT_1)|$ with $|f(\mathbf{x} + \mathbf{d}_f(\mathbf{x}, nT_2), (l+1)T_1) - f(\mathbf{x} - \mathbf{d}_b(\mathbf{x}, nT_2), lT_1)|$ either at a single point or over a small window. Then, the interpolated value can either be computed by the method suspected to be better or by an appropriate weighted combination of the two. A more elaborate measure of motion vector reliability giving good results, based on the *a posteriori* probability of motion vectors, can be found in [7].

Occlusions pose a particular problem since the pixel to be interpolated may be visible only in the previous frame (newly covered area) or in the subsequent frame (newly exposed area). In particular, if $|f(\mathbf{x} + \mathbf{d}_f(\mathbf{x}, nT_2), (l+1)T_1) - f(\mathbf{x} - \mathbf{d}_b(\mathbf{x}, nT_2), lT_1)|$ is relatively large, this may signal that $\mathbf{x}$ lies in an occlusion area. In this case, we may wish to use zero-order hold interpolation based on either the frame at lT_1 or at $(l+1)T_1$,

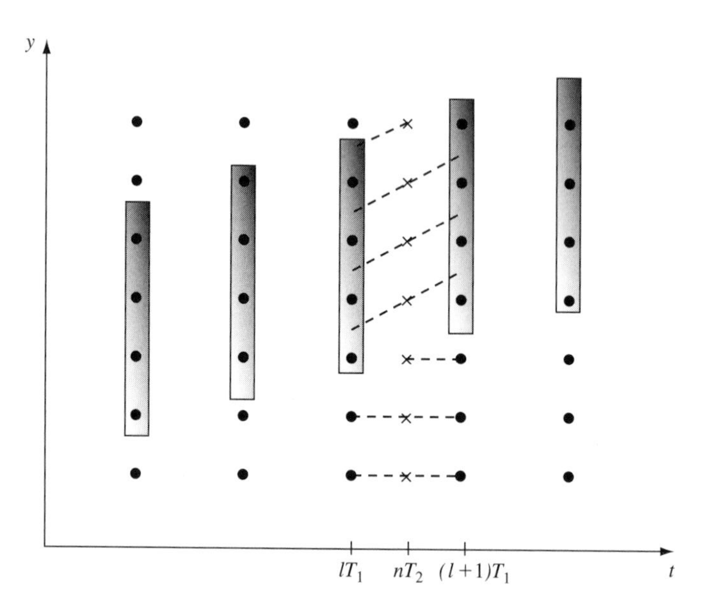

FIGURE 2.10

Example of motion-compensated temporal interpolation including occlusion handling.

according to some local analysis. Figure 2.10 depicts the motion-compensated interpolation of a frame midway between lT_1 and $(l+1)T_1$ including occlusion processing, where we assume that a single object is moving upward.

2.4.2 Spatiotemporal Sampling Structure Conversion

In this section, we consider the case where both the spatial and the temporal sampling structures are changed, and when one or both of the input and output sampling structures is not separable in space and time (usually because of interlace). If the input sampling structure Λ_1 is separable in space and time (as in Eq. (2.18)) and spatial aliasing is minimal, then the methods of the previous section can be combined with pure spatial interpolation. If we want to interpolate a sample at a time mT_1, we can use any suitable spatial interpolation. To interpolate at a sample at a time t that is not a multiple of T_1, the methods of the previous section can be applied.

The difficulties in spatiotemporal interpolation mainly arise when the input sampling structure Λ_1 is not separable in space and time, which is generally the case of interlace. This encompasses both interlaced-to-interlaced conversion, such as in conversion between NTSC and PAL television systems, and interlaced-to-progressive conversion (also called deinterlacing). The reason this introduces problems is that individual fields are undersampled, contrary to the assumption in all the previously discussed methods. Furthermore, as we have seen, there may also be significant aliasing in the spatiotemporal frequency domain due to vertical motion. Thus, a great deal of the research on spatiotemporal interpolation has been addressing these problems due to interlace, and a wide variety of techniques have been proposed, many of them very empirical in nature.

2.4.2.1 Deinterlacing

Deinterlacing generally refers to a 2:1 interpolation from an interlaced grid to a progressive grid with sampling lattices

$$\begin{bmatrix} X & 0 & 0 \\ 0 & Y & 0 \\ 0 & T/2 & T \end{bmatrix} \quad \text{and} \quad \begin{bmatrix} X & 0 & 0 \\ 0 & Y & 0 \\ 0 & 0 & T/2 \end{bmatrix},$$

respectively (see Fig. 2.11). Both input and output lattices consist of fields at time instants $mT/2$. However, because each input field is vertically undersampled, spatial interpolation alone is inadequate. Similarly, because of possible spatiotemporal aliasing and difficulties with motion estimation, motion-compensated interpolation alone is inadequate. Thus, the most successful methods use a nonlinear combination of spatially and temporally interpolated values, according to local measures of which is most reliable. For example, in Fig. 2.11, sample A might best be reconstructed using spatial interpolation, sample B with pure temporal interpolation, and sample C with motion-compensated temporal interpolation. Another sample like D may be reconstructed using a combination of spatial and motion-compensated temporal interpolation. See [8] for a detailed presentation and discussion of a wide variety of deinterlacing methods. It is shown there that some adaptive motion-compensated methods can give reasonably good deinterlacing results on a wide variety of moving and fixed imagery.

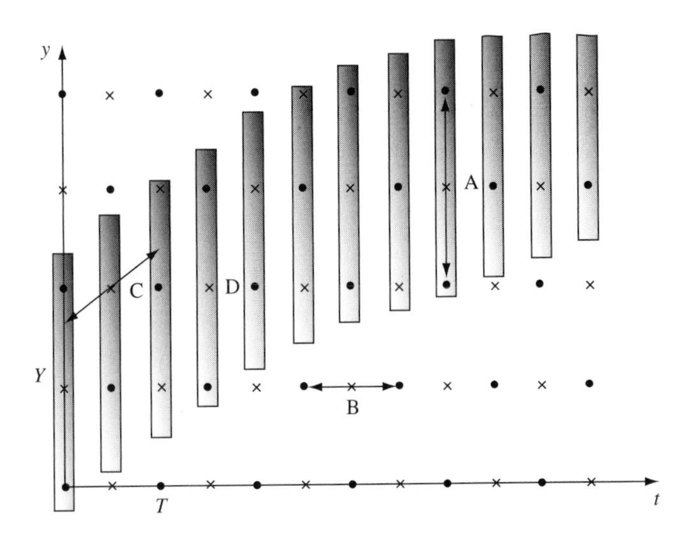

FIGURE 2.11

Input and output sampling structures for deinterlacing.

2.5 CONCLUSION

This chapter has provided an overview of the basic theory related to sampling and interpolation of time-varying imagery. In contrast to other types of signals, it has been shown that it is *not* desirable to limit the spectrum of the continuous signal to a *fixed* three-dimensional frequency band prior to sampling since this leads to excessive loss of spatial resolution. It is sufficient to ensure that the replicated spectra due to sampling do not overlap. However, optimal reconstruction requires the use of motion-compensated temporal interpolation.

The interlaced scanning structure that is widely used in video systems has a fundamental problem whereby aliasing in the presence of vertical motion is inevitable. This makes operations, such as motion estimation, coding, and so on, more difficult to accomplish. Thus, it is likely that interlaced scanning will gradually disappear as camera technology improves and the full spatial resolution desired can be obtained with frame rates of 50-60 Hz and above.

Spatiotemporal interpolation will remain an important technology to convert between the wide variety of scanning standards in both new and archival material. Research will continue into robust, low-complexity methods for motion-compensated temporal interpolation that can be incorporated into any receiver. Further work is also required to fully exploit the model of Fig. 2.8 or similar models in the video sampling structure conversion problem in ways similar to what has been done for still images [9].

REFERENCES

[1] E. Dubois. The sampling and reconstruction of time-varying imagery with application in video systems. *Proc. IEEE*, 73:502–522, 1985.

[2] T. Kalker. On multidimensional sampling. In *The Digital Signal Processing Handbook*, V. K. Madisetti and D. B. Williams, eds. chapter 4, pp. 4–1–4–21. CRC Press, Boca Raton, FL, 1998.

[3] H. A. Aly and E. Dubois. Design of optimal camera apertures adapted to display devices over arbitrary sampling lattices. *IEEE Signal Process. Lett.*, 11:443–445, 2004.

[4] E. Dubois. Motion-compensated filtering of time-varying images. *Multidimensional Syst. Signal Process.*, 3:211–239, 1992.

[5] B. Girod and R. Thoma. Motion-compensating field interpolation from interlaced and non-interlaced grids. In *Proc. SPIE Image Coding*, 594:186–193, 1985.

[6] E. Dubois and J. Konrad. Estimation of 2-D motion fields from image sequences with application to motion-compensated processing. In *Motion Analysis and Image Sequence Processing*, M. Sezan and R. Lagendijk, eds. chapter 3, pp. 53–87. Kluwer Academic Publishers, N1993.

[7] D. Wang, A. Vincent, and P. Blanchfield. Hybrid de-interlacing algorithm based on motion vector reliability. *IEEE Trans. Circuits Syst. Video Technol.*, 15:1019–1025, 2005.

[8] G. de Haan. Deinterlacing-an overview. *Proc. IEEE*, 86:1839–1857, 1998.

[9] H. A. Aly and E. Dubois. Image up-sampling using total-variation regularization with a new observation model. *IEEE Trans. Image Process.*, 14:1647–1659, 2005.

[10] P. Mertz and F. Gray. A theory of scanning and its relation to the characteristics of the transmitted signal in telephotography and television. *Bell Syst. Tech. J.*, 13:464–515, 1934.

[11] D. P. Petersen and D. Middleton. Sampling and reconstruction of wave-number-limited functions in n-dimensional euclidean spaces. *Inf. Control.*, 5:279–323, 1962.

[12] M. A. Isnardi. Modeling the television process. Technical Report 515, Research Laboratory of Electronics, Massachusetts Institute of Technology, Cambridge, MA, 1986.

[13] E. Dubois, G. de Haan, and T. Kurita, "Special issue on motion estimation and compensation technologies for standards conversion." *Signal Process: Image Commu.*, vol. 6, June 1994.

FURTHER INFORMATION

The classic paper on television scanning is [10]. The use of lattices for the study of spatiotemporal sampling was introduced in [11]. A detailed study of camera and display aperture models for television can be found in [12]. Research papers on spatiotemporal interpolation can be found regularly in the IEEE Transactions on Image Processing and IEEE Transactions on Circuits and Systems for Video Technology and Signal Processing and Image Communication. See [13] for a special issue on motion estimation and compensation for standards conversion.

CHAPTER

Motion Detection and Estimation

3

Janusz Konrad

Boston University, Department of Electrical and Computer Engineering, Boston

3.1 INTRODUCTION

Although a still image is a rich source of visual information, it is a sequence of such images that has captured the imagination of cinematographers, home video enthusiasts, and, more recently, video loggers (vloggers). The success of video as a medium is primarily due to the capture of motion; a single image provides snapshot of a scene, whereas a sequence of images also records scene's dynamics. The recorded motion is a very strong cue for human vision; we can easily recognize objects as soon as they move even if they are inconspicuous when still. Motion is equally important for video processing and compression for two reasons. First, motion carries information about spatiotemporal relationships between objects in the field of view of a camera. This information can be used in applications such as traffic monitoring or security surveillance (Chapter 19), for example, to identify objects that move or those entering/leaving the scene. Second, image properties, such as intensity or color, have a very high correlation in the direction of motion, that is, they do not change significantly when tracked over time (the color of a car does not change as it moves across the camera field of view). This can be used for the removal of temporal redundancy in video coding (Chapters 9–11). Motion can also be used in temporal filtering of video (Chapters 2 and 4); one-dimensional filter applied along a motion trajectory can reduce noise without spatially blurring a frame.

The above applications require that image points be identified as moving or stationary (surveillance), or that it be measured how they move (compression, filtering). The first task is often referred to as motion detection, whereas the latter as motion estimation. The goal of this chapter is to present today's most promising approaches to solving both. Note that only two-dimensional (2D) motion of intensity patterns in the image plane, often referred to as *apparent motion*, will be considered. Three-dimensional (3D) motion of objects will not be treated here. *Motion segmentation*, that is, the identification of groups of image points moving similarly, is discussed in Chapter 6.

The discussion of motion in this chapter will be carried out from the point of view of video processing and compression. Necessarily, the scope of methods reported will not be complete. To present the methods in a consistent fashion, a classification will be made based on models, estimation criteria, and search strategies used. This classification will be introduced for two reasons. First, it is essential for the understanding of methods described here and elsewhere in the literature. Second, it should help the reader in the development of his/her own motion detection or estimation methods.

In the next section, the notation is established, followed by a brief review of some tools needed. Then, in Section 3.3, motion detection is formulated as hypothesis testing, maximum a posteriori probability (MAP) estimation, and variational problem. In Section 3.4, motion estimation is described in two parts. First, models, estimation criteria, and search strategies are discussed. Then, five motion estimation algorithms are described in more detail, of which three are based on models supported by the current video compression standards. Both motion detection and estimation are illustrated by experimental results.

3.2 NOTATION AND PRELIMINARIES

Let $I : \Omega \times \mathcal{T} \rightarrow R^+$ be the intensity of image sequence defined over spatial domain Ω and temporal domain $\mathcal{T}$. Let $\boldsymbol{x} = (x_1, x_2)^T \in \Omega$ and $t \in \mathcal{T}$ denote spatial and temporal positions of a point in this sequence, respectively. In this chapter, both continuous and discrete representations of motion and images will be used. Let $I(\boldsymbol{x}, t)$ denote a continuous-coordinate representation of intensity at $(\boldsymbol{x}, t)$, and let I_t denote a complete image at time t. In the process of sampling, spatial pixel position $\boldsymbol{x}$ is approximated by $\boldsymbol{n} = (n_1, n_2)^T$, whereas temporal position t is approximated by t_k or k. With this notation, $I[\boldsymbol{n}, k]$ denotes a discrete-coordinate representation of $I(\boldsymbol{x}, t)$. The triplet $(n_1, n_2, k)^T$ belongs to a 3D sampling grid, for example a 3D lattice (Chapter 2). It is assumed here that images are either continuous or discrete simultaneously in position and in amplitude. Consequently, the same symbol I will be used for continuous and quantized intensities; the nature of I can be inferred from its argument (continuous-valued for $\boldsymbol{x}, t$ and quantized for $\boldsymbol{n}, k$).

Motion in continuous images can be described by *velocity* vector $\boldsymbol{v} = (v_1, v_2)^T$. While $\boldsymbol{v}(\boldsymbol{x})$ is a velocity at spatial position $\boldsymbol{x}$, $\boldsymbol{v}_t$ will denote a velocity field or motion field, that is, the set of all velocity vectors within the image, at time t. Often, the computation of this *dense* representation is replaced by the computation of a small number of motion parameters $\boldsymbol{b}_t$ with the benefit of reduced computational complexity. Then, $\boldsymbol{v}_t$ is approximated by $\boldsymbol{b}_t$ through a known transformation. For discrete images, the notion of velocity is replaced by *displacement* $\boldsymbol{d}$.

3.2.1 Binary Hypothesis Testing

Let y be an observation and let Y be the associated random variable. Suppose that there are two hypotheses H_0 and H_1 with corresponding probability distributions $P(Y = y|H_0)$

and $P(Y = y|H_1)$, respectively. The goal is to decide from which of the two distributions a given y is more likely to have been drawn. Clearly, four possibilities exist (true hypothesis/decision): H_0/H_0, H_0/H_1, H_1/H_0, and H_1/H_1. Although H_0/H_0 and H_1/H_1 correspond to correct choices, H_0/H_1 and H_1/H_0 are erroneous. To make a decision, a decision criterion is needed that attaches some relative importance to the four possible scenarios.

Under the Bayes criterion, two a priori probabilities π_0 and $\pi_1 = 1 - \pi_0$ are assigned to the two hypotheses H_0 and H_1, respectively, and a cost is assigned to each of the four scenarios listed above. Naturally, one would like to design a decision rule so that on average the cost associated with making a decision based on y is minimal. By computing an average risk, and by assuming that costs associated with erroneous decisions are higher than those associated with the corresponding correct decisions, it can be shown that an optimal decision can be made according to the following rule [1, Chapter 2]:

$$\frac{P(Y = y|H_1)}{P(Y = y|H_0)} \underset{H_0}{\overset{H_1}{\gtrless}} \vartheta \frac{\pi_0}{\pi_1}. \tag{3.1}$$

The quantity on the left is called the *likelihood ratio* and ϑ is a constant dependent on the costs of the four scenarios. Since these costs are determined in advance, ϑ is a fixed constant. If π_0 and π_1 are predetermined as well, the above hypothesis test compares the likelihood ratio with a fixed threshold. Alternatively, the prior probabilities can be made variable; variable-threshold hypothesis testing results. To simplify notation, we will subsequently use $P_0(y)$, and $P_1(y)$ to denote $P(Y = y|H_0)$ and $P(Y = y|H_1)$, respectively.

3.2.2 Markov Random Fields

A Markov random field (MRF) is a multidimensional random process, which generalizes the notion of a 1D Markov process. Below, some essential properties of MRFs are described; for a more detailed account the reader is referred to the literature (e.g., [2] and references therein).

Let Λ be a sampling grid in R^N and let $\eta(\boldsymbol{n})$ be a neighborhood of $\boldsymbol{n} \in \Lambda$, that is, a set of such $\boldsymbol{n}$'s that $\boldsymbol{n} \notin \eta(\boldsymbol{n})$ and $\boldsymbol{n} \in \eta(\boldsymbol{l}) \Leftrightarrow \boldsymbol{l} \in \eta(\boldsymbol{n})$. The first-order neighborhood consists of immediate top, bottom, left, and right neighbors of $\boldsymbol{n}$. Let $\mathcal{N}$ be a neighborhood system, that is, a collection of neighborhoods of all $\boldsymbol{n} \in \Lambda$.

A random field Υ over Λ is a multidimensional random process where each site $\boldsymbol{n} \in \Lambda$ is assigned a random variable. A random field Υ with the following properties:

1. $P(\Upsilon = \upsilon) > 0, \quad \forall \upsilon \in \Gamma$, and
2. $P(\Upsilon_{\boldsymbol{n}} = \upsilon_{\boldsymbol{n}}|\Upsilon_{\boldsymbol{l}} = \upsilon_{\boldsymbol{l}}, \forall \boldsymbol{l} \neq \boldsymbol{n}) = P(\Upsilon_{\boldsymbol{n}} = \upsilon_{\boldsymbol{n}}|\Upsilon_{\boldsymbol{l}} = \upsilon_{\boldsymbol{l}}, \forall \boldsymbol{l} \in \eta(\boldsymbol{n})), \ \forall \boldsymbol{n} \in \Lambda, \forall \upsilon \in \Gamma,$

where P is a probability measure, is called a Markov random field with state space Γ.

To define the Gibbs distribution, the concepts of clique and potential function are needed. A *clique* c defined over Λ with respect to $\mathcal{N}$ is a subset of Λ such that either c consists of a single site or every pair of sites in c are neighbors, that is, belong to η. The set of all cliques is denoted by C. Examples of a two-element spatial clique $\{\boldsymbol{n}, \boldsymbol{l}\}$ are two immediate horizontal, vertical or diagonal neighbors. Gibbs distribution with respect to Λ and $\mathcal{N}$ is a probability measure π on Γ such that

$$\pi(\Upsilon = v) = \frac{1}{Z} e^{-U(v)/T}, \tag{3.2}$$

where the constants Z and T are called the partition function and temperature, respectively, and the *energy function* U is of the form

$$U(v) = \sum_{c \in C} V(v, c).$$

where $V(v, c)$ is called a potential function, and depends only on the value of v at sites that belong to the clique c. For two-element cliques $\{\boldsymbol{n}, \boldsymbol{l}\}$ considered here, we will explicitly write $V(v[\boldsymbol{n}], v[\boldsymbol{l}])$.

The equivalence between Markov random fields and Gibbs distributions is provided through the important Hammersley-Clifford theorem, which states that Υ is a MRF on Λ with respect to $\mathcal{N}$ if and only if its probability distribution is a Gibbs distribution with respect to Λ and $\mathcal{N}$. The equivalence between MRFs and Gibbs distributions results in a straightforward relationship between qualitative properties of a MRF and its parameters through the potential functions V. Extension of the Hammersley-Clifford theorem to vector MRFs is straightforward (new definition of a state is needed).

3.2.3 MAP Estimation

Let Y be a random field of observations, and let Υ be a random field modeling the quantity we want to estimate based on Y. Let y, v be their respective realizations. For example, y could be a difference between two images, and v could be a field of motion detection labels. To compute v based on y, a powerful tool is the (MAP) estimation, expressed as follows:

$$\hat{v} = \arg\max_{v} P(\Upsilon = v|y) = \arg\max_{v} P(Y = y|v) \cdot P(\Upsilon = v), \tag{3.3}$$

where $\max_v P(\Upsilon = v|y)$ denotes the maximum of the posterior probability $P(\Upsilon = v|y)$ with respect to v, and arg denotes the argument of this maximum, that is, such $\hat{v}$ that $P(\Upsilon = \hat{v}|y) \geq P(\Upsilon = v|y)$ for any v. Above, the Bayes rule was used, and $P(Y = y)$ was omitted because it does not depend on v. If $P(\Upsilon = v)$ is the same for all realizations v, then only the likelihood $P(Y = y|v)$ is maximized, resulting in the *maximum likelihood* (ML) estimation.

3.2.4 Variational Formulations

In hypothesis testing and MRF models, moving regions to be estimated are defined explicitly in discrete state space (pixel labels), whereas region boundaries are implicit (label differences). Alternatively, these boundaries can be considered continuous, thus leading to functionals instead of sets of variables.

Let the continuous boundary one wishes to estimate be modeled by a closed parameterized planar curve $\vec{\gamma}$, oriented counterclockwise. If $\mathcal{R}$ is a region enclosed by $\vec{\gamma}$ and

s is a variable moving along $\vec{\gamma}$, then a simple formulation that seeks partitioning of the image domain Ω into $\mathcal{R}$ and its complement $\mathcal{R}_c = \Omega \backslash \mathcal{R}$ can be written as follows:

$$\min_{\vec{s}} \iint_{\mathcal{R}} f(\delta I(\boldsymbol{x}))d\boldsymbol{x} + \lambda \int_{\vec{\gamma}} ds, \tag{3.4}$$

where $f(\cdot)$ is a strictly decreasing function, δI is an observation derived from the data, for example image gradient, and ds is the Euclidean length element. While the first term is related to intensity variation within $\mathcal{R}$, the second term is proportional to the boundary length of $\mathcal{R}$. This formulation may seek, for example, a minimal-length boundary estimate that surrounds large intensity gradients.

A solution to (3.4) can be found by computing and solving Euler-Lagrange equations [3]. At any $\boldsymbol{x}$, this results in the following evolution equation [4]:

$$\frac{\partial \vec{\gamma}}{\partial \tau} = F\vec{n} = [f(\delta I) + \lambda\kappa]\vec{n}, \tag{3.5}$$

where τ is the evolution time (unlike the true time t, which is the third coordinate in the image sequence definition), F denotes the contour evolution force, κ is the Euclidean curvature, and $\vec{n}$ is the inward unit normal to $\vec{\gamma}$. The term $\lambda\kappa\vec{n}$ smoothes out the contour by reducing curvature, whereas the term $f(\delta I)\vec{n}$ is a constant "balloon" force that pushes the contour towards large-gradient image areas. The evolution force vanishes for $\delta I \to \infty$ ($f(\delta I) \to 0$), and $\kappa \to 0$, that is, ideal straight edge. In practice, the evolution equation is stopped if the curve fails to evolve significantly between two iterations.

The active-contour evolution Eq. (3.5) suffers from topology invariance (contours cannot be added/removed), as well as potential convergence problems (as the contour evolves, sampling density along the contour changes, and the contour needs to be resampled). Both issues can be avoided by embedding the active contour $\vec{\gamma}$ into a surface u, which leads to the following level-set evolution equation [5]:

$$\frac{\partial u}{\partial \tau} = F\|\nabla u\| = [f(\delta I) + \lambda\kappa]\|\nabla u\|. \tag{3.6}$$

Again κ is the curvature computed from u ($\kappa = div(\nabla u/|\nabla u|)$). This equation can be implemented iteratively using standard discretization [5]. In each iteration, the force F is calculated at zero level-set points ($\boldsymbol{x}$ such that $u(\boldsymbol{x}) = 0$), and then extended to other positions $\boldsymbol{x}$ using, for example, the fast marching algorithm by solving $\nabla u \cdot \nabla F = 0$ for F. Then, the surface u is updated according to (3.6), and periodically re-initialized using the fast marching algorithm by solving $\|\nabla u\| = 1$ (signed distance).

3.3 MOTION DETECTION

Motion detection is, arguably, the simplest of the three motion-related tasks, that is, detection, estimation and segmentation. Its goal is to identify which image points, or, more generally, which regions of the image, have moved. As such, motion detection applies to images acquired with a static camera. However, if camera motion is counteracted,

For example, by global motion estimation and compensation, then the method equally applies to images acquired with a moving camera [6, Chapter 8].

It is essential to realize that motion of image points is not perceived directly but rather through intensity changes. However, such intensity changes over time may be also induced by camera noise or illumination variations. Differentiating intensity changes due to object motion from those due to camera noise or illumination variations is far from trivial, especially when objects have little texture (e.g., uniform luminance/color).

3.3.1 Hypothesis Testing with Fixed Threshold

Fixed-threshold hypothesis testing belongs to the simplest motion detection algorithms as it requires few arithmetic operations. Several early motion detection methods belong to this class, although originally they were not developed as such.

Let $H_{\mathcal{S}}$ and $H_{\mathcal{M}}$ be two hypotheses, declaring an image point at $\boldsymbol{n}$ as stationary ($\mathcal{S}$) and moving ($\mathcal{M}$), respectively. In the context of Section 3.2.1, the state $\mathcal{S}$ means 0, whereas the state $\mathcal{M}$ means 1. Let's assume that $I_k[\boldsymbol{n}] = I_{k-1}[\boldsymbol{n}] + q$, and that q is a noise term, zero-mean Gaussian with variance σ^2 in stationary areas and uniformly distributed in range $[-L, L]$ in moving areas. Clearly, $P_{\mathcal{S}}$ is assumed Gaussian, while $P_{\mathcal{M}}$ is assumed uniform. The motivation is that in stationary areas only camera noise will distinguish same-position pixels at t_{k-1} and t_k, whereas in moving areas this difference is attributed to motion and therefore unpredictable. Let

$$\rho_k[\boldsymbol{n}] = I_k[\boldsymbol{n}] - I_{k-1}[\boldsymbol{n}]$$

be an observation, upon which we intend to select one of the two hypotheses. With the above assumptions, and after taking the natural logarithm of both sides of (3.1) the hypothesis test can be written as follows:

$$\rho_k^2[\boldsymbol{n}] \underset{\mathcal{S}}{\overset{\mathcal{M}}{\gtrless}} \theta, \tag{3.7}$$

where $\theta = 2\sigma^2 \ln(\vartheta \cdot 2L \cdot \pi_{\mathcal{S}}/(\sqrt{2\pi\sigma^2} \cdot \pi_{\mathcal{M}})$. A similar test can be derived for a Laplacian-distributed noise term q; in (3.7) $\rho_k^2[\boldsymbol{n}]$ is replaced by $|\rho_k[\boldsymbol{n}]|$ and θ is computed accordingly. Such a test was used in some early motion detection algorithms. Note that both the Laplacian and Gaussian models are equivalent under an appropriate selection of θ. Although θ includes the prior probabilities, they are usually fixed as is the noise variance, and thus, ρ_k^2 is compared to a constant.

The above pixel-based hypothesis test is not robust to noise in the image; for small θ's "noisy" detection masks result (many isolated small regions), whereas for large θ's only object boundaries and its most textured parts are detected. To attenuate the impact of noise, the method can be extended by averaging the observations over an N-point spatial window $\mathcal{W}_{\boldsymbol{n}}$ centered at $\boldsymbol{n}$:

$$\frac{1}{N} \sum_{\boldsymbol{m} \in \mathcal{W}_{\boldsymbol{n}}} \rho_k^2[\boldsymbol{m}] \underset{\mathcal{S}}{\overset{\mathcal{M}}{\gtrless}} \theta.$$

This approach exploits the fact that data captured in a typical camera can be closely approximated by an additive white noise model; by averaging over $\mathcal{W}_{\boldsymbol{n}}$ the noise impact can be significantly reduced. Still, the method is not very robust and is usually followed by some postprocessing (e.g., median filtering, suppression of small regions). Moreover, since the classification at position $\boldsymbol{n}$ is performed based on all pixels within $\mathcal{W}_{\boldsymbol{n}}$, the resolution of the method is reduced; a moving image point affects the decision of many of its neighbors.

Motion detection based on frame differences, as described above, does not perform well for large, untextured objects (e.g., a large, uniformly colored truck). Only pixels $\boldsymbol{n}$ where $|I_k[\boldsymbol{n}] - I_{k-1}[\boldsymbol{n}]|$ is sufficiently large can be reliably detected. Such pixels concentrate in narrow areas close to moving boundaries where object intensity is distinct from the background in the previous frame. This leads to excessive false negatives, as shown in Fig. 3.1(b). This deficiency can be addressed to a degree by comparing the current intensity $I_k[\boldsymbol{n}]$ to background intensity $B_k[\boldsymbol{n}]$ instead of the previous frame $I_{k-1}[\boldsymbol{n}]$, that is, by defining $\rho_k[\boldsymbol{n}] = I_k[\boldsymbol{n}] - B_k[\boldsymbol{n}]$. Perhaps the simplest approach to estimating $B_k[\boldsymbol{n}]$ is by means of temporal averaging or median filtering the intensity at each $\boldsymbol{n}$. Median filtering is particularly attractive since for a sufficiently large support (temporal window), it can suppress intensities associated with moving objects.

Although temporal median filtering is a fast and quite effective approach to background modeling, it often fails in the presence of parasitic motion, such as fluttering leaves or waves on water surface. To account for such motion, richer statistical models have been proposed. For example, instead of modeling $I_k[\boldsymbol{n}]$ as a Gaussian random variable with mean $I_{k-1}[\boldsymbol{n}]$ or $B_k[\boldsymbol{n}]$, mixture-of-Gaussians models [7] and non-parametric distributions [8] have been successfully used.

Let us briefly describe the latter approach for its simplicity and good performance. At each location $\boldsymbol{n}$ of frame k, an estimate of the stationary (background) probability distribution is computed from K recent frames as follows:

$$P_{\mathcal{S}}(I_k[\boldsymbol{n}]) = \frac{1}{K}\sum_{i=1}^{K}\mathcal{K}(I_k[\boldsymbol{n}] - I_{k-i}[\boldsymbol{n}]), \tag{3.8}$$

where $\mathcal{K}$ is a zero-mean Gaussian with variance σ^2 that, for simplicity, we consider constant throughout the sequence. Assuming uniformly distributed intensity in range $[-L, L]$ for moving pixels (no knowledge about moving areas), the hypothesis test (3.1) now becomes

$$P_{\mathcal{S}}(I_k[\boldsymbol{n}]) \underset{\mathcal{M}}{\overset{\mathcal{S}}{\gtrless}} \theta, \tag{3.9}$$

where $\theta = \pi_{\mathcal{M}}/(2L\vartheta\pi_{\mathcal{S}})$. Note that, unlike in (3.7), the thresholding takes place now in the space of probabilities rather than intensities. An intensity at location $\boldsymbol{n}$ in frame k is deemed stationary only if it is likely to have been drawn from $P_{\mathcal{S}}$. This improves robustness of the detection to small parasitic movements that are accounted for in $P_{\mathcal{S}}$. Also, since the $P_{\mathcal{S}}$ model is based on K recent frames, it adapts to slow background changes such as illumination variations. Note, that in order to avoid model contamination intensities from moving areas in previous frames need to be excluded from the summation in (3.8), for example based on previous detections.

3.3.2 Hypothesis Testing with Adaptive Threshold

The motion detection methods presented thus far assumed no knowledge about intensities of the moving object; $P_{\mathcal{M}}$ was considered uniform. However, one may hope for a performance gain should a more accurate model of moving pixels be allowed. In order to estimate $P_{\mathcal{M}}$, one could use a similar strategy to the one used for background pixels (3.8). However, unlike in background modeling, the location of moving pixels is time-varying and Eq. (3.8) does not apply. Although one can limit the number of frames used (K) so that intensity $I_k[\boldsymbol{n}]$ remains within a moving object for $k = 1,\ldots,K$ [9], the accuracy of this approach strongly depends on moving object size and temporal window size K. Note, however, that as the object travels through location $\boldsymbol{n}$ in the image, intensities $I_1[\boldsymbol{n}]$, $I_2[\boldsymbol{n}],\ldots$ are, effectively, samples derived from this object. This sampling is, in general, non-uniform, and depends on the movement of the object. Instead of using a temporal history to derive object samples, one can directly use object samples in the current frame (spatial history). This can be thought of as the assumption of ergodicity in spatial coordinates instead of ergodicity in time. It has been recently demonstrated that using spatial ergodicity is not detrimental to motion detection but, in fact, reduces memory requirements (no need for multi-frame buffer) [10]; local-in-time and local-in-space models produce very similar results.

To build the local-in-space model at position $\boldsymbol{n}$, ideally, one needs to identify all pixels of a moving object to which $\boldsymbol{n}$ belongs. Clearly, this labeling is not known in advance but since a typical motion detection algorithm is both causal and iterative, one can use motion labels from previous frame or previous iteration. Below the latter approach is outlined.

Let $e_k[\boldsymbol{n}]$ be a change label detected at position $\boldsymbol{n}$ in frame k ($e_k[\boldsymbol{n}] = \mathcal{S}$ or $e_k[\boldsymbol{n}] = \mathcal{M}$). Using the naïve uniform foreground model and the simplified likelihood ratio test (3.9), one first finds an initial detection mask, denoted $e_k^0[\boldsymbol{n}]$. Although using a connected-component analysis one could identify all pixels belonging to a moving object that $\boldsymbol{n}$ is part of, a much simpler approach, that works well, is to consider only a small neighborhood of $\boldsymbol{n}$. For each $\boldsymbol{n}$, one can define a set of neighbors belonging to a moving object as follows:

$$\mathcal{N}_{\mathcal{M}_i}(\boldsymbol{n}) = \{\boldsymbol{m} \in \mathcal{N}(\boldsymbol{n}) : e_k^{i-1}[\boldsymbol{m}] = \mathcal{M}\},$$

where i is the iteration number and $\mathcal{N}(\boldsymbol{n})$ is a small neighborhood around $\boldsymbol{n}$. The moving-pixel probability is then calculated using the same kernel-based method as in the case of $P_{\mathcal{S}}$ (3.8), except that local-in-space samples are used in place of local-in-time samples:

$$P_{\mathcal{M}_i}(I_k[\boldsymbol{n}]) = \frac{1}{|\mathcal{N}_{\mathcal{M}_i}(\boldsymbol{n})|} \sum_{\boldsymbol{m}\in\mathcal{N}_{\mathcal{M}_i}(\boldsymbol{n})} \mathcal{K}(I_k[\boldsymbol{n}] - I_k[\boldsymbol{m}]). \tag{3.10}$$

The estimated moving-pixel probability can be used in a refined likelihood ratio test:

$$\frac{P_{\mathcal{S}}(I_k[\boldsymbol{n}])}{P_{\mathcal{M}_i}(I_k[\boldsymbol{n}])} \underset{\mathcal{M}}{\overset{\mathcal{S}}{\gtrless}} \theta, \tag{3.11}$$

where $\theta = \pi_{\mathcal{M}}/(\vartheta \pi_{\mathcal{S}})$, to produce a new label $e_k^i[\boldsymbol{n}]$. This test is iterated with the new label field defining a new neighborhood $\mathcal{N}_{\mathcal{M}_{i+1}}$ and new PDF $P_{\mathcal{M}_{i+1}}$ which, in turn, produce a new estimate e_k^{i+1}. If there are no detected pixels in the neighborhood of $\boldsymbol{n}$ ($\mathcal{N}_{\mathcal{M}_i}(\boldsymbol{n}) = \emptyset$), there is presumably no moving object at $\boldsymbol{n}$, and one can revert to the naïve assumption that $P_{\mathcal{M}}$ is uniform.

Note that $P_{\mathcal{M}_i}(I_k[\boldsymbol{n}])$ in (3.11) can, effectively, be considered as a scale for the threshold θ. If for a particular intensity $I_k[\boldsymbol{n}]$ its foreground probability $P_{\mathcal{M}_i}(I_k[\boldsymbol{n}])$ is lower, then the effective threshold $\theta \cdot P_{\mathcal{M}_i}(I_k[\boldsymbol{n}])$ is reduced, thus encouraging assignment of the stationary label. To the contrary, a higher value of $P_{\mathcal{M}_i}(I_k[\boldsymbol{n}])$ will encourage assignment of the moving label. The introduction of a foreground model can be interpreted as spatial threshold adaptation.

The above methods are based solely on image intensities and make no *a priori* assumptions about the nature of moving areas. However, moving 3D objects usually create compact, closed boundaries in the image plane, that is, if an image point is declared moving, it is likely that its neighbors are moving as well, and the boundary is smooth rather than rough. To take advantage of this *a priori* information, hypothesis testing can be combined with Markov random field models.

Let E_k be a MRF of all labels assigned at time t_k whose realization is e_k. Let's assume for the time being that $e_k[\boldsymbol{l}]$ is known for all $\boldsymbol{l}$ except $\boldsymbol{n}$. Since motion detection is often iterative, this assumption is not unreasonable; previous estimates are known at $\boldsymbol{l} \neq \boldsymbol{n}$. Thus, the estimation process is reduced to deciding between $e_k[\boldsymbol{n}] = \mathcal{S}$ and $e_k[\boldsymbol{n}] = \mathcal{M}$. Let the label field such that $e_k[\boldsymbol{n}] = \mathcal{S}$ be denoted by $e_k^{\mathcal{S}}$, and let the one with $e_k[\boldsymbol{n}] = \mathcal{M}$ be $e_k^{\mathcal{M}}$. Then, based on (3.1), the decision rule for $e_k[\boldsymbol{n}]$ can be written as follows:

$$\frac{P(R_k = \rho_k | e_k^{\mathcal{M}})}{P(R_k = \rho_k | e_k^{\mathcal{S}})} \underset{\mathcal{S}}{\overset{\mathcal{M}}{\gtrless}} \vartheta \frac{\pi(E_k = e_k^{\mathcal{S}})}{\pi(E_k = e_k^{\mathcal{M}})}, \tag{3.12}$$

where R_k is a random field modeling temporal frame differences, and $\pi(E_k = e_k)$ is a Gibbs distribution governing E_k (Section 3.2.2). By making the simplifying assumption that the temporal differences $\rho_k[\boldsymbol{l}]$ are conditionally independent, given e_k, that is, $P(R_k = \rho_k | e_k) = \prod_{\boldsymbol{l}} P(R_k[\boldsymbol{l}] = \rho_k[\boldsymbol{l}] | e_k[\boldsymbol{l}])$, Eq. (3.12) can be further rewritten as follows:

$$\frac{P_{\mathcal{M}}(\rho_k[\boldsymbol{n}])}{P_{\mathcal{S}}(\rho_k[\boldsymbol{n}])} \underset{\mathcal{S}}{\overset{\mathcal{M}}{\gtrless}} \vartheta \frac{\pi(E_k = e_k^{\mathcal{S}})}{\pi(E_k = e_k^{\mathcal{M}})}. \tag{3.13}$$

This simplified form is due to the fact that with the assumed conditional independence all constituent probabilities on the left-hand side cancel out except for those at $\boldsymbol{n}$ ($e_k^{\mathcal{M}}$ and $e_k^{\mathcal{S}}$ differ only at $\boldsymbol{n}$). Although the conditional independence assumption is reasonable in stationary areas (temporal differences are mostly due to camera noise), it is less so in moving areas. However, a convincing argument based on experimental results can be made in favor of such independence [11].

For the Gaussian $P_{\mathcal{S}}$ and uniform $P_{\mathcal{M}}$ defined earlier this generalizes the hypothesis test (3.7) to:

$$\rho_k^2[\boldsymbol{n}] \underset{\mathcal{S}}{\overset{\mathcal{M}}{\gtrless}} 2\sigma^2 \left(\ln \frac{\vartheta \cdot 2L}{\sqrt{2\pi\sigma^2}} + \ln \frac{\pi(E_k = e_k^{\mathcal{S}})}{\pi(E_k = e_k^{\mathcal{M}})} \right). \tag{3.14}$$

By suitably defining the *a priori* probabilities, one can adapt the threshold in response to the properties of e_k. Because the required properties are object compactness and smoothness of its boundaries, a simple MRF model supported on the second-order neighborhood with two-element cliques $c = \{\boldsymbol{n}, \boldsymbol{l}\}$ and the Ising potential function [2]:

$$V(e_k[\boldsymbol{n}], e_k[\boldsymbol{l}]) = \begin{cases} 0 & \text{if } e_k[\boldsymbol{n}] = e_k[\boldsymbol{l}], \\ \beta & \text{if } e_k[\boldsymbol{n}] \neq e_k[\boldsymbol{l}], \end{cases} \tag{3.15}$$

is appropriate. Whenever a neighbor of $\boldsymbol{n}$ has different label than $e_k[\boldsymbol{n}]$, a penalty $\beta > 0$ is incurred; summed over the whole field it is proportional to the length of the moving mask boundary. Thus, the resulting prior (Gibbs) probability $\pi(E_k = e_k)$ (3.2) will increase for configurations with smooth boundaries and will decrease for those with rough boundaries.

Note that the MRF model facilitates threshold adaptation: if $\pi(E_k = e_k^{\mathcal{S}}) > \pi(E_k = e_k^{\mathcal{M}})$, the overall threshold in (3.14) increases, thus biasing the decision toward a static label. Conversely, for $\pi(E_k = e_k^{\mathcal{S}}) < \pi(E_k = e_k^{\mathcal{M}})$ the bias is in favor of a moving label.

The same Markov model can be applied to the intensity-based hypothesis test (3.11), resulting in the following decision rule:

$$\frac{P(\mathcal{I}_k = I_k | e_k^{\mathcal{M}})}{P(\mathcal{I}_k = I_k | e_k^{\mathcal{S}})} \underset{\mathcal{S}}{\overset{\mathcal{M}}{\gtrless}} \vartheta \frac{\pi(E_k = e_k^{\mathcal{S}})}{\pi(E_k = e_k^{\mathcal{M}})},$$

where $\mathcal{I}_k$ is a random field modeling the image at time t_k. If the intensities, while dependent on the label field, are mutually independent spatially, that is $P(\mathcal{I}_k = I_k | e_k) = \prod_{\boldsymbol{m}} P(\mathcal{I}_k[\boldsymbol{m}] = I_k[\boldsymbol{m}] | e_k[\boldsymbol{m}])$, the following pixel-by-pixel hypothesis test results:

$$\frac{P_{\mathcal{M}}(I_k[\boldsymbol{n}])}{P_{\mathcal{S}}(I_k[\boldsymbol{n}])} \underset{\mathcal{S}}{\overset{\mathcal{M}}{\gtrless}} \vartheta \frac{\pi(E_k = e_k^{\mathcal{S}})}{\pi(E_k = e_k^{\mathcal{M}})}.$$

For the second-order Markov model with Ising potential (3.15) this simplifies to

$$\frac{P_{\mathcal{M}}(I_k[\boldsymbol{n}])}{P_{\mathcal{S}}(I_k[\boldsymbol{n}])} \underset{\mathcal{S}}{\overset{\mathcal{M}}{\gtrless}} \vartheta \left(\frac{\beta}{T} (Q_{\mathcal{S}}[\boldsymbol{n}] - Q_{\mathcal{M}}[\boldsymbol{n}]) \right),$$

where T is the natural temperature of Gibbs distribution, whereas $Q_{\mathcal{S}}[\boldsymbol{n}]$ and $Q_{\mathcal{M}}[\boldsymbol{n}]$ denote the number of stationary and moving neighbors of $\boldsymbol{n}$, respectively (between 0 and 8). The effect of incorporating the prior into the likelihood ratio test is quite apparent. The detected moving neighbors reduce the effective threshold toward declaring $\mathcal{M}$, whereas stationary neighbors increase this threshold toward declaring $\mathcal{S}$. The constant $1/T$ controls the nonlinear behavior of the threshold as the function of $Q_{\mathcal{S}}[\boldsymbol{n}] - Q_{\mathcal{M}}[\boldsymbol{n}]$. If the moving-pixel probability is estimated from local-in-space samples, then $P_{\mathcal{M}_i}(I_k[\boldsymbol{n}])$ (3.10) should replace $P_{\mathcal{M}}(I_k[\boldsymbol{n}])$ above.

3.3.3 MAP MRF Formulation

The MRF label model introduced in the previous section can be also incorporated into the MAP criterion (Section 3.2.3). To find a MAP estimate of the random field E_k, the posterior probability $P(E_k = e_k|\rho_k)$, or its Bayes equivalent $P(R_k = \rho_k|e_k) \cdot \pi(E_k = e_k)$, needs to be maximized.

Let's consider the likelihood $P(R_k = \rho_k|e_k)$. One of the questionable assumptions made in the previous section was the conditional independence of ρ_k given e_k (3.13). To alleviate this problem, let $|I_k[\boldsymbol{n}] - I_{k-1}[\boldsymbol{n}]|$ be an observation modeled as $\rho_k[\boldsymbol{n}] = \xi(e_k[\boldsymbol{n}]) + q[\boldsymbol{n}]$, where q is zero-mean uncorrelated Gaussian noise with variance σ^2 and

$$\xi(e_k[\boldsymbol{n}]) = \begin{cases} 0 & \text{if } e_k[\boldsymbol{n}] = \mathcal{S}, \\ \alpha & \text{if } e_k[\boldsymbol{n}] = \mathcal{M}. \end{cases}$$

Above, α is considered to be an average of the observations in moving areas. For example, α could be computed as an average temporal intensity difference based on previous-iteration moving labels e_k^{i-1} or previous-time moving labels e_{k-1}. Clearly, ξ attempts to closely model the observations since for a static image point it is zero, whereas for a moving point it tracks average temporal intensity mismatch; the uncorrelated q should be a better approximation here than in (3.13).

Under the uncorrelated Gaussian assumption for the likelihood $P(R_k = \rho_k|e_k)$ and a Gibbs distribution for the *a priori* probability $\pi(E_k = e_k)$, the overall energy function can be written as follows:

$$U(\rho_k, e_{k-1}, e_k) = \frac{1}{2\sigma^2}\sum_{\boldsymbol{n}}\Big((\rho_k[\boldsymbol{n}] - \xi(e_k[\boldsymbol{n}]))^2 + \sum_{\{\boldsymbol{n},\boldsymbol{l}\}\in C} V_s(e_k[\boldsymbol{n}], e_k[\boldsymbol{l}]) + \sum_{\{t_{k-1},t_k\}} V_t(e_{k-1}[\boldsymbol{n}], e_k[\boldsymbol{n}])\Big).$$

The first term measures how well each label at $\boldsymbol{n}$ explains the observation $\rho_k[\boldsymbol{n}]$. The other terms measure how contiguous the labels are in the image plane (V_s) and in time (V_t). Both V_s and V_t can be specified similarly to (3.15), thus favoring spatial and temporal similarity of the labels [12]. This basic model can be enhanced by a more flexible likelihood [9, 13] or a more complete prior model including spatiotemporal, as opposed to purely spatial and temporal, cliques [14]. The above cost function can be optimized using various approaches, such as those discussed in Section 3.4.3, namely simulated annealing, iterated conditional modes (ICMs) or highest confidence first (HCF). The latter method, based on an adaptive selection of visited labels according to their impact on the energy U (most influential visited first), gives the best compromise between performance (final energy value) and computing time.

3.3.4 MAP Variational Formulation

So far, the motion detection problem has been formulated in discrete domain; pixels were explicitly labeled as moving or stationary. Alternatively, as mentioned before, moving areas can be defined implicitly by closed contours; the problem can be formulated and solved in continuous domain, and the final solution–discretized. One possible approach is through variational formulation (Section 3.2.4).

To formulate the problem in this fashion, a model for boundaries of moving areas is needed. One popular class of such models have been active contours. Let $\vec{\gamma}$ be a closed parameterized planar curve, oriented counterclockwise, $\mathcal{R}$–region enclosed by $\vec{\gamma}$, and $\mathcal{R}_c = \Omega \backslash \mathcal{R}$–its complement. The problem of detecting moving areas between images $I_{t-\Delta t}$ and I_t can be formulated as follows [15]:

$$\min_{\vec{\gamma}} \iint_{\mathcal{R}} \alpha d\boldsymbol{x} + \iint_{\mathcal{R}_c} |I(\boldsymbol{x},t) - I(\boldsymbol{x},t-\Delta t)| d\boldsymbol{x} + \lambda \int_{\vec{\gamma}} ds. \tag{3.16}$$

The first term assigns cost to the moving regions that is proportional to the area of $\mathcal{R}$. The second term assigns cost to the stationary regions but the cost at each location $\boldsymbol{x}$ is proportional to the temporal intensity change. The third term measures length of the boundary of $\mathcal{R}$, and can be considered prior information. Clearly, a minimum will be achieved if the boundary is smooth and all large temporal intensity differences (moving areas) are included in $\mathcal{R}$, whereas small differences (stationary areas) are in $\mathcal{R}_c$. This simple formulation can be viewed as an example of MAP estimation; the first two terms relate the unknown contour $\vec{\gamma}$ to the data $\{I_{t-\Delta t}, I_t\}$, whereas the last term measures the length of moving-area boundary. Details of a MAP derivation in this context can be found in [16], where also more advanced data terms based on motion compensation have been proposed. Note that minimization (3.16) can be also viewed as region competition [17] since regions $\mathcal{R}$ and $\mathcal{R}_c$ both compete for the membership of pixel at $\boldsymbol{x}$ (is it less expensive to assign at $\boldsymbol{x}$ the cost α or $|I(\boldsymbol{x},t) - I(\boldsymbol{x},t-\Delta t)|$?).

The contour evolution for (3.16) obtained by solving Euler-Lagrange equations is:

$$\frac{\partial \vec{\gamma}}{\partial \tau} = [\alpha - |I(\boldsymbol{x},t) - I(\boldsymbol{x},t-\Delta t)| + \lambda\kappa]\vec{n}. \tag{3.17}$$

Ignoring the curvature κ, $\alpha > |I(\boldsymbol{x},t) - I(\boldsymbol{x},t-\Delta t)|$ will result in the contour shrinking and thus relinquishing the point $\vec{\gamma}(\boldsymbol{x})$, whereas $\alpha < |I(\boldsymbol{x},t) - I(\boldsymbol{x},t-\Delta t)|$ will cause the contour to expand thus englobing this point. Clearly, there will be a competition between two forces, one related to $\mathcal{R}$ and the other related to $\mathcal{R}_c$ that will claim or relinquish image points on and around the curve $\vec{\gamma}$. The curvature κ plays the role of a smoothing filter with respect to curve-point coordinates. For sufficiently large λ, the curvature term will assure smooth boundaries of the detected moving areas. The corresponding level-set evolution equation:

$$\frac{\partial u}{\partial \tau} = F\|\nabla u\| = [\alpha - |I(\mathbf{x},t) - I(\mathbf{x},t-\Delta t)| + \lambda\kappa]\|\nabla u\|. \tag{3.18}$$

can be implemented iteratively using standard discretization [5].

In the formulation above, motion detection performed on one image pair is independent of a detection performed on a neighboring image pair. An interesting, alternative approach is the joint motion detection (or segmentation) over multiple images [18], leading to the concept of object tunnel. The object boundary model across time becomes now a 3D parameterized surface $\vec{\varsigma}$. If $\mathcal{V}$ is a volume enclosed by $\vec{\varsigma}$, and $\mathcal{V}_c = (\Omega \times \mathcal{T})\backslash\mathcal{V}$ is its complement, the problem of joint multi-image motion detection can be formulated as follows:

$$\min_{\vec{\varsigma}} \iiint_{\mathcal{V}} \alpha d\mathbf{x}dt + \iiint_{\mathcal{V}_c} |I(\mathbf{x},t) - I(\mathbf{x},t-\Delta t)|d\mathbf{x}dt + \lambda \iint_{\vec{\varsigma}} d\vec{s}.$$

The first two terms have similar meaning as in (3.16) except that both are evaluated over 3D volumes rather than 2D regions. The third term measures the surface area of the volume $\mathcal{V}$ ($d\vec{s}$ is the Euclidean area element). Solving for $\vec{\varsigma}$, leads to the following surface evolution equation:

$$\frac{\partial \vec{\varsigma}}{\partial \tau} = [\alpha - |I(\mathbf{x},t) - I(\mathbf{x},t-\Delta t)| + \lambda\kappa_m]\vec{n}.$$

This equation is very similar to the contour evolution Eq. (3.17) except for dimensionality of the normal vector $\vec{n}$ and the nature of the curvature (mean curvature κ_m is used here). The level-set evolution equation for 4D surface u is identical to (3.18), again except for the curvature, and can be solved through similar discretization.

3.3.5 Experimental Comparison of Motion Detection Methods

Figure 3.1 shows motion detection results on a typical urban surveillance video for the variational formulation (Fig. 3.1(a), bottom), frame-difference hypothesis test (Fig. 3.1(b)), stationary-only hypothesis test (Fig. 3.1(c)), and stationary/moving hypothesis test (Fig. 3.1(d)). The latter three results are shown without and with Markov model (bottom). Note the significant rate of misses on car bodies in the variational and frame-difference results. This is due to the frame difference ρ_k used as the observation. The other two models rectify this problem, and the only difference between them is in further reduced misses in Fig. 3.1(d) because of the inclusion of the moving-pixel model $P_{\mathcal{M}}$. In each case, the addition of Markov prior clearly improves the detection accuracy by reducing both false positives and misses. The object tunnels shown in Fig. 3.1(e) confirm the accuracy of detections and also illustrate the dynamic evolution of individual objects' masks.

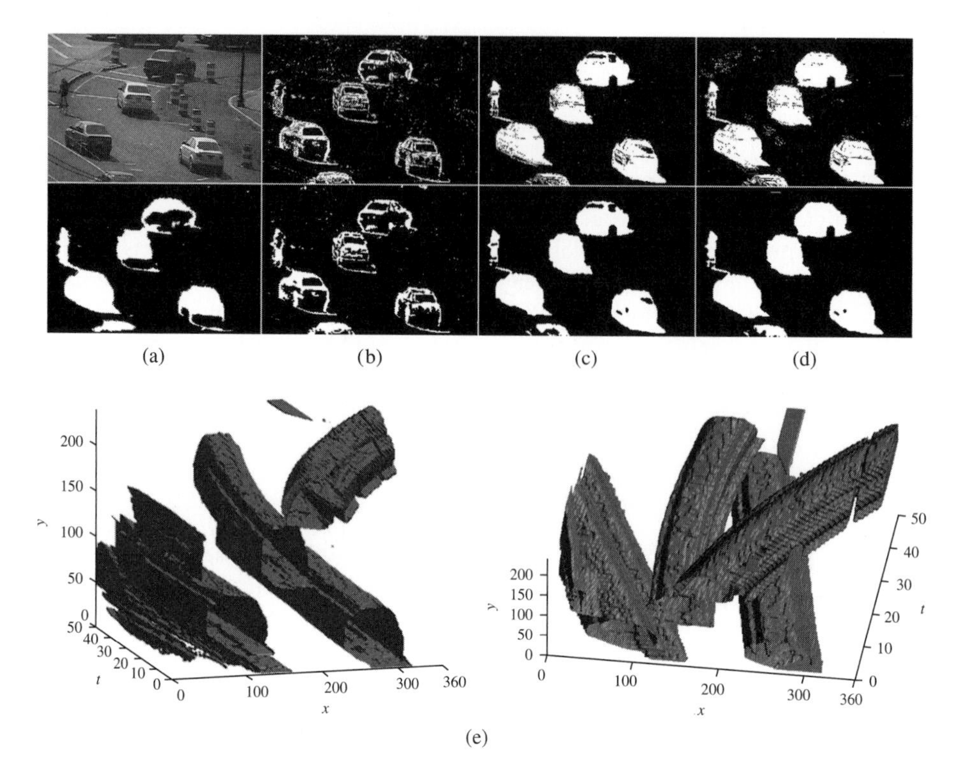

FIGURE 3.1

Motion detection results for a 360 × 240-pixel road traffic video: (a) original frame and active-surface detection result (bottom); (b) frame-difference result (3.7); (c) stationary-only hypothesis test result (3.9); and (d) stationary/moving hypothesis test result (3.11). Bottom results in (b–d) include MRF label model. (e) Two views of object tunnels, that is, surfaces "wrapped" around each moving object's mask, for the case of stationary/moving hypothesis test with Markov prior.

3.4 MOTION ESTIMATION

As mentioned in the Introduction, the knowledge of motion is essential for both the compression and processing of image sequences. Although compression is often considered to be encompassed by processing, a clear distinction between these two terms will be made here. Methods explicitly reducing the number of bits needed to represent a video sequence will be classified as video compression techniques. For example, motion-compensated hybrid (predictive/DCT) coding is exploited today in all video compression standards (Chapters 9 and 10). However, methods that transform a video sequence, for example, to improve quality rather than reduce bit rate, will be considered part of video processing, with such examples as motion-compensated noise reduction (Chapter 4), motion-compensated interpolation (Chapter 2), and motion-based video segmentation (Chapter 6).

The above classification is important from the point of view of the goals of motion estimation that, in turn, influence the choice of models and estimation criteria. In the case of video compression, the estimated motion parameters should lead to the highest compression ratio possible (for a given video quality). Therefore, the computed motion need not resemble the true motion of image points as long as some minimum bit rate is achieved. In video processing, however, it is the true motion of image points that is sought. For example, in motion-compensated temporal interpolation (Fig. 3.2) the task is to compute new images located between existing images of a video sequence (e.g., video frame rate conversion between NTSC and PAL scanning standards). In order that the new images be consistent with the existing ones, image points belonging to moving objects must be displaced according to the true motion as otherwise "jerky" motion of objects would result. This is a very important difference that influences the design of motion estimation algorithms and, most importantly, that usually precludes a good performance of compression-optimized motion estimation in video processing and vice versa.

To develop a motion estimation algorithm, three important elements need to be considered: models, estimation criteria, and search strategies. They will be discussed next, but no attempt will be made to include an exhaustive list pertaining to each of them. Clearly, this cannot be considered a universal classification scheme of motion estimation algorithms, but it is very useful in understanding the properties and merits of various approaches. Then, five practical motion estimation algorithms will be discussed in more detail.

3.4.1 Motion Models

There exist two fundamental models in motion estimation: a motion model, that is, how to represent motion in an image sequence, and a model relating motion parameters to

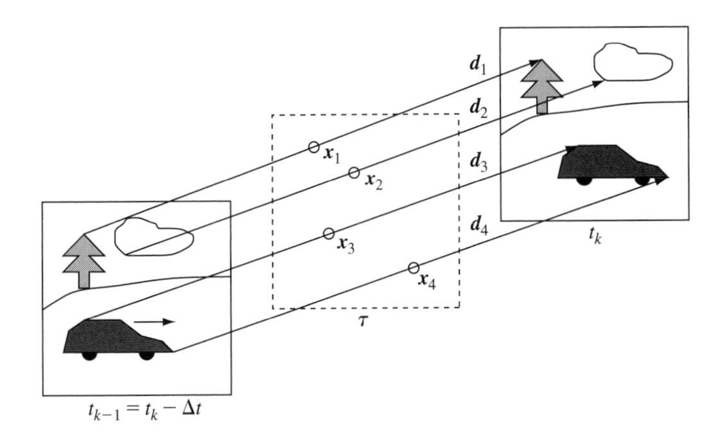

FIGURE 3.2

Motion-compensated interpolation between images at time $t_k - \Delta t$ and t_k. Motion compensation is essential for smooth rendition of moving objects. Shown are four motion vectors that map the corresponding image points at time $t_k - \Delta t$ and t_k onto image at time τ.

image intensities, called an observation model. The latter model is needed since, as was mentioned before, the computation of motion is carried out indirectly by examining intensity changes in time.

3.4.1.1 ***Spatial Motion Models***

The goal is to estimate the motion of image points, that is, the *2D* motion or apparent motion. Such motion is a combination of projections of the motion of objects in a 3D scene and of 3D camera motion. Although camera motion affects the movement of all or almost all image points, the motion of 3D objects only affects a subset of image points corresponding to objects' projections. Since, in principle, the camera-induced motion can be compensated for by either estimating it (Section 3.5.1) or by physically measuring it at the camera, we need to consider only the object-induced motion. This type of motion depends on the following:

1. image formation model, for example, perspective, orthographic projection [19],
2. motion model of 3D object, for example, rigid-body with 3D translation and rotation, 3D affine motion,
3. surface model of 3D object, for example, planar, parabolic.

Although the dependence of 2D motion on the above models is complex in general, two cases are relatively simple and have been used extensively in practice. For an orthographic projection and arbitrary 3D surface undergoing 3D translation, the resulting 2D instantaneous velocity at position $\boldsymbol{x}$ in the image plane is described by a 2D vector:

$$\boldsymbol{v}(\boldsymbol{x}) = \begin{pmatrix} b_1 \\ b_2 \end{pmatrix}, \tag{3.19}$$

where parameters $\boldsymbol{b} = (b_1, b_2)^T = (v_1, v_2)^T$ depend on camera geometry and 3D translation parameters. This 2D translational model has proved very powerful in practice, especially in video compression because locally it provides a close approximation for most natural images.

The second powerful, yet simple, parametric model is that of orthographic projection combined with 3D affine motion of a planar surface. It leads to the following six-parameter affine model [20, Chapter 6]:

$$\boldsymbol{v}(\boldsymbol{x}) = \begin{pmatrix} b_1 \\ b_2 \end{pmatrix} + \begin{pmatrix} b_3 & b_4 \\ b_5 & b_6 \end{pmatrix} \boldsymbol{x}, \tag{3.20}$$

where, again, $\boldsymbol{b} = (b_1, \ldots, b_6)^T$ is a vector of parameters related to the camera as well as 3D surface and motion parameters. Clearly, the translational model (3.19) is a special case of the affine model (3.20). More complex models have been proposed as well but, depending on application, they do not always improve the precision of estimated motion fields. In general, the higher the number of motion parameters, the more precise the description of motion. However, an excessive number of parameters may be detrimental to the performance. This depends on the number of degrees of freedom, that is, model complexity (size of $\boldsymbol{b}$ and the functional dependence of $\boldsymbol{v}$ on x, y) versus the size of the region of support (see below). A complex model applied to a small region of support may lead to an actual increase in the estimation error compared to a simpler model such as one described by (3.20).

3.4.1.2 Temporal Motion Models

The trajectories of individual image points drawn in the (x, y, t) space of an image sequence can be fairly arbitrary since they depend on object motion. In the simplest case, trajectories are linear, such as the ones shown in Fig. 3.2. Assuming that the velocity $\boldsymbol{v}_t(\boldsymbol{x})$ is constant between $t = t_{k-1}$ and τ $(\tau > t)$, a linear trajectory can be expressed as follows:

$$\boldsymbol{x}(\tau) = \boldsymbol{x}(t) + \boldsymbol{v}_t(\boldsymbol{x}) \cdot (\tau - t) = \boldsymbol{x}(t) + \boldsymbol{d}_{t,\tau}(\boldsymbol{x}), \tag{3.21}$$

where $\boldsymbol{d}_{t,\tau}(\boldsymbol{x}) = \boldsymbol{v}_t(\boldsymbol{x}) \cdot (\tau - t)$ is a displacement vector[1] measured in the positive direction of time, that is, from t to τ. Consequently, for linear motion the task is to find the two components of velocity or displacement at each $\boldsymbol{x}$. This simple motion model embedding the two-parameter spatial model (3.19) has proved a powerful motion estimation tool in practice.

A natural extension of the linear model is a quadratic trajectory model, accounting for acceleration of image points, which can be described by

$$\boldsymbol{x}(\tau) = \boldsymbol{x}(t) + \boldsymbol{v}_t(\boldsymbol{x}) \cdot (\tau - t) + \frac{1}{2} \cdot \boldsymbol{a}_t(\boldsymbol{x}) \cdot (\tau - t)^2. \tag{3.22}$$

The model is based on two velocity (linear) variables and two acceleration (quadratic) variables $\boldsymbol{a} = (a_1, a_2)^T$, thus accounting for second-order effects. This model has been demonstrated to greatly benefit such motion-critical tasks as frame rate conversion [21] because of its improved handling of variable-speed motion present in typical videoconferencing images (e.g., hand gestures, facial expressions).

The models above require two (3.21) or four (3.22) parameters at each position $\boldsymbol{x}$. To reduce the computational burden, parametric (spatial) motion models can be combined with the temporal models above. For example, the affine model (3.20) can be used to replace $\boldsymbol{v}_t$ in (3.21) within a suitable region of support. This approach has been successfully used in various region-based motion estimation algorithms. A similar parametric extension of the quadratic trajectory model ($\boldsymbol{v}_t$ and $\boldsymbol{a}_t$ replaced by affine expressions) has been proposed as well [6, Chapter 4] but its practical importance remains to be verified.

---

[1] In the sequel, the dependence of $\boldsymbol{d}$ on t and τ will be dropped whenever it is clear between what time instants $\boldsymbol{d}$ applies.

3.4.1.3 Region of Support

The set of points $\boldsymbol{x}$ to which spatial and temporal motion models apply is called the region of support, denoted by $\mathcal{R}$. The selection of a motion model and region of support is one of the major factors determining the precision of the resulting motion parameter estimates. Usually, for a given motion model, the smaller the region of support $\mathcal{R}$, the better the approximation of motion. This is due to the fact that over a larger area motion may be more complicated, and thus, may require a more complex model. For example, the translational model (3.19) can fairly well describe motion of one car in a highway scene, whereas this model would be quite poor for a scene with many cars. Typically, the region of support for a motion model belongs to one of the four types listed below. Fig. 3.3 shows schematically each type of region.

1. $\mathcal{R}$ = the whole image
   A single motion model applies to all image points. This model is suitable for the estimation of camera-induced motion in a simple static scene as very few parameters can approximate the motion of all image points. This is the most constrained model (relatively small number of motion fields can be represented), but with the fewest parameters to estimate.

2. $\mathcal{R}$ = one pixel
   This model applies to a single image point. Typically, the translational spatial model (3.19) is used jointly with the linear (3.21) or quadratic temporal model (3.22). This pixel-based or *dense* motion representation is the least constrained one because at least two parameters describe the movement of each image point. Consequently, a very large number of motion fields can be represented by all possible combinations of parameter values, but computational complexity is, in general, high.

3. $\mathcal{R}$ = rectangular block of pixels
   This motion model applies to a rectangular (or square) block of image points. In the simplest case, the blocks are disjoint and their union covers the whole image.

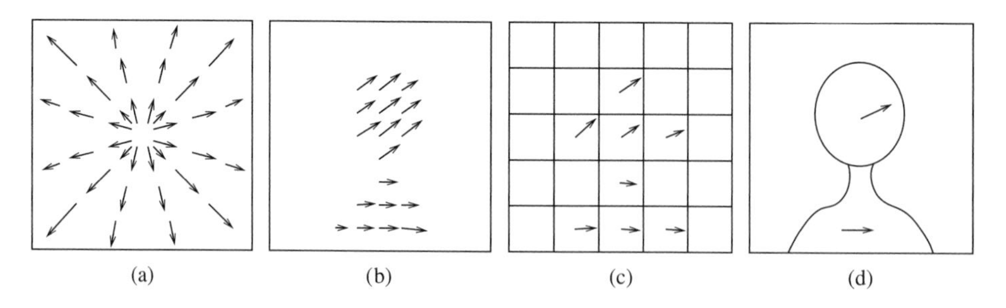

FIGURE 3.3

Schematic representation of motion for the four regions of support $\mathcal{R}$: (a) whole image, (b) pixel, (c) block, and (d) arbitrarily-shaped region. The implicit underlying scene is "head-and-shoulders" as captured by the region-based model (d).

A spatially translational (3.19) and temporally linear (3.21) motion of a square block of pixels has proved to be a very powerful model and is used today in all digital video compression standards (Chapters 9 and 10). It can be also argued that a spatially translational but temporally quadratic (3.22) motion is implicitly used in Moving Picture Experts Group (MPEG) since two motion vectors (backward and forward) defined for blocks in B frames can describe both velocity and acceleration. Although very successful in hardware implementations, due to its simplicity, the translational model lacks precision for images with rotation, zoom, deformation, and is often replaced by the affine model (3.20).

4. $\mathcal{R}$ = irregularly-shaped region
This model applies to all pixels in region $\mathcal{R}$ of arbitrary shape. The motivation is that for objects with sufficiently smooth 3D surface and 3D motion, the induced 2D motion can be closely approximated by the affine model (3.20) applied linearly over time (3.21) to the image area arising from object projection. This is the most advanced motion model found in compression standards; a square block divided into arbitrarily shaped parts, each with independent translational motion, is used in MPEG-4 (Chapter 10).

3.4.1.4 Observation Models

Since the goal is to estimate motion based on intensity variations in time, the relationship between motion parameters and image intensities plays a very important role. The usual, and reasonable, assumption made in this context is that objects do not change their appearance as they move, that is, image intensity remains constant along motion trajectory. Assuming for now that image intensity I is a continuous function and letting s be a variable along motion trajectory, the constant-intensity assumption is equivalent to the directional derivative of I being zero:

$$\frac{dI}{ds} = 0. \tag{3.23}$$

Using the chain rule, the above equation can be written as the well-known motion constraint equation [22]

$$\frac{\partial I}{\partial x}v_1 + \frac{\partial I}{\partial y}v_2 + \frac{\partial I}{\partial t} = (\nabla I)^T \boldsymbol{v} + \frac{\partial I}{\partial t} = 0, \tag{3.24}$$

where $\nabla = (\partial/\partial x, \partial/\partial y)^T$ denotes the spatial gradient and $\boldsymbol{v} = (v_1, v_2)^T$ is the velocity to be estimated. The above constraint equation has served as the basis for many motion estimation algorithms. However, note that applied at single position (x, y) Eq. (3.24) is underconstrained (one equation, two unknowns) and allows to determine only the component of $\boldsymbol{v}$ in the direction of image gradient ∇I [22]. Thus, additional constraints are needed in order to uniquely solve for $\boldsymbol{v}$ [22]. Moreover, Eq. (3.24) does not hold exactly for real images, and usually a minimization of some function of $(\nabla I)^T\boldsymbol{v} + \frac{\partial I}{\partial t}$ is required.

For I sampled in time, the constant-intensity assumption means that $I_{t_k}(\boldsymbol{x}(t_k)) = I_{t_{k-1}}(\boldsymbol{x}(t_{k-1}))$. Furthermore, assuming spatial sampling of intensities on lattice Λ and using the relationship (3.21) with $t = t_{k-1}$ and $\tau = t_k$, this condition can be expressed as follows:

$$I_k[\boldsymbol{n}] - I_{k-1}[\boldsymbol{n} - \boldsymbol{d}[\boldsymbol{n}]] = 0, \quad \forall \boldsymbol{n} \in \Lambda. \tag{3.25}$$

Again, similarly to (3.24), the above relationship is not enough to solve for $\boldsymbol{d}$ (two unknowns) and additional constraints are required. Moreover, it does not hold exactly due to noise, aliasing, illumination variations, etc., and a minimization of some function of $I_k[\boldsymbol{n}] - I_{k-1}[\boldsymbol{n} - \boldsymbol{d}[\boldsymbol{n}]]$ is needed. Note that $I_k[\boldsymbol{n}] - I_{k-1}[\boldsymbol{n} - \boldsymbol{d}[\boldsymbol{n}]]$ is a finite-difference approximation of the directional derivative dI/ds for discrete intensity I. It has been successfully used in video compression since it yields small motion-compensated prediction error.

The assumption about intensity constancy is violated when scene illumination changes. In this case, a constraint based on the spatial gradient's constancy in the direction of motion can be used [23]

$$\frac{d\nabla I}{ds} = \vec{0}.$$

This equation can be re-written as follows:

$$\begin{bmatrix} \partial^2 I/\partial x^2 & \partial^2 I/\partial x \partial y \\ \partial^2 I/\partial x \partial y & \partial^2 I/\partial y^2 \end{bmatrix} \boldsymbol{v} + \frac{\partial(\nabla I)}{\partial t} = \vec{0}. \tag{3.26}$$

It relaxes the constant-intensity assumption but requires that the amount of dilation/contraction, and rotation in the image be negligible,[2] a limitation often satisfied in practice. Although both vector equations above are linear with two unknowns, in practice they do not lend themselves to the direct computation of motion, but need to be further constrained by a motion model. The primary reason for this is that both hold only approximately for real data. Furthermore, they are based on second-order image derivatives that are difficult to compute reliably due to the high-pass nature of the operator; usually image smoothing must be performed first.

Since color is a very important attribute of images, a possible extension of the above models would include chromatic image components. The assumption is that in areas of uniform intensity but substantial color detail, the inclusion of a color-based constraint could prove beneficial. In such a case, Eqs. (3.24), (3.25), and (3.26) would hold with a multicomponent (vector) function replacing I. In video compression, the small gains from color-based motion estimation do not justify the substantial increase in complexity. However, motion estimation using color data is useful in video processing tasks (e.g., motion-compensated filtering, resampling), where motion errors may result in visible

[2]Even when the constant-intensity assumption is valid, the intensity gradient changes its amplitude under dilation/contraction, and its direction under rotation.

distortions. Moreover, a multicomponent motion constraint is interesting for estimating motion from multiple data sources (e.g., range/intensity data).

3.4.2 Estimation Criteria

The models discussed need to be incorporated into an estimation criterion that will be subsequently optimized. There is no unique criterion for motion estimation because its choice depends on the task at hand. For example, in compression an average performance (prediction error) of a motion estimator is important, whereas in motion-compensated interpolation the worst case performance (maximum interpolation error) may be of concern. Moreover, the selection of a criterion may be guided by the processor capabilities on which the motion estimation will be implemented.

3.4.2.1 Pixel-Domain Criteria

Most of the criteria arising from the discrete version of the constant-intensity assumption (3.25) aim at the minimization of a function (e.g., absolute value) of the following error

$$\varepsilon_k[\boldsymbol{n}] = I_k[\boldsymbol{n}] - \tilde{I}_k[\boldsymbol{n}], \qquad \forall \boldsymbol{n} \in \Lambda \tag{3.27}$$

where $\tilde{I}_k[\boldsymbol{n}] = I_{k-1}[\boldsymbol{n} - \boldsymbol{d}[\boldsymbol{n}]]$ is called motion-compensated prediction of $I_k[\boldsymbol{n}]$. Since, in general, $\boldsymbol{d}$ is real-valued, intensities at positions $\boldsymbol{n} - \boldsymbol{d}[\boldsymbol{n}]$ outside of the sampling grid Λ must be recovered by suitable interpolation. For estimation methods that apply intensity matching, $I_{k-1}[\boldsymbol{n} - \boldsymbol{d}[\boldsymbol{n}]]$ must be evaluated for all $\boldsymbol{n}$, and thus, C^0 interpolators that assure continuous interpolated intensity (e.g., bilinear) are sufficient. However, for methods applying gradient descent a derivative of $I_{k-1}[\boldsymbol{n} - \boldsymbol{d}[\boldsymbol{n}]]$ with respect to $\boldsymbol{d}[\boldsymbol{n}]$ must be computed at each $\boldsymbol{n}$ and thus C^1 interpolators giving both continuous intensity and its derivative are preferable for stability reasons.

A common choice for the estimation criterion is the following sum

$$\mathcal{E}(\boldsymbol{d}) = \sum_{\boldsymbol{n} \in \mathcal{R}} \Phi(I_k[\boldsymbol{n}] - \tilde{I}_k[\boldsymbol{n}]) \tag{3.28}$$

where Φ is a nonnegative real-valued function. The often-used quadratic function $\Phi(\varepsilon) = \varepsilon^2$ is not a good choice since a single large error ε (an outlier) overcontributes to $\mathcal{E}$ and biases the estimate of $\boldsymbol{d}$. A more robust function is the absolute value $\Phi(\varepsilon) = \alpha|\varepsilon|$ since the cost grows linearly with error (Fig. 3.4.a.). Since it does not require multiplications, the absolute value criterion is often used in video encoders today. An even more robust criterion is based on the Lorentzian function $\Phi(\varepsilon) = \log(1 + \varepsilon^2/2\omega^2)$ that grows slower than $|x|$ for large errors. The growth of the cost for increasing ε can be adjusted by the parameter ω as shown in Fig. 3.4.a.

Since for algorithms based on intensity matching the continuity of Φ is not important (no gradient computations), noncontinuous functions based on the concept of the truncated quadratic:

$$\Phi_{tq}(\varepsilon, \theta, \beta) = \begin{cases} \varepsilon^2 & |\varepsilon| < \theta, \\ \beta & \text{otherwise,} \end{cases} \tag{3.29}$$

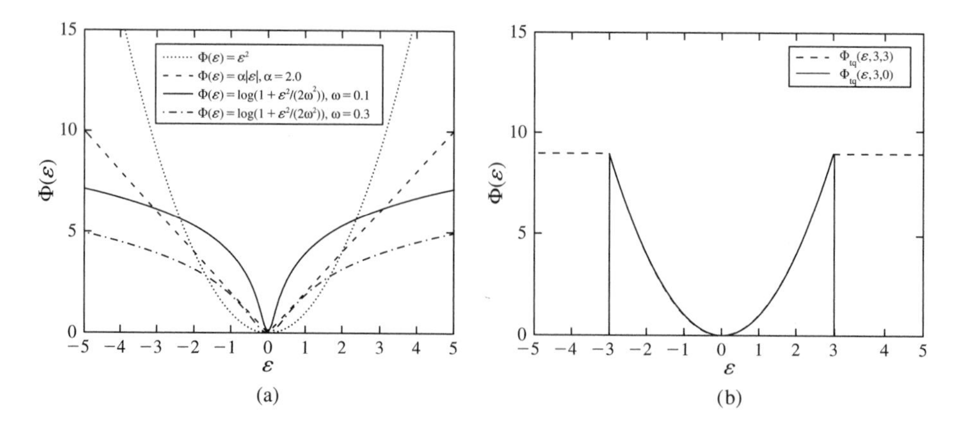

FIGURE 3.4

Comparison of estimation criteria: (a) quadratic, absolute value, Lorentzian functions (two different ω's); and (b) truncated-quadratic functions.

are often used (Fig. 3.4.b). If $\beta = \theta^2$, the usual truncated quadratic results, fixing the cost of outliers at θ^2. An alternative is to set $\beta = 0$ with the consequence that the outliers have zero cost and do not contribute to the overall criterion $\mathcal{E}$. In other words, the criterion is defined only for nonoutlier pixels, and therefore, the estimate of $\boldsymbol{d}$ will be computed solely on the basis of reliable pixels.

The similarity between $I_k[\boldsymbol{n}]$ and its prediction $\tilde{I}_k[\boldsymbol{n}]$ can be also measured by the following cross-correlation function:

$$\mathcal{C}(\boldsymbol{d}) = \sum_{\boldsymbol{n}} I_k[\boldsymbol{n}] I_{k-1}[\boldsymbol{n} - \boldsymbol{d}[\boldsymbol{n}]]. \tag{3.30}$$

Although more complex computationally than the absolute-value criterion due to multiplications involved, this criterion is an interesting and practical alternative to the prediction error-based criteria (Section 3.5.3). Note that a cross-correlation criterion requires maximization unlike the prediction-based criteria.

For a detailed discussion of robust estimation criteria in the context of motion estimation, the reader is referred to the literature (e.g., [24] and references therein).

3.4.2.2 Frequency-Domain Criteria

Although frequency-domain criteria are less used in practice today than the space/time-domain methods, they form an important alternative. Let $\widehat{I_k}(\boldsymbol{u}) = \mathcal{F}[I_k[\boldsymbol{n}]]$ be a spatial (2D) Fourier transform of the intensity signal $I_k[\boldsymbol{n}]$, where $\boldsymbol{u} = (u, v)^T$ is a 2D frequency. Suppose that the image I_{k-1} has been uniformly shifted to create the image I_k, i.e., that $I_k[\boldsymbol{n}] = I_{k-1}[\boldsymbol{n} - \boldsymbol{z}]$. This means that only translational global motion exists in the image, and all boundary effects are neglected. Then, by the shift property of the Fourier transform

$$\mathcal{F}[I_{k-1}[\boldsymbol{n} - \boldsymbol{z}]] = \widehat{I}_{k-1}(\boldsymbol{u}) e^{-j2\pi \boldsymbol{u}^T \boldsymbol{z}}, \tag{3.31}$$

where $\boldsymbol{u}^T$ denotes a transposed vector $\boldsymbol{u}$. Since the amplitudes of both Fourier transforms are independent of $\boldsymbol{z}$, whereas the argument difference

$$\arg\{\mathcal{F}[I_k[\boldsymbol{n}]]\} - \arg\{\mathcal{F}[I_{k-1}[\boldsymbol{n}]]\} = -2\pi \boldsymbol{u}^T \boldsymbol{z}$$

depends linearly on $\boldsymbol{z}$, global motion can be recovered by evaluating the phase difference over a number of frequencies and solving the resulting over-constrained system of linear equations. In practice, this method will work only for single objects moving across a uniform background. Moreover, the positions of image points to which the estimated displacement $\boldsymbol{z}$ applies are not known; this assignment must be performed in some other way. Also, care must be taken of the nonuniqueness of the Fourier phase function which is periodic.

A Fourier-domain representation is particularly interesting for the cross-correlation criterion (3.30). Based on the Fourier transform properties and under the assumption that the intensity function I is real-valued, it is easy to show that

$$\mathcal{F}[\mathcal{C}(\boldsymbol{d})] = \mathcal{F}\left[\sum_{\boldsymbol{n}} I_k[\boldsymbol{n}] I_{k-1}[\boldsymbol{n}-\boldsymbol{d}]\right] = \widehat{I}_k(\boldsymbol{u})\widehat{I}^*_{k-1}(\boldsymbol{u}), \tag{3.32}$$

where the transform is applied in spatial coordinates, $\widehat{I}^*$ is the complex conjugate of $\widehat{I}$, and $\boldsymbol{d}$ is assumed constant across the image (independent of $\boldsymbol{n}$). This equation expresses spatial cross-correlation in the Fourier domain, where it can be efficiently evaluated using the Discrete Fourier Transform.

3.4.2.3 Regularization

The criteria described thus far deal with the underconstrained nature of Eqs. (3.24) and (3.25) by applying a motion measurement to either a region, such as a block of pixels, or to the whole image (frequency-domain criteria). In consequence, resolution of the computed motion may suffer.

To maintain motion resolution at the level of original images, the pixel-wise motion constraint Eq. (3.24) can be used but, to address its underconstrained nature, it needs to be combined with another constraint. In typical real-world images, moving objects are close to being rigid. Upon projection onto the image plane this induces very similar motion of neighboring image points within the object's projection area. In other words, the motion field is locally smooth. Therefore, a motion field $\boldsymbol{v}_t$ must be sought that satisfies the motion constraint (3.24) as close as possible and simultaneously is as smooth as possible. Since gradient is a good measure of local smoothness, this may be achieved by minimizing the following criterion [22]:

$$\mathcal{E}(\boldsymbol{v}) = \int_{\mathcal{D}} \left(\nabla^T I(\boldsymbol{x})\boldsymbol{v}(\boldsymbol{x}) + \frac{\partial I(\boldsymbol{x})}{\partial t}\right)^2 + \lambda\left(\|\nabla(v_1(\boldsymbol{x}))\|^2 + \|\nabla(v_2(\boldsymbol{x}))\|^2\right) d\boldsymbol{x}, \tag{3.33}$$

where $\mathcal{D}$ is the domain of the image. This formulation is often referred to as *regularization* [23]. Note that the smoothness constraint may be also viewed as an alternative spatial motion model to those described in Section 3.4.1.

3.4.2.4 Bayesian Criteria

Bayesian criteria form a very powerful probabilistic alternative to the deterministic criteria described, thus far. If motion field $\boldsymbol{d}_k$ is a realization of vector random field $\boldsymbol{D}_k$ with a given a priori probability distribution, and image I_k is a realization of scalar random field $\mathcal{I}_k$, then the MAP estimate of $\boldsymbol{d}_k$ (Section 3.2.3) can be computed as follows [25]:

$$\begin{aligned}\widehat{\boldsymbol{d}}_k &= \arg\max_{\boldsymbol{d}} P(\boldsymbol{D}_k = \boldsymbol{d}_k | \mathcal{I}_k = I_k; I_{k-1}) \\ &= \arg\max_{\boldsymbol{d}} P(\mathcal{I}_k = I_k | \boldsymbol{D}_k = d_k; I_{k-1}) \cdot P(\boldsymbol{D}_k = \boldsymbol{d}_k; I_{k-1}).\end{aligned} \tag{3.34}$$

In this notation, the semicolon indicates that subsequent variables are only deterministic parameters. The first (conditional) probability distribution denotes the likelihood of image I_k given displacement field $\boldsymbol{d}_k$ and the previous image I_{k-1} and therefore is closely related to the observation model. In other words, this term quantifies how well a motion field $\boldsymbol{d}_k$ explains the intensity change between I_k and I_{k-1}. The second probability $P(\boldsymbol{D}_k = \boldsymbol{d}_k; I_{k-1})$ embodies the prior knowledge about the random field $\boldsymbol{D}_k$, such as its spatial smoothness and therefore can be thought of as a motion model. It becomes particularly interesting when $\boldsymbol{D}_k$ is a MRF. By maximizing the product of the likelihood and the prior probabilities one attempts to strike a balance between motion fields that give a small prediction error and those that are smooth.

3.4.3 Search Strategies

Once models have been identified and incorporated into an estimation criterion, the last step is to develop an efficient (complexity) and effective (solution quality) strategy for finding an estimate of motion parameters.

For a small number of motion parameters and a small state space for each of them, the most common search strategy when minimizing a prediction error, like (3.27), is matching. In this approach, motion-compensated predictions $\tilde{I}_k[\boldsymbol{n}] = I_{k-1}[\boldsymbol{n} - \boldsymbol{d}[\boldsymbol{n}]]$ for various motion candidates $\boldsymbol{d}$ are compared (matched) with $I_k[\boldsymbol{n}]$ within the region of support of the motion model (pixel, block, etc.). The candidate yielding the best match for a given criterion becomes the optimal estimate. For small state spaces, as is the case in block-constant motion models used in today's video coding standards, the full state space of each motion vector can be examined (exhaustive search) but partial search often gives almost as good results (Section 3.5.2).

As opposed to matching, *gradient-based techniques* require an estimation criterion $\mathcal{E}$ that is differentiable. Since this criterion depends on motion parameters through the image function, as in $I_{k-1}[\boldsymbol{n} - \boldsymbol{d}[\boldsymbol{n}]]$, to avoid nonlinear optimization I is usually linearized using Taylor expansion with respect to $\boldsymbol{d}[\boldsymbol{n}]$. Due to the Taylor approximation, the model is applicable only in a small vicinity of the initial $\boldsymbol{d}$. Since the initial motion is usually assumed to be zero, it comes as no surprise that gradient-based estimation yields accurate results only in regions of small motion; the approach fails if motion is large. This deficiency is usually compensated for by a hierarchical or multiresolution implementation [26, Chapter 1]. An example of hierarchical gradient-based method is reported in Section 3.5.1.

For algorithms using a spatial noncausal motion model, such as one based on MRFs, simultaneous optimization of thousands of parameters may be computationally prohibitive.[3] Therefore, relaxation techniques are usually employed to construct a series of estimates such that consecutive estimates differ in one variable at most. In case of estimating motion field $\boldsymbol{d}$, a series of motion fields $\boldsymbol{d}^{(0)}, \boldsymbol{d}^{(1)}, \ldots$ is constructed so that any two consecutive estimates $\boldsymbol{d}^{(k-1)}, \boldsymbol{d}^{(k)}$ differ at most at a single site $\boldsymbol{n}$. At each step of the relaxation procedure the motion vector at a single site is computed; vectors at other sites remain unchanged. Repeating this process results in propagation of motion properties, such as smoothness, that are embedded into the estimation criterion. Relaxation techniques are most often used in dense motion field estimation, but they equally apply to block-based methods.

In deterministic relaxation, such as Jacobi or Gauss-Seidel, each motion vector is computed with probability 1, that is, there is no uncertainty in the computation process. For example, a new local estimate is computed by minimizing the given criterion; variables are updated one after another and the criterion is monotonically improved step by step. Deterministic relaxation techniques are capable of correcting spurious motion vectors in the initial state $\boldsymbol{d}^{(0)}$ but they often get trapped in a local optimum near $\boldsymbol{d}^{(0)}$. Therefore, the availability of a good initial state is crucial.

The HCF algorithm [27] is an interesting variant of deterministic relaxation that is insensitive to the initial state. The distinguishing characteristic of the method is its site visiting schedule that is not fixed but driven by the input data. Without going into details, the HCF algorithm initially selects motion vectors that have the largest potential for reducing the estimation criterion $\mathcal{E}$. Usually, these are vectors in highly textured parts of an image. Later, the algorithm includes more and more motion vectors from low-texture areas, thus building on the neighborhood information of sites already estimated. By the algorithm's construction, the final estimate is independent of the initial state. The HCF is capable of finding close to optimal MAP estimates at a fraction of the computational cost of the globally optimal methods.

A deterministic algorithm specifically developed to deal with MRF formulations is called ICMs [28]. Although it does not maximize the a posteriori probability, it finds solutions that are reasonably close. The method is based on the division of sites of a random field into N sets such that each random variable associated with a site is independent of other random variables in the same set. The number of sets and their geometry depend on the selected cliques of the MRF. For example, for the first-order neighborhood system (Section 3.2.2), N equals 2 and the two sets look like a chess board. First, all the sites of one set are updated to find the optimal solution. Then, the sites of the other set are examined with the state of the first set already known. The procedure is repeated until a convergence criterion is met. Although the method converges quickly, it does not lead to as good solutions as the HCF approach. However, for each of the N

[3]There exist methods based on causal motion models that are computationally inexpensive, for example, pel-recursive motion estimation, but their accuracy is usually lower than that of methods based on noncausal motion models.

sets, the method can be implemented in parallel on a single-instruction multiple-data architecture.

The dependence on a good initial state is eliminated in stochastic relaxation. In contrast to the deterministic relaxation, the motion vector $\boldsymbol{v}$ under consideration is selected randomly (both its location $\boldsymbol{x}$ and parameters $\boldsymbol{b}$), thus allowing (with a small probability) a momentary deterioration of the estimation criterion [25]. In the context of minimization, such as in simulated annealing [2], this allows the algorithm to "climb" out of local minima and reach the global minimum. Stochastic relaxation methods, are easy to implement and capable of finding excellent solutions, but are slow to converge.

3.5 PRACTICAL MOTION ESTIMATION ALGORITHMS

3.5.1 Global Motion Estimation

As discussed in Section 3.4.1, camera movement induces motion of all image points and, therefore, is often an obstacle to solving various video processing problems. For example, to detect motion in video captured by a mobile camera, camera motion must be compensated first [6, Chapter 8]. Global motion compensation (GMC) plays also an important role in video compression because only a few motion parameters are sufficient to greatly reduce the prediction error when images to be encoded are acquired, for example, by a panning camera. GMC has been included in version 2 of the MPEG-4 video compression standard (Chapter 10).

Since camera motion is limited to translation and rotation, and affects all image points, a spatially parametric (e.g., affine (3.20)) and temporally linear (3.21) motion model supported on the whole image is appropriate. Note, that the spatially parametric motion models are accurate only for specific image formation, and object motion/surface models (Section 3.4.1). Under the constant-intensity observation model (3.25), the pixel-based quadratic criterion (3.28) leads to the following minimization:

$$\min_{\boldsymbol{b}} \mathcal{E}(\boldsymbol{v}), \qquad \mathcal{E}(\boldsymbol{v}) = \sum_{\boldsymbol{n}} \varepsilon^2[\boldsymbol{n}], \qquad \varepsilon[\boldsymbol{n}] = I_k[\boldsymbol{n}] - I_{k-1}[\boldsymbol{n} - \boldsymbol{v}(\boldsymbol{n}) \cdot (t_k - t_{k-1})], \tag{3.35}$$

where the dependence of $\boldsymbol{v}$ on $\boldsymbol{b}$ is implicit (3.20) and $t_k - t_{k-1}$ is usually assumed to equal 1. To perform the above minimization, gradient descent can be used. However, since this method gets easily trapped in a local minimum, an initial search for approximate translation components b_1 and b_2 (3.20), that can be quite large, needs to be performed. This search can be executed, for example, using the three-step block matching (Section 3.5.2).

Since the dependence of the cost function $\mathcal{E}$ on $\boldsymbol{b}$ is nonlinear, an iterative minimization procedure is typically used:

$$\boldsymbol{b}^{n+1} = \boldsymbol{b}^n + \boldsymbol{H}^{-1}\boldsymbol{c},$$

where $\boldsymbol{b}^n$ is the parameter vector $\boldsymbol{b}$ at iteration n, $\boldsymbol{H}$ is a $K \times K$ matrix equal to 1/2 of the Hessian matrix of $\mathcal{E}$ (i.e., matrix with elements $\partial^2\mathcal{E}/\partial b_k \partial b_l$), $\boldsymbol{c}$ is a K-dimensional vector equal to $-1/2$ of $\nabla\mathcal{E}$, and K is the number of parameters in the motion model (6 for affine). The above equation can be equivalently written as $\sum_{l=1}^{K} H_{kl}\Delta b_l = c_k, k = 1,\ldots,K$, where $\Delta\boldsymbol{b} = \boldsymbol{b}^{n+1} - \boldsymbol{b}^n$ and

$$H_{kl} = \frac{1}{2}\sum_{\boldsymbol{n}} \frac{\partial^2 \varepsilon^2[\boldsymbol{n}]}{\partial b_k \partial b_l} = \sum_{\boldsymbol{n}} \left(\frac{\partial \varepsilon[\boldsymbol{n}]}{\partial b_k} \frac{\partial \varepsilon[\boldsymbol{n}]}{\partial b_l} + \varepsilon[\boldsymbol{n}] \frac{\partial^2 \varepsilon[\boldsymbol{n}]}{\partial b_k \partial b_l} \right) \approx \sum_{\boldsymbol{n}} \left(\frac{\partial \varepsilon[\boldsymbol{n}]}{\partial b_k} \frac{\partial \varepsilon[\boldsymbol{n}]}{\partial b_l} \right),$$

$$c_k = -\frac{1}{2}\sum_{\boldsymbol{n}} \frac{\partial \varepsilon^2[\boldsymbol{n}]}{\partial b_k} = -\sum_{\boldsymbol{n}} \varepsilon[\boldsymbol{n}] \frac{\partial \varepsilon[\boldsymbol{n}]}{\partial b_k}.$$

The approximation of H_{kl} is due to dropping the second-order derivatives (see [29, page 683] for justification).

To handle large velocities and to speed up computations, the method needs to be implemented hierarchically. Thus, an image pyramid is built with spatial prefiltering and subsampling (usually by 2) applied between each two levels. The computation starts at the top level of the pyramid (lowest resolution) with b_1 and b_2 estimated in the initial step, and the other parameters set to zero. Then, gradient descent is performed by solving for $\Delta\boldsymbol{b}$, for example, using singular value decomposition, and updating $\boldsymbol{b}^{n+1} = \boldsymbol{b}^n + \Delta\boldsymbol{b}$ until a convergence criterion is met. The resulting motion parameters are projected onto a lower level of the pyramid[4] and the gradient descent is repeated. This cycle is repeated until the bottom of the pyramid is reached.

Since the global motion model applies to all image points, it cannot account for local motion. Thus, points moving independently of the global motion may generate large errors $\varepsilon[\boldsymbol{n}]$ and thus bias an estimate of the global motion parameters. The corresponding pixels are called *outliers* and, ideally, should be eliminated from the minimization (3.35). This can be achieved by using a robust criterion (Fig. 3.4) instead of the quadratic. For example, a Lorentzian function or a truncated quadratic can be used, but both provide a nonzero cost for outliers. This reduces the impact of outliers on the estimation but does not eliminate it completely. To exclude the impact of outliers altogether, a modified truncated quadratic should be used such as $\Phi_{tq}(\varepsilon,\theta,0)$ defined in (3.29). This criterion effectively limits the summation in (3.35) to the nonoutlier pixels and is used only during the gradient descent. The threshold θ can be fixed or it can be made adaptive, for example, by limiting the false alarm rate.

Figure 3.5 shows outlier pixels for two images "Foreman" and "Coastguard" declared using the above method based on the eight-parameter perspective motion model [20, Chapter 6]. Note the clear identification of outliers on the moving head, boats, and water. The outliers tend to appear at intensity transitions since it is there that any inaccuracy in global motion caused by a local (inconsistent) motion will induce large error ε; in uniform intensity areas undergoing local motion the error ε remains small. By excluding the outliers from the estimation, the accuracy of computed motion parameters is improved. Since the true camera motion is not known for these two sequences,

[4]The projection is performed by scaling the translation parameters b_1 and b_2 by two and leaving the other four parameters unchanged.

FIGURE 3.5

(a,b) Original images from CIF sequences "Foreman" and "Coastguard," and (c,d) pixels declared as outliers (black) according to a global motion estimate. Results from Dufaux and Konrad [30]. Reproduced with kind permission of the Institute of Electrical and Electronics Engineers (©2000 IEEE).

the improvement was measured in the context of the GMC mode of MPEG-4 compression.[5] In comparison with nonrobust global motion estimation ($\Phi_{tq}(\varepsilon,\infty,\cdot)$), the robust method ($\Phi_{tq}(\varepsilon,\theta,0)$) resulted in bit rate reduction of 8% and 15% for "Foreman" and "Coastguard," respectively [30].

[5]MPEG-4 encoder (version 2) can send parameters of global motion for each frame. Consequently, for each macroblock it can make a decision as to whether to perform a temporal prediction based on global motion parameters or local macroblock motion. The benefit of GMC is that only few motion parameters (e.g., 8) are sent for the whole frame. The GMC mode is beneficial for sequences with camera motion or zoom.

3.5.2 Block Matching

Block matching is the simplest algorithm for the estimation of local motion. It uses a spatially constant (3.19) and temporally linear (3.21) motion model over a rectangular region of support. Although as explained in Section 3.4.1, the translational 2D motion is only valid for the orthographic projection and 3D object translation, this model applied locally to a small block of pixels can be quite accurate for a large variety of 3D motions. It has proved accurate enough to serve as the basis for most of the practical motion estimation algorithms used today. Due to its simplicity and regularity (the same operations are performed for each block of the image), block matching can be relatively easily implemented in VLSI and, therefore, is used today in real-time encoders for *all* video compression standards (see Chapters 9 and 10).

In video compression, motion vectors $\boldsymbol{d}$ are used to eliminate temporal video redundancy *via* motion-compensated prediction (3.27). Hence, the goal is to achieve as low an amplitude of the prediction error $\varepsilon_k[\boldsymbol{n}]$ as possible. By applying a criterion Φ at each pixel and accumulating the results over a block, the method can be described by the following minimization:

$$\min_{\boldsymbol{d}[\boldsymbol{m}]\in\mathcal{P}} \mathcal{E}(\boldsymbol{d}[\boldsymbol{m}]), \qquad \mathcal{E}(\boldsymbol{d}[\boldsymbol{m}]) = \sum_{\boldsymbol{n}\in\mathcal{B}_{\boldsymbol{m}}} \Phi(I_k[\boldsymbol{n}] - I_{k-1}[\boldsymbol{n} - \boldsymbol{d}[\boldsymbol{m}]]) \qquad \forall \boldsymbol{m} \tag{3.36}$$

where $\mathcal{P}$ is the search area to which $\boldsymbol{d}[\boldsymbol{m}]$ belongs, defined as follows:

$$\mathcal{P} = \{\boldsymbol{n} = (n_1, n_2) : -P \leq n_1 \leq P, -P \leq n_2 \leq P\},$$

and $\mathcal{B}_{\boldsymbol{m}}$ is an $M \times N$ block of pixels with the top left corner coordinate at $\boldsymbol{m} = (m_1, m_2)$. The goal is to find the best, in the sense of the criterion Φ, displacement vector $\boldsymbol{d}[\boldsymbol{m}]$ for each block $\mathcal{B}_{\boldsymbol{m}}$. This is illustrated graphically in Fig. 3.6(a); a block is sought within image I_{k-1} that best matches the current block in I_k.

3.5.2.1 Estimation Criterion

Although an average error is used in (3.36), other measures are possible, such as maximum error (min-max estimation). To fully define the estimation criterion, the function Φ must be specified. Originally, $\Phi(x) = x^2$ was often used in block matching, but it was replaced by the absolute error criterion $\Phi(x) = |x|$ for its simplicity (no multiplications) and robustness in the presence of outliers. Other improved criteria have been proposed, such as based on the median of squared errors; however, their computational complexity is significantly higher.

Also, simplified criteria have been proposed to speed up the computations, for example, based on adaptive quantization to 2 bits or pixel subsampling [31]. Usually, a simplification of the original criterion Φ leads to suboptimal performance. However, with an adaptive adjustment of the criterion's parameters (e.g., quantization levels, decimation patterns) a close-to-optimal performance can be achieved at significantly reduced complexity.

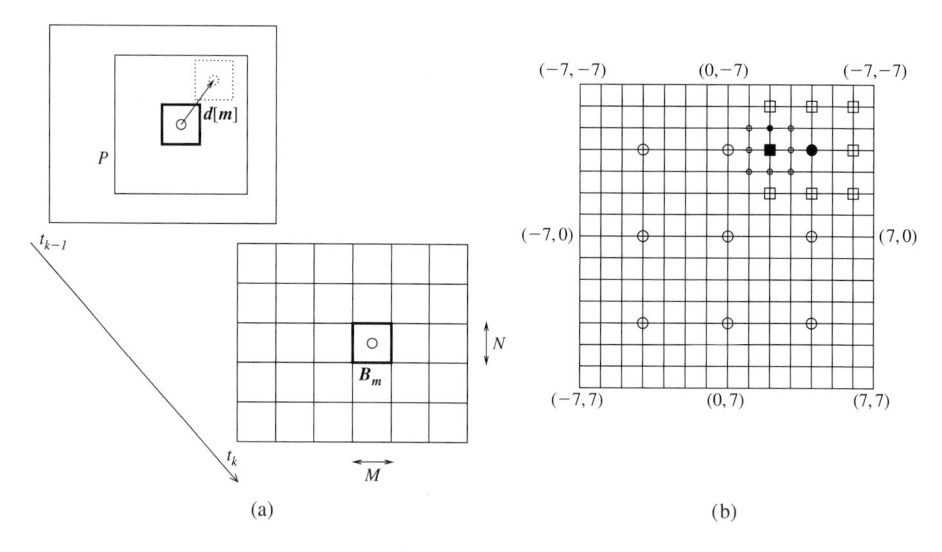

FIGURE 3.6

(a) Block matching between block $\mathcal{B}_m$ at time t_k (current image) and all possible blocks in the search area $\mathcal{P}$ at time t_{k-1}. (b) Three-step search method. Large circles denote level 1 ($\mathcal{P}_1$), squares denote level 2 ($\mathcal{P}_2$) and small circles denote level 3 ($\mathcal{P}_3$). The filled-in elements denote the best match found at each level. The final vector is (2, −5).

3.5.2.2 *Search Methods*

An exhaustive search for $\boldsymbol{d}[\boldsymbol{m}] \in \mathcal{P}$ that gives the lowest error $\mathcal{E}$ is computationally costly. An "intelligent" search, whereby only the more likely candidates from $\mathcal{P}$ are evaluated, usually results in substantial computational savings. One popular technique for reducing the number of candidates is the logarithmic search. Assuming that $P = 2^k - 1$ and denoting $P_l = (P+1)/2^l$, where k and l are integers, the new reduced-size search area is established as follows:

$$\mathcal{P}_l = \{\boldsymbol{n} : \boldsymbol{n} = (\pm P_l, \pm P_l) \text{ or } \boldsymbol{n} = (\pm P_l, 0) \text{ or } \boldsymbol{n} = (0, \pm P_l) \text{ or } \boldsymbol{n} = (0,0)\},$$

that is, $\mathcal{P}_l$ is reduced to the vertexes, midway points between vertexes and the central point of the half-sized original rectangle $\mathcal{P}$. For example, for $P = 7$, $\mathcal{P}_1$ consists of the following candidates: $(-4,-4)$, $(-4,4)$, $(4,-4)$, $(4,4)$, $(-4,0)$, $(4,0)$, $(0,-4)$, $(0,4)$, and $(0,0)$. The search starts with the candidates from $\mathcal{P}_1$. Once the best match is found, the new search area $\mathcal{P}_2$ is centered around this match, and the procedure is repeated for candidates from $\mathcal{P}_2$. Note that the error $\mathcal{E}$ does not have to be evaluated for the $(0,0)$ candidate since it had been evaluated at the previous level. The procedure is continued with subsequently reduced search spaces. Since typically only 3 levels ($l = 1,2,3$) are used, such a method is often referred to as the *three-step search* (Fig. 3.6(b)).

In the logarithmic search above, at each step a 2D search is performed. An alternative approach is to perform 1D searches only, usually in orthogonal directions. Examples of block matching algorithms based on 1D search methods are as follows:

1. *One-at-a-time search* [32]: In this method, first a minimum of $\mathcal{E}$ is sought in one, for example horizontal, direction. Then, given the horizontal estimate, a vertical search is performed. Subsequently, horizontal search is performed given the previous vertical estimate, and so on. In the original proposal, only the 1D minima closest to the origin were examined, but later a 1D fullsearch was used to avoid the problem of getting trapped close to the origin. Note that the searches are not independent since each relies on the result of the previous one.

2. *Parallel hierarchical one-dimensional search* [31]: This method also performs 1D searches in orthogonal directions (usually horizontal and vertical) but independently of each other, that is, the horizontal search is performed simultaneously with the vertical search since it does not depend on the outcome of the latter. In addition, the 1D search is implemented hierarchically. First, every K-th location from $\mathcal{P}$ is taken as a candidate for the 1D search. Once the minima in both directions are identified, new 1D searches begin with every $K/2$-th location from $\mathcal{P}$, and so on. Typically, horizontal and vertical searches using every eighth, fourth, second and finally every pixel (within the limits of $\mathcal{P}$) are performed.

A remark is in order at this point. All the fast search methods are based on the assumption that the error $\mathcal{E}$ has a single minimum for all $\boldsymbol{d}[\boldsymbol{m}] \in \mathcal{P}$ or, in other words, that $\mathcal{E}$ increases monotonically when moving away from the best-match position. In practice, this is rarely true since $\mathcal{E}$ depends on $\boldsymbol{d}[\boldsymbol{m}]$ through the intensity I, which can be arbitrary. Therefore, multiple local minima often exist within $\mathcal{P}$, and a fast search method can be easily trapped in any one of them, whereas an exhaustive search will always find the "deepest" minimum. This is not a very serious problem in video coding since a suboptimal motion estimate translates into an increased prediction error (3.27) that will be entropy coded and, at most, will result in rate increase. It is a serious problem, however, in video processing where the *true* motion is sought and any motion errors may result in uncorrectable distortions. A good review of block matching algorithms can be found in [31].

3.5.3 Phase Correlation

As discussed above, block matching can precisely estimate local displacement but must examine all possible candidates (exhaustive search). At the same time, methods based on the frequency-domain criteria (Section 3.4.2.2) are capable of identifying global motion but cannot localize it in space-time. By combining the two approaches, phase correlation [33] is able to exploit advantages of both. First, likely candidates are computed using a frequency-domain approach and then they are assigned a spatial location by local block matching.

Recall the cross-correlation criterion $C(\boldsymbol{d})$ expressed in the Fourier domain (3.32). By normalizing $\mathcal{F}[C(\boldsymbol{d})]$ and taking the inverse transform one obtains

$$\Psi_{k-1,k}(\boldsymbol{n}) = \mathcal{F}^{-1}\left\{\frac{\widehat{I}_k(\boldsymbol{u})\widehat{I}^*_{k-1}(\boldsymbol{u})}{|\widehat{I}_k(\boldsymbol{u})\widehat{I}^*_{k-1}(\boldsymbol{u})|}\right\}. \tag{3.37}$$

$\Psi_{k-1,k}(\boldsymbol{n})$ is a normalized correlation computed between images I_k and I_{k-1}. In the special case of global translation ($I_k[\boldsymbol{n}] = I_{k-1}[\boldsymbol{n}-\boldsymbol{z}]$), by using the transform (3.31) it can be easily shown that this correlation becomes a Kronecker delta function ($\delta(\boldsymbol{x})$ equals 0 for $\boldsymbol{x} \neq 0$ and 1 for $\boldsymbol{x} = 0$):

$$\Psi_{k-1,k}(\boldsymbol{n})|_{I_k[\boldsymbol{n}]=I_{k-1}[\boldsymbol{n}-\boldsymbol{z}]} = \mathcal{F}\{e^{-j2\pi\boldsymbol{u}\cdot\boldsymbol{z}}\} = \delta(\boldsymbol{n}-\boldsymbol{z}).$$

In practice, when global translation and intensity constancy hold only approximately, $\Psi_{k-1,k}$ has a more complicated shape, potentially with numerous peaks buried in noise. These peaks correspond to dominant displacements between I_{k-1} and I_k, and, if identified, are very good candidates for fine-tuning by, for example, block matching. Note that no explicit motion model has been used, thus far, while the observation model, as usual, is that of constant intensity and the estimation criterion is the cross-correlation. In practice, the method can be implemented as follows [33]:

1. divide I_{k-1} and I_k into large blocks, for example, 64×64 (motion range of $\pm$ 32 pixels), and take the fast Fourier transform (FFT) of each block,
2. compute $\Psi_{k-1,k}$ using same-position blocks in I_{k-1} and I_k,
3. take the inverse FFT of $\Psi_{k-1,k}$ and identify the dominant peaks,
4. use the dominant peak coordinates as the candidate vectors for 16×16-pixel block matching.

The phase correlation method is basically an efficient maximization of a correlation-based error criterion. The shape of the maxima of the correlation surface is weakly dependent on the image content, and the measurement of their locations is relatively independent of illumination changes. This is due, predominantly, to the normalization in (3.37). However, rotations and zooms cannot be easily handled since the peaks in $\Psi_{k-1,k}$ are hard to distinguish due to nonconstancy of the corresponding motion fields.

3.5.4 Optical Flow via Regularization

Recall the regularized estimation criterion (3.33). It uses a translational/linear motion model at each pixel under the constant-intensity observation model and quadratic error criterion. To find the continuous functions v_1 and v_2, implicitly dependent on $\boldsymbol{x}$, the functional in (3.33) needs to be minimized, which is a problem in the calculus of variations. The Euler-Lagrange equations yield [22]:

$$\lambda\nabla^2 v_1 = \left(\frac{\partial I}{\partial x}v_1 + \frac{\partial I}{\partial y}v_2 + \frac{\partial I}{\partial t}\right)\frac{\partial I}{\partial x},$$

$$\lambda \nabla^2 v_2 = \left(\frac{\partial I}{\partial x} v_1 + \frac{\partial I}{\partial y} v_2 + \frac{\partial I}{\partial t} \right) \frac{\partial I}{\partial y},$$

where $\nabla^2 = \partial^2/\partial x^2 + \partial^2/\partial y^2$ is the Laplacian operator. This pair of elliptic partial differential equations can be solved iteratively using finite-difference or finite-element discretization.

An alternative is to formulate the problem directly in the discrete domain. Then, the integral in (3.33) is replaced by a summation, whereas the derivatives are replaced by finite differences. In [22], for example, an average of first-order differences computed over a $2 \times 2 \times 2$ cube was used. By differentiating this discrete cost function, a system of equations can be computed and subsequently solved by Jacobi or Gauss-Seidel relaxation. This discrete approach to regularization is a special case of the MAP estimation presented next.

3.5.5 MAP Estimation of Dense Motion

The MAP formulation (3.34) is very general and requires further assumptions. The likelihood $P(\mathcal{I}_k = I_k | \boldsymbol{D}_k = d_k; I_{k-1})$ relates one image to the other through $\boldsymbol{d}_k$. Since $I_k[\boldsymbol{n}] = I_{k-1}[\boldsymbol{n} - \boldsymbol{d}_k[\boldsymbol{n}]] + \varepsilon_k[\boldsymbol{n}]$ (3.27), the characteristics of this likelihood reside in ε_k. For accurate displacement estimates, the prediction error ε_k is expected to behave like a noise term. In fact, it has been shown that the statistics of ε_k at specific $\boldsymbol{n}$ are reasonably close to those of a zero-mean Gaussian distribution, although a generalized Gaussian is a better fit [34]. Therefore, assuming no correlation among $\varepsilon_k[\boldsymbol{n}]$ for different $\boldsymbol{n}$, $P(\mathcal{I}_k = I_k | \boldsymbol{D}_k = d_k; I_{k-1})$ can be fairly accurately modeled by a product of zero-mean Gaussian distributions.

The prior probability is particularly flexible when $\boldsymbol{D}_k$ is assumed to be a MRF. Then, $P(\boldsymbol{D}_k = \boldsymbol{d}_k; I_{k-1})$ is a Gibbs distribution (Section 3.2.2) uniquely specified by cliques and a potential function. For example, for two-element cliques $\{\boldsymbol{n}, \boldsymbol{l}\}$ the smoothness of $\boldsymbol{D}_k$ can be expressed using the following potential function:

$$V_s(\boldsymbol{d}_k[\boldsymbol{n}], \boldsymbol{d}_k[\boldsymbol{l}]) = \|\boldsymbol{d}_k[\boldsymbol{n}] - \boldsymbol{d}_k[\boldsymbol{l}]\|^2, \qquad \forall \{\boldsymbol{n}, \boldsymbol{l}\} \in C,$$

where $\|\cdot\|$ denotes the Euclidean norm. Clearly, for similar $\boldsymbol{d}_k[\boldsymbol{n}]$ and $\boldsymbol{d}_k[\boldsymbol{l}]$, the potential V_s is small, and thus, the prior probability is high, whereas for dissimilar vectors this probability is small.

Since both likelihood and prior probability distributions are exponential in this case, the MAP estimation (3.34) can be rewritten as energy minimization:

$$\widehat{\boldsymbol{d}}_k = \arg\min_{\boldsymbol{d}} \left(\frac{1}{2\sigma^2} \sum_{\boldsymbol{n}} \left[(I_k[\boldsymbol{n}] - I_{k-1}[\boldsymbol{n} - \boldsymbol{d}[\boldsymbol{n}]])^2 + \sum_{\{\boldsymbol{n}, \boldsymbol{l}\} \in C} \|\boldsymbol{d}[\boldsymbol{n}] - \boldsymbol{d}[\boldsymbol{l}]\|^2 \right] \right). \tag{3.38}$$

The above energy can be minimized in various ways. To attain the global minimum, simulated annealing (Section 3.4.3) should be used. Given sufficiently many iterations, the method is theoretically capable of finding the global minimum, however, at considerable computational cost. However, the method is easy to implement [25]. A faster alternative is the deterministic ICM method that does not find a true MAP estimate, although usually finds a close enough solution in a fraction of time taken by simulated annealing. Even more effective is the HCF method, although its implementation is a little bit more complex.

It is worth noting that the formulation (3.38) comprises, as a special case, the discrete formulation of the optical flow computation mentioned in Section 3.5.4. Consider the constraint (3.24). Multiplying both sides by $\Delta t = t_k - t_{k-1}$ it becomes $I^x d_1 + I^y d_2 + I^t \Delta t = 0$, where $d_1 = v_1 \Delta t$ and $d_2 = v_2 \Delta t$, whereas I^x, I^y, and I^t are discrete approximations to horizontal, vertical, and temporal derivatives, respectively. This constraint is not satisfied exactly for real data; $I^x d_1 + I^y d_2 + I^t \Delta t$ is a noise-like term with characteristics similar to the prediction error ε_k. This is not surprising since both originate from the same constant-intensity hypothesis. By replacing the prediction error in (3.38) with this new term one obtains a cost function equivalent to the discrete formulation of the optical flow problem (Section 3.5.4) [22].

The minimization (3.38) leads to smooth displacement fields $\boldsymbol{d}_k$, also at object boundaries which is undesirable. To relax the smoothness constraint at object boundaries, explicit models of motion discontinuities (line field) [25] or of motion segmentation labels (segmentation field) [34] can be easily incorporated into the MRF formulation, although their estimation is far from trivial.

3.5.6 Experimental Comparison of Motion Estimation Methods

To demonstrate the impact of various motion models, Fig. 3.7 shows results for the QCIF sequence "Carphone." Both the estimated displacements and the resulting motion-compensated prediction errors are shown for pixel-based (dense), block-based, and region-based motion models. The latter motion estimate was obtained by minimizing the mean squared error ($\Phi(\varepsilon) = \varepsilon^2$ in (3.28)) within each region $\mathcal{R}$ from Fig. 3.7(c) for the affine motion model (3.20).

Note the lack of detail in the block-based motion estimate (16×16-pixel blocks), but approximately correct motion of the head. The pixel-based model results in a very smooth estimate with more spatial detail around the moving head but a reduced precision in low-texture areas. The region-based motion estimate shows both better accuracy and detail. Although the associated segmentation (Fig. 3.7(c)) does not correspond exactly to objects as perceived by humans, it nevertheless closely matches object boundaries. The motion of the head and of the upper body is well-captured, but the motion of landscape in the car window is exaggerated due to the lack of image details. As for the prediction error, note the blocking artifacts for the block-based motion model (31.8 dB[6]) but a very small error for the pixel-based model (35.9 dB). The region-based model results in a slightly higher prediction error (35.5 dB) than the pixel-based model, but significantly lower than that of the block model.

[6] The prediction error is defined as follows: $10\log(255^2/\mathcal{E}(\boldsymbol{d}))$ [dB] for $\mathcal{E}$ from (3.28) with quadratic Φ and $\mathcal{R}$ being the whole image.

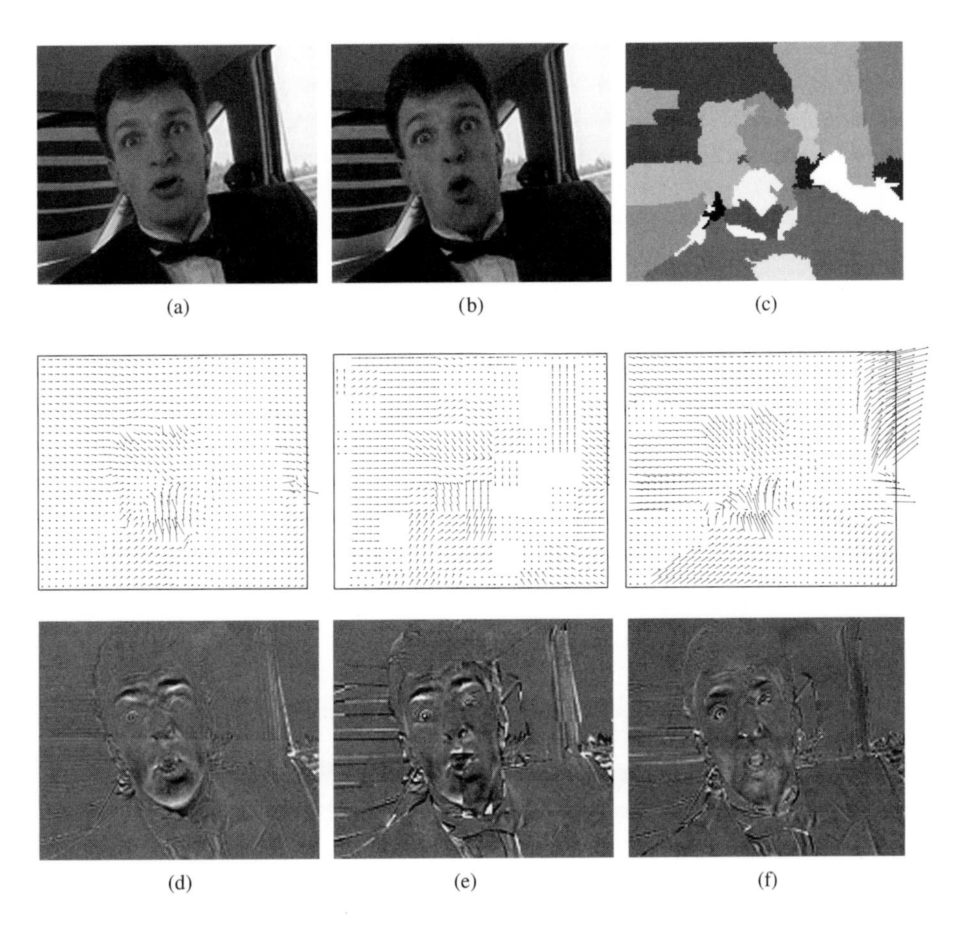

FIGURE 3.7

Original frames (a) #168 and (b) #171 from QCIF sequence "Carphone;" (c) motion-based segmentation of frame #171; motion estimates (subsampled by 4) and the resulting motion-compensated prediction error (magnified by 2) at frame #171 for (d) dense-field MAP estimation; (e) 16×16-pixel block matching; (f) region-based estimation for segments from (c). Results from Konrad and Stiller [6, Chapter 4]. Reproduced with kind permission of Springer Science and Business Media.

3.6 PERSPECTIVES

In the last two decades, motion detection and estimation have moved from research laboratories to specialized products. This has been made possible by two factors. First, enormous advances in VLSI have facilitated practical implementation of CPU-hungry motion detection and estimation algorithms. Second, new models and estimation algorithms have lead to improved reliability and accuracy of the estimated motion. With the continuing advances in VLSI, the complexity constraints plaguing motion algorithms will become less of an issue. This should allow practical implementation of more advanced motion models and estimation criteria, and, in turn, further improve the accuracy of the computed motion. One of the promising approaches studied today is the joint motion segmentation and estimation that effectively combines the detection and estimation discussed separately in this chapter.

3.7 ACKNOWLEDGMENTS

The author would like to thank Dr Mirko Ristivojević and Mr Kai Guo for their generous help with the preparation of some experimental results. Various parts of this work were supported over many years by the Natural Sciences and Engineering Research Council of Canada and, more recently, by the National Science Foundation.

REFERENCES

[1] H. van Trees. *Detection, Estimation and Modulation Theory.* John Wiley and Sons, New York, 1968.

[2] S. Geman and D. Geman. Stochastic relaxation, Gibbs distributions, and the Bayesian restoration of images. *IEEE Trans. Pattern Anal. Mach. Intell.*, 6:721–741, 1984.

[3] R. Weinstock. *Calculus of Variations.* Dover Publications, New York, 1974.

[4] V. Caselles, R. Kimmel, and G. Sapiro. Geodesic active contours. *Int. J. Comput. Vis.*, 22(1): 61–79, 1997.

[5] J. Sethian. *Level Set Methods.* Cambridge University Press, Cambridge, UK, 1996.

[6] H. Li, S. Sun, and H. Derin editors. *Video Compression for Multimedia computing–Statistically Based and Biologically Inspired Techniques.* Kluwer Academic Publishers, New York, NY, 1997.

[7] C. Stauffer and E. Grimson. Learning patterns of activity using real-time tracking. *IEEE Trans. Pattern Anal. Mach. Intell.*, 22(8):747–757, 2000.

[8] A. Elgammal, R. Duraiswami, D. Harwood, and L. Davis. Background and foreground modeling using nonparametric kernel density for visual surveillance. *Proc. IEEE*, 90:1151–1163, 2002.

[9] Y. Sheikh and M. Shah. Bayesian modeling of dynamic scenes for object detection. *IEEE Trans. Pattern Anal. Mach. Intell.*, 27(11):1778–1792, 2005.

[10] P.-M. Jodoin, M. Mignotte, and J. Konrad. Statistical background subtraction using spatial cues. *IEEE Trans. Circuits Syst. Video Technol.*, 17:1758–1763, 2007.

[11] T. Aach and A. Kaup. Bayesian algorithms for adaptive change detection in image sequences using Markov random fields. *Signal Process. Image Commun.*, 7: 147–160, 1995.

[12] P. Bouthemy and P. Lalande. Recovery of moving object masks in an image sequence using local spatiotemporal contextual information. *Opt. Eng.*, 32(6): 1205–1212, 1993.

[13] J. Migdal and E. L. Grimson. Background subtraction using markov thresholds. In *Proc. IEEE Workshop Motion Video Comput. WACV/MOTIONS'05*, Vol. 2, 58–65, 2005.

[14] F. Luthon, A. Caplier, and M. Liévin. Spatiotemporal MRF approach with application to motion detection and lip segmentation in video sequences. *Signal Process.*, 76:61–80, 1999.

[15] S. Jehan-Besson, M. Barlaud, and G. Aubert. Detection and tracking of moving objects using a new level set based method. In *Proc. Int. Conf. Pattern, Recognit.*, 1112–1117, September 2000.

[16] A.-R. Mansouri and J. Konrad. Multiple motion segmentation with level sets. *IEEE Trans. Image Process.*, 12:201–220, 2003.

[17] S. Zhu and A. Yuille. Region competition: Unifying snakes, region growing, and Bayes/MDL for multiband image segmentation. *IEEE Trans. Pattern Anal. Machine Intell.*, 18:884–900, 1996.

[18] M. Ristivojević and J. Konrad. Space-time image sequence analysis: object tunnels and occlusion volumes. *IEEE Trans. Image Process.*, 15:364–376, 2006.

[19] E. Trucco and A. Verri. *Introductory Techniques for 3-D Computer Vision.* Prentice Hall, Upper Saddle River, NJ, 1998.

[20] L. Torres and M. Kunt, editors. *Video Coding: Second Generation Approach.* Kluwer Academic Publishers, New York, NY, 1996.

[21] M. Chahine and J. Konrad. Estimation and compensation of accelerated motion for temporal sequence interpolation. *Signal Process. Image Commun.*, 7: 503–527, 1995.

[22] B. Horn. *Robot Vision.* MIT Press, Cambridge, MA, 1986.

[23] M. Bertero, T. Poggio, and V. Torre. Ill-posed problems in early vision. *Proc. IEEE*, 76: 869–889, 1988.

[24] M. Black. *Robust incremental optical flow.* PhD thesis, Yale University, Department of Computer Science, September 1992.

[25] J. Konrad and E. Dubois. Bayesian estimation of motion vector fields. *IEEE Trans. Pattern Anal. Mach. Intell.*, 14:910–927, 1992.

[26] I. Sezan and R. Lagendijk, editors. *Motion Analysis and Image Sequence Processing.* Kluwer Academic Publishers, New York, NY, 1993.

[27] P. Chou and C. Brown. The theory and practice of Bayesian image labelling. *Int. J. Comput. Vis.*, 4: 185–210, 1990.

[28] J. Besag. On the statistical analysis of dirty pictures. *J. R. Stat. Soc.* Ser. B 48: 259–279, 1986.

[29] W. Press, S. Teukolsky, W. Vetterling, and B. Flannery. *Numerical Recipes in C: The Art of Scientific Computing*, 2nd ed. Cambridge University Press, Cambridge, UK, 1992.

[30] F. Dufaux and J. Konrad. Robust, efficient and fast global motion estimation for video coding. *IEEE Trans. Image Process.*, 9: 497–501, 2000.

[31] V. Bhaskaran and K. Konstantinides. *Image and Video Compression Standards: Algorithms and Architectures.* Kluwer Academic Publishers, New York, NY, 1997.

[32] R. Srinivasan and K. Rao. Predictive coding based on efficient motion estimation. *IEEE Trans. Commun.*, 33:888–896, 1985.

[33] A. Tekalp. *Digital Video Processing.* Prentice Hall PTR, Upper Saddle River, NJ, 1995.

[34] C. Stiller. Object-based estimation of dense motion fields. *IEEE Trans. Image Process.*, 6:234–250, 1997.

CHAPTER

Video Enhancement and Restoration

4

Reginald L. Lagendijk[1], Jan Biemond[1], Andrei Rareş[2], and Marcel J. T. Reinders[1]

[1] *Information and Communication Theory Group, Department of Mediamatics, Faculty of Electrical Engineering, Mathematics and Computer Science, Delft University of Technology, The Netherlands*

[2] *Division of Image Processing (LKEB), Department of Radiology, Leiden University Medical Center, The Netherlands*

4.1 INTRODUCTION

Even with the advancing camera and digital recording technology, there are many situations in which recorded image sequences—or video for short—may suffer from severe degradations. The poor quality of recorded image sequences may be due to, for instance, the imperfect or uncontrollable recording conditions, such as one encounters in astronomy, forensic sciences, and medical imaging. Video enhancement and restoration has always been important in these application areas not only to improve the visual quality but also to increase the performance of subsequent tasks such as analysis and interpretation.

The available bandwidth for transmission or storage can also affect the quality of digital video data. In general, lossy compression is applied using industry compression standards, such as JPEG, H.26x, and MPEG-x, which may result in visible coding artifacts, such as blocking, ringing, and mosquito noise. Then, the challenge is to design noise or artifact reduction filters—so-called video enhancement filters—that provide a graceful trade-off between the amount of noise reduction and the resulting loss of perceptual picture quality.

Another important application of video enhancement and restoration is preserving motion pictures and video tapes recorded over the last century. These unique records of historic, artistic, and cultural developments are deteriorating rapidly due to aging effects of the physical reels of film and magnetic tapes that carry the information. The preservation of these fragile archives is of interest not only to professional archivists but also to broadcasters as a cheap alternative to fill the many television channels that have come available with digital broadcasting. However, reusing old film and video material is only feasible if the visual quality meets the today's standards. First, the archived film and video is transferred from the original film reels or magnetic tape to digital media. Then, all kinds of degradations are removed from the digitized image sequences, in this way increasing the visual quality and commercial value. Because the objective of restoration is to remove irrelevant information such as noise and blotches, it restores the original spatial and temporal correlation structure of digital image sequences. Consequently, restoration may also improve the efficiency of the subsequent MPEG compression of image sequences.

An important difference between the enhancement and restoration of 2D images and video is the amount of data to be processed. While for the quality improvement of important images elaborate processing is still feasible, this is no longer true for the absolutely huge amounts of pictorial information encountered in medical sequences and film/video archives. Consequently, enhancement and restoration methods for image sequences should have a manageable complexity and should be semiautomatic. The term semiautomatic indicates that in the end professional operators control the visual quality of the restored image sequences by selecting values for some of the critical restoration parameters.

The most common artifact encountered in the above-mentioned applications is noise. Over the last two decades, an enormous amount of research has focused on the problem of enhancing and restoring 2D images. Clearly, the resulting spatial methods are also applicable to image sequences, but such an approach implicitly assumes that the individual pictures of the image sequence, or frames, are temporally independent. By ignoring the temporal correlation that exists, suboptimal results may be obtained, and the spatial intraframe filters tend to introduce temporal artifacts in the restored image sequences. In this chapter, we focus our attention specifically on exploiting temporal dependencies, yielding interframe methods. In this respect, the material offered in this chapter is complementary to that on image enhancement in the Chapters 10–13 of *The Essential Guide to Image Processing* [1]. The resulting enhancement and restoration techniques operate in the temporal dimension by definition, but often have a spatial filtering component as well. For this reason, video enhancement and restoration techniques are sometimes referred to as spatiotemporal filters or 3D filters. Section 4.2 of this chapter presents three important classes of noise filters for video frames, namely linear temporal filters, order-statistic (OS) filters, and multiresolution filters. Section 4.3 focuses on coding artifact reduction, in particular on blockiness reduction due to the lossy Discrete Cosine Transform (DCT).

In forensic sciences and film and video archives, a large variety of artifacts are encountered. Besides noise, we discuss the removal of other important impairments that rely on spatial or temporal processing algorithms, namely blotches (Section 4.4), vinegar syndrome (Section 4.5), intensity flicker (Section 4.6), kinescope moiré (Section 4.7), and scratches (Section 4.8). Blotches are dark and bright spots that are often visible in damaged film image sequences. The removal of blotches is essentially a temporal detection and interpolation problem. In certain circumstances, their removal needs to be done by means of spatial algorithms, as will be shown later in this chapter. Vinegar syndrome represents a special type of impairment related to film, and it may have various appearances (e.g., partial loss of color, blur). In some cases, blotch removal algorithms can be applied for blotch removal. In general, however, the particular properties of their appearance

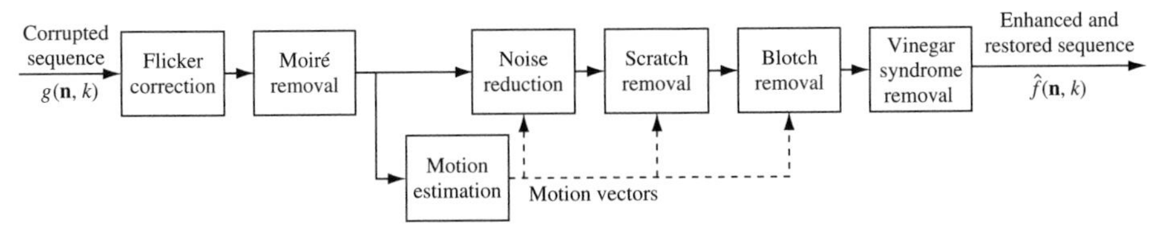

FIGURE 4.1

Some processing steps in the removal of various video artifacts.

need to be taken into account. Intensity flicker refers to variations in intensity in time, caused by aging of film, by copying and format conversion (e.g., from film to video), and—in case of earlier film—by variations in shutter time. While blotches are spatially highly localized artifacts in video frames, intensity flicker is usually a spatially global, but not stationary, artifact. The kinescope moiré phenomenon appears during film-to-video transfer using telecine devices. It is caused by the superposition of (semi-)periodical signals from the film contents and the scan pattern used by the telecine device. Because of its (semi-)periodical nature, spectrum analysis is generally used for its removal. Film scratches are either bright or dark vertical lines spanning the entire frame. They appear approximately at the same place in consecutive frames.

It becomes apparent that image sequences may be degraded by multiple artifacts. For practical reasons, restoration systems follow a sequential procedure, where artifacts are removed one by one. As an example, Fig. 4.1 illustrates the order in which the removal of flicker, moiré, noise, scratches, blotches, and vinegar syndrome takes place. The reasons for this modular approach are the necessity to judge the success of the individual steps (e.g., by an operator) and the algorithmic and implementation complexity. The coding artifacts removal has not been displayed because it usually takes place in a preprocessing step.

As already suggested in Fig. 4.1, most temporal filtering techniques require an estimate of the motion in the image sequence. Motion estimation has been discussed in detail in Chapters 3, 8, and 11 of this book. However, the estimation of motion from degraded image sequences is problematic. We are faced with the problem that the impairments of the video disturb the motion estimator, but at the same time correct motion estimates are assumed in developing enhancement and restoration algorithms. In this chapter, we do not discuss the design of new motion estimators [2–4] that are robust to the various artifacts, but we assume that existing motion estimators can be modified appropriately such that sufficiently correct and smooth motion fields are obtained. The reason for this approach is that even under ideal conditions motion estimates are never perfect. Usually, incorrect or unreliable motion vectors are dealt with in a few special ways. First, clearly incorrect or unreliable motion vectors can be repaired. Second, the enhancement and restoration algorithms should be robust against a limited amount of incorrect or unreliable motion vectors. Third, areas with wrong motion vectors that are impossible to repair can be protected against temporal restoration to avoid an outcome, which is visually more objectionable than the input sequence itself. In such a case, the unavailability of temporal information makes the spatial-only restoration more suitable than the temporal one.

4.2 SPATIOTEMPORAL NOISE FILTERING

Any recorded signal is affected by noise, no matter how precise the recording equipment. The sources of noise that can corrupt an image sequence are numerous (see Chapter 7 of *The Essential Guide to Image Processing* [1]). Examples of the more prevalent ones include camera noise, shot noise originating in electronic hardware and the storage on magnetic tape, thermal noise, and granular noise on film. Most recorded and digitized image sequences contain a mixture of noise contributions, and often the (combined) effects of the noise are nonlinear of nature. In practice, however, the aggregated effect of noise is modeled as an additive white (sometimes Gaussian) process with zero mean and variance σ_w^2 that is independent of the ideal uncorrupted image sequence $f(\mathbf{n},k)$. The recorded image sequence $g(\mathbf{n},k)$ corrupted by noise $w(\mathbf{n},k)$ is then given as follows:

$$g(\mathbf{n},k) = f(\mathbf{n},k) + w(\mathbf{n},k), \tag{4.1}$$

where $\mathbf{n} = (n_1, n_2)$ refers to the spatial coordinates and k refers to the frame number in the image sequence. More accurate models are often much more complex but lead to little gain compared to the added complexity.

The objective of noise reduction is to make an estimate $\hat{f}(\mathbf{n},k)$ of the original image sequence given only the observed noisy image sequence $g(\mathbf{n},k)$. Many different approaches toward noise reduction are known, including optimal linear filtering, nonlinear filtering, scale-space processing, and Bayesian techniques. In this section, we discuss successively the class of linear image sequence filters, OS filters, and multiresolution filters. In all cases, the emphasis is on the temporal filtering aspects. More rigorous reviews of noise filtering for image sequences are given in [5–7].

4.2.1 Linear Filters

4.2.1.1 *Temporally Averaging Filters*

The simplest temporal filter carries out a weighted averaging of successive frames. That is, the restored image sequence is obtained by [2, 8]:

$$\hat{f}(\mathbf{n},k) = \sum_{l=-K}^{K} h(l) g(\mathbf{n}, k-l), \tag{4.2}$$

where $h(l)$ are the temporal filter coefficients used to weight $2K+1$ consecutive frames. In case the frames are considered equally important, we have $h(l) = 1/(2K+1)$. Alternatively, the filter coefficients can be optimized in a minimum mean squared error fashion

$$h(l) \leftarrow \min_{h(l)} \mathrm{E}\left[\left(f(\mathbf{n},k) - \hat{f}(\mathbf{n},k)\right)^2\right], \tag{4.3}$$

yielding the well-known temporal Wiener filtering solution:

$$\begin{pmatrix} h(-K) \\ \vdots \\ h(0) \\ h(1) \\ \vdots \\ h(K) \end{pmatrix} = \begin{pmatrix} R_{gg}(0) & \cdots & R_{gg}(-K) & \cdots & \cdots & R_{gg}(-2K) \\ \vdots & \ddots & & & & \vdots \\ R_{gg}(K) & & R_{gg}(0) & & & \vdots \\ \vdots & & & R_{gg}(0) & & \vdots \\ \vdots & & & & \ddots & \vdots \\ R_{gg}(2K) & \cdots & \cdots & \cdots & \cdots & R_{gg}(0) \end{pmatrix}^{-1} \begin{pmatrix} R_{fg}(-K) \\ \vdots \\ R_{fg}(0) \\ R_{fg}(1) \\ \vdots \\ R_{fg}(K) \end{pmatrix}, \tag{4.4}$$

where $R_{gg}(m)$ is the temporal auto-correlation function defined as $R_{gg}(m) = E\left[g(\mathbf{n},k)g(\mathbf{n},k-m)\right]$, and $R_{fg}(m)$ the temporal cross-correlation function defined as $R_{fg}(m) = E\left[f(\mathbf{n},k)g(\mathbf{n},k-m)\right]$. The temporal window length, that is, the parameter K, determines the maximum degree by which the noise power can be reduced. The larger the window, the greater the reduction of the noise, at the same time, however, the more visually noticeable the artifacts resulting from motion between the video frames. A dominant artifact is blur of moving objects due to the averaging of object and background information.

The motion artifacts can be greatly reduced by operating the filter (4.2) along the picture elements (pixels) that lie on the same motion trajectory [9, 10]. Equation (4.2) then becomes a motion-compensated temporal filter (see Fig. 4.2):

$$\hat{f}(\mathbf{n},k) = \sum_{l=-K}^{K} h(l)\, g\left(n_1 - d_x(n_1,n_2;k,l), n_2 - d_y(n_1,n_2;k,l), k-l\right), \tag{4.5}$$

where $\mathbf{d}(\mathbf{n};k,l) = \left(d_x(n_1,n_2;k,l), d_y(n_1,n_2;k,l)\right)$ is the motion vector for spatial coordinate (n_1,n_2) estimated between the frames k and l. It is pointed out here that the problems of noise reduction and motion estimation are inversely related as far as the temporal

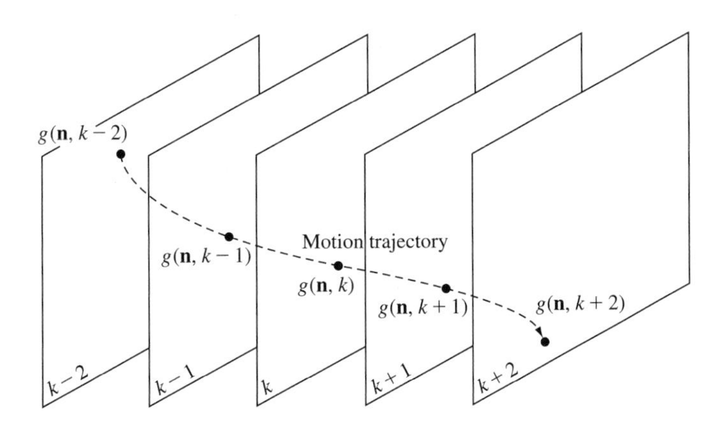

FIGURE 4.2

Noise filter operating along the motion trajectory of the picture element $(\mathbf{n},k)$.

window length K is concerned. That is, as the length of the filter is increased temporally, the noise reduction potential increases but so are the artifacts due to incorrectly estimated motion between frames that are temporally far apart.

To avoid the explicit estimation of motion, which might be problematic at high noise levels, two alternatives are available that turn (4.2) into a motion-adaptive filter. First, in areas where motion is detected (but not explicitly estimated) the averaging of frames should be kept to a minimum. Different ways exist to realize this. For instance, the temporal filter (4.2) can locally be switched off entirely, or can locally be limited to use only future or past frames, depending on the temporal direction in which motion was detected. Basically, the filter coefficients $h(l)$ are spatially adapted as a function of detected motion between frames. Second, the filter (4.2) can be operated along M a priori selected motion directions at each spatial coordinate. The finally estimated value $\hat{f}(\mathbf{n},k)$ is subsequently chosen from the M partial results according to some selection criterion, for instance, as the median [2, 8]:

$$\hat{f}_{i,j}(\mathbf{n},k) = \frac{1}{3}\left(g\left(n_1 - i, n_2 - j, k-1\right) + g\left(n_1, n_2, k\right) + g\left(n_1 + i, n_2 + j, k+1\right)\right), \tag{4.6a}$$

$$\hat{f}(\mathbf{n},k) = \text{median}\left(\hat{f}_{-1,-1}(\mathbf{n},k), \hat{f}_{1,-1}(\mathbf{n},k), \hat{f}_{-1,1}(\mathbf{n},k), \hat{f}_{1,1}(\mathbf{n},k), \hat{f}_{0,0}(\mathbf{n},k), g(\mathbf{n},k)\right). \tag{4.6b}$$

Clearly, cascading (4.6a) and (4.6b) turns the overall estimation procedure into a nonlinear one, but the partial estimation results are still obtained by the linear filter operation (4.6a).

It is easy to see that the filter (4.2) can be extended with a spatial filtering part. There exist many variations to this concept, basically as many as there are spatial restoration techniques for noise reduction. The most straightforward extension of (4.2) is the following 3D weighted averaging filter [7]:

$$\hat{f}(\mathbf{n},k) = \sum_{(\mathbf{m},l)\in S} h(\mathbf{m},l) g(\mathbf{n}-\mathbf{m}, k-l), \tag{4.7}$$

where S is the spatiotemporal support or *window* of the 3D filter (see Fig. 4.3). The filter coefficients $h(\mathbf{m},l)$ can be chosen to be all equal, but a performance improvement is obtained if they are adapted to the image sequence being filtered, for instance, by optimizing them in the mean squared error sense (4.3). In the latter case, (4.7) becomes the theoretically optimal 3D Wiener filter.

However, there are two disadvantages with the 3D Wiener filter. The first is the requirement that the 3D autocorrelation function for the original image sequence is known a priori. The second is the 3D wide-sense stationarity assumptions, which are virtually never true because of moving objects and scene changes. These requirements are detrimental to the performance of the 3D Wiener filter in practical situations of interest. For these reasons, simpler ways of choosing the 3D filter coefficients are usually preferred, provided that they allow for adapting the filter coefficients. One such choice for adaptive filter coefficients is the following [11]:

$$h(\mathbf{m},l;\mathbf{n},k) = \frac{c}{1 + \max\left(\alpha, \left(g(\mathbf{n},k) - g(\mathbf{n}-\mathbf{m}, k-l)\right)^2\right)}, \tag{4.8}$$

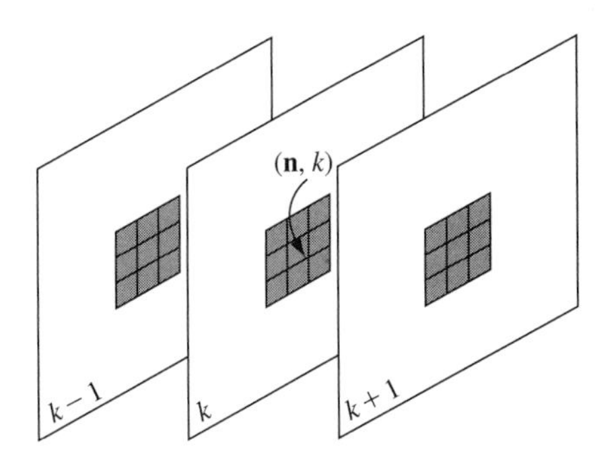

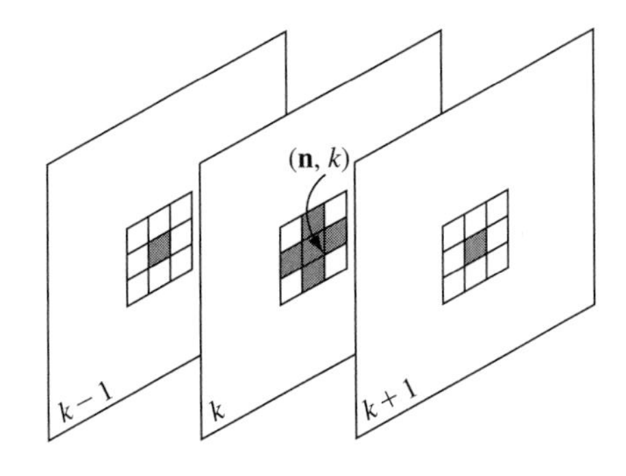

FIGURE 4.3

Examples of spatiotemporal windows to collect data for noise filtering of the picture element $(\mathbf{n}, k)$.

where $h(\mathbf{m}, l; \mathbf{n}, k)$ weights the intensity at spatial location $\mathbf{n} - \mathbf{m}$ in frame $k - l$ for the estimation of the intensity $\hat{f}(\mathbf{n}, k)$. The adaptive nature of the resulting filter can be seen immediately from (4.8). If the difference between the pixel intensity $g(\mathbf{n},k)$ being filtered and the intensity $g(\mathbf{n} - \mathbf{m}, k - l)$ for which the filter coefficient is calculated is less than α, this pixel is included in the filtering with weight $c/(1 + \alpha)$, otherwise it is weighted with a much smaller factor. In this way, pixel intensities that seem to deviate too much from $g(\mathbf{n},k)$—for instance due to moving objects within the spatiotemporal window S—are excluded from (4.7). As with the temporal filter (4.2), the spatiotemporal filter (4.7) can be carried out in a motion-compensated way by arranging the window S along the estimated motion trajectory.

4.2.1.2 Temporally Recursive Filters

A disadvantage of the temporal filter (4.2) and spatiotemporal filter (4.7) is that they need to buffer several frames of an image sequence. Alternatively, a recursive filter structure can be used that generally needs to buffer fewer (usually only one) frames. Furthermore, these filters are easier to adapt because there are fewer parameters to control. The general form of a recursive temporal filter is as follows:

$$\hat{f}(\mathbf{n}, k) = \hat{f}_b(\mathbf{n}, k) + \alpha(\mathbf{n}, k)\left[g(\mathbf{n}, k) - \hat{f}_b(\mathbf{n}, k)\right], \tag{4.9}$$

where $\hat{f}_b(\mathbf{n}, k)$ is the prediction of the original kth frame on the basis of previously filtered frames and $\alpha(\mathbf{n}, k)$ is the filter gain for updating this prediction with the observed kth frame. Observe that for $\alpha(\mathbf{n}, k) = 1$ the filter is switched off, that is, $\hat{f}_b(\mathbf{n}, k) = g(\mathbf{n}, k)$. Clearly, a number of different algorithms can be derived from (4.9) depending on the way the predicted frame $\hat{f}_b(\mathbf{n}, k)$ is obtained and the gain $\alpha(\mathbf{n}, k)$ is computed. A popular choice for the prediction $\hat{f}_b(\mathbf{n}, k)$ is the previously restored frame, either in direct form

$$\hat{f}_b(\mathbf{n}, k) = \hat{f}(\mathbf{n}, k - 1) \tag{4.10a}$$

or in motion-compensated form:

$$\hat{f}_b(\mathbf{n}, k) = \hat{f}(\mathbf{n} - \mathbf{d}(\mathbf{n}; k, k - 1), k - 1). \tag{4.10b}$$

More elaborate variations of (4.10) make use of a local estimate of the signal's mean within a spatiotemporal neighborhood. Furthermore, Eq. (4.9) can also be cast into a formal 3D motion-compensated Kalman estimator structure [12, 13]. In this case, the prediction $\hat{f}_b(\mathbf{n}, k)$ depends directly on the dynamic spatiotemporal state-space equations used for modeling the image sequence.

The simplest case for selecting $\alpha(\mathbf{n}, k)$ is by using a globally fixed value. As with the filter structures (4.2) and (4.7), it is generally necessary to adapt $\alpha(\mathbf{n}, k)$ to the presence or correctness of the motion to avoid filtering artifacts. Typical artifacts of recursive filters are "comet-tails" that moving objects leave behind.

A switching filter is obtained if the gain takes on the values α and 1 depending on the difference between the prediction $\hat{f}_b(\mathbf{n}, k)$ and the actually observed signal value $g(\mathbf{n}, k)$:

$$\alpha(\mathbf{n},k) = \begin{cases} 1 & \textit{if } \left| g(\mathbf{n},k) - \hat{f}_b(\mathbf{n},k) \right| > \varepsilon \\ \alpha & \textit{if } \left| g(\mathbf{n},k) - \hat{f}_b(\mathbf{n},k) \right| \le \varepsilon \end{cases}. \tag{4.11}$$

For areas that have a lot of motion (if the prediction [4.10a] is used) or for which the motion has been estimated incorrectly (if the prediction [4.10b] is used), the difference between the predicted intensity value and the noisy intensity value is large, causing the filter to switch off. For the areas that are stationary or for which the motion has been estimated correctly, the prediction differences are small yielding the value α for the filter coefficient.

A finer adaptation is obtained if the prediction gain is optimized to minimize the mean squared restoration error (4.3), yielding:

$$\alpha(\mathbf{n},k) = \max\left(1 - \frac{\sigma_W^2}{\sigma_g^2(n,k)}, 0\right), \tag{4.12}$$

where $\sigma_g^2(\mathbf{n}, k)$ is an estimate of the image sequence variance in a local spatiotemporal neighborhood of $(\mathbf{n}, k)$. If this variance is high, it indicates large motion or incorrectly estimated motion, causing the noise filter to switch off, that is, $\alpha(\mathbf{n}, k) = 1$. If $\sigma_g^2(\mathbf{n}, k)$ is in the same order of magnitude as the noise variance σ_w^2, the observed noisy image sequence is obviously very unreliable so that the predicted intensities are used without updating it, that is, $\alpha(\mathbf{n}, k) = 0$. The resulting estimator is known as the local linear minimum mean squared error (LLMMSE) estimator. A drawback of (4.12), as with any noise filter that requires the calculation of $\sigma_g^2(\mathbf{n}, k)$, is that outliers in the windows used to calculate this variance may cause the filter to switch off. Order-statistic filters are more suitable for handling data in which outliers are likely to occur.

4.2.2 Order-Statistic Filters

Order-statistic filters are nonlinear variants of weighted-averaging filters. The distinction is that in OS filters the observed noisy data—usually taken from a small spatiotemporal window—is ordered before being used. Because of the ordering operation, correlation information is ignored in favor of magnitude information. Examples of simple OS filters are the minimum operator, maximum operator, and median operator. OS filters are often applied in directional filtering. In directional filtering, different filter directions are considered corresponding to different spatiotemporal edge orientations. Effectively, this means that the filtering operation takes place along the spatiotemporal edges, avoiding the blurring of moving objects. The directional filtering approach may be superior to adaptive or switching filters because noise around spatiotemporal edges can be effectively eliminated by filtering along those edges, as opposed to turning off the filter in the vicinity of edges [7].

The general structure of an OS restoration filter is as follows:

$$\hat{f}(\mathbf{n},k) = \sum_{r=1}^{|S|} h_{(r)}(\mathbf{n},k) g_{(r)}(\mathbf{n},k), \tag{4.13}$$

where $g_{(r)}(\mathbf{n},k)$ are the ordered intensities, or *ranks*, of the corrupted image sequence, taken from a spatiotemporal window S with finite extent centered around $(\mathbf{n},k)$ (see Fig. 4.3). The number of intensities in this window are denoted by $|S|$. As with linear filters, the objective is to choose appropriate filter coefficients $h_{(r)}(\mathbf{n},k)$ for the ranks.

The most simple OS filter is a straightforward temporal median, for instance, taken over three frames:

$$\hat{f}(\mathbf{n},k) = \text{median}\left(g(\mathbf{n},k-1), g(\mathbf{n},k), g(\mathbf{n},k+1)\right). \tag{4.14}$$

Filters of this type are very suitable for removing shot noise. To avoid artifacts at the edges of moving objects, Eq. (4.14) is normally applied in a motion-compensated way. A more elaborate OS filter is the multistage median filter (MMF) [14, 15]. In the MMF, the outputs of basic median filters with different spatiotemporal support are combined. An example of the spatiotemporal supports has been shown in Fig. 4.4. The outputs of these intermediate median filter results are then combined as follows:

$$\hat{f}(\mathbf{n},k) = \text{median}\left(g(\mathbf{n},k), \max\left(\hat{f}_1(\mathbf{n},k), \ldots, \hat{f}_9(\mathbf{n},k)\right), \min\left(\hat{f}_1(\mathbf{n},k), \ldots, \hat{f}_9(\mathbf{n},k)\right)\right). \tag{4.15}$$

The advantage of this class of filters is that although it does not incorporate motion estimation explicitly, artifacts on edges of moving objects are significantly reduced. Nevertheless, the intermediate medians can also be computed in a motion-compensated way by positioning the spatiotemporal windows in Fig. 4.4 along the motion trajectories.

The filter coefficients $h_{(r)}(\mathbf{n},k)$ in (4.13) can also be statistically designed, as described in Chapter 7 of *The Essential Guide to Image Processing* [1]. If the coefficients are optimized in the mean squared error sense, the following general solution for the restored image sequence is obtained [16]:

$$\begin{pmatrix} \hat{f}(\mathbf{n},k) \\ \hat{\sigma}_w^2(\mathbf{n},k) \end{pmatrix} = \begin{pmatrix} h_{(1)}(\mathbf{n},k) & h_{(2)}(\mathbf{n},k) & \cdots & h_{(|S|)}(\mathbf{n},k) \\ n_{(1)}(\mathbf{n},k) & n_{(2)}(\mathbf{n},k) & \cdots & n_{(|S|)}(\mathbf{n},k) \end{pmatrix} \begin{pmatrix} g_{(1)}(\mathbf{n},k) \\ \vdots \\ g_{(|S|)}(\mathbf{n},k) \end{pmatrix}$$

$$= \left(\mathbf{A}^t \mathbf{C}_{(w)}^{-1} \mathbf{A}\right)^{-1} \mathbf{A}^t \mathbf{C}_{(w)}^{-1} \begin{pmatrix} g_{(1)}(\mathbf{n},k) \\ \vdots \\ g_{(|S|)}(\mathbf{n},k) \end{pmatrix}. \tag{4.16a}$$

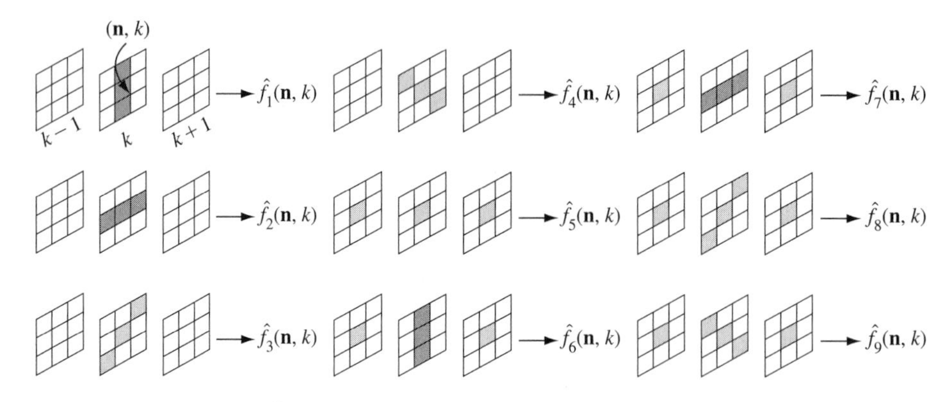

FIGURE 4.4

Spatiotemporal windows used in the multistage median filter.

This expression formulates the optimal filter coefficients $h_{(r)}(\mathbf{n},k)$ in terms of a matrix product involving the $|S| \times |S|$ auto-covariance matrix of the ranks of the noise, denoted by $\mathbf{C}_{(w)}$, and a matrix $\mathbf{A}$ defined as

$$\mathbf{A} = \begin{pmatrix} 1 & E\left[w_{(1)}(\mathbf{n},k)\right] \\ 1 & E\left[w_{(2)}(\mathbf{n},k)\right] \\ \vdots & \vdots \\ 1 & E\left[w_{(|S|)}(\mathbf{n},k)\right] \end{pmatrix}, \tag{4.16b}$$

where $E[w_{(r)}(\mathbf{n},k)]$ denotes the expectation of the ranks of the noise. The result in (4.16a) not only gives an estimate of the filtered image sequence but also for the local noise variance. This quantity is of use by itself in various noise filters to regulate the noise reduction strength. To calculate $E[w_{(r)}(\mathbf{n},k)]$ and $\mathbf{C}_{(w)}$, the probability density function of the noise has to be assumed known. In case the noise $w(\mathbf{n},k)$ is uniformly distributed, (4.16a) becomes the average of the minimum and maximum observed intensity. For Gaussian distributed noise, (4.16a) degenerates to (4.2) with equal weighting coefficients.

An additional advantage of ordering the noisy observation prior to filtering is that outliers can easily be detected. For instance, with a statistical test—such as the rank order test [16]—the observed noisy values within the spatiotemporal window S that are significantly different from the intensity $g(\mathbf{n},k)$ can be detected. These significantly different values originate usually from different objects in the image sequence, for instance, due to motion. By letting the statistical test reject these values, the filters (4.13) and (4.16) use locally only data from the observed noisy image sequence that is close—in intensity—to $g(\mathbf{n},k)$. This further reduces the sensitivity of the noise filter (4.13) to outliers due to motion or incorrectly compensated motion.

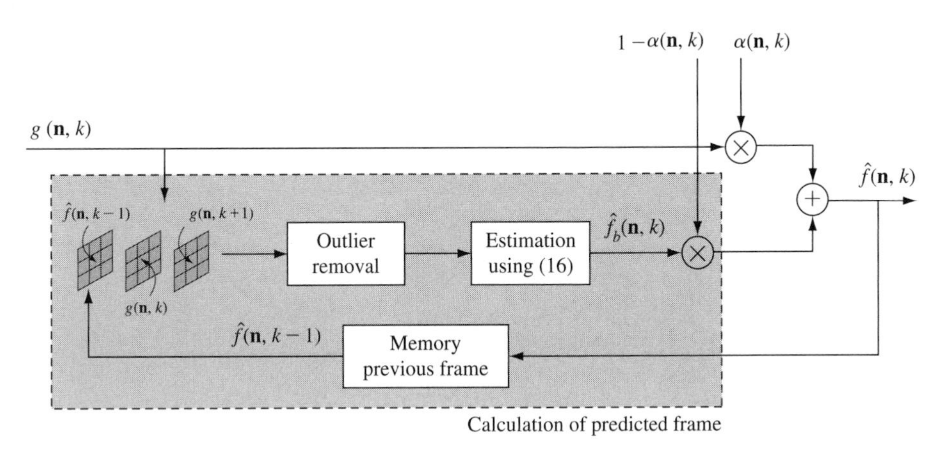

FIGURE 4.5

Overall filtering structure combining (4.9), (4.16), and an outlier removing rank order test.

The estimator (4.16) can also be used in a recursive structure such as the one in Eq. (4.9). Essentially (4.16) is then interpreted as an estimate for the local mean of the image sequence, and the filtered value resulting from (4.16) is used as the predicted value $\hat{f}_b(\mathbf{n},k)$ in (4.9). Furthermore, instead of using only noisy observations in the estimator, previously filtered frames can be used by extending the spatiotemporal window S over the current noisy frame $g(\mathbf{n},k)$ and the previously filtered frame $\hat{f}(\mathbf{n},k-1)$. The overall filter structure is shown in Fig. 4.5.

4.2.3 Multiresolution Filters

The multiresolution representation of 2D images has become quite popular for analysis and compression purposes [17]. This signal representation is also useful for image sequence restoration. The fundamental idea is that if an appropriate decomposition into bands of different spatial and temporal resolutions and orientations is carried out, the energy of the structured signal will locally be concentrated in selected bands, whereas the noise is spread out over all bands. The noise can therefore effectively be removed by mapping all small (noise) components in all bands to zero, while leaving the remaining larger components relatively unaffected. Such an operation on signals is also known as coring [18, 19]. Figure 4.6 shows two coring functions, namely soft-thresholding and hard-thresholding. Chapter 11 of *The Essential Guide to Image Processing* [1] discusses 2D wavelet-based thresholding methods for image enhancement.

The discrete wavelet transform has been widely used for decomposing one-dimensional and multidimensional signals into bands. However, a problem with this

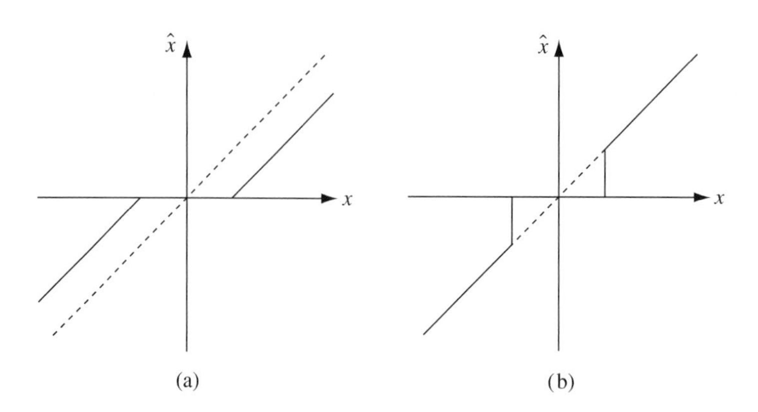

FIGURE 4.6

Coring functions: (a) Soft-thresholding. (b) Hard-thresholding. Here x is an signal amplitude taken from one of the spatiotemporal bands (which carry different resolution and orientation information), and $\hat{x}$ is the resulting signal amplitude after coring.

transform for image sequence restoration is that the decomposition is not shift-invariant. Slightly shifting the input image sequence in spatial or temporal sense can cause significantly different decomposition results. For this reason, in [20] a shift-invariant, but over-complete, decomposition was proposed known as the Simoncelli pyramid. Figure 4.7(a) shows the 2D Simoncelli pyramid decomposition scheme. The filters $L_i(\omega)$ and $H_i(\omega)$ are linear phase low-pass and high-pass filters, respectively. The filters $F_i(\omega)$ are fan filters that decompose the signal into four directional bands. The resulting spectral decomposition is shown in Fig. 4.7(b). From this spectral tessellation, the different resolutions and orientations of the spatial bands obtained by Fig. 4.7(a) can be inferred. The radial bands have a bandwidth of 1 octave.

The Simoncelli pyramid gives a spatial decomposition of each frame into bands of different resolution and orientation. The extension to temporal dimension is obtained by temporally decomposing each of the spatial resolution and orientation bands using a regular wavelet transform. The low-pass and high-pass filters are operated along the motion trajectory to avoid blurring of moving objects. The resulting motion-compensated spatiotemporal wavelet coefficients are filtered by one of the coring functions, followed by the reconstruction of the video frame by an inverse wavelet transformation and Simoncelli pyramid reconstruction. Figure 4.8 shows the overall scheme.

Though multiresolution approaches have been shown to outperform the filtering techniques described in Sections 4.2.1 and 4.2.2 for some types of noise, they generally require much more processing power due to the spatial and temporal decomposition, and—depending on the temporal wavelet decomposition—they require a significant number of frame stores.

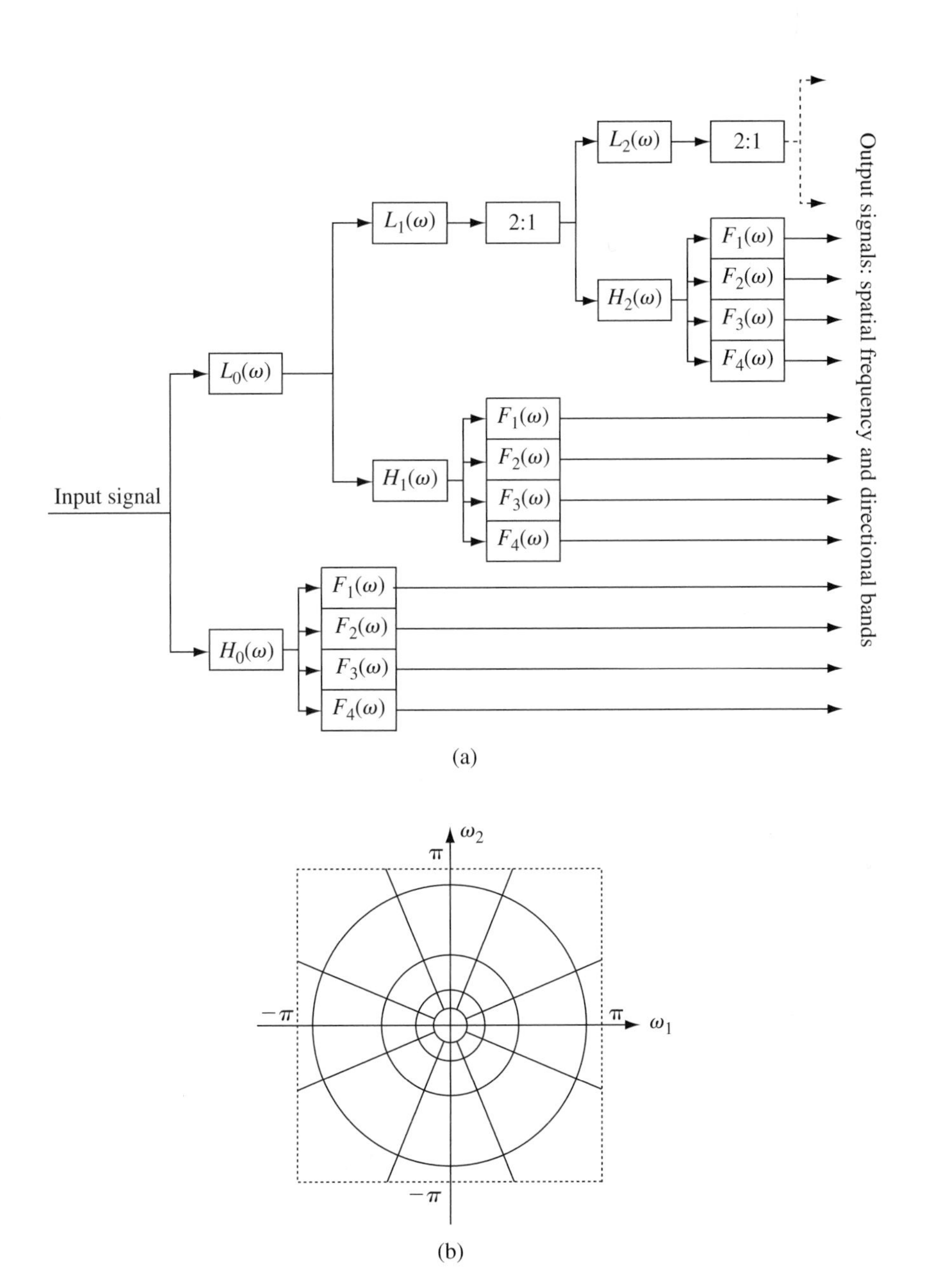

FIGURE 4.7

(a) Simoncelli pyramid decomposition scheme. (b) Resulting spectral decomposition, illustrating the spectral contents carried by the different resolution and directional bands.

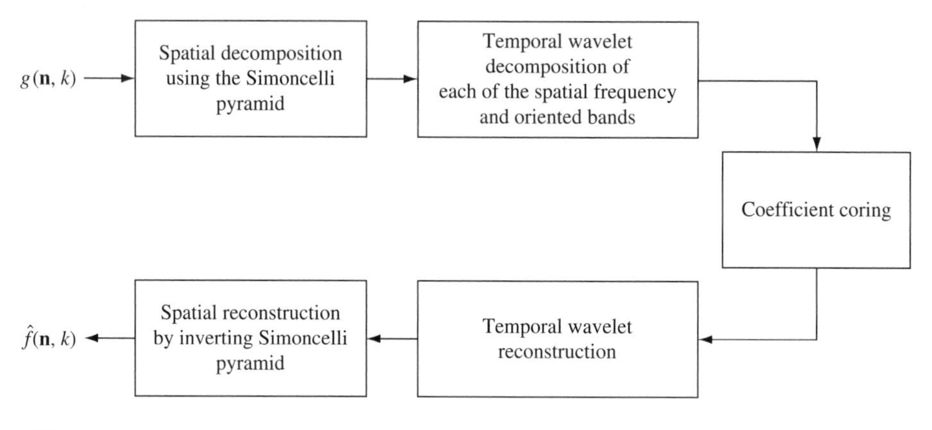

FIGURE 4.8

Overall spatiotemporal multiresolution filtering using coring.

4.3 CODING ARTIFACT REDUCTION

The call for coding artifact reduction started with the introduction of digital transmission and storage. Theoretically, digital transmission can be lossless, but in practice, the available channel bandwidth makes lossy compression necessary, which results in visible artifacts.

The block-based DCT is among the most popular transform techniques and has been widely used in image and video compression. It is the core of several industry standards of image and video compression, such as JPEG, H.26x, and MPEG-x. But these block-based codecs suffer from annoying blocking artifacts when they are applied in low bit rate coding to obtain a reasonable compression ratio.

The quantization is the lossy stage because the original values of the DCT coefficients are permanently lost. Because each block is encoded separately, there may emerge visible discontinuities along block boundaries of the decoded image frame, commonly defined as blocking artifacts. Also ringing, blurring, and mosquito noise can be introduced as a consequence of omitting the high frequencies by truncating the DCT coefficients [21]. Generally speaking, increasing coding bit rate can improve the quality of the reconstructed image/video, but it is limited by channel bandwidth or storage capacity.

There is an increasing demand among consumers for high-quality digital video content, resulting in technologies such as high-definition broadcast channels and storage devices. Interesting challenges arise for content creators while processing video for high-quality applications. The challenge then is to design noise reduction filters that provide a graceful trade-off between the amount of noise reduction and the resulting loss of perceptual picture quality.

Because the blocking artifact is the most visible one, we focus here solely on this artifact. In the literature, very many approaches for blocking artifact reduction have been published. Among them, we can distinguish two main classes, the methods that operate in the spatial domain and the methods defined in the frequency domain. The methods in the spatial domain are the most popular as they operate directly on the received (decoded) spatial data (pixels) [22]. In other words, they do not require access to the DCT coefficients, which are usually not available for postprocessing.

4.3.1 Artifact Reduction in the Spatial Domain

Due to horizontal and vertical "edges" or lines demarking the block grid, there are additional high frequencies in the spectrum of the decoded video signal. The simplest remedy would be the application of low-pass filtering to suppress those frequencies. Initially, Gaussian low-pass filtering with a high-frequency emphasis has been proposed, or Gaussian filtering only applied to pixels near the block boundaries. The general drawback of these early methods is the loss of high frequencies and excessive blurring of true edges.

To overcome these drawbacks, adaptive deblocking algorithms have been proposed. In general, adaptive filtering requires a classification step to determine whether a block is a monotone one or contains an edge, followed by a (non) linear filtering step. If a block is monotone, a 2D filtering is applied; otherwise 1D directional filtering is applied [23, 24].

Adaptive filtering can also be used for simultaneous blocking artifacts reduction and sharpness enhancement. In [25], a method has been proposed that does not require grid position detection. Instead, multiple filter coefficient classes are distinguished, based on an off-line training process using both the original image data and compressed blurred versions of the original images. Based on image pattern classification, the algorithm effectively discriminates between coding artifacts and image edges, thus, significantly increasing perceptual quality of the processed images by removing artifacts and enhancing image detail.

Finally, we mention the so-called projection onto convex sets (POCS)-based methods for blocking artifacts reduction, originating from image restoration. One tries to use a priori information about the original image data to iteratively restore a degraded (sub)image. The convergence behavior can be explained by the theory of POCS. In blockiness reduction, the general assumption is that that the input image is highly correlated so that similar frequency characteristics are maintained between adjacent blocks. If we are able to detect the high-frequency components that are not present within blocks, we can consider them the result of the blocking artifact. Some of the popular constraints such as the intensity and the smoothness constraint as well as other information about the POCS technique can be found in [26]. The proposed method [27] shows that blocking artifacts can be significantly alleviated, while the original high-frequency components, such as edges, are faithfully preserved. Although quite effective for still pictures, these POCS-based methods are less practical for real-time video postprocessing implementations due to their computationally demanding iterative nature.

4.3.2 Artifact Reduction in the Frequency Domain

We have seen that spatial domain methods apply low-pass filtering and as such result in blurring of the picture. To alleviate this problem, we have discussed a few advanced methods as well. Relatively speaking, only a few approaches in the literature have tackled the problem of blockiness reduction in the frequency domain. These methods can either use coefficients available in the bit stream or re-compute DCT coefficients in the postprocessing.

An early method for removing blocking effects is proposed in [28]. It exploits the correlation between intensity values of boundary pixels of two neighboring blocks. Specifically, it is based on the empirical observation that quantization of the DCT coefficients of two neighboring blocks increases the mean squared difference of slope (MSDS) between the neighboring pixels on their boundaries. Therefore, among all permissible inverse quantized coefficients, the set that minimizes this MSDS is most likely to decrease the blocking effect. This minimization problem can be formulated as a computationally intensive quadratic programming (QP) problem.

In [29] a method has been described to exploit the periodic nature of the blocking artifact due to the grid structure. The harmonics generated by the regular lattice pattern can be measured easily in the frequency domain and will give vital information for blockiness estimation. The amplitude of the harmonics is proportional to the degree of blockiness, while the phase of the harmonics can be used to verify that the harmonics are not due to contextual details in the picture. Examining both the amplitude and the phase information of the harmonics leads to an accurate blockiness detector that needs no reference pictures.

The smoothing constraint, part of the POCS technique can also be applied in the frequency domain [27]. The authors propose to remove the high frequencies in the DCT domain by looking into the 8-point and 16-point DCTs. If a high frequency is found in the result of the 16-point DCT and not in the 8-point DCT, they conclude that this high frequency is a consequence of the block boundary. They remove the frequency component by zeroing the appropriate coefficient(s) and performing the inverse direct cosine transform (IDCT).

4.4 BLOTCH DETECTION AND REMOVAL

Blotches are artifacts that are typically related to film. Dirt particles covering the film introduce bright or dark spots on the frames, and the mishandling or aging of film causes loss of gelatin covering the film. Figure 4.9(a) shows a film frame containing dark and bright spots: the blotches. A model for this artifact is the following [2, 30]:

$$g(\mathbf{n},k) = (1 - b(\mathbf{n},k))f(\mathbf{n},k) + b(\mathbf{n},k)c(\mathbf{n},k) + w(\mathbf{n},k), \tag{4.17}$$

where $b(\mathbf{n},k)$ is a binary mask that indicates for each spatial location in each frame whether or not it is part of a blotch. The (more or less constant) intensity values at the corrupted spatial locations are given by $c(\mathbf{n},k)$. Though noise is not considered to be the dominant degrading factor in the section, it is still included in (4.17) as the term $w(\mathbf{n},k)$. The removal of blotches is a two-step procedure. In the first step, the blotches need to be detected, that is, an estimate for the mask $b(\mathbf{n},k)$ is made [31]. In the second step, the incorrect intensities $c(\mathbf{n},k)$ at the corrupted locations are spatiotemporally interpolated

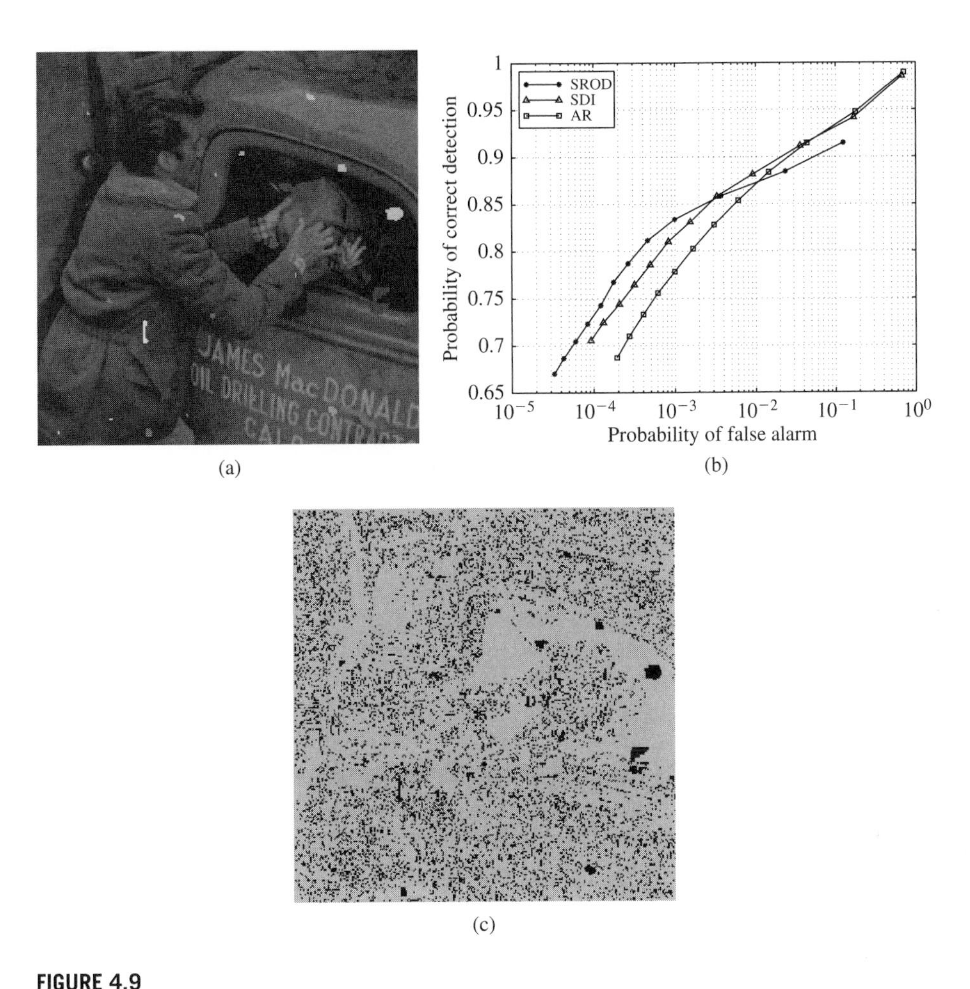

FIGURE 4.9

(a) Video frame with blotches. (b) Correct detection versus false detection for three different blotch detectors. (c) Blotch detection mask using the sROD ($T = 0$).

[32]. In case a motion-compensated interpolation is carried out, the second step also involves the local repair of motion vectors estimated from the blotched frames. The overall blotch detection and removal scheme is shown in Fig. 4.10.

4.4.1 Blotch Detection

Blotches have three characteristic properties that are exploited by blotch detection algorithms. First, blotches are temporally independent and therefore hardly ever occur at the same spatial location in successive frames. Second, the intensity of a blotch is significantly different from its neighboring uncorrupted intensities. Finally, blotches form coherent regions in a frame, as opposed to, for instance, spatiotemporal shot noise.

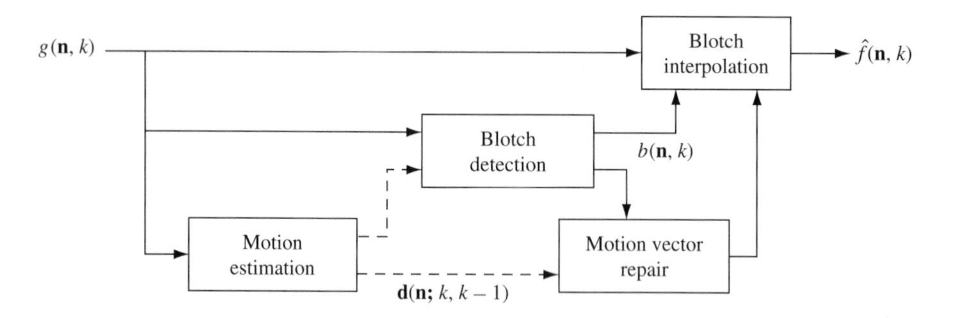

FIGURE 4.10

Blotch detection and removal system.

There are various blotch detectors that exploit these characteristics. The first is a pixel-based blotch detector known as the spike-detector index (SDI). This method detects temporal discontinuities by comparing pixel intensities in the current frame with motion-compensated reference intensities in the previous and following frames:

$$\mathrm{SDI}(\mathbf{n},k) = \min\left(\left(g(\mathbf{n},k) - g\left(\mathbf{n} - \mathbf{d}(\mathbf{n};k,k-1),k-1\right)\right)^2, \left(g(\mathbf{n},k) - g\left(\mathbf{n} - \mathbf{d}(\mathbf{n};k,k+1),k+1\right)\right)^2\right). \tag{4.18}$$

Because blotch detectors are pixel oriented, the motion field $\mathbf{d}(\mathbf{n};k;l)$ should have a motion vector per pixel, that is, the motion field is dense. Observe that any motion-compensation procedure must be robust against the presence of intensity spikes: this will be discussed later in this section. A blotch pixel is detected if $\mathrm{SDI}(\mathbf{n},k)$ exceeds a threshold:

$$b(\mathbf{n},k) = \begin{cases} 1 & \text{if} \quad \mathrm{SDI}(\mathbf{n},k) > T \\ 0 & \quad\text{otherwise} \end{cases}. \tag{4.19}$$

Because blotch detectors are essentially searching for outliers, OS-based detectors usually perform better. The rank order difference (ROD) detector is one such method. It takes $|S|$ reference pixel intensities: $\boldsymbol{r} = \{r_i | i = 1,2,\ldots,|S|\}$ from a motion-compensated spatiotemporal window S (e.g., see the grayed pixels in Fig. 4.11), and ranks them by intensity value: $r_1 \leq r_2 \leq \ldots \leq r_{|S|}$. It then finds the deviation between the pixel intensity $g(\mathbf{n},k)$ and the reference pixels ranked by intensity value as follows:

$$\mathrm{ROD}_i(\mathbf{n},k) = \begin{cases} r_i - g(\mathbf{n},k) & \text{if } g(\mathbf{n},k) \leq \mathrm{medium}(\mathbf{r}) \\ g(\mathbf{n},k) - r_{|S|-i} & \text{if } g(\mathbf{n},k) > \mathrm{medium}(\mathbf{r}) \end{cases} \quad \text{for } i = 1,2,\ldots,\frac{|S|}{2} \tag{4.20}$$

A blotch pixel is detected if any of the RODs exceeds a specific threshold T_i:

$$b(\mathbf{n},k) = \begin{cases} 1 & \text{if } \exists\, i \text{ such that } \mathrm{ROD}_i(\mathbf{n},k) > T_i \\ 0 & \quad\text{otherwise} \end{cases}. \tag{4.21}$$

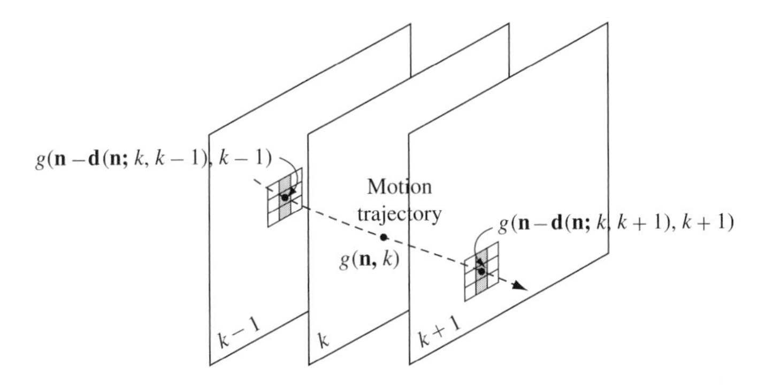

FIGURE 4.11

Example of motion-compensated spatiotemporal window for obtaining reference intensities in the ROD detector.

More complicated blotch detectors explicitly incorporate a model for the uncorrupted frames, such as a two- or three-dimensional autoregressive model or a Markov Random Field to develop the maximum a posteriori detector for the blotch mask $b(\mathbf{n},k)$. Figure 4.9(b) illustrates the detection probability versus the false detection probability of three different detectors on a sequence of which a representative blotched frame is shown in Fig. 4.9(a). These results indicate that for reasonable detection probabilities the false detection probability is fairly high. False detections are detrimental to the restoration process because the interpolation process itself is fallible and may introduce disturbing artifacts that were not present in the blotched image sequence.

The blotch detectors described so far are essentially pixel-based detectors. They do not incorporate the spatial coherency of the detected blotches. The effect is illustrated by Fig. 4.9(c), which shows the detected blotch mask $b(\mathbf{n},k)$ using a simplified version of the ROD detector (with $T_i \rightarrow \infty, i \geq 2$). The simplified ROD (sROD) is given by the following relations [30]:

$$\text{sROD}(\mathbf{n},k) = \begin{cases} \min(\mathbf{r}) - g(\mathbf{n},k) & \text{if } g(\mathbf{n},k) < \min(\mathbf{r}) \\ g(\mathbf{n},k) - \max(\mathbf{r}) & \text{if } g(\mathbf{n},k) > \max(\mathbf{r}) \\ 0 & \text{elsewhere} \end{cases} \tag{4.22a}$$

$$b(\mathbf{n},k) = \begin{cases} 1 & \text{if } \text{sROD}(\mathbf{n},k) > T \\ 0 & \text{otherwise} \end{cases}. \tag{4.22b}$$

The sROD basically looks at the range of the reference pixel intensities obtained from the motion-compensated window, and compares it with the pixel intensity under investigation. A blotch pixel is detected if the intensity of the current pixel $g(\mathbf{n},k)$ lies far enough outside that range.

The performance of even this simple pixel-based blotch detector can be improved significantly by exploiting the spatial coherence of blotches. This is done by postprocessing the blotch mask in Fig. 4.9(c) in two ways, namely by removing small blotches and by completing partially detected blotches. We first discuss the removal of small blotches.

The detector output (4.22a) is not only sensitive to intensity changes due to blotches corrupting the image sequence but also to noise. If the probability density function of the noise—denoted by $f_W(w)$—is known, the probability of false detection for a single pixel can be calculated. Namely, if the sROD uses $|S|$ reference intensities in evaluating (4.22a), the probability that sROD($\mathbf{n}, k$) for a single pixel is larger than T due to noise only is as follows [30]:

$$
\begin{aligned}
P\left(\text{sROD}(\mathbf{n},k) > T|\text{no blotch}\right) &= P\left(g(\mathbf{n},k) - \max(\mathbf{r}) > T|\text{no bl.}\right) + P\left(\min(\mathbf{r}) - g(\mathbf{n},k) > T|\text{no bl.}\right) \\
&= 2\int_{-\infty}^{\infty}\left[\int_{-\infty}^{u-T} f_W(w)dw\right]^{S} f_W(u)du. \qquad (4.23)
\end{aligned}
$$

In the detection mask $b(\mathbf{n}, k)$, blotches may consist of single pixels or multiple connected pixels. A set of connected pixels that are all detected as (being part of a) blotch is called a spatially coherent blotch. If a coherent blotch consists of N connected pixels, the probability that this blotch is due to noise only is as follows:

$$
P\left(\text{sROD}(\mathbf{n},k) > T \text{ for } N \text{ connected pixels} \mid \text{no blotch}\right) = \left(P\left(\text{sROD}(\mathbf{n},k) > T|\text{no blotch}\right)\right)^{N}. \qquad (4.24)
$$

By bounding this false detection probability to a certain maximum, the minimum number of pixels identified by the sROD detector as being part of a blotch can be computed. Consequently, coherent blotches consisting of fewer pixels than this minimum are removed from the blotch mask $b(\mathbf{n}, k)$.

A second postprocessing technique for improving the detector performance is hysteresis thresholding. First, a blotch mask is computed using a very low detection threshold T, for instance, $T = 0$. From the detection mask, the small blotches are removed as described above, yielding the mask $b_0(\mathbf{n}, k)$. Nevertheless, due to the low detection threshold, this mask still contains many false detections. Then, a second detection mask $b_1(\mathbf{n}, k)$ is obtained by using a much higher detection threshold. This mask contains fewer detected blotches and the false detection rate in this mask is small. The second detection mask is now used to validate the detected blotches in the first mask: only those spatially coherent blotches in $b_0(\mathbf{n}, k)$ that have a corresponding blotch in $b_1(\mathbf{n}, k)$ are preserved, all others are removed. The result of the above two postprocessing techniques on the frame shown in Fig. 4.9(a) is shown in Fig. 4.12(a). In Fig. 4.12(b), the detection and false detection probabilities are shown.

4.4.2 Motion Vector Repair and Interpolating Corrupted Intensities

Block-based motion estimators will generally find the correct motion vectors even in the presence of blotches, provided that the blotches are small enough. The disturbing effect of blotches is usually confined to small areas of the frames. Hierarchical motion estimators will experience little influence of the blotches at the lower resolution levels. At higher resolution levels, blotches covering larger parts of (at those levels) small blocks will significantly influence the motion estimation result. If the blotch mask $b(\mathbf{n}, k)$ has been estimated, it is also known which estimated motion vectors are unreliable.

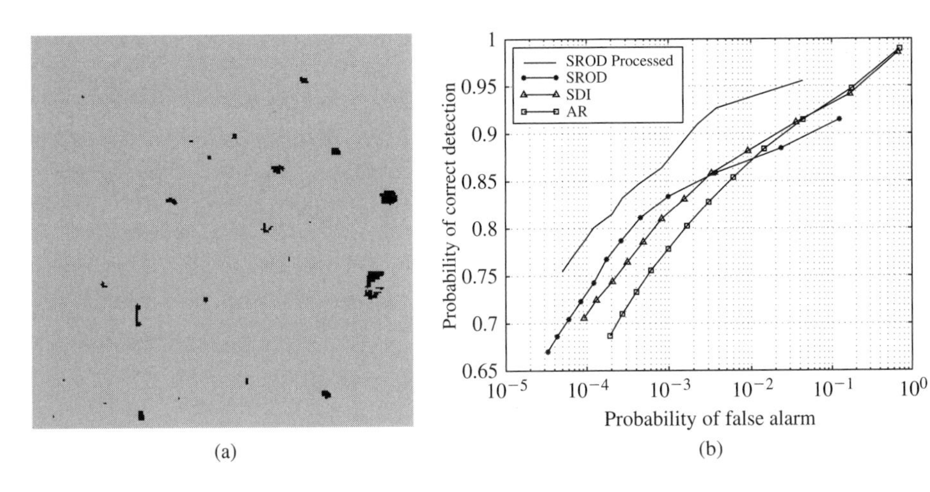

FIGURE 4.12

(a) Blotch detection mask after postprocessing. (b) Correct detection versus false detections obtained for sROD with postprocessing (top curve), compared to results from Fig. 4.9(b).

There are two strategies in recovering motion vectors that are known to be unreliable. The first approach is to take an average of surrounding motion vectors. This process—known as *motion vector interpolation* or *motion vector repair*—can be realized using, for instance, the median or average of the motion vectors of uncorrupted regions adjacent to the corrupted blotch. Though simple, the disadvantages of averaging are that motion vectors may be created that are not present in the uncorrupted part of the image and that no validation of the selected motion vector on the actual frame intensities takes place.

The second—more elaborate—approach circumvents this disadvantage by validating the corrected motion vectors using intensity information directly neighboring the blotched area. As a validation criterion, the motion-compensated mean squared intensity difference can be used [6]. Candidates for the corrected motion vector can be obtained either from motion vectors taken from adjacent regions or by motion re-estimation using a spatial window containing only uncorrupted data such as the pixels directly bordering the blotch.

The estimation of the frame intensities labeled by the mask as being part of a blotch can be done either by a spatial or temporal interpolation, or by a combination of both. We concentrate on spatiotemporal interpolation. Once the motion vector for a blotched area has been repaired, the correct temporally neighboring intensities can be obtained. In a multistage median interpolation filter, five interpolated results are computed using the (motion-compensated) spatiotemporal neighborhoods shown in Fig. 4.13. Each of the five interpolated results is computed as the median over the corresponding neighborhood S_i:

$$\hat{f}_i(\mathbf{n},k) = \text{median}\left\{\left[f(\mathbf{n},k-1)\,|\mathbf{n}\in S_i^{k-1}\right]\cup\left[f(\mathbf{n},k)|\mathbf{n}\in S_i^{k}\right]\cup\left[f(\mathbf{n},k+1)|\mathbf{n}\in S_i^{k+1}\right]\right\}. \tag{4.25}$$

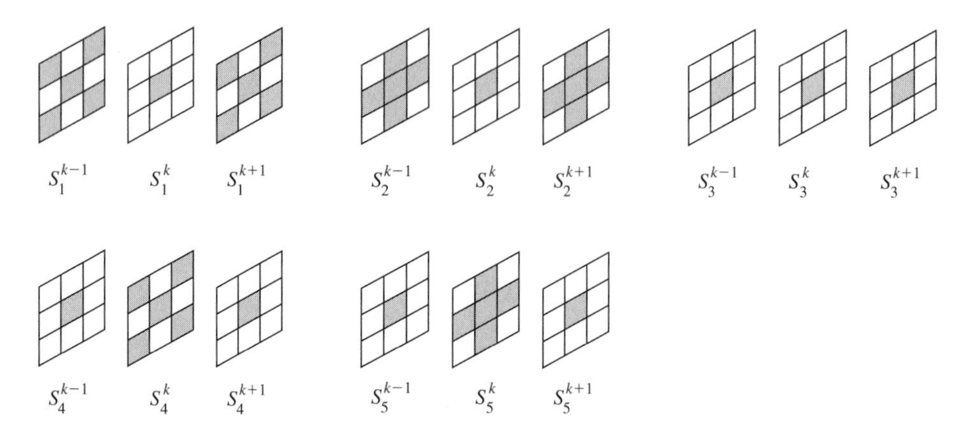

FIGURE 4.13

Five spatiotemporal windows used to compute the partial results in Eq. (4.25).

The final result is computed as the median over the five intermediate results:

$$\hat{f}(\mathbf{n},k) = \text{median}\left(\hat{f}_1(\mathbf{n},k), \hat{f}_2(\mathbf{n},k), \hat{f}_3(\mathbf{n},k), \hat{f}_4(\mathbf{n},k), \hat{f}_5(\mathbf{n},k)\right). \tag{4.26}$$

The MMF does not rely on any model for the image sequence. Though simple, this is at the same time a drawback of median filters. If a model for the original image sequence can be assumed, it is possible to find statistically optimal values for the missing intensities. For the sake of completeness, we mention here that if one assumes the popular Markov Random Field, the following complicated expression needs to be optimized:

$$P\left(\hat{f}(\mathbf{n},k)|f(\mathbf{n}-\mathbf{d}(\mathbf{n};k,k-1),k-1), f(\mathbf{n},k), f(\mathbf{n}-\mathbf{d}(\mathbf{n};k,k+1),k+1)\right)$$

$$\times \propto \exp\left(-\sum_{\mathbf{m}:d(\mathbf{m},k)=1} \gamma(\mathbf{m})\right),$$

$$\gamma(\mathbf{m}) = \sum_{\mathbf{s}\in S^k}\left(\hat{f}(\mathbf{m},k)-\hat{f}(\mathbf{s},k)\right)^2 + \lambda\left(\sum_{\mathbf{s}\in S^{k-1}}\left(\hat{f}(\mathbf{m},k)-f(\mathbf{n}-\mathbf{d}(\mathbf{n};k,k-1),k-1)\right)^2\right.$$

$$\left. + \sum_{\mathbf{s}\in S^{k+1}}\left(\hat{f}(\mathbf{m},k)-f(\mathbf{n}-\mathbf{d}(\mathbf{n};k,k+1),k+1)\right)^2\right) \tag{4.27}$$

The first term of γ on the right-hand side of (4.27) forces the interpolated intensities to be spatially smooth, whereas the second and third terms enforce temporal smoothness. The sets S^{k-1}, S^k, and S^{k+1} denote appropriately chosen spatial windows in the frames $k-1$, k, and $k+1$. The temporal smoothness is calculated along the motion trajectory using the repaired motion vectors. The optimization of (4.27) requires an *iterative optimization*

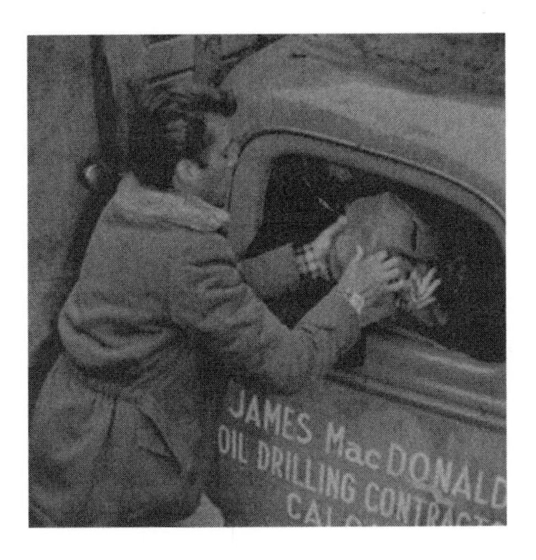

FIGURE 4.14

Blotch-corrected frame resulting from Figure 4.9a.

technique. If a simpler 3D autoregressive model for the image sequence is assumed, the interpolated values can be calculated by solving a set of linear equations.

Instead of interpolating the corrupted intensities, it is also possible to directly copy and paste intensities from past or future frames. The simple copy-and-paste operation instead of a full spatiotemporal data regeneration is motivated by the observation that—at least on local and motion-compensated basis—image sequences are heavily correlated. Furthermore, straightforward interpolation is not desirable in situations where part of the information in the past and future frames itself is unreliable, for instance, if it was part of a blotch itself or if it is situated in an occluded area. The objective is now to determine—for each pixel being part of a detected blotch—whether intensity information from the previous or next frames should be used. This decision procedure can be again cast into a statistical framework [30, 33]. As an illustration, Fig. 4.14 shows the interpolated result of the blotched frame in Fig. 4.9(a).

4.4.3 Video Inpainting

In the last decade, several types of blotch removal methods were proposed that are able to fill in large missing areas in still images [34–38] Initially, these spatial restoration algorithms could be classified mainly in two categories: smoothness-preserving and texture synthesis. The first category comprises various inpainting approaches which usually propagate isophotes (i.e., level lines), gradients, or curvatures inside the artifact by means of variational models [34, 39–41], or approaches which reconstruct the explicit structure of objects based on edge reconnection [38]. In the second category, parametric and nonparametric approaches are employed for texture synthesis based on stochastic models such as Markov Random Fields [35, 36, 42], Bayesian autoregressive models [43], projection onto convex sets [44], and other models [45].

As their naming suggests, texture synthesis methods do a better reconstruction of the textural content. However, they do not preserve object edges properly. This is done better with smoothness-preserving methods, which, on their turn, do not reproduce textural content. To combine the advantages of both methods, a third category of algorithms has appeared comprising hybrid methods that combine structure and texture reconstruction [37, 38, 46, 47]. A valuable survey of the different inpainting approaches can be found in [48].

A part of the aforementioned methods were later extended to 2D+time or 3D data [49–53]. In [49], a method is presented for video inpainting that extends previous 2D solutions [36] and [42] to 2D+time. Each frame is segmented into static background and moving foreground, which will be restored separately. To this purpose, the motion field inside the missing areas is first recovered by means of *motion inpainting*, which is similar to the classic inpainting: healthy motion vectors are copied from areas with similar appearance in neighboring frames. The gaps in the motion field are filled inward, starting from the patch borders. The foreground is then iteratively reconstructed such that the later reconstructed areas benefit from the already recovered ones. In the end, the background is reconstructed by simply copying temporally nearest areas. If the background that needs to be reconstructed is always occluded in the available frames, the reconstruction falls back to a simple texture synthesis.

Another approach, focusing on video stabilization, is presented in [50]. The areas to be inpainted lie therefore on the border of the frames, which remain empty in the presence of film/video unsteadiness. In this approach, motion inpainting is used again to first repair the motion field, which is then employed for the reconstruction of the missing areas. To avoid unsharp reconstructions, an image deblurring algorithm is also introduced that copies a sharper look of the same objects from the neighboring frames.

4.4.4 Restoration in Conditions of Difficult Object Motion

Due to various types of complicated object movements [54], wrong motion vectors are sometimes extracted from the sequence. As a result, the spatiotemporal restoration process that follows may introduce unnecessary errors that are visually more disturbing than the blotches themselves. The extracted temporal information becomes unreliable, and a source of errors by itself. This triggered research on situations in which the motion vectors cannot be used.

An initial solution to the problem was to detect areas of "pathological" motion and protect them against any restoration [54]. This solution, however, preserves the artifacts which happen to lie in those areas. To restore these artifacts, too, several solutions have been proposed, which discard the temporal information and use only spatial information coming from the same frame.

Figure 4.15 presents an example of an artificially degraded image, which was restored by means of a hybrid method. The restoration consisted of a texture synthesis constrained by edge-based structure reconstruction [38]. The main steps are 1) edge detection and feature extraction; 2) image structure reconstruction; and 3) texture synthesis. In the first step, edges are detected around the artifact, based on the contours that result from a

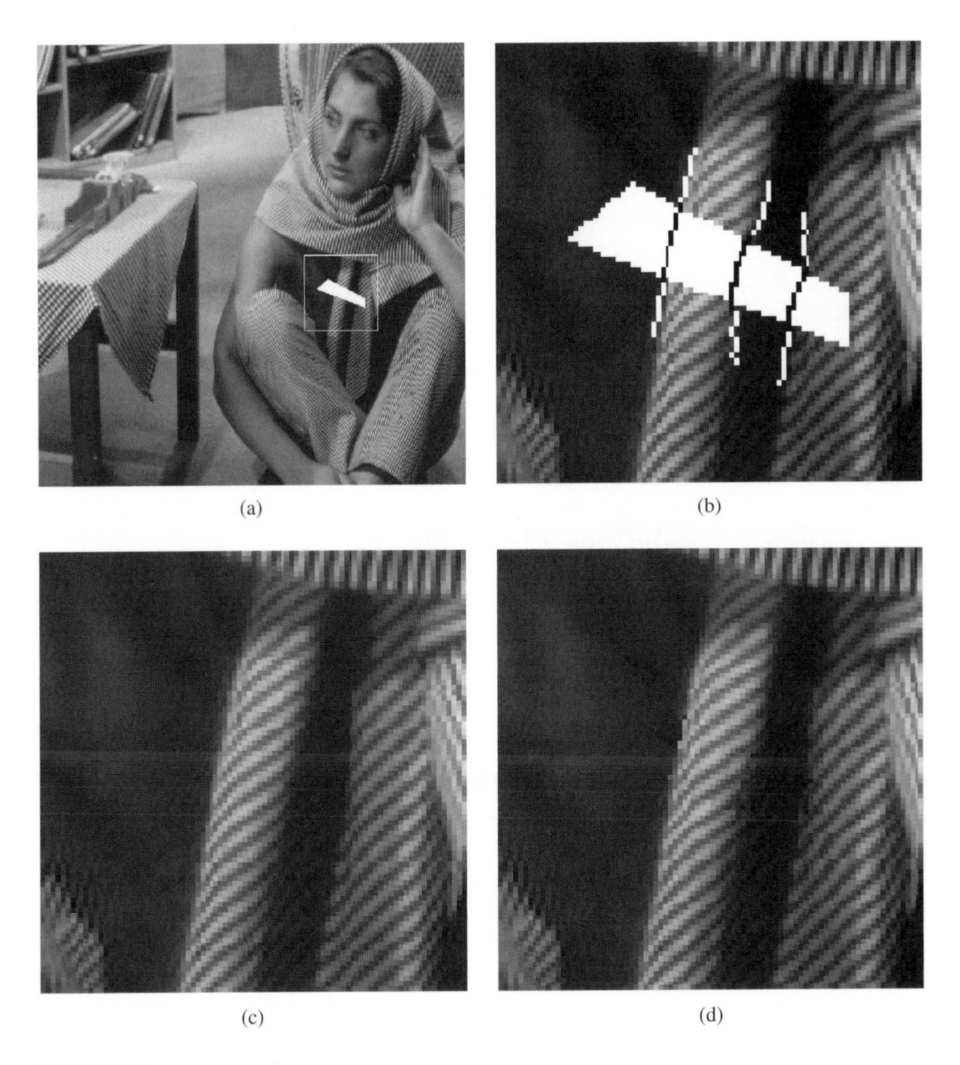

FIGURE 4.15

Constrained texture synthesis. (a) Original image, with artificial artifact surrounded by a white box. (b) Structure. (c) Original content of the artifact. (d) Restoration result.

segmentation procedure. Features are extracted for each edge, such as histograms of the luminance and gradient angles on both sides of the edge. In the second step, the missing object edges within the artifact area are recovered by pairing the edges extracted around the artifact. The pairing procedure assesses the similarity of the paired edges based on the features extracted in the previous step, on an estimate of how well one edge continuous into another one (assuming, for example, local circularity), and on the spatial order of the edges with respect to each other. All these factors contribute to a global consistency score of the resulting structure, and the best scoring configuration is chosen to represent the final image structure. The edges that do not belong to any pair are considered to be T-junctions (coming from edge occlusions). Finally, in the third step, the artifact is restored by texture synthesis, confined by the structure reconstructed in the previous step.

In image sequences, this restoration method can be automatically switched on to repair single frames spatially when the extracted temporal information is not reliable. This is realized by means of a detector of complex events that supervises the output of the motion estimator. The task of the complex event detector is actually twofold: besides detecting the frame areas where the motion estimator fails, it has to discover artifacts which may lie in such areas. The latter task is performed by means of spatial algorithms, because the common artifact detectors would fail due to the unreliability of the temporal information [38, 54].

4.5 VINEGAR SYNDROME REMOVAL

Content stored on acetate-based film rolls faces deterioration at a progressive pace. These rolls can be affected by dozens of types of film artifacts, each of them having its own properties and showing up in particular circumstances. One of the major problems of all film archives is the vinegar syndrome [38]. This syndrome appears when, in the course of their chemical breakdown, the acetate-based film bases start to release acetic-acid, giving a characteristic vinegar smell. It is an irreversible process, and from a certain moment on, it becomes auto-catalytic, progressively fuelling itself in the course of time.

The vinegar syndrome has various appearances. It may show up as a partial loss of color, bright or dark tree-like branches, nonuniformly blurred images, etc [38]. Here, we only focus on one type of vinegar syndrome, namely the partial loss of color. This manifests itself as a localized total loss of information in some of the film dye layers (e.g., green and blue). Thanks to the sandwiched structure of the film (sketched in Fig. 4.16), the inner layer (red) is more protected and still preserves some of the original information. This type of vinegar syndrome may be accompanied by emulsion melting (a dissolution of the dye layers). Figure 4.17(a) shows an example of vinegar syndrome combined with emulsion melting.

The detection and correction of this type of vinegar syndrome can be performed with the normal spatiotemporal techniques, as long as it has a temporally impulsive appearance, like normal blotches. However, there are cases when the vinegar syndrome happens in areas of pathological motion, or when it appears in the same place in consecutive frames. In both cases, the information from the surrounding frames is of little use and one has to rely only on the information from the current frame which lies in and around the artifact. Moreover, the blotch removal algorithms do not exploit the information that exists inside the artifact, in the red layer.

For the detection step, one may use the fact that the loss of color makes the artifact look very bright mainly in the green and blue layers [38]. To avoid the influence of irregularities inside the artifact, a hysteresis thresholding is performed as described in

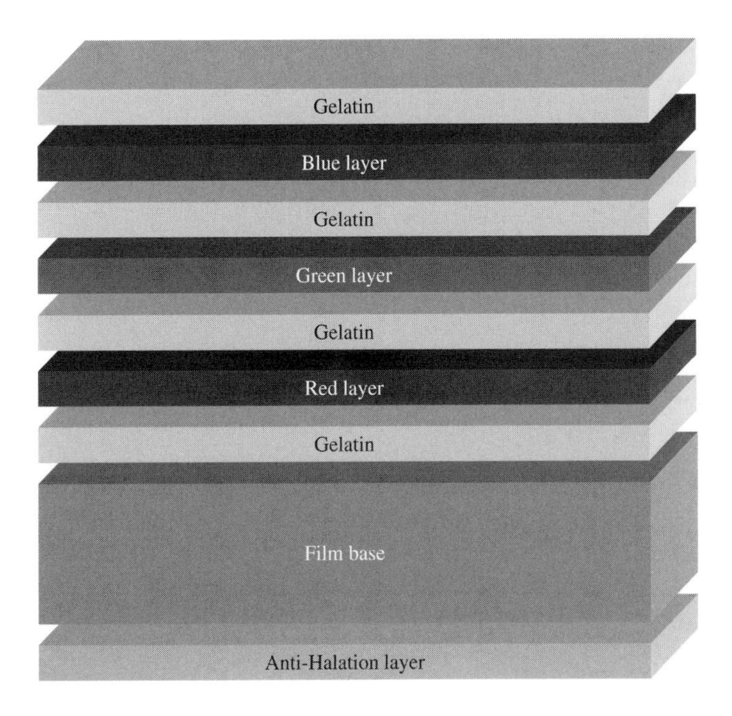

FIGURE 4.16

The layered structure of a film.

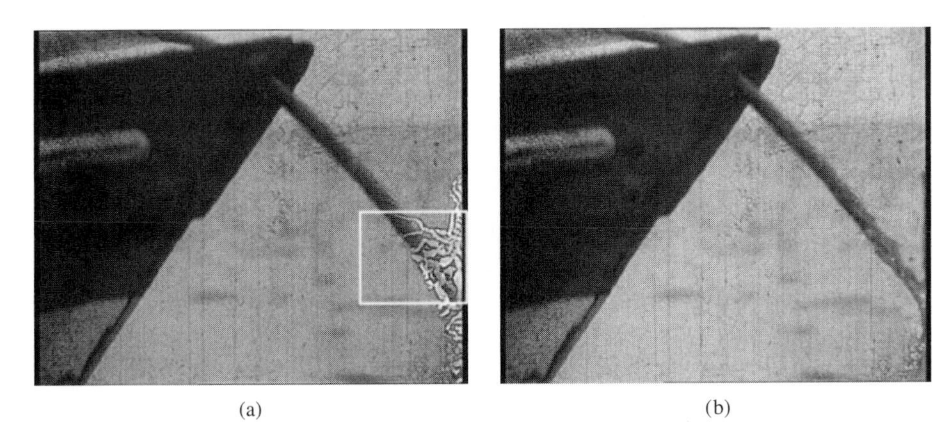

FIGURE 4.17

Restoration example (sequence courtesy of RTP-Radiotelevisão Portuguesa). (a) Original frame, with artifact surrounded by a white box. (b) Restored frame.

Section 4.4.1 of this chapter, with different thresholds for each color layer and in each of the two thresholding phases. In the end, a conditional dilation makes sure that the dark border surrounding the artifact is also added to the artifact mask.

As opposed to conventional spatial restoration techniques, in the restoration step under consideration one may use the information that still exists inside the artifact in the red layer [38]. This information may also be partially affected by the artifact, so a Gaussian smoothing is first performed on the red layer. The isophotes that result from this smoothing are presumed to follow the shapes of the isophotes from the original, undegraded image. These isophotes are then used to "propagate" information inside the artifact from its surrounding area. The general scheme is presented in Fig. 4.18.

An artifact pixel p^i with smoothed red value r_s^i will be overwritten with the nonsmoothed RGB values of a pixel $p^k = (r^k, g^k, b^k)$ lying in the neighborhood of the artifact and representing the "closest" pixel to p^i, (in terms of physical distance and color resemblance). Pixel p^k is selected from a set of pixels representing the connected neighborhood of pixel p^i and having values in $[r_s^i - \Delta_s \ldots r_s^i + \Delta_s]$. p^k is found using Eq. (4.28):

$$k = \arg\min_j \sqrt{\left(r_s^i - r_s^j\right) + \left(\frac{d(p^i, p^j)}{d(p^i, p^j) + 1}\right)^2}, \tag{4.28}$$

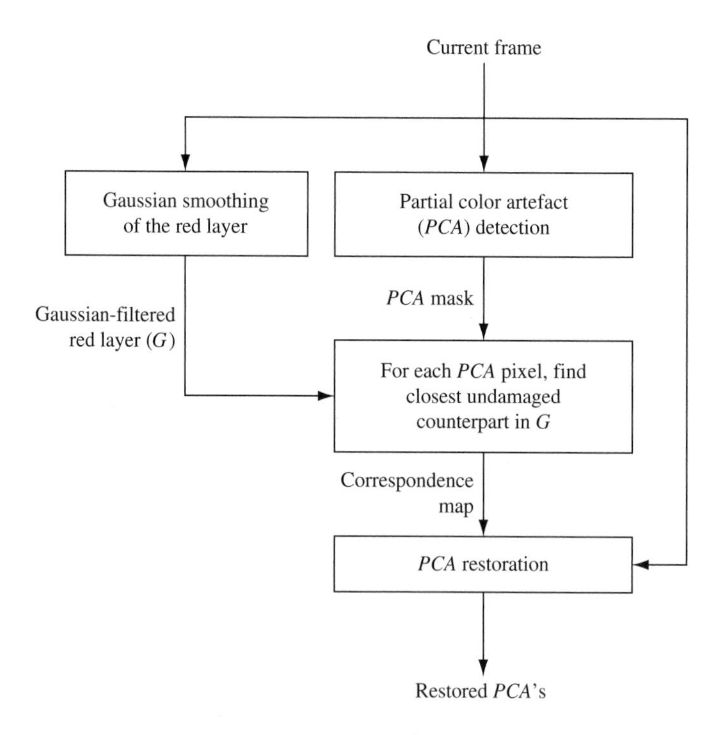

FIGURE 4.18

Restoration scheme for the vinegar syndrome.

where $d(p^i, p^j)$ is the physical distance between pixels p^i and p^j, and is normalized to values in $[0\ldots 1]$ to have the same range as the color distance (all pixel values are in the $[0\ldots 1]$ interval).

Figure 4.17 presents the recovery of portions of a film sequence that were presumed to have been completely lost. A value $\Delta_s \approx 4\%$ of the overall gray value range was used for the connected neighborhood calculation, and a standard deviation $\sigma = 7$ for the Gaussian smoothing step. The result largely reflects the image structure that has been recovered in the smoothed red layer.

4.6 INTENSITY FLICKER CORRECTION

Intensity flicker is defined as unnatural temporal fluctuations of frame intensities that do not originate from the original scene. Intensity flicker is a spatially localized effect that occurs in regions of substantial size. Figure 4.19 shows three successive frames from a sequence containing flicker. A model describing the intensity flicker is the following:

$$g(\mathbf{n}, k) = \alpha(\mathbf{n}, k) f(\mathbf{n}, k) + \beta(\mathbf{n}, k) + w(\mathbf{n}, k), \tag{4.29}$$

where, $\alpha(\mathbf{n}, k)$ and $\beta(\mathbf{n}, k)$ are the multiplicative and additive unknown flicker parameters, that locally scale the intensities of the original frame. The model includes a noise term $w(\mathbf{n}, k)$ that is assumed to be flicker-independent. In the absence of flicker, we have $\alpha(\mathbf{n}, k) = 1$ and $\beta(\mathbf{n}, k) = 0$. The objective of flicker correction is the estimation of the flicker parameters, followed by the inversion of Eq. (4.29). Because flicker always affects fairly large areas of a frame in the same way, the flicker parameters $\alpha(\mathbf{n}, k)$ and $\beta(\mathbf{n}, k)$ are assumed to be spatially smooth functions. Temporally the flicker parameters in one frame may not be correlated at all with those in a subsequent frame.

The earliest attempts to remove flicker from image sequences applied intensity histogram equalization or mean-equalization on frames. These methods do not form a general solution to the problem of intensity flicker correction because they ignore changes in scene contents, and do not appreciate that intensity flicker is a localized effect. In Section 4.6.1, we show how the flicker parameters can be estimated on stationary image

FIGURE 4.19

Three successive frames that contain intensity flicker.

sequences. Section 4.6.2 addresses the more realistic case of parameter estimation on image sequences with motion [55].

4.6.1 Flicker Parameter Estimation

When removing intensity flicker from an image sequence, we essentially make an estimate of the original intensities, given the observed image sequence. Note that the undoing of intensity flicker is only relevant for image sequences, because flicker is a temporal effect by definition. From a single frame, intensity flicker cannot be observed or be corrected.

If the flicker parameters were known, then one can form an estimate of the original intensity from a corrupted intensity using the following straightforward linear estimator:

$$\hat{f}(\mathbf{n},k) = h_1(\mathbf{n},k)g(\mathbf{n},k) + h_0(\mathbf{n},k) \tag{4.30}$$

To obtain estimates for the coefficients $h_i(\mathbf{n},k)$, the mean squared error between $f(\mathbf{n},k)$ and $\hat{f}(n,k)$ is minimized, yielding the following optimal solution:

$$h_0(\mathbf{n},k) = -\frac{1}{\alpha(\mathbf{n},k)}\left(\beta(\mathbf{n},k) + \frac{\sigma_w^2(\mathbf{n},k)}{\sigma_g^2(\mathbf{n},k)}E\left[g(\mathbf{n},k)\right]\right), \tag{4.31a}$$

$$h_1(\mathbf{n},k) = \frac{1}{\alpha(\mathbf{n},k)}\frac{\sigma_g^2(\mathbf{n},k) - \sigma_w^2(\mathbf{n},k)}{\sigma_g^2(\mathbf{n},k)}. \tag{4.31b}$$

If the observed image sequence does not contain any noise, then (4.31) degenerates to the obvious solution:

$$\begin{aligned} h_0(\mathbf{n},k) &= -\frac{\beta(\mathbf{n},k)}{\alpha(\mathbf{n},k)} \\ h_1(\mathbf{n},k) &= \frac{1}{\alpha(\mathbf{n},k)}. \end{aligned} \tag{4.32}$$

In the extreme situation that the variance of the corrupted image sequence is equal to the noise variance, the combination of (4.30) and (4.31) shows that the estimated intensity is equal to the expected value of the original intensities $E[f(\mathbf{n},k)]$.

In practice, the true values for the intensity flicker parameters $\alpha(\mathbf{n},k)$ and $\beta(\mathbf{n},k)$ are unknown and need to be estimated from the corrupted image sequence itself. Because the flicker parameters are spatially smooth functions, we assume that they are locally constant:

$$\begin{cases} \alpha(\mathbf{n},k) = \alpha_{\mathbf{m}}(k) \\ \beta(\mathbf{n},k) = \beta_{\mathbf{m}}(k) \end{cases} \quad \forall \mathbf{n} \in S_{\mathbf{m}}, \tag{4.33}$$

where $S_{\mathbf{m}}$ indicates a small frame region. This region can, in principle, be arbitrarily shaped, but in practice rectangular blocks are chosen. By computing the averages and variances of both sides of Eq. (4.29), the following analytical expressions for the estimates of $\alpha_{\mathbf{m}}(k)$ and $\beta_{\mathbf{m}}(k)$ can be obtained:

$$\hat{\alpha}_{\mathbf{m}}(k) = \sqrt{\frac{\sigma_g^2(\mathbf{n},k) - \sigma_w^2(\mathbf{n},k)}{\sigma_f^2(\mathbf{n},k)}},$$
$$\hat{\beta}_{\mathbf{m}}(k) = E\left[g(\mathbf{n},k)\right] - \hat{\alpha}_{\mathbf{m}}(k) E\left[f(\mathbf{n},k)\right]. \tag{4.34}$$

To solve (4.34) in a practical situation, the mean and variance of $g(\mathbf{n},k)$ are estimated within the region $S_{\mathbf{m}}$. The only quantities that remain to be estimated are the mean and variance of the original image sequence $f(\mathbf{n},k)$. If we assume that the flicker correction is done frame-by-frame, we can estimate these values from the previous corrected frame $k-1$ in the temporally corresponding frame region $S_{\mathbf{m}}$:

$$E\left[f(\mathbf{n},k)\right] \approx \frac{1}{|S_{\mathbf{m}}|} \sum_{\mathbf{m} \in S_{\mathbf{m}}} \hat{f}(\mathbf{m},k-1),$$
$$\sigma_f^2(\mathbf{n},k) \approx \frac{1}{|S_{\mathbf{m}}|} \sum_{\mathbf{m} \in S_{\mathbf{m}}} \left(\hat{f}(\mathbf{m},k-1) - E\left[f(\mathbf{n},k)\right]\right)^2. \tag{4.35}$$

There are situations in which the above estimates are unreliable. The first case is that of uniform intensity areas. For any original image intensity in a uniform region, there are an infinite number of combinations of $\alpha_{\mathbf{m}}(k)$ and $\beta_{\mathbf{m}}(k)$ that lead to the observed intensity. The estimated flicker parameters are also potentially unreliable because of ignoring the noise $w(\mathbf{n},k)$ in (4.34) and (4.35). The reliability of the estimated flicker parameters can be assessed by the following measure:

$$W_{\mathbf{m}}(k) = \begin{cases} 0 & \text{if } \sigma_g^2(\mathbf{m},k) < T \\ \sqrt{\dfrac{\sigma_g^2(\mathbf{m},k) - T}{T}} & \text{otherwise} \end{cases}. \tag{4.36}$$

The threshold T depends on the noise variance. Large values of $W_{\mathbf{m}}(k)$ indicate reliable estimates, while for the most unreliable estimates $W_{\mathbf{m}}(k) = 0$.

4.6.2 Estimation on Sequences with Motion

The results (4.34) and (4.35) assume that the image sequence intensities do not change significantly over time. Clearly, this is an incorrect assumption if motion occurs. The estimation of motion on image sequences that contain flicker is, however, problematic because virtually all motion estimators are based on the constant luminance constraint. Because of the intensity flicker, this assumption is violated heavily. The only motion that can be estimated with sufficient reliability is global motion such as camera panning or zooming. In the following, we assume that in the evaluation of (4.35) and (4.36) possible global motion is compensated for. At that point, we still need to detect areas with any remaining—and uncompensated—motion, and areas that were previously occluded. For both of these cases, the approximation in (4.35) leads to incorrect estimates, which in turn lead to visible artifacts in the corrected frames.

There are various approaches for detecting local motion. One possibility is the detection of large differences between the current and previously (corrected) frames. If local motion occurs, the frame differences will be large. Another possibility to detect local

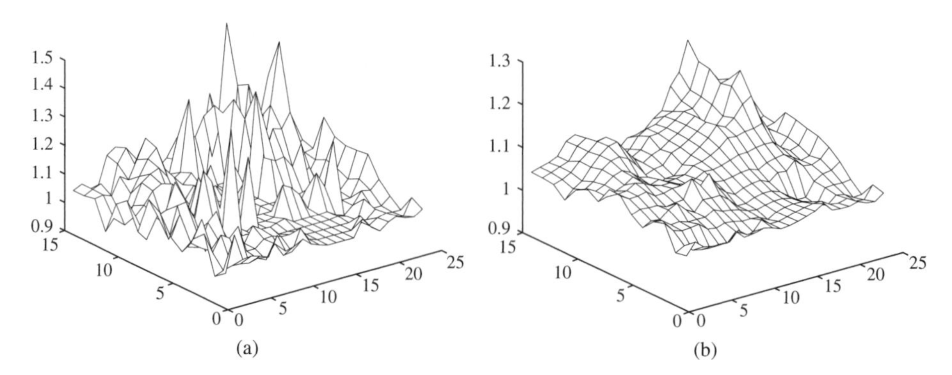

FIGURE 4.20

(a) Estimated intensity flicker parameter $\alpha_{\mathbf{m}}(k)$ using (4.34) and local motion detection. (b) Smoothed and interpolated $\alpha_{\mathbf{m}}(k)$ using SOR.

motion is to compare the estimated intensity flicker parameters to threshold values. If disagreeing temporal information has been used for computing (4.35), we will locally find flicker parameters that do not correspond with its spatial neighbors or with the a priori expectations of the range of the flicker parameters. An outlier detector can be used to localize these incorrectly estimated parameters.

For frame regions $S_{\mathbf{m}}$ where the flicker parameters could not be estimated reliably from the observed image sequence, the parameters are estimated on the basis of the results in spatially neighboring regions. At the same time, for the regions in which the flicker parameters could be estimated, a smoothing postprocessing step needs to be applied to avoid sudden parameter changes that lead to visible artifacts in the corrected image sequence. Such an interpolation and smoothing postprocessing step may exploit the reliability of the estimated parameters, as for instance given by Eq. (4.36). Furthermore, in those frame regions where insufficient information was available for reliably estimating the flicker parameters, the flicker correction should switchoff itself. Therefore, smoothed and interpolated parameters are biased toward $\alpha_{\mathbf{m}}(k) = 1$ and $\beta_{\mathbf{m}}(k) = 0$.

In Fig. 4.20, an example of smoothing and interpolating the estimated flicker parameter for $\alpha_{\mathbf{m}}(k)$ is shown as a 2D matrix [55]. Each entry in this matrix corresponds to a 30×30 pixels region $\Omega_{\mathbf{m}}$ in the frame shown in Fig. 4.19. The interpolation technique used is successive over-relaxation (SOR). Successive over-relaxation is a well-known iterative interpolation technique based on repeated low-pass filtering. Starting off with an initial estimate $\alpha^0_{\mathbf{m}}(k)$ found by solving (4.34), at each iteration a new estimate is formed as follows:

$$
\begin{aligned}
r^{i+1}_{\mathbf{m}}(k) &= W_{\mathbf{m}}(k)\left(\alpha^i_{\mathbf{m}}(k) - \alpha^0_{\mathbf{m}}(k)\right) + \lambda C\left(\alpha^i_{\mathbf{m}}(k)\right), \\
\alpha^{i+1}_{\mathbf{m}}(k) &= \alpha^i_{\mathbf{m}}(k) + \omega \frac{r^{i+1}_{\mathbf{m}}(k)}{W_{\mathbf{m}}(k) + \lambda},
\end{aligned} \tag{4.37}
$$

where $W_{\mathbf{m}}(k)$ is the reliability measure, computed by (4.36), and $C\left(\alpha_{\mathbf{m}}(k)\right)$ is a function that measures the spatial smoothness of the solution $\alpha_{\mathbf{m}}(k)$. The convergence of the

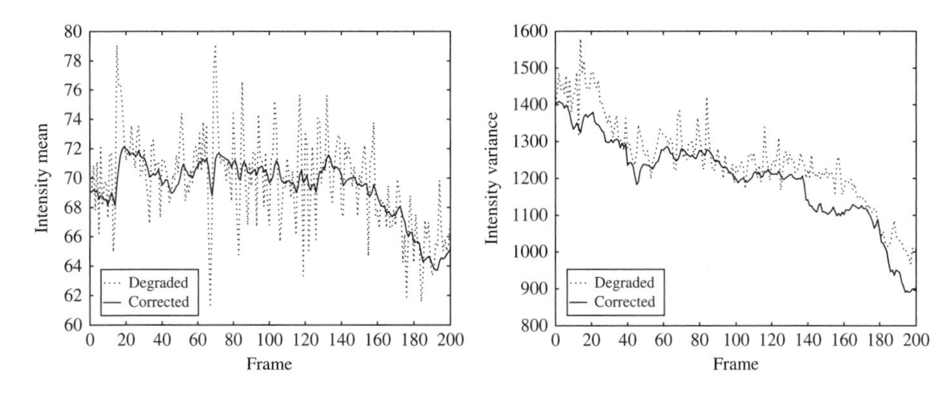

FIGURE 4.21

(a) Mean of the corrupted and corrected image sequence. (b) Variance of the corrupted and corrected image sequence.

iteration (4.37) is determined by the parameter ω, while the smoothness is determined by the parameter λ. For those estimates that have a high reliability, the initial estimates $\alpha_{\mathbf{m}}^{0}(k)$ are emphasized, while for the initial estimates that are deemed less reliable, that is, $\lambda >> W_{\mathbf{m}}(k)$, emphasis is on achieving a smooth solution. Other smoothing and interpolation techniques include dilation and 2D polynomial interpolation. The smoothing and interpolation need to be applied not only to multiplicative parameter $\alpha_{\mathbf{m}}(k)$ but also to the additive parameter $\beta_{\mathbf{m}}(k)$.

As an example, Fig. 4.21 shows the mean and variance as a function of the frame index k of the corrupted and corrected image sequence "Tunnel." Clearly the temporal fluctuations of the mean and variance have been greatly reduced, indicating the suppression of flicker artifacts. An assessment of the resulting visual quality, as with most results of video processing algorithms, has been done by actually viewing the corrected image sequences. Although the original sequence cannot be recovered, the flicker-corrected sequences have a much higher visual quality and they are virtually without any remaining visible flicker.

4.7 KINESCOPE MOIRÉ REMOVAL

The moiré phenomenon represents an interference of (semi-)periodical structures, which gives rise to patterns that did not exist in the original signal. It appears in scanned or printed images [56], in video sequences [57, 58] etc. Moiré phenomena are caused in general by the sampling process (due to the aliasing phenomenon), or a superposition of structures. Kinescope moiré is caused by the process of transferring old TV archives to magnetic tapes. In the early times of the television, broadcasted programs were stored on films that were recorded directly from a TV monitor. The scan lines of the monitor were thus recorded as well. The transfer from film to magnetic tapes took place years later by means of a telecine device. This machine scans the film with a flying spot that

(a)

(b)

FIGURE 4.22

(a) Image affected by moiré (With permission from Sidorov and Kokaram [58]. Courtesy of INA (Institut National de L'Audiovisuel and BBC). (b) Restored image.

moves in horizontal lines. These scan lines and those of the initial TV monitor do not always coincide, giving rise to an aliasing phenomenon. The resulting moiré shows up as a beating pattern of horizontal lines (see Fig. 4.22(a)). Because the initial TV monitors were not always flat, some curvilinear distortion also takes place in the recorded signal, which further complicates the modeling and removal of the moiré.

Currently, the most successful approaches work in the spectral domain. The algorithm presented in [58] replaces peaks of vertical frequency from the Fourier domain with a noise floor, as explained below. The algorithm steps are as follows:

1. After applying the 2D Fourier transform, a range of vertical frequencies around the origin in the spectral domain: $[-\Omega \ldots +\Omega]$ is selected. This horizontal band from the spectrum will be protected against any changes.

2. A binary mask $\boldsymbol{B}$ is created that will indicate which frequencies will be changed:

$$\mathbf{B}\left(\omega_h,\omega_v\right) = \begin{cases} 1 & \text{if } \left(\left|\mathbf{F}\left(\omega_h,\omega_v\right)\right| > \varepsilon\Delta\right) \text{ AND } \left(\left|\omega_v\right| > \Omega\right) \\ 0 & \text{Otherwise} \end{cases}, \tag{4.38}$$

   where $|\mathbf{F}(\omega_h,\omega_v)|$ represents the spectrum magnitude at frequency (ω_h,ω_v), and $\Delta = \frac{\max|\mathbf{F}(\omega_h,\omega_v)| - \min|\mathbf{F}(\omega_h,\omega_v)|}{2} + \min|\mathbf{F}(\omega_h,\omega_v)|$.

3. The magnitudes of the spectral components selected with mask $\boldsymbol{B}$ are replaced with a noise floor value, as shown in Fig. 4.23. This value represents the median magnitude of the frequencies lying outside the band defined in step 1.

4. The resulting magnitude is combined with the original phase extracted in step 1, and the inverse 2D Fourier transform is applied to obtain the restored image.

Figure 4.22 presents the results of the moiré removal algorithm. The parameter values used in this case were $\varepsilon = 0.1, \Delta = 94.4$, and $\Omega = 25$. The moiré pattern was largely

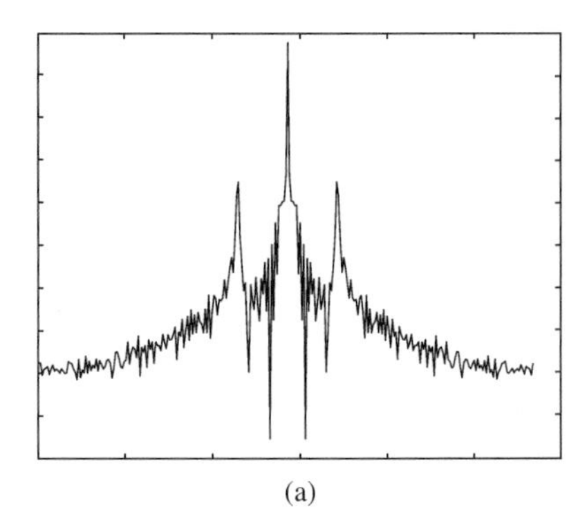

(a)

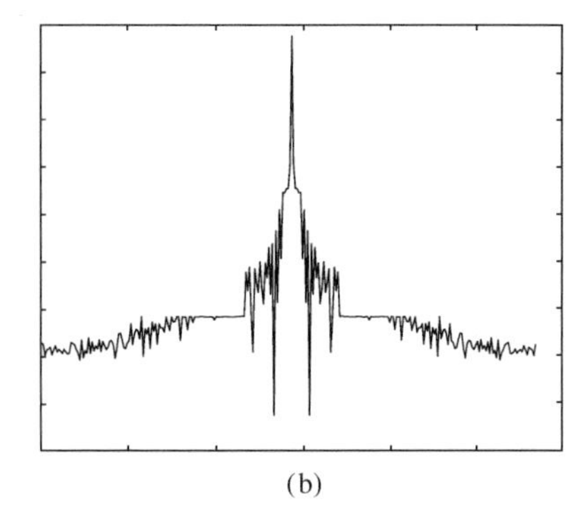

(b)

FIGURE 4.23

An example of spectrum magnitude at zero horizontal frequency (With permission from Sidorov and Kokaram [58]). (a) Degraded image; (b) Restored image.

removed. Some ringing effect still persists due to the discontinuities inserted in the Fourier domain by the magnitude replacement operation shown in Fig. 4.23(b).

The kinescope moiré phenomenon remains an open research track. The nonlinear distortions associated with it (in particular, the curved geometry) are difficult to model. Until now, the algorithms working in the frequency domain were more successful than those working in the spatial domain. More insight into moiré phenomenon is expected to be gained from its temporal analysis.

4.8 SCRATCH REMOVAL

Scratches appear on film rolls mainly because of the particles that accumulate in the projection mechanisms. These particles literally scratch the surface of the film for many frames in a row. This is why scratches are usually vertical and are located around the same position in consecutive frames. Often, they present an "undershoot" or an "overshoot" on their sides, depending on whether they took place on the positive or the negative prints. For simplicity, in the following we consider that the scratches are always bright, with slightly darker sides.

The detection of scratches takes advantage of their particular properties. In [59], the Radon projection for detection of straight lines is used along the y-axis, after which the local maxima are detected. Then, the spatially detected scratches are tracked temporally by means of a Kalman filter. In another approach, the cross-section of the scratch is modeled with a damped sinusoid [2]. This method takes advantage of both the intense lightness of a scratch, as well as the undershoot that is present along the scratch. The scratch influence on the original image is considered to be purely additive. In [60], the above method is generalized to account for the partial loss of data inside a scratch. In [61], a multiframe hypothesis tree is built based on local 1D extrema from each frame.

(a)

(b)

FIGURE 4.24

(a) Image affected by a scratch (Courtesy of Bernard Bessere, Université La Rochelle and CNC - Archives Françaises du Film). (b) Restored image [21].

This hypothesis tree is then sorted based on an energy function. The resulted tree is finally used for tracking the scratches across multiple frames. Other approaches include wavelet-based methods [62], joint detection, and removal of line scratches [63], etc.

The quality of detection may suffer due to either false positives or false negatives, in the presence of long, vertical structures in the image: a pole, the edge of a building, etc. [2]. These are usually vertical, long, and stationary across consecutive frames, so they resemble the scratches very well. A solution to this problem would be to scan not only the film frames but also their blank boundaries [38], as the scratches would appear in the boundaries as well, whereas the long, vertical objects would only appear inside the frames.

The removal of scratches was also subject to various approaches, from 1D and 2D autoregression models [2], wavelets [62], image or video inpainting [63], etc. Each approach has its own advantages and limitations. Many of these approaches are 2D interpolations. The main drawback of the 2D methods is that the results may look very good in each separate frame, but when playing the entire sequence, the scratch correction becomes visible as a "transparent" band that hovers over the image and introduces a small but often noticeable distortion.

In Fig. 4.24, results are shown for a method [64] that extends a 2D exemplar-based algorithm to the temporal domain. While this algorithm was demonstrated for scratch removal, it actually flags a trend of convergence between the scratch removal methods and the video inpainting ones, used for the concealment of blotches (see subsection 4.4.3).

4.9 CONCLUSIONS

This chapter has described methods for enhancing and restoring corrupted video and film sequences. The material that was offered in this chapter is complementary to the spatial enhancement and restoration techniques described in *The Essential Guide to Image Processing* [1]. For this reason, the algorithmic details concentrated mostly on the temporal processing aspects of image sequences or on solutions for cases where the temporal information is not usable. Although the focus has been on the detection and removal of a limited number of impairments, namely noise, coding artifacts, blotches, scratches, vinegar syndrome, flicker and moiré, the approaches and tools described in this chapter are of a more general nature, and they can be used for developing enhancement and restoration methods for other types of degradation.

ACKNOWLEDGEMENTS

We would like to acknowledge the contribution of Peter M.B. van Roosmalen to the previous editions of this chapter.

REFERENCES

[1] A. C. Bovik, editor. *The Essential Guide to Image Processing*. Elsevier, Burlington, MA, 2009.

[2] A. C. Kokaram. *Motion Picture Restoration: Digital Algorithms for Artifact Suppression in Degraded Motion Picture Film and Video*. Springer Verlag, New York, NY, 1998.

[3] M. I. Sezan and R. L. Lagendijk, editors. *Motion Analysis and Image Sequence Processing*. Kluwer Academic Publishers, Boston, MA, 1993.

[4] C. Stiller and J. Konrad. Estimating motion in image sequences. *IEEE Signal Process. Mag.*, 16(4): 70–91, 1999.

[5] J. C. Brailean, R. P. Kleihorst, S. Efstratiadis, A. K. Katsaggelos, and R. L. Lagendijk. Noise reduction filters for dynamic image sequences: A review. *Proc. IEEE*, 83(9):1272–1291, 1995.

[6] M. J. Chen, L. G. Chen, and R. Weng. Error concealment of lost motion vectors with overlapped motion compensation. *IEEE Trans. Circuits Syst. Video Technol.*, 7(3): 560–563, 1997.

[7] A. M. Tekalp. *Digital Video Processing*. Prentice Hall, Upper Saddle River, NJ, 1995.

[8] T. S. Huang, editor. *Image Sequence Analysis*. Springer Verlag, Berlin, Germany, 1991.

[9] E. Dubois and S. Sabri. Noise reduction using motion compensated temporal filtering. *IEEE Trans. Commun.*, 32:826–831, 1984.

[10] A. C. Kokaram and S. Godsill. Joint noise reduction, motion estimation, missing data reconstruction, and model parameter estimation for degraded motion pictures. *Proc. SPIE*, San Diego, 1998.

[11] M. K. Ozkan, M. I. Sezan, and A. M. Tekalp. Adaptive motion-compensated filtering of noisy image sequences. *IEEE Trans. Circuit Syst. Video Technol.*, 3(4):277–290, 1993.

[12] J. W. Woods and J. Kim. Motion-compensated spatio-temporal Kalman filter. In M. I. Sezan and R. L. Lagendijk, editors, *Motion Analysis and Image Sequence Processing*, Kluwer Academic Publishers, Boston, MA, 1993.

[13] J. Kim and J. W. Woods. Spatio-temporal adaptive 3-D Kalman filter for video. *IEEE Trans. Image Process.*, 6(3):414–424, 1997.

[14] G. R. Arce. Multistage order statistic filters for image sequence processing. *IEEE Trans. Signal Process.*, 39(5):1147–1163, 1991.

[15] N. Balakrishnan and C. R. Rao, editors. *Handbook of Statistics 16: Order Statistics: Theory & Methods.* North-Holland, Burlington, MA, 1998.

[16] R. P. Kleihorst, R. L. Lagendijk, and J. Biemond. Noise reduction of image sequences using motion compensation and signal decomposition. *IEEE Trans. Image Process.*, 4(3):274–284, 1995.

[17] E. P. Simoncelli. Bayesian denoising of visual images in the wavelet domain. In P. Müller and B. Vidakovic, editors, *Bayesian Inference in Wavelet Based Models,* Springer-Verlag, Lecture Notes in Statistics 141, 1999.

[18] D. L. Donoho and I. M. Johnstone. Ideal spatial adaptation via wavelet shrinkage. *Biometrika,* 81:425–455, 1994.

[19] P. M. B. van Roosmalen, R. L. Lagendijk, and J. Biemond. Embedded coring in MPEG video compression. *IEEE Trans. Circuits Syst. Video Technol.*, 3:205–211, 2002.

[20] E. P. Simoncelli, W. T. Freeman, E. H. Adelson, and D. J. Heeger. Shiftable multiscale transform. *IEEE Trans. Inf. Theory,* 38(2):587–607, 1992.

[21] G. de Haan. *Video Processing for Multimedia Systems.* University Press Eindhoven, The Netherlands, 2001, ISBN 90-9014015-8.

[22] A. Beric. *Video Post Processing Architectures.* PhD thesis, TU Eindhoven, The Netherlands, 2008, ISBN 978-90-386-1844-9.

[23] B. Ramarmurthi and A. Gersho. Nonlinear space-variant postprocessing of block coded images. *IEEE Int. Conf. Acoust. Speech Signal Process. (ICASSP),* 34(5):1258–1268, 1986.

[24] Y. L. Lee, H. C. Kim, and H. W. Park. Blocking effect reduction of JPEG images by signal adaptive filtering. *IEEE Trans. Image Process.*, 7(2):229–234, 1998.

[25] H. Hu and G. de Haan. Simultaneous coding artifact reduction and sharpness enhancement. *IEEE Int. Conf. Consum. Electron. (ICCE),* 213–214, 2007.

[26] Y. Yang, N. P. Galatsanos, and A. K. Katsaggelos. Projection-based spatially adaptive reconstruction of block-transform compressed images. *IEEE Trans. Circuits Syst. Video Technol.*, 4(7):896–908, 1995.

[27] H. Paek, R.-C. Kim, and S.-U. Lee. On the POCS-based postprocessing technique to reduce the blocking artifacts in transform coded images. *IEEE Trans. Circuits Syst Video Technol.*, 8(3):358–367, 1998.

[28] S. Minami and A. Zakhor. An optimization approach for removing blocking effects in transform coding. *IEEE Trans. Circuits Syst. Video Technol.*, 5(2):74–82, 1995.

[29] K. T. Tan and M. Ghanbari. Frequency domain measurement of blockiness in MPEG-2 coded video. *IEEE Int. Conf. Image Process., ICIP,* 3:977–980, 2000.

[30] P. M. B. Roosmalen. *Restoration of Archived Film and Video.* PhD thesis, Delft University of Technology, The Netherlands, 1999.

[31] A. C. Kokaram, R. D. Morris, W. J. Fitzgerald, and P. J. W. Rayner. Detection of missing data in image sequences. *IEEE Trans. Image Process.*, 4(11):1496–1508, 1995.

[32] A. C. Kokaram, R. D. Morris, W. J. Fitzgerald, and P. J. W. Rayner. Interpolation of missing data in image sequences. *IEEE Trans. Image Process.*, 4(11):1509–1519, 1995.

[33] A. C. Kokaram. On missing data treatment for degraded video and film archives: A survey and a New Bayesian approach. *IEEE Trans. Image Process.*, 13(3):397–415, 2004.

[34] M. Bertalmio, G. Sapiro, V. Caselles, and C. Ballester. *Image Inpainting. Proc. SIGGRAPH 2000*, New Orleans, LA, 2000.

[35] R. Bornard. *Probabilistic Approaches for the Digital Restoration of Television Archives.* PhD thesis, École Centrale, Paris, France, 2002.

[36] A. A. Efros and T. K. Leung. Texture synthesis by Non-parametric sampling. *Proc. ICCV 1999*, Corfu, Greece, September, 1999.

[37] J. Jia and C.-K. Tang. Inference of segmented color and texture description by Tensor voting. *IEEE Trans. Pattern. Anal. Mach. Intell.*, 26(6):771–786, 2004.

[38] A. Rareş. *Archived Film Analysis and Restoration.* PhD thesis, Delft University of Technology, The Netherlands, 2004.

[39] C. Ballester, M. Bertalmio, V. Caselles, G. Sapiro, and J. Verdera. Filling-in by joint interpolation of vector fields and gray levels. *IEEE Trans. Image Process.*, 10(8), 2001.

[40] T. Chan and J. Shen. Mathematical models for local non-texture inpainting. *SIAM J. Appl. Math.*, 62(3):1019–1043, 2001.

[41] S. Masnou. Disocclusion: A variational approach using level lines. *IEEE Trans. Image Process.*, 11(2), 2002.

[42] A. Criminisi, P. Perez, and K. Toyama. Region filling and object removal by exemplar-based inpainting. *IEEE Trans. Image Process.*, 9:1200–1212, 2004.

[43] A. C. Kokaram. Parametric texture synthesis using stochastic sampling. *Proc. ICIP 2002* (IEEE), New York, USA, 2002.

[44] A. N. Hirani and T. Totsuka. Combining frequency and spatial domain information for fast interactive image noise removal. *ACM SIGGRAPH*, 269–276, 1996.

[45] S. T. Acton, D. P. Mukherjee, J. P. Havlicek, and A. C. Bovik. Oriented texture completion by AM-FM Reaction-Diffusion. *IEEE Trans. Image Process.*, 10(6), 2001.

[46] M. Bertalmio, L. Vese, G. Sapiro, and S. Osher. Simultaneous structure and texture image inpainting. *IEEE Trans. Image Process.*, 12(8), 2003.

[47] S. Rane, M. Bertalmio, and G. Sapiro. Structure and texture filling-in of missing image blocks for wireless transmission and compression applications. *IEEE Trans. Image Process.*, 2002.

[48] Z. Tauber, Z.-N. Li, and M. S. Drew. Review and preview: Disocclusion by inpainting for image-based rendering. *IEEE Trans. Syst., Man, and Cybern. C Appl. Rev.*, 37(4):527–540, 2007.

[49] K. A. Patwardhan, G. Sapiro, and M. Bertalmio. Video inpainting under constrained camera motion. *IEEE Trans. Image Process.*, 16(2), 2007.

[50] Y. Matsushita, E. Ofek, W. Ge, X. Tang, and H.-Y. Shum. Full-frame video stabilization with motion inpainting. *IEEE Trans. Pattern. Anal. Mach. Intell.*, 28(7):1150–1163, 2006.

[51] T. Korah and C. Rasmussen. Spatiotemporal inpainting for recovering texture maps of occluded building facades. *IEEE Trans. Image Process.*, 16(9):2262–2271, 2007.

[52] H. Wang, H. Li, and B. Li. Video inpainting for largely occluded moving human. *IEEE Int. Conf. Multimedia Expo*, 1719–1722, 2007.

[53] J. Verdera, V. Caselles, M. Bertalmio, and G. Sapiro. Inpainting surface holes. *Int. Conf. Image Process. (ICIP)*, 2:II-903–II-906, 2003.

[54] A. Rareş, M. J. T. Reinders, and J. Biemond. Statistical analysis of pathological motion areas. *IEE Semin. Digit. Restor. Film Video Arch.*, 2001.

[55] P. M. B. van Roosmalen, R. L. Lagendijk, and J. Biemond. Correction of intensity flicker in old film sequences. *IEEE Trans. Circuits Syst. Video Technol.*, 9(7):1013–1019, 1999.

[56] I. Amidror. *The Theory of the Moiré Phenomenon.* Kluwer Academic Publ., 2000.

[57] C. Hentschel. Video moiré cancellation filter for high-resolution CRTs. *IEEE Trans. Consum. Electron.*, 47(1):16–24, 2001.

[58] D. N. Sidorov and A. C. Kokaram. Suppression of moiré patterns via spectral analysis. In C.-C. Jay Kuo, editor, *Visual Communications and Image Processing*, Vol. 4671, 895–906. *Proc. SPIE*, 2002.

[59] L. Joyeux, O. Buisson, B. Besserer, and S. Boukir. Detection and removal of line scratches in motion picture films. *IEEE Comput. Soc. Conf. Comput. Vis Pattern Recognit.*, 1:548–553, 1999.

[60] V. Bruni and D. Vitulano. A generalized model for scratch detection. *IEEE Trans. Image Process.*, 13(1):44–50, 2004.

[61] B. Besserer and C. Thiré. Detection and tracking scheme for line scratch removal in an image sequence. *8th European Conference on Computer Vision*, 11–14 May, Part III, 264–275, 2004.

[62] X. Jin, G. Jinghuo, W. Xingdong, S. Jun, Z. Guangtao, and L. Zhengguo. An OWE-based algorithm for line scratches restoration in old movies. *IEEE Int. Symp. Circuits Syst.*, 3431–3434, 2007.

[63] S.-W. Kim and K.-H. Ko. Efficient optimization of inpainting scheme and line scratch detection for old film restoration. In Q. Yang and G. Webb, editors, *Lecture Notes in Computer Science*, Springer, Berlin/Heidelberg, 2006, ISBN 978-3-540-36667-6.

[64] G. Forbin, B. Besserer, J. Boldyš, and D. Tschumperlé. Temporal extension to exemplar-based inpainting applied to scratch correction in damaged image sequences. *Visual., Imaging, Image Process.*, 2005.

CHAPTER

Video Stabilization and Mosaicing

5

Mahesh Ramachandran, Ashok Veeraraghavan, and Rama Chellappa

Center for Automation Research,
Department of Electrical and Computer Engineering,
University of Maryland, College Park, Maryland

5.1 INTRODUCTION

A sequence of temporal images acquired by a single sensor adds a whole new dimension to two-dimensional (2D) image data. Availability of an image sequence permits the measurement of quantities, such as subpixel intensities, camera motion and depth, and detection and tracking of moving objects. In turn, the processing of image sequences necessitates the development of sophisticated techniques to extract this information. With the recent availability of powerful yet inexpensive computers, data storage systems, and image acquisition devices, image sequence analysis has transitioned from an esoteric research domain to a practical area with significant applications in many areas.

Temporal variation in the image luminance field is caused by several factors including camera motion, rigid object motion, nonrigid deformation, sensor noise, illumination, and reflectance change. In several situations, it can be assumed that the imaged scene is rigid, and temporal variation in the image sequence is only due to camera and object motion. Classic motion estimation characterizes the local shifts in the image luminance patterns. The global motion that occurs across the entire image frame is typically a result of camera motion and can often be described in terms of a low-order model whose parameters are the unknowns. Global motion analysis is the estimation of these model parameters.

The computation of global motion has seldom attained the center stage of research due to the (often incorrect) assumption that it is a linear or otherwise well-conditioned problem. In practice, an image sequence displays phenomena that voids the assumption of Gaussian noise in the motion field data. The presence of moving foreground objects or occlusion locally invalidates the global motion model, giving rise to outliers. Robustness to such outliers is required of global motion estimators. Researchers have formulated solutions to global motion problems, usually with an application perspective. These can be broadly classified as feature-based and flow-based techniques. Feature-based methods extract and match discrete features between frames, and trajectories of these features are fit to a global motion model. In flow-based algorithms, the optical flow of the image sequence is an intermediate quantity that is used in determining the global motion.

Electronic stabilization of video, creating mosaics from image sequences, and performing motion super-resolution are examples of global motion problems. Applications of these processes are often encountered in surveillance, navigation, tele-operation of vehicles, automatic target recognition (ATR), and forensic science. Reliable motion estimation is critical to these tasks, which is particularly challenging when the sequences display random as well as highly structured systematic errors. The former is primarily a result of sensor noise, atmospheric turbulence, and lossy compression, while the latter is caused by occlusion, shadows, and independently moving foreground objects.

5.1.1 Video Stabilization

Video stabilization refers to the compensation of the motion of pixels on the image plane when a video sequence is captured from a moving camera. Since several practical applications just require video stabilization as opposed to complete estimation of camera trajectory and scene structure, this is an important subproblem that has received much attention. Depending on the type of scenario and the type of motion involved, we have different algorithms to achieve stabilization.

- *Presence of a dominant plane in the scene:* If a dominant plane is present, then we can register all the frames using a planar perspective transformation (homography) corresponding to that plane. For pixels that do not lie on the plane, we need to warp them appropriately depending on the amount of parallax. This is a very common assumption for aerial videos and surveillance cameras monitoring a scene with a dominant ground-plane.

- *Derotation of the image sequence:* In some applications [1, 2], we may want to estimate and remove the motion due to only the 3D rotation of the camera. This corresponds to derotation of the image sequence.

- *Mosaic construction:* We may need to build an extended field-of-view mosaic of the scene using images in the sequence. In this case, we need to accurately register and blend the various images onto the mosaicing surface.

- *Presence of moving objects:* One objective of stabilization is to register the video frames and segment out the moving objects from the scene. This involves detecting independent motion that is different from ego-motion.

5.1.2 Outline

The rest of this chapter is organized as follows. In the next section, we will discuss how simple biological agents, such as insects, are able to perform tasks, such as stabilization and ego-motion estimation, and this section will serve as a motivation for some of the algorithms that we describe in later sections. To measure image motion, we use models that approximate the geometry of projection of a camera. We also use simple models to describe the motion induced in images due to camera motion. Section 5.3 discusses these commonly used models and also discusses the use of image features for measuring the image motion. An alternative way to measure image motion in the absence of features is dense optical flow, and this is discussed in depth in Section 5.4. Then, we describe the various types of image mosaics in Section 5.5 and also discuss how to build mosaics from video sequences. Section 5.6 discusses how we can use measurements from additional sensors along with an image sequence for video stabilization. Section 5.7 discusses motion super-resolution which is an application of stabilization, and Section 5.8 briefly reviews approaches to three-dimensional stabilization. Finally, we conclude the chapter.

5.2 BIOLOGICAL MOTIVATION: INSECT NAVIGATION

Insects are able to fly and navigate in this complex visual world. Despite their relatively small nervous system with very few neurons when compared to the human brain, they are still capable of complex tasks, such as safe landing, obstacle avoidance, and dead reckoning. Behavioral research with insects suggest that insects primarily use visual information: Specifically image motion induced due to ego motion for a number of these navigational tasks. The visual system of insects differs significantly from the visual system of humans. These differences have profound consequences regarding the nature of visual tasks that insects are adept at. Insects have immobile eyes with fixed focal length. Moreover, they do not possess stereoscopic vision [3]. Insect eyes possess inferior spatial acuity but their eyes sample the world at a significantly higher rate than human eyes do. Moreover, their eyes are also placed very close to each other. Because of these differences, insects have evolved to use very different strategies for a range of visual tasks. Srinivasan [4] says "Vision in insects is a very active process in which perception and action are tightly coupled." In this section, we will study this coupling.

A study of how animal visual systems have evolved to do tasks, such as computing optical flow, stabilization, flight control, and ego-motion estimation can help us in two significant ways. The study can serve as a pure motivational tool indicating that such complex tasks, such as stabilization, can be performed real-time, with the accuracy desired. Second, this study can lead us into the paradigm of "active vision" or "purposive vision" [5], where cameras and algorithms are designed with the purpose of estimating certain parameters of interest. In fact, several researchers have used such biologically inspired mechanisms for flight control and obstacle avoidance [6, 7].

5.2.1 Centering Behavior and Collision Avoidance

Bees that fly through holes tend to fly through the center of these holes. Bees, like most other insects, cannot measure distances from surfaces by using stereoscopic vision [3]. Therefore, it is surprising that they are still able to orient themselves at the center of openings. Recent experiments have indicated that bees balance the image motion on the lateral portion of their two eyes as they fly through openings [8]. Bees were trained to fly in narrow tunnels with certain patterns on the side walls of the tunnels. It was shown in [8] that bees tended to fly at the center of this tunnel when the patterns on the side walls were stationary. If one of these patterned side walls was moved in the direction of the bee's flight, thereby reducing the image motion experienced by the bee on that side, then the bees moved closer to that side wall. Similarly, when one of the patterned side walls was moved in the direction opposite to the direction of the bee's flight, the bee moved away from the moving wall. This indicates that bees use image motion for centering behavior, by balancing the image flow on both eyes.

Collision avoidance is another task that is visually driven in most insects. When an insect approaches an obstacle, its image expands on it's eyes. Insects are sensitive to this image expansion and turn away from the direction in which the image expansion occurs [4], thereby avoiding collision with obstacles.

5.2.2 Control of Flight Speed and Stabilization

If some insects are indeed capable of measuring image motion, specifically the angular velocity of the image, then questions arise as to whether they use this information to control other aspects of their flight. In fact, experiments in [8] and [9] indicate that insects do use estimates of image motion to alter and control their speed of flight. Ref. [8] indicates visual control of flight speed in bees is achieved by monitoring and regulating the apparent image motion, specifically the angular velocity of the image. This modulation of flight speed depending on the image motion has some distinct advantages. This vision based control of flight speed enables bees to slow down while encountering narrow passages. Bees also owe their ability of flawless landing on surfaces to this. When a bee is landing on a surface, the bee steadily reduces it's forward speed as it approaches the surface. Experiments in [8] indicate that forward speed is proportional to altitude which ensures that the angular velocity of the image of the surface is maintained constant as the bee approaches the surface. This technique ensures that the flight speed is close to zero at touchdown and helps in a bee's flawless landing.

Insects also use the image motion in order to achieve stabilization of their flight. If an insect moving straight is pushed left due to the wind, then elementary motion detectors [10] indicate that the image on the front of the retina has moved to the right. The insects generate a counteractive torque to bring them back on course. Studies have indicated that insects use such visual cues in conjunction with other nonvisual sensors like halteres (that act as gyroscopes) in order to stabilize their flight.

5.2.3 Measuring Distance by Integrating Optical Flow

Some social insects like honeybees, for example, are able to use visual cues to navigate accurately and repeatedly to food sources that are far away. Moreover, during this flight they are also able to infer the direction and the distance of their food source and reliably communicate it to other bees in their hive. Recent research has indicated that the odometer of the bees is "visually driven" [4]. Experiments have indicated that bees use the extent of image motion in their eyes as a measure of the distance of their flight [8, 11]. This system that estimates distance flown by integrating image motion seems relatively robust to variations in the texture of the environment. Furthermore, it has long been known [12] that honeybees use celestial landmarks to determine the direction of their flight. Interestingly, foraging honeybees are also able to accurately communicate the distance and direction of food sources to other "recruits" through a waggle dance [12].

Thus, insects in general and honeybees in particular possess exceptional navigational abilities that are primarily driven by visual feedback from the environment. Moreover, insects seem to prefer image motion based computations to feature-based methods. This preference to image motion based methods can be attributed to the fact that while their eyes possess very little spatial acuity (ability to identify or extract features), they sample the world at very high rates. Therefore, they are able to estimate image motion and image velocity better.

We have motivated the study of image motion in video sequences by describing how insects perform motion computations and navigate in the world. In video stabilization, we need to analyze the image motion and obtain models for the global motion in image sequences. In the next section, we will describe computational models for characterizing image motion. In the succeeding sections, we will describe some optical flow based algorithms for measuring the image motion and computing the motion-model parameters. We will describe how this is useful for stabilization and mosaicing.

5.3 CAMERA MODEL AND IMAGE MOTION MODEL

Prior to discussing models for the global motion problem, it is worthwhile to verify whether the apparent motion on the image induced by the camera motion can indeed be approximated by a global model. This study takes into consideration an analytic model for the camera as a projective device, the 3D structure of the scene being viewed, and its corresponding image. We describe the model for a projective camera and study how the image of a world point moves as the camera undergoes general motion (three translations and three rotations).

5.3.1 Camera Model

The imaging geometry of a perspective camera is shown in Fig. 5.1. The origin of the 3D coordinate system (X, Y, Z) lies at the optical center C of the camera. The *retinal plane* or *image plane* is normal to the optical axis Z and is offset from C by the focal length f. Images of unoccluded 3D objects in front of the camera are formed on the image plane. The 2D image plane coordinate system (x, y) is centered at the *principal point*, which is the intersection of the optical axis with the image plane. The orientation of (x, y) is flipped with respect to (X, Y) in Fig. 5.1 due to inversion caused by simple transmissive optics. For this system, the image plane coordinate (x_i, y_i) of the image of the unoccluded 3D point (X_i, Y_i, Z_i) is given by

$$x_i = f\frac{X_i}{Z_i}, y_i = f\frac{Y_i}{Z_i}. \tag{5.1}$$

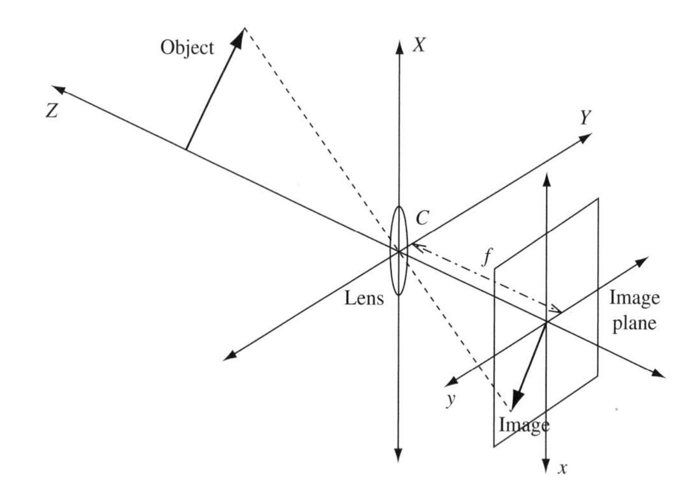

FIGURE 5.1
3D imaging geometry.

The projective relation (5.1) assumes a rectilinear system, with an isotropic optical element. In practice, the plane containing the sensor elements may be misaligned from the image plane, and the camera lens may suffer from optical distortions including nonisotropy. However, these effects can be compensated by calibrating the camera and/or remapping the image. In the remainder of this chapter, it is assumed that the linear dimensions are normalized with reference to the focal length, that is, $f = 1$.

5.3.2 Effect of Camera Motion

The effect of camera motion can be computed using projective geometry [13, 14]. Assume that an arbitrary point in the 3D scene lies at (X_0, Y_0, Z_0) in the reference frame of the first camera and moves to (X_1, Y_1, Z_1) in the second. The effect of camera motion relates the two coordinate systems according to

$$\begin{pmatrix} X_1 \\ Y_1 \\ Z_1 \end{pmatrix} = \begin{pmatrix} r_{xx} & r_{xy} & r_{xz} \\ r_{yx} & r_{yy} & r_{yz} \\ r_{zx} & r_{zy} & r_{zz} \end{pmatrix} \begin{pmatrix} X_0 \\ Y_0 \\ Z_0 \end{pmatrix} + \begin{pmatrix} t_x \\ t_y \\ t_z \end{pmatrix}, \tag{5.2}$$

where the rotation matrix $\left[r_{ij}\right]$ is a function of ω. Combining (5.1) and (5.2) permits the expression of the projection of the point in the second image in terms of that in the first as

$$x_1 = \frac{r_{xx}x_0 + r_{xy}y_0 + r_{xz} + t_x/Z_0}{r_{zx}x_0 + r_{zy}y_0 + r_{zz} + t_z/Z_0} \quad \text{and}$$
$$y_1 = \frac{r_{yx}x_0 + r_{yy}y_0 + r_{yz} + t_y/Z_0}{r_{zx}x_0 + r_{zy}y_0 + r_{zz} + t_z/Z_0}. \tag{5.3}$$

Assuming either that 1) points are distant compared to the interframe translation, that is, neglecting the effect of translation, or 2) a planar embedding of the real world (5.16), the *perspective* transformation is obtained

$$x_1 = \frac{p_{xx}x_0 + p_{xy}y_0 + p_{xz}}{p_{zx}x_0 + p_{zy}y_0 + p_{zz}} \quad \text{and}$$

$$y_1 = \frac{p_{yx}x_0 + p_{yy}y_0 + p_{yz}}{p_{zx}x_0 + p_{zy}y_0 + p_{zz}}. \tag{5.4}$$

Other popular global deformations mapping the projection of a point between two frames are the similarity and affine transformations, which are given by

$$\begin{pmatrix} x_1 \\ y_1 \end{pmatrix} = s\begin{pmatrix} \cos\theta & \sin\theta \\ -\sin\theta & \cos\theta \end{pmatrix}\begin{pmatrix} x_0 \\ y_0 \end{pmatrix} + \begin{pmatrix} b_0 \\ b_1 \end{pmatrix} \tag{5.5}$$

and

$$\begin{pmatrix} x_1 \\ y_1 \end{pmatrix} = \begin{pmatrix} a_0 & a_1 \\ a_2 & a_3 \end{pmatrix}\begin{pmatrix} x_0 \\ y_0 \end{pmatrix} + \begin{pmatrix} b_0 \\ b_1 \end{pmatrix}, \tag{5.6}$$

respectively. Free parameters for the similarity model are the scale factor s, image plane rotation θ, and translation (b_0, b_1). The affine transformation is a superset of the similarity operator, and incorporates shear and skew as well. The perspective operator is a superset of the affine, as can be readily verified by setting $p_{zx} = p_{zy} = 0$ in (5.4).

Next, we discuss how to extract features from images and how they can be used for computing the image motion using the models described earlier.

5.3.3 Image Features

The basic goal in feature-based motion estimation is to use features to find maps that relate the images taken from different view-points. These maps are then used to estimate the image motion by computing the parameters of a motion model. Consider the case of pure rotation. Here, the camera center is fixed and the image plane is moved to another position. The image of a point in the real world is formed by the intersection on the image plane of the ray joining the camera center and the world point. The resulting images formed on the image planes are quite different, but they are related in an interesting way.

Though various lengths, ratios, and angles formed on the images are all different, the *cross ratio* remains the same [15]. Given four collinear points A, B, C, and D on an image, the cross ratio is $\frac{AC}{CB} \div \frac{AD}{DB}$, and it remains constant. In other words, $\frac{AC}{CB} \div \frac{AD}{DB} = \frac{\acute{A}\acute{C}}{\acute{C}\acute{B}} \div \frac{\acute{A}\acute{D}}{\acute{D}\acute{B}}$, where $\acute{A}$, $\acute{B}$, $\acute{C}$, and $\acute{D}$ are the corresponding points in the second image (formed after rotating the camera about its axis).

Looking carefully, we can see that this intuition leads to a map relating the two images. Given four corresponding points in general position in the two images, we can map any point from one image to the other. Suppose we know that A maps to $\acute{A}$, B to $\acute{B}$, C to $\acute{C}$, and D to $\acute{D}$. Then, the point of intersection of AB and CD (say E) will map to the point of intersection of $\acute{A}\acute{B}$ and $\acute{C}\acute{D}$ (say $\acute{E}$). Now, any point F on ABE will map to point $\acute{F}$ such that the cross ratio $\frac{AE}{EB} \div \frac{AF}{FB}$ is preserved. This way one can map each point on one image to the other image. Such a map is called *homography*. As mentioned before, such a map is defined by four corresponding points in general position. So, if x maps to $\acute{x}$ by homography H, $x' = Hx$. Note that such a map exist only in case of pure rotation.

However, for planar scenes, homography relating the two views exist irrespective of the motion involved. In the case of planar scene, there exist a homography relating the first image to the real-world plane and another one mapping the real-world plane to the second image plane, that is,

$$x_1 = H_1 x_p, \tag{5.7}$$

$$x_p = H_2 x_2, \quad \text{and} \tag{5.8}$$

$$\Rightarrow x_1 = H_1 H_2 x_2 = H x_2, \tag{5.9}$$

where H_1 maps x_1, a point on first image plane to x_p, the corresponding point on the real plane, while H_2 maps x_p to x_2, the corresponding point on the second image plane. Thus, homography $H = H_1 H_2$ maps points from one image plane to the other. Such a homography exists, no matter what the underlying motion between the two camera positions is. This happens because the images formed by camera rotation (or in the case of planar scenes) do not depend on the scene structure. On the other hand, when there are depth variations in the scene, such a homography doesn't exist between images formed by camera translation.

In the case of depth variations, we can use structure from motion (SFM) approaches to estimate the motion of the camera. The estimated camera motion can be used to selectively stabilize the image sequence for the camera motion (such as compensation of rotation, or sideways translation etc.) The next section discusses about SFM and describes an algorithm for SFM using image feature points.

5.3.4 Structure from Motion

Structure from motion refers to the task of inferring the camera motion and scene structure from an image sequence taken from a moving camera, using either image features or flow. We describe an illustrative algorithm for SFM using feature point tracks in a sequence. Consider an image sequence with N images. Feature points are detected and tracked throughout the sequence. Suppose the scene has M features and their projections in each image are denoted by $x_{i,j} = (u_{i,j}, v_{i,j})^T$, where $i \in \{1,\ldots,M\}$ denotes the feature index and $j \in \{1,\ldots,N\}$ denotes the frame index. For the sake of simplicity, assume that all features are visible in all frames. The structure-from-motion problem involves solving for the camera locations and the 3D coordinates of the feature points in the world coordinate system.

The camera poses are specified by a rotation matrix R_j and a translation vector T_j for $j = 1, \ldots, N$. The coordinates of a point in the camera system and world system are related by $P_c = R_i P_w + T_i$, where P_c denotes the coordinates of a point in the camera coordinate system, and P_w denotes the coordinates of the same point in the world coordinate system. The 3D coordinates of the world landmarks are denoted by $\underline{X}_i = (X_i, Y_i, Z_i)^T$ for $i = 1, \ldots, M$. We assume an orthographic projection model for the camera. Landmarks are projected onto the image plane according to the following equation:

$$\begin{pmatrix} u_{i,j} \\ v_{i,j} \end{pmatrix} = K \cdot \begin{bmatrix} R_j & T_j \end{bmatrix} \underline{X}_i. \tag{5.10}$$

In Eq. (5.10), K denotes the 2×3 camera matrix. Let the centroid of the $3D$ points be C and the centroid of the image projections of all features in each frame be c_j. We can eliminate the translations from these equations by subtracting out C from all world point locations and c_j from the image projections of all features in the jth frame. Let $\hat{x}_{i,j} = x_{i,j} - c_j$ and $\hat{X}_j = \underline{X}_j - C$. The projection equation can be rewritten as

$$\hat{x}_{i,j} = P_j \cdot \hat{X}_i. \tag{5.11}$$

In Eq. (5.11), $P_j = K \cdot R_j$ denotes the 2×3 projection matrix of the camera.

We can stack up the image coordinates of the all the feature points in all the frames and write it in the factorization format as follows:

$$\begin{bmatrix} \hat{x}_{1,1} & \cdots & \hat{x}_{M,1} \\ \vdots & \ddots & \vdots \\ \hat{x}_{1,N} & \cdots & \hat{x}_{M,N} \end{bmatrix} = \begin{bmatrix} P_1 \\ \vdots \\ P_N \end{bmatrix} \begin{bmatrix} \hat{X}_1 & \cdots & \hat{X}_M \end{bmatrix}. \tag{5.12}$$

The matrix on the left hand side of Eq. (5.12) is the measurement matrix, and it has been written as a product of a $2N \times 3$ projection matrix and a $3 \times M$ structure matrix. This factorization implies that the measurement matrix is of rank 3. This observation leads to a factorization algorithm in solving for the projection matrices P_j and the $3D$ point locations X_i. The details of the algorithm are described in detail in [16].

5.3.5 Feature based Algorithms

We have seen that a homography can be used to map one image to the other in the case of pure camera rotation or a planar scene. If such a homography exists between the images, four points are sufficient to specify it precisely. In practice, we extract a number of features in each image and use feature matching algorithms [17] to establish correspondence between the images. The resulting set of feature matches between two images usually have a subset of wrong ("outlier") matches due to errors in the feature extraction and matching process. We handle these outliers in a RANSAC framework [18] which attempts to find the motion parameters by first identifying the set of inlier feature matches. If we have an image sequence, we use feature tracking algorithms like KLT to track a set of features through an image sequence. The correspondences specified by these tracks are used to compute the motion model parameters.

Usually, neither the scene being viewed is planar nor the motion a pure rotation. In such cases, there is no linear map that relates one image to the other unless one neglects the effect of translation (similar to assumption made in Eq. (5.4)). In such cases, researchers either make simplifying assumptions based on the domain knowledge or include additional constraints involving more views to take care of the limitations of the geometric approach. In [19], Morimoto *et al.* demonstrate real time image stabilization that can handle large image displacements based on a two-dimensional multiresolution technique. [20] propose an operation called *threading* that connects two consecutive fundamental matrices using the trifocal tensor as the thread. This makes sure that consecutive camera matrices are consistent with the 3D scene without explicitly recovering it.

All feature-based methods assume that there are features in the images which can reliably be extracted and matched across frames. There are cases when there are hardly any features like in aerial imagery. For such situations, flow-based methods are more suitable. The next section will discuss flow-based approaches.

5.4 FLOW-BASED APPROACHES

Optical flow is the apparent motion of image pixels caused by the actual relative motion between the scene and the camera. When a 3D scene is imaged by a moving camera, with translation $t = (t_x, t_y, t_z)$ and rotation $\omega = (\omega_x, \omega_y, \omega_z)$, the optical flow of the scene (Chapter 3.8) is given by

$$\begin{aligned} u(x,y) &= (-t_x + xt_z)g(x,y) + xy\omega_x - (1+x^2)\omega_y + y\omega_z \quad \text{and} \\ v(x,y) &= (-t_y + yt_z)g(x,y) + (1+y^2)\omega_x - xy\omega_y - x\omega_z \end{aligned} \tag{5.13}$$

for small ω. Here, $g(x,y) = 1/Z(x,y)$ is the inverse scene depth, and all linear dimensions are normalized in terms of the focal length f of the camera.

5.4.1 Global Flow Models

Clearly, the optical flow field can be arbitrarily complex and does not necessarily obey a low-order global motion model. However, several approximations to (5.13) exist that reduce the dimensionality of the flow field. One possible approximation is to assume that the translations are small compared to the distance of the objects in the scene from the camera. In this situation, image motion is caused purely by camera rotation and is given by

$$\begin{aligned} u(x,y) &= xy\omega_x - (1+x^2)\omega_y + y\omega_z \quad \text{and} \\ v(x,y) &= (1+y^2)\omega_x - xy\omega_y - x\omega_z. \end{aligned} \tag{5.14}$$

Equation (5.14) represents a true global motion model, with three degrees of freedom $(\omega_x, \omega_y, \omega_z)$. When the field of view (FOV) of the camera is small, that is, when $|x|, |y| \ll 1$, the second order terms can be neglected, giving a further simplified three parameter global motion model

$$u(x,y) = \omega_y + y\omega_z \quad \text{and}$$
$$v(x,y) = \omega_x - x\omega_z. \tag{5.15}$$

Alternatively, the 3D world being imaged can be assumed to be approximately planar. It can be shown that the inverse scene depth for an arbitrarily oriented planar surface is a planar function of the image coordinates (x, y)

$$g(x,y) = ax + by + c. \tag{5.16}$$

Substituting (5.16) into (5.13) gives the eight parameter global motion model

$$u(x,y) = a_0 + a_1x + a_2y + a_6x^2 + a_7xy \quad \text{and}$$
$$v(x,y) = a_3 + a_4x + a_5y + a_6xy + a_7y^2, \tag{5.17}$$

for appropriately computed $\{a_i, i = 0\ldots7\}$. Eq. (5.17) is called the *pseudo-perspective* model or transformation.

Further, let us consider the perspective transformation model introduced for features in Eq. 5.6. The flow field (u, v) is the difference between image plane coordinates $(x_1 - x_0, y_1 - y_0)$ across the entire image. When the FOV is small, we can assume $|p_{zx}x_0|, |p_{zy}y_0| \ll 1$. Under this assumption, the flow field, as a function of image coordinate, is given by

$$u(x,y) = \frac{(p_{xx} - p_{zz})x + p_{xy}y + p_{xz}}{p_{zx}x + p_{zy}y + p_{zz}} \quad \text{and}$$
$$v(x,y) = \frac{p_{yx}x + (p_{yy} - p_{zz})y + p_{yz}}{p_{zx}x + p_{zy}y + p_{zz}}, \tag{5.18}$$

which is also a perspective transformation albeit with different parameters. Without loss of generality, we can set $p_{zz} = 1$, giving eight degrees of freedom for the perspective model.

5.4.2 Flow-Based Algorithm

The computation of optical flow from an image sequence is based on the luminance constancy assumption. Under this assumption the flow-field can be related to the image derivatives using the gradient constraint equation

$$\frac{\partial\psi}{\partial t} + u\frac{\partial\psi}{\partial x} + v\frac{\partial\psi}{\partial y} = 0 \qquad \forall x, y, t, \tag{5.19}$$

where $\psi(x,y,t)$ is the image luminance pattern. The flow field (u,v) is a function of location (x,y). For smooth motion fields encountered in typical global motion problems, it is meaningful to model (u,v) as a weighted sum of basis functions:

$$u = \sum_{k=0}^{K-1} u_k \phi_k, \quad v = \sum_{k=0}^{K-1} v_k \phi_k. \tag{5.20}$$

The basis function $\phi_k(x,y)$ is typically a locally supported interpolator generated by shifts of a prototype function $\phi_0(x,y)$ along a square grid of spacing w. An example of linear basis function modeling in 1D is shown in Fig. 5.2. Additional requirements are imposed on ϕ_0, to ensure computational ease and an intuitive appeal for modeling a flow field. These are

1. Separability: $\phi_0(x,y) = \phi_0(x)\phi_0(y)$.
2. Differentiability: $\frac{d\phi_0(x)}{dx}$ exists $\forall x$.
3. Symmetry about the origin: $\phi_0(x) = \phi_0(-x)$.
4. Peak at the origin: $|\phi_0(x)| \leq \phi_0(0) = 1$.
5. Compact support: $\phi_0(x) = 0 \;\; \forall |x| > w$.

The cosine window

$$\phi_0(x) = \frac{1}{2}\left[1 + \cos\left(\frac{\pi x}{w}\right)\right], \qquad x \in [-w, w], \tag{5.21}$$

is one such choice of basis that has been shown to accurately model typical optical flow fields associated with global motion problems. A useful range for w is between 8 and 32.

It can be shown that an unbiased estimate for the basis function model parameters $\{u_k, v_k\}$ is obtained by solving the following $2K$ equations [21]

$$\sum_k u_k \int \frac{\partial \phi_k \phi_l \frac{\partial \hat{\psi}}{\partial x}}{\partial x} \psi + \sum_k v_k \int \frac{\partial \phi_k \phi_l \frac{\partial \hat{\psi}}{\partial x}}{\partial y} \psi = \int \phi_l \frac{\partial \hat{\psi}}{\partial t} \frac{\partial \hat{\psi}}{\partial x} \quad \text{and}$$

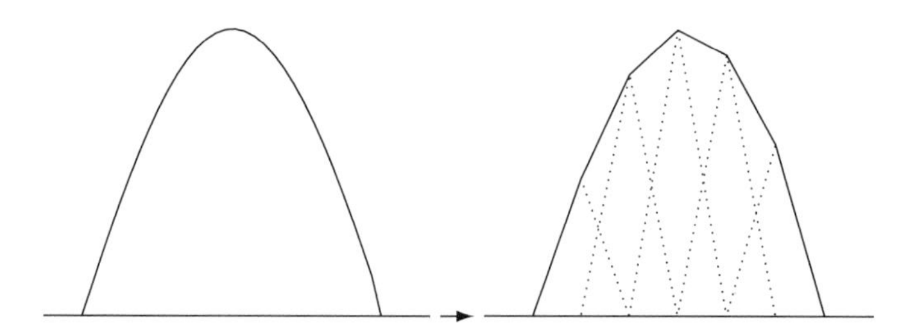

FIGURE 5.2

A function (left) and its modeled version (right). The model used here is the linear interpolator or triangle function. The contribution of each model basis function is denoted by the dotted curves.

$$\sum_k u_k \int \frac{\partial \phi_k \phi_l \frac{\partial \hat{\psi}}{\partial y}}{\partial x} \psi + \sum_k v_k \int \frac{\partial \phi_k \phi_l \frac{\partial \hat{\psi}}{\partial y}}{\partial y} \psi = \int \phi_l \frac{\partial \hat{\psi}}{\partial t} \frac{\partial \hat{\psi}}{\partial y}, \forall\, l = 0, 1 \ldots K - 1. \quad (5.22)$$

Each pair of equations of the type (5.22) characterizes the solution around the image area covered by the basis function ϕ_l. The dominant unknowns, which are the corresponding model weights, are u_l, v_l. The finite support requirement on basis function ϕ_l ensures that only the center weights u_l, v_l and their immediate neighbors in the cardinal and diagonal directions enter each equation. In practice, sampled differentiations and integrations are performed on the sequence. Each equation pair is computed as follows:

1. First, the X, Y, and temporal gradients are computed for the observed frame of the sequence. Smoothing is performed prior to gradient estimation if the images are dominated by sharp edges.
2. Three templates each of size $2w \times 2w$ are formed. The first template is the prototype function ϕ_0, with its support coincident with the template. The other two are its X and Y gradients. Knowledge of the analytical expression for ϕ_0 means that its gradients can be determined with no error.
3. Next, a square tile of size $2w \times 2w$ of the original and spatiotemporal gradient images, coincident with the support of ϕ_l is extracted.
4. The eighteen left hand side terms of each equation and one right hand side term are computed by overlaying the templates as necessary and by computing the sum of products.
5. Steps 3 and 4 are repeated for all K basis functions.
6. Since the interactions are only between spatially adjacent basis function weights, the resulting matrix is sparse, block tridiagonal, with tridiagonal submatrices, each entry of which is a 2×2 matrix. This permits convenient storage of the left hand side matrix.
7. The resulting sparse system is solved rapidly using the Preconditioned Biconjugate Gradients algorithm [22, 23].

The procedure described above produces a set of model parameters $\{u_k, v_k\}$ that largely conforms to the appropriate global motion model, where one exists. In the second phase, these parameters are simultaneously fit to the global motion model while outliers are identified, using the iterated weighted least squares technique outlined below:

1. *Initialization*
   (a) All flow field parameters whose support regions show sufficiently large high frequency energy (quantified in terms of the determinant and condition number of the covariance matrix of the local spatial gradient) are flagged as valid data points.
   (b) A suitable global motion model is specified.

2. *Model fitting*
   (a) If there are an insufficient number of valid data points, the algorithm signals an inability to compute the global motion. In this event, a more restrictive motion model must be specified.

   (b) If there are sufficient data points, model parameters are computed to be the least squares (LS) solution of the linear system relating observed model parameters with the global motion model of choice.

   (c) When a certain number of iterations of this step are complete, the LS solution of valid data points is output as the global motion model solution.

3. *Model consistency check*
   (a) The compliance of the global motion model to the overlapped basis flow vectors is computed at all grid points flagged as valid, using a suitable error metric.

   (b) The mean error $\bar{\epsilon}$ is computed. For a suitable multiplier f, all grid points with errors larger than $f\bar{\epsilon}$ are declared invalid.

   (c) Step 2 is repeated.

Typically, three to four iterations are sufficient. Since this system is open-loop, small errors do tend to build up over time. It is also conceivable to use a similar approach to refine the global motion estimate by registering the current image with a suitably transformed origin frame.

5.5 STABILIZATION AND MOSAICING

Stabilization is the process of estimating and compensating for the background image motion occurring due to the ego-motion of the camera. Mosaicing is the process of compositing or piecing together successive frames of the stabilized image sequence so as to virtually increase the field of view of the camera [24]. This process is especially important for remote surveillance, teleoperation of unmanned vehicles, rapid browsing in large digital libraries, and in video compression. Mosaics are commonly defined only for scenes viewed by a pan/tilt camera, for which the images can be related by a projective transformation. However, recent studies look into qualitative representations, nonplanar embeddings [25, 26], and layered models [27]. The newer techniques permit camera translation and gracefully handle the associated parallax. These techniques compute a "parallax image" [28] and warp the off-planar image pixels on the mosaic using the corresponding values in the parallax image. Mosaics represent the real world in 2D, on a plane or other manifold like the surface of a sphere or "pipe." Mosaics that are built on spherical or cylindrical surfaces belong to the class of panoramic mosaics [29, 30]. For general camera motion, there are techniques to construct a mosaic on an adaptive surface depending on the camera motion. Such mosaics, called manifold mosaics, are described in [26, 31]. Mosaics that are not true projections of the 3D world yet present extended information on a plane are referred to as *qualitative* mosaics.

Several options are available while building a mosaic. A *simple* mosaic is obtained by compositing several views of a static 3D scene from the same view point and different view angles. Two alternatives exist, when the imaged scene has moving objects, or when there is camera translation. The *static* mosaic is generated by aligning successive images with respect to the first frame of a batch and performing a temporal filtering operation on the stack of aligned images. Typical filters are pixelwise mean or median over the batch of images, which have the effect of blurring out moving foreground objects. In addition, the edges in the mosaic are smoothed, and sharp features are lost. Alternatively the mosaic image can be populated with the first available information in the batch.

Unlike the static mosaic, the *dynamic* mosaic is not a batch operation. Successive images of a sequence are registered to either a fixed or a changing origin, referred to as the *backward* and *forward stabilized* mosaics, respectively. At any time instant, the mosaic contains all the new information visible in the most recent input frame. The fixed coordinate system generated by a backward stabilized dynamic mosaic literally provides a snapshot into the transitive behavior of objects in the scene. This finds use in representing video sequences using still frames. The forward stabilized dynamic mosaic evolves over time, providing a view port with the latest past information supplementing the current image. This procedure is useful for virtual field of view enlargement in the remote operation of unmanned vehicles.

In order to generate a mosaic, the global motion of the scene is first estimated. This information is then used to rewarp each incoming image to a chosen frame of reference. Rewarped frames are combined in a manner suitable to the end application. The techniques presented in earlier Sections 5.3 and 5.4 are used to efficiently compute the global motion model parameters. Illustrative mosaics obtained using these algorithms are presented in the following examples. Example 5.1 shows a mosaic obtained using the feature-based method. In the absence of prominent features that can be reliably

EXAMPLE 5.1

This shows a mosaic formed by a feature-based method. Features are extracted in each frame and tracked through the image sequence. Since this video consists of a set of frames when the camera is undergoing pure rotation, the motion of the feature points on the image-plane can be described via a homography. The homography is computed with respect to a common frame of reference and the resulting mosaic is shown above.

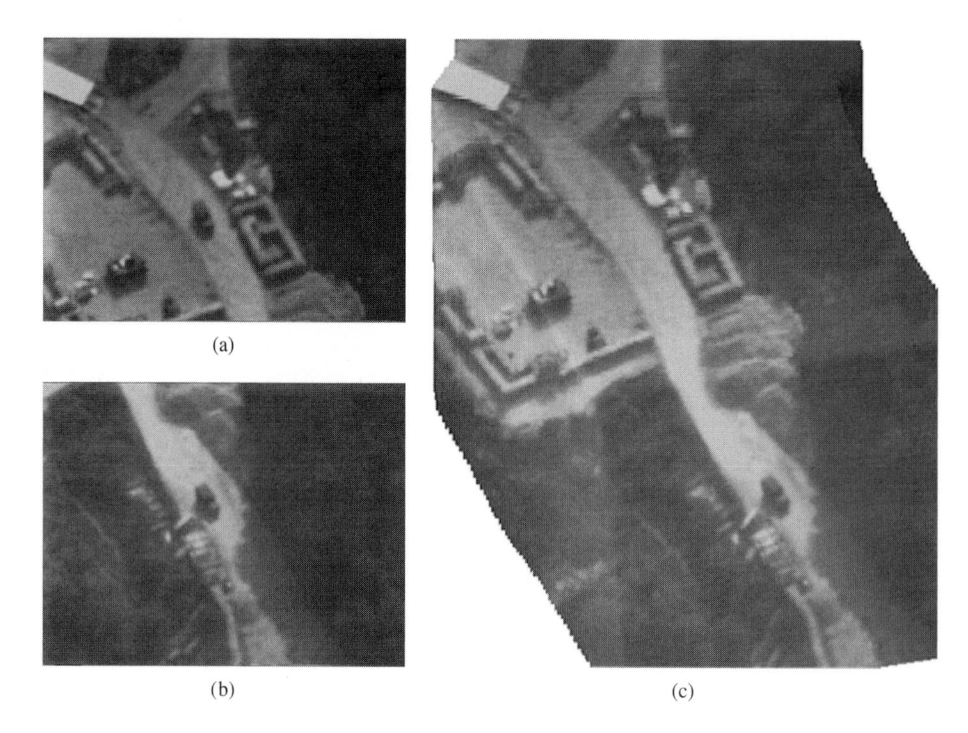

(a)

(b)

(c)

EXAMPLE 5.2

Images (a) and (b) show the first and 180th frames of the *Predator F* sequence. The vehicle near the center moves as the camera pans across the scene in the same general direction. Poor contrast is evident in the top right of (a) and in most of (b). The use of basis functions for computing optical flow pools together information across large areas of the sequence, thereby mitigating the effect of poor contrast. Likewise, the iterative process for obtaining model parameters successfully eliminates outliers caused by the moving vehicle. The mosaic constructed from this sequence is shown in (c).

tracked, correspondence establishment becomes difficult. Examples 5.2 and 5.3 illustrate the effectiveness of flow-based methods in such *featureless* scenes.

5.5.1 Video Mosaicing

Constructing mosaics out of video sequences is an important application for cameras deployed on MAVs and UAVs, where the sensors are of low resolution ("strawhole sensors"). Building mosaics of these sequences will give an enhanced field of view of the scene for an operator viewing the video. The noise level in the images due to the low camera quality needs to be addressed while mosaicing video sequences. Because of the changing noise-level, the error in the estimates of the parameters of the image motion model varies for different images. During the process of developing the mosaics, only

EXAMPLE 5.3

This example demonstrates that mosaics can be used to visualize dynamic information. This is a 2200-frames mosaic obtained from predator imagery. Red marks show the path of moving vehicles. The mosaic was created using an iterative variant [32] of the optical-flow based algorithm described in Section 5.4. The paths were obtained by applying KLT tracker [33] on the stabilized sequence.

EXAMPLE 5.4

This figure illustrates a sample mosaic of a sequence collected from an aerial platform. The video sequence has low SNR, and its resolution is 230×310. The motion of the camera is erratic, and the displacement between consecutive pairs of images can be quite high. The mosaic is built by registering each incoming image to the mosaic and updating the mosaic with the information in each image. Finally, the mosaic is blended at the seams in order to remove visual artifacts.

those frames must be stitched whose image motion parameters lead to good registration of the images.

The following criterion can be used as a metric for measuring the registration quality [34].

$$\begin{aligned} R(I_M, I_t) &= D(I_M, I_t) + G(I_M, I_t), \\ D(I_M, I_t) &= \sum_{r \in R} \left[I_M(r) - I_t(p(r;m)) \right]^2, \quad \text{and} \\ G(I_M, I_t) &= \sum_{r \in R} \left[\nabla I_M(r) - \nabla I_t(p(r;m)) \right]^2. \end{aligned} \tag{5.23}$$

The above criterion (5.23) $R(\cdot,\cdot)$ depends on the image difference error $D(\cdot,\cdot)$ as well as the image-gradient difference error $G(\cdot,\cdot)$. I_t is the current frame and I_M is the current mosaic image. The region R denotes the region of overlap between I_t and I_M. The second error measure $G(I_M, I_t)$ is derived from gradient-domain image registration methods [35–38]. It primarily measures the mis-registration of high-gradient pixels in both the images. The reason for adding this extra term is because, in low-quality images, of the image difference error (that works on raw intensities) by itself does not accurately reflect the mis-registration between two images. The gradient error term (applied on smoothed images) measures the mis-registration between prominent edges in the image.

We start off with an empty mosaic image and sequentially fill the pixels in the mosaic with the information in each frame. The error measure (5.23) is computed for each frame and the mosaic is updated with information from a frame only if its registration error is below a chosen threshold. There are two ways in which the pixels of the current frame are used to update the mosaic: either as reference frame pixels or as nonreference frame pixels. Reference frames are directly pasted onto the mosaic whereas nonreference frames update only unfilled pixels on the mosaic.

- If the most recent reference frame is more than k frames away from the current frame in the sequence (for an appropriately chosen k), then the current frame is incorporated as a reference frame in the current frame as a reference frame.
- If the overlap between the current frame warped and the last reference frame (when warped on the mosaic) falls below a threshold, we incorporate the current frame as a reference frame.
- If neither of the above two conditions are satisfied, the current frame is incorporated as a nonreference frame.

5.6 STABILIZATION AND MOSAICING WITH ADDITIONAL INFORMATION

An aerial platform typically hosts a multitude of other sensors apart from a video camera. Some examples of these sensors are Global Positioning Systems (GPS) [39], three-axis gyrometers, three-axis accelerometers, magnetic sensors, acoustic sensors, sonar rangefinders, etc. Gyros and accelerometers are the main sensing components in an Inertial Measurement Unit (IMU) [39, 40]. IMUs are popular sensors that are collocated with video cameras in order to aid motion estimation.

In this section, we discuss how additional information available along with an image sequence can be used for the purpose of video stabilization. We demonstrate results on the Video Verification and Identification (VIVID) dataset (which consists of image sequences and metadata).

5.6.1 VIVID Metadata

Metadata refers to the additional information that is collected and made available along with a video sequence. The VIVID metadata contains the following information: platform roll, pitch and heading, and sensor azimuth, elevation and twist. In addition, it also has the platform latitude, longitude, and altitude. Using this information, it is possible to compute the position and orientation of the camera in terms of a rotation matrix and a translation vector with respect to the world coordinate system [41].

The projection matrix relating the image plane to the world plane is given by

$$P = K \cdot R \cdot \begin{bmatrix} I & -T \end{bmatrix}, \tag{5.24}$$

where K is the camera calibration matrix, R and T are the rotation and translation between the camera and the world coordinate systems, and I is the identity matrix.

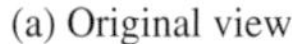

(a) Original view

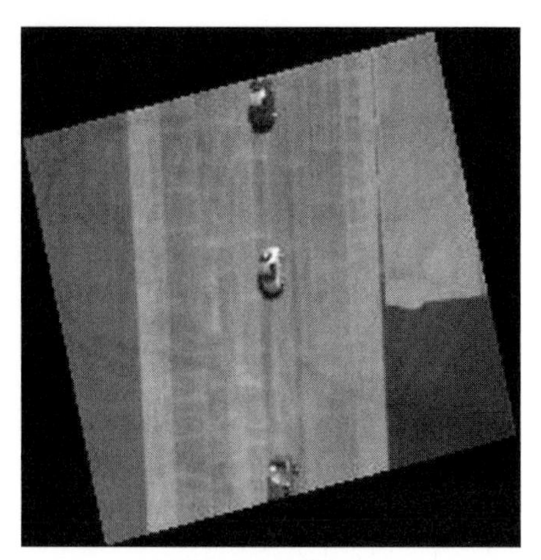

(b) Fronto-parallel view

EXAMPLE 5.5

(a) An image of the sequence and (b) The corresponding fronto-parallel view obtained using the metadata.

This projection matrix (5.24) can be used to transform each image of the sequence into a frontoparallel view as shown in Example 5.5.

5.6.2 Stabilization with Metadata

Metadata can be used for image motion estimation. As noted in Section 5.6.1, each image can be transformed into the frontoparallel view using the rotations derived from the metadata. The registration problem can then be transformed into a problem which involves measuring the image-plane translation between the view-normalized images. For every consecutive pair of frames, we compute the optical flow and robustly fit a two-parameter translation model to the flow.

Let I_{i-1} and I_i be two consecutive images, and I_M be the current mosaic. Assume that we have registered I_{i-1} onto the mosaic, and let the homography that warps I_{i-1} onto I_M be H_{i-1}. The coordinates of corresponding feature points on the mosaic and image are related by $P_M = H_{i-1} \cdot P_{i-1}$, where P_M denotes the coordinates of a point on the mosaic and P_{i-1} denotes the coordinates of the corresponding point on the image I_{i-1}. Suppose we have an estimate of the homography F_i that transforms the current image onto a frontoparallel view and approximately places I_i on the mosaic. To register I_i on the mosaic, we need to solve for the pure-translation (t_x, t_y) that registers the frontoparallel view on the mosaic. We can represent the translation in a matrix transformation form as below:

$$T_i = \begin{bmatrix} 1 & 0 & t_x \\ 0 & 1 & t_y \\ 0 & 0 & 1 \end{bmatrix}. \tag{5.25}$$

The homography between the image I_i and the mosaic can now be represented as $P_M = H_i \cdot P_i$, where $H_i = T_i F_i$.

We solve for the translation (t_x, t_y) by computing the flow between two consecutive view-normalized images and fitting a two-parameter motion model to the flow. We observe that during the stabilization of long sequences, fitting a pure-translation model is much more robust than fitting a full projective model, especially when the presence of homogeneous texture regions in the image do not permit accurate flow estimation over the entire image.

Example 5.6 illustrates two parts of a big mosaic constructed from a minute-long video sequence in the VIVID dataset (using the available metadata). The sequence is

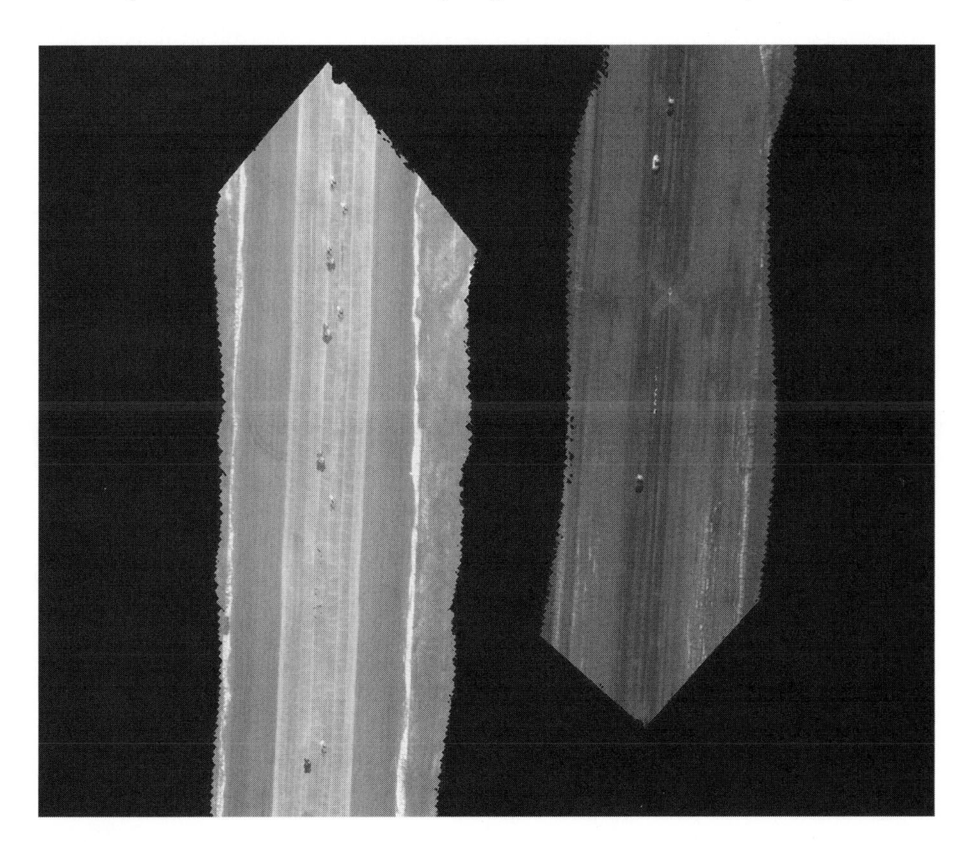

EXAMPLE 5.6

This figure illustrates two sample mosaics of a VIVID sequence that were constructed using the metadata. The sequence is one minute long and had 1750 frames. Metadata was available for approximately every third frame. Since we solve only for the translation part of the motion from the images and obtain the rotational part of the transformation from the metadata, we are able to chain together the inter-image displacements over long sequences. The entire mosaic is 8000 pixels long and is too big to be displayed. We display two parts of the mosaic side by side. The moving vehicles are displayed on the mosaic to illustrate the trajectory.

1750 frames long and metadata was available for around 580 frames of the sequence. In the absence of metadata, mosaicing of long sequences becomes more challenging as illustrated in Example 5.7. In this example, the image motion parameters were solved by fitting a full projective model to the the estimated flow, and these parameters were used to mosaic the image sequence. It is observed that towards the end of the sequence, the projective distortion becomes significant enough and the width of the road keeps increasing. This suggests that metadata can aid the mosaicing long sequences.

5.6.3 Inertial Measurements

An inertial sensor measures the acceleration and angular velocity of an object along three mutually perpendicular axes. IMUs measure these quantities based on the physical laws of motion (i.e., by indirectly measuring specific forces). For an IMU moving along a localized environment near the ground, we can write the mechanization equations as follows [39]:

$$\begin{aligned} \dot{p}^e &= v^e, \\ \dot{v}^e &= C_b^e f^b + g^e, \quad \text{and} \\ \dot{C}_b^e &= C_b^e \Omega_{ib}^b. \end{aligned} \tag{5.26}$$

In (5.26), superscripts denote the coordinate systems with respect to which the parameters are measured: b means body coordinate system, i denotes inertial system, and e represents the earth fixed coordinate system. f^b represents the force measured by the accelerometer, Ω_{ib}^b represents the angular velocity of the body relative to the inertial frame, C_b^e is the rotation matrix from the body to earth-fixed coordinate system (ECS), p^e and v^e are the position and velocity of the body (or sensor), and g^e represents the gravity.

The rate measurements from an IMU can be numerically integrated [39] to obtain estimates of the position and orientation of the object. However, the measurement errors lead to drift in these estimates, and it is necessary to correct for this drift by fusing IMU rates with measurements from another sensor. GPS has traditionally been used along with an IMU for motion estimation. Recently, researchers have investigated [42, 43] the fusion of IMU data with images from a camera. The next section briefly describes the fusion of inertial measurements with feature point trajectories from an image sequence in order to estimate the camera motion.

5.6.4 Stabilization with Inertial Measurements

An IMU that is fixed to a camera measures the camera's acceleration and angular velocity. Rate measurements from an IMU can be used for video stabilization. IMU measurements can be fused with the image measurements (feature point trajectories) in order to obtain the camera motion. The estimated camera motion can in turn be used to stabilize the video sequence.

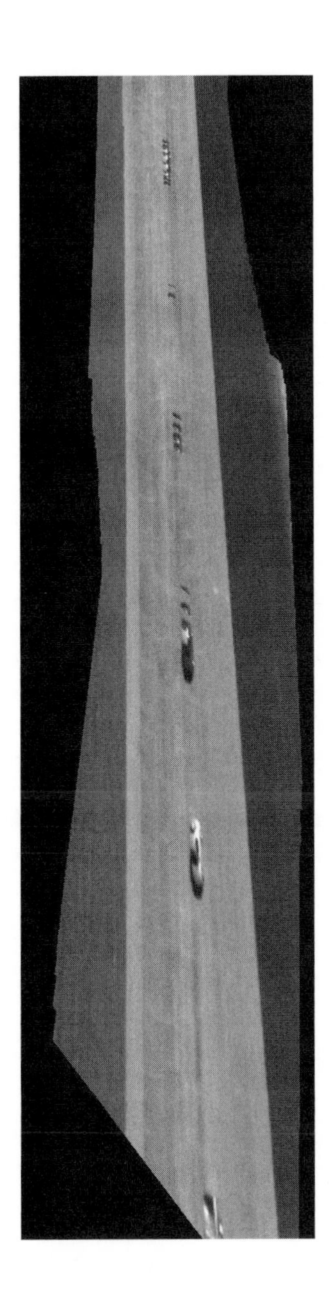

EXAMPLE 5.7

This figure illustrates a sample mosaic of a VIVID sequence that was constructed without using the metadata (images only). As the registration progresses, the projective distortion starts increasing, and towards the bottom of the mosaic, there is significant distortion. The width of the road keeps increasing along the length of the mosaic. These kinds of distortions are absent in a metadata-based stabilization system.

An estimation problem can be formulated where the rotations and translations of the camera need to be estimated for every image in a sequence. Suppose there are N images in the sequence and M feature point trajectories tracked through the sequence. Let the feature point locations be denoted by $x_{i,j}$, where $i \in \{1,\ldots,M\}$ denotes the feature point index and $j \in \{1,\ldots,N\}$ denotes the frame index. Let X_i denote the world coordinates of the ith feature point. Let the position and orientation of the camera location corresponding to the jth image in the sequence be specified by the translation vector T_j and the rotation matrix R_j. Let the 3×3 camera calibration matrix be denoted by K. For a point P whose world coordinates are specified by P_w and camera coordinates are specified by P_c, the relation between the coordinate systems is as follows: $P_c = R_j \cdot P_w + T_j$. A point X_i on the world projects on the image at the jth camera location according to the relation:

$$x_{i,j} = K \cdot (R_j X_j + T_j). \tag{5.27}$$

An estimation problem is formulated in a state-space approach [44], with the state consisting of the rotation R_j and translation T_j of the camera, the $3D$ locations X_j of all feature points, and additional variables for the parameters of the error model of the inertial measurements. The problem is solved in a prediction-correction framework using an Extended Kalman Filter. Given the state of the system at the instant of the current frame, the pose at the time instant of the next image in the sequence is predicted by integrating the inertial measurements in between. The image features in the next image are used to correct the pose estimate of the camera linearizing the image projection Eq. (5.27) around the predicted camera pose estimates. The mathematical equation and details of this approach are outside of the scope of this survey chapter, and interested readers may refer to the papers listed in the bibliography.

Stabilization of a sequence is achieved by computing the homographies between the images induced by a plane in the scene. The homographies are indirectly computed using the estimates of the camera poses using the following equation:

$$H = R\left(I - \frac{T \cdot n^T}{d}\right). \tag{5.28}$$

In the above, H is the homography relating the current frame to a reference frame, I is a 3×3 identity matrix, R and T denote the rotation and orientation of the current camera pose with respect to the reference pose, n is the plane normal vector of the scene plane in the coordinate system of the current camera pose, and d is the perpendicular distance of the camera center to the scene plane. The homographies are computed for each frame using the state estimates at the instant of all frames, and these homographies are used to stabilize the video sequence.

Example 5.8 illustrates a result of stabilization using inertial sensors of a video sequence obtained from a camera on a robotic arm, with synchronized inertial measurements. We show the stabilized results of the first and the thirty-fifth frames in the sequence, and also an illustration of the registration between the two frames.

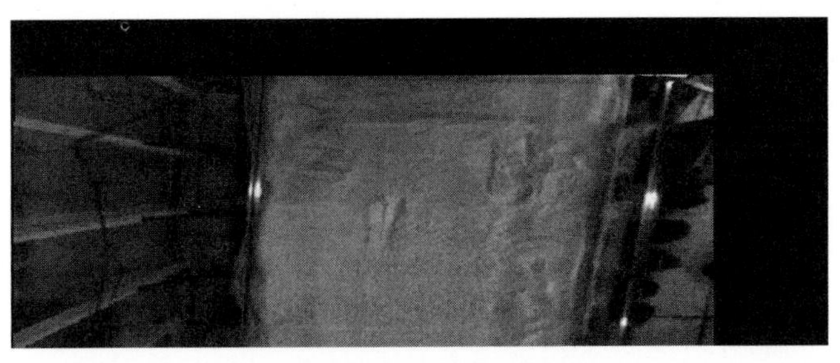

(a) First image in the sequence

(b) Thirty-fifth image in the sequence

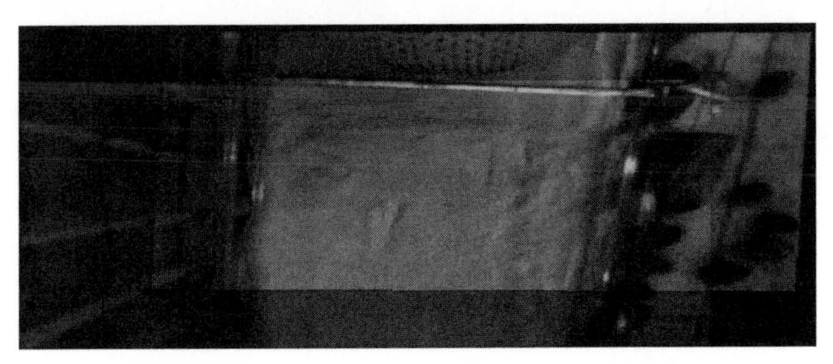

(c) Illustration of registration of the two images above

EXAMPLE 5.8

This example illustrates the result of stabilization on a video sequence captured from a robotic arm. Synchronized inertial measurements were available along with the video sequence. The data was provided by Dennis Strelow [43]. Feature points were detected and tracked through the sequence. The state of the system (camera positions, orientations, velocities etc) was estimated for each image in the sequence using an extended kalman filter. The inertial measurements were integrated between two time instants to predict the predict the camera location at the time-instant of the next image. This predicted position was then corrected using the image observations. Figure (a) shows the first image of the sequence, and Figure (b) shows the 35th image. Both images are stabilized onto a common reference plane. Figure (c) shows an overlay of (a) and (b) in different color channels to illustrate the registration. The registration of the ground plane in the images can be observed in the overlap area. The registration is good, since the overlapping area does not have features in two different colors. The parallax pixels are not registered, and therefore appear twice in the image (similar to a ghost-like image).

5.6.5 Motion Segmentation

Motion Segmentation is the task of identifying the independently moving objects (pixels) in the video and separating them from the background motion. If the background consists of a plane, then we can register the various frames onto a common frame perfectly, using projective transformations. The areas of the image that do not register well belong to the moving objects. If the registration of all frames is perfect, we can take the image difference of two registered images. The pixels that have high intensity difference can be classified as moving objects. However, this simple scheme has a lot of false alarms since registration is not always perfect. In the presence of slight misregistration, the pixels near prominent edges usually flag up as moving pixels. In order to reduce the false alarms, we resort to a detection scheme that combines evidences from (1) image differences, and (2) optical flow discrepancy.

Example 5.9 shows a frame marking the moving object pixels in the video sequence detected in that particular frame.

(a) Original image

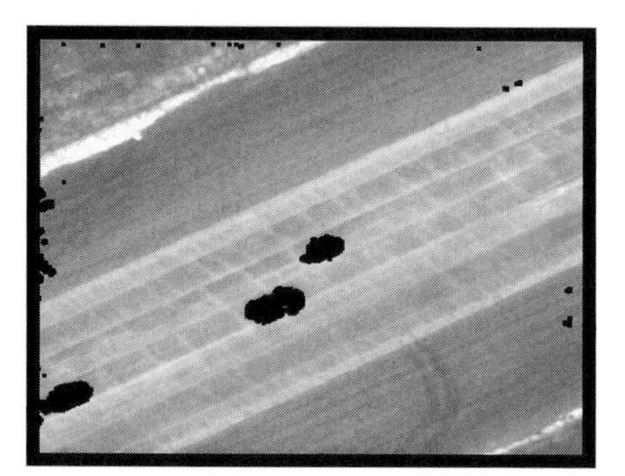

(b) Motion segmentation result

EXAMPLE 5.9

(a) An image from the sequence and (b) The result of the motion segmentation algorithm, where the pixels belonging to moving vehicles are detected and removed from the image.

The next section discusses an application of video stabilization which is motion super-resolution.

5.7 MOTION SUPER-RESOLUTION

Besides being used to eliminate foreground objects, data redundancy in a video sequence can be exploited for enhancing the resolution of an image mosaic, especially when the overlap between the frames is significant. This process is known as motion super-resolution [45, 46]. Each frame of the image sequence is assumed to represent a warped subsampling of the underlying high resolution original. In addition, blur and noise effects can be incorporated into the image degradation model. Let ψ_u represent the underlying image, and $K(x_u, y_u, x, y)$ be a multirate kernel that incorporates the effect of global deformation, subsampling, and blur. The observed low resolution image ψ is given by

$$\psi(x,y) = \sum_{x_u,y_u} \psi_u(x_u,y_u)K(x_u,y_u,x,y) + \eta(x,y), \tag{5.29}$$

where η is a noise process.

To illustrate the operation of (5.29), consider a simple example. Let the observed image be a 4:1 downsampled representation of the original, with a global translation of $(2,-3)$ pixels and no noise. Also assume that the downsampling kernel is a perfect antialiasing filter. The observed image ψ formed by this process is given by

$$\psi_4(x_u,y_u) = \psi_u(x_u,y_u) * K_0(4x-2, 4y+3),$$

$$\mathcal{F}(K_0)(\omega_x,\omega_y) = \begin{cases} 1 & |\omega_x|,|\omega_y| < \frac{\pi}{4} \\ 0 & \text{otherwise} \end{cases}, \quad \text{and}$$

$$\psi(x,y) = \psi_4(4x,4y), \tag{5.30}$$

with K_0 being the antialiasing filter and $\mathcal{F}(K_0)$ its Fourier transform. The process defined in (5.30) represents, in some ways, the worst case scenario. For this case, it can be shown that the original high pass frequencies can never be estimated since they are perfectly filtered out in the image degradation process. Thus, multiple high resolution images produce the same low resolution images after (5.30). On the other hand, when the kernel K is a finite support filter, the high frequency information is attenuated but not eliminated. In theory, it is now possible to restore the original image content, at almost all frequencies, given sufficient low resolution frames.

Motion super-resolution algorithms usually comprise three distinct stages of processing, viz. 1) registration, 2) blur estimation, and 3) refinement. Registration is the process of computing and compensating for image motion. More often than not, the blur is assumed to be known, although in theory the motion super-resolution problem can be formulated to perform blind deconvolution. The kernel K is specified given the motion and blur. The process of reconstructing the original image from this information and the image sequence data is termed as refinement. Often, these stages are performed iteratively and the high resolution image estimate evolves over time.

The global motion estimation algorithms outlined in Sections 5.3 and 5.4 can be used to perform rapid super-resolution. It can be shown that super-resolution can be approximated by first constructing an upsampled static mosaic, followed by some form of inverse filtering to compensate for blur. This approximation is valid when the filter K has a high attenuation over its stopband and thereby minimizes aliasing. Moreover, such a procedure is highly efficient to implement and provides reasonably detailed super-resolved frames. Looking into the techniques used in mosaicing, the median filter emerges

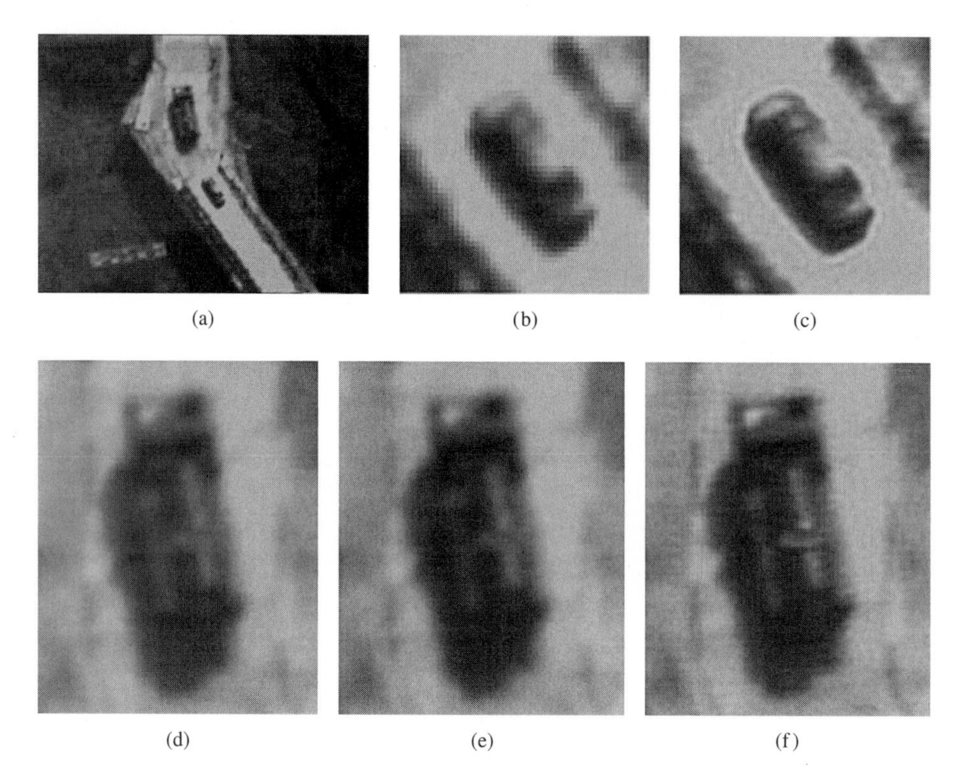

(a) (b) (c)

(d) (e) (f)

EXAMPLE 5.10

A demonstration of the ability of this relatively simple approach for performing motion super-resolution is presented here. The *Predator B* sequence data is gathered from an aerial platform (the predator unmanned air vehicle), and compressed with loss. One frame of this sequence is shown in (a). Forty images of this sequence are co-registered using an affine global motion model, upsampled by a factor of 4, combined and sharpened to generate the super-resolved image. (b) and (d) show the car and truck present in the scene, at the original resolution, while (e) shows the truck image upsampled by a factor of 4, using a bilinear interpolator. The super-resolved images of the car and truck are shown in (c) and (f) respectively. The significant improvement in visual quality is evident. It must be mentioned here that for noisy input imagery, much of the data redundancy is expended in combating compression noise. More dramatic results can be expected when noise-free input data is available to the algorithm.

as an excellent procedure for robustly combining a sequence of images prone to outliers. The super-resolution process is defined in terms of the following steps:

1. Compute the global motion for the image sequence.
2. For an upsampling factor M, scale up the relevant global motion parameters.

3. Using a suitable interpolation kernel and scaled motion parameters, generate a stabilized, upsampled sequence.
4. Build a static mosaic using a robust temporal operator like the median filter.
5. Apply a suitable sharpening operator to the static mosaic.

5.8 THREE-DIMENSIONAL STABILIZATION

3D stabilization is the process of compensating an image sequence for the true 3D rotation of the camera. Extracting the rotation parameters for the image sequence under general conditions involves solving the structure from motion (SFM) problem, which is the simultaneous recovery of full 3D camera motion and scene structure. Mathematical analysis of SFM shows the nonlinear interdependence of structure and motion given observations on the image plane. Solutions to SFM are based on elimination of the depth field by cross multiplication [13, 47–51], differentiation of flow fields [52, 53], nonlinear optimization [54, 55], and other approaches. For a comprehensive discussion of SFM algorithms, the reader is encouraged to refer to [13, 14, 21, 56]. Alternatively, camera rotation can be measured using transducers.

Upon computing the three rotation angles, viz. the pitch, roll, and yaw of the camera, the original sequence can be rewarped to compensate for these effects. Alternatively, one can perform selective stabilization by compensating the sequence for only one or two of these components. Extending this concept, one can selectively stabilize for certain frequencies of motion so as to eliminate handheld jitter, while preserving deliberate camera pan.

5.9 SUMMARY

This chapter discussed the topics of video stabilization, registration, and mosaicing. Stabilization involves compensating for the image motion occurring due to camera ego-motion. Mosaicing is the process of stitching together all the frames of a sequence onto a common surface, to build an extended field-of-view image of the scene. We presented some schemes to perform video stabilization using images alone, and with metadata. Such additional information available along with a video sequence will help in the estimation of the parameters describing the image transformation. Results are presented on a variety of sequences demonstrating the working of the algorithms. We also briefly discussed about motion super-resolution which is an application of stabilization. We briefly discussed the concept of three-dimensional stabilization. The following bibliography provides a useful set of references on stabilization, optical flow, mosaicing, registration, etc. for interested readers.

ACKNOWLEDGEMENTS

This work was partially funded thanks to Army Research Office MURI ARMY-W911NF0410176 under the technical monitorship of Dr. Tom Doligalski. We also thank Dr. Gaurav Aggarwal and Dr. Sridhar Srinivasan for contributions to earlier versions of this book chapter.

REFERENCES

[1] Y. S. Yao and R. Chellappa. Selective stabilization of images acquired by unmanned ground vehicles. *IEEE Trans. Robot. Autom.*, 13(5):693–708, 1997.

[2] C. Morimoto and R. Chellappa. Fast 3d stabilization and mosaic construction. *Proc. IEEE Int. Conf. Comput. Vis. Pattern Recognit.*, 660–667, 1998.

[3] T. Collett and L. Harkness. Depth vision in animals. *Anal. Vis. Behav.*, 111–176, 1982.

[4] M. Srinivasan and S. Zhang. Visual motor computations in insects. *Annu. Rev. Neurosci.*, 27: 679–696, 2004.

[5] Y. Aloimonos. Purposive and active vision. In *CVGIP: Image Understanding*, 840–850, 1992.

[6] F. Mura and N. Franceschini. Visual control of altitude and speed in a flight agent. *Proc. 3rd Int. Conf. Simulation of Adaptive Behaviour: From Animal to Animats*, 91–99, 1994.

[7] T. Neumann and H. Bulthoff. Insect inspired visual control of translatory flight. *Proc. of the 6th European Conf. on Artificial Life ECAL 2001*, 627–636, 2001.

[8] M. Srinivasan, S. Zhang, M. Lehrer, and T. Collett. Honeybee navigation en route to the goal: visual flight control and odometry. *J. Exp. Biol.*, 199:237–244, 1996.

[9] C. David. Compensation for height in the control of groundspeed by drosophila in a new, 'barbers's pole' wind tunnel. *J. comput. physiol.*, A147:485–493, 1982.

[10] W. Reichardt. Movement perception in insects. In W. Reichardt, editor, *Processing of Optical Data by Organisms and by Machines,* 465–493, Academic Press, Burlington, MA, 1969.

[11] H. Esch and J. Burns. Distance estimation by foraging honeybees. *J. Exp. Biol.*, 199:155–162, 1996.

[12] V. Frisch. *The Dance Language and orientation of bees.* Harvard University Press, Cambridge MA, 1993.

[13] A. Mitiche. *Computational Analysis of Visual Motion.* Plenum, New York, 1994.

[14] O. Faugeras. *Three-Dimensional Computer Vision.* MIT Press, Cambridge, MA, 1993.

[15] R. Hartley and A. Zisserman. *Multiple View Geometry in computer vision.* Cambridge University Press, Cambridge, UK, 2000.

[16] C. Tomasi and T. Kanade. Shape and motion from image streams under orthography: a factorization method. *Int. J. Comput. Vis.*, 9(2):137–154, 1992.

[17] D. Lowe. Distinctive image features from scale-invariant keypoints. *Int. J. Comput. Vis.*, 60(2): 91–110, 2004.

[18] M. A. Fischler and R. C. Bolles. Random sample consensus: A paradigm for model fitting with applications to image analysis and automated cartography. *Commun. ACM*, 24:381–395, 1981.

[19] C. Morimoto and R. Chellappa. Fast electronic digital image stabilization for off-road navigation. *Real-time Imaging*, 2:285–296, 1996.

[20] S. Avidan and S. A. Threading fundamental matrices. *IEEE Trans. Pattern Anal. Mach. Intell.*, 23(1):73–77, 2001.

[21] S. Srinivasan. *Image Sequence Analysis: Estimation of Optical Flow and Focus of Expansion, with Applications.* PhD thesis, University of Maryland, College Park, MD, 1999.

[22] O. Axelsson. *Iterated Solution Methods.* Cambridge University Press, Cambridge, UK, 1994.

[23] W. H. Press, S. A. Teukolsky, W. T. Vetterling, and B. P. Flannery. *Numerical Recipes in C, Second Edition.* Cambridge University Press, Cambridge, UK, 1992.

[24] M. Irani, P. Anandan, and S. Hsu. Mosaic based representations of video sequences and their applications. *Int. Conf. Comput. Vis., Cambridge, MA*, 605–611, 1995.

[25] B. Rousso, S. Peleg, I. Finci, and A. Rav-Acha. Universal mosaicing using pipe projection. *Int. Conf. Comput. Vis., Mumbai, India*, 945–952, 1998.

[26] S. Peleg, B. Rousso, A. Rav-Acha, and A. Zomet. Mosaicing on adaptive manifolds. *IEEE Trans. Pattern Anal. Mach. Intell.*, 22(10):1144–1154, 2000.

[27] J. Wang and E. Adelson. Representing moving images with layers. *IEEE Trans. Image Process.*, 3(5):625–638, 1994.

[28] R. Kumar, P. Anandan, and K. Hanna. Shape recovery from multiple views: a parallax based approach. *ARPA Image Understanding Workshop, Monterey, CA, Nov. 1994*, Morgan Kauffmann Publishers, November 1994. San Mateo, California 94403o.

[29] H.-Y. Shum and R. Szeliski. Construction of panoramic mosaics with global and local alignment. *Int. J. Comput. Vis.*, 36(2):101–130, 2000.

[30] R. Szeliski. *Image Alignment and Stitching: A Tutorial.* Now Publishers Inc., Boston, MA, 2006.

[31] S. Peleg and J. Herman. Panoramic mosaics by manifold projection. *Proc. IEEE Int. Conf. Comput. Vis. Pattern Recognit.*, 338–343, 1997.

[32] F. R. Frigole. *Robust Stabilization and Mosaicking for Micro Air Vehicle Imagery.* PhD thesis, University of Maryland, College Park, MD, 2002.

[33] J. Shi and C. Tomasi. Good features to track. *IEEE Conf. Comput. Vis. Pattern Recognit.*, 593–600, 1994.

[34] M. Ramachandran and R. Chellappa. Stabilization and mosaicing of aerial videos. *Proc. IEEE Int. Conf. Image Process.*, 345–348, 2006.

[35] M. Irani and P. Anandan. Robust multi-sensor image alignment. *Proc. IEEE Int. Conf. Comput. Vis.*, 959–966, 1998.

[36] A. Zomet, A. Levin, S. Peleg, and Y. Weiss. Seamless image stitching by minimizing false edges. *IEEE Trans. Image Process.*, 15(4):969–977, 2006.

[37] Y. Keller and A. Averbuch. Multisensor image registration via implicit similarity. *IEEE Trans. Pattern Anal. Mach. Intell.*, 28(5):794–801, 2006.

[38] J. Pluim, J. Maintz, and M. Viergever. Image registration by maximization of combined mutual information and gradient information. *IEEE Trans. Med. Imaging*, 19(8):809–814, 2000.

[39] J. Farrell and M. Barth. *The Global Positioning System and Inertial Navigation.* McGraw-Hill, New York, 2004.

[40] D. H. Titterton and J. L. Weston. *Strapdown Inertial Navigation Technology.* The Institution of Electrical Engineers, United Kingdom, 2004.

[41] C. Shekhar. Semi-automatic video-to-site registration for aerial monitoring. *Proc. IEEE Int. Conf. Pattern Recognit.*, 4:959–966, Barcelona, 1998.

[42] J. Lobo and J. Dias. Vision and inertial sensor cooperation using gravity as a vertical reference. *IEEE Trans. Pattern Anal. Mach. Intell.*, 25(12):1597–1608, 2003.

[43] D. Strelow. *Motion estimation from image and inertial measurements.* PhD thesis, Carnegie Mellon University, Pittsburg, 2004.

[44] G. Qian and R. Chellappa. Structure from motion using sequential monte carlo methods. *Int. J. Comput. Vis.*, 59(1):5–31, 2004.

[45] A. Zomet, A. Rav-Acha, and S. Peleg. Robust super-resolution. *Proc. IEEE Comput. Soc. Conf. Comput. Vis. Pattern Recognit.*, 1:645–650, 2001.

[46] G. Chantas, N. Galatsanos, and N. Woods. Super-resolution based on fast registration and maximum a posteriori reconstruction. *IEEE Trans. Image Process.*, 16(7):1821–1830, 2007.

[47] N. Gupta and L. Kanal. 3-d motion estimation from motion field. *Artificial Intelligence*, 78(1-2): 45–86, 1995.

[48] R. Tsai and T. Huang. Estimating 3-d motion parameters of a rigid planar patch i. *IEEE Trans. Acoustics, Speech Signal Process.*, 29(12):1147–1152, 1981.

[49] X. Zhuang, T. Huang, N. Ahuja, and R. Haralick. A simplified linear optical flow-motion algorithm. *Comput. Vis. Graphics Image Process.*, 42(3):334–344, 1988.

[50] X. Zhuang, T. Huang, N. Ahuja, and R. Haralick. Rigid body motion and the optic flow image. *First IEEE Conf. AI Applications*, 366–375, 1984.

[51] A. Waxman, B. Kamgar-Parsi, and M. Subbarao. Closed-form solutions to image flow equations for 3d structure and motion. *Int. J. Comput. Vis.*, 1(3):239–258, 1987.

[52] H. Longuet-Higgins and K. Prazdny. The interpretation of a moving retinal image. *Proc. R. Soc. Lond. B*, B-208:385–397, 1980.

[53] A. Waxman and S. Ullman. Surface structure and three-dimensional motion fromimage flow kinematics. *Int. J. Robot. Res.*, 4(3):72–94, 1985.

[54] G. Adiv. Determining 3-d motion and structure from optical flow generated by several moving objects. *IEEE Trans. Pattern Anal. Mach. Intell.*, 7(4):384–401, 1985.

[55] A. Bruss and B. Horn. Passive navigation. *Comput. Vis. Graphics Image Process.*, 21(1):3–20, 1983.

[56] J. Weng, T. S. Hwang, and N. Ahuja. *Motion and Structure from Image Sequences.* Springer-Verlag, Berlin, 1991.

CHAPTER

Video Segmentation

A. Murat Tekalp

College of Engineering, Koc University,
Istanbul, Turkey

6.1 INTRODUCTION

Video segmentation refers to partitioning video into spatial, temporal, or spatiotemporal regions that are homogeneous in some feature space [4]. It is an integral part of many video analysis and coding problems, including (i) video summarization, indexing, and retrieval, (ii) advanced video coding, (iii) video authoring and editing, (iv) improved motion (optical flow) estimation, (v) 3D motion and structure estimation with multiple moving objects [1–3], and (vi) video surveillance/understanding. The first three applications concern multimedia services, which require temporal segmentation of video into shots or groups of pictures (GoP). They may also benefit from spatiotemporal segmentation of video into objects for better content description, object-based video editing, and rate allocation. The latter three are computer vision applications, where spatiotemporal segmentation helps to identify foreground and background objects, as well as optical flow boundaries (motion edges) and occlusion regions.

As with any segmentation problem, effective video segmentation requires proper feature selection and an appropriate distance measure. Different features and homogeneity criteria generally lead to different segmentations of the same video, for example, color, texture, or motion segmentation. Furthermore, there is no guarantee that any of the resulting automatic segmentations will be semantically meaningful, since a semantically meaningful region may have multiple colors, multiple textures, and/or multiple motion. Although semantic objects can be computed automatically in some well-constrained settings, for example, in video surveillance systems [16], where objects can be extracted by change detection and background subtraction when the camera is stationary, in general, semantic object segmentation requires specialized capture methods (chroma-keying) or user interaction. Specific video segmentation methods should be considered in the context of the requirements of the application in which they are used. Factors that affect the choice of a specific segmentation method include [5] the following:

- *Real-time performance:* If segmentation must be performed in real time, for example, for rate control in videotelephony, then simple algorithms that are fully automatic must be used. On the other hand, one can employ semiautomatic, interactive algorithms for off-line applications such as video indexing or off-line video coding to obtain semantically meaningful segmentations [6].

- *Precision of segmentation:* If segmentation is employed to improve the compression efficiency or rate control, then certain misalignment between segmentation results and actual object borders may not be of concern. On the other hand, if segmentation is needed for object-based video authoring/editing or shape similarity matching, then it is of utmost importance that the estimated boundaries align with actual object boundaries perfectly, where even a single pixel error may not be tolerable.
- *Scene complexity:* Complexity of video content can be modeled in terms of amount of camera motion, color and texture uniformity within objects, contrast between objects, smoothness of motion of objects, objects entering and leaving the scene, regularity of object shape along the temporal dimension, frequency of cuts and special effects, etc. Clearly, more complex scenes require more sophisticated segmentation algorithms. For example, it is easier to detect cuts than special effects such as wipes or fades.

This chapter provides an overview of some video segmentation methods ranging from simple shot boundary detection and change detection techniques to more sophisticated motion segmentation and interactive video object segmentation and tracking methods. Although multimodal signal processing methods have been shown to be effective in specific applications [7], we cover only video modality in this chapter. We start our discussion with temporal video segmentation methods in Section 6.2. Change detection methods, where we study both two-frame methods and methods with memory, are covered in Section 6.3. Motion segmentation methods are discussed in Section 6.4. We first introduce the "dominant motion" approach, which aims to label independently moving regions sequentially (one at a time). We then present the alternative "multiple motion segmentation" approach, including clustering in the motion parameter space, maximum likelihood (ML) segmentation, maximum a posteriori probability (MAP) segmentation, and region labeling methods. Section 6.5 addresses simultaneous optical flow estimation and segmentation method, since the accuracy of segmentation results depends on the accuracy of the estimated motion field and vice versa. Finally, Section 6.6 deals with semantically meaningful object segmentation with emphasis on chroma-keying and semiautomatic (interactive) object tracking methods.

6.2 SCENE CHANGE DETECTION

Scene change or shot boundary detection is a relatively easy segmentation problem since it is one dimensional, along the temporal dimension. Shot boundary detection methods locate temporal discontinuities, that is, frames across which large differences are observed in some feature space, usually a combination of color and motion [8–13].

Temporal discontinuities may be abrupt (cuts) or gradual (special effects, such as wipes and fades). It is easier to detect cuts than special effects. The simplest approach for detecting temporal discontinuities is to quantify frame differences in the pixel intensity domain. If a predetermined number of pixels exhibit differences larger than a threshold value, then a cut can be declared. Clearly, this method is sensitive to presence of noise, camera motion, and compression artifacts in the video. A slightly more robust approach may be to divide each frame into rectangular blocks, compute statistics of each block such as the mean and variance independently, and then check the count of blocks with changing statistics against a set threshold. Applying low-pass filtering to each frame prior to computing frame differences or block statistics should also improve robustness.

A more robust alternative is to consider frame histogram differences instead of pixel-wise or blockwise frame intensity differences. To this effect, we compute n-bin color histogram, $h_k(i)$, $i = 1,\ldots,n$, for each frame k. Various measures and tests have been developed to quantify similarity or dissimilarity of histograms. These include the histogram intersection measure, χ^2 test, and Kolmogorov–Smirnov test [11]. A closely related approach is to detect changes in the counts of edge-pixels in successive frames, that is, similarity of edge histograms. Although they are effective to detect cuts and fades, neither histogram differences nor intensity differences can usually differentiate between wipes and camera motion, such pans and zooms. Detection of these special effects requires combination of histogram difference and camera motion estimation. Global motion can be estimated and frames are motion compensated before computation of the features [14]. Another approach to detect gradual changes is the so-called twin comparison method [15], which can be used with different features. A lower threshold is used to detect abrupt scene changes, whereas a higher threshold is used to detect the actual position of gradual ones.

There also exist shot boundary detection algorithms for specific domains, such as surveillance video [16], sports video [17], and movies [18, 19]. Sports video is arguably one of the most challenging domains for robust shot boundary detection due to 1) existence of a strong color correlation between successive shots, since a single dominant color background, such as the soccer field, may be seen in successive shots; 2) existence of large camera and object motions; 3) existence of many gradual transitions, such as wipes and dissolves. Ekin et al. [17] observed that gradual transitions in sports video are not accurately detected by generic algorithms, and proposed using two features, the absolute difference between two frames of the ratio of dominant colored pixels to total number of pixels and color histogram dissimilarity, measured by histogram intersection, for reliable shot boundary detection.

Videos are almost always stored and transmitted in compressed form. Detection of scene changes in real time poses a challenge in many applications since decompressing and processing video data sequentially requires significant computational resources. Hence, the need for scene segmentation algorithms in the compressed domain (without completely decoding the bit stream) [21]. DC images that are spatially reduced versions of the original video frames can be constructed from the DC coefficient of each 8×8 block [20]. Successful results have been obtained for detection of both abrupt and gradual scene changes using DC images [2].

6.3 SPATIOTEMPORAL CHANGE DETECTION

Change detection methods segment each frame into two regions, namely changed and unchanged regions in the case of a static camera or global and local motion regions in the case of a moving camera [22]. This section deals only with the former case, where unchanged regions correspond to the background (null hypothesis) and changed regions to the foreground object(s) or uncovered (occlusion) areas. The case of moving camera is identical to the former, once the global motion between successive frames due to camera motion is estimated and compensated [23]. However, accurate estimation of the camera motion may require scene segmentation resulting in a chicken-egg problem. Fortunately, the dominant motion segmentation approach, presented in the next section, offers a solution to the estimation of the camera motion without prior scene segmentation. Hence, the discussion of the case of moving camera is deferred until Section 6.4.1.

Various change detection methods in the literature differ according to (i) what features and background model are used, (ii) what distance metrics are used, and (iii) what kind of threshold and background model adaptation rules are used. In the following, we first discuss change detection using two frames. Temporal integration (using more than two frames) and combination of spatial and temporal segmentation are also studied to obtain spatially and temporally coherent regions.

6.3.1 Spatial Change Detection Using Two Frames

The simplest method to detect changes between two registered frames would be to analyze the frame difference (FD) image, which is given by

$$\mathrm{FD}_{k,r}(\mathbf{x}) = s(\mathbf{x}, k) - s(\mathbf{x}, r), \tag{6.1}$$

where $\mathbf{x} = (x_1, x_2)$ denotes pixel location and $s(\mathbf{x}, k)$ stands for the intensity value at pixel $\mathbf{x}$ in frame k. FD image shows the pixel-by-pixel difference between the current image k and the reference image r. The reference image r may be taken as the previous image $k-1$ (successive FD) or an image at a fixed time. Methods using a fixed reference frame, which may be updated for global illumination changes, are called background subtraction methods [26]. For example, if we are interested in monitoring a hallway using a fixed camera, an image of the hallway when it is empty may be used as a fixed reference image. Assuming that we have a static camera and the illumination remains more or less constant between the frames, the pixel locations where $\mathrm{FD}_{k,r}(\mathbf{x})$ differs from zero indicate regions "changed" due to local motion. To distinguish the nonzero differences that are due to noise from those that are due to local motion, segmentation can be achieved by thresholding the FD as

$$z_{k,r}(\mathbf{x}) = \begin{cases} 1 & \text{if } |\mathrm{FD}_{k,r}(\mathbf{x})| > T \\ 0 & \text{otherwise} \end{cases}, \tag{6.2}$$

where T is an appropriate threshold. Here, $z_{k,r}(\mathbf{x})$ is called a segmentation label field, which is equal to "1" for changed regions and "0" otherwise. The value of the threshold T

can be chosen by an optimal threshold determination algorithm. This pixelwise thresholding is generally followed by one or more postprocessing steps to eliminate isolated labels. Postprocessing operations include forming 4- or 8-connected regions and discarding labels with less than a predetermined number of entries and morphological filtering of the changed and unchanged region masks.

In practice, a simple FD image analysis is not satisfactory for two reasons: First, a uniform intensity region may be interpreted as stationary even if it is moving (aperture problem). It may be possible to avoid the aperture problem using a multiresolution decision procedure, since uniform intensity regions are smaller at lower resolution levels. Second, the intensity difference due to motion is affected by the magnitude of the spatial gradient in the direction of motion. This problem can be addressed by considering a locally normalized FD function [24] or locally adaptive thresholding [25]. An improved change detection algorithm that addresses both concerns can be summarized as follows:

(i) Construct a Gaussian pyramid where each frame is represented in multiple resolutions. Start processing at the lowest resolution level.

(ii) For each pixel at the present resolution level, compute the normalized FD given by [24]

$$\mathrm{FDN}_{k,r}(\mathbf{x}) = \frac{\sum_{\mathbf{x}\in\mathcal{N}} |s(\mathbf{x},k) - s(\mathbf{x},r)|\,|\nabla s(\mathbf{x},r)|}{\sum_{\mathbf{x}\in\mathcal{N}} |\nabla s(\mathbf{x},r)|^2 + c}, \tag{6.3}$$

where $\mathcal{N}$ denotes a local neighborhood of the pixel $\mathbf{x}$, $\nabla s(\mathbf{x},r)$ denotes the gradient of image intensity at pixel $\mathbf{x}$, and c is a constant to avoid numerical instability. If the normalized difference is high (indicating that the pixel is moving), replace the normalized difference from the previous resolution level at that pixel with the new value. Otherwise, retain the value from the previous resolution level.

(iii) Repeat step ii) for all resolution levels.

Finally, we threshold the normalized motion detection function at the highest resolution level.

6.3.2 Temporal Integration

An important consideration is to add memory to the motion detection process to ensure both spatial and temporal continuity of the changed regions at each frame. This can be achieved in a number of different ways, including temporal filtering (integration) of the intensity values across multiple frames before thresholding and postprocessing of labels after thresholding.

A variation of the successive FD and normalized FD is the FD with memory $\mathrm{FDM}_k(\mathbf{x})$, which is defined as the difference between the present frame $s(\mathbf{x},k)$ and a weighted average of past frames $\bar{s}(\mathbf{x},k)$, given by [24]

$$\mathrm{FDM}_k(\mathbf{x}) = s(\mathbf{x},k) - \bar{s}(\mathbf{x},k), \tag{6.4}$$

where

$$\bar{s}(\mathbf{x},k) = (1-\alpha)s(\mathbf{x},k) + \alpha\bar{s}(\mathbf{x},k-1), \quad k = 1,\ldots \tag{6.5}$$

and

$$\bar{s}(\mathbf{x},0) = s(\mathbf{x},0).$$

Here, $0 < \alpha < 1$ is a constant. After processing a few frames, the unchanged regions in $\bar{s}(\mathbf{x},k)$ maintain their sharpness with a reduced level of noise, whereas the changed regions are blurred. The function $\text{FDM}_k(\mathbf{x})$ is thresholded either by a global or a spatially adaptive threshold as in the case of two-frame methods. The temporal integration increases the likelihood of eliminating spurious labels, thus, resulting in spatially contiguous regions.

Accumulative differences can be employed when detecting changes between a sequence of images and a fixed reference image (as opposed to successive frame differences). Let $s(\mathbf{x},k)$, $s(\mathbf{x},k-1),\ldots,s(\mathbf{x},k-N)$ be a sequence of N frames, and let $s(\mathbf{x},r)$ be a reference image. An accumulative difference image is formed by comparing every frame in the sequence with this reference image. For every pixel location, the accumulative image is incremented if the difference between the reference image and the current image in the sequence at that pixel location is bigger than a threshold. Thus, pixels with higher counter values are more likely to correspond to changed regions.

An alternative procedure that was adopted by MPEG-4 as a non-normative tool considers postprocessing of labels [25]. First, scene changes are detected. Within each scene (shot), an initial change detection mask is estimated between successive pairs of frames by global thresholding of the FD function. Next, the boundary of the changed regions is smoothed by a relaxation method using local adaptive thresholds [22]. Then, memory is incorporated by relabeling unchanged pixels which correspond to changed locations in one of the last L frames. This step ensures temporal continuity of changed regions from frame to frame. The depth of the memory L may be adapted to scene content to limit error propagation. Finally, postprocessing to obtain the final changed and unchanged masks eliminates small regions.

6.3.3 Combination with Spatial Segmentation

Another consideration is to enforce consistency of the boundaries of the changed regions with spatial edge locations at each frame. This may be accomplished by first segmenting each frame into uniform color and/or texture regions. Next, each region resulting from the spatial segmentation is labeled as changed or unchanged as a whole as opposed to labeling each pixel independently. Region labeling decisions may be based on the number of changed and unchanged pixels within each region or thresholding the average value of the frame differences within each region [27].

6.4 MOTION SEGMENTATION

Motion segmentation (also known as optical flow segmentation) methods label pixels (or optical flow vectors) at each frame that are associated with independently moving part of a scene. The region boundaries may or may not be pixel-accurate or semantically meaningful. For example, a single object with articulated motion may be segmented into multiple regions. The occlusion and aperture problems are mainly responsible for misalignment of motion and actual object boundaries. Furthermore, model misfit possibly due to deviation of the surface structure from a plane generally leads to oversegmentation of the motion field. Although it is possible to achieve fully automatic motion segmentation with some limited accuracy for certain content domains, semantically meaningful video object segmentation generally requires user interaction to define the object of interest in at least some key frames as discussed in Section 6.6.

Motion segmentation is closely related to two other problems, motion (change) detection and motion estimation. Change detection is a special case of motion segmentation with only two regions, namely changed and unchanged regions (in the case of a static camera) or global and local motion regions (in the case of a moving camera). An important distinction between change detection and motion segmentation is that the former can be achieved without motion estimation if the scene is recorded with a static camera. Change detection in the case of a moving camera and general motion segmentation, on the other hand, requires some sort of global and/or local motion estimation either explicitly or implicitly. Motion detection and segmentation are also plagued with the same two fundamental limitations associated with motion estimation: occlusion and aperture problems [4]. For example, pixels in a flat image region may appear stationary even if they are moving due to the aperture problem (hence the need for hierarchical methods); and/or erroneous labels may be assigned to pixels in covered or uncovered image regions due to the occlusion problem.

In general, application of standard image segmentation methods directly to estimated optical flow vectors may not yield meaningful results, since an object moving in 3D usually generates a spatially varying optical flow field [28]. For example, in the case of a rotating object, there is no flow at the center of the rotation, and the magnitude of the flow vectors grows as we move away from the center of rotation. Therefore, a parametric model-based approach, where we assume that the motion field can be described by a set of K parametric models, is usually adopted. In parametric motion segmentation, the model parameters are the motion features. Then, motion segmentation algorithms aim to determine the number of motion models that can adequately describe a scene, type/complexity of these motion models, and the spatial support of each motion model. Most commonly used types of parametric models are affine, perspective, and quadratic mappings, which assume a 3D planar surface in motion. In the case of a nonplanar object, the resulting optical flow can be modeled by a piecewise affine, perspective, or quadratic flow field if we approximate the object surface by a union of a small number of planar patches. Since each independently moving object and/or planar patch will best fit a different parametric model, the parametric approach may lead to oversegmentation of motion in the case of nonplanar objects.

6.4.1 Dominant Motion Segmentation

Segmentation by dominant motion analysis refers to extracting one object (with the dominant motion) from the scene at a time [24, 29, 30, 38]. Dominant motion segmentation can be considered a hierarchically structured top-down approach, which starts by fitting a single parametric motion model to the entire frame and then partitions the frame into two regions, those pixels which are well represented by this dominant motion model and those that are not. The process converges to the dominant motion model in a few iterations, each time fitting a new model to only those pixels that are well represented by the motion model in the previous iteration. The dominant motion may correspond to the camera (background) motion or a foreground object motion, whichever occupies a larger area in the frame. The dominant motion approach may also handle separation of individually moving objects. Once the first dominant object is segmented, it is excluded from the region of analysis, and the entire process is repeated to define the next dominant object. This is unlike the multiple motion segmentation approaches that are discussed in the next section, which start with an initial segmentation mask (usually with many small regions) and refine them according to some criterion function to form the final mask. It is worth noting that the dominant motion approach is a direct method that is based on spatiotemporal image intensity gradient information. This is in contrast to first estimating the optical flow field between two frames and then segmenting the image based on the estimated optical flow field.

6.4.1.1 *Segmentation Using Two Frames*

Motion estimation in the presence of more than one moving objects with unknown supports is a difficult problem. It was Burt et al. [29] who first showed that the motion of a 2D translating object can be accurately estimated using a multiresolution iterative approach even in the presence of other independently moving objects without prior knowledge of their supports. This is, however, not always possible with more sophisticated motion models (e.g., affine and perspective), which are more sensitive to presence of other moving objects in the region of analysis.

To this effect, Irani et al. [24] proposed multistage parametric modeling of dominant motion. In this approach, first a translational motion model is employed over the whole image to obtain a rough estimate of the support of the dominant motion. The complexity of the model is then gradually increased to affine and projective models with refinement of the support of the object in between. The parameters of each model are estimated only over the support of the object based on the previously used model. The procedure can be summarized as follows:

(i) Compute the dominant 2D translation vector (d_x, d_y) over the whole frame as the solution of

$$\begin{bmatrix} \sum I_x^2 & \sum I_x I_y \\ \sum I_x I_y & \sum I_y^2 \end{bmatrix} \begin{bmatrix} d_x \\ d_y \end{bmatrix} = \begin{bmatrix} -\sum I_x I_t \\ -\sum I_y I_t \end{bmatrix}, \tag{6.6}$$

where I_x, I_y, and I_t denote partials of image intensity with respect to x, y, and t. In case the dominant motion is not a translation, the estimated translation becomes a first-order approximation of the dominant motion.

(ii) Label all pixels that correspond to the estimated dominant motion as follows:

(a) Register the two images using the estimated dominant motion model. The dominant object appears stationary between the registered images, whereas other parts of the image are not.

(b) Then, the problem reduces to labeling stationary regions between the registered images, which can be solved by the multiresolution change detection algorithm given in Section 6.3.1.

(c) Here, in addition to the normalized FD (6.3), define a motion reliability measure as the reciprocal of the condition number of the coefficient matrix in (6.6), given by [24]

$$R(\mathbf{x}, k) = \frac{\lambda_{\min}}{\lambda_{\max}}, \tag{6.7}$$

where $\lambda_{\min}$ and $\lambda_{\max}$ are the smallest and largest eigenvalue of the coefficient matrix. A pixel is classified as stationary at a resolution level if its normalized FD is low and its motion reliability is high. This step defines the new region of analysis.

(iii) Estimate the parameters of a higher order motion model (affine, perspective, or quadratic) over the new region of analysis as in [24]. Iterate over steps ii) and iii) until a satisfactory segmentation is attained.

6.4.1.2 Temporal Integration

Temporal continuity of the estimated dominant objects can be facilitated by extending the temporal integration scheme introduced in Section 6.3.2. To this effect, we define an internal representation image [24]

$$\bar{s}(\mathbf{x}, k) = (1 - \alpha)s(\mathbf{x}, k) + \alpha\ warp(\bar{s}(\mathbf{x}, k-1), s(\mathbf{x}, k)), \quad k = 1, \ldots, \tag{6.8}$$

where

$$\bar{s}(\mathbf{x}, 0) = s(\mathbf{x}, 0)$$

and $warp(A, B)$ denotes warping image A toward image B according to the dominant motion parameters estimated between images A and B and $0 < \alpha < 1$. As in the case of change detection, the unchanged regions in $\bar{s}(\mathbf{x}, k)$ maintain their sharpness with a reduced level of noise, whereas the changed regions are blurred after processing a few frames.

The algorithm to track the dominant object across multiple frames can be summarized as follows [24]: For each frame,

(i) compute the dominant motion parameters between the internal representation image $\bar{s}(\mathbf{x}, k)$ and the new frame $s(\mathbf{x}, k)$ within the support M_{k-1} of the dominant object at the previous frame;

(ii) warp the internal representation image at frame $k-1$ toward the new frame according to the computed motion parameters;

(iii) detect the stationary regions between the registered images as described in Section 6.4.1.1 using M_{k-1} as an initial estimate to compute the new mask M_k;

(iv) Update the internal representation image using (6.8).

Comparing each new frame with the internal representation image as opposed to the previous frame allows the method to track the same object. This is because the noise is significantly filtered in the internal representation image of the tracked object, and the image gradients outside the tracked object are lowered due to blurring. Note that there is no temporal motion constancy assumption in this tracking scheme.

6.4.1.3 *Multiple Motions*

Multiple object segmentation can be achieved by repeating the same procedure on the residual image after each object is extracted. Once the first dominant object is segmented and tracked, the procedure can be repeated recursively to segment and track the next dominant object after excluding all pixels belonging to the first object from the region of analysis. Hence, the method is capable of segmenting multiple moving objects in a top-down fashion if a dominant motion exists at each stage.

Some difficulties with the dominant motion approach were reported when there is no overwhelmingly dominant motion. Then, in the absence of competing motion models, the dominant motion approach can lead to arbitrary decisions (relying upon absolute threshold values) that are irrevocable, especially when the motion measure indicates unreliable motion vectors (in low spatial gradient regions). Sawhney et al. [32] proposed using robust estimators to partially alleviate this problem.

6.4.2 Multiple Motion Segmentation

Multiple motion segmentation methods let multiple motion models compete against each other at each decision site. They consist of three basic steps, which are strongly interrelated: estimation of the number K of independent motions, estimation of model parameters for each motion, and determination of support of each model (segmentation labels). If we assume that we know the number K of motions and the K sets of motion parameters, then we can determine the support of each model. The segmentation procedure then assigns the label of the parametric motion vector that is closest to the estimated flow vector at each site. Alternatively, if we assume that we know the value of K and a segmentation map consisting of K regions, the parameters for each model can be computed in the least squares sense (either from estimated flow vectors or from spatiotemporal intensity values) over the support of the respective region. But since both the parameters and supports are unknown in reality, we have a chicken-egg problem; that is, we need to know the motion model parameters to find the segmentation labels, and the segmentation labels are needed to find the motion model parameters.

Various approaches exist in the literature for solving this problem by iterative procedures. They may be grouped as segmentation by clustering in the motion parameter space [28, 37, 42], ML segmentation [38, 40, 44], and MAP segmentation [34], which are covered in Sections 6.4.2.1 – 6.4.2.3, respectively. Pixel-based segmentation methods suffer from the drawback that the resulting segmentation maps may contain isolated labels. Spatial continuity constraints in the form of Gibbs random field (GRF) models have been introduced to overcome this problem within the MAP formulation [34]. However, the computational cost of these algorithms may be prohibitive. Furthermore, they do not guarantee that the estimated motion boundaries coincide with spatial color edges (object boundaries). Section 6.4.2.4 presents an alternative region labeling approach to address this problem.

6.4.2.1 *Clustering in the Motion Parameter Space*

A simple segmentation strategy is to first determine the number K of models (motion hypotheses) that are likely to be observed in a sequence and then perform clustering in the model parameter space (e.g., a six dimensional space for the case of affine models) to find K models representing the motion. In the following, we study two distinct approaches in this class: the K-means method and the Hough transform method.

6.4.2.1.1 K-Means Method

Wang and Adelson (W–A) [37] employed K-means clustering for segmentation in their layered video representation. W–A method starts by partitioning the image into nonoverlapping blocks uniformly distributed over the image and fits an affine model to the estimated motion field (optical flow) within each block. To determine the reliability of the parameter estimates at each block, the sum of squared distances between the synthesized and estimated flow vectors is computed as

$$\bar{\eta}^2 = \sum_{\mathbf{x} \in \mathcal{B}} ||\mathbf{v}(\mathbf{x}) - \tilde{\mathbf{v}}(\mathbf{x})||^2, \tag{6.9}$$

where $\mathcal{B}$ refers to a block of pixels. Obviously, if the flow within the block complies with a single affine model, the residual will be small. On the other hand, if the block falls on the boundary between two distinct motions, the residual will be large. The motion parameters for blocks with acceptably small residuals are selected as the seed models. Then, the seed model parameter vectors are clustered to find the K representative affine motion models. The clustering procedure can be described as follows: given N seed affine parameter vectors $\mathbf{A}_1, \mathbf{A}_2, \ldots, \mathbf{A}_N$, where

$$\mathbf{A}_n = \begin{bmatrix} a_{n,1} \\ a_{n,2} \\ a_{n,3} \\ a_{n,4} \\ a_{n,5} \\ a_{n,6} \end{bmatrix}, \quad n = 1, \ldots, N, \tag{6.10}$$

find K cluster centers $\bar{\mathbf{A}}_1, \bar{\mathbf{A}}_2, \ldots, \bar{\mathbf{A}}_K$, where $K \ll N$, and the label k, $k = 1, \ldots, K$, assigned to each affine parameter vector $\mathbf{A}_n$ which minimizes

$$\sum_{n=1}^{N} \mathcal{D}(\mathbf{A}_n, \bar{\mathbf{A}}_k). \tag{6.11}$$

The distance measure $\mathcal{D}$ between two affine parameter vectors $\mathbf{A}_n$ and $\mathbf{A}_k$ is given by

$$\mathcal{D}(\mathbf{A}_n, \mathbf{A}_k) = \mathbf{A}_n^T \mathbf{M} \mathbf{A}_k, \tag{6.12}$$

where $\mathbf{M}$ is a 6×6 scaling matrix.

The solution to this problem is given by the well-known K-means algorithm, which consists of the following iteration:

(i) Initialize $\bar{\mathbf{A}}_1, \bar{\mathbf{A}}_2, \ldots, \bar{\mathbf{A}}_K$ arbitrarily.

(ii) For each seed block n, $n = 1, \ldots, N$, find k given by

$$k = Arg \min_s \mathcal{D}(\mathbf{A}_n, \bar{\mathbf{A}}_s), \tag{6.13}$$

where s takes values from the set $\{1, 2, \ldots, K\}$. It should be noted that if the minimum distance exceeds a threshold, then the site is not labeled, and the corresponding flow vector is ignored in the parameter update that follows.

(iii) Define $\mathcal{S}_k$ as the set of seed blocks whose affine parameter vector is closest to $\bar{\mathbf{A}}_k$, $k = 1, \ldots, K$. Then, update the class means

$$\bar{\mathbf{A}}_k = \frac{\sum_{n \in \mathcal{S}_k} \mathbf{A}_n}{\sum_{n \in \mathcal{S}_k} 1}. \tag{6.14}$$

(iv) Repeat steps ii) and iii) until the class means $\bar{\mathbf{A}}_k$ do not change by more than a predefined amount between successive iterations.

Statistical tests can be applied to eliminate some parameter vectors that are deemed as outliers.

Once the K cluster centers are determined, a label assignment procedure is employed to assign a segmentation label $z(\mathbf{x})$ to each pixel $\mathbf{x}$ as

$$z(\mathbf{x}) = Arg \min_k \| \mathbf{v}(\mathbf{x}) - \mathcal{P}(\bar{\mathbf{A}}_k; (\mathbf{x})) \|^2, \tag{6.15}$$

where k is from the set $\{1, 2, \ldots, K\}$, the operator $\mathcal{P}$ is defined as

$$\mathcal{P}(\bar{\mathbf{A}}_k; (\mathbf{x})) = \begin{bmatrix} (\bar{a}_{k,1} - 1)x_1 + \bar{a}_{k,2}x_2 + \bar{a}_{k,3} \\ \bar{a}_{k,4}x_1 + (\bar{a}_{k,5} - 1)x_2 + \bar{a}_{k,6} \end{bmatrix}, \tag{6.16}$$

and $\mathbf{v}(\mathbf{x})$ is the dense motion vector at pixel $\mathbf{x}$ given by

$$\mathbf{v}(\mathbf{x}) = \begin{bmatrix} v_1(\mathbf{x}) \\ v_2(\mathbf{x}) \end{bmatrix}, \tag{6.17}$$

where v_1 and v_2 denote the horizontal and vertical components, respectively. All sites without labels are assigned one according to the motion compensation criterion, which assigns the label of the parameter vector that gives the best motion compensation at that site. This feature ensures more robust parameter estimation by eliminating the outlier vectors. Several postprocessing operations may be employed to improve the accuracy of the segmentation map. The procedure can be repeated by estimating new seed model parameters over the regions estimated in the previous iteration. Furthermore, the number of clusters can be varied by splitting or merging of clusters between iterations. The K-means method requires a good initial estimate of the number of classes K. The Hough transform methods do not require this information but are more expensive.

6.4.2.1.2 Hough Transform Methods

The Hough transform is a well-known clustering technique where the data samples "vote" for the most representative feature values in a quantized feature space. In a straightforward application of the Hough transform method to optical flow segmentation using the six-parameter affine flow model (6.16), the six-dimensional feature space $a_1, \ldots, a_6$ would be quantized to certain parameter states after the minimal and maximal values for each parameter are determined. Then, each flow vector $\mathbf{v}(\mathbf{x}) = [v_1(\mathbf{x})\ v_2(\mathbf{x})]^T$ votes for a set of quantized parameters which minimizes

$$\eta^2(\mathbf{x}) \doteq \eta_1^2(\mathbf{x}) + \eta_2^2(\mathbf{x}), \tag{6.18}$$

where $\eta_1(\mathbf{x}) = v_1(\mathbf{x}) - a_1 - a_2x_1 - a_3x_2$ and $\eta_2(\mathbf{x}) = v_2(\mathbf{x}) - a_4 - a_5x_1 - a_6x_2$. The parameter sets that receive at least a predetermined amount of votes are likely to represent candidate motions. The number of classes K and the corresponding parameter sets to be used in labeling individual flow vectors are hence determined. The drawback of this scheme is the significant amount of computation and memory requirements involved.

To keep the computational burden at a reasonable level, several modified Hough methods have been presented. Proposed simplifications to ease the computational load include [28] (i) decomposition of the parameter space into two disjoint subsets $\{a_1, a_2, a_3\} \times \{a_4, a_5, a_6\}$ to perform two 3D Hough transforms, (ii) a multiresolution Hough transform, where at each resolution level the parameter space is quantized around the estimates obtained at the previous level, and (iii) a multipass Hough technique, where the flow vectors that are most consistent with the candidate parameters are grouped first. In the second stage, those components formed in the first stage, which are consistent with the same flow model in the least squares sense, are merged together to form segments. Several merging criteria have been proposed. In the third and final stage, ungrouped flow vectors are assimilated into one of their neighboring segments. Other simplifications that are proposed include the probabilistic Hough transform [41] and randomized Hough transform [42].

Clustering in the parameter space has some drawbacks: (i) both methods rely on precomputed optical flow as an input representation, which is generally blurred at motion boundaries and may contain outliers, (ii) clustering based on distances in the parameter space can lead to clustered parameters, which are not physically meaningful and the results are sensitive to the choice of the weight matrix $\mathbf{M}$ and small errors in the estimation of affine parameters, and (iii) parameter clustering and label assignment procedures are decoupled; hence, ad hoc postprocessing operations that depend on some threshold values are needed to clean up the final segmentation map. The following section proposes a ML segmentation method, which addresses all of these shortcomings.

6.4.2.2 *Maximum Likelihood Segmentation*

Motion segmentation approaches in general are classified as optical flow segmentation methods, which operate on precomputed optical flow estimates as an input representation, and direct methods, which operate on spatiotemporal intensity values. We present here a unified formulation that covers both cases. The ML method finds the segmentation labels that maximize the likelihood function, which models the deviation of the observations (estimated dense motion vectors or observed intensity values) from a parametric description of them (parametric motion vectors or motion-compensated intensity values, respectively) for a given motion model.

We start by defining the log-likelihood function as

$$L(\mathbf{o}|\mathbf{z}) = \log(p(\mathbf{o}|\mathbf{z})), \tag{6.19}$$

where $\mathbf{z}$ denotes the lexicographical ordering of the segmentation labels $\mathbf{z}(\mathbf{x})$, which takes values from the set $1,2,\ldots,K$ at each pixel $\mathbf{x}$. The vector $\mathbf{o}$ stands for the lexicographic ordering of the observations, which are either estimated dense motion (optical flow) vectors or image intensity values. The conditional probability $p(\mathbf{o}|\mathbf{z})$ quantifies how well piecewise parametric motion modeling fits the observations $\mathbf{o}$ given the segmentation labels $\mathbf{z}$. If we model the mismatch between the observations $\mathbf{o}(\mathbf{x})$ and their parametric representations computed by the operator $\mathcal{O}(\mathbf{A}_{z(\mathbf{x})};\mathbf{x})$,

$$\eta = \mathbf{o}(\mathbf{x}) - \mathcal{O}(\mathbf{A}_{z(\mathbf{x})};\mathbf{x}), \tag{6.20}$$

where $\mathbf{A}_k$ denotes the kth parametric motion model, by white, Gaussian noise with zero mean and variance σ^2, then the conditional pdf of the observations given the segmentation labels can be expressed as

$$p(\mathbf{o}|\mathbf{z}) = \frac{1}{(2\pi\sigma^2)^{M/2}} \exp\left\{ -\sum_{i=1}^{M} \eta^2(\mathbf{x}_i)/2\sigma^2 \right\}, \tag{6.21}$$

where M is the number of observations available at the sites $\mathbf{x}_i$. Assuming that the parametric flow model is more or less accurate, this deviation is due to presence of observation noise (given correct segmentation labels). Then, the problem is to find K motion models $\mathbf{A}_1,\mathbf{A}_2,\ldots,\mathbf{A}_K$ and a label field $z(\mathbf{x})$ to maximize the log-likelihood function $L(\mathbf{o}|\mathbf{z})$.

We consider two cases:

I. Precomputed optical flow segmentation: The observation $\mathbf{o}(\mathbf{x})$ stands for the estimated dense motion vectors $\mathbf{v}(\mathbf{x})$ and the operator $\mathcal{O}$ stands for the parametric motion operator $\mathcal{P}$ given by Eq. (6.16) or a higher order model given by

$$\begin{aligned} \tilde{v}_1(\mathbf{x}) &= a_1x_1 + a_2x_2 - a_3 + a_7x_1^2 + a_8x_1x_2 \\ \tilde{v}_2(\mathbf{x}) &= a_4x_1 + a_5x_2 - a_6 + a_7x_1x_2 + a_8x_2^2. \end{aligned} \tag{6.22}$$

Then,

$$\eta^2(\mathbf{x}_i) = (v_1(\mathbf{x}_i) - \tilde{v}_1(\mathbf{x}_i))^2 + (v_2(\mathbf{x}_i) - \tilde{v}_2(\mathbf{x}_i))^2 \tag{6.23}$$

is the norm-squared deviation of the actual flow vectors from what is predicted by the quadratic flow model. This case concerns motion segmentation by motion-vector matching.

II. Direct segmentation: The observation $\mathbf{o}(\mathbf{x})$ stands for the scalar pixel intensities $I_t(\mathbf{x})$ at frame t, and the operator $\mathcal{O}$ is the motion-compensation operator $\mathcal{Q}$, defined by

$$\mathcal{Q}(\mathbf{A}_{z(\mathbf{x})};\mathbf{x}) = \mathbf{I}_{t-1}(\mathbf{x}'), \tag{6.24}$$

where

$$\mathbf{x}' = [x_1'\ x_2']^T = [x_1\ x_2]^T + \mathcal{P}(\mathbf{A}_{z(\mathbf{x})};\mathbf{x}). \tag{6.25}$$

Then,

$$\eta^2(\mathbf{x}_i) = (I_t(\mathbf{x}) - I_{t-1}(\mathbf{x}'))^2. \tag{6.26}$$

This case corresponds to motion segmentation by motion-compensated intensity matching. The motion parameters $\mathbf{A}_k$ are estimated over the support of model k using direct methods (see step (iii) below).

In either case, assuming that the variances for all classes are the same, maximization of the log-likelihood function is equivalent to minimization of the cost function

$$\sum_{x_1,x_2} \| \mathbf{o}(\mathbf{x}) - \mathcal{O}(\mathbf{A}_{z(\mathbf{x})};\mathbf{x}) \|^2 \tag{6.27}$$

or equivalently

$$\sum_{k=1}^{K} \sum_{\mathbf{x}\in\mathcal{Z}_k} \| \mathbf{o}(\mathbf{x}) - \mathcal{O}_k(\mathbf{x}) \|^2, \tag{6.28}$$

where $\mathcal{Z}_k$ is the set of pixels $\mathbf{x}$ with the motion label $z(\mathbf{x}) = k$ and $\mathcal{O}_k(\mathbf{x}) \doteq \mathcal{O}(\mathbf{A}_k;\mathbf{x})$.

A two-step iterative solution to this problem is given as follows:

(i) Initialize $\mathbf{A}_1, \mathbf{A}_2, \ldots, \mathbf{A}_K$.

(ii) Assign a motion label $z(\mathbf{x})$ to each pixel $\mathbf{x}$ as

$$z(\mathbf{x}) = Arg\ \min_k \| \mathbf{o}(\mathbf{x}) - \mathcal{O}(\mathbf{A}_k;\mathbf{x}) \|^2, \tag{6.29}$$

where k takes values from the set $\{1, 2, \ldots, K\}$.

(iii) Update $\mathbf{A}_1, \mathbf{A}_2, \ldots, \mathbf{A}_K$ as

$$\mathbf{A}_k = Arg\ \min_{\mathbf{A}} \sum_{\mathbf{x}\in\mathcal{Z}_k} \| \mathbf{v}(\mathbf{x}) - \mathcal{P}(\mathbf{A};\mathbf{x}) \|^2. \tag{6.30}$$

This minimization is equivalent to least squares estimation of the affine motion model fit to the motion vectors with the label $z(\mathbf{x}) = k$. A closed form solution to this problem can be expressed in terms of a linear matrix equation

$$\begin{bmatrix} x_1 & x_2 & 1 & 0 & 0 & 0 \\ 0 & 0 & 0 & x_1 & x_2 & 1 \end{bmatrix} \begin{bmatrix} a_{k,1} \\ a_{k,2} \\ a_{k,3} \\ a_{k,4} \\ a_{k,5} \\ a_{k,6} \end{bmatrix} = \begin{bmatrix} x_1 + v_1(\mathbf{x}) \\ x_2 + v_2(\mathbf{x}) \end{bmatrix} \tag{6.31}$$

for all $\mathbf{x}$ such that $z(\mathbf{x}) = k$.

(iv) Repeat steps ii) and iii) until the class means $\mathbf{A}_k$ do not change by more than a predefined amount between successive iterations.

This method does not require gradient-based optimization or other numeric search procedures for optimization of a cost function. Thus, it is robust and computationally efficient. Extensions of this formulation using mixture modeling and robust estimators, that require gradient-based optimization, have also been proposed [38].

Motion-vector matching is a good motion segmentation criterion when the estimated motion field is accurate; that is, all outlier motion estimates are properly eliminated. Motion-compensated intensity matching is a more suitable criterion when spatial intensity (color) variations are sufficient and/or a multiresolution labeling procedure is employed. A possible limitation of ML segmentation framework is that it lacks constraints to enforce spatial and temporal continuity of the segmentation labels. Thus, rather ad hoc steps are needed to eliminate small, isolated regions in the segmentation label field. The MAP segmentation strategy promises to impose continuity constraints in an optimization framework.

6.4.2.3 *Maximum A Posteriori Probability Segmentation*

The MAP method is a Bayesian approach that searches for the maximum of the a posteriori pdf of the segmentation labels given the observations (either precomputed optical flow or spatiotemporal intensity data). This pdf is not only a measure of how well the segmentation labels explain the observed data but also how well they conform with our prior expectations. The MAP formulation differs from the ML approach in that it includes smoothness terms to enforce spatial continuity of the output motion segmentation map.

The a posteriori pdf $p(\mathbf{z}|\mathbf{o})$ of the segmentation label field $\mathbf{z}$ given the observed data $\mathbf{o}$ can be expressed, using the Bayes theorem, as

$$p(\mathbf{z}|\mathbf{o}) = \frac{p(\mathbf{o}|\mathbf{z})p(\mathbf{z})}{p(\mathbf{o})}, \tag{6.32}$$

where $p(\mathbf{o}|\mathbf{z})$ is the conditional pdf of the optical flow data given the segmentation $\mathbf{z}$ and $p(\mathbf{z})$ is the a priori pdf of the segmentation. Observe that (i) $\mathbf{z}$ is a discrete-valued random vector with a finite sample space Ω and (ii) $p(\mathbf{o})$ is constant with respect to the segmentation labels, and hence can be ignored for the purpose of computing $\mathbf{z}$. The MAP estimate, then, maximizes the numerator of (6.32) over all possible realizations of the segmentation field $\mathbf{z} = \omega$, $\omega \in \Omega$.

Modeling of the conditional pdf $p(\mathbf{o}|\mathbf{z})$ has been discussed in detail in Section 6.4.2.2 through (6.21) and (6.23) or (6.26). The prior pdf is modeled by a Gibbs distribution, which effectively introduces local constraints on the segmentation. It is given by

$$p(\mathbf{z}) = \frac{1}{Q} \sum_{\omega \in \Omega} \exp\{-U(\mathbf{z})\}\delta(\mathbf{z} - \omega), \tag{6.33}$$

where Ω denotes the sample space of the discrete-valued random vector $\mathbf{z}$, Q is the partition function (normalization constant) given by

$$Q = \sum_{\omega \in \Omega} \exp\{-U(\omega)\} \text{ and} \tag{6.34}$$

$U(\mathbf{z})$ is the potential function given by

$$U(\mathbf{z}) = \sum_{\mathbf{x}_i} \sum_{\mathbf{x}_j \in \mathcal{N}_{\mathbf{x}_i}} V_C(z(\mathbf{x}_i), z(\mathbf{x}_j)), \tag{6.35}$$

which can be expressed as a sum of local clique potential functions, such as

$$V_C(z(\mathbf{x}_i), z(\mathbf{x}_j)) = \begin{cases} -\gamma & \text{if } z(\mathbf{x}_i) = z(\mathbf{x}_j) \\ +\gamma & \text{otherwise} \end{cases}, \tag{6.36}$$

and $\mathcal{N}_{\mathbf{x}_i}$ denotes the neighborhood system for the label field. Prior constraints on the structure of the segmentation labels, such as spatial smoothness, can be specified in terms of the clique potential function. Temporal continuity of the labels can similarly be modeled [34].

Substituting (6.21) and (6.33) into the criterion (6.32) and taking the logarithm of the resulting expression, maximization of the a posteriori probability distribution can be performed by minimizing the cost function

$$E = \frac{1}{2\sigma^2} \sum_{i=1}^{M} \eta^2(\mathbf{x}_i) + U(\omega). \tag{6.37}$$

The first term describes how well the predicted data fit the actual measurements (estimated optical flow vectors or image intensity values) and the second term measures how well the segmentation conforms to our prior expectations.

Because the motion model parameters corresponding to each label are not known a priori, the MAP segmentation must alternate between estimation of the model parameters and assignment of the segmentation labels to optimize the cost function (6.37). Murray and Buxton [34] were the first to propose a MAP segmentation method where the optical flow was modeled by a piecewise quadratic flow field (6.22), and the segmentation labels were assigned based on a simulated annealing (SA) procedure. Given the estimated flow field $\mathbf{v}$ and the number of independent motion models K, the MAP segmentation using the Metropolis algorithm can be summarized as follows:

(i) Start with an initial labeling $\mathbf{z}$ of the optical flow vectors. Calculate the model parameters $\mathbf{a} = [a_1 \ldots a_8]^T$ for each region using least squares fitting (similar to that in Section 6.4.2.2). Set the initial temperature for SA.

(ii) Update the segmentation labels at each site $\mathbf{x}_i$ as follows:

(a) Perturb the label $z_i = z(\mathbf{x}_i)$ randomly.

(b) Decide whether to accept or reject this perturbation, based on the change ΔE in the cost function (6.37),

$$\Delta E = \frac{1}{2\sigma^2}\Delta\eta^2(\mathbf{x}_i) + \sum_{\mathbf{x}_j \in \mathcal{N}_{\mathbf{x}_i}} \Delta V_C(z(\mathbf{x}_i), z(\mathbf{x}_j)), \tag{6.38}$$

where $\mathcal{N}_{\mathbf{x}_i}$ denotes a neighborhood of the site $\mathbf{x}_i$ and $V_C(z(\mathbf{x}_i), z(\mathbf{x}_j))$ is given by Eq. (6.36). The first term indicates whether or not the perturbed label is more consistent with the given flow field determined by the residual (6.23), and the second term reflects whether or not it is in agreement with the prior segmentation field model.

Because the update at each site is dependent on the labels of the neighboring sites, the order in which the sites are visited affects the result of this step.

(iii) After all pixel sites are visited once, re-estimate the mapping parameters for each region based on the new segmentation label configuration.

(iv) Exit if a stopping criterion is satisfied. Otherwise, lower the temperature according to a predefined temperature schedule and go to step ii).

We can make the following observations: (i) The MAP method carries a high computational cost. (ii) The procedure proposed by Murray–Buxton suggests performing step iii) above, the model parameter update, after each and every perturbation. We did not notice a significant difference in performance if motion parameter updates are done after all sites are visited once. (iii) The method can be applied with any parametric motion model, although the original formulation has been developed on the basis of the eight-parameter model.

6.4.2.4 *Region-Based ML Segmentation: Fusion of Color and Motion*

In this section, we extend the ML approach (Section 6.4.2.2) to region-based motion segmentation (as opposed to pixel-based), where the image is first divided into homogeneous color regions, and then, at every iteration, each region is assigned a single motion label. This region-based motion label assignment strategy facilitates obtaining spatially continuous segmentation maps that are closely related to actual object boundaries, without the heavy computational burden of statistical Markov random field (MRF) model-based approaches. The predefined regions should be such that each region has a single motion. It is generally true that motion boundaries coincide with color segment boundaries, but not vice versa; that is, color segments are almost always a subset of motion segments as illustrated in Fig. 6.1. Therefore, one can first perform a color segmentation to obtain a set of candidate motion segments. Other approaches to region definition include mesh-based partitioning of the scene [58] and macro pixels ($N \times N$ blocks) to improve the robustness of the ML motion segmentation. Here, we assume that each frame of video has been subject to a region formation procedure. We let $C(\mathbf{x})$ denote the region map of a frame consisting of M mutually exclusive and exhaustive regions and define $\mathcal{C}_m$ as the set of pixels $\mathbf{x}$ with the region label $C(\mathbf{x}) = m$, $m = 1,\ldots,M$.

We wish to find the motion segmentation map $\mathbf{z}$ (a vector formed by lexicographic ordering of $z(\mathbf{x})$) and the corresponding affine parameter vectors $\mathbf{A}_1, \mathbf{A}_2, \ldots, \mathbf{A}_K$, which best fit the dense motion-vector field, such that [44]

$$\sum_{m=1}^{M} \sum_{\mathbf{x} \in \mathcal{C}_m} \| \mathbf{v}(\mathbf{x}) - \mathcal{P}(\mathbf{A}_{z(m)}; \mathbf{x}) \|^2 \tag{6.39}$$

is minimized. Here, $z(m)$ refers to the motion label of all pixels within $\mathcal{C}_m$ and takes one of the values $1,2,\ldots,K$; $\mathcal{P}$ is an operator defined by Eq. (6.16), and $\mathbf{v}(\mathbf{x})$ is the dense motion vector at pixel $\mathbf{x}$ as defined by Eq. (6.17). The procedure is given as follows:

(i) Initialize the motion segmentation map $\mathbf{z}$ by assigning a single motion label k, $k = 1,\ldots,K$ to each $\mathcal{C}_m$.

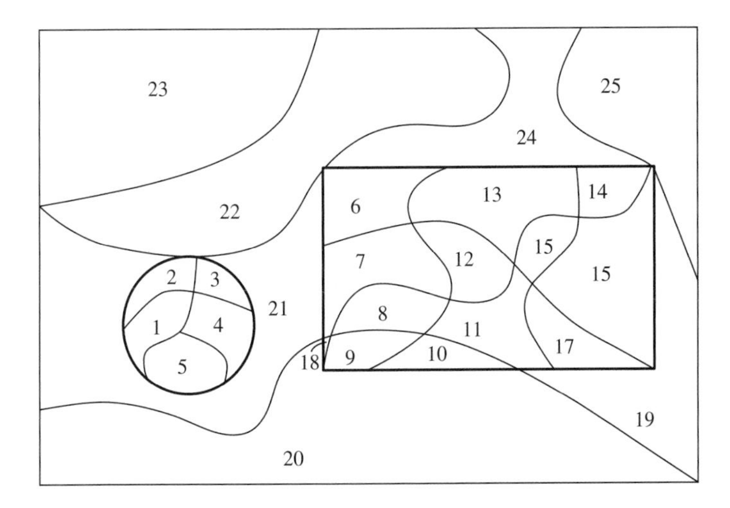

FIGURE 6.1

Illustration of the observation that color segments are generally subsets of motion segments. Here, the bold lines indicate motion segment boundaries, and each motion segment is composed of many color regions.

(ii) Update the parameter vectors $\mathbf{A}_1, \mathbf{A}_2, \ldots, \mathbf{A}_K$ as

$$\mathbf{A}_k = Arg \min_{\mathbf{A}} \sum_{\mathbf{x} \in \mathcal{Z}_k} \| \mathbf{v}(\mathbf{x}) - \mathcal{P}(\mathbf{A}; \mathbf{x}) \|^2, \tag{6.40}$$

where $\mathcal{Z}_k$ is the set of pixels $\mathbf{x}$ with the label $z(\mathbf{x}) = k$. This minimization can be achieved by solving the linear matrix equation

$$\begin{bmatrix} x_1 & x_2 & 1 & 0 & 0 & 0 \\ 0 & 0 & 0 & x_1 & x_2 & 1 \end{bmatrix} \begin{bmatrix} a_{k,1} \\ a_{k,2} \\ a_{k,3} \\ a_{k,4} \\ a_{k,5} \\ a_{k,6} \end{bmatrix} = \begin{bmatrix} v_1(\mathbf{x}) \\ v_2(\mathbf{x}) \end{bmatrix} \tag{6.41}$$

for all $\mathbf{x}$ in $\mathcal{Z}_k$.

(iii) Assign a motion label to each region $\mathcal{C}_m$, $m = 1, 2, \ldots, M$, such that

$$z(\mathcal{C}_m) = Arg \min_{k} \sum_{\mathbf{x} \in \mathcal{C}_m} \| \mathbf{o}(\mathbf{x}) - \mathcal{O}(\mathbf{A}_k; \mathbf{x}) \|^2, \tag{6.42}$$

where $k = 1, 2, \ldots K$ and $\mathbf{o}(\mathbf{x})$ and $\mathcal{O}(\mathbf{x})$ are as defined in Section 6.4.2.2. This allows region-based affine motion segmentation with pixel-based motion-vector or intensity matching.

(iv) Repeat steps ii) and iii) until the class means $\mathbf{A}_k$ do not change by more than a predefined amount between successive iterations.

We note that the pixel-based ML motion segmentation method presented in Section 6.4.2.2 is a special case of this region-based framework. If each region $\mathcal{C}_m$ contains a single pixel, then the iterations are carried over individual pixels, and the motion label assignment is performed at each pixel independently.

We conclude this section by observing that the methods discussed here that used precomputed optical flow as an input representation are limited by the accuracy of the available optical flow estimates. Next, we introduce a framework in which optical flow estimation and segmentation interact in a mutually beneficial manner.

6.5 SIMULTANEOUS MOTION ESTIMATION AND SEGMENTATION

Up to now, we discussed methods to compute the segmentation labels from either precomputed optical flow or directly from intensity values but did not address how to compute an improved dense motion field along with the segmentation map. It is clear that the success of optical flow segmentation is closely related to the accuracy of the estimated optical flow field (in the case of using precomputed flow values) and vice versa.

It follows that optical flow estimation and segmentation should be addressed simultaneously for best results. Here, we present a simultaneous Bayesian approach based on a representation of the motion field as the sum of a parametric field and a residual field. The interdependence of optical flow and segmentation fields is expressed in terms of a Gibbs distribution within the MAP framework. The resulting optimization problem, to find estimates of a dense set of motion vectors, a set of segmentation labels, and a set of mapping parameters, is solved using the highest confidence first (HCF) and iterated conditional mode (ICM) algorithms.

6.5.1 Motion-Field Model and MAP Framework

We model the optical flow field $\mathbf{v}(\mathbf{x})$ as the sum of a parametric flow field $\tilde{\mathbf{v}}(\mathbf{x})$ and a nonparametric residual field $\mathbf{v}_r(\mathbf{x})$, which accounts for local motion and other modeling errors; that is,

$$\mathbf{v}(\mathbf{x}) = \tilde{\mathbf{v}}(\mathbf{x}) + \mathbf{v}_r(\mathbf{x}). \tag{6.43}$$

The parametric component of the motion field clearly depends on the segmentation label $z(\mathbf{x})$, which takes on the values $1, \ldots, K$.

The simultaneous MAP framework aims at maximizing the a posteriori pdf

$$p(\mathbf{v}_1, \mathbf{v}_2, \mathbf{z} \mid \mathbf{g}_k, \mathbf{g}_{k+1}) = \frac{p(\mathbf{g}_{k+1} \mid \mathbf{g}_k, \mathbf{v}_1, \mathbf{v}_2, \mathbf{z}) p(\mathbf{v}_1, \mathbf{v}_2 \mid \mathbf{z}, \mathbf{g}_k) p(\mathbf{z} \mid \mathbf{g}_k)}{p(\mathbf{g}_{k+1} \mid \mathbf{g}_k)} \tag{6.44}$$

with respect to the optical flow $\mathbf{v}_1$, $\mathbf{v}_2$, and the segmentation labels $\mathbf{z}$, where $\mathbf{v}_1$ and $\mathbf{v}_2$ denote the lexicographic ordering of the first and second components of the flow vectors $\mathbf{v}(\mathbf{x}) = [v_1(\mathbf{x})\ v_2(\mathbf{x})]^T$ at each pixel $\mathbf{x}$. Through careful modeling of these pdfs, we can express an interrelated set of constraints that help improve both optical flow and segmentation estimates.

The first conditional pdf $p(\mathbf{g}_{k+1} \mid \mathbf{g}_k, \mathbf{v}_1, \mathbf{v}_2, \mathbf{z})$ provides a measure of how well the present displacement and segmentation estimates conform with the observed frame $k+1$ given frame k. It is modeled by a Gibbs distribution as

$$p(\mathbf{g}_{k+1} \mid \mathbf{g}_k, \mathbf{v}_1, \mathbf{v}_2, \mathbf{z}) = \frac{1}{Q_1} \exp\left\{-U_1(\mathbf{g}_{k+1} \mid \mathbf{g}_k, \mathbf{v}_1, \mathbf{v}_2, \mathbf{z})\right\}, \tag{6.45}$$

where Q_1 is the partition function (normalizing constant) and

$$U_1(\mathbf{g}_{k+1} \mid \mathbf{g}_k, \mathbf{v}_1, \mathbf{v}_2, \mathbf{z}) = \sum_{\mathbf{x}} [g_k(\mathbf{x}) - g_{k+1}(\mathbf{x} + \mathbf{v}(\mathbf{x})\Delta t)]^2 \tag{6.46}$$

is called the Gibbs potential. Here, the Gibbs potential corresponds to the norm square of the displaced FD (DFD) between the frames $\mathbf{g}_k$ and $\mathbf{g}_{k+1}$. Thus, maximization of (6.45) imposes the constraint that $\mathbf{v}(\mathbf{x})$ minimizes the DFD.

The second term in the numerator in (6.44) is the conditional pdf of the displacement field given the motion segmentation and the search image. It is also modeled by a Gibbs distribution

$$p(\mathbf{v}_1, \mathbf{v}_2 \mid \mathbf{z}, \mathbf{g}_k) = p(\mathbf{v}_1, \mathbf{v}_2 \mid \mathbf{z}) = \frac{1}{Q_2} \exp\{-U_2(\mathbf{v}_1, \mathbf{v}_2 \mid \mathbf{z})\}, \tag{6.47}$$

where Q_2 is a constant and

$$U_2(\mathbf{v}_1,\mathbf{v}_2 \mid \mathbf{z}) = \alpha \sum_{\mathbf{x}} ||\mathbf{v}(\mathbf{x}) - \tilde{\mathbf{v}}(\mathbf{x})||^2 + \beta \sum_{\mathbf{x}_i} \sum_{\mathbf{x}_j \in \mathcal{N}_{\mathbf{x}_i}} ||\mathbf{v}(\mathbf{x}_i) - \mathbf{v}(\mathbf{x}_j)||^2 \, \delta(z(\mathbf{x}_i) - z(\mathbf{x}_j)) \tag{6.48}$$

is the corresponding Gibbs potential, $||\cdot||$ denotes the Euclidian distance, and $\mathcal{N}_{\mathbf{x}}$ is the set of neighbors of site $\mathbf{x}$. The first term in (6.48) enforces a minimum norm estimate of the residual motion field $\mathbf{v}_r(\mathbf{x})$; that is, it aims to minimize the deviation of the motion field $\mathbf{v}(\mathbf{x})$ from the parametric motion field $\tilde{\mathbf{v}}(\mathbf{x})$ while minimizing the DFD. Note that the parametric motion field $\tilde{\mathbf{v}}(\mathbf{x})$ is calculated from the set of model parameters $\mathbf{a}_i$, $i = 1,\ldots,K$, which in turn is a function of $\mathbf{v}(\mathbf{x})$ and $z(\mathbf{x})$. The second term in (6.48) imposes a piecewise local smoothness constraint on the optical flow estimates without introducing any extra variables such as line fields. Observe that this term is active only for those pixels in the neighborhood $\mathcal{N}_{\mathbf{x}}$, which share the same segmentation label with the site $\mathbf{x}$. Thus, spatial smoothness is enforced only on the flow vectors generated by a single object. The parameters α and β allow for relative scaling of the two terms.

The third term in (6.44) models the a priori probability of the segmentation field in a manner similar to that in MAP segmentation. It is given by

$$p(\mathbf{z} \mid \mathbf{g}_k) = p(\mathbf{z}) = \frac{1}{Q_3} \sum_{\omega \in \Omega} \exp\{-U_3(\mathbf{z})\}\delta(\mathbf{z} - \omega), \tag{6.49}$$

where Ω denotes the sample space of the discrete-valued random vector $\mathbf{z}$, and Q_3 and $U_3(\mathbf{z})$ are as defined in (6.34) and (6.35), respectively. The dependence of the labels on the image intensity is usually neglected, although region boundaries generally coincide with intensity edges.

6.5.2 Two-Step Iteration Algorithm

Maximizing the a posteriori pdf (6.44) is equivalent to minimizing the cost function,

$$E = U_1(\mathbf{g}_{k+1} \mid \mathbf{g}_k, \mathbf{v}_1, \mathbf{v}_2, \mathbf{z}) + U_2(\mathbf{v}_1, \mathbf{v}_2 \mid \mathbf{z}) + U_3(\mathbf{z}) \tag{6.50}$$

that is composed of the potential functions in Eq. (6.45), (6.47), and (6.49). Direct minimization of (6.50) with respect to all unknowns is an exceedingly difficult problem, because the motion and segmentation fields constitute a large set of unknowns. To this effect, we perform the minimization of (6.50) through the following two-step iterations [48]:

(i) Given the best available estimates of the parameters $\mathbf{a}_i$, $i = 1,\ldots,K$, and $\mathbf{z}$, update the optical flow field $\mathbf{v}_1, \mathbf{v}_2$. This step involves the minimization of a modified cost function

$$E_1 = \sum_{\mathbf{x}} [g_k(\mathbf{x}) - g_{k+1}(\mathbf{x} + \mathbf{v}(\mathbf{x})\Delta t)]^2 + \alpha \sum_{\mathbf{x}} ||\mathbf{v}(\mathbf{x}) - \tilde{\mathbf{v}}(\mathbf{x})||^2$$
$$+ \beta \sum_{\mathbf{x}_i} \sum_{\mathbf{x}_j \in \mathcal{N}_{\mathbf{x}_i}} ||\mathbf{v}(\mathbf{x}_i) - \mathbf{v}(\mathbf{x}_j)||^2 \, \delta(z(\mathbf{x}_i) - z(\mathbf{x}_j)), \tag{6.51}$$

which is composed of all terms in (6.50) that contain $\mathbf{v}(\mathbf{x})$. Although the first term indicates how well $\mathbf{v}(\mathbf{x})$ explains our observations, the second and third terms impose prior constraints on the motion estimates that they should conform with the parametric flow model and that they should vary smoothly within each region. To minimize this energy function, we employ the HCF method recently proposed by Chou and Brown [49]. HCF is a deterministic method designed to efficiently handle the optimization of multivariable problems with neighborhood interactions.

(ii) Update the segmentation field $\mathbf{z}$, assuming that the optical flow field $\mathbf{v}(\mathbf{x})$ is known. This step involves the minimization of all the terms in (6.50), which contain $\mathbf{z}$ as well as $\tilde{\mathbf{v}}(\mathbf{x})$, given by

$$E_2 = \alpha \sum_{\mathbf{x}} ||\mathbf{v}(\mathbf{x}) - \tilde{\mathbf{v}}(\mathbf{x})||^2 + \sum_{\mathbf{x}_i} \sum_{\mathbf{x}_j \in \mathcal{N}_{\mathbf{x}_i}} V_C(z(\mathbf{x}_i), z(\mathbf{x}_j)). \tag{6.52}$$

The first term in (6.52) quantifies the consistency of $\tilde{\mathbf{v}}(\mathbf{x})$ and $\mathbf{v}(\mathbf{x})$. The second term is related to the a priori probability of the present configuration of the segmentation labels. We use an ICM procedure to optimize E_2 [48]. The mapping parameters $\mathbf{a}_i$ are updated by least squares estimation within each region.

An initial estimate of the optical flow field can be found by using the Bayesian approach with a global smoothness constraint. Given this estimate, the segmentation labels can be initialized by a procedure similar to Wang and Adelson's [37]. The determination of the free parameters α, β, and γ is a design problem. One strategy is to choose them to provide a dynamic range correction so that each term in the cost function (6.50) has equal emphasis. However, because the optimization is implemented in two steps, the ratio α/γ also becomes consequent. We recommend to select $1 \leq \alpha/\gamma \leq 5$, depending on how well the motion field can be represented by a piecewise-parametric model and whether we have a sufficient number of classes.

A hierarchical implementation of this algorithm is also possible by forming successive low-pass filtered versions of the images $\mathbf{g}_k$ and $\mathbf{g}_{k+1}$. Thus, the quantities $\mathbf{v}_1, \mathbf{v}_2$, and $\mathbf{z}$ can be estimated at different resolutions. The results of each hierarchy are used to initialize the next lower level. Note that the Gibbsian model for the segmentation labels has been extended to include neighbors in scale by Kato et al. [53].

Several other motion analysis approaches can be formulated as special cases of this framework. If we retain only the first and the third terms in (6.50) and assume that all sites possess the same segmentation label, then we have Bayesian motion estimation with a global smoothness constraint. The motion estimation algorithm proposed by Iu [47] uses the same two terms but replaces the $\delta(\cdot)$ function by a local outlier rejection function. The motion estimation and region labeling algorithm proposed by Stiller [52] involve all terms in (6.50) except the first term in (6.48). Furthermore, the segmentation labels in Stiller's algorithm are used merely as tokens to allow for a piecewise smoothness constraint on the flow field and do not attempt to enforce consistency of the flow vectors with a parametric component. We also note that the motion estimation method of Konrad and Dubois [51], which use line fields, are fundamentally different in that they model discontinuities in the motion field rather than modeling regions that correspond to different parametric motions. On the other hand, the motion segmentation algorithm of Murray and Buxton [34] (Section 6.4.2.3) employs only the second term in (6.48) and third term in (6.50) to model the conditional and prior pdf, respectively. Wang and Adelson [37] relies on the first term in (6.48) to compute the motion segmentation (Section 6.4.2.2). However, they also take the DFD of the parametric motion vectors into consideration when the closest match between the estimated and parametric motion vectors, represented by the second term, exceeds a threshold.

6.6 SEMANTIC VIDEO OBJECT SEGMENTATION

So far, we discussed methods for automatic motion segmentation. However, it is difficult to achieve semantically meaningful object segmentation using fully automatic methods based on low-level features such as motion, color, and texture. This is because a semantic object may contain multiple motions, colors, textures, and so on, and definition of semantic objects may depend on the context, which may not be possible to capture by low-level features. Thus, in this section, we present two approaches that can extract semantically meaningful objects using capture-specific information or user interaction.

6.6.1 Chroma-Keying

Chroma-keying is an object-based video capture technology where each video object is recorded individually in a special studio against a key color. The key color is selected such that it does not appear on the object to be captured. Then, the problem of extracting the object from each frame of video becomes one of color segmentation. Chroma-keyed video capture requires special attention to avoid shadows and other nonuniformity in the key color within a frame; otherwise, segmentation of key color may become a nontrivial problem.

6.6.2 Semiautomatic Segmentation

Because chroma-keying requires special studios and/or equipment to capture video objects, an alternative approach is interactive segmentation using automated tools to aid a human operator. To this effect, we assume that the contour of the first occurrence of the semantic object of interest is marked interactively by a human operator. Although detection of moving regions (by change detection methods) may result in semantically meaningful objects in well-constrained settings, in an unconstrained environment, user interaction is indeed the only way to define a semantically meaningful object unambiguously because only the user can know what is semantically meaningful in the context of an application. For example, if we have the video clip of a person carrying a ball, whether the ball and the person are two separate objects or a single object may depend on the application. Once the boundary of the object of interest is interactively determined in one or more key frames, its boundary in all other frames can be automatically computed by 2D motion tracking until the object exits the field of view.

2D object tracking is closely related to the problem of spatiotemporal segmentation in the sense it provides temporally linked spatial segmentation maps. The general approach can be summarized as projecting the current segmentation map into the next frame using 2D motion information. The projected region can be updated by morphological or other operators using the color and edge information in the next frame. This update step allows fine tuning of the segmentation map to alleviate motion estimation errors as well as including newly uncovered regions in the segmentation map. Object tracking methods can be classified as feature-point-based, contour-based, and region-based tracking methods [4]. Feature points are points on the object (current segmentation map) that can be used as markers, such as corner points. Motion of these points can be found by gradient-based (e.g., Lukas–Kanade) or matching-based (e.g., block matching) methods. Goodness of tracking results at each feature point can be evaluated at each frame and some feature points can be removed and others may be added [54]. In contour-based methods, the tracking step defines a polygonal or spline approximation of the boundary of the video object, which may be further refined automatically or interactively using appropriate software tools [55, 56]. In region-based methods, the object region is repartitioned into color- and/or motion-homogeneous subregions, and each subpartition is projected into the next frame individually with or without using subregion connectivity constraints [57, 58].

6.7 EXAMPLES

Examples are shown for automatic motion segmentation using the pixel-based and region-based ML methods on two MPEG-4 test sequences: "Mother & Daughter" (frames 1 and 2) and "Mobile & Calendar" (frames 136 and 137). The former is an example of a slowly moving object against a still background, where mother's head is rotating while her body, the background, and the child are stationary. The latter is a challenging sequence, with several distinctly moving objects such as a rotating ball, a moving train, and a vertically translating calendar against a background that moves due to camera pan. Figure 6.2(a) and (b) shows the first and second frames of the "Mother & Daughter" sequence, and Fig. 6.2(c) shows the estimated motion field between these frames. Figures 6.3(a–c) shows the corresponding pictures for frames 136 and 137 of the

"Mobile & Calendar" sequence. Motion estimation was performed by using the hierarchical version of the Lucas–Kanade method [4] with three levels of hierarchy. In both cases, region definition by color segmentation is performed on the temporally second frame using the fuzzy c-means technique [43]. Each spatially disconnected piece of the color segmentation map was defined as an individual region. The resulting region maps are shown in Figs. 6.2(d) and 6.3(d).

Figure 6.4 demonstrates the performance of the ML method for foreground/background separation (i.e., $K = 2$) with two different initializations. Figure 6.4(a) and (b) shows two possible initial segmentation maps, where the segmentation map is divided into two horizontal and vertical parts, respectively. Figure 6.4(b) and (c) shows the segmentation maps using pixel-based labeling by motion-vector matching after 10 iterations

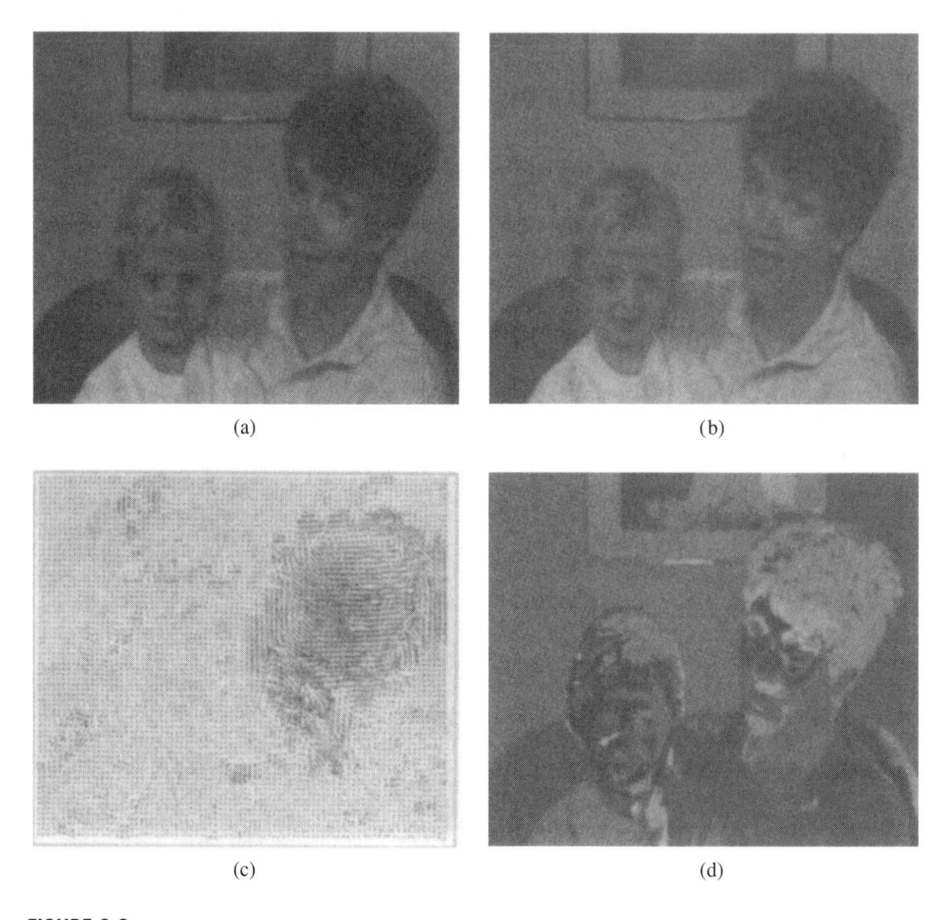

FIGURE 6.2

(a) The first and (b) second frames of the Mother & Daughter sequence. (c) 2D dense motion field from second frame to first frame. (d) Region map obtained by color segmentation.

FIGURE 6.3

(a) The 136th and (b) 137th frames of the "Mobile & Calendar" sequence. (c) 2D dense motion field from 137th to 136th frame. (d) Region map obtained by color segmentation.

starting from Fig. 6.4(a) and (b), respectively. Figure 6.4(d) and (e) show the results of region-labeling by motion vector-matching starting with the affine parameter sets obtained from the maps Fig. 6.4(b) and (c), respectively. Observe that the segmentation maps obtained by pixel-labeling contain many misclassified pixels, whereas the maps obtained by color-region-labeling are more coherent with the moving object in the scene.

Figure 6.5 illustrates the performance of the ML method with different number of initial segments, K. Figure 6.5(a) and (b) shows two initial segmentation maps with $K = 4$ and $K = 6$, respectively. The results of pixel-based labeling by motion-vector matching after 10 iterations for both initializations are depicted in Fig. 6.5(c) and (d), respectively. Figure 6.5(e) shows the result of region-based labeling using the color regions depicted in Fig. 6.3(d) and the affine model parameters initialized by those computed from the map in Fig. 6.5(c) with $K = 4$. We observed that this procedure results in oversegmentation when repeated with $K = 6$. Therefore, we employ motion-compensated intensity matching and region merging to reduce the number of the motion classes if necessary. In this step, a region is merged with another if the latter set of affine parameters gives

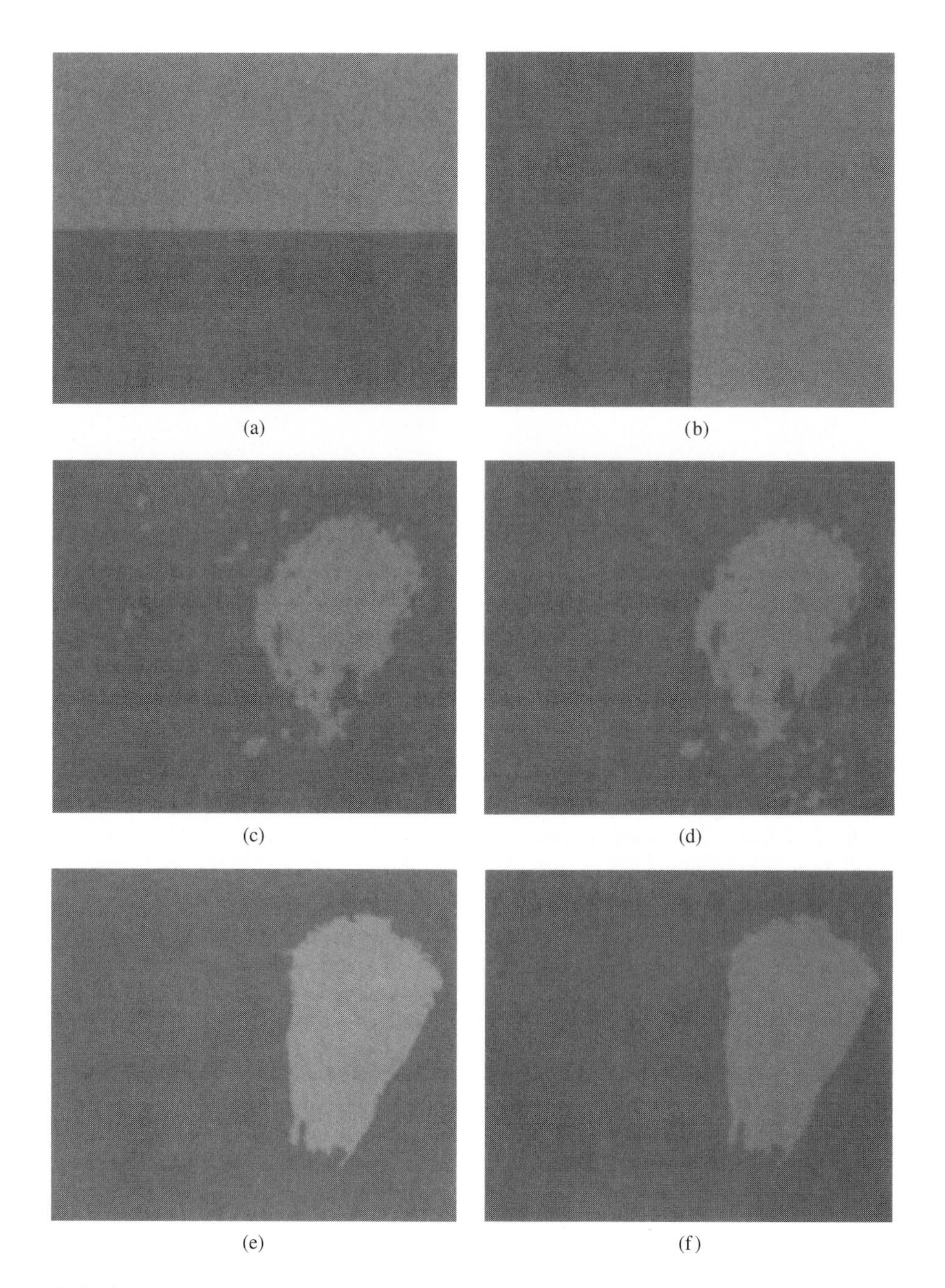

FIGURE 6.4

Results of the ML method with two different initializations: (a) and (b) Initial map. (c) and (d) Pixel-based motion-vector matching. (e) and (f) Region-based motion-vector matching.

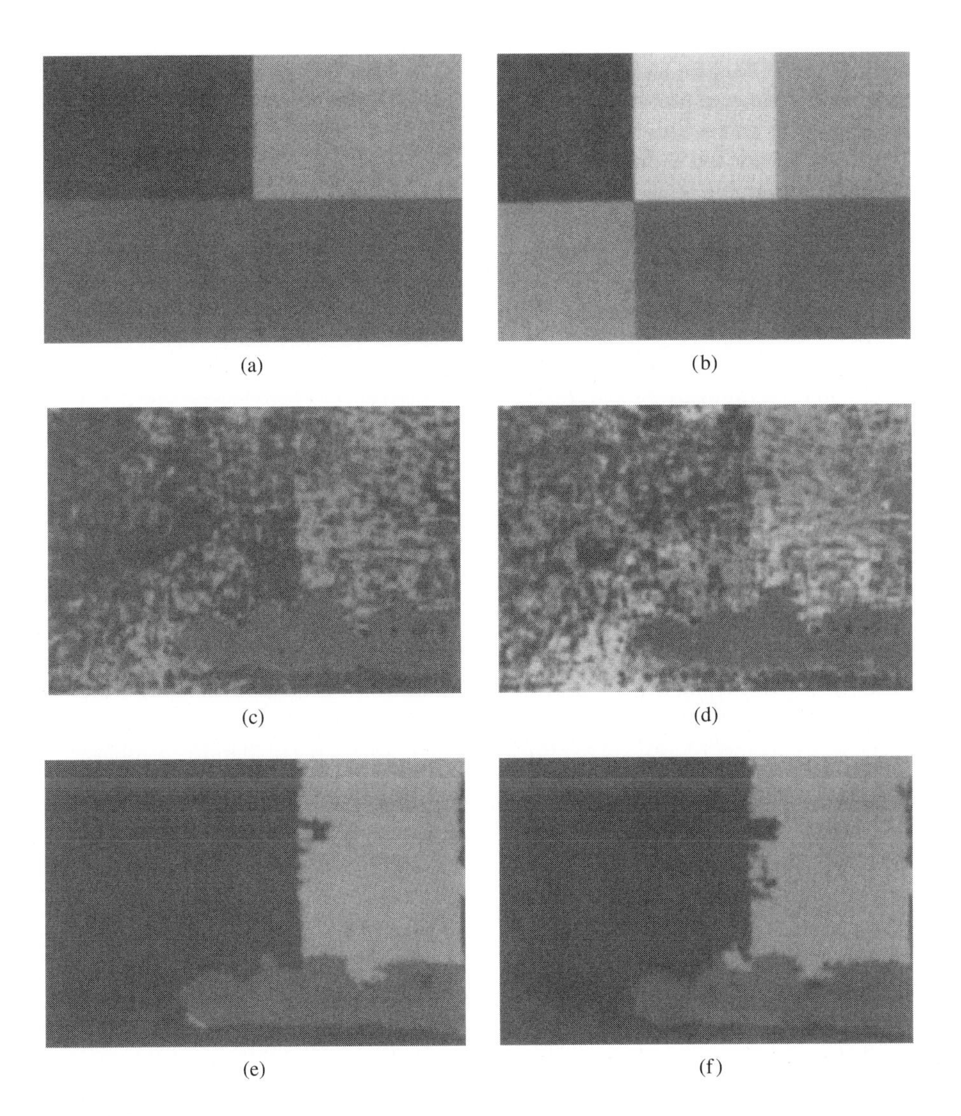

FIGURE 6.5

Results of the ML method: Initial map (a) $K=4$ and (b) $K=6$. Pixel-based labeling (c) $K=4$ and (d) $K=6$. Region-based labeling (e) $K=4$ and (f) $K=6$.

a comparable DFD as the former. The result of this final step is depicted in Fig. 6.5(f) for $K = 6$, where two of the six classes are eliminated by motion-compensated intensity matching. The ML segmentation method is computationally efficient since it does not require gradient-based optimization or any numeric search. It converges within approximately 10 iterations, and each iteration involves solution of only two 3×3 matrix equations.

6.8 PERFORMANCE EVALUATION OF VIDEO SEGMENTATION

Comparative assessment of segmentation results is often based on subjective judgment, which is qualitative and time consuming. Hence, there is a need to associate a generic figure of merit with video segmentation results. A straightforward approach would be to compare the segmentation results with a ground truth (GT), if GT data is available. However, in practical applications, GT data is hardly available. Therefore, measures that do not rely on GT data should be developed and used to evaluate temporal consistency (in case of scene change detection) or spatial and spatiotemporal consistency (in case of spatiotemporal object segmentation) of results.

A set of performance measures that do not rely on GT data have been proposed in [59], which can be grouped as intraobject homogeneity and interobject disparity measures. The intraobject homogeneity measures are based on shape regularity, spatial uniformity, temporal stability, and motion consistency. The interobject disparity measures include local color and motion contrast with neighbor regions. The usefulness of these measures have been demonstrated based on how the results predicted by these measures correlate with judgments of human observers.

Other measures that do not rely on GT data have been proposed in an independent study [60]. These include spatial color contrast along the estimated object boundary, motion difference along the estimated object boundary, and color histogram difference between successive object segmentation masks in the temporal direction. These measures can be computed per object and per frame so that it is possible to identify the objects and frames that are poorly segmented within a long video sequence.

ACKNOWLEDGMENTS

The author acknowledges Yucel Altunbasak and P. Erhan Eren for their contributions to the section on the maximum likelihood segmentation and Michael Chang for his contributions to the section on the maximum a posteriori probability segmentation. This work was supported by grants from National Science Foundation, New York State Science and Technology Foundation, and Eastman Kodak Company.

REFERENCES

[1] N. Dimitrova, H. Zhang, B. Shahraray, I. Sezan, T. Huang, and A. Zakhor. Applications of video content analysis and retrieval. *IEEE Multimed.*, 9:42–55, 2002.

[2] P. Salembier and F. Marques. Region-based representations of images and video: Segmentation tools for multimedia services. *IEEE Trans. Circ. Syst. Video Tech.*, 9(8):1147–1169, 1999.

[3] R. Castagno, T. Ebrahimi, and M. Kunt. Video segmentation based on multiple features for interactive multimedia applications. *IEEE Trans. Circ. Syst. Video Tech.*, 8(5):562–571, 1998.

[4] A. M. Tekalp. *Digital Video Processing.* Prentice Hall, New Jersey, NJ, 1995.

[5] P. L. Correia and F. Pereira. Classification of video segmentation application scenarios. *IEEE Trans. Circ. Syst. Video Tech.*, 14(5):735–741, 2004.

[6] E. Izquierdo and M. Ghanbari. Key components for an advanced segmentation system. *IEEE Trans. Multimed.*, 4(1):97–113, 2002.

[7] Y. Wang, Z. Liu, and J.-C. Huang. Multimedia content analysis using both audio and video clues. *IEEE Signal Process. Mag.*, 17:12–36, 2000.

[8] G. Ahanger and T. D. C. Little. A survey of technologies for parsing and indexing digital video. *J. Visual Comm. Image Repres.*, 7:28–43, 1996.

[9] H. Jiang, A. Helal, A. K. Elmagarmid, and A. Joshi. Scene change detection techniques for video databases. *Multimed. Syst.*, 6:186–195, 1998.

[10] U. Gargi, R. Kasturi, and S. H. Strayer. Performance characterization of video-shot change detection methods. *IEEE Trans. Circ. Syst. Video Tech.*, 10:1–13, 2000.

[11] I. Koprinska and S. Carrato. Temporal video segmentation: A survey. *Signal Process. Image Commun.*, 16(5):477–500, 2001.

[12] R. Lienhart. Reliable transition detection in videos: A survey and practitioner's guide: Int. *J. Image Graph.*, 1:469–486, 2001.

[13] A. Hanjalic. Shot-boundary detection: Unraveled and resolved? *IEEE Trans. Circ. Syst. Video Tech.*, 12:90–105, 2002.

[14] P. Bouthemy, M. Gelgon, and F. Ganansia. A unified approach to shot change detection and camera motion characterization. *IEEE Trans. Circ. Syst. Video Tech.*, 9:1030–1044, 1999.

[15] H. J. Zhang, A. Kankanhalli, and S. W. Smoliar. Automatic partitioning of full-motion video. *Multimed. Syst.*, 1:10–28, 1993.

[16] E. Stringa and C. S. Regazzoni. Real-time video shot detection for scene surveillance applications. *IEEE Trans. Image Process.*, 9(1):69–79, 2000.

[17] A. Ekin, A. M. Tekalp, and R. Mehrotra. Automatic soccer video analysis and summarization. *IEEE Trans. Image Process.*, 12(7):796–807, 2003.

[18] A. Hampapur, R. Jain, and T. E. Weymouth. Production model based digital video segmentation. *Multimed. Tools Appl.*, 1:9–46, 1995.

[19] H. Sundaram and S.-F. Chang. Computable scenes and structures in films. *IEEE Trans. Multimed.*, 4:482–491, 2002.

[20] B. L. Yeo and B. Liu. Rapid scene analysis on compressed videos. *IEEE Trans. Circ. Syst. Video Tech.*, 5(6):533–544, 1995.

[21] D. Lelescu and D. Schonfeld. Statistical sequential analysis for real-time video scene change detection on compressed multimedia stream. *IEEE Trans. Multimed.*, 5(1):106–117, 2003.

[22] T. Aach, A. Kaup, and R. Mester. Statistical model-based change detection in moving video. *Signal Process.*, 31(2):165–180, 1993.

[23] R. Mech and M. Wollborn. A noise robust method for 2D shape estimation of moving objects in video sequences considering a moving camera. *Signal Process.*, (special issue), 66(2):203–217, 1998.

[24] M. Irani, B. Rousso, and S. Peleg. Computing occluding and transparent motions. *Int. J. Comput. Vis.*, 12(1):5–16, 1994.

[25] A. Neri, S. Colonnese, G. Russo, and P. Talone. Automatic moving object and background separation. *Signal Process.*, (special issue), 66(2):219–232, 1998.

[26] S.-Y. Chien, S. Y. Ma, and L.-G. Chen. Efficient moving object segmentation algorithm using background registration technique. *IEEE Trans. Circ. Syst. Video Tech.*, 12(7):577–586, 2002.

[27] C. Gu, T. Ebrahimi, and M. Kunt. Morphological moving object segmentation and tracking for content-based video coding. *Proc. Int. Symp. Multimed. Comm. Video Coding*, New York, NY, October 1995.

[28] G. Adiv. Determining three-dimensional motion and structure from optical flow generated by several moving objects. *IEEE Trans. Patt. Anal. Mach. Intell.*, 7:384–401, 1985.

[29] P. J. Burt, R. Hingorani, and R. Kolczynski. Mechanisms for isolating component patterns in the sequential analysis of multiple motion. In *IEEE Workshop on Visual Motion*, 187–193, Princeton, New Jersey, October 1991.

[30] J. R. Bergen, P. J. Burt, K. Hanna, R. Hingorani, P. Jeanne, and S. Peleg. Dynamic multiple-motion computation. In Y. A. Feldman and A. Bruckstein, editors. *Artificial Intelligence and Computer Vision*, Elsevier, Holland, 147–156, 1991.

[31] J. R. Bergen, P. J. Burt, R. Hingorani, and S. Peleg. A three-frame algorithm for estimating two-component image motion. *IEEE Trans. Patt. Anal. Mach. Intell.*, 14:886–896, 1992.

[32] H. Sawhney, S. Ayer, and M. Gorkani. Model-based 2D and 3D dominant motion estimation for mosaicing and video representation. *IEEE Int. Conf. Comput. Vision*, Cambridge, MA, June 1995.

[33] W. B. Thompson. Combining motion and contrast for segmentation. *IEEE Trans. Patt. Anal. Mach. Intel.*, 2:543–549, 1980.

[34] D. W. Murray and B. F. Buxton. Scene segmentation from visual motion using global optimization. *IEEE Trans. Patt. Anal. Mach. Intel.*, 9(2):220–228, 1987.

[35] M. Hoetter and R. Thoma. Image segmentation based on object oriented mapping parameter estimation. *Signal Process.*, 15:315–334, 1988.

[36] P. Schroeter and S. Ayer. Multi-frame based segmentation of moving objects by combining luminance and motion. *Signal Process. VII, Theories and Appl. Proc. Seventh European Signal Proc. Conf.*, September 1994.

[37] J. Y. A. Wang and E. Adelson. Representing moving images with layers. *IEEE Trans. on Image Process.*, 3:625–638, 1994.

[38] S. Ayer and H. Sawhney. Layered representation of motion video using robust maximum-likelihood estimation of mixture models and MDL coding. *IEEE Int. Conf. Comp. Vision*, Cambridge, MA, June 1995.

[39] J.-M. Odobez and P. Bouthemy. Direct model-based image motion segmentation for dynamic scene analysis. *Proc. Second Asian Conf. Comput. Vision (ACCV)*, December 1995.

[40] Y. Weiss and E. H. Adelson. A unified mixture framework for motion segmentation: incorporating spatial coherence and estimating the number of models. *Proc. IEEE Int. Conf. Comput. Vision & Patt. Recognit.*, June, 1996.

[41] N. Kiryati, Y. Eldar, and A. M. Bruckstein. A probabilistic Hough transform. *Patt. Recognit.*, 24(4):303–316, 1991.

[42] S.-M. Kruse. Scene segmentation from dense displacement vector fields using randomized Hough transform. *Signal Process: Image Comm.*, 9:29–41, 1996.

[43] Y. W. Lim and S. U. Lee. On the color image segmentation algorithm based on the thresholding and the fuzzy c-means techniques. *Patt. Recognit.*, 23(9):935–952, 1990.

[44] Y. Altunbasak, E. Eren, and A. M. Tekalp. Region-based affine motion segmentation using color information. *Graphical Models Image Process.*, 60(1):13–23, 1998.

[45] N. Diehl. Object-oriented motion estimation and segmentation in image sequences. *Signal Process: Image Comm.*, 3:23–56, 1991.

[46] S. F. Wu and J. Kittler. A gradient-based method for general motion estimation and segmentation. *J. Vis. Comm. Image Rep.*, 4(1):25–38, 1993.

[47] S.-L. Iu. Robust estimation of motion vector fields with discontinuity and occlusion using local outliers rejection. *SPIE*, 2094:588–599, 1993.

[48] M. M. Chang, A. M. Tekalp, and M. I. Sezan. Simultaneous motion estimation and segmentation. *IEEE Trans. Image Process.*, 6(9):1326–1333, 1997. (also in Proc. ICASSP'94, Adelaide, Australia)

[49] P. B. Chou and C. M. Brown. The theory and practice of Bayesian image labeling. *Int. J. Comp. Vision*, 4:185–210, 1990.

[50] S. Hsu, P. Anandan, and S. Peleg. Accurate computation of optical flow by using layered motion representations. *Proc. Int. Conf. Patt. Recognit.*, Jerusalem, Israel, 743–746, October 1994.

[51] E. Dubois and J. Konrad. Estimation of 2-D motion fields from image sequences with application to motion-compensated processing. In M. I. Sezan and R. L. Lagendijk, editors. *Motion Analysis and Image Sequence Processing*, Kluwer, Norwell, MA, 1993.

[52] C. Stiller. Object-oriented video coding employing dense motion fields. *Proc. Int. Conf. ASSP*, Adelaide, Australia, 1994.

[53] Z. Kato, M. Berthod, and J. Zerubia. Parallel image classification using multiscale Markov random fields. *Proc. IEEE Int. Conf. ASSP*, Minneapolis, MN, V137–140, 1993.

[54] J. Shi and C. Tomasi. Good features to track. *IEEE Conf. Comp. Vision Patt. Recognit. (CVPR)*, Seattle, WA, June 1994.

[55] Y. Fu, A. T. Erdem, and A. M. Tekalp. Tracking visible boundary of objects using occlusion adaptive motion snake. *IEEE Trans. Image Process.*, 9(12):2051–2060, 2000.

[56] H. W. Park, T. Schoepflin, and Y. Kim. Active contour model with gradient directional information: Directional snake. *IEEE Trans. Circ. Syst. Video Tech.*, 11(2):252–256, 2001.

[57] P. Salembier. Morphological multiscale segmentation for image coding. *Signal Process.*, 38:339–386, 1994.

[58] A. M. Tekalp, P. J. L. van Beek, C. Toklu, and B. Gunsel. 2D mesh-based visual object representation for interactive synthetic/natural video. *Proc. IEEE* (special issue), 86(6):1029–1051, 1998.

[59] P. L. Correia and F. Pereira. Objective evaluation of video segmentation quality. *IEEE Trans. Image Process.*, 12(2):186–200, 2003.

[60] C. Erdem, B. Sankur, and A. M. Tekalp. Performance measures for video object segmentation and tracking. *IEEE Trans. Image Process.*, 13(7):937–951, 2004.

CHAPTER 7

Motion Tracking in Video

Nikos Nikolaidis, Michail Krinidis, Evangelous Loutas, Georgios Stamou, and Ioannis Pitas

Department of Informatics, Aristotle University of Thessaloniki, Box 451, 54124 Thessaloniki, Greece

7.1 INTRODUCTION

Motion tracking in digital video aims at deriving the trajectory over time of moving objects or, in certain cases, the trajectory of the camera. Tracking should be distinguished from object detection, that aims at estimating an object's position and/or orientation in a certain image. However, detection and tracking are not totally unrelated. As a matter of fact, detection is involved in one of the two major approaches that one can adopt to devise a tracking algorithm [1]. According to this approach, object detection is performed on each frame of a video sequence and, subsequently, correspondences between objects detected in successive frames are sought. Thus, the trajectory of each object is established. According to the second approach, that essentially combines the detection and correspondence finding steps, the objects positions, and/or orientations in the next frame(s) are predicted, rather than detected, using information derived from the current (or previous) frames.

The output of an object tracking algorithm depends on the application and the representation used to describe the object that is being tracked over time. Thus, this output can be, for example, the contour (silhouette) of the object, the 2D image coordinates of its center of mass, its 3D position in world coordinates, the posture of an articulated object (i.e., the set of joint angles that define the configuration in space of an articulated structure), and so forth.

Object tracking has received considerable attention in the past few years mainly due to the wide range of its potential applications. One important application domain is advanced human–machine interfaces, where tracking, along with human motion analysis and behavior understanding, plays a role complementary to speech recognition and natural language understanding [2]. In this context, gesture recognition, body and face pose estimation, facial expression analysis and recognition are employed in an effort to enable machines to interact more cleverly with their users and also with their environment. In all these tasks, tracking constitutes an essential part of the overall process.

Another important application domain is smart surveillance, obviously due to the large number of security-sensitive areas such as banks, department stores, parking lots, and so forth, that need to be monitored efficiently. Such a "smart" system does not simply detect motion, which alone might lead to false alarms, but it can also classify the motion (e.g., human or nonhuman motion) and perhaps perform face recognition and tracking for access control purposes, or wide field tracking across multiple-camera. Other systems proceed even further, performing human behavior analysis to detect suspicious behavior (e.g., in a car parking area, or in front of an ATM). In this way, the huge number of cameras already installed in security-sensitive areas could be used as a tool for efficient real-time automated or semiautomated surveillance, for example to alert security officers or human operators. Furthermore, human operators could search archived video for specific activity patterns without actually viewing the video. The benefits of such systems, however, could be counterbalanced by possible drawbacks, such as privacy violations, since sensitive and complex sociological issues are involved.

Virtual reality and computer animation have substantially benefited from motion tracking and analysis. To insert and animate an avatar in a virtual reality environment, one needs to "capture" the motion of the entity (e.g., a person) represented by the avatar in the real environment and use it to drive the avatar. Similarly, animating autonomous human-like characters in virtual reality or games is often performed by synthesizing human movements based on motion data acquired off-line by tracking the body and/or facial expressions of real humans. Another very powerful example of the successful application of the above can be found in the motion pictures industry, for example in movies involving both real and computer-generated actors, where the degree of realism of the virtual actors motion is surprisingly high. Training athletes and analyzing their performance can also benefit from vision-based tracking of human motion [3, 4]. Additionally, medical diagnosis of gait disorders and treatment support can be performed by gait analysis [5, 6].

Video coding is another important application domain of motion tracking. For example, when using a videophone, tracking the user's face in the video would make it possible to code it with more detail than the background and can result in more effective video storage and transmission. Object tracking is also involved in model-based coding. In the videophone application example, tracking can be used to derive pose and deformation parameters of a 3D head model at the transmitter side. These parameters are then sent to the receiver that uses them to animate a similar model. This way, the amount of information to be transmitted is limited to the vector of animation parameters over time. Traffic monitoring, namely road surveillance to detect abnormal situations, calculate statistics regarding traffic flow, and so forth is also largely based on tracking. Moreover, providing a vehicle with automatic path planning and obstacle avoidance capabilities often involves tracking. Finally, content-based querying, indexing, and retrieval in multimedia databases can also benefit from advanced object tracking techniques. For instance, motion path data obtained by tracking and analyzing the motion of players in sports video footage can be used for content-based indexing and retrieval of such data.

Video-based object tracking is just one of the many techniques devised for tracking objects. It belongs, along with microphone array sound source tracking, to the broader class of passive object tracking techniques that rely on measuring signals naturally emitted by the tracked object, such as light or sound [7]. Another broad category of tracking techniques is active object tracking, which involves placing devices (e.g., sensors or transmitters) on the object [8]. Active object tracking techniques include [9, 10]:

- mechanical trackers that are based on a kinematic structure (either serial or parallel), which is attached to the object (usually the human body) and consists of links interconnected with sensorized joints.
- inertia trackers, which are devices that consist of gyroscopes and accelerators and measure the translational acceleration, as well as the angular velocity of an object.
- ultrasonic trackers, where transmission and sensing of ultrasonic waves is used. The time taken for a brief ultrasonic pulse to travel from a stationary transmitter placed in the environment to a receiver attached to the moving object is measured and used to identify the object position or orientation.
- magnetic trackers, which are noncontact devices that use the magnetic field produced by a stationary transmitter to measure the position of a receiver placed on the moving object.
- radio and microwave trackers, where the time-of-flight of the corresponding type of waves from a stationary transmitter to a receiver attached on the object of interest is measured to determine the range of an object.
- hybrid trackers, which employ more than one of the above position/orientation measurement technologies to track objects more accurately than a single technology would allow.

The use of such devices greatly simplifies motion capture. However, active trackers are "intrusive" and mainly suitable for well-controlled environments. Therefore, passive trackers are preferable (but more difficult to devise) to active ones.

Computer vision researchers have been trying to achieve results comparable to active object tracking using video information for a long time, in an effort to produce widely applicable motion tracking systems, free of intrusive devices, able to function in uncontrolled (indoor or outdoor) environments. In some cases, video-based object tracking employing simple markers placed on the object of interest is used. Other systems employ light emitting diodes (LED). They are placed either on the moving object, with multiple stationary infrared or other cameras sensing the transmitted light, or in the environment (e.g., the ceiling) with cameras placed on the moving object sensing the emitted light. However, these two variations can be considered as belonging to the category of active object tracking technologies. We limit our discussion to computer-vision-based object tracking techniques that employ passive sensing without markers.

Various criteria can be used to provide a classification of object tracking algorithms. In the rest of this Section, a number of such classification schemes will be presented. The main features of object tracking algorithms will be described along with the various classification schemes.

Two important criteria which are adopted in this chapter, aiming at the coarse classification of the presented techniques, are the dimensionality of the tracking space and the structure of the object to be tracked. According to the first criterion, the object tracking algorithms are classified as 2D or 3D ones. 2D object tracking aims at recovering the motion in the image plane of the projection of objects that, in the general case, move in the 3D space. 3D object tracking, on the other hand, attempts to estimate the actual 3D object movement using the 2D information conveyed by video data captured by one or more cameras. The structure of the object to be tracked is another characteristic that affects the type of motion that needs to be estimated. Therefore, rigid and deformable object tracking refer to estimating the motion of rigid and deformable objects respectively, whereas articulated object tracking refers to estimating the motion of articulated objects, that is objects composed of rigid parts (links) connected by joints allowing rotational or translational motion in 1, 2, or 3 degrees of freedom. In other words, articulated motion can be defined as piecewise rigid motion, where the rigid parts conform to the rigid motion constraints, but the overall motion is not rigid [11].

A different classification of object tracking methods is based on their mode of operation. That is, tracking can either be performed online or off-line. Trackers in the former category can employ information about the object coming from one or more previous frames to predict its location in the current frame (i.e., information from future frames is not available). Trackers from the latter category can potentially make use of the entire image sequence, prior and posterior to the frame of interest. Off-line object tracking can potentially provide better results because more information is available. However, future frames are often not available. Furthermore, utilizing this extra information comes at the expense of increased computational load.

Tracking over time involves matching objects in consecutive frames using some kind of information. Essentially, object tracking methods attempt to identify coherent relations of image information parameters (position, velocity, color, texture, shape, etc.) between frames. Therefore, an alternative classification of object tracking algorithms can be based on the type of information they employ for matching. Another important characteristic of object tracking algorithms is whether a model (geometric or other) of the object that needs to be tracked (e.g., a human body model when tracking people in video sequences) is used. Consequently, tracking algorithms can be classified to model-free or model-based ones. The decision to use a model, its type (2D image template, 3D volumetric or surface geometry model, color distribution model, etc.), as well as its complexity depend on the application. For example, in surveillance applications [12, 13], 3D geometry models of the object to be tracked are hardly necessary, because the parameters of interest involve only the presence and the spatial position of humans. In contrast, in a motion capture application aiming at obtaining data for the animation of a virtual actor, a detailed 3D face and body model is required.

Particular attention has been paid to motion models, especially for humans. On one hand, humans move in complex and rich patterns. On the other hand, many of the human activities involve highly structured motion patterns. Some motions are repetitive in nature (e.g., walking or running [14]), while others represent "cognitive routines" (e.g., crossing the street by checking for cars to the left and to the right). It is safe to assume that, if such models could be identified in the images, they would provide strong constraints for the tracking process, with image measurements being used to fine tune the estimates of the object motion parameters. However, this is not a trivial task. Human activities are often affected by unforeseen factors that are usually impossible to recover from image data. Also, several activities combine more than one motion models. Attempts have been made to tackle this problem (learning individual motion models, switching models, etc.), but at the expense of computational complexity.

It is often the case that tracking objects in consecutive frames is supported by a prediction scheme. Based on the information extracted from previous frames (and any high-level information that can be obtained), the state of the object (e.g., its location) is predicted and compared with the state of objects identified in the actual frame in question. Regardless of the type and number of parameters used to describe the object state, a model of its evolution in time is required. An excellent framework for prediction is the Kalman filter [15–17], which additionally estimates prediction errors. In complex scenes, however, it is most likely that deriving a single hypothesis for the next state of the object is impossible. For this reason, alternative prediction schemes have been devised that are capable of keeping track of multiple hypotheses, such as the well-known Condensation algorithm [18]. This leads to another possible taxonomy of object tracking algorithms, that is single-hypothesis versus multiple-hypothesis trackers.

Devising a tracking algorithm that is capable of deriving accurate object motion information under all possible conditions and environments is a very challenging task, because it requires tackling a number of difficult issues that include but are not limited to projection ambiguities, occlusion and self-occlusion, unconstrained motion, clutter, poor or varying lighting conditions, use of a single-camera, the deformable clothing of humans (which produces variability in the body shape and appearance), and so forth. These difficulties have led researchers to adopt a number of assumptions or constraints to focus on tackling specific aspects of an overall very complex problem. Assumptions can be either related to the motion of the camera or subjects (fixed camera, single-person scenes, occlusion-free scenes, known motion models, e.g., front-to-parallel movement with respect to the camera, etc.) or refer to the appearance of the environment (constant lighting conditions, uniform or static background, etc.) or the subject(s) (tight clothing, tracking of a specific type of objects, e.g., cars, etc.). Any of the aforementioned assumptions can provide constraints that can greatly facilitate the solution of the tracking problem. The constraints incorporated in an object tracking algorithm can be used for its characterization. Therefore, tracking algorithms can be characterized on the basis of whether they focus on tracking specific objects, such as car tracking or tracking of human body parts (face, hand, etc.) or not, the number of views available (single-view, multiple-view, and omnidirectional view tracking techniques), the state of the camera (moving vs. stationary), the tracking environment (indoors vs. outdoors), the number of tracked objects (single object, multiple objects, groups of objects), and so forth.

An important step prior to applying any tracking algorithm is proper initialization. This can be performed off-line or online and aims at recovering information about the camera and/or the scene and/or the object to be tracked [7]. The first is most often dealt by off-line camera calibration that identifies the intrinsic (focal length, radial distortion, etc.) and the extrinsic (scene geometry) camera parameters. This, however, requires a fixed camera setup. If the camera setup changes, recalibration is necessary. Online (or self) calibration is also possible [19, 20]. Recovering information about the scene can be important for subsequent tracking. For example, in tracking algorithms where the initial object position in the image is found by background subtraction (see Section 7.2.1.1), the initialization step might include capturing the background reference images. Finally, the third initialization goal might include the initialization of the model used in a model-based object tracking algorithm or the estimation of the initial pose and position of the object. Obviously, the scope of the initialization step differs between different tracking algorithms. For example, in a 2D contour-based object tracking method, the initialization could be an object detection step, aiming at finding the contour of the object in the first frame of the video sequence. If the method is a feature point-based one, the initialization should provide the 2D coordinates of all the object feature points that are to be tracked.

Another important feature of a tracking algorithm is its ability to handle occlusions. Occlusion is usually distinguished in partial and total occlusion, where the object of interest is partially or totally occluded by another object (fixed or moving). Self-occlusion is also of particular interest. In this case, parts of the object are occluded by the object itself (e.g., limb occlusion when walking humans are being tracked or face occlusion caused by hand gestures during a conversation). Several object tracking techniques assume no occlusion at all whereas others provide occlusion handling mechanisms. One way of handling occlusion is by means of reinitialization. In this case, the initialization phase (e.g., object detection) is applied periodically or at certain instances to detect (and then continue to track) objects that might have been occluded. This usually comes at a computational expense that may affect the real-time capabilities of the algorithm. Alternatively, object position prediction schemes could be employed for the whole duration of the occlusion.

This chapter aims to describe the basic principles behind 2D and 3D object tracking algorithms and provide a basic literature overview of the subject. However, the literature is very rich. Therefore, certain categories of object tracking methods, such as deformable object tracking (used for example in facial expression tracking), are not thoroughly covered. Additional information and further details about the topics described in this chapter can be found in the excellent books [21, 22] and reviews [1, 2, 7, 11, 23–28] that have appeared in the literature.

7.2 RIGID OBJECT TRACKING

7.2.1 2D Rigid Object Tracking

2D rigid object tracking tries to determine the motion of the projection of one or more rigid objects on the image plane. This motion is induced by the relative motion between the camera and the observed scene. A basic assumption behind 2D rigid motion tracking is that there is only one, rigid, relative motion between the camera and the observed scene [29]. This is the case of for example a moving car. This assumption rules out articulated objects, like a moving human body, or deformable objects like a piece of cloth.

Methods for 2D rigid object tracking can be classified in different categories according to the tools that are used in tracking:

- region-based methods,
- contour-based methods,
- feature point-based methods,
- template-based methods.

2D rigid object tracking methods sometimes constitute the basic building blocks for other categories of tracking algorithms. For example, an articulated object tracking algorithm may include a rigid object tracking module in order to track the rigid parts that make up the articulated structure. In the following, we will review the basic principles of the main categories of 2D rigid object tracking algorithms, describe the Bayesian framework frequently employed in object tracking algorithms and discuss the crucial topic of occlusion handling.

7.2.1.1 Region-Based Object Tracking

This is usually an efficient way to interpret and analyze motion observed in a video sequence. An image region can be defined as a set of pixels having homogeneous characteristics. It can be derived by image segmentation, which can be based on distinctive object features (e.g., color, edges) and/or on the motion observed in the frames of a video sequence. Essentially, a region would be the image area covered by the projection of the object of interest onto the image plane. Alternatively, a region can be the bounding box or the convex hull of the projected object under examination.

Color information proved to be effective in region-based object tracking, because it enables fast processing that can lead to real-time tracking (i.e., 20–30 frames per second) while providing robust results. Color segmentation is the core of color-based object tracking algorithms. If the color of the object to be tracked can be modeled efficiently and distinguished from the color of other objects in the scene and the color of the background, it can be a very useful tracking cue, combined of course with an appropriate color similarity metric. The major issue with color segmentation and tracking is how to provide robustness against illumination changes. This can be achieved for example by controlling the illumination conditions, which is, of course, impossible in real world environments especially for outdoor scenes. Alternatively, illumination invariance or color correction can be employed. The former aims at representing color information in a way that is invariant to illumination changes, whereas the latter attempts to map the color responses of a camera obtained under unknown illumination conditions to illumination independent descriptors. For example, one common way to try to obtain a certain degree of illumination invariance is by using only the chromaticity values in a suitable color space (e.g., the chromaticity components H,S in the Hue-Saturation-Value space). Alternatively, one can normalize the color space. For example, in the RGB space, this would mean that instead of using the R,G,B component values, one can use the values $\frac{R}{R+G+B}$, $\frac{G}{R+G+B}$, and $\frac{B}{R+G+B}$ respectively (which is the widely used normalized RGB space) or the values $\frac{R}{\max\{G,B\}}$, $\frac{G}{\max\{R,B\}}$, $\frac{B}{\max\{R,G\}}$ [30]. For color constancy, a number of well-known algorithms, such as variants of the Grey World Algorithm [31], color by correlation [32], and so forth can be applied.

Many tracking algorithms focus on tracking humans (the whole body or body parts, such as hands, face, etc.). It becomes obvious that, in this case, the distinctive color of the human skin can serve as an appropriate means of locating and tracking people in video sequences (Fig. 7.1). A color segmentation algorithm can be devised in three steps, namely the choice of a suitable color space, the modeling of the skin (or any other object) color distribution over the selected color space, and the method used to classify the individual pixels to object (skin) and nonobject (nonskin) pixels. Methods for color modeling can be roughly classified as parametric (using a single Gaussian or a mixture of Gaussians) and nonparametric (histogram-based, such as lookup tables and Bayes skin probability maps). The selection of an appropriate color space has been proven to be coupled to the model used for the skin color distribution. For parametric techniques, it was determined that spaces normalized with respect to illumination (e.g., normalized RGB) perform better with the single Gaussian model, whereas mixtures of Gaussians can produce comparable results when applied to spaces with no illumination normalization [33]. For nonparametric modeling methods, the HS color spaces (HSV, HSI, HLS, etc.), which are inherently related to the human perception of color, perform better when employed in methods based on lookup tables. On the other hand, the choice of color space does not affect methods based on skin probability maps [34]. Experimental studies have determined that, for a specific camera, the skin color distribution forms a cluster (the so-called skin locus) in various color spaces. The authors in [35] argue that for every color space, there exists an optimum skin detector scheme, such that the performance of all these skin detector schemes is the same, thus rendering the choice of a specific color space irrelevant.

FIGURE 7.1

Skin detection based on simple thresholding in the Hue-Saturation components of the HSV color space.

FIGURE 7.2

A typical environment for color-based object tracking using chroma keying. Image source: SurfCap project, CVSSP, University of Surrey, UK [36].

A simple scenario that allows robust color-based object tracking is chroma-keying. In this case, the background is single-colored (most often blue) and the object colors are very different from the background (Fig. 7.2), or the object is single-colored (e.g., a moving person that wears single-color clothes, most often dark-colored) and this specific color does not appear in the background. In both cases, simple color thresholding can be employed to separate the object from the background and track it. The same technique can be used along with markers of distinctive color that have been placed on the object.

One of the most often used approaches in color-related region-based object tracking is the one that involves color histograms [37–39], which falls in the category of nonparametric techniques. The color histogram of the i-th region A_i in an image is denoted by $\mathbf{O}_i$. In the initialization step, color histograms of all the objects (i.e., regions) of interest in the scene are computed from a number of frames of a video sequence and stored in a database as reference color histograms, denoted by $\mathbf{O}_i^r$. These reference (model) histograms are later used in the matching process. In each new frame of the video sequence, for each of the tracked objects, a color histogram, denoted by $\mathbf{O}_i^t$, is calculated for every candidate object position. Each derived histogram, that is target histogram, is compared against the reference color histogram of the object to determine the best match and find the position of the tracked object in the current frame. Various criteria, which depend on the specific algorithm employed, are used to measure histogram distance or similarity. These include the histogram intersection measure [37] that performs a bin-by-bin comparison between two histograms and returns a relative match score based on the portion of pixels that are found in the same color bin of each histogram. It can be defined as:

$$\bigcap(\mathbf{O}_i^t, \mathbf{O}_i^r) = \sum_{n=1}^{U} \min\left\{\mathbf{O}_{i,n}^t, \mathbf{O}_{i,n}^r\right\}, \tag{7.1}$$

where U is the number of bins in the histogram and n is the corresponding bin index. The sum of squared differences (SSD) can also be employed, that is:

$$\mathrm{SSD}\left(\mathbf{o}_i^t, \mathbf{o}_i^r\right) = \sum_{n=1}^{U} \left(\mathbf{o}_{i,n}^r - \mathbf{o}_{i,n}^t\right)^2. \tag{7.2}$$

It is obvious that the reference histograms described earlier can handle only a fixed color distribution, thus not being able to account for changes in illumination that could potentially lead to tracking drift or failure. To overcome this, the number of bins of the reference histograms as well as the bin content can be dynamically updated at regular intervals during the tracking process, using color information from the frames of the video sequence. This, however, comes at the expense of computational burden. The selected color quantization level can affect the result, that is the accuracy of determining the object position in the frame under examination. Histograms have been shown to be effective only when the number of bins is neither too low nor too high.

In [40], a region-based technique that uses weighted histograms of features to represent a target (region to be tracked) is proposed. Color histograms in the RGB space were used in the implementation presented in this chapter. The main contribution of this work is the use of a spatial isotropic kernel whose values decrease as we move away from the center. This kernel is applied during the evaluation of the color histogram of the pixels in the ellipsoidal image region used to represent the target. By doing so, region pixels away from the center contribute less to the evaluation of the histogram. Assuming that the weighted histogram that models the target (target model) is known and that we have evaluated the target position (x, y) in the current frame, the ellipsoidal region in the next frame whose weighted histogram exhibits maximum similarity with the target model is sought around (x, y). Histogram similarity is evaluated through a distance based on the Bhattacharyya coefficient. Because of the utilization of the kernel, the distance function varies smoothly within the search region and thus finding the position that minimizes this distance can be performed through a gradient optimization method. The mean shift procedure [41] is used for this purpose. It should be noted that the target model can be periodically updated to cope with changes in the color of the target, (for example, due to illumination changes). The proposed real-time tracker successfully coped with camera motion, partial occlusions, clutter and target scale variations.

For the parametric techniques [42–47] that use parametric color reference models, the operating principle is the following: the current frame is searched for a region, namely a window of variable size but fixed shape, whose color content best matches a reference color model, for example a mixture of Gaussians [47]. More specifically, in this case, the color distribution of the object is considered as multimodal and, as such, is approximated by a number of Gaussian functions in some color space, for example the Hue-Saturation color space. Starting from the object location in the previous frame, the method proceeds iteratively at each frame so as to minimize a distance measure to the reference color model. Because object color can often change due to the illumination conditions, (i.e., the same object can be perceived as having two different colors when such conditions change abruptly), the model is adapted to reflect the changing appearance of the tracked object.

A statistical approach is employed in which color distributions are estimated over time by sampling from the object pixels to obtain a new pixel set that is used to update the Gaussian mixture model.

An example of an easily implemented region-based object tracking algorithm that relies on color is the one introduced in [48]. The algorithm tracks colored regions from frame to frame. To cope with color changes due to illumination changes, the color space used is normalized. The object to be tracked is divided into L image regions $R_1, \ldots, R_L$. The regions are assumed to be fixed with respect to size and relative position. When initializing the algorithm, every region R_i is assigned a reference color vector $\bar{\mathbf{g}}_i$, which represents the averaged color of pixels within the region. A color vector $\mathbf{g}_i$ is also computed in a similar manner for each region in every frame of the video sequence. The reference and the computed color vectors are compared using a goodness of fit criterion. For each of the three color components (RGB), the ratio of the component values in the computed and the reference color vectors (e.g., $\frac{r_i}{\bar{r}_i}$ for the red component) is computed. The goodness of fit criterion is then chosen as the ratio of the maximum of these three ratios to the minimum of the ratios. Values close to 1 correspond to good matches.

Another region-based tracking system that employs color as a tracking cue is Pfinder, which was presented in [49]. The system is capable of tracking a single person in scenes with complex background, captured by a fixed camera. It uses blob representations, that is coherent connected regions where the pixels have similar image properties. For each pixel in a blob, its spatial coordinates (x, y), along with its textural (color) components are used to form a feature vector. The corresponding spatial and color distributions are assumed to be independent. In an initialization step, the algorithm builds a model for the background (see the next paragraph for details). When a person enters the scene, large changes in the scene are used to detect it and a statistical blob model is constructed. For subsequent frames, the spatial distribution of each blob is predicted using a Kalman filter. Then, for each pixel, the likelihoods that this pixel is a member of one of the blobs are evaluated and used to assign this pixel to a certain blob or the background. Finally, the statistical models of each blob are updated.

Another popular region-based object tracking approach is based on background subtraction [1, 50–54]. In this class of techniques, and in contrast to the parametric object color models presented earlier, the interest is in building a model of the background rather than of the object. In the most naive implementation of this approach, the model of the scene is a single image that is acquired without the presence of any moving object, during an initialization step. It is then assumed that the background remains static during the acquisition of the video sequence. Each new frame is subtracted from the scene model to segment any foreground (i.e., moving) objects. However, the subtraction itself is not sufficient to obtain accurate information because of the noisy measurements captured by the camera, as well as changes in the scene environment, for example illumination changes, slow, repetitive motion of scene elements such as waves in water, moving clouds, and so forth. This problem is obviously more prominent in outdoors video sequences and various more sophisticated approaches to background subtraction have been proposed. In [49], the color of each pixel of the background is modeled by a different 3D Gaussian whose mean and covariance are evaluated from a number of consecutive frames that depict only the background. Thus a model of the background is built. This model is continuously updated within the actual operation of the tracking system using a simple adaptive filter. For each pixel of a video sequence (captured with the same camera), the likelihood that its color stems from the color distribution of the pixel with the same location in the background model is evaluated. If the likelihood is small the pixel is labeled as foreground pixel. A more complex background model is presented [51]. In this model the color of each pixel in the recent history is modeled as a mixture of three to five Gaussians. Each of these Gaussians can represent either a background or a foreground "pixel process" (i.e., a time series of the values of the pixel over time). Gaussians with a large weight and small variance are considered as modeling the background and the rest are considered as modeling the foreground. The color of a pixel (x,y) of a certain frame is checked against the existing Gaussians of the model for this pixel. If the pixel matches a background (respectively foreground) distribution it is labeled as a background (respectively foreground) one. The parameters (mean and variance) of the matching distribution are updated to account for the new pixel. Also, if the pixel matches no existing distribution a new Gaussian whose mean is equal to this pixel's color is inserted into the mixture, replacing the least probable existing one. Foreground pixels are grouped into regions through a connected components algorithm. A multitude of other approaches that, for example, try to incorporate spatial [55] information along with color, or involve autoregressive moving average ARMA processes to model and predict variations in a scene [56, 57] have been proposed. However, one should have in mind that all background subtraction methods are applicable only in situations where the camera is static and the background is slowly varying.

A simple real-time tracking algorithm for tracking people that is based on background subtraction is presented in [58]. The method uses just one image as the background model and performs no updates. Foreground pixels (i.e., pixels that correspond to humans) are detected in each frame using the luminance contrast between the frame and the background image. These pixels are then grouped into blobs, represented as bounding boxes. Blobs detected in two consecutive frames are matched using their overlap as a matching criterion. Tracking and handling of cases such as blob splitting/merging and blob entering/leaving the scene is performed through the so-called matching matrices that carry information regarding the correspondences between blobs in the two frames.

In [59], tracking is treated as a classification problem where an ensemble of classifiers are constantly trained to distinguish the tracked object (represented as a rectangular image area) from the background. A feature vector is constructed for every pixel of the current frame. This vector consists, for example, from the pixel's R,G,B values and bin values of a histogram of oriented gradients calculated on a window centered at the pixel. An ensemble of weak classifiers is trained using positive and negative examples (i.e., feature vectors from object and background pixels) to separate the two entities (object/background). Then, a strong classifier, formed using Adaboost [60], is used to classify the pixels in the next frame and build a confidence map. The mean shift algorithm [41] is employed to find the peak of the confidence map, which is the new position of the object in the next frame. Then, to take into account the object and background changes over time, the best weak classifiers are kept whereas the rest are replaced by new ones, trained on the new data. This procedure is applied repeatedly.

In a similar vein, a two-class (object/background) support vector machine (SVM) classifier [61] is combined with a tracker based on optic-flow in [62]. The method can track only objects it has been trained for. An example application of this method on the tracking of the rear end of vehicles from sequences captured with a camera mounted on a moving vehicle is presented in this chapter. The SVM is trained off-line with positive examples (edge images of rear-ends of a multitude of different cars) and negative examples (edge images of background scenes). Instead of performing tracking by trying to find the transformation parameters that, when applied on the tracked region on the current frame, minimize the difference between this region and the resulting region in the next frame (as in the case of optic-flow based tracking), the method tries to find the region in the next frame for which the trained SVM classifier provides the largest score for being a member of the object class. The advantage of such an approach is that tracking utilizes all knowledge stored in the SVM during the training procedure.

Most region-based tracking algorithms (including those presented in this Section) use a fixed feature vector (e.g., color) for the background/foreground (object) separation and object tracking. In contrast to this, the authors in [63] proposed a method that selects the best set of features for tracking and continuously adjust this set as the tracking proceeds. The authors argue that the features that best distinguish between the object and its background are the best features for tracking, since, similar to [59] tracking is viewed as a local discrimination/classification problem between the object and the background. A pool of features is used for the selection of the most appropriate ones. For the implementation presented in this chapter, the authors use 49 different linear combinations of the three primary colors R, G, and B (for example R, R-B, R+G+B, etc.). For a given current frame, where tracking has already been performed (i.e., the object has been separated from the background), the histograms of the object and background pixels (i.e., the class conditional sample densities) are evaluated for each feature and their log likelihood ratio is evaluated. This step essentially produces (for each feature) a function F that maps feature values associated with the object to positive values and feature values associated with the background to negative values. This function can be interpreted as a nonlinear transformation of each feature into a new, so-called "tuned" feature that is tailored to discriminate the object from the background in the frame under examination. Then, the two-class variance ratio is used to rank all features according to how well they separate the sample distributions of object and background pixels and the top N features are selected. By applying the F functions of the selected features on the pixels of the next frame, a weight map where object pixels are ideally labeled with positive values and background pixels with negative ones is produced for each of the N selected features. The mean shift algorithm [41] is then applied on each of the N weight maps to obtain N estimates for the location of the object. These estimates are combined to provide the final estimate of the object in the next frame. The same procedure is applied iteratively to all frames.

Many other region-based tracking techniques such as those presented in [64–68] can be found in the corresponding literature.

7.2.1.2 Contour-Based Object Tracking

An alternative way of devising an object tracking algorithm is by representing the object using outline contour information and tracking it over time, thus retrieving both its position and its shape. Such a modeling method is more complicated than modeling entire regions, for example using color. However, contour-based tracking is usually more robust than region-based object tracking algorithms, because it can be adapted to cope with partial occlusions and outline information is, in most cases, more insensitive to illumination variations. Contour tracking has found numerous applications in surveillance, medical diagnosis, and audiovisual speech recognition. Apart from tracking rigid objects, contour-based tracking can also be used for tracking deformable objects.

Active contours, also known as "snakes," have been extensively used by researchers to perform object segmentation and tracking. An active contour algorithm dynamically deforms a contour to "lock" onto image features such as lines, edges, boundaries, and so forth. Active contours were first introduced in [69], for robust segmentation and region tracking. Snakes consist of elastic parametric curves whose deformation is subject to internal forces (contour elastic forces) and external forces (due to image content and other constraints). More formally, an active contour is a collection of n points in the image plane that define a polygonal line:

$$\Psi(\gamma) = \psi(\gamma)_1, \ldots, \psi(\gamma)_n, \tag{7.3}$$

$$\psi(\gamma)_i = (x(\gamma)_i, y(\gamma)_i), \quad i = 1, \ldots, n. \tag{7.4}$$

Because active contour models are a special instance of deformable models [69], we can also define them over a space of allowed contours using a functional to be minimized. The latter represents the energy of the model and, as already mentioned, in its initial formulation, consists of two terms:

$$E_{\text{total}}(\Psi(\gamma)) = E_{\text{int}}(\Psi(\gamma)) + E_{\text{ext}}(\Psi(\gamma)), \tag{7.5}$$

where E_{int} is an energy function dependent on the shape of the contour and E_{ext} is an energy function that depends on the image features of interest and other user-defined constraints. These functions are defined by:

$$E_{\text{int}} = \int_0^1 (\phi_1 \|\Psi'(\gamma)\|^2 + \phi_2 \|\Psi''(\gamma)\|^2) d\gamma \tag{7.6}$$

and

$$E_{\text{ext}} = \int_0^1 F(\Psi(\gamma)) d\gamma, \tag{7.7}$$

where $\Psi'(\gamma)$, $\Psi''(\gamma)$ are the first and second order derivatives of the contour with respect to γ and ϕ_1 and ϕ_2 are constant or dynamically changing parameters associated with the internal energy that control the behavior of the snake. More specifically, ϕ_1 controls the "tension" of the contour (i.e., its ability to resist to stretch), whereas ϕ_2 controls the "stiffness" of the contour (the flexibility and the smoothness of the snake). Also, F in

(7.7) is the potential related to the external forces applied to the snake. It depends on the object features that we are interested in. For example, if edges are the features of interest, it can be defined as [69]:

$$F = -\|\nabla I\|^2, \tag{7.8}$$

where I is the intensity of the image. Such a formulation for F causes the contour to be attracted to edges.

The points on the active contour iteratively approach the boundary of an object through the solution of the energy minimization problem (7.5). Many object tracking algorithms use this concept, for example [70]. When the internal and the external forces applied to the snake counterbalance each other, the snake is said to have reached its equilibrium state. The equation describing the equilibrium state is:

$$\left(\frac{\partial(\phi_1\Psi)}{\partial\gamma} - \frac{\partial^2(\phi_2\Psi)}{\partial\gamma^2}\right) + \nabla F = 0. \tag{7.9}$$

A snake can be made to converge towards specific image features by a suitable choice of F and the internal shape parameters ϕ_1, ϕ_2. To perform contour tracking, the minimization of (7.5) is repeated for successive frames. Because snakes are sensitive to initialization (i.e., the snake has to be initially placed close to the object that needs to be outlined, otherwise it will fail), a temporal prediction module capable of estimating the position of the object outline in the next frame is often used in conjunction with the snake algorithm, thus achieving, at the same time, a reduction in the computational complexity of the snake algorithm.

A robust algorithm for tracking the visible boundary of an object in the presence of occlusion is presented in [70]. First, an initial outline of the object contour is specified by the user and is automatically refined by using intra-energy terms. Then, a number of node points are selected along the contour to define the initial snake. Afterwards, the snake is segmented into nonoverlapping pieces by selecting a set of nodes whose local curvature or color variation exceeds a predefined threshold. For each contour segment, two predicted locations are obtained, the other based on local motion vectors within the object and the other based on motion vectors on the background side of the contour. The next location is the one that results in the smaller prediction energy. Finally, the predicted contour is refined using interframe and intraframe energy terms.

In Wang et al. [71], objects are represented at time t by an image curve $\chi(\mu, t)$ parameterized in terms of B-splines defined by a number of control points:

$$\chi(\mu,t) = (\Theta(\mu)\cdot\Gamma^x(t), \Theta(\mu)\cdot\Gamma^y(t), \text{ for } 0 \le \mu \le L), \tag{7.10}$$

where $\Theta(\mu)$ is a vector $[B_0(\mu),\ldots,B_{N_B-1}(\mu)]^T$ of B-spline basis functions, Γ^x and Γ^y are vectors of control points coordinates, and L is the number of spans. The positions of control points are continuously adjusted to have them evenly distributed and prevent them from crashing together or moving far away from one another. Moreover, the control points are not allowed to cross over during the tracking process. Temporal prediction is based on the movement of the centroid of the object and is performed using a simple prediction filter. Another approach, based on B-splines, is presented in [72], where Bascle et al. integrate the ideas of snake-based contour tracking and region-based motion analysis. They use a snake to track the region outline and perform segmentation. Afterwards, the motion of the extracted region is estimated by a dense analysis of the apparent motion over the region, using spatiotemporal image gradients.

A different approach is presented in [73]. This algorithm uses graph cuts based on active contours to track object contours in video sequences. The minimum cut of a graph G that separates a source set $\{s_1, s_2, \ldots, s_n\}$ and a sink set $\{t_1, t_2, \ldots, t_m\}$ is exactly the $s - t$ minimum cut of the resulting graph after identifying $s_1, s_2, \ldots, s_n$ to a new source s and identifying $t_1, t_2, \ldots, t_m$ to a new sink t [74]. The algorithm separates the area of interest into an inner and an outer boundary, represents the information contained in the region of interest by an adjacency graph, identifies all the nodes on the inner boundary as a single source s and all the nodes on the outer boundary as a single sink t. It then proceeds to compute the $s - t$ minimum cut and repeats the whole procedure until the algorithm converges, that is, given an initial boundary near the object under consideration, the method can iteratively deform to match the desired object boundary. The result in each frame is used to initialize the contour in the next frame. The method utilizes both the intensity information within the current frame and the intensity difference between the current and the previous frame to find the next position of the object contour. An application of the method in object segmentation is described in [74].

A Bayesian framework for contour tracking is presented in [75]. The method employs an energy functional that contains two energy terms. The first term is the image energy that is evaluated in a band around the contour and is based on color and texture features. The latter are obtained from the subbands of a steerable pyramid representation of the image. The aforementioned visual features are modeled by semiparametric models and are fused using the so-called independent opinion polling strategy. The second term is the shape energy that is based on the past contour observations and preserves the shape of the object in case of occlusions. A shape model is learned from nonrigid contour deformations as tracking proceeds. Tracking is achieved by deforming the contour through the minimization of the energy functional using a gradient descent approach. A level set representation of the contour is used, because of its flexibility in splitting and merging events. The experimental results provided by the authors show that the method can indeed handle successfully full and partial occlusions.

Other interesting contour-based tracking techniques are presented in [76–81].

7.2.1.3 *Feature Point-Based Object Tracking*

Feature point-based object tracking can be defined as the attempt to recover the motion parameters of a feature point in a video sequence, more specifically the parameters associated with the planar translation of a point, because points in 2D space neither rotate nor translate with respect to depth. More formally, let $A = \{A_0, A_1, \ldots, A_{N-1}\}$ denote the N frames of a video sequence and $\mathbf{m}_i(x_i, y_i)$, $i = 0 \ldots N-1$ denote the positions of the same feature point in those frames. The task at hand is to determine a motion vector

$\mathbf{d}_i(d_{x,i}, d_{y,i})$ that best determines the position of the feature point in the next frame, $\mathbf{m}_{i+1}(x_{i+1}, y_{i+1})$, that is: $\mathbf{m}_{i+1} = \mathbf{m}_i + \mathbf{d}_i$. The object to be tracked is usually defined by the bounding box or the convex hull of the tracked feature points.

Feature point-based object tracking, though more prone to individual outliers and tracking drift of individual points, can be implemented very efficiently [82]. An important problem in feature point-based tracking is to determine and, consequently, track salient feature points in an image. Potentially good feature points for tracking are those that have distinctive local characteristics, such as brightness, contrast, texture, and so forth. Such points can be, for example, those associated with high local curvature information, that is, corners, edges, and so forth. Robustness of the point neighborhood (or of the corresponding descriptor) to illumination variations and viewpoint changes is another desirable characteristic of good feature points. Many feature point (also known as key or interest point) detectors have been proposed in the literature. These include the Harris detector [83], the SIFT detector [84], and the detector proposed in [85] that will be described in the next paragraph. Alternatively, information originating from the type of object that is to be tracked can be exploited, as in [86], where a combination of the adaptive Hough transform, the block matching algorithm, and active contours is used to extract salient facial features and perform face tracking.

Many feature point-based tracking approaches exist [85, 87–97]. Many of these approaches are parts of a larger system, for example, of a face tracking or pose estimation system. Shi and Tomasi proposed a method for finding good features to track in [85]. For each candidate feature point, the following 2×2 matrix is constructed:

$$\mathbf{Z} = \begin{bmatrix} \sum_W J_x^2 & \sum_W J_x J_y \\ \sum_W J_x J_y & \sum_W J_y^2 \end{bmatrix}, \tag{7.11}$$

where J_x and J_y are the image gradients evaluated on the point under consideration in the x and y direction respectively and W is a $n \times n$ window centered on the candidate feature point. A good feature point is defined to be a point where the minimum eigenvalue of its matrix $\mathbf{Z}$ is larger than a predefined threshold. Such feature points represent corners or salt-and-pepper textures (Fig. 7.3). To measure the quality of image feature points during tracking and make sure that the same points are tracked throughout the video sequence, the authors employ a measure of feature dissimilarity that quantifies the change of appearance of a feature point between the first and the current frame.

The above feature point selection process is tightly coupled with the corresponding object tracking algorithm, described in [87]. In this algorithm, the displacement of a feature point is chosen so as to minimize the dissimilarity defined by the following double integral over the given window W, centered at the pixel under consideration:

$$\epsilon = \iint_W [\mathbf{A}_i(\mathbf{m} - \mathbf{d}) - \mathbf{A}_{i+1}(\mathbf{m})]^2 \, \omega(\mathbf{m}) d\mathbf{m}, \tag{7.12}$$

where $\mathbf{A}_i$ and $\mathbf{A}_{i+1}$ are two successive frames, $\mathbf{d}$ is the displacement vector and $\omega(\mathbf{m})$ is a weighting function (in the simplest case it is set equal to 1). As it was firstly introduced

FIGURE 7.3

Feature point selection based on [85].

in [98], when the interframe motion is sufficiently small, the displacement vector can be written approximately as the solution to a 2×2 linear system of equations:

$$\mathbf{Zd} = \mathbf{e}, \tag{7.13}$$

where

$$\mathbf{e} = 2 \iint_W [A_i(\mathbf{m}) - A_{i+1}(\mathbf{m})] g(\mathbf{m}) \omega(\mathbf{m}) d\mathbf{m}. \tag{7.14}$$

$\mathbf{Z}$ is the matrix (7.11), $\omega(\mathbf{m})$ is the weighting function and

$$g(\mathbf{m}) = \begin{bmatrix} \frac{\partial(A_i(\mathbf{m}) + A_{i+1}(\mathbf{m}))}{\partial x} \\ \frac{\partial(A_i(\mathbf{m}) + A_{i+1}(\mathbf{m}))}{\partial y} \end{bmatrix}. \tag{7.15}$$

A cross-correlation method was used in [96] to track feature points from frame to frame. Assuming relatively small displacements between adjacent frames, the tracking procedure follows the method described in [99]. The algorithm begins with a set of feature points selected on the initial frame. Every n frames, new feature points are selected to maintain the overall point number. For each feature point in the current frame, the algorithm searches (within a window) for its new location in the next frame. The criterion that is used is the maximum cross-correlation of square neighborhoods between the current and the next frame. Consequently, a refinement of the position of each feature point is performed based on the fact that all features belong to the same rigid object and, thus, they should move in a consistent way.

A tracking system that utilizes the pyramid-based Kanade–Lucas–Tomasi feature point tracking algorithm [100] and a foreground color distribution obtained through training is presented in [101]. Feature points are selected on the object to be tracked using the algorithm [85] presented earlier. The feature points are considered to be a flock of features, that is, a global constraint is enforced on the feature point locations that keeps them spatially combined. During the tracking process, the feature points must be close to each other and not far from the median feature point. Two thresholds define the allowable distances in each case. The algorithm is tested on human hand tracking. The color of the human hand is learned by a hand detection scheme that provides a normalized-RGB histogram. The CamShift algorithm [44] operates on this modality. This results in a map that assigns to each pixel a probability that its color is skin-like. This map is used during the selection and tracking of the feature points. Feature point locations with high skin color probability are preferred.

In [95], feature selection is based on Gabor wavelets, because they exhibit a number of desirable properties [102]. The convolution of an image A with Gabor wavelets leads to the Gabor wavelet transform of the image that is:

$$\hat{A}(x,y) = \int A(x',y')g(x-x',y-y')dx'dy', \tag{7.16}$$

where $g(x,y)$ is a 2D Gabor function:

$$g(x,y) = \left(\frac{1}{2\pi\sigma_x\sigma_y}\right)\exp\left[-\frac{1}{2}\left(\frac{x^2}{\sigma_x^2}+\frac{y^2}{\sigma_y^2}\right)+2\pi jWx\right], \tag{7.17}$$

where σ_x and σ_y are the standard derivations of $g(x,y)$ along the corresponding directions. One can obtain Gabor wavelets by scaling and rotating (7.17), essentially using Gabor functions with different frequency centers and orientations. If we use S' different frequencies and T different orientations, each image pixel will be associated with $S' \times T$ coefficients, whose amplitudes form the feature vector for this pixel. The energy of the coefficients is used in [95] to determine the discriminating power of each pixel and select the pixels to be tracked. A mesh is then created using the selected feature points. To track the object in the next frame of a video sequence, the authors employ a 2D golden section algorithm that calculates the best possible translation of the center of the formed mesh. To allow for deformation, they also allow the nodes of the mesh to perturbate locally. In both cases, appropriately defined similarity functions are used to determine the best possible match.

A deformable surface model was utilized to select and track characteristic feature points in [103]. This surface deforms to approximate the local image intensity surface (i.e., the surface obtained when considering the image intensity as the height in a height field defined over image coordinates x, y) around points to be tracked. Modal analysis is used to describe the surface deformation and the so-called generalized displacement vector, which appears in the explicit surface deformation governing equations, is used as the feature vector that characterizes the image neighborhood around a point. The position of the point in the next frame is the one whose generalized displacement vector best matches that of the point in the current frame. The generalized displacement vector is also used for the selection of "good" points on the object to be tracked. This vector can be proven to be a combination of the output of various line and edge detection operators, thus leading to distinct, robust feature points that can be tracked efficiently.

7.2.1.4 *Template-Based Object Tracking*

Template-matching techniques are used by many researchers to perform 2D object tracking. Template-based tracking is closely related to region-based tracking because a template is essentially a model of the image region (and, therefore, of the corresponding object) to be tracked. These approaches follow the same principles with the template matching techniques used in object recognition and detection. The first step (initialization step) is to select the template that will be used. Templates can be acquired prior to tracking in a number of ways. First, a template that is specific for a particular instance of a class of objects can be created. For example, in a face tracking module employed in a videophone application, the user might be asked to face the camera for a period of time, enabling the system to detect its face and use the corresponding image region as a template. A template can be alternatively created off-line by employing statistical methods. For example, in a face tracking application, a generic face template can be created by incorporating information from the various existing face databases, for example by evaluating an "average" face. Such a template can be obtained, for example, through the use of eigenfaces [104] that have been extensively used in face recognition, verification, and tracking [105] tasks. Eigenfaces are essentially the eigenvectors of the high-dimensional vector space of possible faces of humans. To generate a set of eigenfaces, a large set of face images are normalized (i.e., the eyes and mouth are aligned), resampled at the same pixel resolution (e.g., $q_1 \times q_2$ pixels), and then treated as $q_1 q_2$-dimensional vectors whose components are their pixel intensities. The eigenvectors of the covariance matrix of the face image vectors are then extracted. Because the eigenvectors belong to the same vector space as the face images, they can be considered as $q_1 \times q_2$ pixel face images (called eigenfaces). When properly weighted, they can be averaged together into a gray-scale rendering of an average human face (Fig. 7.4) that can be used as a template in face tracking. Instead of representing intensity or color values, the pixels of a template can be also assigned feature vectors that consist of values obtained through an image processing operation (e.g., morphological operations).

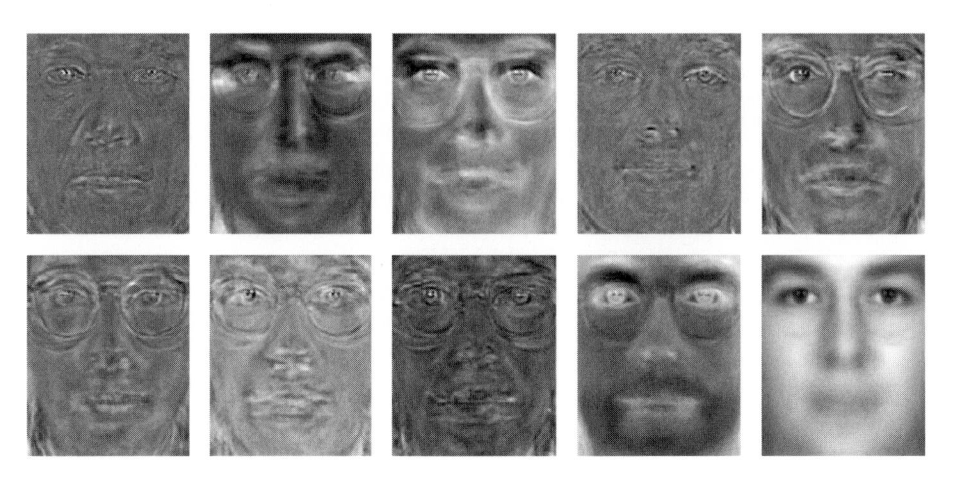

FIGURE 7.4

Sample eigenfaces computed from images of the M2VTS database [106] and the corresponding "average" face (lower right corner).

Template matching can be defined as the process of searching the target image (i.e., the current frame of the video sequence) to determine the image region that resembles the template, based on a similarity or distance measure. Essentially, the template region should undergo a geometrical transformation that would "place" it onto the target image in such a way as to minimize the distance measure used. The goal of a template matching algorithm is to estimate the parameters of such a transformation. More formally, let $\mathrm{Y}(t)$ be the image region corresponding to the object being tracked at time step t. For a rigid object, $\mathrm{Y}(t)$ can be found from the template Y_0 by employing an appropriate geometric transformation $\theta(\mathrm{Y}_0; \mathbf{q}(t))$, the parameters $\mathbf{q}(t)$ of which should be estimated by the algorithm. Affine (rotation, translation, scaling) [107, 108] or quadratic transformations can be employed. The object location in the current frame is determined by the vector $\mathbf{q}(t)$. Estimation of the transformation parameters is performed by identifying the image region that best matches the template. To avoid exhaustive search for the best match in the frame, various techniques can be used. Background subtraction can be employed to determine image regions where motion activity appears and limit the search in these regions. Alternatively, prediction schemes like Kalman filters can be used to estimate the location of the object being tracked in the next frame and use it as the center of a limited-size search region. Finally, prior knowledge of the scene and other constraints (e.g., constraints on the expected object displacement between consecutive frames) can restrict the search area.

Many similarity/distance metrics have been used in the template matching step. More specifically, if T_i is the brightness of the i-th pixel (x_i, y_i) in the template, $I_{i,v}$ is the brightness of the i-th pixel (x_i, y_i) in the image region v, and M is the number of pixels in the template, matching can be performed by finding the image region v that minimizes one of the following distance metrics:

- the Hamming distance (i.e., the number of "different" pixels in the template and the image region),
- the sum of absolute differences (SAD),

$$\sum_{i=1}^{M} |I_{i,v} - T_i|, \tag{7.18}$$

- the sum of squared differences (SSD),

$$\sum_{i=1}^{M} (I_{i,v} - T_i)^2, \tag{7.19}$$

or maximizes one of the following similarity criteria:

- the normalized correlation,

$$\frac{\sum_{i=1}^{M}(T_i - \bar{T})(I_{i,v} - \bar{I}_v)}{\sqrt{\sum_{i=1}^{M}(T_i - \bar{T})^2 \sum_{i=1}^{M}(I_{i,v} - \bar{I}_v)^2}}, \tag{7.20}$$

where $\bar{T}$ is the mean brightness of the template and $\bar{I}_\nu$ is the mean brightness of the image region ν.

- the joint entropy,

$$H(I_\nu, T) = -\sum_{i=1}^{K}\sum_{j=1}^{K} h(i,j)\log(h(i,j)), \tag{7.21}$$

where $h(i,j)$ are the values of the joint histogram of I_ν, T and K are the number of histogram bins in each of the two dimensions.

- the mutual information, which can be expressed as:

$$H(I_\nu) + H(T) - H(I_\nu, T), \tag{7.22}$$

where

$$H(I_\nu) = -\sum_{i=1}^{K} h(i)\log(h(i))$$

is the entropy of the image region I_ν ($h(i)$ being the values of the histogram of I_ν), $H(T)$ is the entropy of the template T and $H(I_\nu, T)$ is the joint entropy (7.21),

- the maximum likelihood [109].

Up to this point, it was assumed that template tracking is performed using a single template. However, due to the nonrigid nature of many natural objects or due to viewpoint changes, template tracking, as described so far, fails to provide satisfactory results in a number of real-world scenarios. For that reason, deformable template tracking methods have been introduced [110–112]. In these methods, prior knowledge of the object shape is used in an energy minimization scheme. Deformable templates are specified by a set of parameters that enable a priori knowledge about the expected shape of the object to guide the template matching process. The deformable templates interact with the image in a dynamic manner. An energy function is defined for the template, consisting of terms attracting the template to salient features (intensity, edges). The template parameters are obtained by a minimization of the energy function. In essence, the deformable template is obtained by allowing an original template to deform using any appropriate deformation function. The result should cover the various instances of the deformable object as much as possible, with minimum computational overhead while maintaining the attributes of the template (smoothness, connectivity, etc.).

An easily implemented template-based face tracking technique is described in [113]. The face template is acquired in a training step where the user is asked to look at the camera, rendering the method useful only for tracking the face of the person it was trained for. Multiple people cannot be tracked. The template is a facial image covering the eyebrows, the outer left and right edges of the eyes, and the bottom of the mouth. To cope with changes in face appearance due to head rotations, eight additional templates that correspond to rotations towards eight major directions (left, right, up, down, etc.) are constructed by warping the initial template. A match function that finds the minimum sum of absolute differences between the value of the red color component of each pixel in the image region and the appropriate face template is used for template matching.

A template-based object tracking algorithm that uses color invariant features (independent of the viewpoint, surface orientation, illumination direction, illumination intensity, and highlights) at each pixel position is presented in [108]. To account for sudden changes in illumination, the template is dynamically updated by means of robust filters like Kalman or particle filters. The filters are also used to estimate the position of the pixels comprising the object being tracked, as well as to handle short-time and partial occlusions. The distance metric used to compare the template and the candidate image regions is the Mahalanobis distance.

In [114], two template sizes are available but both sizes can be magnified by a scale factor in the x-th and y-th direction independently. The template is matched within the search window using the mean absolute error between the image regions and the template.

A deformable template-based tracking technique is introduced in [115]. A hand-drawn prototype template describing the representative contours/edges of the object to be tracked is used. The original template undergoes deformation transformations to obtain a deformed template. The template matching process consists of comparing the template with the candidate image regions with respect to shape similarity, region similarity (using for example color and texture information), interframe motion (the object outline in the template should be attracted to pixels exhibiting large motion), and image gradient (the object outline in the template should be attracted to pixels that have large image gradients). The template is dynamically updated in shape and size from the newly detected object.

7.2.1.5 Bayesian Object Tracking

Many object tracking techniques perform tracking within a Bayesian framework. The latter belongs to the class of state space approaches that attempt to estimate the state of a system over discrete time steps, assuming that noisy measurements are available at these time steps. The state vector contains all data required to describe the system. For example, when tracking a moving object in two dimensions, the state vector would typically consist of the object position (x and y coordinates), as well as its velocity and acceleration along each coordinate (i.e., it would be a six-dimensional vector). The measurement vector contains observations corrupted by noise that are related to the state vector. Its dimension is usually smaller than that of the state vector. In the previous example, the measurement vector would contain "noisy" object positions (x and y coordinates, as measured in the image).

To perform the estimation, one needs a system model that describes the evolution of the object state over time, that is:

$$\mathbf{x}_k = f_k(\mathbf{x}_{k-1}, \mathbf{w}_{k-1}) \tag{7.23}$$

and a measurement model that links the noisy measurements to the state vector:

$$\mathbf{z}_k = h_k(\mathbf{x}_k, \mathbf{v}_k), \tag{7.24}$$

where $f_k : \Re^{n_x} \times \Re^{n_w} \rightarrow \Re^{n_x}$ and $h_k : \Re^{n_x} \times \Re^{n_v} \rightarrow \Re^{n_z}$ are possibly nonlinear functions; $\mathbf{w}_k$ and $\mathbf{v}_k$ are sequences that represent the i.i.d. process noise and measurement noise, respectively; n_x, n_w, n_z, and n_v denote the size of the state, process noise, measurement, and measurement noise vectors, respectively.

To be able to perform object tracking in a Bayesian framework, two remarks should be taken into consideration. First, the system and measurement models should be available in a probabilistic form. Second, in object tracking, an estimate of the object position is required every time a new measurement becomes available. Hence, estimation can be performed recursively. Bayesian object tracking belongs to the class of online methods, that is it produces an estimate at each time step k based only on all past measurements $\mathbf{z}_k$ up to time k. Assuming that the initial pdf of the state vector $p(\mathbf{x}_0|\mathbf{z}_0)$ is known ($\mathbf{z}_0$ is the set containing no measurements), the goal is to obtain the posterior pdf $p(\mathbf{x}_k|\mathbf{z}_k)$ at time step k. More specifically, let $p(\mathbf{x}_k|\mathbf{x}_{k-1})$ be the system model defined by (7.23) and the process noise $\mathbf{w}_k$ statistics in probabilistic form and $p(\mathbf{z}_k|\mathbf{x}_k)$ the measurement model (also known as the likelihood function) defined by (7.24) and the measurement noise $\mathbf{v}_k$ statistics. The estimation process comprises of two steps. During the first step (prediction), the posterior pdf at time step $k-1$, that is $p(\mathbf{x}_{k-1}|\mathbf{z}_{k-1})$, is propagated forward in time, using the system model $p(\mathbf{x}_k|\mathbf{x}_{k-1})$ [116]:

$$p(\mathbf{x}_k|\mathbf{z}_{k-1}) = \int p(\mathbf{x}_k|\mathbf{x}_{k-1})p(\mathbf{x}_{k-1}|\mathbf{z}_{k-1})d\mathbf{x}_{k-1}, \tag{7.25}$$

thus, obtaining the prior pdf $p(\mathbf{x}_k|\mathbf{z}_{k-1})$ at time step k. The second step (update) modifies the propagated pdf by exploiting the latest measurement available. Thus, the desired posterior pdf, $p(\mathbf{x}_k|\mathbf{z}_k)$, can be obtained by employing Bayes' theorem:

$$p(\mathbf{x}_k|\mathbf{z}_k) = \frac{p(\mathbf{z}_k|\mathbf{x}_k)p(\mathbf{x}_k|\mathbf{z}_{k-1})}{p(\mathbf{z}_k|\mathbf{z}_{k-1})}, \tag{7.26}$$

where $p(\mathbf{z}_k|\mathbf{z}_{k-1})$ is used for normalization and is calculated as follows [116]:

$$p(\mathbf{z}_k|\mathbf{z}_{k-1}) = \int p(\mathbf{z}_k|\mathbf{x}_k)p(\mathbf{x}_k|\mathbf{z}_{k-1})d\mathbf{x}_k. \tag{7.27}$$

The optimum solution in the Bayesian sense can be obtained based on (7.25) and (7.26) [116]. Analytical forms of the solution can be obtained either when certain assumptions hold, as in the case of standard Kalman filters or by approximations, as in the case of extended Kalman filters (EKF) and particle filters (PF) [18, 117, 118].

7.2.1.5.1 Kalman Filters and Extended Kalman Filters

The Kalman filter is a special case of the Bayesian filters mentioned earlier and is the best possible estimator, if the posterior pdf is Gaussian and the following conditions hold:

- Functions f and h in (7.23) and (7.24) are linear and known.
- The distributions of the process and measurement noises are Gaussian.

If we wish to provide a formal definition, the Kalman filter is a set of mathematical equations that provides a computationally efficient, recursive solution to the least-squares method [15–17]. Let us assume that the process and measurement noise, denoted by $\mathbf{w}_k$ and $\mathbf{v}_k$ respectively, are independent, white ones with normal probability distributions:

$$p(\mathbf{w}_k) \sim N(0, \mathbf{Q}_k), \tag{7.28}$$

$$p(\mathbf{v}_k) \sim N(0, \mathbf{R}_k), \tag{7.29}$$

where $\mathbf{Q}_k$ and $\mathbf{R}_k$ are the process and measurement noise covariance matrices, respectively, which can be constant or dynamically changing. Linear stochastic difference equations describe both the system model, that is the evolution of the object state over time, and the measurement model $\mathbf{z} \in \Re^m$:

$$\mathbf{x}_k = \mathbf{A}_k\mathbf{x}_{k-1} + \mathbf{B}_k\mathbf{u}_{k-1} + \mathbf{w}_{k-1}, \tag{7.30}$$

$$\mathbf{z}_k = \mathbf{H}_k\mathbf{x}_k + \mathbf{v}_k, \tag{7.31}$$

where $\mathbf{u}_k \in \Re^l$ is an optional control input. We can observe that to propagate the state forward in time, the $n \times n$ matrix $\mathbf{A}_k$ must be defined. If there is an optional control input, then, the $n \times l$ matrix $\mathbf{B}_k$ must also be defined to relate it to the state $\mathbf{x}_k$. Finally, the state and the measurement vectors are linked with the $m \times n$ matrix $\mathbf{H}_k$ in the measurement equation (7.31). All three matrices can be either constant or dynamically changing.

The Kalman filter operates in a two step predictor–corrector manner. During the first step, the current estimate along with an estimate of the error covariance are propagated forward in time. The second stage incorporates a new measurement to modify the propagated current state and error covariance estimates. Let $\hat{\mathbf{x}}_k^- \in \Re^{n_x}$ denote the a priori state estimate at step k (based on all past measurements prior to step k) and $\hat{\mathbf{x}}_k \in \Re^{n_x}$ denote the a posteriori state estimate at step k (as soon as measurement $\mathbf{z}_k$ becomes available). The errors of the two state estimates can be respectively defined as [16]:

$$\mathbf{e}_k^- \equiv \mathbf{x}_k - \hat{\mathbf{x}}_k^-$$

and

$$\mathbf{e}_k \equiv \mathbf{x}_k - \hat{\mathbf{x}}_k.$$

Then, the a priori and a posteriori error covariances can be respectively defined as:

$$\mathbf{P}_k^- = E\left[\mathbf{e}_k^- \mathbf{e}_k^{-T}\right] \tag{7.32}$$

and

$$\mathbf{P}_k = E\left[\mathbf{e}_k\mathbf{e}_k^T\right]. \tag{7.33}$$

The equations of the first step (prediction) are [15, 16]:

$$\hat{\mathbf{x}}_k^- = \mathbf{A}_k\hat{\mathbf{x}}_{k-1} + \mathbf{B}_k\mathbf{u}_{k-1}, \tag{7.34}$$

$$\mathbf{P}_k^- = \mathbf{A}_k\mathbf{P}_{k-1}\mathbf{A}_k^T + \mathbf{Q}_{k-1}. \tag{7.35}$$

The update stage begins by computing the so-called "gain" of the Kalman filter, denoted by $\mathbf{K}_k$. It is chosen so as to minimize the a posteriori error covariance (7.33). One popular form that performs this minimization is [16]:

$$\mathbf{K}_k = \mathbf{P}_k^- \mathbf{H}_k^T \left(\mathbf{H}_k \mathbf{P}_k^- \mathbf{H}_k^T + \mathbf{R}_k \right)^{-1}. \tag{7.36}$$

We can observe from (7.36) that, as $\mathbf{R}_k$ approaches zero, the actual measurement $\mathbf{z}_k$ becomes more trustworthy, whereas the predicted measurement $\mathbf{H}_k \hat{\mathbf{x}}_k^-$ becomes less trustworthy. The exactly opposite occurs when the a priori estimate error covariance $\mathbf{P}_k^-$ approaches zero.

The gain is used, along with a measurement $\mathbf{z}_k$ (when it becomes available) to modify the a priori estimate $\hat{\mathbf{x}}_k^-$, so that the a posteriori state estimate $\hat{\mathbf{x}}_k$ can be computed:

$$\hat{\mathbf{x}}_k = \hat{\mathbf{x}}_k^- + \mathbf{K}_k \left(\mathbf{z}_k - \mathbf{H}_k \hat{\mathbf{x}}_k^- \right). \tag{7.37}$$

The difference in parentheses in (7.37) is called the measurement innovation, or the residual and it reflects the error between the predicted measurement $\mathbf{H}_k \hat{\mathbf{x}}_k^-$ and the actual measurement $\mathbf{z}_k$. The gain is also used to modify the a priori error covariance and obtain an estimate of the a posteriori error covariance:

$$\mathbf{P}_k = (\mathbf{I} - \mathbf{K}_k \mathbf{H}_k) \mathbf{P}_k^-. \tag{7.38}$$

The two steps are repeated using the previous a posteriori estimates to predict the new a priori estimates. In the following, we will provide an example of applying Kalman filtering for object tracking by providing the description of an existing method.

In [119], a Kalman filter is used to perform pupil tracking for subsequently monitoring eyelid movements, determining gaze and estimating face orientation. The state vector at time k is defined as $\mathbf{x}_k = [x_k, y_k, v_{x_k}, v_{y_k}]^T$, where x_k and y_k denote the coordinates of the pupil's centroid pixel position and v_{x_k} and v_{y_k} denote its velocity in the x and y directions, respectively. The system model equation assumes no optional control input $\mathbf{u}_k$:

$$\mathbf{x}_k = \mathbf{A}_k \mathbf{x}_{k-1} + \mathbf{w}_{k-1}.$$

Assuming that the interframe pupil movements are small, the state transition matrix $\mathbf{A}_k$ can be parameterized as:

$$\mathbf{A}_k = \begin{bmatrix} 1 & 0 & 1 & 0 \\ 0 & 1 & 0 & 1 \\ 0 & 0 & 1 & 0 \\ 0 & 0 & 0 & 1 \end{bmatrix}.$$

A pupil detector is employed in each time step to provide measurements, that is $\mathbf{z}_k = [\hat{x}_k \ \hat{y}_k]^T$ at time step k. Because $\mathbf{z}_k$ refers to position only and for simplicity, matrix $\mathbf{H}_k$ in (7.31) is chosen as:

$$\mathbf{H}_k = \begin{bmatrix} 1 & 0 & 0 & 0 \\ 0 & 1 & 0 & 0 \end{bmatrix}.$$

Initialization of the Kalman filter is performed by means of pupil detection. If the latter is successful for two consecutive frames c and $c+1$, the state vector assumes values from the last successful detection, that is $\mathbf{x}_0 = [x_{c+1}, y_{c+1}, v_{x_{c+1}}, v_{y_{c+1}}]^T$. The prediction phase of the Kalman filter involves the propagation of the covariance (7.35), hence an initial covariance matrix must be defined. Because this is updated iteratively when more images are acquired, it can be simply initialized to large values. Assuming that the error of the predicted position and velocity of the pupil's centroid is $\pm d_{\max}$ and $\pm v_{\max}$ in both directions, a suitable initial covariance matrix could then be:

$$\mathbf{P}_0 = \begin{bmatrix} d_{\max}^2 & 0 & 0 & 0 \\ 0 & d_{\max}^2 & 0 & 0 \\ 0 & 0 & v_{\max}^2 & 0 \\ 0 & 0 & 0 & v_{\max}^2 \end{bmatrix}.$$

Additionally, the two covariance matrices associated with the process noise and measurement noise must be defined. These can be constant or changing over time. In [119], they are assumed constant and are empirically determined.

In a number of computer vision problems where Kalman filtering is employed (including object tracking), the state and the measurement models might not be linear. In such cases, the standard Kalman filter cannot be used, unless some kind of linearization is performed. Indeed, a Kalman filter that linearizes about the current mean and covariance is known as the EKF [17] and has been used in the context of object tracking. Let $\mathbf{w}_k$ and $\mathbf{v}_k$ denote the process and measurement noise, respectively, as described previously. The system model that describes the evolution of the object state over time:

$$\mathbf{x}_k = f(\mathbf{x}_{k-1}, \mathbf{u}_{k-1}, \mathbf{w}_{k-1}) \tag{7.39}$$

and the measurement model $\mathbf{z} \in \Re^m$:

$$\mathbf{z}_k = h(\mathbf{x}_k, \mathbf{v}_k) \tag{7.40}$$

are nonlinear. In other words, the formulation remains the same with the standard Kalman filter, the only difference being that the functions f and h are considered nonlinear.

Although the values of the process and measurement noise are not known, we can still approximate the state and measurement vector without them [16], that is:

$$\tilde{\mathbf{x}}_k = f(\hat{\mathbf{x}}_{k-1}, \mathbf{u}_{k-1}, 0), \tag{7.41}$$

$$\tilde{\mathbf{z}}_k = h(\tilde{\mathbf{x}}_k, 0), \tag{7.42}$$

where $\hat{\mathbf{x}}_{k-1}$ denotes the a posteriori estimate of the state from a previous time step. Using the same notation with the standard Kalman filter and following the linearization process and the derivation in [16], the equations of the first step (prediction) are:

$$\tilde{\mathbf{x}}_k = f(\hat{\mathbf{x}}_{k-1}, \mathbf{u}_{k-1}, 0), \tag{7.43}$$

$$\mathbf{P}_k^- = \mathbf{A}_{k-1}\mathbf{P}_{k-1}\mathbf{A}_{k-1}^T + \mathbf{W}_{k-1}\mathbf{Q}_{k-1}\mathbf{W}_{k-1}^T, \tag{7.44}$$

where $\mathbf{A}_k$ is the Jacobian matrix of partial derivatives of $f(\cdot)$ with respect to $\mathbf{x}_k$, $\mathbf{W}_k$ is the Jacobian matrix of partial derivatives of $f(\cdot)$ with respect to $\mathbf{w}_k$, and $\mathbf{Q}_k$ is the process noise covariance matrix.

The second step (update) equations are:

$$\mathbf{K}_k = \mathbf{P}_k^- \mathbf{H}_k^T \left(\mathbf{H}_k \mathbf{P}_k^- \mathbf{H}_k^T + \mathbf{V}_k \mathbf{R}_k \mathbf{V}_k^T\right)^{-1}, \tag{7.45}$$

$$\hat{\mathbf{x}}_k = \tilde{\mathbf{x}}_k + \mathbf{K}_k(\mathbf{z}_k - \mathbf{h}_k(\tilde{\mathbf{x}}_k, 0)), \tag{7.46}$$

$$\mathbf{P}_k = (I - \mathbf{K}_k \mathbf{H}_k)\mathbf{P}_k^-, \tag{7.47}$$

where $\mathbf{H}_k$ is the Jacobian matrix of partial derivatives of $h(\cdot)$ with respect to $\mathbf{x}_k$, $\mathbf{V}_k$ is the Jacobian matrix of partial derivatives of $h(\cdot)$ with respect to $\mathbf{v}_k$, and $\mathbf{R}_k$ is the measurement noise covariance matrix.

Other variations of the Kalman filter have been devised to improve its performance with respect to its application to computer-vision problems. These include the Unscented Kalman Filter (UKF) [120], which is an improvement over the EKF. Although the EKF uses only the first order terms of the Taylor expansion and, consequently, introduces errors, UKF avoids these errors by using the third and higher order terms of the Taylor expansion.

7.2.1.5.2 Particle Filters

If the posterior pdf is not Gaussian, Kalman filters will not perform adequately. In such a case, PF can be used. They are sequential Monte Carlo methods that can be used for object tracking within a Bayesian framework. They come in a variety of names, such as Conditional Density Propagation (or the Condensation algorithm) [18], survival of the fittest [117], interacting particle approximations [118], and so forth and they have been extensively used in tracking objects. The main concept behind PF is to represent the probability distribution of alternative solutions as a set of samples (i.e., particles), each of which carries a weight. Estimates of the posterior distribution are calculated based on these samples and their associated weights. As the number of samples grows, the filter approaches the optimal Bayesian estimate [116]. The ideal scenario would involve sampling directly from the posterior distribution. However, this is hardly the case. Solutions to this would be to use sampling techniques, such as factored sampling or importance sampling. If we cannot sample directly from the posterior pdf, because it is too complex, but we can sample from the prior pdf, a random sampling technique called factored sampling can be employed. Each random sample is assigned a weight, and the weighted set can be the approximation to the posterior density. To improve the results of random sampling, alternative sampling techniques can be employed. For example, importance sampling does not sample from the prior pdf, but from another density function that can "drive" the selection of samples towards areas of the posterior pdf that contain the most information. By doing so, the resulting sample set will describe the posterior pdf more efficiently. Other sampling methods have also been devised, for example [121]. More thorough descriptions of Kalman filters and PF are provided in [15–17] and [116, 122, 123], respectively.

An advantage of PF is that they allow the principled fusion of information from different information sources. This is exploited in [124] where a tracker that is based on PF and fuses color, motion and sound information is proposed. The main source of information is color. Color is fused with either motion/temporal activity information (obtained from frame differencing) for video surveillance or with sound-related information (time delay of arrival between signals arriving in a pair of microphones) in a teleconferencing framework. Other attempts to fuse multiple cues within a particle filtering framework have been also reported [125]. Okuma et al. [126] combine a cascaded Adaboost algorithm [127] with a mixture of PF [128] for multiple-object tracking. The cascaded Adaboost algorithm is utilized to learn models of the objects under examination. These detection models are subsequently used to guide the PF.

Numerous tracking methods that are based on or incorporate PF have been reported in the literature. Examples include approaches that combine PF with appearance-adaptive models [129], utilize Markov chain Monte Carlo-based PF to deal with targets that are influenced by the proximity and/or the behavior of other targets [130], formulate particle filtering within a geometric active contour framework [76], embed an online discriminative feature selection procedure within the particle filtering process [131], apply particle filtering for the tracking of high degrees of freedom articulated structures [132], and so forth.

7.2.1.6 *Occlusion Handling*

An inevitable problem in object tracking is the occlusion or self-occlusion of the target object. In the majority of video sequences, parts or even the entire tracked object are not visible in all the frames of the sequence, either due to the existence of static objects (e.g., walls, trees, obstacles) that occlude it or due to the existence of more than one moving objects (e.g., two people walking and crossing each other). A large number of algorithms in the tracking literature have ignored occlusions and treated them simply as noise in the matching process. However, a number of approaches try to deal with occlusion in more advanced ways.

The simplest way to handle occlusions is to reinitialize the tracker in a frame where the detected level of occlusion is high enough to lead to tracking drift or failure. For example, in a face tracking algorithm, this would employ using a face detection scheme similar to the one used to initialize the tracking process for redetecting a face that has been "lost." Obviously, this comes at the expense of computational burden. On the basis of the observation that the occlusion can be treated as a local effect, another more clever approach could be to segment the object of interest into N parts and perform tracking on the whole set of these parts [133]. This is performed by estimating N possible transformations (e.g., affine ones) and using a voting scheme to select a single transformation representative of the entire object, whereas the performance of all the transformations is evaluated at regular intervals to enhance the tracking process. This can enable tracking even in the presence of substantial occlusion, because a relatively low number of the object parts needs to be visible to successfully track the object of interest under partial occlusion. In a similar manner, Fu et al. in [70] track the boundary of an object by using an occlusion adaptive motion snake that consists of many nonoverlapping contour segments. The method has been described in Section 7.2.1.2.

Finally, prediction schemes, such as Kalman filters can be employed to handle occlusions. For example, in [134], a Kalman filter is used to track occluded feature points in a feature point-based 2D object tracking algorithm. In case of partial occlusion, their coordinates are updated by employing the Kalman filter to predict the movement of the upper left and the lower right point of the object bounding rectangle. In case of total occlusion, the Kalman filter predicts the position of the occluded region based on the velocity estimates of the region corners, obtained from the measurements prior to total occlusion. In [108], the algorithm also employs a Kalman filter as an occlusion handling mechanism. This approach is reported to be robust against short-time partial occlusions. The method cannot handle efficiently severe, complete, and long-time occlusions. A similar approach was used in [135] to track moving people in video sequences. Additionally, algorithms that can handle multiple hypotheses, such as the PF (or the Condensation algorithm) described in Section 7.2.1.5, can be used for the occlusion period until the image measurements can disambiguate with respect to the actual object position in the frame under examination.

All of the aforementioned occlusion handling mechanisms mainly refer to monocular object tracking. In multiple-camera systems, self-occlusions and occlusions between moving objects or between moving and static objects can be treated more efficiently, because an object that is occluded in one view might be fully visible in another. Many promising multiple-camera methods can cope with occlusion. In [136–138], multiple-camera systems were used to track people successfully. These methods combine the information from all cameras to determine the "best" view. When a camera loses the target due to occlusion, information is obtained from other cameras that are also tracking the object. A probabilistic map function aiming at detecting occlusions calculates the probability that a particular location in the image is visible from a specific camera.

A different approach was presented in [139], where a probabilistic weighting scheme for spatial data integration through a simple Bayesian network was employed. The presented tracking scheme firstly extracts a set of measurements (observations) for the estimation of the state vector. Then, a predictor–corrector filter is exploited to filter the results. Finally, a Bayesian network performs spatial data integration using triangulation, perspective projections and Bayesian inference. The input to the network is the set of measurements from the previous time step and the states from the views of the other cameras. The output constitutes the input to a Kalman filter whose goal is to maintain a temporal smoothing on the vector of the 3D trajectories.

Handling occlusion of rigid parts of an articulated object is presented in [140, 141]. Occlusion relationships are defined and handled using the relative motion of the articulated object parts. Simple occlusion relationships between two objects A and B are defined using visibility order: A occludes B, B occludes A, A and B are not occluded. Occlusion graphs can describe more complex occlusion relations when more than two objects are involved. Occlusion between two objects is considered unambiguous, if the convex hull of the union of all possible moves of one object can be partitioned from the other object using a separating plane. Ambiguous occlusion cases are also defined. Matching is performed by registering layered templates that represent different occlusion situations. As this chapter aims at finger tracking, templates depict fingers in a number of occlusion configurations. The minimization of the sum of squared differences is used for template registration.

7.2.2 3D Rigid Object Tracking

3D rigid object tracking can be defined as the estimation of the position and orientation of a rigid object in 3D space from video data obtained from one or more video cameras. The location of a rigid object in the 3D space is determined by the position of its center of mass with respect to a world coordinate system, as well as the relative orientation of a coordinate system attached to its center of mass with respect to the world coordinate system. Thus, 3D rigid object tracking has to estimate a total of six parameters although in certain applications determining only the position (i.e., considering the object as a point mass) or only the orientation parameters might suffice. One of the most important applications of 3D rigid object tracking is 3D head tracking (often referred to as head pose estimation) which is being used as a preprocessing step or a building block in face recognition and verification, facial expression analysis, avatar animation, human-computer interaction, and model-based coding systems. Although the human head is not a rigid object, considering (and tracking) it as such (i.e., considering only the global head motion) is sufficient in a number of cases. Alternatively, head deformations can be taken into account in the tracking algorithm [142]. Other methods focus on tracking the facial features and expressions in two or three dimensions [143–148] but these fall outside the scope of this Section. 3D vehicle tracking is another application of 3D rigid object tracking [149]. Apart from its obvious importance as a stand-alone task, 3D rigid object tracking often constitutes a basic building block of 3D articulated object tracking methods, where one needs to estimate the position of rigid objects (links) that make up the articulated structure.

Certain methods (e.g., [150]) use 2D tracking techniques to derive the 2D image-plane motion of the object of interest and then a Kalman filter for deriving the 3D motion parameters. Another approach for head pose estimation [90] employs tracking in 2D of salient facial features (eye corners and nose). Projective invariance of the cross-ratios of the eye corners and anthropometric statistics are subsequently used to compute orientation relative to the camera plane. A similar approach in [86] locates robust facial features (eyebrows, eyes, nostrils, mouth, cheeks, and chin) and uses the symmetric properties of certain facial features and rules of projective geometry to determine the direction of gaze.

An approach for estimating 3D head orientation (i.e., the pan, tilt, and roll angles) in single-view video sequences is presented in [151]. Following initialization by a face detector, a variant of the tracking technique [103] (see Section 7.2.1.3) is used to track the face in the video sequence. The difference with [103] is that, for face tracking, the deformable surface model is used to approximate the image intensity surface of the entire face area. The generalized displacement vector (which, as already mentioned in Section 7.2.1.3, is involved in the equations that govern the deformation of the surface model) is used for face tracking. This vector is also used for head pose estimation. In more detail, the generalized displacement vector of a certain frame is provided, along with the head angles estimates from the previous frame, as input to three RBF interpolation networks that are trained off-line to estimate the pan, tilt, and roll angles. Training of the RBF networks (i.e., estimation of their parameters) is performed using example image sequences for which the head pose angles of each frame have been evaluated through the use of a magnetic head tracker.

A significant number of techniques are model-based, that is, they involve a 3D geometry model (usually enriched with texture information) of the object of interest to derive 3D information from the 2D projections of the object on the video sequence. A typical model-based 3D rigid object tracking algorithm can be roughly outlined as follows: a 3D geometric model of the object is initialized and enriched with texture information. Then for each video frame the model parameter vector (position and orientation) that best describes the object in the current frame is evaluated. The model parameters derived in the previous frame are used to provide a rough estimate of the parameters in the current frame, which is subsequently refined so as to maximize the similarity between the projection of the model on the image plane and the underlying image content. The rendering capabilities of current graphics hardware can be utilized for the fast projection (rendering) of the 3D model.

The 3D model representation can be either surface-based (usually a triangular mesh) or volumetric (spheres, cylinders, superquadrics, etc.) [2]. The level of detail of the model depends on the desired accuracy and the overall system purpose and architecture. A more detailed account of the various types of models that are used to provide a "flesh" representation of the human body parts (considered as rigid bodies) will be given in Section 7.3.1. In what concerns the human head, different surface models exist [144, 152–154], some of which were obtained using 3D scanning technologies. The use of a cylindrical representation of the human head is very common because of its simplicity, especially when used in conjunction with texture information [155].

Tracking initialization for such methods is not trivial in most cases. During the initialization step, the initial geometric model configuration has to be estimated. Furthermore, the texture of the model (reference texture) should be obtained from the video sequence (Fig. 7.5). Automatic approaches usually involve a step for detecting the object in the first video frame and a subsequent step for registering the 3D model with this projection. In [155] for example, a 2D face detector is used and then the model position is obtained by assuming that the object to be tracked is a head facing upright and towards the camera. If the geometric model is a generic one and does not correspond to the actual object that is to be tracked (i.e., the object depicted in the video sequence), an additional step that deforms it to achieve adaptation to the tracked object might be applied. The fitting of a generic polygonal face model to a human face image is presented in [144]. Although the chapter deals with facial feature and expression tracking, the fitting procedure is characteristic of this type of initialization for face tracking techniques. Fitting is performed by localizing important facial features, such as eyes, forehead line, jaw line, mouth, and so forth on a pair of frontal and profile images. This is achieved through edge detection followed by snake-like techniques. Enhancement of some edges is also performed.

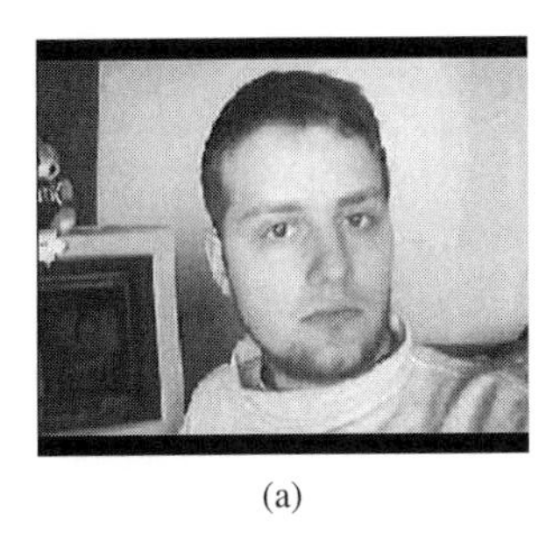

(a)

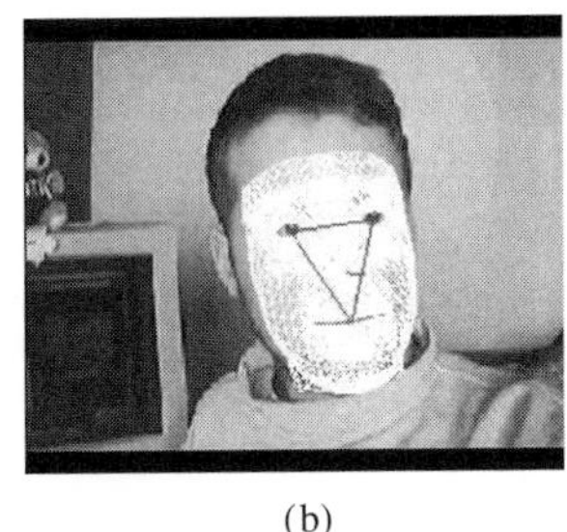

(b)

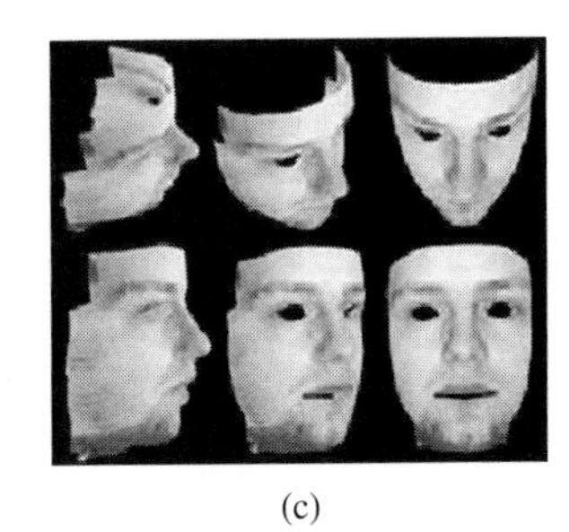

(c)

FIGURE 7.5

(a) A sample frame from a head and shoulders video sequence, (b) a generic face polygon model fitted on the video frame, and (c) the geometric model enriched with texture information.

In [153], the automatic adaptation of the CANDIDE face model [156] to video data is presented. Matching is performed by finding the main facial features (eyes, mouth) using deformable templates. Then, the model is fit using global and local adaptation. The global adaptation step estimates 3D eye and mouth center positions of the face model and uses them to perform scaling, rotation, and translation of the face model in the 3D world. The system presented in [152] also involves finding the main facial features and performing global and local adaptation. Geometrical considerations are used to perform the adaptation. The method involves a geometric model that is more detailed than CANDIDE and does not restrict the initialization images to depict the subject in a certain facial expression (e.g., with mouth closed). The two methods presented earlier are applied to head and shoulders image sequences as they aim at the initialization of coding algorithms for videophone sequences.

Enriching the geometric model with texture information during initialization involves finding a mapping between the 2D image plane coordinate system (u, v) and the model surface, represented in parametric form (s, t) in a 2D parametric space (also known as texture map coordinate system). Finding this mapping requires that the 3D location of the model and the projection matrix associated with the camera are known, which is true for the initialization phase.

Because the projection of the 3D model on the image plane (u, v) is a 2D image, its matching with the actual image content can be performed with the techniques and similarity metrics used in 2D rigid object tracking like color similarity, sum of squared differences, joint entropy, and so forth (see Section 7.2.1.4). Matching using edge information or other features is also used. Especially in the case of head pose estimation, facial feature (eyes, noise, mouth) localization on the video frames followed by registration of the resulting features with the corresponding features of the projected model can be used.

Estimating the model parameters in a certain frame so that the residual error is minimized can be done using an appropriate search strategy. The trivial solution would be to test all possible combinations of model positions and orientations, but this is computationally prohibitive. Therefore, clever ways of reducing the possible solutions are required. Prior knowledge about the tracking environment or the intended application can provide such constraints. For example, in a videophone application, it would be safe to assume that the user's head would stay relatively close to the camera at all times and that it would not rotate too much with respect to the camera. Constraints can also be introduced through appropriate motion models. Similar constraints for articulated object tracking are described in more detail in Section 7.3.1. In some cases (e.g., [142]), and due to the relatively high dimensionality of the matching problem, iterative approaches that aim at minimizing the similarity function over the model parameter space (e.g., standard nonlinear function optimization approaches like the conjugate direction method) are used. Alternatively, a prediction–correction scheme (e.g., the EKF) having error signal as the difference between the model projection and the actual image content can be used to update the initial estimate of the model state (i.e., the model position and orientation parameters).

An alternative to projecting the 3D textured model onto the image plane (u, v), and performing matching there, is to perform matching in the texture map coordinate system (s, t) [155]. Indeed, representing the image-derived texture information in the 2D (s, t) coordinate system corresponds to a warped (flattened) image of this texture (Fig. 7.6). For each model position that has to be tested during the matching procedure, the video frame is warped to the (s, t) coordinate space and compared to the reference texture (i.e., the model texture) represented in the same space. Again, the matching procedure is a 2D one, therefore the similarity metrics outlined earlier and in Section 7.2.1.4 are still applicable.

Texture mapping is used in the 3D head tracking system presented in [155]. The head is modeled as a cylinder. Tracking is initialized using a face detector in the first frame and finding the model position and orientation that fits best with the detected face. The detected face region is mapped as a texture on the cylinder. Tracking is subsequently formulated as an image registration problem, where the previous estimate of the model position and orientation is updated so as to minimize the sum of squared differences between the reference texture (i.e., the texture of the model) and the warped target texture (texture obtained from the current frame). No iterative optimization is involved in the

FIGURE 7.6

Head and shoulders image obtained from the MIT Vision and Modeling Group face database [104], along with the corresponding texture map.

process making the system capable of performing at 15 frames per second. To improve the performance, illumination regularization is applied using illumination templates. The illumination templates are superimposed on the target region using a rough registration procedure. The templates do not depend on the specific person being tracked and are obtained using a training procedure. Geometric head models generated by surface scans have been also used but were found to provide no advantage over the simple cylindrical model.

In [92], a generalized 3D model of an average human face is built using a database of range images. An automatic detection step is employed to detect the position of a face, as well as the position of certain facial features (eyes, nose, mouth) in the first frame of a video sequence. The position of the detected facial features is used to properly align the general 3D model to the 2D image data. Subsequently, texture is attached to the 3D mesh using the underlying image intensity values. The 3D coordinates of the feature points are expressed in terms of their obtained 2D coordinates and they are provided as input (along with other parameters describing the camera) to an EKF. SSD trackers are used to track the feature points in the frames of the video sequence. The output of these trackers becomes the observation vector (see Section 7.2.1.5.1) of the Kalman filter. The final output of the filter is the estimate of the 3D position of the head.

The techniques in [157, 158] are also worth studying.

7.3 ARTICULATED OBJECT TRACKING

A number of physical entities in real-world environments can only be represented using articulated structures, that is, structures composed of rigid parts (links) connected by joints, typically described through a tree-like kinematic hierarchy (Fig. 7.7). Living beings, such as humans or animals, exhibit such attributes. To be able to extract higher level information about the behavior of such entities (e.g., gesture recognition, understanding animal behavior, etc.), precise tracking of the corresponding articulated structures in 3D is necessary. Therefore, most of the articulated tracking algorithms attempt object tracking in a 3D space. Furthermore, even if the goal is to track an articulated object in 2D (i.e., in the frames of a video sequence), the methodology is similar to that of 3D articulated tracking algorithms. Moreover, many 2D articulated object tracking algorithms employ a 3D model of the articulated object. For all these reasons, most of this section covers 3D articulated object tracking, followed by a brief discussion on 2D articulated object tracking.

7.3.1 3D Articulated Object Tracking

3D articulated object tracking approaches can be model-free or model-based. In the former case, no model of the articulated structure is used. Instead, a bottom-up approach is used to combine image information extracted locally (edges, corners, etc.), to create coherent structures, such as the limbs of the human body. Obviously, this approach requires that the structures reconstructed are constantly visible in the images.

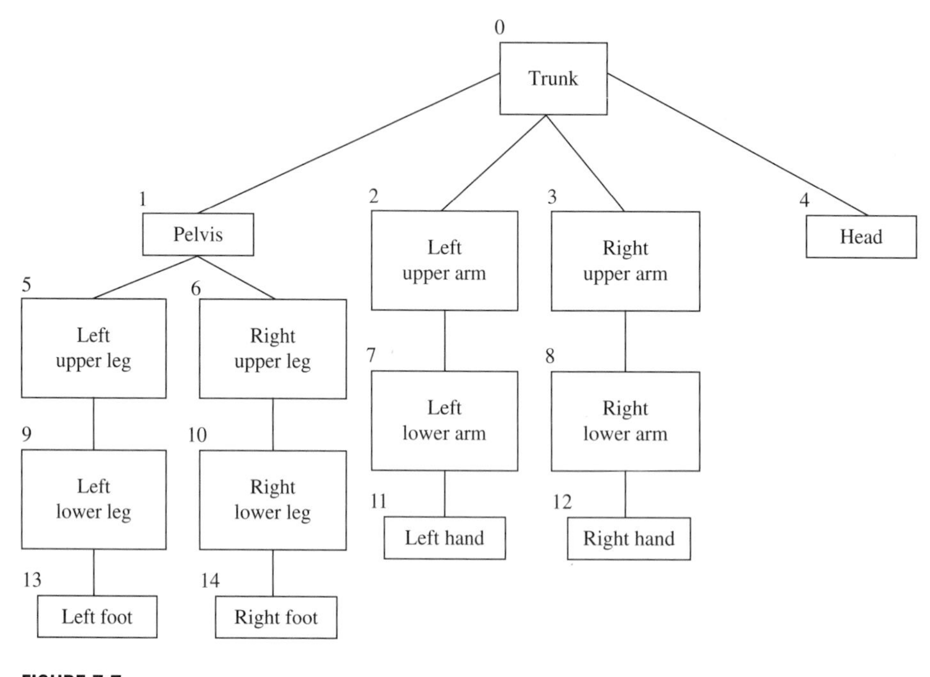

FIGURE 7.7

A tree-like hierarchical representation of the human body.

A subset of model-free approaches for 3D human body (or human body parts) pose estimation and tracking are based on learning [159]. These methods often utilize the fact that the set of typical human poses is considerably smaller than the set of kinematically possible ones. Thus, these approaches try to devise methods for recovering the 3D body pose directly from image observations. A subcategory of learning-based approaches are the so-called example-based methods [160–163]. These methods store in an explicit way a large number of training images for which the corresponding 3D poses of the human body (or parts of it) are known. For given frame of an image sequence, stored images that are similar to this frame are retrieved and the pose in this frame is estimated by interpolating between the poses corresponding to the retrieved images. Some of these methods, for example [162, 163], are actually 3D pose estimation methods applicable to single images. However, one can easily incorporate them in a tracking framework, for example by applying them in each frame of a video sequence.

In model-based approaches, a model of the articulated object is employed. The complexity of the model depends on the accuracy required in a specific application. The human body, for instance, is represented by rigid parts (resembling limbs) connected to each other at joints. Even such a minimal representation has around 30 degrees of freedom (DOF).

A widely used approach tries to predict the model configuration that, when projected onto the image, minimizes the error between the model projection and the actual image data. The 3D model is initialized either manually or automatically, using information extracted from the first frame of a video sequence. The ideal case would then be to perform an exhaustive search over all possible model configurations, attempting to match their projections against the information extracted from the next frame in the sequence. Such an exhaustive search, however, would be possible only for a model of a very low dimensionality. For models having of 20–30 degrees of freedom, "clever" ways of reducing the high-dimensional search space are required. This can be performed by "pruning" the search space using kinematic and other constraints. Even then, the complexity of an exhaustive search remains prohibitively large. Further reduction can be achieved if additional constraints in the model configurations that need to be searched are introduced. The selection of a motion model is a way of introducing such constraints. For instance, if the motion of the object of interest is known to be periodic or if we wish to track the motion of people performing specific actions such as running or walking [14], appropriate motion models can be used to significantly reduce the complexity of the tracking process and, hence, the computational time required.

If one model configuration is to be estimated, single hypothesis methods, such as Kalman filters [49, 164] and least squares [165–167] can be used. It is often the case that, due to the complex nature of 3D articulated object modeling and tracking, a single hypothesis method will result in loss of tracking or severe tracking drift. Methods capable of tracking multiple hypotheses, such as particle filtering or the condensation algorithm [14, 18, 168], described in Section 7.2.1 or others [169] can be used instead. Multiple hypotheses are maintained until the extracted image data can help pinpointing a single model configuration. Regardless of the single or multiple hypotheses maintained in each time step, the choice of the type of image data used to match against the projected model configurations and the method of their extraction plays a crucial role in accurately tracking the articulated object in 3D.

A number of decisions have to be made when devising an algorithm for tracking 3D articulated objects. One such important decision is the number of cameras used to obtain the video sequence(s) in which objects will be tracked. More than one cameras can be used to deal with the inherent depth ambiguities of monocular tracking. Regardless of the number of cameras, camera calibration might need to be performed prior to obtaining the sequences, as described in Section 7.1.

Although multiple-camera can help disambiguate depth problems, monocular tracking is often desirable, because of the technical simplicity of single-camera systems. However, 3D monocular articulated object tracking algorithms should deal with a number of difficult issues. First, depth information is lost when projecting the world into 2D images, classifying 3D articulated object tracking as an ill-posed problem that might produce more than one solutions unless additional constraints or prior knowledge are utilized. This problem is, for example, obvious in the case of symmetric positions of the arm in 3D, close to its neutral position, which results in the same projection. Additionally, self-occlusion often occurs between different parts of the articulated object (e.g., the human body or limbs). Such a situation is considerably more difficult to tackle with a single-camera.

Another difficulty in 3D articulated tracking algorithms is associated with the deformable clothing of humans, where most of the attempts on articulated tracking are focused on. Attempts to provide a solution by cloth simulation and reconstruction techniques require a carefully controlled multicamera and lighting environment [170], and result in severe computational overheads. Attention should also be paid to the fact that the object model is matched against noisy image measurements. Images often contain many distracting features that can be potentially mistaken for the object parts. In addition, illumination changes create problems if image edges or intensities are the features used for tracking. Shadows can produce false edges and varying illumination conditions can cause texture changes that cannot be accounted for by the model used. Also, during rapid motion, blurring occurs at the motion boundaries, which can severely affect methods that rely on static feature extraction, such as the ones that compute the optical flow. Solutions to this problem involve the fusion of image cues that provide complementary information.

In the rest of this section, some of the issues arising when designing a 3D articulated object tracking algorithm (with a bias towards algorithms that incorporate a geometrical model) will be presented in more detail, along with example techniques.

7.3.1.1 3D Modeling

Because various applications such as human–computer interaction, surveillance, and so forth focus on people, many efforts have been targeted on representing highly articulated structures, such as the human body, using 3D models that vary in complexity, depending on the accuracy needed in each specific application. 3D articulated models are also employed in motion capture applications, where the goal of the tracking process is to recover the full-body pose. Because of the complexity of these models, the computational overhead increases.

Human body models generally consist of two components: the skeletal structure and a representation for the flesh surrounding it. The skeletal structure is a collection of segments and joints with various degrees of freedom at the joints, as illustrated in Fig. 7.8. In general, each joint is associated with up to 3 DOFs, while additional DOFs are required for the global position and orientation of the model. In many cases, joints are allocated lesser DOFs than the actual ones to reduce model complexity to a tractable level. The parameters of the skeletal structure are the joint angles. If the individual segments comprising the articulated structure are allowed to deform, the shape parameters of the segments are also included in the set of parameters that parameterize the model. Additionally, the articulated object model should be initialized in a consistent manner, for example for humans, by taking into account the standard humanoid dimensions, as specified in [171, 172].

The flesh can be represented using polygonal meshes (Fig. 7.9) or other surface primitives [144, 152]. Alternatively, volumetric primitives such as spheres, cylinders (Fig. 7.10) [14, 155, 165], cones [164, 168, 173], and superquadrics (generalized ellipsoids with

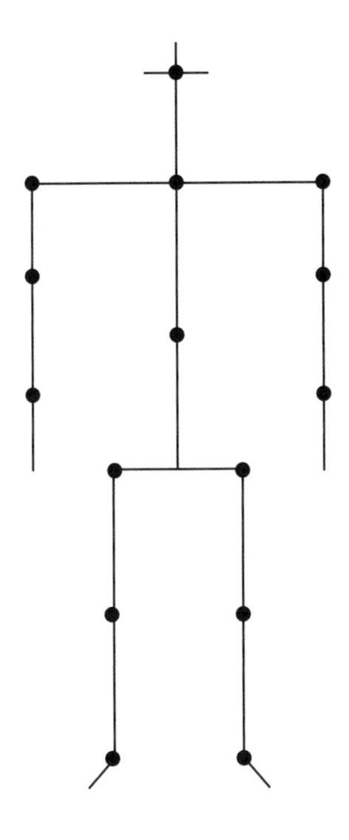

FIGURE 7.8

A example of a stick human body model. Each joint can be associated with 1–3 degrees of freedom (DOF).

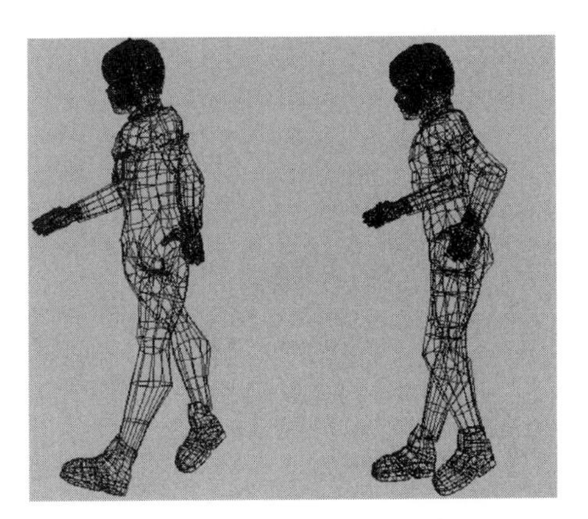

FIGURE 7.9

Polygonal representation of the human body, adapted from [175].

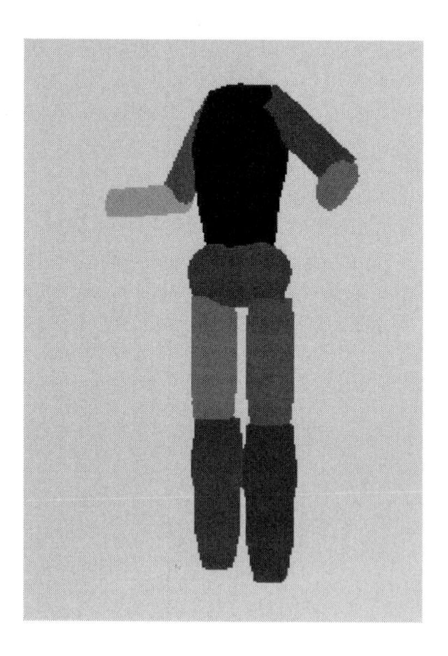

FIGURE 7.10

Cylinder-based volumetric representation of the human body.

additional parameters along each axis, which encode the "squareness") [167, 169, 174] can be used.

The accuracy of the representation depends on the intended application as mentioned earlier. In computer animation, extremely accurate surface models (often obtained through body scans of actual people), which consist of thousands of polygons, are employed. In computer vision, however, the inverse problem we deal with, that is recovering a 3D model from images is much harder and solutions are less accurate. Consequently, coarse volumetric primitives have been preferred because of their lower complexity. These models are subsequently used for tracking human body parts in 3D space.

7.3.1.2 ***Kinematic and Motion Constraints***

Kinematic constraints are distinguished in "hard" and "soft" constraints. The former are usually in the form of angular displacement, velocity, and acceleration limits of the joints of an articulated structure, whereas the latter are probabilistic and associated to previous instances of the object (e.g., human) motion. For example, the range of angles where the human elbow can move is limited and these limits should be taken into account during the tracking process. Enforcing anatomical joint angle limits is relatively straightforward. They can be applied by specifying minimum and maximum allowed values for the joint degrees of freedom [164, 169]. Moreover, the speed attained by different body parts connected on the same joint is different. An example showing the need of imposing constraints on the motion of certain body parts is the human finger, which consists of three parts. The middle part moves to greater extent and more rapidly than the other two. Moreover, the human finger is not allowed to move to all directions. Finally, non interpenetration constraints between the different parts of the articulated structure have to be incorporated [169]. These constraints stem from the fact that two parts cannot occupy the same space simultaneously.

Prior information about the motion to be tracked can also lead to further constraints that can be used to improve the tracking process. For instance, several human activities are repetitive in nature (e.g., walking or running) and the corresponding motion can, therefore, be modeled using specific motion models that can be learned from training data [14]. Alternative methods include example-based motion models such as the one in [176], where given a database of example human motions, the authors construct a low-dimensional model of the motion, and project subsequences in the database onto this low-dimensional representation, obtaining a coefficient vector at each time step. Coefficients are then used to index the database. The same projection is applied to any new motion that is observed (and should be tracked). The problem of tracking then becomes a problem of searching the database efficiently and matching the observed motion with examples from the database. The examples matched are used as samples in a stochastic sampling (i.e., condensation algorithm [18]) framework. Finally, motion can be modeled as a higher-order autoregressive process, as in [168]. Attempts have also been made to perform tracking by switching between multiple motion models [177]. Such an approach, however, severely increases the computational burden. One of the major problems with motion models is that models used so far are not generic enough to be applied in various real-world conditions.

7.3.1.3 Image Cues

As soon as the model of the object has been properly defined, a method for matching the projected model configurations to the image data is required. Many different image features can be utilized for this purpose. They range from low-level (e.g., edges) to high-level cues (e.g., the locations of the joints). Joint locations are difficult to recover directly, because there is no characteristic intensity distribution around them. They are consequently inferred using the adjoining rigid parts of the articulated object, thus making the method very sensitive to the segmentation of the articulated object. These difficulties lead researchers to the use of low-level features for the matching process. Edge information can be employed, because edges are partially invariant to changes in viewpoint and lighting conditions [178], while being easy to detect, especially in humans. There are 3D motion patterns, however, that cannot be detected using edges. For instance, when the human limbs rotate around their 3D symmetry axes, the edge changes in the image are insufficient for the detection of such a motion [169].

To alleviate such problems, intensity can be also used as a cue for image/model matching. This involves a step of acquiring a reference texture, either as part of an initialization stage [140] or from the previous time instance [14, 164, 165, 169]. The reference texture is mapped onto the model surface. The textured model is projected onto the image and matched against the underlying image texture. This, however, requires that image texture can be reliably extracted at all times, otherwise tracking may deteriorate. Complex real-world environments, where intensity may vary significantly, the subjects may wear loose and deformable clothing and texture information may not be easy to extract, can severely affect the performance of such algorithms. Silhouettes can additionally be employed to track objects (especially people). They can be acquired by motion segmentation, contour tracking, or background segmentation [12, 173], as described earlier.

To increase the robustness of the incorporated algorithms, an object tracking system may employ more than one image cues, in an effort to extract as much information as possible from the image. Several variations are possible and include a combination of the above, such as edges and intensity, silhouettes and edges [164, 168], and so forth.

7.3.1.4 *Example Techniques*

In this section, a number of works on 3D articulated object tracking are briefly presented. It should be noted that the field of 3D articulated tracking of humans, especially from single view data, has witnessed a boom during the last few years, with a wealth of new approaches towards this difficult problem. Readers are thus encouraged to consult the recent publications in this field such as Refs. [179, 180] for multiple-camera approaches and Refs. [181–184] for single-camera approaches.

In [185], a 3D tracking system for high DOF articulated structures, that is applied on hand tracking, is presented. The human hand is modeled using an articulated structure consisting of 16 rigid parts (15 rigid parts for the fingers and 1 for the palm). The total hand configuration is represented by a 28-dimensional vector. Each finger part is characterized by its central axis line. The finger tip coordinates are also used. A priori knowledge of the hand kinematics and geometry and exact hand localization is required. Matching is performed by using a prediction mechanism. A correction scheme, formulated as a linear least squares problem, is subsequently applied.

In [174], the authors perform tracking in a hierarchical manner. They use an articulated three-dimensional model. The human body skeletal structure is modeled by a 22 DOF "stick figure" model, whereas for the flesh-shape representation, the class of tapered superquadrics that includes shapes such as cylinders, spheres, ellipsoids, and hyper-rectangles is utilized. Edges are employed as image cues and chamfer matching is used to evaluate the similarity between projected model configurations and actual image data. To deal with depth ambiguities, the authors make use of a ring of four inwards looking calibrated cameras, whereas to increase the accuracy of the method, they use subjects that wear tight clothing with sleeves of contrasting colors. Their hierarchical approach produces solutions for the different parts of the human body in a top-down manner using a kinematic chain, that is the position of the torso is found first and then the other body parts lower down in the hierarchy, such as the individual limbs, are recursively estimated. The proposed technique does not handle occlusion well.

In [165], twists are used to model the kinematic chain, whereas the individual body parts are modeled using cylinders. The authors perform monocular tracking of subjects walking either parallel to the image plane or diagonally towards the camera. Tracking is also performed on the classic Muybridge image sequences, where three different viewpoints are available at each time instance. To reduce the complexity, only half the human body skeleton (19 DOF), corresponding to the visible side of the body, is used.

In [173], a volumetric body model comprising of parallelepipeds, spheres, and truncated cones is used. The authors employed three calibrated cameras and derived 2D spring-like forces between the predicted model and the extracted silhouette. The derived 2D forces from each camera are combined to obtain 3D forces to be applied to the 3D model and align it with the data. The dynamical simulation is embedded in a Kalman filtering framework. Results are reported on subjects running in an indoor environment, wearing unconstrained clothing.

In [164], a shape model built from truncated cones is used to estimate motion in a monocular sequence using articulated kinematics. Edges and intensity are used as image cues in an EKF. In the Kalman filter prediction step, anatomical joint parameter limits are enforced. The results reported involve tracking in an unconstrained environment of a subject moving parallel to the image plane.

Sidenbladh et al. [14] use an articulated 3D model based on cylinders. They perform tracking in monocular cluttered sequences, using intensity as the image cue. The condensation algorithm [18] is employed for tracking. To constrain the tracking process, they use prior cyclic motion models for tracking walking humans. In a more recent work of the same authors, instead of using prior motion models, they employ models learned from training data [176]. Results are reported for tracking the planar motion of an arm and the motion of a walking person.

A single-camera model-free method that is based on learning and regression is proposed in [159]. The method tries to estimate in each frame the 55-D vector $\mathbf{x}$ of angles from 18 major joints of the human body. The image cue (observations) used is the body silhouette shape that is encoded by a 100-D histogram of shape contexts descriptors $\mathbf{z}$. For pose estimation in a single image, given a set of labelled training examples $(\mathbf{x}_i, \mathbf{z}_i)$ a regression model of the form

$$\mathbf{x} = \mathbf{r}(\mathbf{z}) = \sum_k \mathbf{a}_k \phi_k(\mathbf{z}) \tag{7.48}$$

that maps the (not previously seen) observation $\mathbf{z}$ to a pose $\mathbf{x}$ is learnt. In the previous equation, $\phi_k()$ are prespecified scalar basis functions that can be either linear or nonlinear (kernels) and $\mathbf{a}_k$ are the weight vectors that are estimated using the training data. Different regression methods that include relevance vector machine (RVM) regression and support vector machine (SVM) regression have been tested, with similar results. To apply the method on a video sequence, the method described earlier is embedded in a regression-based tracking framework that involves two levels of regression. The first is a dynamical model that, given the pose estimates in the two previous frames, provides a pose estimate $\tilde{\mathbf{x}}$ for the next frame. The second is a likelihood model that uses the rough pose estimate $\tilde{\mathbf{x}}$ as well as the image observations $\mathbf{z}$ derived from this frame for the evaluation of the final pose estimate for this frame. In this regression, the estimate $\tilde{\mathbf{x}}$ is essentially used to select which of the many possible pose estimates derived from observation-only regression is the correct one. The parameters of both regression models are again estimated through training with labeled examples. The additional temporal and dynamic information available in the tracking framework helps resolving ambiguities that often result (for example due to loss of depth information) when the method is applied on single images. The obtained results are very good although the method cannot be used for tracking complicated motions or motions that it has not been trained on.

Another example-based, model-free algorithm for the 3D tracking of the human hand is presented in [161]. This algorithm uses a tree of templates (example images), generated from a 3D geometric hand model. Each template depicts the hand in a different configuration. For each of these templates, the 3D pose parameters (27 degrees of freedom) are known. The templates are stored on the leaves of the tree, whereas each internal node corresponds to a hand configuration that is (roughly speaking) the "average" of the configurations stored in its children. Thus, a hierarchical (coarse-to-fine) organization/-partition of the state (3D pose) space is achieved. The likelihood that the hand depicted in a certain frame is in a certain 3D pose is evaluated by matching it to the templates (through the tree structure). Matching utilizes a cost/distance function that is based on the chamfer distance between edges in the frame and the templates. Additionally, edge orientation and skin color information are used so as to make the matching more robust in cluttered backgrounds. The method is presented in a Bayesian filtering formulation.

In [186] a model-free "tracking as recognition" approach to 3D articulated human body tracking is proposed. Articulated tracking is the last part of a larger system that involves a single calibrated camera. First, a statistical background appearance model is used to detect areas where moving persons might exist. Since the camera is calibrated and because humans are assumed to move on a known ground plane, the derivation of 3D coordinates from image coordinates is possible. A 3D coarse human shape model (an ellipsoid) is used to segment and track the human in 3D. The final 3D articulated tracking step can handle only three motion modes namely walking, running, and standing. This step is based on an articulated motion model organized as a hierarchical (two-level) finite state machine. The higher-level states represent the three modes, whereas each of the lower-level states represents a 3D body posture (i.e., one phase in the cycle of one of the motion modes). Each such state is associated with an optic flow template for the legs, as well as with a vector of joint angles of the corresponding body configuration, both derived from averaged motion capture data. Finding the motion mode that corresponds to an image sequence (i.e., whether it depicts walking, running or standing action) and subsequently identifying for each frame the corresponding phase of the motion cycle (e.g., the instance of the walk cycle) is accomplished by finding the optimal path within the state transition diagram, namely the path that maximizes the likelihood of the observations. The similarity between frames (observations) and lower-level states is evaluated using the speed of the body as well as optic flow derived from the frame. Once the optimal path is found, the 3D pose of the human body in a certain frame is estimated as the joint angles vector of the corresponding state.

An example-based method for single-view 3D articulated body tracking is proposed in [160]. The method is based on the so-called exemplars. Each exemplar is a pose (phase) within a motion sequence and consists of the joint angles vector along with multiple views from different viewing angles (in the form of silhouettes). Exemplars are generated using motion capture data and a 3D rendering software. Tracking is performed using a PF that is coupled to the dynamics of the motion, as represented by the exemplar-based model. Dynamics are modeled by clustering exemplars that correspond to similar 3D poses and by taking into account the sequencing of the poses within each motion sequence. The major disadvantage of this method is that it can handle only a single type of motion, for example walking.

7.3.2 2D Articulated Object Tracking

2D articulated object tracking refers to recovering the position in 2D, that is, on the image plane, of the rigid parts comprising the articulated structure. It follows that 2D articulated object tracking shares characteristics and principles with 2D rigid object tracking, described in Section 7.2.1, as well as 3D articulated object tracking, described in Section 7.3.1. Contour-based tracking presented in Section 7.2.1.2 can be also used for articulated object tracking in two dimensions.

Similarly to 3D algorithms, 2D articulated object tracking can be model-free or model-based. Model-free methods (e.g. [135]) proceed by exploiting image information (edges, intensity, etc.) to create coherent structures that correspond to the rigid parts of the articulated structure (e.g., the upper and lower segments of the human arm). Occlusion can create problems in such methods, due to the lack of visibility or partial visibility of one or more of the rigid parts comprising the articulated object. Alternatively, model-based approaches employ a 2D or 3D model of the articulated object, depending on the application and the precision required. The rigid parts of a 2D model can be represented using geometric primitives, such as sticks (i.e., lines), circles, rectangles, and ellipses or by curves and snakes if the object parts are allowed to deform. Additionally, the appearance of the object can be modeled by including texture on the above-mentioned geometric primitives. If a 3D model is used, this can be similar to the ones described in Section 7.3.1.

In an early work, Lowe [187] uses a 3D model to track an object in video sequences. The Marr–Hildreth edge detector [188] is used to identify edges in the frames of the video sequence. The edges are grouped based on local connectivity. The matching process finds the best match between the extracted image edges and the projected contours of the 3D model. The matches are ranked and the best ones are chosen for minimization. The model is then rotated and translated and the aforementioned procedure is iteratively repeated until the best configuration is found. The computational time depends on the complexity of the model used. The algorithm is robust but rather slow.

Another model-based 2D articulated object tracking algorithm is presented in [189]. The proposed system employs three different models: a 2D articulated model of the object being tracked, a dynamic model, and an appearance model. The 2D articulated model approximates the shape of the object in the image (in this case the human body) and consists of rigid parts that are connected with each other. The displacement vectors and the joint angles are the parameters that describe the motion of the articulated object. The dynamic model is a stochastic linear equation that rules these parameters and is used to predict the model configuration. It consists of a trained linear stochastic equation and a look-up table which contains a list of exceptions (nonlinear configurations). The appearance model is a set of profiles of the object centered at specific feature points on the object. These feature points are detected during the tracking process using template matching. The tracking of the object is achieved through the following steps: first, the dynamic model predicts potential configurations of the model and consequently a template matching technique is used to detect the specific feature points for each configuration. After filtering the results by EKF and evaluating them by measuring the color matchings between the results and the appearance model, the configuration with the largest matching score is selected as the final result for this frame. Other model-based articulated object tracking methods are presented in [141, 190–195].

Two methods for 2D articulated human body tracking are presented in [196]. The methods build an appearance model for each person in a video and then track this person by detecting those models in each frame. The first method builds the appearance model using a bottom-up approach, that is it groups together candidate body parts found throughout a sequence. The second one (that is also described in [197]) follows a top-down approach; it automatically builds the appearance model for the entire body as detected in a convenient key pose in a single frame. The first component of this second method is a person detector that only detects people in typical poses, more specifically in the mid-stance of a lateral walk. The reasoning behind this choice is that such a pose is easier to detect and that the appearance information for the various body parts derived from such a pose is accurate (because, for example, no self occlusion occurs) and can be used to localize the limbs of this person in other arbitrary poses. This per-frame (and single scale) detector is applied on the frames of the sequence and results in at least one detection. Given the estimated limb (head, torso, upper/lower arm, and left/right upper/lower legs) positions from that detection, a quadratic logistic regression classifier for each limb in RGB space (i.e., an appearance model) is learned (using the detected limb pixels as positive and all non-person pixels as negative examples). These classifiers are then applied on all frames of the sequence (both before and after the detection, obviously depicting the person in any possible pose) to obtain limb masks. The masks are searched (using general pose deformations at multiple scales) for candidate limbs arranged in a pictorial structure. Pictorial structures [198] are collections of parts arranged in a deformable configuration. Each part is represented using a simple appearance model and the deformable configuration is represented by spring-like connections between pairs of parts.

Approaches that aim at estimating the 2D configuration of the human body in a single image have been also proposed [199, 200]. Obviously these approaches are of interest to the tracking community because they can be generalized or embedded within a tracking framework.

REFERENCES

[1] A. Yilmaz, O. Javed, and M. Shah. Object tracking: A survey. *ACM Comput. Surv. (CSUR)*, 38(4), 2006.

[2] D. M. Gavrila. The visual analysis of human movement: A survey. *Comput. Vis. Image Underst.*, 73(1):82–98, 1999.

[3] Y. Luo, T. D. Wu, and J. N. Hwang. Object-based analysis and interpretation of human motion in sports video sequences by dynamic bayesian networks. *Comput. Vis. Image Underst.*, 92(2–3):196–216, 2003.

[4] P. Figueroa, N. Leite, R. M. L. Barros, I. Cohen, and G. Medioni. Tracking soccer players using the graph representation. In *Int. Conf. Pattern Recognit. (ICPR2004)*, Vol. IV, 787–790, Cambridge, England, August 2004.

[5] M. Köhle, D. Merkl, and J. Kastner. Clinical gait analysis by neural networks: issues and experiences. In *10th IEEE Symp. Comput. Based Med. Syst. (CBMS97)*, 138–143, Maribor, Slovenia, March 1997.

[6] D. Meyer, J. Denzler, and H. Niemann. Model based extraction of articulated objects in image sequences. In *IEEE Int. Conf. Image Process. (ICIP97)*, Vol. III, 78–81, Washington, DC, USA, October 1997.

[7] T. B. Moeslund and E. Granum. A survey of computer vision-based human motion capture. *Comput. Vis. Image Underst.*, 81(3):231–268, 2001.

[8] T. B. Moeslund. Interacting with a virtual world through motion capture. In L. Qvortrup, editor, Interaction in Virtual Inhabited 3D Worlds, Chap. 11. Springer-Verlag, Berlin/New York, 2000.

[9] G. H. Burdea and P. Coiffet. *Virtual Reality Technology*, 2nd ed. Wiley-Interscience, New York, 2003.

[10] G. Welch and E. Foxlin. Motion tracking: No silver bullet, but a respectable arsenal. *IEEE Comput. Graph. Appl.*, 22(6):24–38, 2002.

[11] J. K. Aggarwal, Q. Cai, W. Liao, and B. Sabata. Nonrigid motion analysis: Articulated and elastic motion. *Comput. Vis. Image Underst.*, 70(2):142–156, 1998.

[12] I. Haritaoglu, R. Cutler, D. Harwoodc, and L. S. Davis. Backpack: Detection of people carrying objects using silhouettes. *Comput. Vis. Image Underst.*, 81(3):385–397, 2001.

[13] M. Isard and J. MacCormick. BraMBLe: A bayesian multiple-blob tracker. In *IEEE Int. Conf. Comput. Vis. (ICCV2001)*, Vol. 2, 34–41, Vancouver, BC, Canada, July 2001.

[14] H. Sidenbladh, M. Black, and D. Fleet. Stochastic tracking of 3D human figures using 2D image motion. In *Eur. Conf. Comput. Vis., (ECCV2000)*, Vol. 2, 720–718, Dublin, Ireland, June 2000.

[15] P. S. Maybeck. *Stochastic Models, Estimation and Control*, Vol. 1. Academic, New York, 1979.

[16] G. Welch and G. Bishop. An introduction to the Kalman filter. *Tech. Rep.* 95-041, University of North Carolina at Chapel Hill, Department of Computer Science, 1995.

[17] P. Zarchan and H. Musoff. *Fundamentals of Kalman Filtering: A Practical Approach.* American Institute of Aeronautics, 2001.

[18] M. Isard and A. Blake. Condensation-conditional density propagation for visual tracking. *Int. J. Comput. Vis.*, 29(1):5–28, 1998.

[19] O. Faugeras, L. Quan, and P. Strum. Self-calibration of a 1D projective camera and its application to the self-calibration of a 2D projective camera. *IEEE Trans. Pattern Anal. Mach. Intell.*, 22(10): 1179–1185, 2000.

[20] E. Malis and R. Cipolla. Camera self-calibration from unknown planar structures enforcing the multiview constraints between collineations. *IEEE Trans. Pattern Anal. Mach. Intell.*, 24(9):1268–1272, 2002.

[21] B. Rosenhahn, R. Klette, and D. Metaxas. *Human Motion: Understanding, Modelling, Capture, and Animation.* Springer-Verlag, New York, 2008.

[22] J. MacCormick. *Stochastic Algorithms for Visual Tracking.* Springer-Verlag, New York, 2002.

[23] L. Wang, W. Hu, and T. Tan. Recent developments in human motion analysis. *Pattern Recognit.*, 36(3):585–601, 2003.

[24] J. K. Aggarwal and Q. Cai. Human motion analysis: A review. *Comput. Vis. Image Underst.*, 73(3):428–440, 1999.

[25] J. J. Wang and S. Singh. Video analysis of human dynamics-a survey. *Real Time Imaging J.*, 9(5):320–345, 2003.

[26] V. Lepetit and P. Fua. Monocular model-based 3d tracking of rigid objects: A survey. *Found. Trends Comput. Graph. Vis.*, 1(1):1–89, 2005.

[27] T. B. Moeslund, A. Hilton, and V. Kruger. A survey of advances in vision-based human motion capture and analysis. *Comput. Vis. Image Underst.*, 104(2–3):90–126, 2006.

[28] W. Hu, T. Tan, L. Wang, and S. Maybank. A survey on visual surveillance of object motion and behaviors. *IEEE Trans. Syst., Man Cybern. C*, 34(3):334–352, 2004.

[29] E. Trucco and A. Verri. *Introductory Techniques for 3D Computer Vision.* Prentice Hall, Upper Saddle River, NJ, 1998.

[30] T. Gevers and A. W. M. Smeulders. Pictoseek: Combining color and shape invariant features for image retrieval. *IEEE Trans. Image Process.*, 9(1):102–119, 2000.

[31] G. Buchsbaum. A spatial processor model for object color perception. *J. Franklin Inst.*, 310(1): 1–26, 1980.

[32] G. Finlayson, S. Hordley, and P. Hubel. Color by correlation: A simple, unifying framework for colour constancy. *IEEE Trans. Pattern Anal. Mach. Intell.*, 23(11):1209–1221, 2001.

[33] B. D. Zarit, B. J. Super, and F. K. H. Quek. Comparison of five color models in skin pixel classification. In *ICCV99 Int. Workshop Recognit., Anal. Track. Faces Gestures Real-Time Syst. (RATFG-RTS99)*, 58–63, Corfu, Greece, September 1999.

[34] C. Terrillon, M. David, and S. Akamatsu. Automatic detection of human faces in natural scene images by use of a skin color model and invariant moments. In *Third IEEE Int. Conf. Automat. Face Gesture Recognit. (AFGR98)*, 112–117, Nara, Japan, April 1998.

[35] A. Albiol, L. Torres, and E. Delp. Optimum color spaces for skin detection. In *IEEE Int. Conf. Image Process. (ICIP2001)*, Vol. 1, 122–124, Thessaloniki, Greece, October 2001.

[36] J. Starck and A. Hilton. Surface capture for performance based animation. *IEEE Comput. Graph. Appl.*, 27(3):21–31, 2007.

[37] M. J. Swain and D. H. Ballard. Color indexing. *Int. J. Comput. Vis.*, 7(1):11–32, 1991.

[38] M. Vezhnevets. Face and facial feature tracking for natural Human-Computer Interface. In *Int. Conf. Comput. Graph. between Eur. Asia (GraphiCon-2002)*, 86–90, Nizhny Novgorod, Russia, September 2002.

[39] K. Schwerdt and J. L. Crowley. Robust face tracking using color. In *Int. Conf. Automat. Face Gesture Recognit. (AFGR2000)*, 90–95, Grenoble, France, March 2000.

[40] D. Comaniciu, V. Ramesh, and P. Meer. Kernel-based object tracking. *IEEE Trans. Pattern Anal. Mach. Intell.*, 25(5):564–577, 2003.

[41] D. Comaniciu and P. Meer. Mean shift: A robust approach toward feature space analysis. *IEEE Trans. Pattern Anal. Mach. Intell.*, 24(5):603–619, 2002.

[42] W. Lu, J. Yang, and A. Waibel. Skin-color modeling and adaptation. In *Third Asian Conf. Comput. Vis. (ACCV98)*, Vol. 2, 687–694, Hong Kong, China, January 1998.

[43] P. Perez, C. Hue, J. Vermaak, and M. Gangnet. Color-based probabilistic tracking. In *Eur. Conf. Comput. Vis. (ECCV2002)*, Vol. 1, 661–675, Copenhagen, Denmark, May-June 2002.

[44] G. R. Bradski. Real time face and object tracking as a component of a perceptual user interface. In *Workshop Appl. Comput. Vis. (WACV98)*, 214–219, Princeton, NJ, USA, October 1998.

[45] Y. Wu and T. S. Huang. Color tracking by transductive learning. In *Int. Conf. Comput. Vis. Pattern Recognit. (CVPR2000)*, Vol. 1, 133–138, Hilton Head, SC, USA, June 2000.

[46] H. Sidenbladh and M. Black. Learning the statistics of people in images and video. *Int. J. Comput. Vis.*, 54(1):181–207, 2003.

[47] S. J. McKenna, Y. Raja, and S. Gong. Tracking and segmenting people in varying lighting conditions using colour. In *Int. Conf. Automat. Face Gesture Recognit. (AFGR98)*, 228–233, Nara, Japan, April 1998.

[48] P. Fieguth and D. Terzopoulos. Color-based tracking of heads and other mobile objects at video frame rates. In *Int. Conf. Comput. Vis. Pattern Recognit. (CVPR97)*, 21–27, San Juan, PR, USA, June 1997.

[49] C. Wren, A. Azarbayejani, T. Darrell, and A. Pentland. PFinder: Real-time tracking of the human body. *IEEE Trans. Pattern Anal. Mach. Intell.*, 19(7):780–785, 1997.

[50] C. Smith, C. Richards, S. A. Brandt, and N. P. Papanikolopoulos. Visual tracking for intelligent vehicle-highway systems. *IEEE Trans. Vehicular Technol.*, 45(4):744–759, 1996.

[51] C. Stauffer and W. E. L. Grimson. Learning patterns of activity using real-time tracking. *IEEE Trans. Pattern Anal. Mach. Intell.*, 22(8):747–757, 2000.

[52] A. Baumberg and D. Hogg. Learning flexible models from image sequences. In *Eur. Conf. Comput. Vis. (ECCV94)*, Vol. 1, 299–308, Stockholm, Sweden, May 1994.

[53] Q. Cai and J. Aggarwal. Tracking human motion using multiple cameras. In *Int. Conf. Pattern Recognit. (ICPR96)*, Vol. 3, 68–72, Vienna, Austria, August 1996.

[54] I. Haritaoglu, D. Harwood, and L. S. Davis. W4: Real-time surveillance of people and their activities. *IEEE Trans. Pattern Anal. Mach. Intell.*, 22(8):809–830, 2000.

[55] A. M. Elgammal, D. Harwood, and L. S. Davis. Non-parametric model for background subtraction. In *Eur. Conf. Comput. Vis. (ECCV00)*, Vol. II, 751–767, 2000.

[56] A. Monnet, A. Mittal, N. Paragios, and V. Ramesh. Background modeling and subtraction of dynamic scenes. In *IEEE Int. Conf. Comput. Vis. (ICCV)*, Vol. 2, 1305–1312, October 2003.

[57] J. Zhong and S. Sclaroff. Segmenting foreground objects from a dynamic textured background via a robust kalman filter. In *IEEE Int. Conf. Comput. Vis. (ICCV)*, Vol. 1, 44–50, October 2003.

[58] L. M. Fuentesa and S. A. Velastin. People tracking in surveillance applications. *Image Vis. Comput.*, 24(11):1165–1171, 2006.

[59] S. Avidan. Ensemble tracking. *IEEE Trans. Pattern Anal. Mach. Intell.*, 29(2):261–271, 2007.

[60] Y. Freund and R. E. Schapire. A decision-theoretic generalization of on-line learning and an application to boosting. In *Second Eur. Conf. Comput. Learn. Theory*, 23–37, 1995.

[61] V. Vapnik. *The nature of statistical learning theory.* Springer Verlag, New York, 1995.

[62] S. Avidan. Support vector tracking. *IEEE Trans. Pattern Anal. Mach. Intell.*, 26(8):1064–1072, 2004.

[63] T. R. Collins, Y. Liu, and M. Leordeanu. Online selection of discriminative tracking features. *IEEE Trans. Pattern Anal. Mach. Intell.*, 27(10):1631–1643, 2005.

[64] A. D. Jepson, D. J. Fleet, and T. F. El-Maraghi. Robust online appearance models for visual tracking. *IEEE Trans. Pattern Anal. Mach. Intell.*, 25(10):1296–1311, 2003.

[65] A.-R. Mansouri. Region tracking via level set pdes without motion computation. *IEEE Trans. Pattern Anal. Mach. Intell.*, 24(7):947–961, 2002.

[66] E. L. Andrade, J. C. Woods, E. Khan, and M. Ghanbari. Region-based analysis and retrieval for tracking of semantic objects and provision of augmented information in interactive sport scenes. *IEEE Trans. Multimed.*, 7(6):1084–1096, 2005.

[67] A. Cavallaro, O. Steiger, and T. Ebrahimi. Tracking video objects in cluttered background. *IEEE Trans. Circuits Syst. Video Technol.*, 15(4):575–584, 2005.

[68] N. Amezquita, R. Alquezar, and F. Serratosa. A new method for object tracking based on regions instead of contours. In *IEEE Conf. Comput. Vis. Pattern Recognit. (CVPR07)*, 1–8, June 2007.

[69] M. Kass, M. Witkin, and A. Terzopoulos. Snakes: Active contour models. *Int. J. Comput. Vis.*, 1(4):321–331, 1988.

[70] Y. Fu, A. T. Erdem, and A. M. Tekalp. Tracking visible boundary of objects using occlusion adaptive motion snake. *IEEE Trans. Image Process.*, 9(12):2051–2060, 2000.

[71] H. Wang, J. Leng, and Z. M. Guo. Adaptive dynamic contour for real-time object tracking. In *Image Vis. Comput. New Zealand (IVCNZ2002)*, Auckland, New Zealand, December 2002.

[72] B. Bascle, P. Bouthemy, R. Deriche, and F. Meyer. Tracking complex primitives in an image sequence. In *Int. Conf. Pattern Recognit. (ICPR94)*, Vol. 1, 426–431, Jerusalem, Israel, October 1994.

[73] N. Xu and N. Ahuja. Object contour tracking using graph cuts based active contours. In *IEEE Int. Conf. Image Process. (ICIP2002)*, Vol. 3, 277–280, Rochester, NY, USA, September 2002.

[74] N. Xu, R. Bansal, and N. Ahuja. Object segmentation using graph cuts based active contours. In *Int. Conf. Comput. Vis. Pattern Recognit. (CVPR2003)*, Vol. 2, 46–53, Madison, WI, USA, June 2003.

[75] A. Yilmaz, X. Li, and M. Shah. Contour-based object tracking with occlusion handling in video acquired using mobile cameras. *IEEE Trans. Pattern Anal. Mach. Intell.*, 26(11):1531–1536, 2004.

[76] Y. Rathi, N. Vaswani, A. Tannenbaum, and A. Yezzi. Particle filtering for geometric active contours with application to tracking moving and deforming objects. In *IEEE Conf. Comput. Vis. Pattern Recognit.*, Vol. 2, 2–9, June 2005.

[77] M. Yokoyama and T. Poggio. A contour-based moving object detection and tracking. In *Joint IEEE Int. Workshop Vis. Surveill. Perform. Eval. Track. Surveill.*, 271–276, Beijing, China, October 2005.

[78] F. P. Nava and A. F. Martel. Condensation-based contour tracking with Sobolev smoothness priors. *Neural Parallel Sci. Comput.*, 10(1):47–56, 2002.

[79] M. Niethammer, A. Tannenbaum, and S. Angenent. Dynamic active contours for visual tracking. *IEEE Trans. Automat. Contr.*, 51(4):562–579, 2006.

[80] J.-H. Jean and R.-Y. Wu. Adaptive visual tracking of moving objects modeled with unknown parameterized shape contour. In *IEEE Int. Conf. Network., Sens. Contr.*, Vol. 1, 76–81, March 2004.

[81] M. Niethammer, P. A. Vela, and A. Tannenbaum. Geometric observers for dynamically evolving curves. *IEEE Trans. Pattern Anal. Mach. Intell.*, 30(6):1093–1108, 2008.

[82] K. Toyama. Prolegomena for robust face tracking. Tech. Rep. MSR-TR-98-65, Microsoft Research, 1998.

[83] C. Harris and M. Stephens. A combined corner and edge detector. In *4th Alvey Vis. Conf.*, 147–151, Manchester, 1998.

[84] D. G. Lowe. Distinctive image features from scale-invariant keypoints. *Int. J. Comput. Vis.*, 60(2): 91–110, 2004.

[85] J. Shi and C. Tomasi. Good features to track. In *IEEE Int. Conf. Comput. Vis. Pattern Recognit. (CVPR94)*, 593–600, Seattle, WA, United States, June 1994.

[86] A. Nikolaidis and I. Pitas. Probabilistic multiple face detection and tracking using entropy measures. *Pattern Recognit.*, 33(11):1783–1791, 2000.

[87] C. Tomasi and T. Kanade. Detection and tracking of point features. *Tech. Rep.* CMU-CS-91-132, School of Computer Science, Carnegie Mellon University, Pittsburgh, 1991.

[88] E. Elagin, J. Steffens, and H. Neven. Automatic pose estimation system for human faces based on bunch graph matching technology. In *Int. Conf. Automat. Face Gesture Recognit. (AFGR98)*, 136–141, Nara, Japan, April 1998.

[89] A. Gee and R. Cipolla. Fast visual tracking by temporal consensus. *Image Vis. Comput.*, 14(2):105–114, 1996.

[90] T. Horprasert, Y. Yacoob, and L. S. Davis. Computing 3-D head orientation from a monocular image sequence. In *Int. Conf. Automat. Face Gesture Recognit. (AFGR96)*, 242–247, Killington, VT, USA, October 1996.

[91] T. Maurer and C. Von Der Malsburg. Tracking and learning graphs and pose on image sequences of faces. In *Int. Conf. Automat. Face Gesture Recognit. (AFGR96)*, 176–181, Killington, VT, USA, October 1996.

[92] T. S. Jebara and A. P. Pentland. Parameterized structure from motion for 3D adaptive feedback tracking of faces. In *Int. Conf. Comput. Vis. Pattern Recognit. (CVPR97)*, 144–150, San Juan, PR, USA, June 1997.

[93] P. Yao, G. Evans, and A. Calway. Face tracking and pose estimation using affine motion parameters. In *12th Scand. Conf. Image Anal.*, 531–536, Norwegian Society for Image Processing and Pattern Recognition, June 2001.

[94] C. Luo, T. S. Chua, and T. K. Ng. Face tracking in video with hybrid of Lucas-Kanade and Condensation algorithm. In *IEEE Int. Conf. Multimed. Expo (ICME2003)*, Baltimore, Maryland, USA, July 2003.

[95] H. Chao, Y. F. Zheng, and S. C. Ahalt. Object tracking using the Gabor wavelet transform and the golden section algorithm. *IEEE Trans. Multimed.*, 4(4):528–538, 2002.

[96] A. Shokurov, A. Khropov, and D. Ivanov. Feature tracking in images and video. In *Int. Conf. Comput. Graph. between Eur. Asia (GraphiCon-2003)*, 177–179, Moscow, Russia, September 2003.

[97] E. Loutas, I. Pitas, and C. Nikou. Probabilistic multiple face detection and tracking using entropy measures. *IEEE Trans. Circuits Syst. Video Technol.*, 14(1):128–135, 2004.

[98] B. D. Lucas and T. Kanade. An iterative image registration technique with an application to stereo vision. In *Int. Joint Conf. Artif. Intell.*, 674–679, Vancouver, BC, Canada, August 1981.

[99] P. Beardsley, P. H. S. Torr, and A. Zisserman. 3D model aquisition from extended image sequences. In *Eur. Conf. Comput. Vis. (ECCV96)*, Vol. 2, 683–695, Cambridge, England, April 1996.

[100] J. Y. Bouguet. Pyramidal implementation of the Lucas Kanade feature tracker. *Tech. Rep.*, Intel Corporation, Microprocessor Research Labs, OpenCV Documents, 1999.

[101] M. Kolsch and M. Turk. Fast 2D hand tracking with flocks of features and multi-cue integration. In *IEEE Workshop Real-Time Vis. Hum. Comput. Interact. (RTV4HCI 2004)*, Washington DC, USA, June 2004.

[102] S. Marcelja. Mathematical description of the responses of simple cortical cells. *J. Opt. Soc. Am.*, 70(11):1297–1300, 1980.

[103] M. Krinidis, N. Nikolaidis, and I. Pitas. 2d feature point selection and tracking using 3d physics-based deformable surfaces. *IEEE Trans. Circuits Syst. Video Technol.*, 17(7):876–888, 2007.

[104] M. Turk and A. Pentland. Eigenfaces for recognition. *J. Cognit. Neurosci.*, 3(1):71–96, 1991.

[105] M. J. Black and A. D. Jepson. Eigentracking: Robust matching and tracking of articulated objects using a view-based representation. *Int. J. Comput. Vis.*, 26(1):63–84, 1998.

[106] S. Pigeon and L. Vandendorpe. The M_2VTS multimodal face database. *In Int. Conf. Audio- Video-Based Biometric Person Authentication (AVBPA97)* 403–409, Crans Montana, Switzerland, March 1997

[107] H. T. Nguyen and A. W. M. Smeulders. Template tracking using color invariant pixel features. In *IEEE Int. Conf. Image Process. (ICIP2000)*, Vol. 1, 569–572, Rochester, NY, USA, September 2000.

[108] H. T. Nguyen and A. W. M. Smeulders. Fast occluded object tracking by a robust appearance filter. *IEEE Trans. Pattern Anal. Mach. Intell.*, 26(8):1099–1104, 2004.

[109] C. F. Olson. Maximum-likelihood image matching. *IEEE Trans. Pattern Anal. Mach. Intell.*, 24(6):853–857, June 2002.

[110] L. V. Tsap, D. B. Goldgof, and S. Sarkar. Fusion of physically-based registration and deformation modeling for nonrigid motion analysis. *IEEE Trans. Image Process.*, 10(11):1659–1669, 2001.

[111] Y. Wang and S. Zhu. Analysis and synthesis of textured motion: Particles and waves. *IEEE Trans. Patern Anal. Mach. Intell.*, 26(10):1348–1363, 2004.

[112] T. Schoepflin, V. Chalana, D. R. Haynor, and Y. Kim. Video object tracking with a sequential hierarchy of template deformations. *IEEE Trans. Circuits Syst. Video Technol.*, 11(11):1171–1182, 2001.

[113] R. Kjeldsen and A. Aner. Improving face tracking with 2D template warping. In *Int. Conf. Face Gesture Recognit. (AFGR2000)*, 129–135, Grenoble, France, March 2000.

[114] J. Heinzmann and A. Zelinsky. Robust real-time face tracking and gesture recognition. In *Int. Joint Conf. Artif. Intell. (IJCAI97)*, 1525–1530, Nagoya, Aichi, Japan, August 1997.

[115] Y. Zhong, A. K. Jain, and M. P. Dubuisson-Jolly. Object tracking using deformable templates. *IEEE Trans. Patern Anal. Mach. Intell.*, 22(5):544–549, 2000.

[116] S. Arulampalam, S. Maskell, N. Gordon, and T. Clapp. A tutorial on particle filters for on-line non-linear/non-Gaussian Bayesian tracking. *IEEE Trans. Signal Process.*, 50(2):174–188, 2002.

[117] K. Kanazawa, D. Koller, and S. Russell. Stochastic simulation algorithms for dynamic probabilistic networks. In *Eleventh Annu. Conf. Uncertain. AI (UAI95)*, 346–351, Montreal, Canada, August 1995.

[118] P. Moral. Non-linear filtering: Interacting particle solution. *Markov Process. Relat. Fields*, 2(4):555–580, 1996.

[119] Q. Ji and X. Yang. Real-time eye, gaze, and face pose tracking for monitoring driver vigilance. *Real Time Imaging*, 8(5):357–377, 2002.

[120] S. J. Julier, J. K. Ulmann, and H. Durrant-Whyte. A new approach for filtering nonlinear systems. In *Am. Contr. Conf.*, 1628–1632, Orlando, April 1995.

[121] G. Kitagawa. Monte Carlo filter and smoother for non-Gaussian non-linear state space models. *J. Comput. Graph. Stat.*, 5(1):1–25, 1996.

[122] A. Doucet, N. De Freitas, and N. Gordon. *Sequential Monte Carlo Methods in Practice.* Springer-Verlag, New York, 2001.

[123] R. Ristic, S. Arulampalam, and N. Gordon. *Beyond the Kalman Filter: Particle Filters for Tracking Applications.* Artech House, Boston, MA, 2004.

[124] P. Perez, J. Vermaak, and A. Blake. Data fusion for visual tracking with particles. *Proc. IEEE*, 92(3):495– 513, 2004.

[125] M. Spengler and B. Schiele. Towards robust multi-cue integration for visual tracking. In *Int. Workshop Comput. Vis. Syst.*, 94–107, Vancouver, Canada, July 2001.

[126] K. Okuma, A. Taleghani, N. De Freitas, J. J. Little, and D. G. Lowe. A boosted particle filter: Multitarget detection and tracking. In *Eur. Conf. Comput. Vis. (ECCV04)*, Vol. I, 28–39, Prague, Czech Republic, May 2004.

[127] P. Viola and M. Jones. Rapid object detection using a boosted cascade of simple features. In *IEEE Conf. Comput. Vis. Pattern Recognit.*, Vol. 1, 511–518, Kauai, USA, December 2001.

[128] J. Vermaak, A. Doucet, and P. Perez. Maintaining multi-modality through mixture tracking. In *Int. Conf. Comput. Vis.*, Vol. 2, 1110–1116, October 2003.

[129] S. K. Zhou, R. Chellappa, and B. Moghaddam. Visual tracking and recognition using appearance-adaptive models in particle filters. *IEEE Trans. Image Process.*, 13(11):1491–1506, 2004.

[130] Z. Khan, T. Balch, and F. Dellaert. MCMC-based particle filtering for tracking a variable number of interacting targets. *IEEE Trans. Pattern Anal. Mach. Intell.*, 27(11):1805–1918, 2005.

[131] J. Wang, X. Chen, and W. Gao. Online selecting discriminative tracking features using particle filter. In *IEEE Conf. Comput. Vis. Pattern Recognit.*, Vol. 2, 1037–1042, June 2005.

[132] M. Bray, E. Koller-Meier, and L. Van Gool. Smart particle filtering for high-dimensional tracking. *Comput. Vis. Image Underst.*, 106(1):116–129, 2007.

[133] C. Gentile, O. Camps, and M. Sznaier. Segmentation for robust tracking in the presence of severe occlusion. *IEEE Trans. Image Process.*, 13(2):166–178, 2004.

[134] E. Loutas, K. I. Diamantaras, and I. Pitas. Occlusion resistant object tracking. In *IEEE Int. Conf. Image Process. (ICIP2001)*, Vol. II, 65–68, Thessaloniki, Greece, October 2001.

[135] Y. Ricquebourg and P. Bouthemy. Real-time tracking of moving persons by exploiting spatio-temporal image slices. *IEEE Trans. Pattern Anal. Mach Intell.*, 22(8):797–808, 2000.

[136] H. Tsutsui, J. Miura, and Y. Shirai. Optical flow-based person tracking by multiple cameras. In *IAPR Workshop Mach. Vis. Appl. (MVA98)*, 418–421, Chiba, Japan, November 1998.

[137] A. Mittal and L. S. Davis. M2Tracker: A multi-view approach to segmenting and tracking people in a cluttered scene using region-based stereo. *Int. J. Comput. Vis.*, 51(3):189–203, 2003.

[138] A. Utsumi, H. Mori, J. Ohya, and M. Yachida. Multiple-human tracking using multiple cameras. In *Int. Conf. Automat. Face Gesture Recognit. (AFGR98)*, 498–503, Nara, Japan, April 1998.

[139] S. L. Dockstader and A. M. Tekalp. Multiple camera tracking of interacting and occluded human motion. *Proc. IEEE*, 89, 1441–1455, 2001.

[140] J. Rehg and T. Kanade. Model-based tracking of self occluding articulated objects. In *IEEE Int. Conf. Comput. Vis. (ICCV95)*, 612–617, Cambridge, MA, USA, June 1995.

[141] D. D. Morris and J. Rehg. Singularity analysis for articulated object tracking. In *Int. Conf. Comput. Vis. Pattern Recognit. (CVPR98)*, 289–296, Santa Barbara, CA, USA, June 1998.

[142] J. Paterson and A. Fitzgibbon. 3D head tracking using non-linear optimization. In *Br. Mach. Vis. Conf.*, Vol. 2, 609–618, Norwich, UK, September 2003.

[143] J. Ahlberg and R. Forchheimer. Face tracking for model-based coding and face animation. *Int. J. Imaging Syst. Technol.*, 13(1):8–22, 2003.

[144] T. Goto, S. Kshirsagar, and N. Magnenat-Thalmann. Automatic face cloning and animation. *IEEE Signal Process. Magaz.*, 18(3):17–25, 2001.

[145] H. Li, P. Roivainen, and R. Forcheimer. 3D motion estimation in model-based facial image coding. *Trans. Pattern Anal. Mach. Intell.*, 15(6):545–555, 1993.

[146] Z. Zhu and Q. Ji. Robust pose invariant facial feature detection and tracking in real-time. In *Int. Conf. Pattern Recognit. (ICPR2006)*, Vol. 1, 1092–1095, 2006.

[147] J. Chen and B. Tiddeman. Robust facial feature tracking under various illuminations. In *IEEE Int. Conf. Image Process.*, 2829–2832, October 2006.

[148] J. Chen and B. Tiddeman. A robust facial feature tracking system. In *IEEE Conf. Adv. Video Signal Based Surveill. (AVSS2005)*, 445–449, September 2005.

[149] W. Hu, X. Xiao, D. Xie, T. Tan, and S. Maybank. Traffic accident prediction using 3D model-based vehicle tracking. *Trans. Vehicular Technol.*, 53(3):677–694, 2004.

[150] A. Azarbayejani, T. Starner, B. Horowitz, and A. Pentland. Visually controlled graphics. *IEEE Trans. Pattern Anal. Mach. Intell.*, 15(6):602–605, 1993.

[151] M. Krinidis, N. Nikolaidis, and I. Pitas. 3d head pose estimation in monocular video sequences using deformable surfaces and radial basis functions. *IEEE Trans. Circuits Syst. Video Technol.*, 2008.

[152] M. Kampmann. Automatic 3d face model adaptation for model-based coding of videophone sequences. *Trans. Circuits Syst. Video Technol.*, 12(3):172–182, 2002.

[153] L. Zhang. Automatic adaptation of a face model using action units for semantic coding of videophone sequences. *IEEE Trans. Circuits Syst. Video Technol.*, 8(3):781–795, 1998.

[154] S.-C. Pei, C.-W. Ko, and M.-S. Su. Global motion estimation in model-based image coding by tracking three-dimensional contour feature points. *IEEE Trans. Circuits Syst. Video Technol.*, 8(2):181–190, 1998.

[155] M. La Cascia, S. Sclaroff, and V. Athitsos. Fast, reliable head tracking under varying illumination: An approach based on registration of texture-mapped 3d models. *IEEE Trans. Medi. Imaging*, 22(4):322–336, 2000.

[156] R. Rydfalk. *CANDIDE*, a parameterised face. *Tech. Rep.* Lith-ISY-I-0866, Univesity of Linköping, Sweden, 1987.

[157] Y. Yoon, A. Kosaka, J. B. Park, and A. C. Kak. A new approach to the use of edge extremities for model-based object tracking. In *IEEE Int. Conf. Rob. Autom.*, 1883–1889, Barcelona, Spain, April 2005.

[158] P. Mittrapiyanuruk, G. N. DeSouza, and A. C. Kak. Accurate 3d tracking of rigid objects with occlusion using active appearance models. In *IEEE Workshop Motion Video Comput. (WACV/MOTION'05)*, Vol. 2, 90–95, 2005.

[159] A. Agarwal and B. Triggs. Recovering 3d human pose from monocular images. *IEEE Trans. Pattern Anal. Mach. Intell.*, 298(1):44–58, 2006.

[160] E.-J. Ong, A. S. Micilotta, R. Bowden, and A. Hilton. Viewpoint invariant exemplar-based 3d human tracking. *Comput. Vis. Image Underst.*, 104(2-3):178–189, 2006.

[161] B. Stenger, A. Thayananthan, P. H. S. Torr, and R. Cipolla. Filtering using a tree-based estimator. In *IEEE Int. Conf. Comput. Vis. (ICCV'03)*, Vol. 2, 1063, Nice, France, October 2003.

[162] G. Shakhnarovich, P. Viola, and T. Darrell. Fast pose estimation with parameter-sensitive hashing. In *IEEE Int. Conf. Comput. Vis. (ICCV'03)*, 750–757, Nice, France, October 2003.

[163] G. Mori and J. Malik. Estimating human body configurations using shape context matching. In *Proc. Eur. Conf. Comput. Vis.*, Vol. 3, 666–680, 2002.

[164] S. Wachter and H. Nagel. Tracking persons in monocular image sequences. *Comput. Vis. Image Underst.*, 74(3):174–192, 1999.

[165] C. Bregler and J. Malik. Tracking people with twists and exponential maps. In *IEEE Int. Conf. Comput. Vis. Pattern Recognit. (CVPR98)*, 8–15, Santa Barbara, CA, USA, June 1998.

[166] D. DiFranco, T. Cham, and J. Rehg. Reconstruction of 3d figure motion from 2d correspondences. In *IEEE Int. Conf. Comput Vis. (ICCV2001)*, Vol. 1, 307–314, Kauai, HI, USA, December 2001.

[167] R. Plankers and P. Fua. Articulated soft objects for multiview shape and motion capture. *IEEE Trans. Pattern Anal. Mach. Intell.*, 25(9):1182–1187, 2003.

[168] J. Deutscher, A. Blake, and I. Reid. Articulated body motion capture by annealed particle filtering. In *IEEE Int. Conf. Comput. Vis. Pattern Recognit. (CVPR2000)*, Vol. 2, 126–133, Hilton Head, SC, USA, June 2000.

[169] C. Sminchisescu and B. Triggs. Estimating articulated human motion with covariance scaled sampling. *Int. J. Rob. Res.*, 22(6):371–393, June 2003.

[170] R.L. Carceroni and K.N. Kutulakos. Multi-view scene capture by surfel sampling: From video streams to non-rigid 3D motion, shape and reflectance. *Int. J. Comput. Vis.*, 49(2–3):175–214, 2002.

[171] Hanim-Humanoid Animation Working Group. Specifications for a standard humanoid. http://www.h-anim.org/Specifications/H-Anim1.1, 2002.

[172] NASA. *Anthropometric source book. Vol. II.: A handbook of anthropometric data.* NASA Reference Publication 1024, 1978.

[173] Q. Delamarre and O. Faugeras. 3D articulated models and multi-view tracking with silhouettes. *Comput. Vis. Image Underst.*, 74(3):174–192, 1999.

[174] D.M. Gavrila and L. S. Davis. 3D model-based tracking of humans in action: a multi-view approach. In *IEEE Int. Conf. Comput. Vis. Pattern Recognit. (CVPR96)*, 73–80, San Fransisco, CA, USA, June 1996.

[175] J. Kundert-Gibbs and P. Lee. *Mastering Maya 3.* Sybex Inc., New York, 2001.

[176] H. Sidenbladh, M. Black, and L. Sigal. Implicit probabilistic models of human motion for synthesis and tracking. In *Eur. Conf. Comput. Vis., (ECCV2002)*, Vol. 1, 784–800, Copenhagen, Denmark, May–June 2002.

[177] V. Pavlovic, J. Rehg, T. Cham, and K. Murphy. A dynamic bayesian approach to figure tracking using learned dynamical models. In *IEEE Int. Conf. Comput. Vis. (ICCV99)*, 94–101, Corfu, Greece, September 1999.

[178] I. Biederman. Recognition-by-components: A theory of human image understanding. *Psychol. Rev.*, 94(2):115–147, 1987.

[179] L. Sigal, S. Bhatia, S. Roth, M. J. Black, and M. Isard. Tracking loose-limbed people. In *IEEE Conf. Comput. Vis. Pattern Recognit.*, Vol. 1, I–421– I–428, June 2004.

[180] S. Hou, A. Galata, F. Caillette, N. Thacker, and P. Bromiley. Real-time body tracking using a gaussian process latent variable model. In *Proc. 11th IEEE Int. Conf. Comput. Vis. (ICCV'07)*, 1–8, Rio de Janeiro, Brazil, October 2007.

[181] R. Urtasun, D. J. Fleet, and P. Fua. 3d people tracking with gaussian process dynamical models. In *IEEE Conf. Comput. Vis. Pattern Recognit.*, Vol. 1, 238–245, New York, NY, USA, June 2006.

[182] F. Guo and G. Qian. Monocular 3d tracking of articulated human motion in silhouette and pose manifolds. *EURASIP J. Image Video Process., 2008 Special Issue on Anthropocentric Video Analy: Tools Appl.*

[183] R. Li, M. H. Yang, S. Sclaroff, and T. P. Tian. Monocular tracking of 3d human motion with a coordinated mixture of factor analyzers. In *ECCV06*, Vol. 2, 137–150, 2006.

[184] C. Sminchisescu, A. Kanaujia, and D. Metaxas. Learning joint top-down and bottom-up processes for 3d visual inference. In *IEEE Conf. Comput. Vis. Pattern Recognit.*, Vol. 1, 1743–1750, New York, NY, USA, June 2006.

[185] J. Rehg and T. Kanade. Visual tracking of high DOF articulated structures: an application to human hand tracking. In *Eur. Conf. Comput. Vis. (ECCV94)*, Vol. 2, 35–46, Stockholm, Sweden, May 1994.

[186] T. Zhao and R. Nevatia. Tracking multiple humans in complex situations. *IEEE Trans. Pattern Anal. Mach. Intell.*, 26(9):1208–1221, 2004.

[187] D. G. Lowe. Fitting parameterized three-dimensional models to images. *IEEE Trans. Pattern Anal. Mach. Intell.*, 13(5):441–450, 1991.

[188] E. Hildreth and D. Marr. Theory of edge detection. In *Proc. Roy. Soc. Lond.*, Vol. 207, 187–217, 1980.

[189] M. Jesus, A. J. Abrantes, and J. S. Marques. Tracking the human body using multiple predictors. In *Proc. Second Int. Workshop Articulated Motion Deformable Objects*, 155–164, London, UK, 2002.

[190] R. Ronfard, C. Schmid, and B. Triggs. Learning to parse pictures of people. In *Eur. Conf. Comput. Vis. (ECCV2002)*, Vol. 4, 700–714, Copenhagen, Denmark, May-June 2002.

[191] S. X. Ju, M. Black, and Y. Yacoob. Cardboard people: A parameterized model of articulated image motion. In *Int. Conf. Automat. Face Gesture Recognit. (AFGR96)*, 38–44, Killington, Vermont, October 1996.

[192] G. McAllister, S. J. McKenna, and I. W. Ricketts. MLESAC tracking with 2D revolute-prismatic articulated models. In *Int. Conf. Pattern Recognit. (ICPR2002)*, Vol. 2, 725–728, Quebec City, Canada, August 2002.

[193] T. J. Cham and J. M. Rehg. A multiple hypothesis approach to figure tracking. In *Int. Conf. Comput. Vis. Pattern Recognit. (CVPR99)*, Vol. 2, 239–245, Fort Collins, CO, USA, June 1999.

[194] H. Moon and R. Chellappa. 3d shape-encoded particle filter for object tracking and its application to human body tracking. *EURASIP J. Image Video Process., 2008 Special Issue on Anthropocentric Video Analy: Tools Appl.*, 2008.

[195] E. Loutas, N. Nikolaidis, and I. Pitas. A mutual information approach to articulated object tracking. In *IEEE Int. Symp. Circuits Syst. (ISCAS2003)*, Vol. II, 672–675, Bangkok, Thailand, May 2003.

[196] D. Ramanan, D. A. Forsyth, and A. Zisserman. Tracking people by learning their appearance. *IEEE Trans. Pattern Anal. Mach. Intell.*, 29(1):65–81, 2007.

[197] D. Ramanan, D. A. Forsyth, and A. Zisserman. Strike a pose: tracking people by finding stylized poses. In *IEEE Conf. Comput. Vis. Pattern Recognit.*, Vol. 1, 271–278, June 2005.

[198] P. F. Felzenszwalb and D. P. Huttenlocher. Efficient matching of pictorial structures. In *IEEE Int. Conf. Comput. Vis. Pattern Recognit. (CVPR2000)*, Vol. 2, 66–73, Hilton Head, SC, USA, June 2000.

[199] G. Hua, M.-H. Yang, and Y. Wu. Learning to estimate human pose with data driven belief propagation. In *IEEE Conf. Comput. Vis. Pattern Recognit.*, Vol. 2, 747–754, June 2005.

[200] G. Mori, X. Ren, A. Efros, and J. Malik. Recovering human body configurations: Combining segmentation and recognition. In *IEEE Conf. Comput. Vis. Pattern Recognit.*, Vol. 2, II–326–II–333, June 2004.

CHAPTER

Basic Transform Video Coding

8

Barry Barnett

IBM Corporation Austin, Texas, USA

The subject of video coding is of fundamental importance to many areas in engineering and the natural and perceptual sciences. Video engineering has become a fundamentally digital discipline although analog TV transmission is still by far the mainstay around the world. Digital transmission of television signals via satellites is commonplace, and widespread High Definition Television (HDTV) terrestrial transmission began in 2000. The FCC has mandated all United States terrestrial TV transmissions to follow the digital standard by February, 2009. Video compression is an absolute requirement for the growth and success for the low or diminished bandwidth transmission and storage of digital video signals. Video encoding is required wherever digital video communications, storage, processing, acquisition, and reproduction occur. The widespread use of digital video encoding has necessitated the development of international standards to guarantee interoperability of the many applications that transmit or exchange digital video information. The current generation of video coding standards has matured over two decades of international cooperation.

The Motion Pictures Expert Group (MPEG) has finalized four well-known encoding standards: MPEG-1, MPEG-2, MPEG-4, and MPEG-7. MPEG-1 and MPEG-2 define methods for the transmission of digital video information for multimedia and television formats. The MPEG standards have been adopted worldwide as the most widely used video coding technology. The MPEG-4 standard, which was adopted in 1999, specifically addresses the transmission of very low bit rate video by introducing the notion of media objects, which are made up of audio, visual, or audiovisual content. It is targeted to satisfy the needs of audiovisual content authors, service providers, and end users. MPEG-4 "Advanced Video Coding" has recently been adopted by both satellite and cable TV providers as the successor to the ubiquitous first-generation MPEG-2 encoder. The MPEG-7 standard, which was adopted in 2001, defines audiovisual content storage and retrieval services (Chapter 15 discusses video storage and retrieval). An aspect central to each of the MPEG standards are the video encoding and decoding algorithms that make digital video applications practical. The MPEG standards are discussed in Chapters 9 and 10.

The International Telecommunications Union (ITU) has similarly finalized four encoding standards: H.261, H.262, H.263, and H.264. H.261 (adopted in 1990) was the first transform-based internationally adopted standard encoder and is the prototype for all subsequent H.26X and MPEG video coders. The H.26X standards have been defined to support the worldwide array of digital formats encountered in key applications such as digital telephony and both terrestrial and satellite digital TV transmission. The H.262 and H.264 encoders are functionally equivalent to the MPEG-2 and MPEG-4 Part 10 encoders. The H.263 encoder was developed to specifically support low bandwidth applications on telephony and data networks.

Video coding not only reduces the storage requirements or transmission bandwidth of digital video applications but also affects many end-to-end system design and performance trade-offs. The design and selection of a video encoder therefore is not only based on its ability to compress information. Issues such as bit rate versus distortion criteria, algorithm complexity, transmission channel characteristics, algorithm symmetry versus asymmetry, video source statistics, fixed versus variable rate coding, and standards compatibility should be considered to make good encoder design decisions.

The growth of digital video applications and technology in the last decade has been explosive, and video compression is playing a central role in this success. Yet, the video coding discipline is relatively young and certainly will evolve and change significantly over the next decade. For instance, perceptual-based video encoding research is still relatively new and promises to be able to significantly influence the future course of the video coding discipline. Research in video coding has great vitality and the body of work is significant. It is apparent that this relevant and important topic will have an immense effect on the future of digital video technologies.

8.1 INTRODUCTION TO VIDEO COMPRESSION

Video or visual communications require significant amounts of information transmission. Video compression as considered here involves the bit rate reduction of the digital video signal carrying visual information. Traditional video-based compression, like other information compression techniques, focuses on eliminating redundancy and unimportant elements of the source. The degree to which the encoder reduces the bit rate is called its *coding efficiency*, or equivalently its inverse is termed the *compression ratio*, that is,

$$\text{coding efficiency} = (\text{compression ratio})^{-1} = \text{encoded bit rate}/\text{decoded bit rate}. \tag{8.1}$$

Compression can be a lossless or lossy operation. Due to the immense volume of video information, lossy operations are a key element used in video compression algorithms. The loss of information or distortion measure is usually evaluated using the mean square error (MSE), mean absolute error (MAE) criteria, or peak signal-to-reconstruction noise (PSNR),

$$\mathrm{MSE} = \frac{1}{MN}\sum_{i=1}^{M}\sum_{j=1}^{N}\left[I(i,j) - \hat{I}(i,j)\right]^2$$

$$\mathrm{MAE} = \frac{1}{MN}\sum_{i=1}^{M}\sum_{j=1}^{N}\left|I(i,j) - \hat{I}(i,j)\right| \tag{8.2}$$

$$\mathrm{PSNR} = 20\log_{10}\left(\frac{2^n}{\mathrm{MSE}^{1/2}}\right),$$

for image I and its reconstructed image $\hat{I}$, pixel indices $1 \leq i \leq M$ and $1 \leq j \leq N$, image size $N \times M$ pixels, and n bits per pixel. The MSE, MAE, and PSNR as described here are global measures and do not necessarily give a good indication of the reconstructed image quality. In the final analysis, the human observer determines the quality of the reconstructed image and video quality. The concept of distortion versus coding efficiency is one of the most fundamental trade-offs in the technical evaluation of video encoders. The topic of perceptual quality assessment of compressed images and video is discussed in Section 8.2.

Video signals contain information in three dimensions. These dimensions are modeled as *spatial* and *temporal* domains for video encoding purposes. Digital video compression methods generally seek to minimize information redundancy independently in each domain. The major international video compression standards (MPEG-1, MPEG-2, MPEG-4, H.261, H262, H.263, and H.264) use this approach. Figure 8.1 schematically depicts a generalized video compression system that implements the spatial and temporal encoding of a digital image sequence. Each image in the sequence I_k is defined as in Eq. 8.2. The spatial encoder operates on image *blocks*, typically on the order of 8×8 pixels each. The temporal encoder generally operates on 16×16 pixel image blocks. The system is designed for two modes of operation: the *intraframe* mode and the *interframe* mode.

The single layer feedback structure of this generalized model is representative of the encoders that are recommended by the International Standards Organization (ISO) and ITU video coding standards MPEG-1, MPEG-2/H.262, MPEG-4, MPEG-7, H.261, H263, and H.264 [1–6]. The feedback loop is used in the interframe mode of operation and generates a *prediction error* between the image blocks of the current frame and the current prediction frame. The prediction is generated by the *motion compensator.* The *motion estimation* unit creates *motion vectors* for each 16×16 block. The motion vectors and previously reconstructed frame are fed to the motion compensator to create the prediction.

The intraframe mode spatially encodes an entire current frame on a periodic basis, such as every 15 frames, to ensure that systematic errors do not continuously propagate. Intraframe mode will also be used to spatially encode a block whenever the interframe

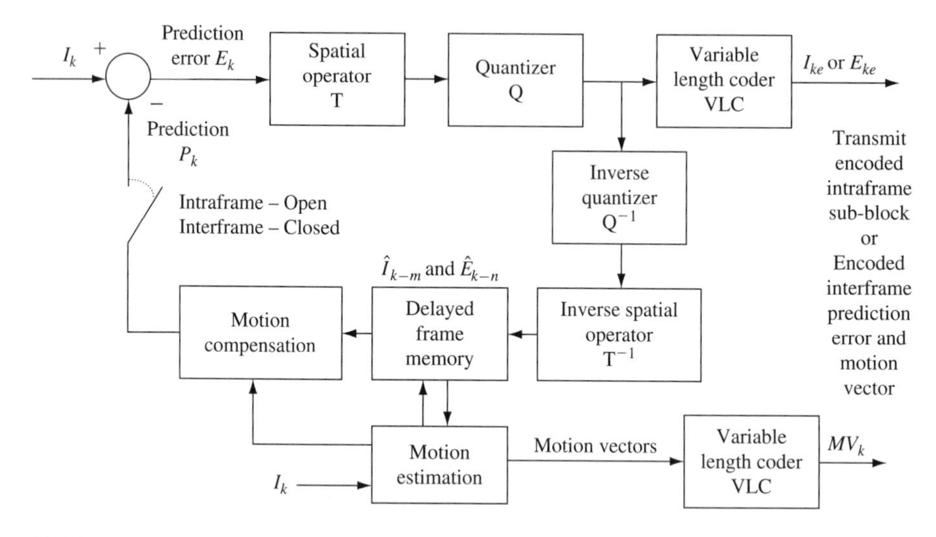

FIGURE 8.1

Generalized transform-based video compression system.

encoding mode cannot meet its performance threshold. The intraframe versus interframe mode selection algorithm is not included in this diagram. It is responsible for controlling the selection of the encoding functions, data flows, and output data streams for each mode.

The intraframe encoding mode does not receive any input from the feedback loop. I_k is spatially encoded, and subsequently encoded by the variable length coder (VLC) forming I_{ke}, which is transmitted to the decoder. The receiver decodes I_{ke} producing the reconstructed image subblock $\hat{I}_k$. During the interframe coding mode, the current frame prediction P_k is subtracted from the current frame input I_k to form the current prediction error E_k. The prediction error is then spatially and VLC encoded to form E_{ke} and is transmitted along with the VLC encoded motion vectors MV_k. The decoder can reconstruct the current frame $\hat{I}_k$ using the previously reconstructed frame $\hat{I}_{k-1}$ (stored in the decoder), the current frame motion vectors, and the prediction error. The motions vectors MV_k operate on $\hat{I}_{k-1}$ to generate the current prediction frame P_k. The encoded prediction error E_{ke} is decoded to produce the reconstructed prediction error $\hat{E}_k$. The prediction error is added to the prediction to form the current frame $\hat{I}_k$. The functional elements of the generalized model are described here in detail.

- Spatial operator—This element is generally a unitary two-dimensional (2D) linear transform, but in principle can be any unitary operator that can distribute most of the signal energy into a small number of coefficients, that is, decorrelate the signal data. Spatial transformations are successively applied to small image blocks to take advantage of the high degree of data correlation in adjacent image pixels. The most widely used spatial operator for image and video coding is the discrete cosine transform (DCT). It is applied to 8×8 pixel image blocks, is well suited for image transformations because it uses real computations with fast implementations, provides excellent decorrelation of signal components, and avoids generation of spurious components between the edges of adjacent image blocks.

- Quantizer—The spatial or transform operator is applied to the input to arrange the signal into a more suitable format for subsequent lossy and lossless coding operations. The quantizer operates on the transform-generated coefficients. This is a lossy operation that can result in a significant reduction in the bit rate. The quantization method used in this kind of video encoder is usually scalar and nonuniform. The scalar quantizer simplifies the complexity of the operation compared with vector quantization (VQ). The nonuniform quantization interval is sized according to the distribution of the transform coefficients (TCOEFF) to minimize the bit rate and the distortion created by the quantization process. Alternatively, the quantization interval size can be adjusted based on the performance of the Human Visual System (HVS). The Joint Pictures Expert Group (JPEG) standard includes two (luminance and color difference) HVS sensitivity weighted quantization matrices in its "Examples and Guidelines" annex.
- VLC—The lossless VLC is used to effectively exploit the "symbolic" redundancies contained in each block of quantized TCOEFF. This step is termed "entropy coding" to designate that the encoder is designed to minimize the source entropy. The VLC is applied to a serial bit stream that is generated by scanning the TCOEFF block. The scanning pattern should be chosen with the objective of maximizing the performance of the VLC. The MPEG encoder for instance describes a zigzag scanning pattern that is intended to maximize transform zero coefficient run lengths. Alternatively, the H.261 VLC is designed to encode these run lengths using a variable length *Huffman* code.

The feedback loop sequentially reconstructs the encoded spatial and prediction error frames and stores the results to create a prediction for the current image subblock. The elements required to do this are the Inverse Quantizer, Inverse Spatial Operator, Delayed Frame Memory, Motion Estimator, and Motion Compensator.

- Inverse operators—The inverse operators Q^{-1} and T^{-1} are applied to the encoded current frame I_{ke} or the current prediction error E_{ke} to reconstruct and store the encoded frame for the motion estimator and motion compensator to generate the next prediction frame.
- Delayed frame memory—Both current and previous frames must be available to the motion estimator and motion compensator to generate a prediction frame. The number of previous frames stored in memory can vary based on the requirements of the encoding algorithm. MPEG-1 defines a B-frame which is a bidirectional encoding that requires that motion prediction be performed in both the forward and backward directions. This necessitates storage of multiple frames in memory.
- Motion estimation—The temporal encoding aspect of this system relies on the assumption that rigid body motion is responsible for the differences between two

or more successive frames. The objective of the motion estimator is to estimate the rigid body motion between two frames. The motion estimator operates on all current frame 16×16 image blocks and generates the pixel displacement or motion vector for each block. The technique used to generate motion vectors is called *block-matching motion estimation* and is discussed further in Section 8.4.3. The method uses the current frame I_k and the previous reconstructed frame $\hat{I}_{k-1}$ as input. Each block in the previous frame is assumed to have a displacement that can be found by searching for it in the current frame. The search is usually constrained to be within a reasonable neighborhood so as to minimize the complexity of the operation. Search matching is usually based on a minimum MSE or MAE criteria. When a match is found, the pixel displacement is used to encode the particular block. If a search does not meet a minimum MSE or MAE threshold criteria, the motion compensator will indicate that the current block is to be spatially encoded using intraframe mode.

- Motion compensation (MC)—The motion compensator makes use of the current frame motion estimates MV_k and the previously reconstructed frame $\hat{I}_{k-1}$ to generate the current frame prediction P_k. The current frame prediction is constructed by placing the previous frame blocks into the current frame according to the motion estimate pixel displacement. The motion compensator uses the threshold criteria to decide which blocks will be encoded as prediction error blocks using motion vectors and which blocks will only be spatially encoded.

The generalized model does not address some video compression system details such as the bit-stream syntax (which supports different application requirements) or the specifics of the encoding algorithms. These issues are dependent on the video compression system design.

Alternative video encoding models have also been the focus of current research. Three-dimensional (3D) video information can be compressed directly using VQ or 3D *wavelet* encoding models. VQ encodes a 3D block of pixels as a codebook index that denotes its "closest or nearest neighbor" in the minimum squared or absolute error sense. However, the VQ codebook size grows on the order as the number of possible inputs. Searching the codebook space for the nearest neighbor is generally very computationally complex, but structured search techniques can provide good bit rates, quality, and computational performance. Tree-structured VQ (TSVQ) [7] reduces the search complexity from codebook size N to $\log N$, with a corresponding loss in average distortion. The simplicity of the VQ decoder (it only requires a simple table lookup for the transmitted codebook index) and its bit-rate distortion performance make it an attractive alternative for specialized applications. The complexity of the codebook search generally limits the use of VQ in real-time applications. VQ quantizers have also been proposed for interframe, variable bit rate, and subband video compression methods [8].

3D wavelet encoding is a topic of recent interest. This video encoding method is based on the *Discrete Wavelet Transform* methods and is discussed in Chapter 11. The Wavelet Transform decomposes a bandwidth limited signal into a *multiresolution* representation.

The multiresolution decomposition makes the Wavelet Transform an excellent signal analysis tool because signal characteristics can be viewed in a variety of time-frequency or space-frequency scales. The Wavelet Transform is implemented in practice via the use of multiresolution or subband filterbanks [9]. The wavelet filterbank is well suited for video encoding because of its ability to adapt to the multiresolution characteristics of video signals. Wavelet Transform encodings are naturally hierarchical in their time-frequency representation and easily adaptable for *progressive transmission* [10]. They have also been shown to possess excellent bit-rate distortion characteristics.

Direct 3D video compression systems suffer from a major drawback for real-time encoding and transmission. To encode a sequence of images in one operation, the sequence must be buffered. This introduces a buffering and computational delay that can be very noticeable in the case of real-time video communications.

Video compression techniques treating visual information in accordance with HVS models have recently been introduced. These methods are termed "Second-Generation or Object-Based" and attempt to achieve very large compression ratios by imitating the operations of the HVS. The HVS model has also been incorporated into more traditional video compression techniques by reflecting visual perception into various aspects of the coding algorithm. HVS weightings have been designed for the DCT AC coefficients quantizer used in the MPEG encoder. A discussion of these techniques can be found in Chapter12.

Digital video compression is currently enjoying tremendous growth in part due to the great advances in VLSI, ASIC, and microcomputer technology in the last decade. The real-time nature of video communications necessitates the use of general purpose and specialized high-performance hardware devices. In the near future, advances in design and manufacturing technologies will create hardware devices that will allow greater adaptability, interactivity, and interoperability of video applications. For instance, MPEG-7 has defined format free operations for the storage and retrieval of audiovisual information that is being used by digital cable TV vendors for "on demand" delivery of digital video content.

8.2 VIDEO COMPRESSION APPLICATION REQUIREMENTS

A wide variety of digital video applications currently exist. They range from simple low-resolution and bandwidth applications (multimedia, PicturePhone) to very high-resolution and bandwidth (HDTV) demands. This section will present requirements of current and future digital video applications and the demands they place on the video compression system.

To demonstrate the importance of video compression, the transmission of digital video television signals is presented. The bandwidth required by a digital television signal is approximately one-half the number of picture elements (pixels) displayed per second. The analog video monitor pixel size in the vertical dimension is the distance between scanning lines, and the horizontal dimension is the distance the scanning spot moves during half cycle of the highest video signal transmission frequency. The video signal bandwidth is given by Eq. 8.3,

$$\begin{aligned} B_W &= (\text{cycles/frame})(F_R) \\ &= (\text{cycles/line})(N_L)(F_R) \\ &= \frac{(0.5)(\text{aspect ratio})(F_R)(N_L)(R_H)}{0.84}, \\ &= (0.8)(F_R)(N_L)(R_H) \end{aligned} \tag{8.3}$$

where B_W = video signal system bandwidth, F_R = number of frames transmitted per second (fps), N_L = number of scanning lines per frame, and R_H = horizontal resolution (lines), proportional to pixel resolution.

The National Television Systems Committee (NTSC) picture aspect ratio is 4/3, the constant 0.5 is the ratio of the number of cycles to the number of lines, and the factor 0.84 is the fraction of the horizontal scanning interval that is devoted to signal transmission.

The NTSC transmission standard used for television broadcasts in the United States has the following parameter values: $F_R = 29.97$ fps, $N_L = 525$ lines, and $R_H = 340$ lines.

This yields an analog video system bandwidth B_W of 4.2 MHz for the NTSC broadcast system. To transmit a color digital video signal, the digital pixel format must be defined. The digital color pixel is made of three components: one luminance (Y) component occupying 8 bits and two color difference components (U and V) each requiring 8 bits. The NTSC picture frame has $720 \times 480 \times 2$ total luminance and color pixels. To transmit this information for an NTSC broadcast system at 29.97 fps, the following bandwidth is required:

$$\begin{aligned} &\text{Digital BW} \simeq 1/2 \text{ bit rate} = 1/2(29.97 \text{ fps}) \\ &\qquad \times (24 \text{ bits/pixel}) \times (720 \times 480 \times 2 \text{ pixels/frame}) = 249\,\text{MHz}. \end{aligned}$$

This represents an increase of approximately 59 times the required NTSC system bandwidth and about 41 times the full transmission channel bandwidth (6 MHz) for current NTSC signals. HDTV picture resolution requires up to three times more raw bandwidth than this example! (Two transmission channels totaling 12 MHz are allocated for terrestrial HDTV transmissions.) It is clear from this example that terrestrial television broadcast systems have to use digital transmission and digital video compression to achieve the overall bit rate reduction and image quality required for HDTV signals.

The example not only points out the significant system bandwidth requirements for digital video information but also indirectly brings up the issue of digital video quality requirements. The trade-off between bit rate and quality or distortion is a fundamental issue facing the design of video compression systems. To this end, it is important to fully characterize an application's video communications requirements before designing or selecting an appropriate video compression system. Factors that should be considered in the design and selection of a video compression system include the following items:

- Video characteristics—Video parameters such as the dynamic range, source statistics, pixel resolution, and noise content can affect the performance of the compression system.
- Transmission requirements—Transmission bit rate requirements determine the power of the compression system. Very high-transmission bandwidth, storage capacity, or quality requirements may necessitate lossless compression. Conversely, extremely low bit rate requirements may dictate compression systems that trade-off image quality for a large compression ratio. *Progressive transmission* is a key issue for selection of the compression system. It is generally used when the transmission bandwidth exceeds the compressed video bandwidth. Progressive coding refers to a multiresolution, hierarchical, or subband encoding of the video information. It allows for transmission and reconstruction of each resolution independently from low to high resolution.

  Channel errors affect system performance and the quality of the reconstructed video. Channel errors can affect the bit stream randomly or in burst fashion. The channel error characteristics can have different effects on different encoders and can range from local to global anomalies. In general, transmission error correction codes (ECCs) are used to mitigate the effect of channel errors, but awareness and knowledge of this issue is important.
- Compression system characteristics and performance—The nature of video applications makes many demands on the video compression system. Interactive video applications such as videoconferencing demand that the video compression systems have symmetric capabilities. That is, each participant in the interactive video session must have the same video encoding and decoding capabilities, and that the system performance requirements must be met by both the encoder and decoder. On the other hand, television broadcast video has significantly greater performance requirements at the transmitter because it has the responsibility of providing real-time high-quality compressed video that meets the transmission channel capacity.

Digital video system implementation requirements can vary significantly. Desktop televideo conferencing can be implemented using software encoding and decoding or may require specialized hardware and transmission capabilities to provide high-quality performance. The characteristics of the application will dictate the suitability of the video compression algorithm for particular system implementations. The importance of the encoder and system implementation decision cannot be overstated; system architectures and performance capabilities are changing at a rapid pace, and the choice of the best solution requires careful analysis of the all possible system and encoder alternatives.

- Rate-distortion requirements—The rate-distortion requirement is a basic consideration in the selection of the video encoder. The video encoder must be able to provide the bit rate(s) and video fidelity (or range of video fidelity) required by the application. Otherwise, any aspect of the system may not meet specifications. For example, if the bit rate specification is exceeded to support a lower MSE, a larger than expected transmission error rate may cause a catastrophic system failure.

- Standards requirements—Video encoder compatibility with existing and future standards is an important consideration if the digital video system is required to inter-operate with existing and/or future systems. A good example is that of a desktop videoconferencing application supporting a number of legacy video compression standards. This requires support of the older video encoding standards on new equipment designed for a newer incompatible standard. Videoconferencing equipment not supporting the old standards would not be capable or as capable to work in environments supporting older standards.

These factors are shown in Table 8.1 to demonstrate video compression system requirements for some common video communications applications. The video

TABLE 8.1 Digital video application requirements.

Application	Bit rate requirement	Distortion requirements	Transmission requirements	Computational requirements	Standards requirements
Network video on demand	1.5 Mbps, 10 Mbps	High, medium	Internet, 100 Mbps LAN	MPEG-1, MPEG-2/4	MPEG-1, MPEG-2/4, MPEG-7
Video phone	64 kbps	High distortion	ISDN p × 64	H.261 encoder, H.261 decoder	H.261
Desktop multimedia video CDROM	1.5 Mbps	High distortion to medium	PC channel	MPEG-1 decoder	MPEG-1, MPEG-2, MPEG-7
Desktop LAN videoconference	10 Mbps	Medium distortion	Fast Ethernet 100 Mbps	Hardware encoders decoders	MPEG-2/4, H.261
Desktop WAN videoconference	1.5 Mbps	High distortion	Ethernet	Hardware encoders decoders	MPEG-1, MPEG-4, H.263
Desktop dial-up videoconference	64 kbps	Very high distortion	POTS and Internet	Software encoder decoder	MPEG-4, H.263
Digital satellite television	10 Mbps	Low distortion	Fixed service satellites FSS	MPEG-2 decoder	MPEG-2
HDTV	20 Mbps	Low distortion	12 MHz terrestrial link	MPEG-2/4 encoder decoder	MPEG-2, MPEG-4
HD DVD, DVD	36.5 Mbps 20 Mbps	Low distortion	PC channel	H.264, MPEG-2 decoder	H.264, MPEG-2

compression system designer as a minimum should consider these factors in making a determination about the choice of video encoding algorithms and technology to implement.

8.3 DIGITAL VIDEO SIGNALS AND FORMATS

Video compression techniques make use of signal models to be able to use the body of digital signal analysis/processing theory and techniques that have been developed over the past 50 or so years. The design of a video compression system as represented by the generalized model introduced in Section 8.1 requires knowledge of the signal characteristics and the digital processes that are used to create the digital video signal. It is also highly desirable to understand video display systems and the behavior of the HVS.

8.3.1 Sampling of Analog Video Signals

Digital video information is generated by sampling the *intensity* of the original continuous analog video signal $I(x, y, t)$ in three dimensions. The spatial component of the video signal is sampled in the horizontal and vertical dimensions (x, y), and the temporal component is sampled in the time dimension (t). This generates a series of digital images or image sequence $I(i, j, k)$. Video signals that contain colorized information are usually decomposed into three parameters (YC_rC_b, YUV, RGB, etc.) whose intensities are likewise sampled in three dimensions. The sampling process inherently quantizes the video signal due to the digital word precision used to represent the intensity values. Therefore, the original analog signal can never be reproduced exactly, but for all intents and purposes, a high-quality digital video representation can be reproduced with arbitrary closeness to the original analog video signal. The topic of video sampling and interpolation is discussed in Chapter 2.

An important result of sampling theory is the *Nyquist Sampling Theorem.* This theorem defines the conditions under which sampled analog signals can be "perfectly" reconstructed. If these conditions are not met, the resulting digital signal will contain *aliased* components, which introduce artifacts into the reconstruction. The Nyquist conditions are depicted graphically for the one-dimensional (1D) case in Fig. 8.2.

The 1D signal l is sampled at rate f_s. It is band limited (as are all real world signals) in the frequency domain with an upper frequency bound of f_B. According to the Nyquist Sampling Theorem, if a band-limited signal is sampled, the resulting *Fourier* spectrum is made up of the original signal spectrum $|L|$ plus replicates of the original spectrum spaced at integer multiples of the sampling frequency f_s. Figure 8.2(a) depicts the magnitude $|L|$ of the Fourier spectrum for l. The magnitude of the Fourier spectrum $|L_s|$ for the sampled signal l_s is shown for two cases. Figure 8.2(b) presents the case where the original signal l can be reconstructed by recovering the central spectral island. Figure 8.2(c) shows the case where the Nyquist sampling criteria has not been met and spectral overlap occurs. The spectral overlap is termed *aliasing* and occurs when $f_s < 2f_B$. When $f_s > 2f_B$, the original signal can be reconstructed by using a low-pass digital filter whose pass band is

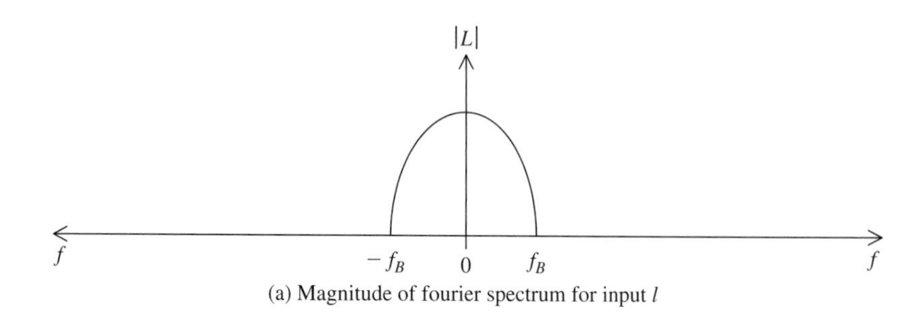

(a) Magnitude of fourier spectrum for input l

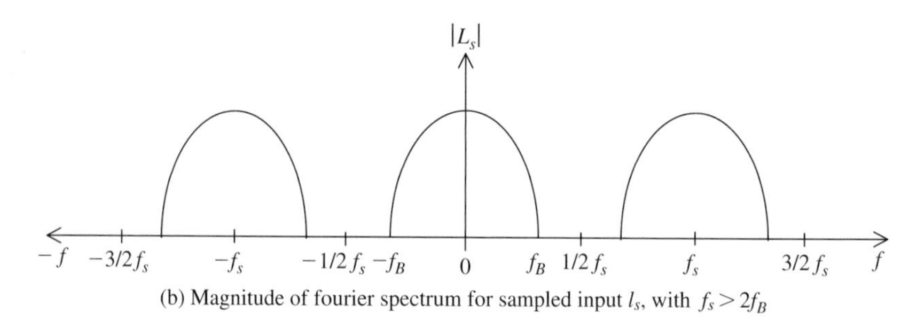

(b) Magnitude of fourier spectrum for sampled input l_s, with $f_s > 2f_B$

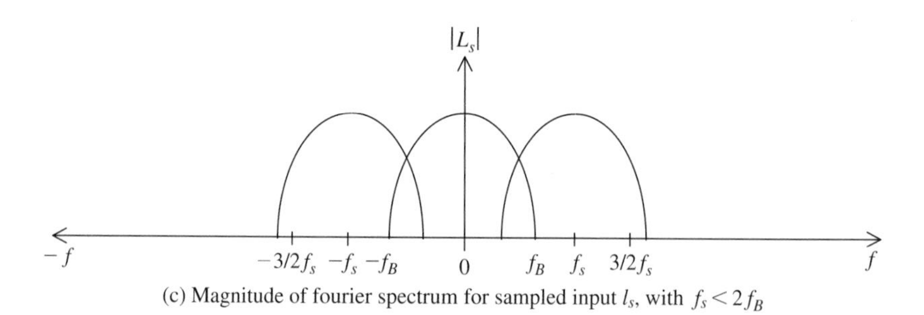

(c) Magnitude of fourier spectrum for sampled input l_s, with $f_s < 2f_B$

FIGURE 8.2

Nyquist sampling theorem.

designed to recover $|L|$. These relationships provide a basic framework for the analysis and design of digital signal processing systems.

2D or spatial sampling is a simple extension of the 1D case. The Nyquist criteria have to be obeyed in both dimensions, that is, the sampling rate in the horizontal direction must be two times greater than the upper frequency bound in the horizontal direction, and the sampling rate in the vertical direction must be two times greater than the upper frequency bound in the vertical direction. In practice, spatial sampling grids are square so that an equal number of samples per unit length in each direction are collected. Charge coupled devices (CCDs) are typically used to spatially sample analog imagery and video.

The sampling grid spacing of these devices is more than sufficient to meet the Nyquist criteria for most resolution and application requirements. The electrical characteristics of CCDs have a greater effect on the image or video quality than its sampling grid size.

Temporal sampling of video signals is accomplished by capturing a spatial or image frame in the time dimension. The temporal samples are captured at a uniform rate of about 60 fields per second for NTSC television and 24 fps for a motion film recording. These sampling rates are significantly less than the spatial sampling rate. The maximum temporal frequency that can be reconstructed according to the Nyquist frequency criteria is 30 Hz in the case of television broadcast. Therefore, any rapid intensity change (caused for instance by a moving edge) between two successive frames will cause aliasing because the harmonic frequency content of such a step-like function exceeds the Nyquist frequency. Temporal aliasing of this kind can be greatly mitigated in CCDs by the use of low-pass temporal filtering to remove the high-frequency content. *Photoconductor Storage Tubes* are used for recording broadcast television signals. They are analog scanning devices whose electrical characteristics filter the high-frequency temporal content and minimize temporal aliasing. Indeed, motion picture film also introduces low-pass filtering when capturing image frames. The exposure speed and the response speed of the photo chemical film combine to mitigate high-frequency content and temporal aliasing. These factors cannot completely stop temporal aliasing, and so intelligent use of video recording devices is still warranted, for example, the main reason movie camera panning is done very slowly is to minimize temporal aliasing.

In many cases where fast motions or moving edges are not well resolved due to temporal aliasing, the HVS will interpolate such motion and provide its own perceived reconstruction. The HVS is very tolerant of temporal aliasing by using its own knowledge of natural motion to provide motion estimation and compensation to the image sequences generated by temporal sampling. The combination of temporal filtering in sampling systems and the mechanisms of human visual perception reduce the effects of temporal aliasing such that temporal under sampling (sub-Nyquist sampling) is acceptable in the generation of typical image sequences intended for general purpose use.

8.3.2 Digital Video Formats

Sampling is the process used to create the image sequences used for video and digital video applications. Spatial sampling and quantization of a natural video signal digitizes the image plane into a 2D set of digital pixels that define a digital image. Temporal sampling of a natural video signal creates a sequence image frames typically used for motion pictures and television. The combination of spatial and temporal sampling creates a sequence of digital images termed digital video. As described earlier, the digital video signal intensity is defined as $I(i,j,k)$, where $0 \leq i \leq M, 0 \leq j \leq N$ are the horizontal and vertical spatial coordinates, and $0 \leq k$ is the temporal coordinate.

The standard digital video formats introduced here are used in the broadcast for both analog and digital television, as well as computer video applications. Composite television signal digital broadcasting formats are included here due to their use in video compression standards, digital broadcasting, and standards format conversion applications. Knowledge of these digital video formats provides background for understanding the international video compression standards developed by the ITU and the ISO. These standards contain specific recommendations for use of the digital video formats described here.

Composite television digital video formats are used for the digital broadcasting, Society of Motion Picture and Television Engineers (SMPTE) digital recording, and conversion of television broadcasting formats. Table 8.2 contains both analog and digital system parameters for the NTSC and Phase Alternating Lines (PAL) composite broadcast formats.

Component television signal digital video formats have been defined by the International Consultative Committee for Radio (CCIR) Recommendation 601. It is based on component video with one luminance (Y) and two color difference signals (C_r and C_b). The raw bit rate for the CCIR 601 format is 162 Mbps. Table 8.3 contains important systems parameters of the CCIR 601 digital video studio component recommendation for both NTSC and PAL/Sequentiel Couleur avec Memoire (SECAM).

The ITU Specialist Group (SGXV) has recommended three formats that are used in the ITU H.261, H.263, H.264, and ISO MPEG video compression standards. They are the Standard Input Format (SIF), Common Interchange Format (CIF), and the low bit rate version of CIF called Quarter CIF (QCIF). Together, these formats describe a comprehensive set of digital video formats that are widely used in current digital video applications. CIF and QCIF support the NTSC and PAL video formats using the same parameters. The SIF format defines different vertical resolution values for NTSC and PAL. The CIF and QCIF formats also support the H.261-modified parameters. The modified parameters are integer multiples of 8 in order to support the 8×8 pixel 2D DCT operation. Table 8.4 lists this set of digital video standard formats. The modified H.26X parameters are listed in parenthesis.

TABLE 8.2 Digital composite television parameters.

Description	NTSC	PAL
Analog video bandwidth (MHz)	4.2	5.0
Aspect ratio, hor size/vert size	4/3	4/3
Frames per second	29.97	25
Lines per second	525	625
Interlace ratio, fields:frames	2:1	2:1
Subcarrier frequency (MHz)	3.58	4.43
Sampling frequency (MHz)	14.4	17.7
Samples per active Line	757	939
Bit rate (Mbps)	114.5	141.9

TABLE 8.3 Digital video component television parameters for CCIR 601.

Description	NTSC	PAL/SECAM
Luminance channel		
Analog video bandwidth (MHz)	5.5	5.5
Sampling frequency (MHz)	13.5	13.5
Samples per active line	710	716
Bit rate (Mbps)	108	108
Two color difference channels		
Analog video bandwidth (MHz)	2.2	2.2
Sampling frequency (MHz)	6.75	6.75
Samples per active line	355	358
Bit rate (Mbps)	54	54

TABLE 8.4 SIF, CIF, and QCIF digital video formats.

Description	SIF NTSC/PAL	CIF	QCIF
Horizontal resolution (Y) pixels	352	360(352)	180(176)
Vertical resolution (Y) pixels	240/288	288	144
Horizontal resolution (C_r, C_b) pixels	176	180(176)	90(88)
Vertical resolution (C_r, C_b) pixels	120/144	144	72
Bits per pixel (bpp)	8	8	8
Interlace fields:frames	1:1	1:1	1:1
Frame rate (fps)	30	30, 15, 10, 7.5	30, 15, 10, 7.5
Aspect ratio hor size/vert size	4:3	4:3	4:3
Bit rate (Y) Mbps @ 30 fps	20.3	24.9	6.2
Bit rate (U, V) Mbps @ 30 fps	10.1	12.4	3.1

8.4 VIDEO COMPRESSION TECHNIQUES

Video compression systems generally comprise two modes that reduce information redundancy in the spatial and the temporal domains. Spatial compression and quantization operates on a single image block, making use of the local image characteristics to reduce the bit rate. The spatial encoder also includes a VLC inserted after the quantization stage. The VLC stage generates a lossless encoding of the quantized image block. Temporal domain compression makes use of optical flow models (generally in the form of block-matching motion estimation methods) to identify and mitigate temporal redundancy.

This section presents an overview of some widely accepted encoding techniques used in video compression systems. *Entropy Encoders* are lossless encoders that are used in the VLC stage of a video compression system. They are best used for information sources that are *memoryless* (sources in which each value is independently generated), and try to minimize the bit rate by assigning variable length codes for the input values according to the input probability density function (pdf). *Predictive Coders* are suited to information sources that have memory, that is, a source in which each value has a statistical dependency on some number of previous and/or adjacent values. Predictive coders can produce a new source pdf with significantly less statistical variation and entropy than the original. The transformed source can then be fed to a VLC to reduce the bit rate. Entropy and predictive coding are good examples for presenting the basic concepts of statistical coding theory.

Block transformations are the major technique for representing spatial information in a format that is highly conducive to quantization and VLC encoding. Block transforms can provide a coding gain by packing most of the block energy into a fewer number of coefficients. The *quantization* stage of the video encoder is the central factor in determining the rate-distortion characteristics of a video compression system. It quantizes the block TCOEFF according to the bit rate and distortion specifications. MC takes advantage of the significant information redundancy in the temporal domain by creating current frame predictions based on block matching motion estimates between the current and previous image frames. MC generally achieves a significant increase in the video coding efficiency over pure spatial encoding.

8.4.1 Entropy and Predictive Coding

Entropy coding is an excellent starting point in the discussion of coding techniques because it makes use of many of the basic concepts introduced in the discipline of *Information Theory* or *Statistical Communications Theory* [11]. The discussion of VLC and predictive coders requires the use of *information source* models to lay the statistical foundation for the development of this class of encoder. An information source can be viewed as a process that generates a sequence of symbols from a finite alphabet. Video sources are generated from a sequence of image blocks that are generated from a "pixel" alphabet. The number of possible pixels that can be generated is 2^n, when n is the number of bits per pixel. The order in which the image symbols are generated depends on how the image block is arranged or scanned into a sequence of symbols. Spatial encoders transform the statistical nature of the original image so that the resulting coefficient matrix can be scanned in a manner such that the resulting source or sequence of symbols contains significantly less information content.

Two useful information sources are used in modeling video encoders: the Discrete Memoryless Source (DMS) and Markov sources. VLC coding is based on the DMS model, and the predictive coders are based on the Markov source models. The DMS is simply a source in which each symbol is generated independently. The symbols are *statistically independent* and the source is completely defined by its symbols/events and the set of probabilities for the occurrence for each symbol, that is, $E = \{e_1, e_2, \ldots, e_n\}$ and the set $\{p(e_1), p(e_2), \ldots, p(e_n)\}$, where n is the number of symbols in the alphabet. It is useful to introduce the concept of entropy at this point. Entropy is defined as the average information content of the information source. The information content of a single event or symbol is defined as

$$I(e_i) = \log \frac{1}{p(e_i)}. \tag{8.4}$$

The base of the logarithm is determined by the number of states used to represent the information source. Digital information sources use base 2 to define the information content using the number of bits per symbol or bit rate. The entropy of a digital source is further defined as the average information content of the source, that is,

$$H(E) = \sum_{i=1}^{n} p(e_i) \log_2 \frac{1}{p(e_i)} = -\sum_{i=1}^{n} p(e_i) \log_2 p(e_i) \text{bits/symbol}. \tag{8.5}$$

This relationship suggests that the average number of bits per symbol required to represent the information content of the source is the entropy. The *Noiseless Source Coding Theorem* states that a source can be encoded with an average number of bits per source symbol that is arbitrarily close to the source entropy. So-called entropy encoders seek to find codes that perform close to the entropy of the source. *Huffman* and *Arithmetic* encoders are examples of entropy encoders.

Modified Huffman coding [12] is commonly used in the image and video compression standards. It produces good performing variable length codes without significant computational complexity. The traditional Huffman algorithm is a two-step process that first creates a table of source symbol probabilities and then constructs code words whose lengths grow according to the decreasing probability of a symbol's occurrence. Modified versions of the traditional algorithm are used in the current generation of image and video encoders. The MPEG-2 encoder uses two sets of static Huffman code words (one each for AC and DC DCT coefficients). A set of 32 code words is used for encoding the AC coefficients. The zigzag scanned coefficients are classified according to the zero coefficient run-length and first nonzero coefficient value. A simple table lookup is all that is then required to assign the code word for each classified pair.

Markov and *Random Field* source models are well suited to describing the source characteristics of natural images. A Markov source has memory of some number of preceding or adjacent events. In a natural image block, the value of the current pixel is dependent on the values of some the surrounding pixels because they are part of the same object, texture, contour, etc. This can be modeled as an mth order Markov source, in which the probability of source symbol e_i depends on the last m source symbols. This dependence is expressed as the probability of occurrence of event e_i conditioned on the occurrence of the last m events, that is, $p(e_i|e_{i-1}, e_{i-2}, \ldots, e_{i-m})$. The Markov source is made up of all possible n^m states, where n is the number of symbols in the alphabet. Each state contains a set of up to n conditional probabilities for the possible transitions between the current symbol and the next symbol. The differential pulse code modulation

(DPCM) predictive coder makes use of the Markov source model. DPCM is used in the MPEG-1 and H.261 standards to encode the set of quantized DC coefficients generated by the DCTs.

The DPCM predictive encoder modifies the use of the Markov source model considerably to reduce its complexity. It does not rely on the actual Markov source statistics at all and simply creates a linear weighting of the last m symbols (mth order) to predict the next state. This significantly reduces the complexity of using Markov source prediction at the expense of an increase in the bit rate. DPCM encodes the *differential signal* d between the actual value and the predicted value, that is, $d = e - \hat{e}$, where the prediction $\hat{e}$ is a linear weighting of m previous values. The resulting differential signal d generally has reduced entropy as compared to the original source. DPCM is used in conjunction with a VLC encoder to reduce the bit rate. The simplicity and entropy reduction capability of DPCM makes it a good choice for use in real-time compression systems. Third-order predictors ($m = 3$) have been shown to provide good performance on natural images [13].

8.4.2 Block Transform Coding—The DCT

Block transform coding is widely used in image and video compression systems. The transforms used in video encoders are *unitary*, which means that the transform operation has an inverse operation that uniquely reconstructs the original input. The DCT successively operates on 8×8 image blocks and is used in the H.261, H.262, H.263, MPEG-1, MPEG-2, and MPEG-4 standards. Block transforms make use of the high degree of correlation between adjacent image pixels to provide *energy compaction* or coding gain in the transformed domain. The *block transform coding gain* G_{TC} is defined as the logarithmic ratio of the arithmetic and geometric means of the transformed block variances (VAR), that is,

$$G_{TC} = 10\log_{10}\left[\frac{\frac{1}{N}\sum_{i=0}^{N-1}\sigma_i^2}{\left(\prod_{i=0}^{N-1}\sigma_i^2\right)^{1/N}}\right], \tag{8.6}$$

where the transformed image block is divided into N subbands, and σ_i^2 is the variance of each subband i, for $0 \leq i \leq N-1$. G_{TC} also measures the gain of block transform coding over pulse code modulation coding. The coding gain generated by a block transform is realized by packing most the original signal energy content into a small number of TCOEFF. This results in a lossless representation of the original signal that is more suitable for quantization. That is, there may be many TCOEFF containing little or no energy that can be completely eliminated. Spatial transforms should also be orthonormal, that is, generate uncorrelated coefficients, so that simple scalar quantization can be used to quantize the coefficients independently.

The Karhunen-Loeve Transform (KLT) creates uncorrelated coefficients and is optimal in the energy packing sense. But the KLT is not widely used in practice. It requires the calculation of the image block covariance matrix so that its unitary orthonormal eigenvector matrix can be used to generate the KLT coefficients. This calculation (for which no fast algorithms exist) and the transmission of the eigenvector matrix are required for every transformed image block.

The DCT is the most widely used block transform for digital image and video encoding. It is an orthonormal transform and has been found to perform close to the KLT [14] for first-order Markov sources. The DCT is defined on an 8×8 array of pixels,

$$F(u,v) = \frac{1}{4} C_u C_v \sum_{i=0}^{7} \sum_{j=0}^{7} f(i,j) \cos\left(\frac{(2i+1)u\pi}{16}\right) \cos\left(\frac{(2j+1)v\pi}{16}\right) \tag{8.7}$$

and the inverse IDCT is defined as,

$$f(i,j) = C_u C_v \sum_{u=0}^{7} \sum_{v=0}^{7} F(u,v) \cos\left(\frac{(2i+1)u\pi}{16}\right) \cos\left(\frac{(2j+1)v\pi}{16}\right), \tag{8.8}$$

where

$$C_u = \frac{1}{\sqrt{2}} \text{ for } u = 0, \quad C_u = 1 \text{ otherwise}$$

$$C_v = \frac{1}{\sqrt{2}} \text{ for } v = 0, \quad C_v = 1 \text{ otherwise}$$

where i and j are the horizontal and vertical indices of the 8×8 spatial array, and u and v are the horizontal and vertical indices of the 8×8 coefficient array. The DCT is the chosen method for image transforms for a couple of important reasons. The DCT has fast $O(n \log n)$ implementations using real calculations. It is even simpler to compute than the DFT because it does not require the use of complex numbers.

The second reason for its success is that the reconstructed input of the Inverse DCT (IDCT) tends not to produce any significant discontinuities at the block edges. Finite discrete transforms create a reconstructed signal that is periodic. Periodicity in the reconstructed signal can produce discontinuities at the periodic edges of the signal or pixel block. The DCT is not as susceptible to this behavior as the Discrete Fourier Transform (DFT). Since the cosine function is real and even, that is, $\cos(x) = \cos(-x)$, and the input $F(u,v)$ is real, the IDCT generates a function that is even and periodic in $2n$, where n is the length of the original sequence. On the other hand, the Inverse DFT (IDFT) produces a reconstruction that is periodic in n, but and necessarily even. This phenomenon is illustrated in Fig. 8.3 for the 1D signal $f(i)$.

The original finite sequence $f(i)$ depicted in Fig 8.3(a) is transformed and reconstructed in Fig. 8.3(b) using the DFT-IDFT transform pairs and in Fig. 8.3(c) using the DCT-IDCT transform pairs. The periodicity of the IDFT in Fig. 8.3(b) is five samples long and illustrates the discontinuity introduced by the discrete transform. The periodicity of the IDCT in Fig. 8.3(c) is 10 samples long due to the evenness of the DCT operation. Discontinuities introduced by the DCT are generally less severe than the DFT. The importance of this property of the DCT is that reconstruction errors and blocking

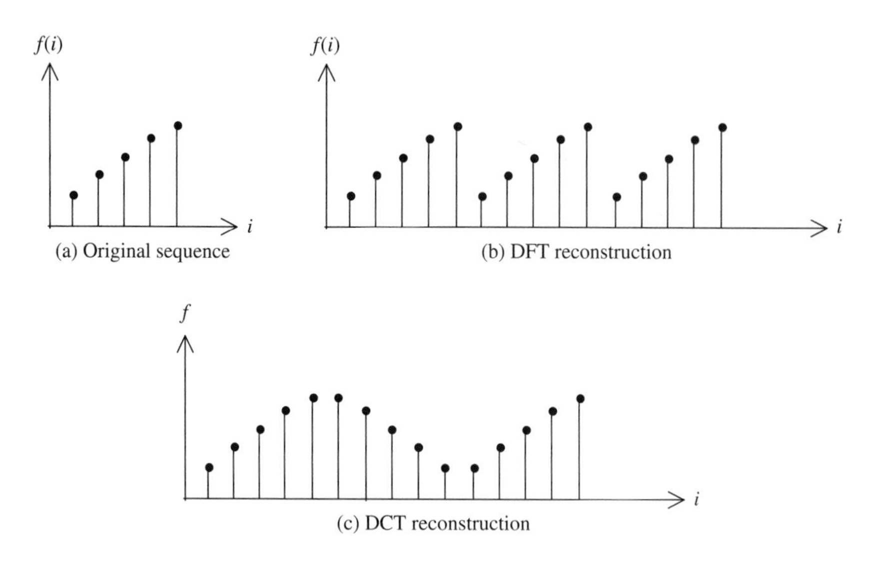

FIGURE 8.3

Reconstruction periodicity of DFT versus DCT.

artifacts are less severe in comparison to the DFT. Blocking artifacts are visually striking and occur due to the loss of high-frequency components that are either quantized or eliminated from the DCT coefficient array. The DCT minimizes blocking artifacts compared with the DFT because it does not introduce the same level of reconstruction discontinuities at the block edges. Figure 8.4 depicts blocking artifacts introduced by gross quantization of the DCT coefficients.

This section ends with an example of the energy packing capability of the DCT. Figure 8.5 depicts the DCT transform operation. The original 8×8 image subblock from the Lena image is displayed in part Fig. 8.5(a), and the DCT transformed coefficient array is displayed in part Fig. 8.5(b).

The original image subblock in Fig. 8.5(a) contains large values in every position is not very suitable for spatial compression in this format. The coefficient matrix Fig. 8.5(b) concentrates most of the signal energy in the top left quadrant. The signal frequency coordinates $(u, v) = (0,0)$ start at the upper left position. The DC component equals 1255 and contains the vast majority of the signal energy by itself. This dynamic range and concentration of energy should yield a significant reduction in nonzero values and bit rate after the coefficients are quantized.

8.4.3 Quantization

The quantization stage of the video encoder creates a lossy representation of the input. The input as discussed earlier should be conditioned with a particular method of quantization in mind. And vice versa, the quantizer should be well matched to the characteristics of the input to meet or exceed the rate-distortion performance requirements. As always

(a) Original

(b) Reconstructed

FIGURE 8.4

Severe blocking artifacts introduced by gross quantization of DCT coefficients.

$$f(i,j) = \begin{bmatrix} 136 & 141 & 143 & 153 & 152 & 154 & 154 & 156 \\ 143 & 150 & 153 & 156 & 160 & 156 & 155 & 155 \\ 149 & 155 & 161 & 163 & 158 & 155 & 156 & 155 \\ 158 & 161 & 162 & 161 & 160 & 158 & 160 & 157 \\ 157 & 161 & 160 & 162 & 161 & 157 & 154 & 155 \\ 160 & 160 & 161 & 160 & 160 & 156 & 156 & 156 \\ 160 & 161 & 160 & 161 & 161 & 157 & 157 & 156 \\ 162 & 162 & 161 & 161 & 162 & 157 & 157 & 157 \end{bmatrix}$$

(a) Original lena 8 × 8 image sub-block

$$F(u,v) = \begin{bmatrix} 1255 & -8 & -9 & -6 & 1 & -1 & -3 & 1 \\ -26 & -20 & -5 & 4 & -1 & 1 & 0 & 1 \\ -9 & -5 & 1 & -1 & 0 & 0 & -1 & 0 \\ -6 & -2 & 0 & 1 & -1 & 0 & 0 & 0 \\ 1 & 0 & 1 & 2 & 0 & -1 & -1 & 0 \\ -2 & 1 & 2 & 0 & 1 & 1 & 0 & -1 \\ -1 & 0 & 0 & -2 & 0 & 0 & 1 & -1 \\ 1 & 0 & -1 & -2 & 0 & 1 & -1 & 0 \end{bmatrix}$$

(b) DCT coefficients

FIGURE 8.5

The 8 × 8 discrete cosine transform.

is the case, the quantizer has an effect on system performance that must be taken under consideration. Simple scalar versus VQ implementations can have significant system performance implications.

Scalar and vector represent the two major types of quantizers. These can be further classified as memoryless or containing memory and symmetric or nonsymmetric. Scalar quantizers control the values taken by a single variable. The quantizer defined by the MPEG-1 encoder scales the DCT TCOEFF. Vector quantizers operate on multiple variables, that is, a vector of variables, and become very complex as the number of variables increases. This discussion will introduce the reader to the basic scalar and vector quantizer concepts that are relevant to image and video encoding.

The Uniform Scalar Quantizer is the most fundamental scalar quantizer. It possesses a nonlinear staircase input-output characteristic that divides the input range into output levels of equal size. For the quantizer to effectively reduce the bit rate, the number of output values should be much less than the number of input values. The reconstruction values are chosen to be at the midpoint of the output levels. This choice is expected to minimize the reconstruction MSE when the quantization errors are uniformly distributed. The quantizers specified in the H.261, H.263, MPEG-1, and MPEG-2 video coders are nearly uniform. They have constant step sizes except for the larger *dead zone* area (the input range for which the output is zero).

Nonuniform quantization is typically used for nonuniform input distributions, such as natural image sources. The scalar quantizer that produces the minimum MSE for a nonuniform input distribution will have nonuniform steps. Compared with the uniform quantizer, the nonuniform quantizer has increasingly better MSE performance as the number of quantization steps increases. The Lloyd-Max [15] is a scalar quantizer design that uses the input distribution to minimize the MSE for a given number of output levels. The Lloyd-Max places the reconstruction levels at the centroids of the adjacent input quantization steps. This minimizes the total absolute error within each quantization step based on the input distribution.

Vector quantizers decompose the input sequence into length n vectors. An image for instance can be divided into $M \times N$ blocks of n pixels each, or the image block can be transformed into a block of TCOEFF. The resulting vector is created by scanning the 2D block elements into a vector of length n. A vector $\mathbf{X}$ is quantized by choosing a codebook vector representation $\hat{\mathbf{X}}$ that is its "closest match." The closest match selection can be made by minimizing an error measure, that is, choose $\hat{\mathbf{X}} = \hat{\mathbf{X}}_i$ such that the MSE over all codebook vectors is minimized,

$$\hat{\mathbf{X}} = \hat{\mathbf{X}}_i : \min_i \text{MSE}(\mathbf{X}, \hat{\mathbf{X}}_i) = \min_i \frac{1}{n} \sum_{j=1}^{n} (x_j - \hat{x}_j)^2. \tag{8.9}$$

The index i of the vector $\hat{\mathbf{X}}_i$ denotes the codebook entry that is used by the receiver to decode the vector. Obviously the complexity of the decoder is much simpler than the encoder. The size of the codebook dictates both the coding efficiency and reconstruction quality. The raw bit rate of a vector quantizer is

$$\text{bit rate}_{VQ} = \frac{\log_2 m}{n} \text{bits/pixel}, \tag{8.10}$$

where $\log_2 m$ is the number of bits required to transmit the index i of the codebook vector $\hat{\mathbf{X}}_i$. The codebook construction includes two important issues that are pertinent to the performance of the video coder. The set of vectors that are included in the codebook determine the bit rate and distortion characteristics of the reconstructed image sequence. The codebook size and structure determines the search complexity to find the minimum error solution for Eq. 8.9. Two important VQ codebook designs are the *Linde-Buzo-Gray* (LBG) [16] and Tree Search VQ (TSVQ) [7]. The LBG design is based on the Lloyd-Max scalar quantizer algorithm. It is widely used because the system parameters can be generated via the use of an input "training set" instead of the true source statistics. The TSVQ design reduces VQ codebook search time by using m-ary tree structures and searching techniques.

8.4.4 MC and Estimation

MC [17] is a technique created in the 1960s, which is used to increase the efficiency of video encoders. Motion compensated video encoders are implemented in three stages. The first stage estimates objective motion (motion estimation) between the previously reconstructed frame and the current frame. The second stage creates the current frame prediction (MC) using the motion estimates and the previously reconstructed frame. The final stage differentially encodes the prediction and the actual current frame as the prediction error. Therefore, the receiver reconstructs the current image only using the VLC encoded motion estimates and the spatially and VLC encoded prediction error.

Motion estimation and compensation are common techniques used to encode the temporal aspect of a video signal. As discussed earlier, block-based MC and motion estimation techniques used in video compression systems are capable of the largest reduction in the raw signal bit rate. Typical implementations generally out-perform pure spatial encodings by a factor of three or more. The interframe redundancy contained in the temporal dimension of a digital image sequence accounts for the impressive signal compression capability that can be achieved by video encoders. Interframe redundancy can be simply modeled as static backgrounds and moving foregrounds to illustrate the potential temporal compression that can be realized. Over a short period of time, image sequences can be described as a static background with moving objects in the foreground. If the background does not change between two frames, their difference is zero, and the two background frames can essentially be encoded as one. Therefore, the compression ratio increase is proportional to two times the spatial compression achieved in the first frame. In general, unchanging or static backgrounds can realize additive coding gains, that is,

$$\text{Static Background Coding Gain} \propto N \bullet (\text{Spatial Compression Ratio of Background Frame}), \tag{8.11}$$

where N is the number of static background frames being encoded. Static backgrounds occupy a great deal of the image area and are typical of both natural and animated image sequences. Some variation in the background always occurs due to random and systematic fluctuations. This tends to reduce the achievable background coding gain.

Moving foregrounds are modeled as nonrotational rigid objects that move independent of the background. Moving objects can be detected by matching the foreground object between two frames. A perfect match results in zero difference between the two frames. In theory, moving foregrounds can also achieve additive coding gain. In practice, moving objects are subject to occlusion, rotational and nonrigid motion, and illumination variations that reduce the achievable coding gain. MC systems that make use of motion estimation methods leverage both background and foreground coding gain. They provide pure interframe differential encoding when two backgrounds are static, that is, the computed motion vector is (0,0). And the motion estimate computed in the case of moving foregrounds generates the minimum distortion prediction.

Motion estimation is an interframe prediction process falling in two general categories: pel-recursive algorithms [18] and block-matching algorithms (BMA) [19]. The pel-recursive methods are very complex and inaccurate, which restrict their use in video encoders. Natural digital image sequences generally display ambiguous object motions that adversely affect the convergence properties of pel-recursive algorithms. This has led to the introduction of block-matching motion estimation, which is tailored for encoding image sequences. Block-matching motion estimation assumes that the objective motion being predicted is rigid and nonrotational. The block size of the BMA for the MPEG, H.261, and H.263 encoders is defined as 16×16 luminance pixels. MPEG-2 also supports 16×8 pixel blocks.

BMAs predict the motion of a block of pixels between two frames in an image sequence. The prediction generates a pixel displacement or motion vector whose size is constrained by the search neighborhood. The search neighborhood determines the complexity of the algorithm. The search for the best prediction ends when the best block match is determined within the search neighborhood. The best match can be chosen as the minimum MSE, which for a full search is computed for each block in the search neighborhood, that is,

$$\text{Best Match}_{\text{MSE}} = \min_{m,n} \frac{1}{N^2} \sum_{i=1}^{M} \sum_{j=1}^{N} \left[I^k(i,j) - I^{k-1}(i+m, j+n) \right]^2, \tag{8.12}$$

where k is the frame index, l is the temporal displacement in frames, M is the number of pixels in the horizontal direction, N is the number of pixels in the vertical direction of the image block, i and j are the pixel indices within the image block, and m and n are the indices of the search neighborhood in the horizontal and vertical directions. Therefore, the best match motion vector estimate $MV(m = x, n = y)$ is the pixel displacement between the block $I^k(i, j)$ in frame k and the best matched block $I^{k-1}(i + x, j + y)$ in the displaced frame $k - l$. The best match is depicted in Fig. 8.6.

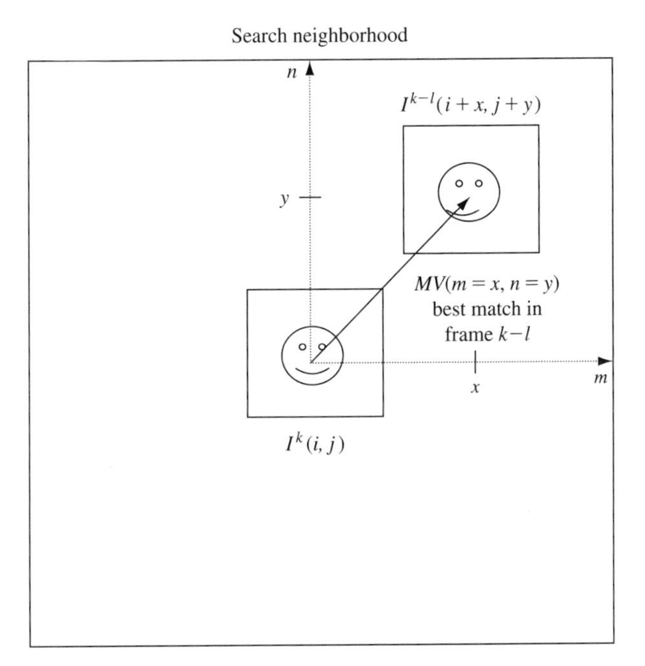

FIGURE 8.6

Best match motion estimate.

In cases where the block motion is not uniform or if the scene changes, the motion estimate may in fact increase the bit rate over the corresponding spatial encoding of the block. In the case where the motion estimate is not effective, the video encoder does not use the motion estimate and encodes the block using the spatial encoder.

The search space size determines the complexity of the motion estimation algorithm. Full search methods are costly and are not generally implemented in real-time video encoders. Fast searching techniques can considerably reduce computational complexity while maintaining good accuracy. These algorithms reduce the search process to a few sequential steps in which each subsequent search direction is based on the results of the current step. The procedures are designed to find local optimal solutions and cannot guarantee selection of the global optimal solution within the search neighborhood. The logarithmic search [20] algorithm proceeds in the direction of minimum distortion until the final optimal value is found. Logarithmic searching has been implemented in some MPEG encoders. The *Three-step* search [5] is a very simple technique that proceeds along a best match path in three steps in which the search neighborhood is reduced for each successive step. Figure 8.7 depicts the Three-step search algorithm.

A 14×14 pixel search neighborhood is depicted. The search area sizes for each step are chosen so that the total search neighborhood can be covered in finding the local minimum. The search areas are square. The length of the sides of the search area for Step 1 is chosen to be larger than or equal to half the length of the range of the search

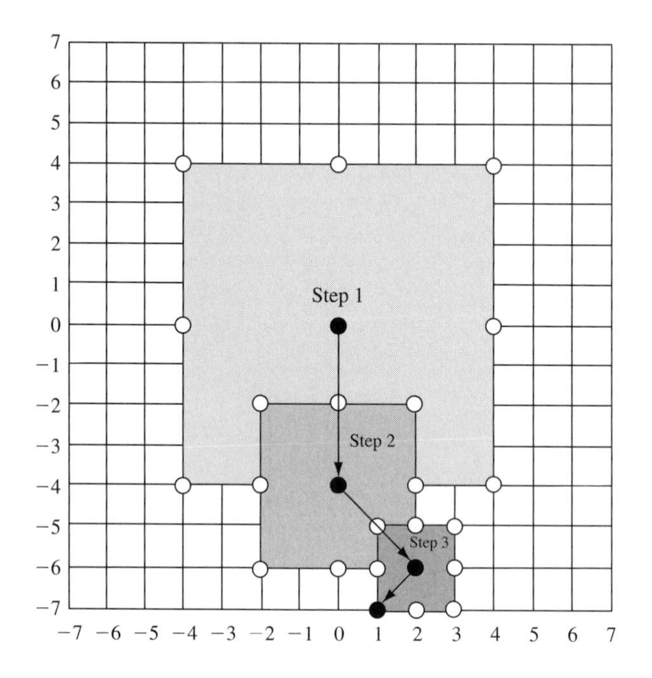

FIGURE 8.7

Three-step search algorithm pictorial.

neighborhood (in this example the search area is 8 × 8). The length of the sides is successively reduced by half after each of the first two steps is completed. Nine points for each step are compared using the matching criteria. These consist of the central point and eight equally spaced points along the perimeter of the search area. The search area for Step 1 is centered on the search neighborhood. The search proceeds in by centering the search area for the next step over the best match from the previous step. The overall best match is the pixel displacement chosen to minimize the matching criteria in Step 3. The total number of required comparisons for the Three-step algorithm is 25. That represents an 87% reduction in complexity versus the full search method for a 14 × 14 pixel search neighborhood.

8.5 TRANSFORM CODING: INTRODUCTION TO THE VIDEO ENCODING STANDARDS

The major internationally recognized video compression standards have been developed by the ISO, the International Electrotechnical Commission (IEC), and the ITU standards organizations. The Moving Pictures Experts Group (MPEG) is a working group operating within ISO and IEC. Since starting its activity in 1988, MPEG has produced ISO/IEC 11172 (MPEG-1, 1992), ISO/IEC 13818 (MPEG-2, 1994), ISO/IEC 14496 (MPEG-4, 1999), ISO/IEC 15938 (MPEG-7, 2001), and ISO/IEC 21000 (MPEG-21, 2002). The ITU adopted the original CCITT Recommendation H.261: "Video Codec for Audio Visual Services at $p \times 64$ kbps," in 1990, followed by the ITU-T SG 15 WP 15/1 Draft Recommendation H.262 (*Infrastructure of audiovisual services—Coding of moving video*) 1995, ITU-T SG 15 WP 15/1 Draft Recommendation H.263 (*Video coding for low bit rate communications*) 1995, and lastly the latest ITU Recommendation H.264: "Advanced Video Coding (AVC)," in 2002. H.264 and MPEG-4 Part 10 are equivalent video coding specifications as are H.262 and MPEG-2.

The MPEG-1 specification was motivated by T1 network transmission speeds, the CD-ROM, and the early multimedia capabilities of the desktop computer. It is intended for video coding up to the rate of 1.5 Mbps and is composed of five sections: System Configurations, Video Coding, Audio Coding, Compliance Testing, and Software for MPEG-1 Coding. The standard does not specify the actual video coding process, but only the syntax and semantics of the bit stream, and the video generation at the receiver. It does not accommodate interlaced video and only supports CIF quality format at 25 or 30 fps.

Activity for MPEG-2 was started in 1991. It was targeted for higher bit rates, broadcast video, and a variety of consumer and telecommunications video and audio applications. The syntax and technical contents of the standard were frozen in 1993. It is composed of four parts: Systems, Video, Audio, and Conformance Testing. MPEG-2 was also recommended by the ITU as H.262.

The MPEG-4 project was initially targeted to enable content developers and users to achieve various forms of interactivity with the audiovisual content of a scene and to mix synthetic and natural audio and video information in a seamless way. MPEG-4 technology comprises two major parts: a set of coding tools for audiovisual objects and a syntactic language to describe both the coding tools and the coded objects. From a technical viewpoint, the most notable departure from traditional coding standards is the ability for a receiver to download the description of the syntax used to represent the audiovisual information. The visual information is not restricted to have the format of conventional video, that is, it may not necessarily be frame-based, but can incorporate audio and/or video foreground and background objects, which can produce significant improvements in both encoder efficiency and functionality.

MPEG-7 [21] is formally named "Multimedia Content Description Interface" and is a common way of describing multimedia content data that is used to access and interpret content by a computer program. Since much of the value of multimedia content can be derived from its accessibility, MPEG-7 strives to define common data access methods to maximize the value of multimedia information regardless of the technologies encompassed by the source and destination or the specific applications using its services. To meet these requirements, MPEG-7 has created a hierarchical framework that can handle many levels of description. In addition, other types of descriptive data are defined, such as coding formats, data access conditions, parental ratings, relevant links, and the overall context. MPEG-7 is made up of three main elements that include description tools, storage and transmission system tools, and a language to define the MPEG-7 syntax. These elements provide the flexibility to meet the stated requirements.

The MPEG-21 [22] standard was adopted in 2002. It was started as an extension to MPEG-7 and is focused on defining the common content and user access model addressing the vast proliferation of new and old multimedia distribution and reception technologies. MPEG-21 specifically looks to define the technology needed to support users to exchange, access, consume, trade, and otherwise manipulate Digital Items in an efficient, transparent, and interoperable way. Digital Items are defined to be the fundamental unit of distribution and transaction, that is, content (web page, picture, movie, etc.).

The precursor to the MPEG video encoding standards development is the H.261 encoder, which contains many video of the coding methods and techniques later adopted by MPEG. The ITU Recommendation H.261 was adopted in 1990 and specifies a video encoding standard for videoconferencing and videophone services for transmission over Integrated Services Digital Network (ISDN) at $p \times 64$ kbps, $p = 1, \ldots, 30$. H.261 describes the video compression methods that were later adopted by the MPEG standards and is presented as an example of a transform-based standardized video coder in the following section. The ITU Experts Group for Very Low Bit Rate Video Telephony (LBC) has produced the H.263 Recommendation for Public Switched Telephone Networks (PSTN), which was finalized in December 1995 [5]. H.263 is an extended version of H.261 supporting bidirectional MC and sub-QCIF formats. The H.263 encoder is based on hybrid DPCM/DCT coding refinements and improvements targeted to generate bit rates of less than 64 kbps. The Joint Video Team (JVT) of the ISO/IEC and ITU Experts Group produced the MPEG-4 Part 10 and H.264 standard for the coding of natural video images in 2002 [6]. The H.264 AVC encoder incorporates a number of transform, quantization, and MC improvements that achieve significantly better rate-distortion performance than the H.262/MPEG-2 encoders at the expense of greater computational requirements.

8.5.1 Transform Coding Standard Example: The H.261 Video Encoder

A brief description of the H261 video coding standard is offered in this section as an introduction to the techniques used in the transform-based video coding standards. The H.261 recommendation [4] is targeted at the videophone and videoconferencing application market running on connection-based ISDN at $p \times 64$ kbps, $p = 1, \ldots, 30$. It explicitly defines the encoded bit-stream syntax and decoder, while leaving the encoder design to be compatible with the decoder specification. The video encoder is required to carry a delay of less than 150 ms so that it can operate in real-time bidirectional videoconferencing applications. H.261 is part of a group of related ITU Recommendations that define Visual Telephony Systems. This group includes as follows:

H.221 – Defines the frame structure for an audiovisual channel supporting 64–1920 kbps.

H.230 – Defines frame control signals for audiovisual systems.

H.242 – Defines audiovisual communications protocol for channels supporting up to 2 Mbps.

H.261 – Defines the video encoder/decoder for audiovisual services at $p \times 64$ kbps.

H.320 – Defines narrow-band audiovisual terminal equipment for $p \times 64$ kbps transmission.

The H.261 encoder block diagrams are depicted in Fig. 8.8(a) and (b). An H.261 Source Coder implementation is depicted in Fig. 8.8(c). The source coder implements the video encoding algorithm that includes the spatial encoder, the quantizer, the temporal prediction encoder, and the VLC. The spatial encoder is defined to use the 2D 8×8 pixel block DCT and a nearly uniform scalar quantizer using a possible 31 step sizes to scale the AC and interframe DC coefficients. The resulting quantized coefficient matrix is zigzag scanned into a vector that is VLC coded using a hybrid modified run length and Huffman coder. MC is optional. Motion estimation is only defined in the forward direction because H.261 is limited to real-time videophone and videoconferencing. The recommendation does not specify the motion estimation algorithm or the conditions for the use of intraframe versus interframe encoding.

The video multiplex coder creates a H.261 bit stream that is based on the data hierarchy described below. The transmission buffer is chosen not to exceed the maximum coding delay of 150 ms and is used to regulate the transmission bit rate via the coding controller. The transmission coder embeds an ECC into the video bit stream that provides error resilience, error concealment, and video synchronization.

H.261 supports most of the internationally accepted digital video formats. These include CCIR 601, SIF, CIF, and QCIF. These formats are defined for both NTSC and PAL broadcast signals. The CIF and QCIF formats were adopted in 1984 by H.261 to support 525-line NTSC and 625-line PAL/SECAM video formats. The CIF and QCIF operating parameters can be found in Table 8.4. The raw data rate for 30 fps CIF is 37.3 and 9.35 Mbps for QCIF. CIF is defined for use in channels in which $p \geq 6$ so that the required compression ratio for 30 fps is less than 98:1. CIF and QCIF formats support frame rates of 30, 15, 10, and 7.5 fps, which allows the H.261 encoder to achieve greater coding efficiency by skipping the encoding and transmission of whole frames. H.261 allows 0, 1, 2, 3, or more frames to be skipped between transmitted frames.

H.261 specifies a set of encoder protocols and decoder operations that every compatible system must follow. The H.261 *Video Multiplex* defines the data structure hierarchy that the decoder can interpret unambiguously. The video data hierarchy defined in H.261 is depicted in Fig. 8.9. They are the Picture layer, Group of Block (GOB) layer, Macroblock (MB) layer, and the basic (8×8) block layer. Each layer is built from the previous or lower layers and contains its associated data payload and header that describes the parameters used to generate the bit stream. The basic 8×8

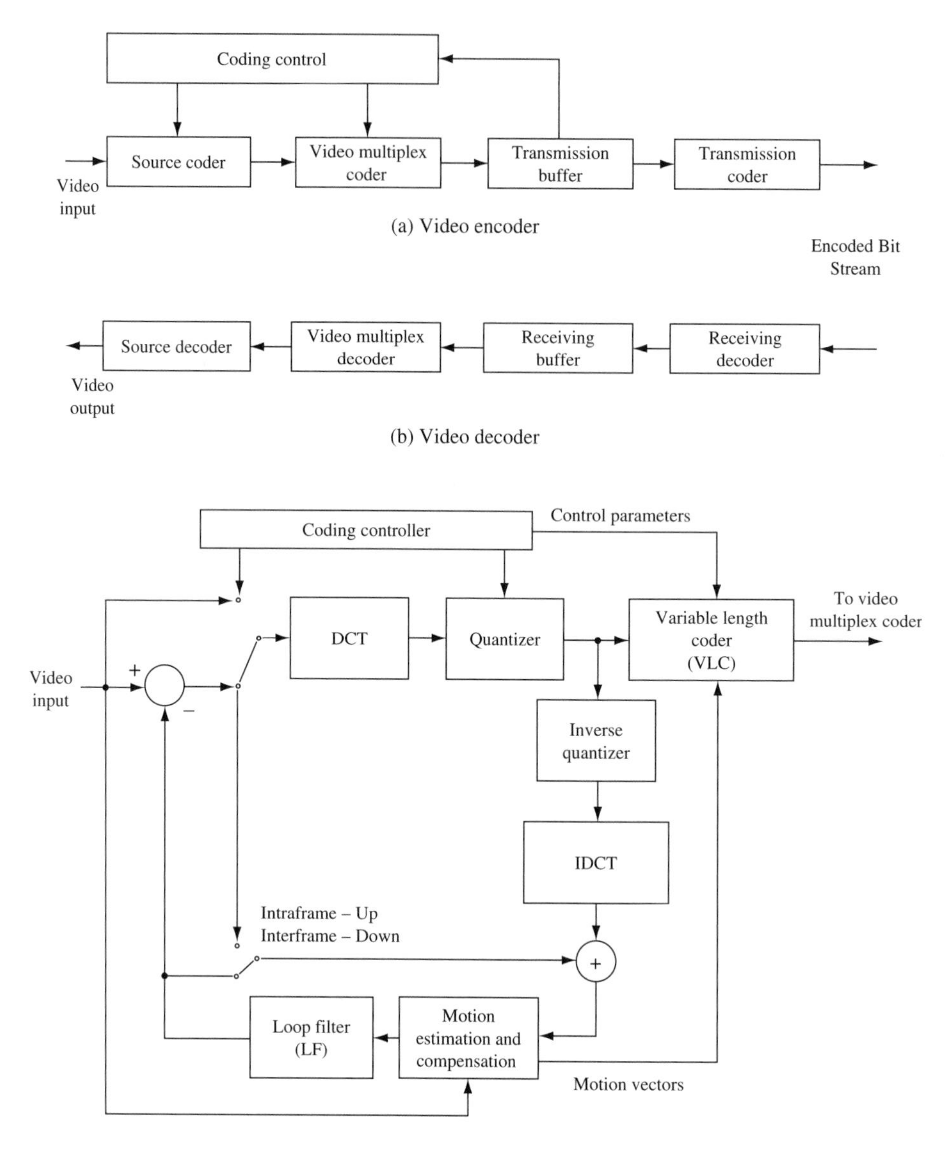

FIGURE 8.8

ITU-T H.261 block diagrams.

block is used in intraframe DCT encoding. The MB is the smallest unit for selecting intraframe or interframe encoding modes. It is made up of four adjacent 8×8 luminance blocks and two subsampled 8×8 color difference blocks (C_B and C_R as defined in Table 8.4) corresponding to the luminance blocks. The GOB is made up of 176×48

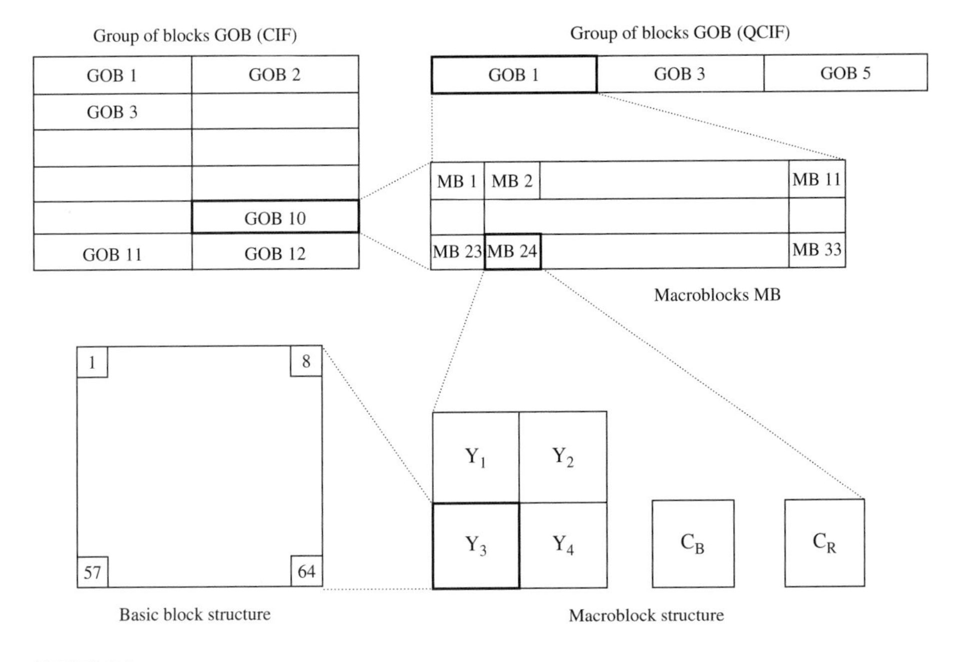

FIGURE 8.9

H.261 block hierarchy.

pixels (33 MBs) and is used to construct the 352 × 288 pixel CIF or 176 × 144 pixel QCIF Picture layer.

The headers for the GOB and Picture layers contain start codes so that the decoder can resynchronize when errors occur. They also contain other relevant information required to reconstruct the image sequence. The following parameters used in the headers of the data hierarchy complete the H.261 Video Multiplex:

Picture layer

- Picture start code (PSC), 20-bit synchronization pattern (0000 0000 0000 0001 0000)
- Temporal reference (TR), 5-bit input frame number
- Type information (PTYPE), which indicates source format, CIF = 1 QCIF = 0, and other controls
- User-inserted bits

GOB layer

- GOBs start code (GBSC), 16-bit synchronization code (0000 0000 0000 0001)
- Group number (GN), 4-bit address representing the 12 GOBs per CIF frame

- Quantizer information (GQUANT), which indicates one of 31 quantizer step sizes to be used in a GOB unless overridden by MB MQUANT parameter
- User-inserted bits

MB layer

- MB address (MBA), which is the position of a MB within a GOB
- Type information (MTYPE), for one of 10 encoding modes used for the MB (This includes permutations of intraframe, interframe, MC, and loop filtering (LF). A prespecified VLC is used to encode these modes.)
- Quantizer (MQUANT), 5-bit normalized quantizer step size from 1 to 31
- Motion vector data (MVD), up to 11-bit VLC describing the differential displacement
- Coded block pattern (CBP), up to 9-bit VLC indicating the location of the encoded blocks in the MB

Block Layer

- TCOEFF, which are zigzag scanned and can be 8-bit fixed or up to 13-bit VLC
- End of block (EOB), symbol

The H.261 bit stream also specifies transmission synchronization and error code correction using a BCH code [23] that is capable of correcting 2-bit errors in every 511-bit block. It inserts 18 parity bits for every 493 data bits. A synchronization bit is added to every code word to be able to detect the BCH code-word boundaries. The transmission synchronization and encoding also operates on the audio and control information specified by the ITU H.320 Recommendation.

The H.261 video compression algorithm depicted in Fig. 8.7 is specified to operate in intraframe and interframe encoding modes. The intraframe mode provides spatial encoding of the 8×8 block and uses the 2D DCT. Interframe mode encodes the prediction error, with MC being optional. The prediction error is optionally DCT encoded. Both modes provide options that affect the performance and video quality of the system. The motion estimate method, mode selection criteria, and block transmission criteria are not specified although the ITU has published reference models [24, 25] that make particular implementation recommendations. The coding algorithm used in the ITU-T Reference Model 8 (RM8) [5] is summarized in three steps and is followed by an explanation of its important encoding elements.

1. The motion estimator creates a displacement vector for each MB. The motion estimator generally operates on the 16×16 pixel luminance MB. The displacement vector is an integer value between $\pm$ 15, which is the maximum size of the search neighborhood. The motion estimate is scaled by a factor of 2 and applied to the C_R and C_B component MBs.

2. The compression mode for each MB is selected using a minimum error criteria that is based on the Displaced Macroblock Difference (DMD),

$$\mathrm{DMD}(i,j,k) = b(i,j,k) - b(i - d_1, j - d_2, k - 1), \tag{8.13}$$

where b is a 16×16 MB, i and j are its spatial pixel indices, k is the frame index, and d_1 and d_2 are the pixel displacements of the MB in the previous frame. The displacements range from $-15 \leq d_1, d_2 \leq +15$. When d_1 and d_2 are set to zero, the DMD becomes the Macroblock Difference (MD). The compression mode determines the operational encoder elements that are used for the current frame. The H.261 compression modes are depicted in Table 8.5.

3. The *Video Multiplex Coder* processes each MB to generate the H.261 video bit stream whose elements are discussed above.

There are five basic MTYPE encoding mode decisions that are carried out in Step 2. These are as follows:

- Use intraframe or interframe mode?
- Use MC?
- Use a CBP?
- Use loop-filtering?
- Change quantization step size MQUANT?

To select the MB compression mode, the VAR of the input MB, the MD, and the DMD (as determined by the best motion estimate) are compared as follows:

TABLE 8.5 H.261 macroblock video compression modes.

Mode	MQUANT	MVD	CBP	TCOEFF
Intra				3
Intra	3			3
Inter			3	3
Inter	3		3	3
Inter+MC		3		
Inter+MC		3	3	3
Inter+MC	3	3	3	3
Inter+MC+LF		3		
Inter+MC+LF		3	3	3
Inter+MC+LF	3	3	3	3

1. If VAR(DBD) < VAR(MD), then interframe + MC (Inter + MC) coding is selected. In this case, the MVD is transmitted. Table 8.5 indicates that there are three Inter + MC modes that allow for the transmission of the prediction error (DMD) with or without DCT encoding of some or all of the four 8 × 8 basic blocks.
2. "VAR input" is defined as the variance of the input MB. If VAR input < VAR(DMD) and VAR input < VAR(MD), then the intraframe mode (Intra) is selected. Intraframe mode uses DCT encoding of all four 8 × 8 basic blocks.
3. If VAR(MD) < VAR(DMD), then interframe mode (Inter) is selected. This mode indicates that the motion vector is zero, and that some or all of the 8 × 8 prediction error (MD) blocks can be DCT encoded.

The transform coefficient CBP parameter is used to indicate whether a basic block is reconstructed using the corresponding basic block from the previous frame or whether it is encoded and transmitted. In other words, no basic block encoding is used when the block content does not change significantly. The CPB parameter encodes 63 combinations of the four luminance blocks and two color difference blocks using a variable length code. The conditions for using CBP are not specified in the H.261 recommendation.

Motion compensated blocks can be chosen to be low-pass filtered before the prediction error is generated by the feedback loop. This mode is denoted as Inter+MC+LF in Table 8.5. The low-pass filter is intended to reduce the quantization noise in the feedback loop, as well as the high-frequency noise and artifacts introduced by the motion compensator. H.261 defines loop filtering as optional and recommends a separable 2D spatial filter design, which is implemented by cascading two identical 1D Finite Impulse Response (FIR) filters. The coefficients of the 1D filter are [1, 2, 1] for pixels inside the block and [0, 1, 0] (no filtering) for pixels on the block boundary.

The MQUANT parameter is controlled by the state of the transmission buffer to prevent overflow or underflow conditions. The dynamic range of the DCT MB coefficients extends between [−2047,...,2047]. They are quantized to the range [−127,...,127] using one of the 31 quantizer step sizes as determined by the GQUANT parameter. The step size is an even integer in the range of [2,...,62]. GQUANT can be overridden at the MB layer by MQUANT to clip or expand the range prescribed by GQUANT so that the transmission buffer is better used. The ITU-T RM8 *liquid level control model* specifies the inspection of 64 Kb transmission buffers after encoding 11 MBs. The step size of the quantizer should be increased (decreasing the bit rate) if the buffer is full, and vice versa, the step size should be decreased (increasing the bit rate) if the buffer is empty. The actual design of the rate control algorithm is not specified.

The DCT MB coefficients are subjected to variable thresholding before quantization. The threshold is designed to increase the number of zero-valued coefficients, which in turn increases the number of the zero run-lengths and VLC coding efficiency. The ITU-T RM8 provides an example thresholding algorithm for the H.261 encoder. Nearly uniform scalar quantization using a dead zone is applied after the thresholding process. All the coefficients in the luminance and chrominance MBs are subjected to the same quantizer except for the intraframe DC coefficient. The intraframe DC coefficient is quantized using a uniform scalar quantizer whose step size is 8. The quantizer decision levels are not specified, but the reconstruction levels are defined in H.261 as follows:

For case QUANT odd
REC_LEVEL = QUANT 5 (2 5 COEFF_VALUE + 1), for COEFF_LEVEL > 0,
REC_LEVEL = QUANT 5 (2 5 COEFF_VALUE − 1), for COEFF_LEVEL < 0,
For case QUANT even
REC_LEVEL = QUANT 5 (2 5 COEFF_VALUE + 1) − 1, for COEFF_LEVEL > 0,
REC_LEVEL = QUANT 5 (2 5 COEFF_VALUE − 1) + 1, for COEFF_LEVEL < 0,
If COEFF_VALUE = 0, then REC_LEVEL = 0,

where REC_LEVEL is the reconstruction value, QUANT is half the MB quantization step size ranging from 1 to 31, and COEFF_VALUE is the quantized DCT coefficient.

To increase the coding efficiency, lossless VLC is applied to the quantized DCT coefficients. The coefficient matrix is scanned in a zig-zag manner to maximize the number of zero coefficient run-lengths. The VLC encodes events defined as the combination of a run-length of zero coefficients preceding a nonzero coefficient and the value of the nonzero coefficient, that is, EVENT = (RUN, VALUE). The VLC EVENT tables are defined in Ref. [4].

8.6 CLOSING REMARKS

Digital video compression, although relatively recently becoming an internationally standardized technology, is strongly based on the information coding technologies researched over the last 50 years. The large variety of bandwidth and video quality requirements for the transmission and storage of digital video information has demanded that a variety of video compression techniques and standards be developed. The major international standards recommended by ISO and the ITU make use of common spatial and video coding methods. The generalized digital video encoder introduced in Section 8.1 illustrates the spatial transform and temporal video compression elements that are central to the current MPEG-1, MPEG-2/H.262, MPEG-4, H.261, H.263, and H.264 standards that have been adopted over the past two decades. They address a vast landscape of application requirements from low to high bit rate environments, as well as stored video and multimedia to real-time videoconferencing and high-quality broadcast television.

REFERENCES

[1] ISO/IEC 11172 *Information Technology: Coding of Moving Pictures and Associated Audio for Digital Storage Media at up to about 1.5 Mbit/s.* 1993.

[2] ISO/IEC JTC1/SC29/WG11. *CD 13818: Generic Coding of Moving Pictures and Associated Audio.* 1993.

[3] ISO/IEC JTC1/SC29/WG11. *CD 14496: Coding of Audio-Visual Objects.* 1999.

[4] CCITT Recommendation H.261: "Video Codec for Audio Visual Services at *p* x 64 kbits/s," COM XV-R 37-E, 1990.

[5] ITU-T SG 15 WP 15/1, *Draft Recommendation H.263* (*Video coding for low bitrate communications*), Document LBC-95-251, October 1995.

[6] ITU-T Recommendation H.264, ISO/IEC 11496-10, *Advanced Video Coding*, Final Committee Draft, Document JVT-F100, December 2002.

[7] W. H. Equitz. A new vector quantization clustering algorithm. *IEEE Trans. Acoust. Speech Signal Process.*, ASSP-37(10):1568–1575, 1989.

[8] H. Hseuh-Ming and J. W. Woods. *Handbook of Visual Communications*, Chapter 6. Academic Press Inc., San Diego, CA, 1995.

[9] J. W. Woods. *Subband Image Coding*. Kluwer Academic Publishers, Norwell, MA, 1991.

[10] L. Wang and M. Goldberg. Progressive image transmission using vector quantization on images in pyramid form. *IEEE Trans. Commun.*, 1339–1349, 1989.

[11] C. E. Shannon. A mathematical theory of communication. *Bell Syst. Tech. J.*, 27:379–423, 623–656, 1948.

[12] D. Huffman. A method for the construction of minimum redundancy codes. *Proc. IRE*, 40:1098–1101, 1952.

[13] P. W. Jones and M. Rabbani. *Digital Image Compression Techniques*. SPIE Optical Engineering Press, Bellingham, WA, 60, 1991.

[14] N. Ahmed, T. R. Natarajan, and K. R. Rao. On image processing and a discrete cosine transform. *IEEE Trans. Comput.*, IT-23:90–93, 1974.

[15] J. J. Hwang and K. R. Rao. *Techniques and Standards For Image, Video, and Audio Coding*. Prentice Hall, Upper Saddle River, NJ, 22, 1996.

[16] R. M. Gray. Vector quantization. *IEEE ASSP Mag.*, IT-1:4–29, 1984.

[17] B. G. Haskell and J. O. Limb. *Predictive Video Encoding using Measured Subjective Velocity*. U.S. Patent No. 3,632,865, 1972.

[18] A. N. Netravali and J. D. Robbins. Motion—compensated television coding: Part I. *Bell Syst. Tech. J.*, 58:631–670, 1979.

[19] J. R. Jain and A. K. Jain. Displacement measurement and its application in interframe image coding. *IEEE Trans. Commun.*, COM-29:1799–1808, 1981.

[20] T. Koga et al. Motion compensated interframe coding for video conferencing. *NTC '81, National Telecommun. Conf.*, G5.3.1–G5.3.5, New Orleans, LA, November 1981.

[21] ISO/IEC JTC1/SC29/WG11 Recommendation MPEG-7. *Coding of Moving Pictures and Audio*, Document N6828. 2004

[22] ISO/IEC JTC1/SC29/WG11 Recommendation MPEG-21. *Coding of Moving Pictures and Audio*. 2002

[23] M. Roser et al. Extrapolation of a MPEG-1 video-coding scheme for low-it-rate applications. *SPIE Video Commun. And PACS for Medical Appl.*, vol. 1977, Berlin, Germany, 1993, pp. 180–187.

[24] CCITT SG XV WP/1/Q4 Specialist Group on Coding for Visual Telephony, *Description of Ref. Model 6* (RM6), Document 396, October 1988.

[25] CCITT SG XV WP/1/Q4 Specialist Group on Coding for Visual Telephony, *Description of Ref. Model 8* (RM8), Document 525, June 1989.

CHAPTER

9 MPEG-1 and MPEG-2 Video Standards

Supavadee Aramvith[1] and Ming-Ting Sun[2]

[1] *Department of Electrical Engineering, Faculty of Engineering, Chulalongkorn University, Bangkok, Thailand*
[2] *Information Processing Laboratory, Department of Electrical Engineering, University of Washington, Seattle, Washington, USA*

9.1 MPEG-1 VIDEO CODING STANDARD

9.1.1 Introduction

9.1.1.1 Background and Structure of MPEG-1 Standards Activities

The development of digital video technology in the 1980s has made it possible to use digital video compression in various kinds of applications. The effort to develop standards for coded representation of moving pictures, audio, and their combination is carried out in the Moving Picture Experts Group (MPEG). MPEG is a group formed under the auspices of the International Organization for Standardization (ISO) and the International Electrotechnical Commission (IEC). It operates in the framework of the Joint ISO/IEC Technical Committee 1 (JTC 1) on Information Technology, which was formally Working Group 11 (WG11) of SubCommittee 29 (SC29). The premise is to set the standard for coding moving pictures and the associated audio for digital storage media at about 1.5 Mbps so that a movie can be compressed and stored in a video compact disc (VCD). The resultant standard is the international standard for moving picture compression, ISO/IEC 11172 or MPEG-1 (MPEG – Phase 1). MPEG-1 standards consist of five parts, including systems (11172-1), video (11172-2), audio (11172-3), conformance testing (11172-4), and software simulation (11172-5). In this chapter, we will focus only on the video part.

The activity of the MPEG committee started in 1988 based on the work of ISO Joint Photographic Experts Group (JPEG) [1] and ITU-T (formerly CCITT) Recommendation H.261: Video Codec for Audiovisual Services at px64 kb/s [2]. Thus, the MPEG-1 standard has much in common with the JPEG and H.261 standards. The MPEG development methodology is similar to that of H.261 and is divided into three phases: requirements, competition, and convergence [3]. The purpose of the requirements phase is to precisely set the focus of the effort and determine the rule for the competition phase. The document of this phase is a proposal package description [4] and a test methodology [5]. The next step is the competition phase in which the goal is to obtain the state of the art technology from the best of academic and industrial research. The criteria are based on the technical merits and the trade-off between the video quality and the cost of implementation [5]. After the competition phase, various ideas and techniques are integrated into one solution in the convergence phase. The solution results in a simulation model, which implements a reference encoder and a decoder. The simulation model is used to carry out simulations to optimize the performance of the coding scheme [6]. A series of fully documented experiments called core experiments are then carried out. The MPEG committee reached the Committee Draft (CD) status in September 1990, and the CD 11172 was approved in December 1991. International Standard (IS) 11172 for the first three parts was established in November 1992. The IS for the last two parts was finalized in November 1994.

9.1.1.2 MPEG-1 Target Applications and Requirements

The MPEG standard is a generic standard, which means that it is not limited to a particular application. A variety of digital storage media applications of MPEG-1 have been proposed based on the assumptions that acceptable video and audio quality can be obtained for a total bandwidth of about 1.5 Mbps. Typical storage media for these applications include VCD, digital audio tape (DAT), Winchester-type computer disks, and writable optical disks. The target applications are asymmetric applications where the compression process is performed once and the decompression process is required often. Examples of the asymmetric applications include VCD, video on demand (VOD), and video games. In these asymmetric applications, the encoding delay is not a concern. The encoders are needed only in small quantities while the decoders are needed in large volumes. Thus, the encoder complexity is not a concern while the decoder complexity needs to be low to result in low-cost decoders.

The requirements for compressed video in digital storage media mandate several important features of the MPEG-1 compression algorithm. The important features include normal playback, frame-based random access and editing of video, reverse playback, fast forward/reverse play, encoding high-resolution still frames, robustness to uncorrectable errors, etc. The applications also require MPEG-1 to support flexible picture sizes and frame rates. Another requirement is that the encoding process can be performed in reasonable speed using existing hardware technologies and the decoder can be implemented in low cost.

Since the MPEG-1 video coding algorithm was developed based on H.261, in the following sections, we will focus only on those parts that are different from H.261.

9.1.2 MPEG-1 Video Coding versus H.261

9.1.2.1 Bidirectional Motion-Compensated Prediction

In H.261, only the previous video frame is used as the reference frame for the motion-compensated prediction (forward prediction). MPEG-1 allows the future frame to be used as the reference frame for the motion-compensated prediction (backward prediction),

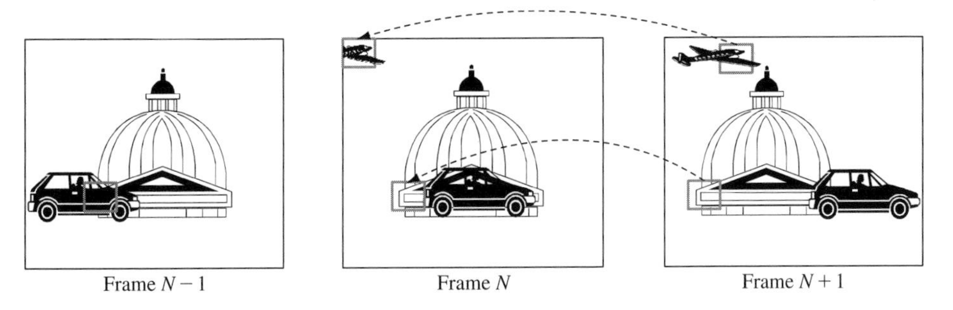

FIGURE 9.1

A video sequence showing the benefits of bidirectional prediction.

which can provide better prediction. For example, as shown in Fig. 9.1, if there are moving objects, and if only the forward prediction is used, there will be uncovered areas (such as the block behind the car in frame N) for which we may not be able to find a good matching block from the previous reference picture (frame $N-1$). On the other hand, the backward prediction can properly predict these uncovered areas since they are available in the future reference picture, that is, frame $N+1$ in this example. Also as shown in the figure, if there are objects moving into the picture (the airplane in the figure), these new objects cannot be predicted from the previous picture, but can be predicted from the future picture. In fact, all the information in the second picture is available from the first and the third picture. Another major advantage of the B-pictures is the denoising capability. In practical situations, the pixel values of an object may not be same, due to various noise effects from lighting changes, shadows, sampling effects, and other noises. Bidirectional prediction could reduce the noise effects due to averaging or simply due to the fact that there is an extra choice that could provide a better matching.

9.1.2.2 *Motion-Compensated Prediction with Half-Pixel Accuracy*

The motion estimation in H.261 is restricted to only integer-pixel accuracy. However, a moving object often moves to a position that is not on the pixel grid but between the pixels. MPEG-1 allows half-pixel accuracy motion vectors. By estimating the displacement at a finer resolution, we can expect improved prediction and, thus, better performance than motion estimation with integer-pixel accuracy. As shown in Fig. 9.2, since there is no pixel value at the half-pixel locations, interpolation is required to produce the pixel values at the half-pixel positions. Bilinear interpolation is used in MPEG-1 for its simplicity. As in H.261, the motion estimation is performed only on luminance blocks. The resulting motion vector is scaled by 2 and applied to the chrominance blocks. Motion vectors are differentially encoded with respect to the motion vector in the preceding adjacent macroblock. The reason is that the motion vectors of adjacent regions are highly correlated, as it is quite common to have relatively uniform motion over areas of the picture.

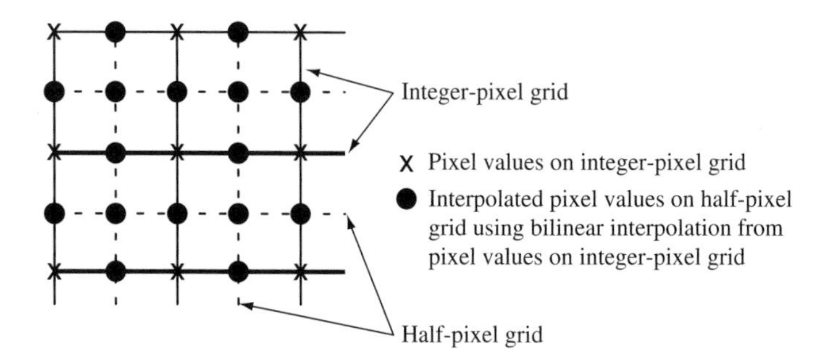

FIGURE 9.2

Half-pixel motion estimation.

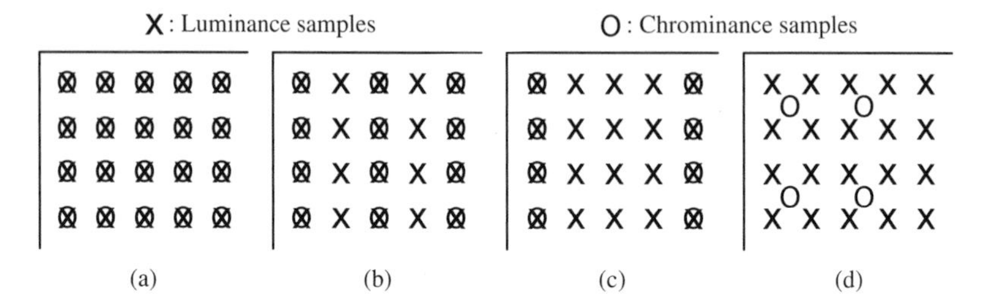

FIGURE 9.3

Luminance and chrominance samples in (a) 4:4:4 format (b) 4:2:2 format (c) 4:1:1 format (d) 4:2:0 format.

9.1.3 MPEG-1 Video Structure

9.1.3.1 Source Input Format

The typical MPEG-1 input format is the Source Input Format (SIF). SIF is derived from ITU-R BT 601, formerly CCIR601, a worldwide standard for digital TV studio. ITU-R BT 601 specifies the Y Cb Cr color coordinate where Y is the luminance component (black and white information) and Cb and Cr are two color difference signals (chrominance components). A luminance sampling frequency of 13.5 MHz was adopted. There are several Y Cb Cr sampling formats, such as 4:4:4, 4:2:2, 4:1:1, and 4:2:0. In 4:4:4, the sampling rates for Y, Cb, and Cr are the same. In 4:2:2, the sampling rates of Cb and Cr are half of that of Y. In 4:1:1 and 4:2:0, the sampling rates of Cb and Cr are one quarter of that of Y. The positions of Y Cb Cr samples for 4:4:4, 4:2:2, 4:1:1, and 4:2:0 are shown in Fig. 9.3.

Converting an analog TV signal to digital video with the 13.5 MHz sampling rate of ITU-R BT 601 results in 720 active pixels per line (576 active lines for PAL and 480 active lines for National Television System Committee [NTSC]). This results in a 720 × 480 resolution for NTSC and a 720 × 576 resolution for PAL. With 4:2:2, the uncompressed bit rate for transmitting ITU-R BT 601 at 30 fps is then about 166 Mbps. Since it is difficult to compress an ITU-R BT 601 video to 1.5 Mbps with good video quality, in MPEG-1, typically the source video resolution is decimated to a quarter of the ITU-R BT 601 resolution by filtering and subsampling. The resultant format is called SIF, which has a 360×240 resolution for NTSC and a 360×288 resolution for PAL. Since in the video coding algorithm, the block size of 16×16 is used for motion-compensated prediction, the number of pixels in both the horizontal and the vertical dimensions should be multiples of 16. Thus, the four leftmost and four rightmost pixels are discarded to give a 352×240 resolution for NTSC systems (30 fps) and a 352×288 resolution for PAL systems (25 fps). The chrominance signals have half of the above resolutions in both the horizontal and vertical dimensions (4:2:0, 176×120 for NTSC and 176×144 for PAL). The uncompressed bit rate for SIF (NTSC) at 30 fps is about 30.4 Mbps.

9.1.3.2 Group of Pictures and I-B-P-Pictures

In MPEG, each video sequence is divided into one or more groups of pictures (GOPs). There are four types of pictures defined in MPEG-1: I-, P-, B-, and D-pictures of which the first three are shown in Fig. 9.4. Each GOP is composed of one or more pictures; one of these pictures must be an I-picture. Usually, the spacing between two anchor frames (I- or P-pictures) is referred to as M and the spacing between two successive I-pictures is referred to as N. In Fig. 9.4, $M = 3$ and $N = 9$. However, it should be noted that a GOP does not need to use a periodical structure.

Intracoded pictures (I-pictures) are coded independently with no reference to other pictures. I-pictures provide random access points in the compressed video data, since the I-pictures can be decoded independently without referencing to other pictures. With I-pictures, an MPEG bit stream is more editable. Also, error propagation due to transmission errors in previous pictures will be terminated by an I-picture since the I-picture does not reference to the previous pictures. Since I-pictures use only transform coding without motion-compensated predictive coding, it provides only moderate compression.

Predictive-coded pictures (P-pictures) are coded using the forward motion-compensated prediction similar to that in H.261 from the preceding I- or P-picture. P-pictures provide more compression than the I-pictures by virtue of motion-compensated prediction. They also serve as references for B-pictures and future

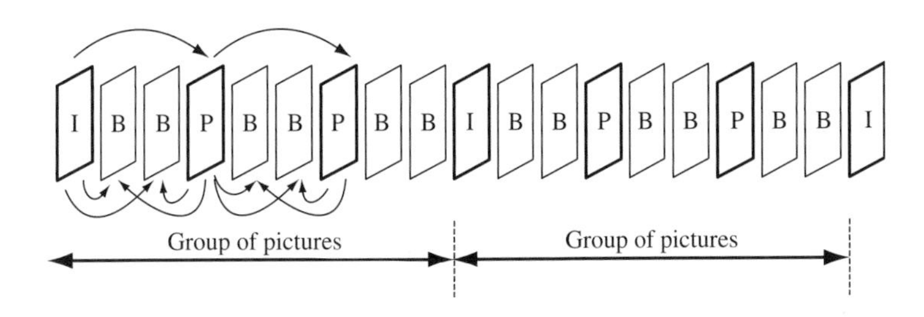

FIGURE 9.4

MPEG group of pictures.

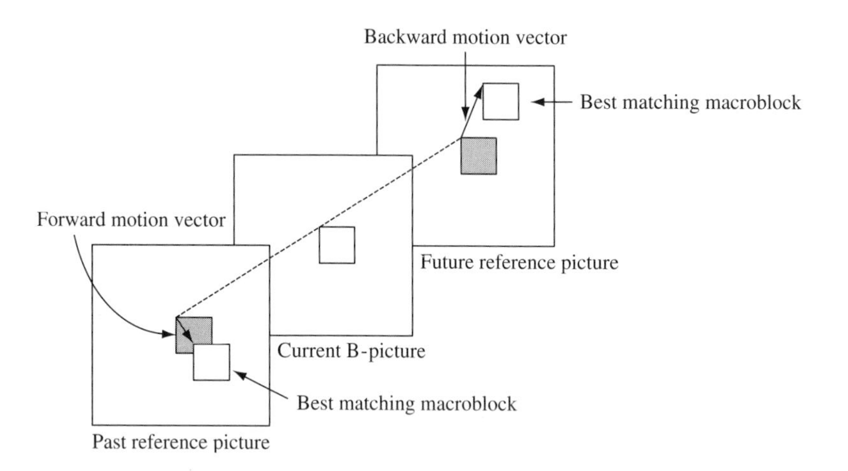

FIGURE 9.5

Bidirectional motion estimation.

P-pictures. Transmission errors in the I-pictures and P-pictures can propagate to the succeeding pictures since the I-pictures and P-pictures are used to predict the succeeding pictures.

Bidirectional-predicted pictures (B-pictures) allow macroblocks to be coded using bidirectional motion-compensated prediction from both the past and future reference I- or P-pictures. In the B-pictures, each bidirectional motion-compensated macroblock can have two motion vectors: a forward motion vector, which references to a best matching block in the previous I- or P-pictures, and a backward motion vector, which references to a best matching block in the next I- or P-pictures as shown in Fig. 9.5. The motion-compensated prediction can be formed by the average of the two referenced motion-compensated blocks. By averaging between the past and the future reference blocks, the effect of noise can be decreased. B-pictures provide the best compression compared to I- and P-pictures. I- and P-pictures are used as reference pictures for predicting B-pictures. To keep the structure simple, the B-pictures are not used as reference pictures. Hence, B-pictures do not propagate errors.

DC pictures (D-pictures) are low-resolution pictures obtained by decoding only the DC coefficient of the discrete cosine transform (DCT) coefficients of each macroblock. They are not used in combination with I-, P-, or B-pictures. D-pictures are rarely used but are defined to allow fast searches on sequential digital storage media.

The trade-off of having frequent B-pictures is that it decreases the correlation between the previous I- or P-picture and the next reference P- or I-picture. It also causes coding delay and increases the encoder complexity. With the example shown in Figs. 9.4 and 9.6, at the encoder, if the order of the incoming pictures is $1,2,3,4,5,6,7,\ldots$, the order of coding the pictures at the encoder will be $1,4,2,3,7,5,6,\ldots$. At the decoder, the order of the decoded pictures will also be $1,4,2,3,7,5,6,\ldots$. However, the display order after the decoder should be $1,2,3,4,5,6,7$. Thus, frame memories have to be used to put the

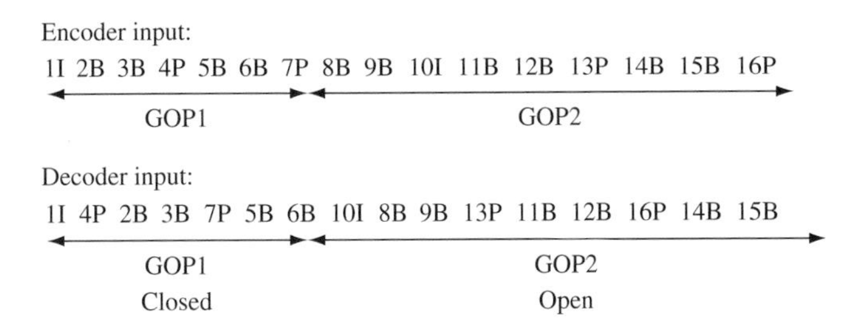

FIGURE 9.6

Frame reordering.

pictures in the correct order. This picture reordering causes delay. The computation and extra memory requirement of bidirectional motion estimation and the picture-reordering frame memories increase the encoder complexity.

In Fig. 9.6, two types of GOPs are shown. GOP1 can be decoded without referencing other GOPs. It is called a Closed GOP. In GOP2, to decode the eighth B- and ninth B-pictures, the seventh P-picture in GOP1 is needed. GOP2 is called an Open GOP, which means the decoding of this GOP needs to reference other GOPs.

9.1.3.3 Slice, Macroblock, and Block Structures

An MPEG picture consists of slices. A slice consists of a contiguous sequence of macroblocks in a raster scan order (from left to right and from top to bottom). In an MPEG-coded bit stream, each slice starts with a slice header, which is a clear codeword (a clear codeword is a unique bit pattern, which can be identified without decoding the variable-length codes [VLCs] in the bit stream). Due to the clear codeword slice header, slices are the lowest level of units that can be accessed in an MPEG-coded bit stream without decoding the VLCs. Slices are important in the handling of channel errors. If a bit stream contains a bit error, the error may cause error propagation due to the variable-length coding. The decoder can regain synchronization at the start of the next slice. Having more slices in a bit stream allows better error termination but the overhead will increase.

A macroblock consists of a 16×16 block of luminance samples and two 8×8 blocks of corresponding chrominance samples as shown in Fig. 9.7. A macroblock thus consists of four 8×8 Y blocks, one 8×8 Cb block, and one 8×8 Cr block. Each coded macroblock contains motion-compensated prediction information (coded motion vectors and the prediction errors). There are four types of macroblocks: intrapredicted, forward-predicted, backward-predicted, and averaged macroblocks. The motion information consists of one motion vector for forward- and backward-predicted macroblocks and two motion vectors for bidirectionally predicted (or averaged) macroblocks. P-pictures can have intrapredicted and forward-predicted macroblocks. B-pictures can have all four types of macroblocks. The first and last macroblocks in a slice must always be coded. A macroblock is designated as a skipped macroblock when its motion vector

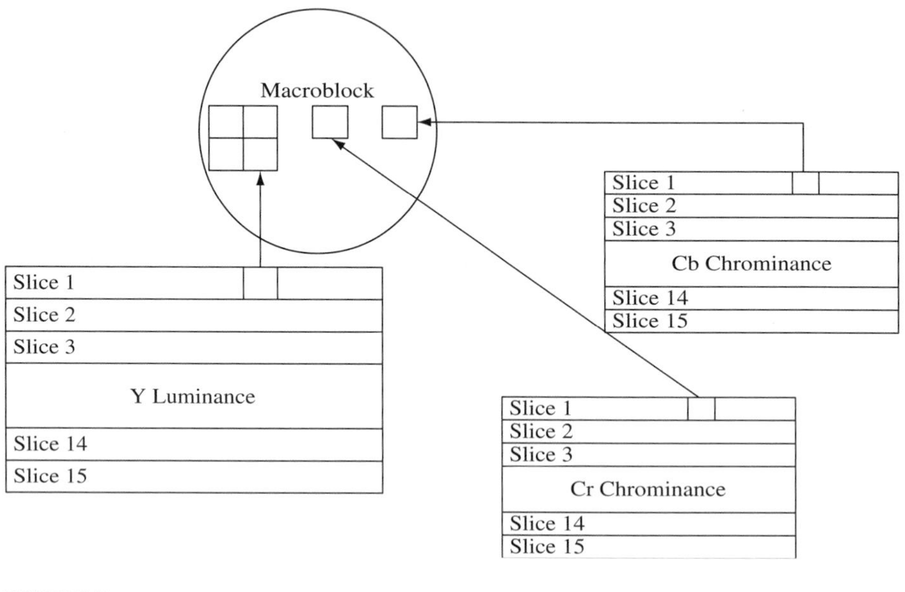

FIGURE 9.7

Macroblock and slice structures.

is zero, and all the quantized DCT coefficients are zero. Skipped macroblocks are not allowed in I-pictures. Nonintracoded macroblocks in P- and B-pictures can be skipped. For a skipped macroblock, the decoder just copies the pixel values of the macroblock from the previous picture.

9.1.4 Summary of the Major Differences between MPEG-1 Video and H.261

As compared to H.261, MPEG-1 video differs in the following aspects:

- MPEG-1 allows bidirectional motion-compensated predictive coding with half-pixel accuracy while H.261 has no bidirectional prediction (B-pictures) and the motion vectors are always in integer-pixel accuracy.
- MPEG-1 supports the maximum motion vector range of -512 to $+511.5$ pixels for half-pixel motion vectors and -1024 to $+1023$ for integer-pixel motion vectors while H.261 has a maximum range of only $\pm$ 15 pixels.
- MPEG-1 uses visually weighted quantization based on the fact that the human eye is more sensitive to quantization errors related to low spatial frequencies than to high spatial frequencies. MPEG-1 defines a default 64-element quantization matrix but also allows custom matrices appropriate for different applications. H.261 has only one quantizer for the intra DC coefficient and 31 quantizers for all other coefficients.

- H.261 only specifies two source formats: Common Intermediate Format (CIF, 352×288 pixels) and Quarter CIF (QCIF, 176×144 pixels). In MPEG-1, the typical source format is SIF (352×240 for NTSC and 352×288 for PAL). However, the users can specify other formats. The picture size can be as large as $4\text{k} \times 4\text{k}$ pixels. There are certain parameters in the bit streams that are left flexible, such as the number of lines per picture (less than 4096), the number of pels per line (less than 4096), picture rate (24, 25, and 30 fps), and 14 choices of pel aspect ratios.
- In MPEG-1, I-, P-, and B-pictures are organized as a flexible GOPs.
- MPEG-1 uses a flexible slice structure instead of group of blocks (GOBs) as defined in H.261.
- MPEG-1 has D-pictures to allow the fast-search option.
- To allow cost effective implementation of user terminals, MPEG-1 defines a constrained parameter set that lays down specific constraints, as listed in Table 9.1.

9.1.5 Simulation Model

Similar to H.261, MPEG-1 specifies only the syntax and the decoder. Many detailed coding options such as the rate control strategy, the quantization decision levels, the motion estimation schemes, and coding modes for each macroblock are not specified. This allows future technology improvement and product differentiation. To have a reference MPEG-1 video quality, simulation models were developed in MPEG-1. A simulation model contains a specific reference implementation of the MPEG-1 encoder and decoder including all the details that are not specified in the standard. The final version of the MPEG-1 simulation model is Simulation Model 3 (SM3) [7]. In SM3, the motion estimation technique uses one forward and/or one backward motion vector per macroblock with half-pixel accuracy. A two-step motion estimation scheme that consists of a full-search in the range of ± 7 pixels with the integer-pixel precision, followed by a search in the eight neighboring half-pixel positions, is used. The decision of the coding mode for each macroblock (whether or not it will use motion-compensated prediction and intracoding/intercoding), the quantizer decision levels, and the rate-control algorithm are all specified in the simulation model.

TABLE 9.1 MPEG-1 constrained parameter set.

- Horizontal size $\leq$ 720 pels
- Vertical size $\leq$ 576 pels
- Total number of macroblocks/picture $\leq$ 396
- Total number of macroblocks/second $\leq 396 \times 25 = 330 \times 30$
- Picture rate $\leq$ 30 fps
- Bit rate $\leq$ 1.86 Mbps
- Decoder buffer $\leq$ 376832 bits

9.1.6 MPEG-1 Video Bit-Stream Structures

As shown in Fig. 9.8, there are six layers in the MPEG-1 video bit stream: the video sequence, GOPs, picture, slice, macroblock, and block layers.

- A video sequence layer consists of a sequence header, one or more GOPs, and an end-of-sequence code. It contains the setting of the following parameters: the picture size (horizontal and vertical sizes), pel aspect ratio, picture rate, bit rate, the minimum decoder buffer size (video buffer verifier size), constraint parameters flag (this flag is set only when the picture size, picture rate, decoder buffer size, bit rate, and motion parameters satisfy the constraints bound in Table 9.1), the control for the loading of 64 eight-bit values for intra and nonintra quantization tables, and the user data.
- The GOP layer consists of a set of pictures that are in a continuous display order. It contains the setting of the following parameters: the time code that gives the hours-minutes-seconds time interval from the start of the sequence, the closed GOP flag that indicates whether the decoding operation needs pictures from the previous GOP for motion compensation, the broken link flag that indicated whether the previous GOP can be used to decode the current GOP, and the user data.
- The picture layer acts as a primary coding unit. It contains the setting of the following parameters: the temporal reference that is the picture number in the sequence and is used to determine the display order, the picture types (I/P/B/D), the decoder buffer initial occupancy that gives the number of bits that must be in the compressed video buffer before the idealized decoder model defined by MPEG decodes the picture (it is used to prevent the decoder buffer overflow and underflow), the forward motion vector resolution and range for P- and B-pictures, the backward motion vector resolution and range for B-pictures, and the user data.

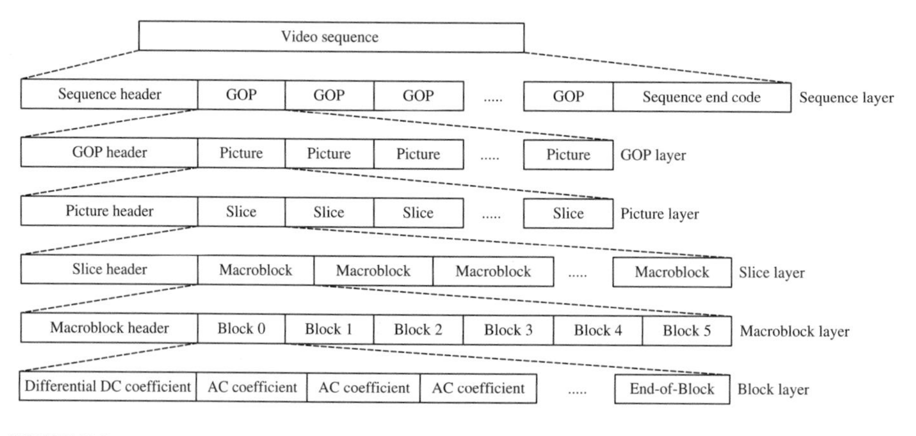

FIGURE 9.8

MPEG-1 bit-stream syntax layers.

- The slice layer acts as a resynchronization unit. It contains the slice vertical position where the slice starts and the quantizer scale that is used in the coding of the current slice.
- The macroblock layer acts as a motion compensation unit. It contains the setting of the following parameters: the optional stuffing bits, the macroblock address increment, the macroblock type, quantizer scale, motion vector, and the coded block pattern that defines the coding patterns of the six blocks in the macroblock.
- The block layer is the lowest layer of the video sequence and consists of coded 8×8 DCT coefficients. When a macroblock is encoded in the intramode, the DC coefficient is encoded similar to that in JPEG (the DC coefficient of the current macroblock is predicted from the DC coefficient of the previous macroblock). At the beginning of each slice, predictions for DC coefficients for luminance and chrominance blocks are reset to 1024. The differential DC values are categorized according to their absolute values and the category information is encoded using VLC. The category information indicates the number of additional bits following the VLC to represent the prediction residual. The AC coefficients are encoded similar to that in H.261 using a VLC to represent the zero-run-length and the value of the nonzero coefficient. When a macroblock is encoded in nonintra modes, both the DC and AC coefficients are encoded similar to that in H.261.

Above the video sequence layer, there is a system layer in which the video sequence is packetized. The video and audio bit streams are then multiplexed into an integrated data stream. These are defined in the systems part.

9.1.7 Summary

MPEG-1 is mainly for storage media applications. Due to the use of B-pictures, it may result in long end-to-end delay. The MPEG-1 encoder is much more expensive than the decoder due to the large search range, the half-pixel accuracy in motion estimation, and the use of the bidirectional motion estimation. The MPEG-1 syntax can support a variety of frame rates and formats for various storage media applications. Similar to other video coding standards, MPEG-1 does not specify every coding option (motion estimation, rate control, coding modes, quantization, preprocessing, postprocessing, etc.). This allows continuing technology improvement and product differentiation.

9.2 MPEG-2 VIDEO CODING STANDARD

9.2.1 Introduction

9.2.1.1 Background and Structure of MPEG-2 Standards Activities

The MPEG-2 standard represents the continuing efforts of the MPEG committee to develop generic video and audio coding standards after their development of MPEG-1. The idea of this second phase of MPEG work came from the fact that MPEG-1 is optimized for applications at about 1.5 Mbps with input source in SIF, which is a relatively low-resolution progressive format. Many higher quality higher bit-rate applications require a higher resolution digital video source such as ITU-R BT 601, which is an interlaced format. New techniques can be developed to code the interlaced video better.

The MPEG-2 committee started working in late 1990 after the completion of the technical work of MPEG-1. The competitive tests of video algorithms were held in November 1991, followed by the collaborative phase. The CD for the video part was achieved in November 1993. The MPEG-2 standard (ISO/IEC 13818) [8] consists of nine parts. The first five parts are organized in the same fashion as MPEG-1: systems, video, audio, conformance testing, and simulation software technical report. The first three parts of MPEG-2 reached IS status in November 1994. Parts 4 and 5 were approved in March 1996. Part 6 of the MPEG-2 standard specifies a full set of digital storage media control commands (DSM-CC). Part 7 is the specification of Advanced Audio Coding (AAC), which is the successor of former MPEG audio standard MP3 (MPEG-1 Layer 3 audio) with more efficient coding but with much higher complexity and nonbackward compatibility. AAC is also defined in MPEG-4 Part 3. Part 8 was originally planned to be the coding of 10-bit video for studio application but was discontinued due to lack of interest for further standard adoption from industry. Part 9 is the specification of real-time interface (RTI) to transport stream decoders, which may be utilized for adaptation to all appropriate networks carrying MPEG-2 transport streams. Part 10 is the specification of conformance testing part of DSM-CC. Part 6 and Part 9 have been approved as ISs in July 1996. Part 7 and Part 10 have been later approved as ISs in April 1997 and July 1999, respectively. Like the MPEG-1 video standard, MPEG-2 video coding standard specifies only the bit-stream syntax and the semantics of the decoding process. Many encoding options were left unspecified to encourage continuing technology improvement and product differentiation.

MPEG-3, which was originally intended for high-definition television (HDTV) at higher bit rates, was merged with MPEG-2. Hence, there is no MPEG-3. MPEG-2 video coding standard (ISO/IEC 13818-2) was also adopted by ITU-T as ITU-T Recommendation H.262 [9].

9.2.1.2 ***Target Applications and Requirements***

MPEG-2 is primarily targeted at coding high-quality video at 4–15 Mbps for VOD, standard definition (SD) and high-definition (HD) digital television broadcasting, and digital storage media such as digital versatile disc (DVD).

The requirements from MPEG-2 applications mandate several important features of the compression algorithm. Regarding picture quality, MPEG-2 needs to be able to provide good NTSC quality video at a bit rate of about 4–6 Mbps and transparent NTSC quality video at a bit rate of about 8–10 Mbps. It also needs to provide the capability of random access and quick channel switching by means of I-pictures in GOPs. Low-delay mode is specified for delay-sensitive visual communications applications. MPEG-2 has scalable coding modes to support multiple grades of video quality, spatial resolutions, and frame rates for various applications. Error resilience options include intramotion vector, data partitioning, and scalable coding. Compatibility with the existing MPEG-1 video standard is another prominent feature provided by MPEG-2. For example, MPEG-2 decoders should be able to decode MPEG-1 bit streams. If scalable coding is used, the base layer of MPEG-2 signals can be decoded by a MPEG-1 decoder. Finally, it should allow reasonable complexity encoders and low-cost decoders be built with mature technology. Since MPEG-2 video is based heavily on MPEG-1, in the following sections, we will focus only on those features which are different from MPEG-1 video.

9.2.2 MPEG-2 Profiles and Levels

MPEG-2 standard is designed to cover a wide range of applications. However, features needed for some applications may not be needed for other applications. If we put all the features into one single standard, it may result in an overly expensive system for many applications. It is desirable for an application to implement only the necessary features to lower the cost of the system. To meet this need, MPEG-2 classified the groups of features for important applications into profiles. A profile is defined as a specific subset of the MPEG-2 bit-stream syntax and functionality to support a class of applications (e.g., low-delay video conferencing applications or storage media applications). Within each profile, levels are defined to support applications that have different quality requirements (e.g., different resolutions). Levels are specified as a set of restrictions on some of the parameters (or their combination) such as sampling rates, frame resolutions, and bit rates in a profile. Applications are implemented in the allowed range of values of a particular profile at a particular level.

Table 9.2 shows the combination of profiles and levels that are defined in MPEG-2. MPEG-2 defines seven distinct profiles: simple, main, signal-to-noise ratio (SNR) scalable, spatially scalable, high, 4:2:2, and multiview. The first five profiles were developed with the final approval of MPEG-2 video in November 1994. The last two profiles, that is, 4:2:2 and multiview, were later developed and finalized in January 1996 and September 1996, respectively. Simple profile is defined for low-delay video conferencing applications using only I- and P-pictures. Main profile is the most important and widely used profile for general high-quality digital video applications such as VOD, DVD, digital TV, and HDTV. SNR scalable profile supports multiple grades of video quality. Spatially scalable profile supports multiple grades of resolutions. High profile supports multiple grades of quality, resolution, and chroma formats. Four levels are defined within the profiles: low (for SIF resolution pictures), main (for ITU-R BT 601 resolution pictures), high-1440 (for European HDTV resolution pictures), and high (for North America HDTV resolution pictures). The 11 combinations of profiles and levels in Table 9.2 define the MPEG-2 conformance points, which cover most practical MPEG-2 target applications. The numbers in each conformance point indicate the maximum bound of the parameters. The number in the first line indicates the luminance-rate in samples/s. The number in the second line indicates bit rate in bits/second. Each conformance point is a subset of the conformance point at the right or above. For example, a main-profile main-level decoder should also decode simple-profile main-level and main-profile low-level bit streams. Among the defined profiles and levels, main-profile at main-level (MP@ML) is

TABLE 9.2 Profiles and levels.

Level	Profile				
	Simple 4:2:0	Main 4:2:0	SNR Scalable 4:2:0	Spatially Scalable 4:2:0	High 4:2:0 or 4:2:2
High 1920 × 1152 (60 fps)		62.7 Ms/s, 80 Mbps			100 Mbps for 3 layers
High 1440 1440 × 1152 (60 fps)		47 Ms/s, 60 Mbps		47 Ms/s, 60 Mbps for 3 layers	80 Mbps for 3 layers
Main 720 × 576 (30 fps)	10.4 Ms/s, 15 Mbps	10.4 Ms/s, 15 Mbps	10.4 Ms/s, 15 Mbps for 2 layers		20 Mbps for 3 layers
Low 352 × 288 (30 fps)		3.04 Ms/s, 4 Mbps	3.04 Ms/s, 4 Mbps for 2 layers		

used for digital television broadcast in ITU-R BT 601 resolution and DVD-video. The main-profile at high-level (MP@HL) is used for HDTV. The 4:2:2 profile at main level (422P@ML) is defined for professional video production environments, which supports a higher bit rate of up to 50 Mbps with 4:2:2 color subsampling, and higher precision in DCT coding. Although the high profile supports 4:2:2, also a high-profile codec needs to support SNR scalable profile and spatially scalable profile. This makes the high-profile codec expensive. The 4:2:2 profile does not need to support the scalabilities and thus will be much cheaper to implement. Multiview profile is defined to support the efficient encoding for the applications involving two video sequences from two cameras shooting the same scene with a small angle between them.

9.2.3 MPEG-2 Video Input Resolutions and Formats

Although the main concern of the MPEG-2 committee is to support the ITU-R BT 601 resolution which is the digital TV resolution, MPEG-2 allows a maximum picture size of 16k × 16k pixels. It also supports the frame rates of 23.976, 24, 25, 29.97, 30, 50, 59.94, and 60 Hz as in MPEG-1. MPEG-2 is suitable for coding the progressive video format as well as the interlaced video format. As for the color subsampling formats, MPEG-2 supports 4:2:0, 4:2:2, and 4:4:4. MPEG-2 uses the 4:2:0 format as in MPEG-1 except that there is a difference in the positions of the chrominance samples as shown in Figs. 9.9(a) and (b).

In MPEG-1, a slice can cross macroblock row boundaries. Therefore, a single slice in MPEG-1 can be defined to cover the entire picture. On the other hand, slices in

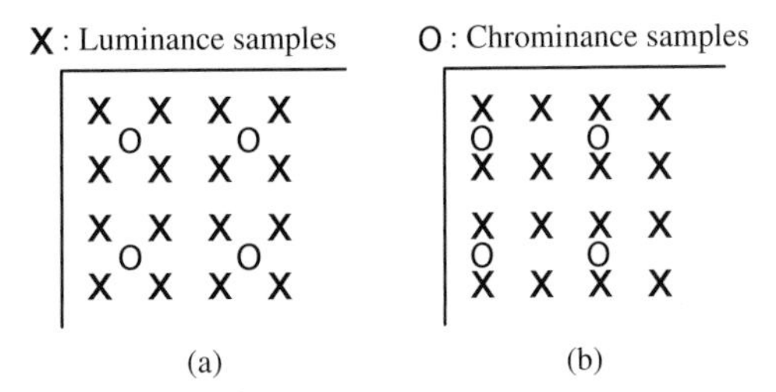

FIGURE 9.9

The position of luminance and chrominance samples for 4:2:0 format in (a) MPEG-1. (b) MPEG-2.

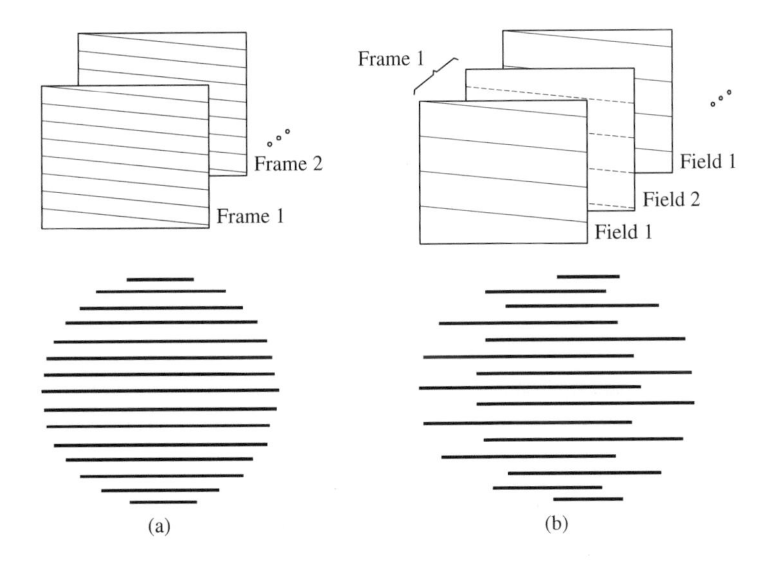

FIGURE 9.10

(a) Progressive scan. (b) Interlaced scan.

MPEG-2 begin and end in the same horizontal row of macroblocks. There are two types of slice structure in MPEG-2: the general and the restricted slice structures. In the general slice structure, MPEG-2 slices need not cover the entire picture. Thus, only the regions enclosed in the slices are encoded. In the restricted slice structure, every macroblock in the picture shall be enclosed in a slice.

9.2.4 MPEG-2 Video Coding Standard Compared to MPEG-1

9.2.4.1 Interlaced versus Progressive Video

Figure 9.10 shows the progressive and interlaced video scan. In the interlaced video, each displayed frame consists of two interlaced fields. For example, Frame 1 consists of Field 1 and Field 2, with the scanning lines in Field 1 located between the lines of Field 2. On the contrary, the progressive video has all the lines of a picture displayed in one frame. There are no fields or half pictures as with the interlaced scan. Thus, progressive video in general requires a higher picture rate than the frame rate of an interlaced video to avoid a flickery display. The main disadvantage of the interlaced format is that when there are object movements, the moving object may appear distorted when we merge two fields into a frame. For example, Fig. 9.10 shows a moving ball. In the interlaced format, since the moving ball will be at different locations in the two fields, if we put the two fields into a frame, the ball will look distorted. Using MPEG-1 to encode the distorted objects in the frames of the interlaced video will not produce the optimal results. On the other hand, if we use MPEG-1 to encode each field separately, the result is also not optimal, since there is less correlation among the pixels due to the larger separation between the scan lines in each field.

9.2.4.2 *Interlaced Video Coding*

Figure 9.11 shows the interlaced video format. As explained earlier, an interlaced frame is composed of two fields. From the figure, the top field (Field 1) occurs earlier in time than the bottom field (Field 2). Both fields together form a frame. In MPEG-2, pictures are coded as I-, P-, and B-pictures like in MPEG-1. To optimally encode the interlaced video, MPEG-2 can encode a picture either as a field picture or a frame picture. In the field-picture mode, the two fields in the frame are encoded separately. If the first field in a picture is an I-picture, the second field in the picture can be either I- or P-pictures as the second field can use the first field as a reference picture. However, if the first field in a picture is a P- or B-field picture, the second field has to be the same type of picture. In a frame picture, two fields are interleaved into a picture and coded together as one picture. In MPEG-2, a video sequence is a collection of frame pictures and field pictures.

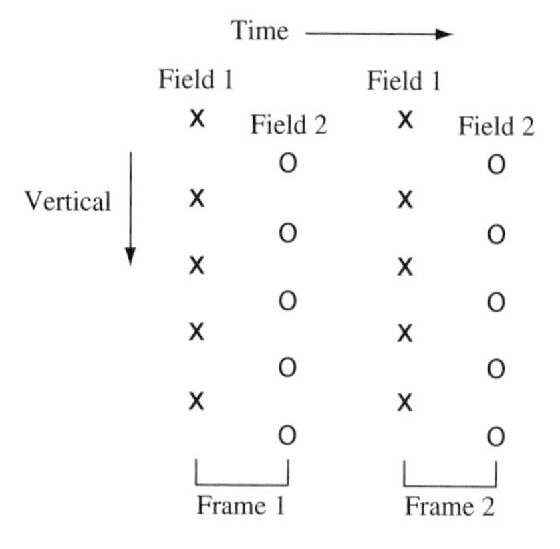

FIGURE 9.11

Interlaced video format.

9.2.4.2.1 Frame-Based and Field-Based Motion-Compensated Prediction

In MPEG-2, an interlaced picture can be encoded as a frame picture or as field pictures. MPEG-2 defines two different motion-compensated prediction types: frame-based and field-based motion-compensated prediction. Frame-based prediction forms a prediction based on the reference frames. Field-based prediction is made based on reference fields. For the simple profile where the bidirectional prediction cannot be used, MPEG-2 introduced a dual-prime motion-compensated prediction to explore the temporal redundancies between fields. Figure 9.12 shows the three types of motion-compensated predictions. Note that all motion vectors in MPEG-2 are specified with a half-pixel resolution.

In the frame-based prediction for frame pictures, as shown in Fig. 9.12(a), the whole interlaced frame is considered as a single picture. It uses the same motion-compensated predictive coding method used in MPEG-1. Each 16 × 16 macroblock can have only one motion vector for each forward or backward prediction. Two motion vectors are allowed in the case of the bidirectional prediction.

The field-based prediction in frame pictures considers each frame picture as two separate field pictures. Separate predictions are formed for each 16 × 8 block of the macroblock as shown in Fig. 9.13. Thus, the field-based prediction in a frame picture needs two sets of motion vectors. A total of four motion vectors are allowed in the case of bidirectional prediction. Each field prediction may select either the Field 1 or the Field 2 of the reference frame.

In field-based prediction for field pictures, the prediction is formed from the two most recently decoded fields. The predictions are made from reference fields, independently for each field, with each field considered as an independent picture. The block size of prediction is 16 × 16; however, it should be noted that the 16 × 16 block in the field picture corresponds to a 16 × 32 pixel-area in the frame picture. Field-based prediction in field picture needs only one motion vector for each forward prediction or

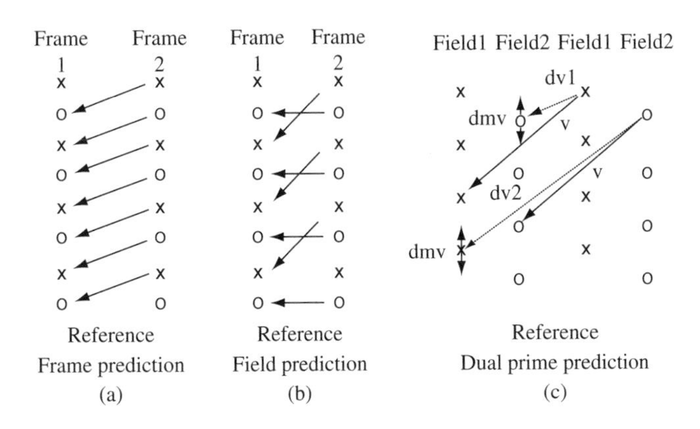

FIGURE 9.12

Three types of motion-compensated prediction.

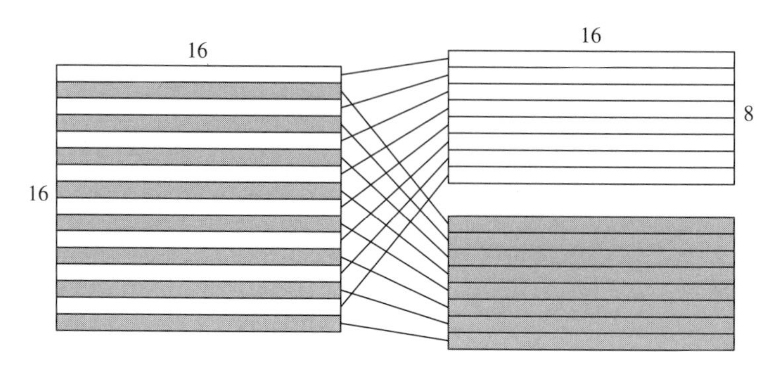

FIGURE 9.13

Blocks for frame-/field-based prediction.

backward prediction. Two motion vectors are allowed in the case of the bidirectional prediction.

In 16×8 prediction for field pictures, two motion vectors are used for each macroblock. The first motion vector is applied to the 16×8 block in Field 1 and the second motion vector is applied to the 16×8 block in Field 2. A total of four motion vectors are allowed in the case of bidirectional prediction.

Dual-prime motion-compensated prediction can be used only in P-pictures. As shown in Fig. 9.12(c), once the motion vector "v" for a macroblock in a field of given parity (Field 1 or Field 2) is known relative to a reference field of the same parity, it is extrapolated or interpolated to obtain a prediction of the motion vector for the opposite parity reference field. In addition, a small correction is also made to the vertical component of the motion vectors to reflect the vertical shift between lines of the Field 1 and Field 2. These derived motion vectors are denoted as dv1 and dv2 (represented by dash lines) in Fig. 9.12(c). Next, a small refinement motion vector, called "dmv," is added. The choice of dmv values $(-1, 0, +1)$ is determined by the encoder. The motion vector "v" and its corresponding "dmv" value are included in the bit stream so that the decoder can also derive dv1 and dv2. In calculating the pixel values of the prediction, the motion-compensated predictions from the two reference fields are averaged, which tends to reduce the noise in the data.

Dual-prime prediction is mainly for low-delay coding applications such as videophone and video conferencing. For low-delay coding using simple profile, B-pictures should not be used. Without using bidirectional prediction, dual-prime prediction is developed for P-pictures to provide a better prediction than the forward prediction.

9.2.4.2.2 Frame/Field DCT

MPEG-2 has two DCT modes: frame-based and field-based DCT as shown in Fig. 9.14. In the frame-based DCT mode, a 16×16-pixel macroblock is divided into four 8×8 DCT blocks. This mode is suitable for the blocks in the background that have little motion because these blocks have high correlation between pixel values from adjacent scan lines. In the field-based DCT mode, a macroblock is divided into four DCT blocks where the

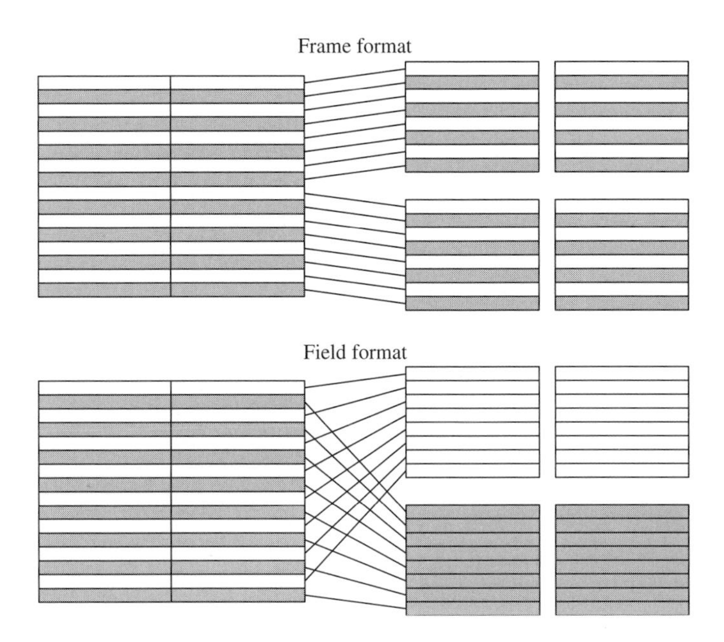

FIGURE 9.14

Frame/field format block for DCT.

pixels from the same field are grouped together into one block. This mode is suitable for the blocks that have motion because as explained, motion causes distortion and may introduce high-frequency noises into the interlaced frame.

9.2.4.2.3 Alternate Scan

MPEG-2 defines two different zigzag scanning orders for scanning the DCT coefficients: zigzag and alternate scans as shown in Fig. 9.15. The zigzag scan used in MPEG-1 is suitable for progressive images where the frequency components have equal importance in each horizontal and vertical direction. In MPEG-2, an alternate scan is introduced based on the fact that interlaced images tend to have relatively higher frequency components in the vertical direction. Thus, the scanning order weighs more on the higher vertical frequencies than the same horizontal frequencies. In MPEG-2, the selection between these two zigzag scan orders can be made on a picture basis.

9.2.4.3 Quantization

In MPEG-2, for intra DC coefficients, 8–11 bits of precision is allowed after quantization, whereas in MPEG-1, only 8-bit precision is allowed. In MPEG-2, intra AC coefficients and all nonintra DC and AC coefficients can be quantized to $[-2048, 2047]$, whereas in MPEG-1, the range is $[-256, 255]$. The finer quantization can reduce the quantization error to improve the reconstructed video quality.

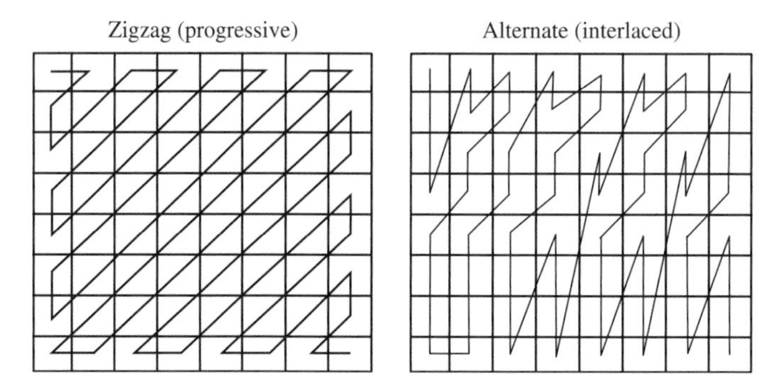

FIGURE 9.15

Progressive/interlaced scan.

9.2.5 Scalable Coding

Scalable coding is also called layered coding. In scalable coding, the video is coded in a base layer and several enhancement layers. If only the base layer is decoded, basic video quality can be obtained. If the enhancement layers are also decoded, enhanced video quality (e.g., higher SNR, higher resolution, higher frame rate) can be achieved. Scalable coding is useful for transmission over noisy channel since the more important layers (e.g., the base layer) can be better protected and sent over a channel with better error performance. Scalable coding is also used in video transport over variable-bit rate channels. When the channel bandwidth is reduced, the less important enhancement layers may not be transmitted. It is also useful for progressive transmission, which means the users can get rough representations of the video fast with the base layer and then the video quality will be refined as more enhancement data arrive. Progressive transmission is useful for database browsing and image transmission over the Internet.

MPEG-2 supports three types of scalability modes: SNR, spatial, and temporal scalability. Different scalable modes can be combined into hybrid coding schemes such as hybrid spatial-temporal and hybrid spatial-SNR scalability. In a basic MPEG-2 scalability mode, there can be two layers of video: lower and enhancement layers. The hybrid scalability allows up to three layers.

9.2.5.1 *SNR Scalability*

MPEG-2 SNR scalability provides two different video qualities from a single video source while maintaining the same spatial and temporal resolutions. A block diagram of the two-layer SNR scalable encoder and decoder is shown in Fig. 9.16(a) and (b), respectively. In the base layer, the DCT coefficients are coarsely quantized and the coded bit stream is transmitted with moderate quality at a lower bit rate. In the enhancement layer, the difference between the nonquantized DCT coefficients and the coarsely quantized DCT coefficients from the lower layer is encoded with finer quantization step sizes. By doing this, the moderate video quality can be achieved by decoding only the lower layer bit streams while the higher video quality can be achieved by decoding both layers.

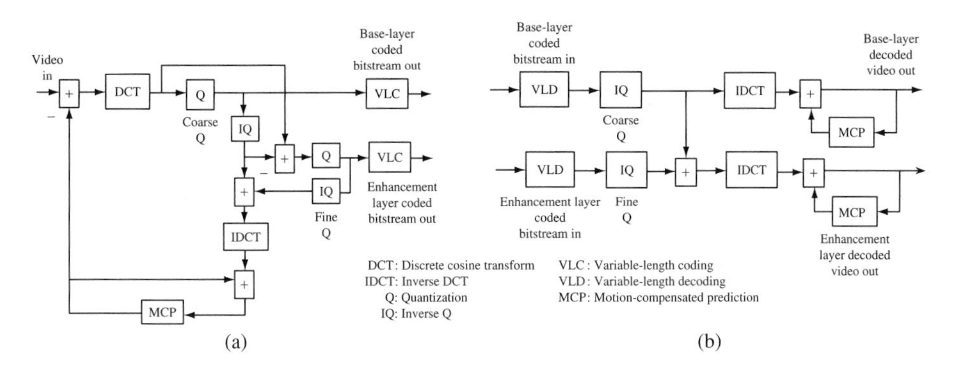

FIGURE 9.16

(a) SNR scalable encoder. (b) SNR scalable decoder.

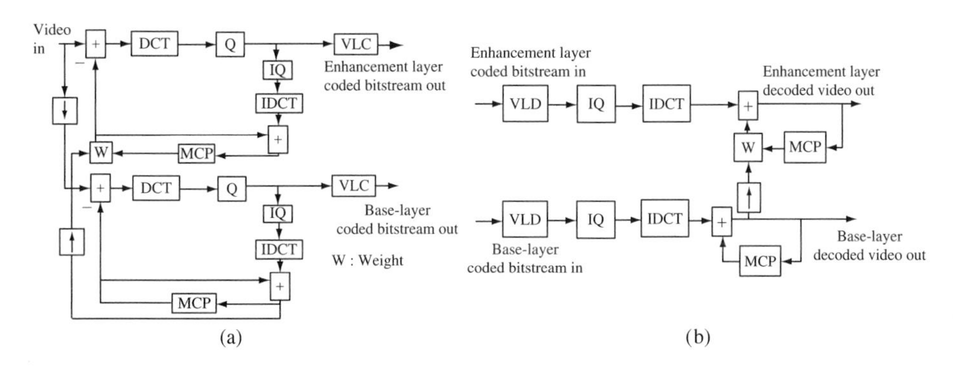

FIGURE 9.17

(a) Spatial scalable encoder. (b) Spatial scalable decoder.

9.2.5.2 Spatial Scalability

With spatial scalability, the applications can support users with different resolution terminals. For example, the compatibility between standard definition TV (SDTV) and HDTV can be achieved with the SDTV being coded as the base layer. With the enhancement layer, the overall bit stream can provide the HDTV resolution. The input to the base layer usually is created by downsampling the original video to create a low-resolution video for providing the basic spatial resolution. The choice of video formats such as frame sizes, frame rates, or chrominance formats is flexible in each layer.

A block diagram of the two-layer spatial scalable encoder and decoder is shown in Figs. 9.17(a) and (b), respectively. In the base layer, the input video signal is downsampled by spatial decimation. To generate a prediction for the enhancement layer video signal input, the decoded lower layer video signal is upsampled by spatial interpolation and is weighted and combined with the motion-compensated prediction from the enhancement layer. The selection of weights is done on a macroblock basis and the selection information is sent as a part of the enhancement-layer bit stream.

The base- and enhancement-layer-coded bit streams are then transmitted over the channel. At the decoder, the lower layer bit streams are decoded to obtain the lower resolution video. The lower resolution video is interpolated and then weighted and added to the motion-compensated prediction from the enhancement layer. In the MPEG-2 video standard, the spatial interpolator is defined as a linear interpolation or a simple averaging for missing samples.

9.2.5.3 Temporal Scalability

The temporal scalability is designed for video services that require different temporal resolutions or frame rates. The target applications include video over wireless channel where the video frame rate may need to be dropped when the channel condition is poor. It is also intended for stereoscopic video and coding of future HDTV formats in which the baseline is to make the migration from the lower temporal resolution systems to the higher temporal resolution systems possible. In temporal scalable coding, the base layer is coded at a lower frame rate. The decoded base layer pictures provide motion-compensated predictions for encoding the enhancement layer.

9.2.5.4 Hybrid Scalability

Two different scalable modes from the three scalability types, SNR, spatial, and temporal, can be combined into hybrid scalable coding schemes. Thus, it results in three combinations: hybrid of SNR and spatial, hybrid of spatial and temporal, and hybrid of SNR and temporal. Hybrid scalability supports up to three layers: the base layer, Enhancement Layer 1, and Enhancement Layer 2. The first combination, hybrid of SNR and spatial scalabilities, is targeted at applications such as HDTV/SDTV or SDTV/videophone at two different quality levels. The second combination, hybrid spatial and temporal scalability, can be used for applications such as high temporal resolution progressive HDTV with basic interlaced HDTV and SDTV. The last combination, hybrid SNR and temporal scalable mode, can be used for applications such as enhanced progressive HDTV with basic progressive HDTV at two different quality levels.

9.2.6 Data Partitioning

Data partitioning is designed to provide more robust transmission in an error-prone environment. Data partitioning splits the block of 64 quantized transform coefficients into partitions. The lower partitions contain more critical information such as low-frequency DCT coefficients. To provide more robust transmission, the lower partitions should be better protected or transmitted with a high priority channel with low probability of error while the upper partitions can be transmitted with a lower priority. This scheme has not been formally standardized in MPEG-2 but was specified in the information annex of the MPEG-2 DIS document [7]. One thing to note is that the partitioned data is not backward compatible with other MPEG-2 bit streams. Therefore, it requires a decoder that supports the decoding of data partitioning. Using the scalable coding and data partitioning may result in mismatch of reconstructed pictures in the encoder and the decoder and thus cause drift in video quality. In MPEG-2, since there are I-pictures that can terminate error propagation, depending on the application requirements, it may not be a severe problem.

9.2.7 Other Tools for Error Resilience

The effect of bit errors in MPEG-2-coded sequences varies depending on the location of the errors in the bit stream. Errors occurring in the sequence header, picture header, and slice header can make it impossible for the decoder to decode the sequence, the picture, or the slice. Errors in the slice data that contain important information such as macroblock header, DCT coefficients, and motion vectors can cause the decoder to lose synchronization or cause spatial and temporal error propagation. There are several techniques to reduce the effects of errors besides the scalable coding. These include concealment motion vectors, the slice structure, and temporal localization by the use of intrapictures/slices/macroblocks.

The basic idea of concealment motion vector is to transmit motion vectors with the intramacroblocks. Since the intramacroblocks are used for future predictions, they may cause severe video quality degradations if they are lost or corrupted by transmission errors. With a concealment motion vector, a decoder can use the best-matching block indicated by the concealment motion vector to replace the corrupted intramacroblock. This improves the concealment performance of the decoder.

In MPEG, each slice starts with a slice header that is a unique bit pattern that can be found without decoding the VLCs. These slice headers represent possible resynchronization markers after a transmission error. A small slice size, that is, a small number of macroblocks in a slice, can be chosen to increase the frequency of synchronization points, thus reducing the effects of the spatial propagation of each error in a picture. However, this can lead to a reduction in coding efficiency as the slice-header overhead information is increased.

The temporal localization is used to minimize the extent of error propagation from picture to picture in a video sequence, for example, by using intracoding modes. For the temporal error propagation in an MPEG video sequence, the error from an I- or a P-picture will stop propagating when the next error-free I-picture occurs. Therefore, increasing the number of I-pictures/slices/macroblocks in the coded sequence can reduce the distortion caused by the temporal error propagation. However, more I-pictures/slices/macroblocks will result in reduction of coding efficiency and it is more likely that errors will occur in the I-pictures, which will cause error propagation.

9.2.8 Test Model

Similar to other video coding standards such as H.261 and MPEG-1, MPEG-2 only specifies the syntax and the decoder. Many detailed coding options are not specified. To have a reference MPEG-2 video quality, test models were developed in MPEG-2. The final test model of MPEG-2 is called Test Model 5 (TM5) [10]. TM5 was defined only for main profile experiments. The motion-compensated prediction techniques involve frame, field, dual-prime prediction and have forward and backward motion vectors as in MPEG-1. The dual prime was kept in main profile but restricted to P-pictures with no intervening B-pictures. Two-step search, which consists of an integer-pixel full-search followed by a half-pixel search, is used for motion estimation. The mode decision (intracoding/intercoding) is also specified. Main profile was restricted to only two quantization matrices: the default table specified in MPEG-1 and the nonlinear quantizer tables. The traditional zigzag scan is used for intercoding while the alternate scan is used for intracoding. The rate-control algorithm in TMN5 consists of three layers operating at the GOP, the picture, and the macroblock levels. A bit allocation per picture is determined at the GOP layer and updated based on the buffer fullness and the complexity of the pictures.

9.2.9 MPEG-2 Video and System Bit-Stream Structures

A high-level structure of the MPEG-2 video bit stream is shown in Fig. 9.18. Every MPEG-2 sequence starts with a sequence header and ends with an end-of-sequence. MPEG-2 syntax is a superset of the MPEG-1 syntax. The MPEG-2 bit stream is based on the basic structure of MPEG-1 (refer to Fig. 9.8). There are two bit-stream syntax allowed: ISO/IEC 11172-2 video sequence syntax or ISO/IEC 13818-2 (MPEG-2) video sequence syntax.

If the sequence header is not followed by the sequence extension, the MPEG-1 bit-stream syntax is used. Otherwise, the MPEG-2 syntax is used, which accommodates more features but at the expense of higher complexity. The sequence extension includes a profile/level indication, a progressive/interlaced indicator, a display extension including choices of chroma formats and horizontal/vertical display sizes, and choices of scalable modes. The GOP header is located next in the bit-stream syntax with at least one picture following each GOP header. The picture header is always followed by the picture coding extension, the optional extension and user data fields, and picture data. The picture coding extension includes several important parameters such as the indication of intra DC precision, picture structures (choices of the first/second fields or frame pictures), intra

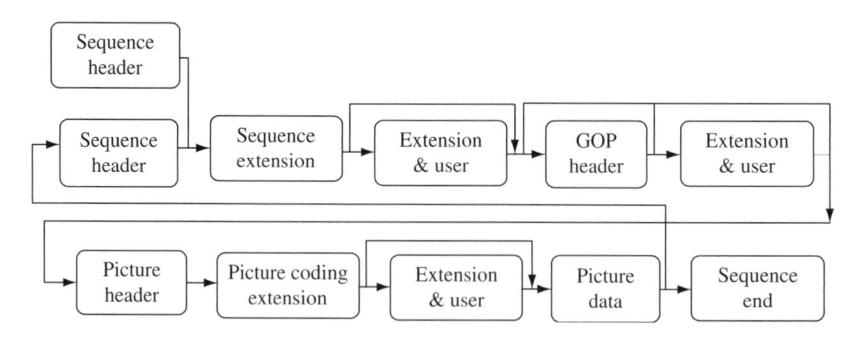

FIGURE 9.18

MPEG-2 data structure and syntax.

VLC format, alternate scan, choices of updated quantization matrix, picture display size, display size of the base layer in the case of the spatial scalability extension, and indicator of the forward/backward reference picture in the base layer in the case of the temporal scalability extension. The picture data consist of slices, macroblocks, and data for the coded DCT blocks. MPEG-2 defines six layers as MPEG-1. However, the specification of some data elements is different. The details of MPEG-2 syntax specification are documented in [8].

9.2.10 Summary

MPEG-2 is mainly targeted at general high-quality video applications at bit rates greater than 2 Mbps. It is suitable for coding both progressive and interlaced video. MPEG-2 uses frame/field adaptive motion-compensated predictive coding and DCT. Dual-prime motion compensation for P-pictures is allowed for low-delay applications with no B-picture. In addition to the default quantization table, MPEG-2 defines a nonlinear quantization table with increased accuracy for small values. Alternate scan and new VLC tables are defined for DCT coefficient coding. MPEG-2 also supports compatibility and scalability with the MPEG-1 standard. MPEG-2 syntax is a superset of MPEG-1 syntax and can support a variety of rates and formats for various applications. Similar to other video coding standards, MPEG-2 defines only syntax and semantics. It does not specify every encoding options (preprocessing, motion estimation, quantizer, rate-quality control, and other coding options) and decoding options (postprocessing and error concealment) to allow continuing technology improvement and product differentiation. It is important to keep in mind that different implementations may lead to different quality, bit rate, delay, and complexity trade-offs with different cost factors. An MPEG-2 encoder is much more expensive than an MPEG-2 decoder, since it needs to perform many more operations (e.g., motion estimation, coding-mode decisions, and rate control). An MPEG-2 encoder is also much more expensive than an H.261 or an MPEG-1 encoder due to the higher resolution and more complicated motion estimations (e.g., larger search range, frame/field bidirectional motion estimation). MPEG-1 has been successfully used in VCD. MPEG-2 is today's predominant video coding standard used in DVD, digital television including SDTV and HDTV broadcasting over terrestrial, satellites, or cable TV (CATV) networks. They have been used in high-quality video networking and streaming applications for transmitting live and prerecorded streams over broadband Internet, corporate intranets, and virtual private networks. The advent of newer video compression standard, H.264/MPEG-4 Part 10 Advanced Video Coding (AVC) [11], delivers much higher compression ratios than MPEG-2, that is, up to 2–3 times more efficient than MPEG-2. However, more efficiency comes with higher computational requirements. H.264 is developed to support applications such as next generation digital TV over terrestrial, satellite, cable, and Internet protocol television (IPTV) over cable or digital subscriber line (DSL), high-definition optical discs such as Blu-Ray disc and HD-DVD, and mobile TV. Currently, there are large amount of contents encoded in the MPEG-2 format. As the conversion to digital high-definition video formats is taking place, the transition from the MPEG-2 format to the H.264 format is making headway and there is an increasing demand for encoder and decoder chips that support both standards as well as both the SD and HD formats. There is also a need for MPEG-2 to H.264 and H.264 to MPEG-2 transcoder during these MPEG-2/H.264 coexisting periods. References [12–33] provide further information on the related MPEG-1 and MPEG-2 topics.

REFERENCES

[1] ISO/IEC JTC1 CD 10918. Digital compression and coding of continuous-tone still images. International Organization for Standardization (ISO). 1993.

[2] ITU-T Recommendation H.261. Line transmission of non-telephone signals. Video codec for audio visual services at px64 kbits/s. 1993.

[3] S. Okubo. Reference model methodology—A tool for the collaborative creation of video coding standards. *Proc. IEEE*, 83(2):139–150, 1995.

[4] MPEG proposal package description. Document ISO/WG8/MPEG/89-128. 1989.

[5] T. Hidaka, K. Ozawa. Subjective assessment of redundancy-reduced moving images for interactive applications: Test methodology and report. *Signal Process.: Image Commun.*, 2:201–219, 1990.

[6] ISO/IEC JTC1 CD 11172. Coding of moving pictures and associated audio for digital storage media up to 1.5 Mbits/s. International Organization for Standardization (ISO). 1992.

[7] ISO/IEC JTC1/SC2/WG11. *MPEG Video Simulation Model Three (SM3)*. MPEG 90/041. 1990.

[8] ISO/IEC JTC1/SC29/WG11. MPEG-2: Generic coding of moving pictures and associated audio information. ISO/IEC International Standard. 2000.

[9] ISO/IEC 13818-2-ITU-T Rec. H.262. Generic coding of moving pictures and associated audio information: Video. 1995.

[10] ISO/IEC JTC1/SC29/WG11. Test Model 5. MPEG 93/457, Document AVC-491. 1993.

[11] ISO/IEC 14496-10/ITU-T Recommendation H.264. Advance video coding for generic audiovisual services. 2005.

[12] M. L. Liou. Visual telephony as an ISDN application. *IEEE Commun. Mag.*, 28:30–38, 1990.

[13] A. Tabatabai, M. Mills, and M. L. Liou. A review of CCITT px64 kbps video coding and related standards. *Int. Electronic Imaging Exposition and Conf.*, 58–61, October 1990.

[14] D. J. Le Gall. MPEG: A video compression standard for multimedia applications. *Commun. ACM*, 34:47–58, 1991.

[15] D. J. Le Gall. The MPEG video compression algorithm. *Signal Process.:Image Commun.*, 4:129–140, 1992.

[16] L. Chiariglione. Standardization of moving picture coding for interactive applications. *GLOBECOM'89*, 559–563, November 1989.

[17] A. Puri. Video coding using the MPEG-1 compression standard. *Soc. Inf. Disp. Int. Sym.*, 123–126, Boston, MA, May 1992.

[18] A. Puri. Video coding using the MPEG-2 compression standard. *SPIE/VCIP*, 2094:1701–1713, Cambridge, MA, November 1993.

[19] S. Okubo, K. McCann, and A. Lippman. MPEG-2 requirements, profiles, and performance verification. *Proc. Int. Workshop on HDTV'93*, Ottawa, Canada, October 1993.

[20] A. Puri, R. Aravind, and B. Haskell. Adaptive frame/field motion compensated video coding. *Signal Process.: Image Commun.*, 5:39–58, 1993.

[21] T. Naveen et al. MPEG 4:2:2 profile: High-quality video for studio applications. *Photonics East*, SPIE, vol. CR60, Philadelphia, PA, October 1995.

[22] A. Puri. Compression of stereoscopic video using MPEG-2. *Photonics East*, SPIE, vol. CR60, Philadelphia, PA, October 1995.

[23] R. J. Clarke. *Digital Compression of Still Images and Video*. Academic Press, Burlington, MA, 1995.

[24] V. Bhaskaran and K. Konstantinides. *Image and Video Compression Standards: Algorithms and Architectures*. Kluwer Academic Publishers, Boston, MA, 1995.

[25] J. L. Mitchell, W. B. Pennebaker, and D. J. Le Gall. *The MPEG Digital Video Compression Standard*. Van Nostrand Reinhold, New York, NY, 1996.

[26] K. R. Rao and J. J. Hwang. *Techniques and Standards for Image, Video, and Audio Coding*. Prentice-Hall, Upper Saddle River, NJ, 1996.

[27] K. Konstantinides, C.-T. Chen, T.-C. Chen, H. Cheng, and F.-T. Jeng. Design of an MPEG-2 codec. *IEEE Signal Process. Mag.*, 19(4):32–41, 2002.

[28] Y. Wu, S. Hirakawa, H. Katoh, U. Reimers, and J. Whitaker. Special issue on global digital television: Technology and emerging services. *Proc. IEEE*, 94(1):1–2, 2006.

[29] J. Bennett and A. Bock. In-depth review of advanced coding technologies for low bit rate broadcast applications. *SMPTE Motion Imaging J.*, 113:413–418, 2004.

[30] J. Xin, C.-W. Lin, and M.-T. Sun. Digital video transcoding. *Proc. IEEE*, 93(1):84–97, 2005.

[31] I. Ahmad, X. Wei, Y. Sun, and Y.-Q. Zhang. Video transcoding: An overview of various techniques and research issues. *IEEE Trans. Multimed.*, 7(5):793–804, 2005.

[32] P. Tseng, Y. Chang, Y. Huang, H. Fang, C. Huang, and L. Chen. Advances in hardware architectures for image and video coding—A survey. *Proc. IEEE*, 93(1):184–197, 2005.

[33] J. Probell. Architecture considerations for multi-format programmable video processors. *J. Signal Process. Syst.*, 50(1):33–39, 2008.

CHAPTER

10 MPEG-4 Visual and H.264/AVC: Standards for Modern Digital Video

Berna Erol[1], Faouzi Kossentini[2], Anthony Joch[3], Gary J. Sullivan[4], and Lowell Winger[5]

[1] *Ricoh Innovations, California Research Center, USA*
[2] *Digital Media Networks, Scientific Atlanta, A Cisco Company, Lawrenceville, GA, USA*
[3] *Avvasi Inc., Waterloo, Canada*
[4] *Microsoft Corporation, Redmond, WA, USA*
[5] *Magnum Semiconductor, Waterloo, Canada*

10.1 INTRODUCTION

The last two decades have witnessed an increasing diversity of digital video applications in many areas, including communications, education, medicine, and entertainment. The vast amount of digital data that is associated with video applications makes the representation, exchange, storage, access, and manipulation of this data a challenging task. To enable interoperability between transmitting and receiving (or recording and playback) products made by different manufacturers, there is a need to standardize the representation of, and access to, these data. There has been significant work in the fields of efficient representation of video by means of compression, storage, and transmission [1–4]. More information on the MPEG-1 and MPEG-2 standards can be found in Chapter 9.

In this chapter, we describe the digital video coding parts of a recent suite of International Standards: MPEG-4. The MPEG-4 suite of standards addresses system issues such as management and composition of audiovisual data in MPEG-4 Part 1 [5], the representation and decoding of visual data in Part 2 [6] and Part 10 [7], the representation and decoding of audio data in Part 3 [8], and various other associated technology in a number of other "parts." Each part can be considered as a distinct standard on its own, or the various parts can be combined for use together for example, for encoding both video and audio data and storing them together in a single file for synchronized playback. In this chapter, we will focus on the visual parts of the standard, that is, MPEG-4 Part 2 (MPEG-4 Visual) and Part 10 (H.264/AVC).

MPEG-4 version 1 became an International Standard in 1999. The second version of the standard, which also included the entire technical content of the first version of the standard, became an International Standard in 2000. After these two versions, more parts were developed, and new tools and profiles were introduced. But among the most important of these was the development of Part 10, better known as H.264 or Advanced Video Coding (AVC), which represented a substantial advance in video compression capability and additional features. H.264/AVC was developed by a Joint Video Team (JVT) that was formed as a partnership project between the ISO, IEC, and ITU-T after earlier design efforts that began in the ITU-T community in a project known as "H.26L" at the time. The first version of H.264/AVC became an International Standard in 2003. Subsequently, a set of extensions to the original standard known as the Fidelity Range Extensions (FRExt) was completed in 2004. A set of further "professional" capabilities and a set of Scalable Video Coding (SVC) extensions were completed in 2007. Further work is under way to extend the design to support multiview video for 3D applications.

One primary objective of MPEG-4 Part 2 was to enable object-based coding that would allow access and manipulation of individual objects in a video scene [9]. However, thus far, the object-based representation has not been demonstrated to be practical for several reasons. One reason is simply that the segmentation of video content into arbitrarily shaped and semantically meaningful video objects (VOs) remains a very difficult task. Another is that adding object handling capability to decoders increases the computational resource requirements of the decoder, although typical existing video applications may not ordinarily substantially benefit from the extra capability. As a result, these features of the MPEG-4 Part 2 design have not been widely deployed in products and remain primarily only of interest for academic and research purposes today. However, some other sets of MPEG-4 Part 2 capabilities (specifically, the Simple Profile and Advanced Simple Profile, and to a lesser extent the Simple Studio Profile) have achieved widespread deployment in mobile applications and Internet-based video for PC-based viewing.

Once MPEG-4 Part 10 (H.264/AVC) was later developed, it was universally recognized as a major advance of video technology and became widely adopted in essentially all digital video applications, because it offers high compression and error resilience capabilities—making the representation of video much more efficient and robust.

In the next section, we first define the terminology used in this chapter, then provide a technical description of the two visual parts of the MPEG-4 standard, followed by a discussion of MPEG-4 video applications. We conclude this chapter with the authors' assessment of the impact of the MPEG-4 standard on the fast-evolving digital video industry.

10.2 TERMINOLOGY

In this section, we describe some of the terminology used in this chapter. Throughout this chapter, we will use the term *picture* for a frame or a field that is coded or processed as a distinct unit for compression or other processing. A picture consists of a number of arrays of samples. Typically, there are three such arrays, one for each of the axes of the color space used in the picture. For example, the color axes may be Y (also called *luma* and representing monochrome brightness), Cb (the color deviation from monochrome towards pure blue), and Cr (the color deviation from monochrome towards pure red).

Each sample consists of the intensity of light along the dimension of one color space axis in one elementally small area of a sampling grid. When using a Y, Cb, Cr color space, the resolution (the size of the array in width and height) of the Cb and Cr arrays is typically lower by a factor of four (i.e., the array width and height are each smaller by a factor of 2) than for the corresponding Y array for the same picture. This practice is known as 4:2:0 chroma sampling. Other important chroma sampling structures include 4:2:2 sampling, in which the width of the chroma arrays is lower but the height is the same as for the luma array, and 4:4:4 sampling, in which the width and height of all three arrays are the same. Note that the term *sample* is used here for precision, rather than the widely used term *pixel.* The former is simply an integer value associated with a spatial location and a single axis of the color space, whereas the latter is typically defined as a point in the image that has a color associated with it (the color typically being specified in terms of three or more color space axes). Therefore, a discrete cosine transform (DCT), for instance, cannot strictly be computed using pixels, but can be computed using luma samples or chroma samples. The term *pixel* is particularly problematic when the resolution of the picture is not the same for all of the axes of the color space, resulting in a lack of a clearly specified color at the location of each sample. In casual usage, the two terms are often interchangeable.

A frame contains two fields. A field consists of half of the samples in a frame, based on the lines (i.e., the rows of the arrays) in which the samples are found in each array. A top field consists of the even numbered lines of samples in a frame (counting line 0 as the top line of the frame) and a bottom field consists of the odd numbered lines of samples. Typically, fields are only discussed in the context of interlaced video, in which the two fields of each frame are sampled at different instants in time.

Areas of video pictures can typically be coded in one of two ways. One of these is to use the values of the samples in some previously decoded reference picture to form a prediction of the values of the samples in the new picture and then, as needed, to add a representation of the residual difference signal between the predicted values of the samples and the actual values of the samples in the picture being encoded. This technique is known as interpicture prediction or interpicture coding, and is referred to more succinctly as *intercoding.* Encoding a picture area without using such interpicture prediction (either without forming a prediction at all or by using a prediction that is formed only from the sample values within the same picture that is being encoded) is referred to as intrapicture coding, or simply as *intracoding.* This usage of "intra" and "inter" as adjectives rather than as prefixes of other words is a widespread characteristic jargon in the video coding community—the suffixing of these adjectives with "picture coding" is implicit.

10.3 MPEG-4 PART 2

MPEG-4 Part 2, officially known as ISO/IEC 14496-2 [6], standardizes an efficient object-based representation of video. Such representation is achieved by defining visual objects and encoding them into separate bitstream segments [5, 9]. Although MPEG-4 Part 2 defines only the bitstream syntax and the decoding process, the precise definitions of some example encoding algorithms are presented in two verification models: one for natural video [10], and the other for synthetic and natural hybrid video [11].

MPEG-4 Part 2 allows four different types of coding tools: video object coding for the coding of natural and/or synthetic generated, rectangular or arbitrarily shaped VOs, mesh object coding for the coding of visual objects represented with mesh structures, model-based coding for the coding of a synthetic representation and animation of the human face and body, and still texture coding for the wavelet coding of still textures.

In the following sections, we first describe the object-based representation and each of the MPEG-4 Part 2 coding tools. Next, we discuss the scalability and the error resilience tools, followed by a presentation of the MPEG-4 Part 2 profiles.

10.3.1 Object-based Representation

As stated above, object-based representation capability is a major feature of the MPEG-4 Part 2 design. Although this capability is not widely implemented in products, this feature of MPEG-4 Part 2 remains a significant milestone in the technology history of compressed representation of visual information.

The object-based representation in MPEG-4 Part 2 is based on the concept of the audiovisual object (AVO). An AVO consists of a visual object component, an audio object component, or a combination of these components. The characteristics of the audio and visual components of the individual AVOs can vary, such that the audio component can be 1) synthetic or natural, 2) mono, stereo or multichannel (e.g., surround sound), and the visual component can be natural or synthetic. Some examples of AVOs include object-based representations of a person recorded by a video camera, a sound clip recorded with a microphone, and a 3D image with text overlay.

MPEG-4 supports the composition of a set of AVOs into a scene, also referred to as an *audiovisual scene.* To allow interactivity with individual AVOs within a scene, it is essential to transmit the information that describes each AVO's spatial and temporal coordinates. This information is referred to as the *scene description information* and is transmitted as a separate stream and multiplexed with AVO elementary bitstreams so that the scene can be composed at the user's end. This functionality makes it possible to change the composition of AVOs without having to change their content.

An example of an audiovisual scene, which is composed of natural and synthetic audio and visual objects, is presented in Fig. 10.1. AVOs can be organized in a hierarchical fashion. Elementary AVOs, such as the blue head and the associated voice, can be combined together to form a compound AVO, that is, a talking head. It is possible to change the position of the AVOs, delete them or make them visible, or manipulate them in a number of ways depending on their characteristics. For example, a visual object can

FIGURE 10.1

An audiovisual scene.

be zoomed and rotated by the user. Even more, the quality, spatial resolution, and temporal resolution of the individual AVOs can be modified. For example, in a mobile video telephony application, the user can request a higher frame rate and/or spatial resolution for the talking person than those of the background objects.

10.3.2 Video Object Coding

A VO is an arbitrarily shaped video segment that has a semantic meaning. A 2D snapshot of a VO at a particular time instant is called a video object plane (VOP). A VOP is defined by its texture (luma and chroma values) and its shape. MPEG-4 Part 2 allows object-based access to the VOs, as well as temporal instances of the VOs, that is, VOPs. To enable access to an arbitrarily shaped object, a separation of the object from the background and the other objects must be performed. This process, known as segmentation, is not standardized in MPEG-4. However, automatic and semiautomatic tools [12], and techniques such as chroma keying [13], although not always effective, can be employed for VO segmentation.

As illustrated in Fig. 10.2, a basic VOP encoder consists mainly of two blocks: a hybrid of a motion compensated predictor and a DCT-based coder, and a shape coder. Similar to MPEG-1 and MPEG-2, MPEG-4 supports intracoded (I-), temporally predicted (P-), and bidirectionally predicted (B-) VOPs, all of which are illustrated in Fig. 10.3. Except for I-VOPs, motion estimation and compensation are applied. Next, the difference between the motion compensated data and the original data is DCT transformed, quantized, and then variable length coded. Motion information is also encoded using variable-length codings (VLCs). Since the shape of a VOP may not change significantly between consecutive VOPs, predictive coding is employed to reduce temporal redundancies. Thus,

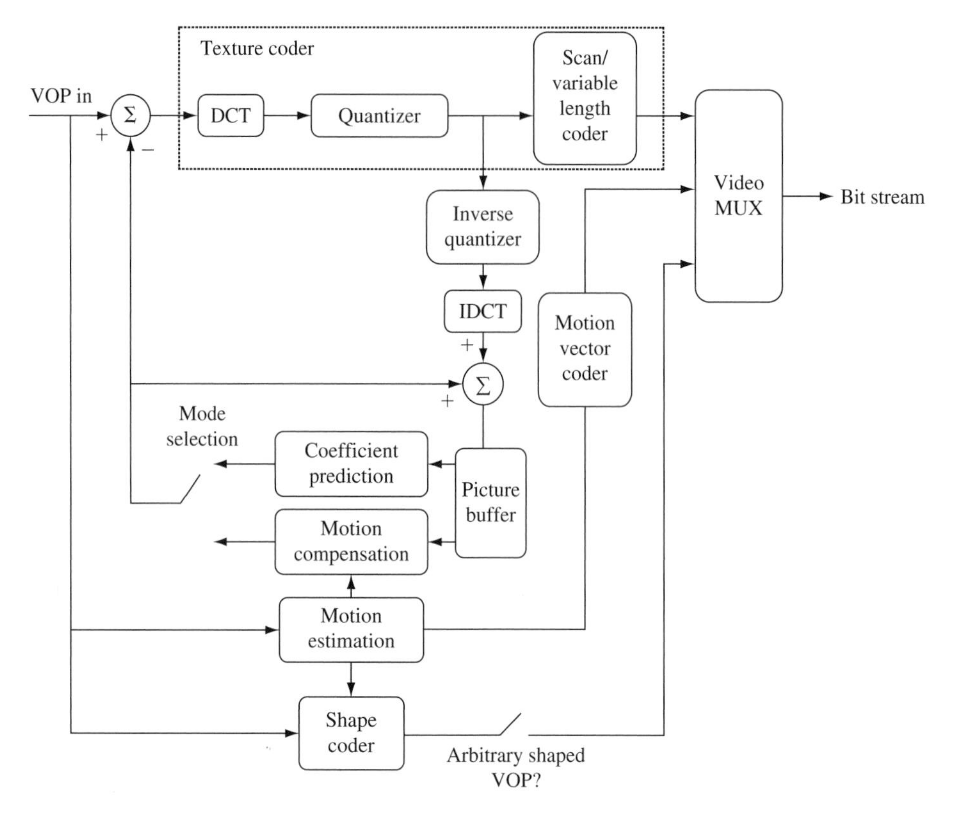

FIGURE 10.2

A basic block diagram of an MPEG-4 Part 2 video encoder.

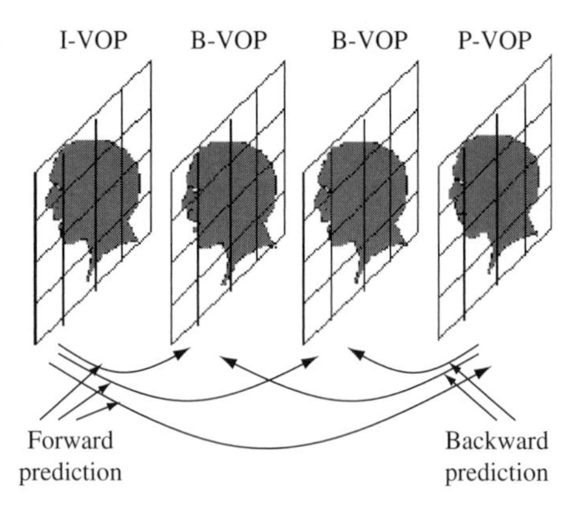

FIGURE 10.3

Prediction types for a video object plane (VOP).

motion estimation and compensation are also applied to the shape of the VOP. Finally, the motion, texture, and shape information is multiplexed with the headers to form the coded VOP bitstream. At the decoder end, the VOP is reconstructed by combining motion, texture, and shape data decoded from the bitstream.

As noted above, most actual usage of MPEG-4 Part 2 in products does not support object coding functionality. When object coding is not used, the VOP is simply a rectangular video frame, and MPEG-4 video coding becomes quite similar to that specified in MPEG-1/MPEG-2 [1, 2] and H.263 [3]. Many aspects are particularly similar to those of H.263, which formed much of the basis of the MPEG-4 Part 2 design. In fact, an MPEG-4 conforming decoder must be able to decode all the bitstreams generated by encoders that use the H.263 "Baseline" format. (In fact, H.263 Baseline profile encoders can also be referred to as MPEG-4 Part 2 encoders for this reason.) The H.263 Baseline format is the format used by H.263 encoders when optional enhanced features are not employed, and it is the most common form of H.263 bitstreams. The most commonly implemented forms of MPEG-4 Part 2 video are those of its Simple Profile (SP) and its Advanced Simple Profile (ASP). SP bitstreams are very similar to H.263 bitstreams, with some enhanced capabilities added for somewhat improved compression capability and loss/error robustness. The ASP adds some extra compression capability improvements.

10.3.2.1 Motion Vector Coding

Motion vectors (MVs) are predicted using a spatial neighborhood of three MVs, and the resulting prediction error is variable length coded. Motion vectors are transmitted only for P-VOPs and B-VOPs. MPEG-4 Part 2 uses a variety of motion compensation techniques, such as the use of unrestricted MVs (motion vectors that are allowed to point outside the coded area of a reference VOP), and the use of four MVs per macroblock. In addition, some new profiles (such as the ASP) that were added in version 2 of MPEG-4 Part 2 support global motion compensation and quarter-sample motion vector precision.

10.3.2.2 Motion Texture Coding

The texture information of motion video data in MPEG-4 Part 2 is encoded in a similar manner as in MPEG-1, MPEG-2, (described in Chapter 9) and H.263. DCT based coding is employed to reduce spatial redundancies. Intra blocks, as well as motion compensation prediction error blocks, are texture coded. Each VOP is divided into macroblocks as illustrated in Fig. 10.4. Each macroblock consists of a 16×16 array of luma samples and two corresponding 8×8 arrays of chroma samples. These arrays are partitioned into 8×8 blocks for DCT processing. DCT coding is applied to the four 8×8 luma and two 8×8 chroma blocks of each macroblock. If a macroblock lies on the boundary of an arbitrarily shaped VOP, then the samples that are outside the VOP are padded before DCT coding. As an alternative, a shape-adaptive DCT coder can be used for coding boundary macroblocks of intra VOPs. This generally results in higher compression performance, at the expense of an increased implementation complexity. Macroblocks that are completely inside the VOP are DCT transformed as in MPEG-1, MPEG-2, and H.263. DCT transformation of the blocks is followed by quantization, zig-zag coefficient

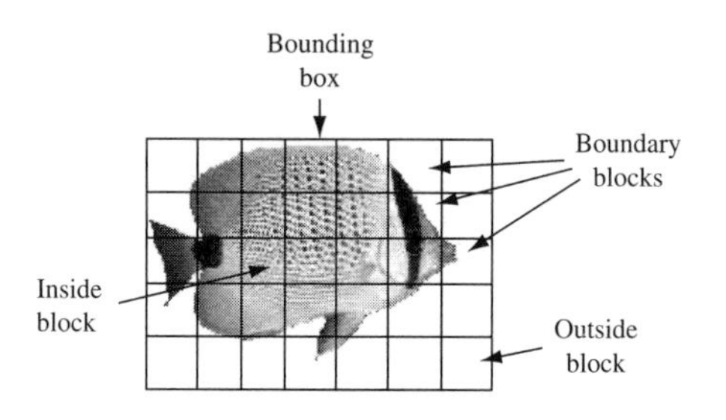

FIGURE 10.4

A VOP enclosed in a rectangular bounding box and divided into macroblocks.

scanning, and variable-length coding (VLC). Adaptive DC/AC prediction methods and alternate scan techniques can be employed for the encoding of the DCT coefficients of intra blocks.

10.3.2.3 Shape Coding

Besides H.263+, which provides some limited shape coding support through its chroma-keying coding technique, MPEG-4 Part 2 was the first video coding standard[1] that supported shape coding. As opposed to the approach taken in H.263+, MPEG-4 Part 2 adopted an alpha-plane based shape coding method because of its ability to clearly distinguish between shape and texture information while retaining high compression performance and reasonable complexity. In this form of shape coding, the shape and transparency of a VOP are defined by a binary (bitmap) or grayscale alpha plane. Each bit of a binary alpha plane indicates whether or not a sample belongs to a VOP. Each sample of a grayscale alpha plane indicates the transparency of a location within a VOP. Grayscale transparency sample values can range from 0 (transparent) to 255 (opaque). If all of the samples in a VOP block are indicated to be opaque or transparent, then no additional transparency information needs to be transmitted for that block.

MPEG-4 Part 2 provides tools for both lossless and lossy coding of binary and grayscale alpha planes. Furthermore, both intra and intershape coding are supported. Binary alpha planes are divided into 16×16 blocks as illustrated in Fig. 10.5. The blocks that are inside the VOP are signaled as opaque blocks, and the blocks that are outside the VOP are signaled as transparent blocks. The samples in boundary blocks (i.e., blocks that contain samples both inside and outside the VOP) are scanned in a raster scan order and coded using context-based arithmetic coding. Grayscale alpha planes, which represent transparency information, are divided into 16×16 blocks and coded the same way as the texture in the luma blocks.

In intrashape coding using binary alpha planes, a context is computed for each sample using 10 neighbouring samples (shown in Fig. 10.6 [a]), and the equation $C = \sum_k c_k 2^k$,

[1]Although not discussed in detail here, MPEG-4 Part 10 (H.264/AVC) now also supports grayscale alpha plane shape coding, using a feature known as "auxiliary pictures", and it supports a wider range of transparency value precisions. This capability was added in the FRExt extension project.

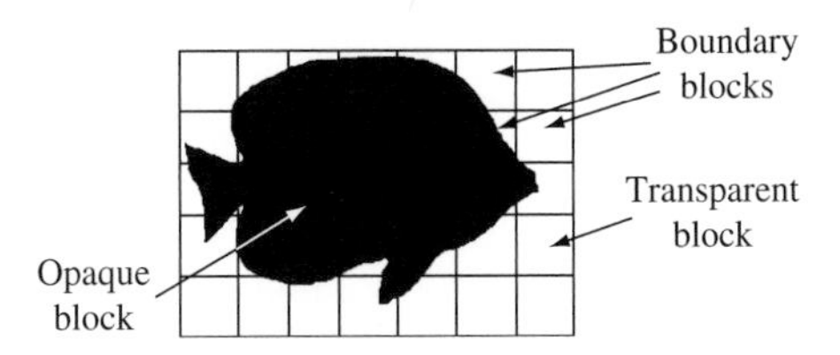

FIGURE 10.5

Binary alpha plane.

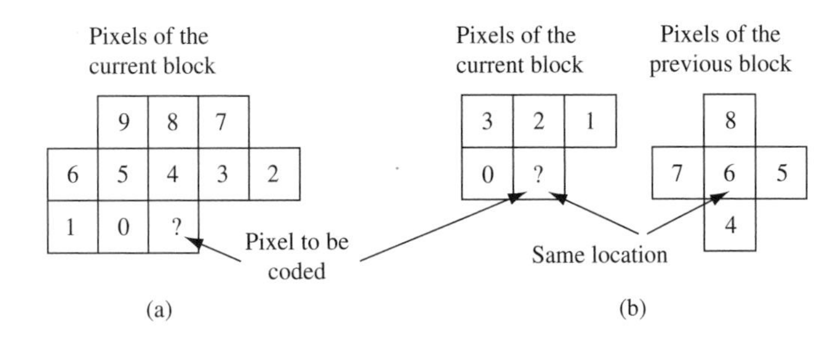

FIGURE 10.6

Template samples that form the context of an arithmetic coder for (a) intra- and (b) intercoded shape blocks.

where k is the sample index, c_k is "0" for transparent samples and "1" for opaque samples. If the context samples fall outside the current block, then samples from neighbouring blocks are used to build the context. The computed context is then used to access the table of probabilities. The selected probability is used to determine the appropriate code space for arithmetic coding. For each boundary block, the arithmetic encoding process is also applied to the transposed version of the block. The representation that results in the least coding bits can be selected by the encoder to be conveyed in the bitstream.

In intershape coding using binary alpha planes, the shape of the current block is first predicted from the shape of the temporally previous VOP by performing motion estimation and compensation using integer sample accuracy. The shape motion vector is then coded predictively. Next, the difference between the current and the predicted shape block is arithmetically coded. The context for an intercoded shape block is computed using a template of nine samples from both the current and temporally previous VOP shape blocks, as shown in Fig. 10.6 (b).

In both intra- and intershape coding, lossy coding of the binary shape is achieved by either not transmitting the difference between the current and the predicted shape block, or subsampling the binary alpha plane by a factor of two or four prior to arithmetic encoding. To reduce the blocky appearance of the decoded shape caused by lossy coding, an upsampling filter is employed during the reconstruction.

10.3.2.4 *Sprite Coding*

In MPEG-4 Part 2, sprite coding is used for representation of VOs that are static throughout a video scene, or are modified such that their changes can be approximated by warping the original object planes [6, 14]. Sprites are generally used for transmitting the background in video sequences. As shown in the example of Fig. 10.7, a sprite may consist of a panoramic image of the background, including the samples that are occluded by other VOs. Such a representation can increase compression capability, since the background image would need to be encoded only once at the beginning of the video segment, and the camera motion, such as panning and zooming, can be represented by only a few global motion parameters.

10.3.3 Mesh Object Coding

A mesh is a tessellation (partitioning) of an image into polygonal patches. Mesh representations have been successfully used in computer graphics for efficient modeling and rendering of 3D objects. To benefit from functionalities provided by such representations, MPEG-4 Part 2 supports 2D and 3D mesh representations of natural and synthetic visual objects, and still texture objects, with triangular patches [6, 15]. The vertices of the triangular mesh elements are called node points, and they can be used to track the motion of a VO, as depicted in Fig. 10.8. Motion compensation is performed by spatially piecewise warping of the texture maps that correspond to the triangular patches. Mesh modeling can efficiently represent continuous motion, resulting in less blocking artifacts at low bit rates as compared to the block-based modeling. It also enables object-based retrieval of VOs by providing accurate object trajectory information and syntax for vertex-based object shape representation.

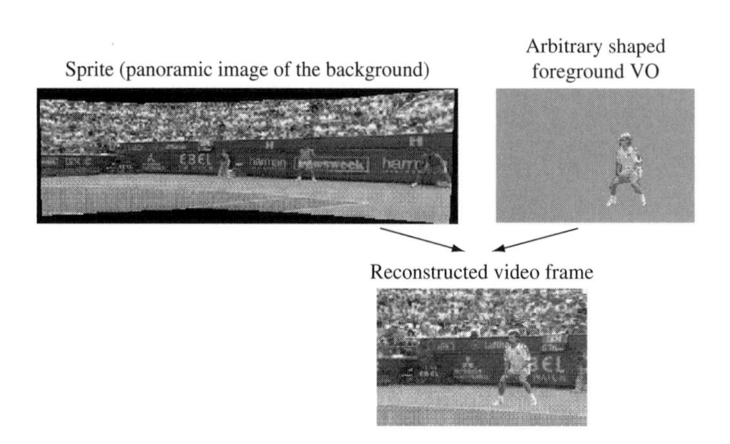

FIGURE 10.7

Sprite coding of a video sequence (courtesy of Dr Thomas Sikora, Technical University of Berlin, [9]).

FIGURE 10.8

Mesh representation of a VO with triangular patches (courtesy of Dr A. Murat Tekalp, University of Rochester, [9]).

10.3.4 Model-based Coding

Model-based representation enables very low bit rate video coding applications by providing the syntax for the transmission of the parameters that describe the behavior of a human being, rather than transmission of the video frames. MPEG-4 Part 2 supports the coding of two types of models: a face object model, which is a synthetic representation of the human face with 3D polygon meshes that can be animated to have visual manifestations of speech and facial expressions, and a body object model, which is a virtual human body model represented with 3D polygon meshes that can be rendered to simulate body movements [6, 16, 17].

10.3.4.1 Face Animation

It is required that every MPEG-4 Part 2 decoder that supports face object decoding have a default face model, which can be replaced by downloading a new face model. Either model can be customized to have a different visual appearance by transmitting facial definition parameters (FDPs). FDPs can determine the shape (i.e., head geometry) and texture of the face model.

The face animation parameters are coded by applying quantization followed by arithmetic coding. The quantization is performed by taking into consideration the limited movements of the facial features. Alternatively, DCT coding can be applied to a vector of 16 temporal instances of the FAP. This solution improves compression capability at the expense of a higher delay.

A face object consists of a collection of nodes, also called feature points, which are used to animate synthetic faces. The animation is controlled by face animation parameters (FAPs) that manipulate the displacements of feature points and angles of face features and expressions. The standard defines a set of 68 low-level animations, such as head and eye

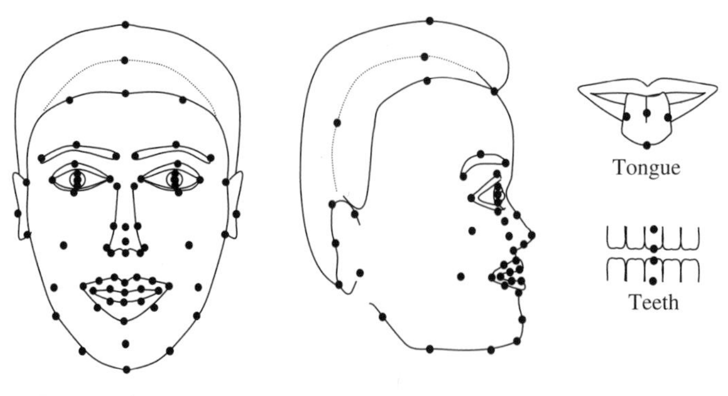

FIGURE 10.9

Feature points used for animation.

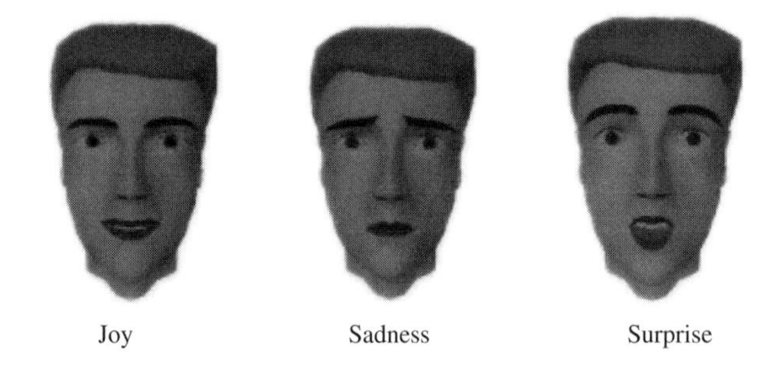

FIGURE 10.10

Examples of face expressions coded with FAPs (courtesy of Dr Joern Ostermann, University of Hannover, [18]).

rotations, as well as motion of a total of 82 feature points for the jaw, lips, eye, eyebrow, cheek, tongue, hair, teeth, nose, and ear. These feature points are shown in Fig. 10.9. High level expressions, such as joy, sadness, fear and surprise, and mouth movements are defined by sets of low-level expressions. For example, the expression of joy is defined by relaxed eyebrows, and an open mouth with the mouth corners pulled back towards ears. Figure 10.10 illustrates several video scenes that are constructed using face animation parameters.

10.3.4.2 ***Body Animation***

Body animation was standardized in the version 2 of MPEG-4 Part 2. Similar to the case of a face object, two sets of parameters are defined for a body object body definition parameters (BDPs), which define the body through its dimensions, surface, and texture, and body animation parameters (BAPs), which define the posture and animation of a given body model.

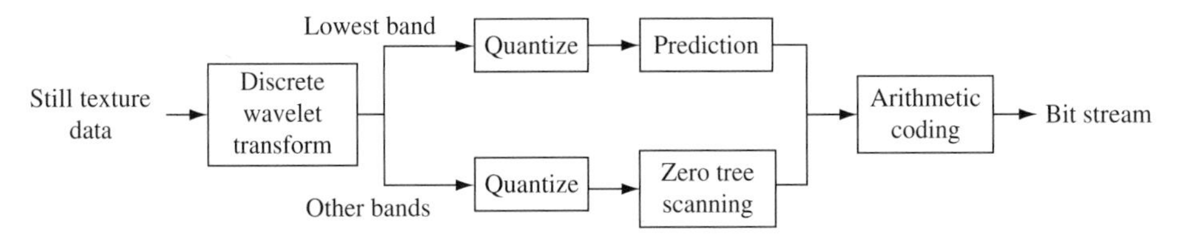

FIGURE 10.11

Block diagram of the still texture coder.

10.3.5 Still Texture Coding

The block diagram of an MPEG-4 Part 2 still texture coder is shown in Fig. 10.11. As illustrated in this figure, the texture is first decomposed using a 2D separable wavelet transform, employing a Daubechies biorthogonal filter bank [6]. The discrete wavelet transform is performed using either integer or floating point operations. For coding of an arbitrarily shaped texture, a shape adaptive wavelet transform can be used.

The DPCM coding method is applied to the coefficient values of the lowest frequency subband. A multiscale zerotree coding method [19] is applied to the coefficients of the remaining subbands. Zerotree modeling is used for encoding the location of nonzero wavelet coefficients by taking advantage of the fact that, if a wavelet coefficient is quantized to zero, then all wavelet coefficients with the same orientation and the same spatial location at finer wavelet scales are also likely to be quantized to zero. Two different zerotree scanning methods are used to achieve spatial and quality (SNR) scalability. After DPCM coding of the coefficients of the lowest frequency subband, and zerotree scanning of the remaining subbands, the resulting data are coded using an adaptive arithmetic coder.

10.3.6 Scalability

In addition to the video coding tools discussed so far, MPEG-4 Part 2 provides scalability tools that allow organization of the bitstream into base and enhancement layers. The enhancement layers are transmitted and decoded depending on the bit rate, display resolution, network throughput, and decoder complexity constraints. Temporal, spatial, quality, complexity, and object-based scalabilities are supported in MPEG-4 Part 2. The first three types of scalability were discussed in Chapter 6.4. Complexity scalability is the scaling of the processing tasks in such a way that the reconstruction quality of the bitstream is adaptable to the processing power of the decoder. Object-based scalability allows the addition or removal of VOs, as well as the prioritization of the objects within a scene. Using this functionality, it is possible to represent the objects of interest with higher spatial and/or temporal resolution, whereas allocating less bandwidth and computational resources to the objects that are less important. All of the above forms of scalability allow prioritized transmission of data, thereby also improving the bitstream's error resilience.

A special case of MPEG-4 Part 2 scalability support, which allows encoding of a base layer and up to 11 enhancement layers, is known as Fine Granularity Scalability (FGS). In FGS coding, the DCT coefficients in the enhancement layers are coded using bit-plane coding instead of the traditional run-level coding [20]. As a result, the FGS coded streams allow for smoother transitions between different quality levels, thereby allowing adaptation to varying network conditions [9, 16, 20]. However, the FGS design specified in MPEG-4 Part 2 had some problems in terms of its compression capability and was ultimately deprecated by the MPEG standardization committee. Today, it serves primarily only as a subject of historical interest.

10.3.7 Error Resilience

To ensure robust operation over error-prone channels, MPEG-4 Part 2 offers a suite of error resilience tools that can be divided into three groups: resynchronization, data partitioning, and data recovery [6, 21]. Resynchronization is enabled by the MPEG-4 Part 2 syntax, which supports a video packet structure that contains resynchronization markers and information such as macroblock number and quantizer in the header. All of these are necessary to restart the decoding operation in case an error is encountered. Data partitioning allows the separation between the motion and texture data, along with additional resynchronization markers in the bitstream to improve the ability to localize the errors. This technique provides enhanced concealment capabilities. For example, if texture information is lost, motion information can be used to conceal the errors. Data recovery is supported in MPEG-4 Part 2 by reversible variable length codes for DCT coefficients and a technique known as NEWPRED. Reversible variable length codes for DCT coefficients can be decoded in both the forward and backward directions. Thus, if part of a bitstream cannot be decoded in the forward direction due to errors, some of the coefficient values can be recovered by decoding the coefficient part of the bitstream in the backward direction. NEWPRED, which is a method intended for real-time encoding applications, makes use of an upstream channel from decoder to encoder, where the encoder dynamically replaces the reference pictures according to the error conditions and feedback received from the decoder [9].

10.3.8 MPEG-4 Part 2 Profiles

Since the MPEG-4 Part 2 syntax is designed to be generic, and includes many tools to enable a variety of video applications, the implementation of an MPEG-4 Part 2 decoder that supports the full syntax is often impractical. Therefore, the standard defines a number of subsets of the syntax, referred to as *profiles*, each of which targets a specific group of applications. For instance, the SP targets low-complexity and low-delay applications such as mobile video communications, whereas the Main Profile is intended primarily for interactive broadcast and DVD applications. A complete list of the MPEG-4 Part 2 version 1 and version 2 profiles is given in Table 10.1. The subsequent versions added more profiles to the ones defined in version 1 and version 2, increasing the total number of profiles in MPEG-4 Part 2 to approximately 20. Examples of newer profiles include the ASP targeting more efficient coding of ordinary rectangular video, the SP Profile targeting studio editing applications and supporting 4:4:4 and 4:2:2 chroma sampling, and the Fine Granularity Scalability Profile targeting web casting and wireless communication applications.

TABLE 10.1 MPEG-4 Part 2 visual profiles.

Profile group	Profile name	Supported functionalities
Profiles for natural video content	Simple	Error resilient coding of rectangular VOs
	Simple scalable	SP + Frame-based temporal and spatial scalability
	Core	SP + Coding of arbitrarily shaped objects
	Main	Core profile + Interlaced video coding + Transparency coding + Sprite coding
	N-bit	Core profile + Coding VOs with sample bit depths between 4 and 12 bits
	Advanced real time simple(2)	Improved error resilient coding of rectangular video with low buffering delay
	Core scalable(2)	Core profile + Temporal and spatial scalability of arbitrarily shaped objects
	Advanced coding efficiency(2)	Coding of rectangular and arbitrarily shaped objects with improved compression capability
Profiles for synthetic and synthetic/natural/ hybrid video content	Simple face animation	Basic coding of simple face animation
	Scalable texture	Spatially scalable coding of still texture objects
	Simple basic animated 2D texture	Simple face animation + Spatial and quality scalability + Mesh-based representation of still texture objects
	Hybrid	Coding of arbitrarily shaped objects + Temporal scalability + Face object coding + Mesh coding of animated still texture objects
	Advanced scalable texture(2)	Scalable coding of arbitrarily shaped texture, shape, wavelet tiling, and error resilience
	Advanced core(2)	Core visual profile + Advanced scalable texture visual profile
	Simple face and body animation(2)	Simple face animation profile + Body animation

The index (2) marks profiles that became available in version 2 of MPEG-4 Part 2. The rest of the profiles shown are available in both version 1 and 2 of MPEG-4 Part 2. Profiles specified after version 2 (such as the Simple Studio Profile) are not listed here.

In terms of actual widespread use in products, the SP and ASP are the dominant adopted configurations.

10.4 MPEG-4 PART 10: H.264/AVC

H.264/AVC provides significantly higher compresion capability over the prior video coding standards. One of the key goals in the development of H.264/AVC was to address the needs of the many different video applications and delivery networks that would be used to carry the coded video data. To facilitate this, the standard is conceptually divided into a Video Coding Layer (VCL) and a Network Abstraction Layer (NAL). The VCL defines a decoding process that can provide an efficient representation of the video, whereas the NAL provides appropriate header and system information for each particular network or storage media. The NAL enables network friendliness by mapping VCL data to a variety of transport layers such as RTP/IP, MPEG-2 systems, and H.323. A more detailed description of the NAL concepts and the error resilience properties of H.264/AVC are provided in [22] and [23]. In this chapter, we provide an overview of the H.264/AVC video coding tools that comprise the VCL.

10.4.1 H.264/AVC Video Coding Layer: Technical Overview

Although H.264/AVC is similar to preceding standards (e.g., MPEG-2, H.263, MPEG-4 Part 2), in that it defines only the bitstream syntax and video decoding process, in this study, for completeness, we discuss both the encoding and decoding processes. H.264/AVC employs a hybrid block-based video compression technique, similar to those defined in earlier video coding standards, which is based on combining interpicture prediction to exploit temporal redundancy and transform-based coding of the prediction errors to exploit spatial redundancy. A generalized block diagram of an H.264/AVC encoder is provided in Fig. 10.12. Although it is based on the same hybrid coding framework, H.264/AVC features a number of significant components that distinguish it from its predecessors. These features include spatial directional prediction; an advanced motion compensation model using variable block size prediction, quarter-sample accurate motion compensation, multiple reference picture prediction, and weighted prediction; an in-loop deblocking filter; a choice of either a large (8×8) or small (4×4) block size integer transform and two context-adaptive entropy coding modes.

A number of special features are provided to enable more flexible use or to provide resilience against losses or errors in the video data. The sequence and picture header information are placed into structures known as parameter sets, which can be transmitted in a highly flexible fashion to either help reduce bit rate overhead or add resilience against header data losses. Pictures are composed of slices that can be highly flexible in shape, and each slice of a picture is coded completely independently of the other slices in the same picture, in order to enable enhanced loss/error resilience. Further loss/error resilience is enabled by providing capabilities for data partitioning (DP) of the slice

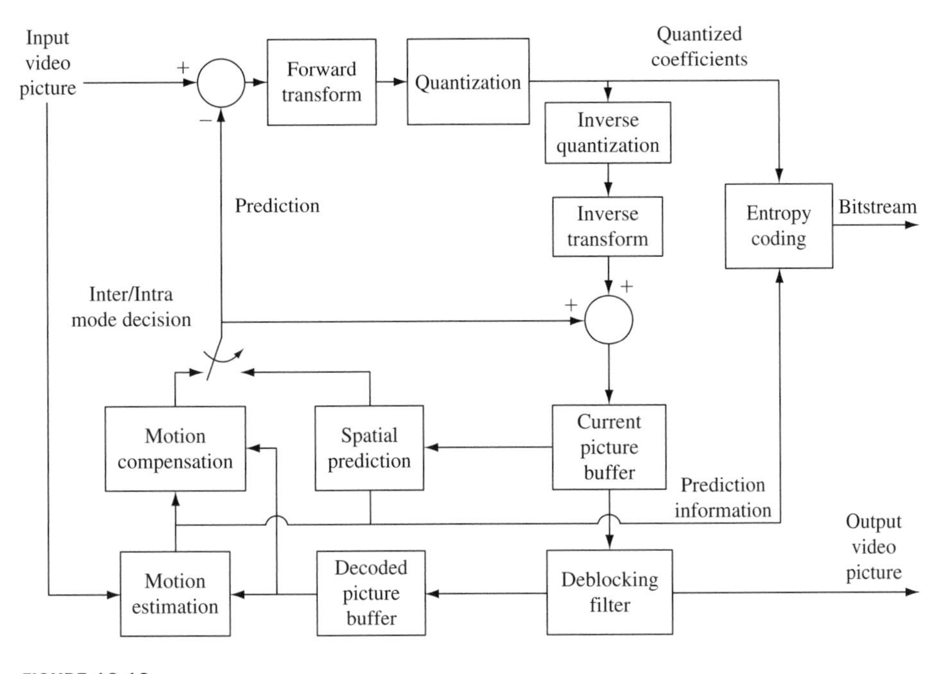

FIGURE 10.12

Block diagram of an H.264/AVC encoder.

data and allowing redundant picture (RP) slice representations to be sent. Robustness against variable network delay is provided by allowing arbitrary slice order (ASO) in the compressed bitstream. A new type of coded slice known as a synchronization slice for intra (SI) or inter (SP) coding is supported that enables efficient switching between streams or switching between different parts of the same stream (e.g., to change the server bit rate of preencoded streaming video or for loss/error robustness). Synchronization slices or subsequences[2] can also be used for trick-mode playback;[3] however, nonreference pictures are more commonly used to help decoders support smooth trickmodes. Subsequences, picture display and bitstream timing information, may be signalled with Supplemental Enhancement Information (SEI) messages. Several H.264/AVC SEI messages and some H.264/AVC video usability information (VUI) parameters, while not required for decoded picture construction, are required by applications standards to aid with normative requirements for features such as video display processing. Such application standards incorporating H.264/AVC include broadcast services such as Digital Video Broadcast (DVB) in Europe, Integrated Services Digital Broadcasting (ISDB) in Japan,

[2] Subsequences can also provide temporal scalability by enabling optional frame rate reduced decoding.

[3] Trick-mode playback includes capabilities such as fast forward and reverse, slow forward and reverse, jumping from section to section, and other forms of unusual navigation through a coded video stream.

Advanced Television Systems Committee (ATSC) in the United States, and high-definition optical disk formats such as Blu-ray Disk (BD).

10.4.1.1 Slices and Slice Groups

As in previous standards, pictures in H.264/AVC are partitioned into macroblocks, which are the fundamental coding units, each consisting of a block of 16×16 luma samples, and two blocks of corresponding 8×8 chroma samples. Also, each picture can be divided into a number of independently decodable slices, where a slice consists of one or more macroblocks.

In contrast with previous standards in which the picture type (I, P, or B) determined the macroblock prediction modes available, it is the slice type in H.264/AVC that determines which prediction modes are available for the macroblocks. For example, I slices contain only intra predicted macroblocks, and P slices can contain interpredicted (motion compensated) macroblocks in addition to the types allowed in I slices. Slices of different types can be mixed within a single picture.

The partitioning of the picture into slices can, in some usage scenarios, be done in a much more flexible way in H.264/AVC than in previous standards. In the previous standards, the shape of a slice was often highly constrained, and the macroblocks within the same slice were always consecutive in the raster-scan order of the picture or of a rectangle within the picture. The greater flexibility of the partitioning allowed in H.264/AVC is due to a concept known as slice groups. The slice group concept, informally known as the Flexible Macroblock Ordering (FMO) feature, is a new feature unique to H.264/AVC. Although this feature is not supported in many application scenarios,[4] it can be used to enable a variety of capabilities such as enhanced loss/error robustness, enhanced Region of Interest (ROI) coding customization, and low-delay real-time construction of composited image scenes from multiple lower-resolution video bitstreams for applications such as "continuous-presence multipoint" video conferencing.

Using the slice group concept, the allocation of macroblocks into slices can in principle be made completely flexible, through the specification of slice groups and macroblock allocation maps in the picture parameter set. Through the FMO feature, a single slice may contain all of the data necessary for decoding a number of macroblocks that are scattered throughout the picture, which can be useful for recovery from errors due to packet loss [22]. An example is given in Fig. 10.13, which illustrates one method of allocating each macroblock in a picture to one of three slice groups. In this example, the bitstream data for every third macroblock in raster-scan order is contained within the same slice group. A macroblock allocation map is specified that categorizes each macroblock of the picture into a distinct slice group. Each slice group is partitioned into one or more slices, where each slice consists of an integer number of macroblocks in raster-scan order within its slice group. Thus, if a single slice is lost, and the remaining slices are successfully reconstructed, several macroblocks that are adjacent to the corrupted macroblocks are available to assist error concealment.

[4]It is supported only in the Baseline, Extended, and Scalable Baseline profiles of the standard.

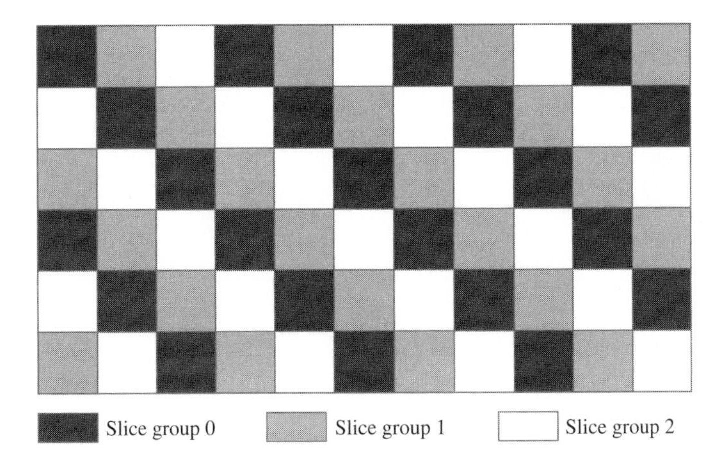

FIGURE 10.13

Example of slice groups.

10.4.1.2 Spatial Directional Intra Prediction

The compression performance of a block-based video codec depends fundamentally on the effective use of prediction of sample blocks in order to minimize the residual prediction error. H.264/AVC provides a very powerful and flexible model for the prediction of each block of samples from previously encoded and reconstructed sample values. This includes spatial directional prediction within the same picture (intraprediction), and flexible multiple reference picture motion compensation from a set of previously reconstructed and stored pictures (interprediction). As stated above, the prediction modes that are available are dependent upon the type of each slice (intra, predictive, or bi-predictive, for I, P, and B slices, respectively).

Intra prediction requires data from only within the current picture. Unlike the previous video standards, in which prediction of intracoded blocks was achieved by predicting only the DC value or a single row or column of transform coefficients from the above or left block, H.264/AVC uses spatial directional prediction, in which individual sample values are predicted based on neighbouring sample values that have already been decoded and fully reconstructed. Three different modes are supported, selectable on a per-macroblock basis: 4×4, 8×8 and 16×16. In the 4×4 intracoding mode, each 4×4 luma block within a macroblock can use a different prediction mode. There are nine possible modes DC and eight directional prediction modes. For example, in the horizontal prediction mode, the prediction is formed by copying the samples immediately to the left of the block across the rows of the block. As illustrated in Fig. 10.14, diagonal modes operate similarly, with the prediction of each sample based on a weighting of the previously reconstructed samples adjacent to the predicted block.

The 8×8 intra prediction mode operates similarly, except that each 8×8 luma block within a macroblock can (after low-pass filtering) use one of eight directional modes or a DC mode. The 16×16 intra prediction mode also operates similarly, except that the

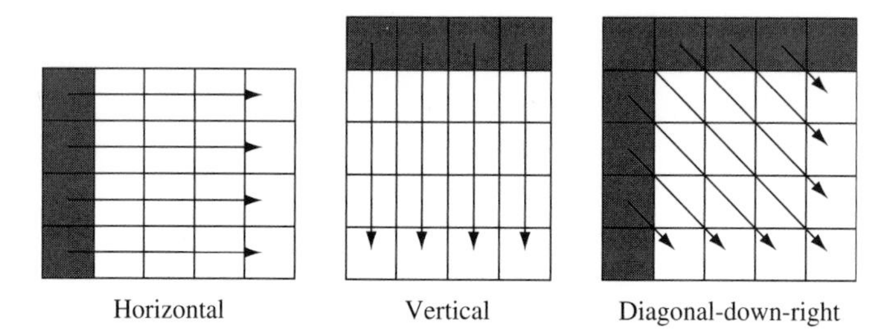

FIGURE 10.14

Intraprediction in H.264/AVC.

entire luma macroblock is predicted at once, based on the samples above and to the left of the macroblock. Also, in this mode there are only four modes available for prediction DC, vertical, horizontal, and planar. The 16 × 16 intra mode is most useful in relatively smooth picture areas.

10.4.1.3 Enhanced Motion Compensation Prediction Model

Motion-compensated prediction (interprediction) in H.264/AVC, similar to that in the previous standards, is based primarily on the translational block-based motion model, in which blocks of samples from previously reconstructed reference pictures are used to predict current blocks through transmission of motion vectors. However, the motion compensation model defined in H.264/AVC [24, 25] is more powerful and flexible than those defined in earlier standards, and provides a much larger number of options in the search for block matches to minimize the residual error. The H.264/AVC motion model includes seven partition sizes for motion compensation, quarter-sample accurate motion vectors, a generalized multiple reference picture buffer, and weighted prediction. For an encoder to take full advantage of the larger number of prediction options, the prediction selection in H.264/AVC is more computationally intense than in some earlier standards that have simpler models, such as MPEG-2.

Previous standards typically allowed motion compensation block sizes of only 16 × 16 or possibly 8 × 8 luma samples. In H.264/AVC, the partitioning of each macroblock into blocks for motion compensation is much more flexible, allowing seven different block sizes, as illustrated in Fig. 10.15. Each macroblock can be partitioned in one of the four ways as illustrated in the top row of the figure. If the 8 × 8 partitioning is chosen, each 8 × 8 block can be further partitioned in one of the four ways as shown in the bottom row of the figure.

All motion vectors in H.264/AVC are transmitted with quarter luma sample accuracy, providing improved opportunities for prediction compared to most previous standards, which allowed only half-sample accurate motion compensation. For motion vectors at fractional sample locations, the sample values are interpolated from the reference picture samples at integer-sample locations. Luma component predictions at half-sample locations are interpolated using a 6-tap FIR filter. Predictions at quarter-sample locations

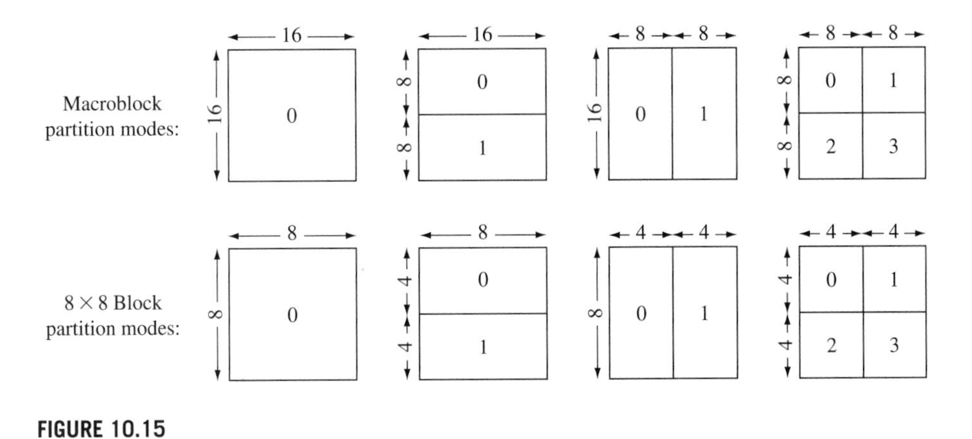

FIGURE 10.15

Illustration of macroblock partitioning into blocks of different sizes.

are computed through a bilinear interpolation of the values for two neighbouring integer- or half-sample locations. Chroma fractional-sample location values are computed by linear interpolation between integer-location values.

H.264/AVC provides great flexibility in terms of which pictures can be used as references to generate motion compensated predictions for subsequent coded pictures. This is achieved through a flexible multiple reference picture buffer. In earlier standards, the availability of reference pictures was generally fixed and limited to a single temporally previous picture for predicting P pictures, and two pictures—one temporally previous and one temporally subsequent—for predicting B pictures. However, in H.264/AVC, a multiple reference picture buffer is available and it may contain up to 16 reference frames or 32 reference fields (depending on the profile, level, and picture resolution specified for the bitstream). The assignment and removal of pictures entering and exiting the buffer can be explicitly controlled by the encoder by transmitting buffer control commands as side information in the slice header, or the buffer can operate on a first-in first-out basis in decoding order. The motion compensated predictions for each macroblock can be derived from one or more of the reference pictures within the buffer by including reference picture selection syntax elements in conjunction with the motion vectors. With the generalized multiple reference picture buffer, the temporal constraints that are imposed on reference picture usage are greatly relaxed. For example, as illustrated in Fig. 10.16, it is possible that a reference picture used in a P slice is located temporally subsequent to the current picture, or that both references for predicting a macroblock in a B slice be located in the same temporal direction (either forward or backward in temporal order). To convey the generality, the terms forward and backward prediction are not used in H.264/AVC. Instead, the terms List 0 prediction and List 1 prediction are used, reflecting the organization of pictures in the reference buffer into two (possibly overlapping) lists of pictures without regard to temporal order. In addition, with the generalized multiple reference picture buffer, pictures coded using B slices can be referenced for the prediction of other pictures, if desired by the encoder [24]. This extremely flexible assignment of

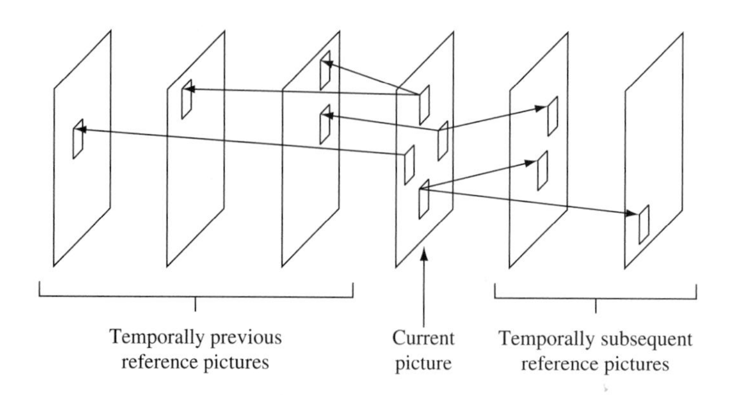

FIGURE 10.16

Multireference picture prediction in H.264/AVC.

reference and nonreference pictures enables video encoders to implement new, powerful, and flexible picture type controllers. Sometimes also called dynamic[5] or adaptive[6] Group of Picture (GOP) controllers, these modules can enable significantly improved compression capability. For example, the use of a flexible GOP[7] structure can greatly reduce intra beating artifacts that may otherwise occur with periodically introduced intra slices.[8] Flexible GOP controllers can also provide substantial perceptual gain for camera flashes and other abrupt lighting changes, which were very difficult to represent efficiently with previous standards.

In P slices, each motion compensated prediction block is derived by quarter-sample interpolation of a prediction from a single block within the set of reference pictures. In B slices, an additional option is provided, allowing each prediction block to be derived from a pair of locations within the set of reference pictures. By default, the final prediction in a bipredictive block is computed using a sample-wise ordinary average of the two interpolated reference blocks. However, H.264/AVC also includes a weighted prediction feature, in which each prediction block can be assigned a relative weighting value rather than using

[5]Broadly used, the term dynamic GOP has become an ambiguous term, often referring to a variable number of interpictures between intra refresh pictures, or to the practice of promotion of interpictures to intra pictures within an otherwise fixed pattern of intra, inter, reference, and non-reference pictures. Such a promotion would typically occur at scene changes.

[6]Also broadly used, the term adaptive GOP often refers to the promotion of nonreference frames to reference frames within an otherwise fixed pattern of frame types. For MPEG-2 this term has frequently been used to refer to a variable number of B-frames.

[7]Here we use the term flexible GOP to refer to varying the location, number, and temporal reference structure of reference, and nonreference frames. A significant example is variable use of hierarchical picture structures between successive I or P pictures. Such a hierarchical picture structure may be composed of a number of tiers. Each tier is a set of pictures that depends for reference upon all sets of pictures assigned to lower tiers. For example, the lowest tier may be composed of I and P pictures, whereas middle tiers are composed of reference B pictures and the highest tier of nonreference B pictures.

[8]Periodic refresh pictures composed of intra slices are often used for random stream access (e.g., channel change) and trick mode support.

an ordinary average value, and an offset can be added to the prediction. These parameters can either be transmitted explicitly by the encoder, or computed implicitly, based on a time-related relationship between the current picture and the reference picture(s). Weighted prediction can provide substantial coding gains in scenes containing fades or cross fades, which are very difficult to code efficiently with previous video standards that do not include this feature.

10.4.1.4 In-Loop Deblocking Filter

Block-based prediction and transform coding, including quantization of transform coefficients, can lead to visible and subjectively objectionable changes in intensity at coded block boundaries, referred to as blocking artifacts. In previous video codecs, such as MPEG-2, the visibility of these artifacts could optionally be reduced to improve subjective quality by applying a deblocking filter after decoding and before display. The goal of such a filter is to reduce the visibility of subjectively annoying artifacts while avoiding excessive smoothing that would result in loss of detail in the image. However, because the process was optional for decoders and was not normatively specified in these standards, the filtering was required to take place outside of the motion compensation loop, and large blocking artifacts would still exist in reference pictures that would be used for predicting subsequent pictures. This reduced the effectiveness of the motion compensation process, and allowed blocking artifacts to be propagated into the interior of subsequent motion compensated blocks, causing the subsequent picture predictions to be less effective and making the removal of the artifacts by filtering more challenging.

To improve upon this, H.264/AVC defines a normative deblocking filtering process. This process is performed identically in both the encoder and decoder, in order to maintain an identical set of reference pictures. This filtering leads to both objective and subjective improvements in quality, as shown in Fig. 10.17, due to the improved

(a)

(b)

FIGURE 10.17

A decoded frame of the sequence Foreman (a) without the in-loop deblocking filtering applied and (b) with the in-loop deblocking filtering (The original sequence Foreman is courtesy of Siemens AG).

prediction and the reduction in visible blocking artifacts. Note that the second version of the ITU-T H.263 standard also included an in-loop deblocking filter as an optional feature, but this feature was not supported in the most widely deployed Baseline profile of that standard, and the design had some problems with inverse-transform rounding error effects when used at high fidelities.

The deblocking filter defined in H.264/AVC operates on a 4×4 grid, which is the smallest basis for the block transform and interprediction. Both luma and chroma samples are filtered. The filter is highly adaptive in order to remove as many artifacts as possible without excessive smoothing. For the line of samples across each horizontal or vertical block edge (such as that illustrated in Fig. 10.18), a filtering strength parameter is determined based on the coding parameters on both sides of the edge. When the coding parameters indicate that large artifacts are more likely to be generated (e.g., intra prediction or coding of nonzero transform coefficients), larger strength values are assigned. This results in stronger filtering being applied. Additionally, sample values (i.e., a_0–a_3 and b_0–b_3 in Fig. 10.18) along each line of samples to be (potentially) filtered are checked against several conditions that are based on the quantization step size employed on either side of the edge in order to distinguish between discontinuities that are introduced by quantization, and those that are true edges that should not be filtered to avoid loss of detail [25]. For instance, in the example shown in Fig. 10.18, a significant lack of smoothness can be detected between the sample values on each side of the block edge. When the quantization step size is large, this lack of smoothness is considered an undesirable artifact and it is filtered. When the quantization step size is small, the lack of smoothness is considered to be the result of actual details in the scene being depicted by the video and it is not altered.

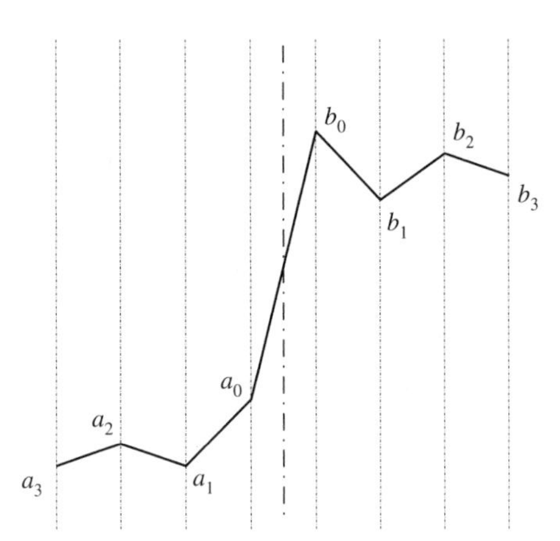

FIGURE 10.18

Example of an edge profile. Notations a_0, a_1, a_2, a_3, and b_0, b_1, b_2, b_3 stand for sample values on each side of the edge.

10.4.1.5 Transform, Quantization, and Scanning

As in earlier standards, H.264/AVC uses block transform coding to efficiently represent the prediction residual signal. H.264/AVC provides a transform based on blocks of 8 × 8 samples; however, unlike previous standards, an H.264/AVC encoder may also choose to use a small transform based on blocks of 4 × 4 samples. Furthermore, instead of performing an approximation of the floating-point DCT (i.e., by specifying that each implementation must choose its own approximation of the ideal IDCT equations), the transforms employed have properties similar to that of the 4 × 4 and 8 × 8 DCTs, but are completely specified in integer operations. This eliminates the problem of mismatch between the inverse transforms performed in the encoder and decoder, enabling an exact specification of the decoded sample values and significantly reducing the complexity of the IDCT computations. In particular, the IDCTs can be computed easily using only 16-bit fixed-point computations (including intermediate values), which would have been very difficult for the IDCT defined in all previous standards. Additionally, the basis of the 4 × 4 transform is extended with additional second-stage transforms that are applied to the 2 × 2 chroma DC coefficients of a macroblock, and the 4 × 4 luma DC coefficients of a macroblock predicted in the 16 × 16 intra mode.

Scalar quantization is employed on the transform coefficients [26], but without the extra wide dead-zone used in all other video standards. The 52 quantization parameter values are designed so that the quantization step size increases by approximately 12.2% for each increment of one in the quantization parameter (such that the step size exactly doubles if the quantization parameter is increased by six). A number of quantization scaling matrices may also be specified to extend the granularity and range of the quantization step size by a further factor of 256. These matrices may customize the step size of each frequency coefficient of the 4 × 4 and 8 × 8 transforms, to optimize for human visual perception in each luma and chroma channel, separately for intra- and interpredicted macroblocks. As in earlier standards, each block of quantized coefficients is zig-zag scanned for ordering the quantized coefficient representations in the entropy coding process. The decoder performs an approximate inversion of the quantization process, by multiplying the quantized coefficient values by the step size that was used in their quantization. This inverse quantization process is referred to as scaling in the H.264/AVC standard, since it consists essentially of just a multiplication by a scale factor. In slice types other than SP/SI, the decoder then performs an inverse transform of the scaled coefficients and adds this approximate residual block to the spatial-domain prediction block to form the final picture reconstruction before the deblocking filter process.

In SP and SI synchronization slices, the decoder performs an additional forward transform on the prediction block and then quantizes the transform-domain sum of the prediction and the scaled residual coefficients. This additional forward transform and quantization operation can be used by the encoder to discard any differences between the details of the prediction values obtained when using somewhat different reference pictures in the prediction process (e.g., differences arising from the reference pictures being from an encoding of the same video content at a different bit rate). The reconstructed result in this case is then formed by an inverse transform of the quantized transform-domain sum, rather than the ordinary case where the reconstructed result is the sum of a spatial-domain prediction and a spatial-domain scaled residual.

10.4.1.6 *Entropy Coding*

Two methods of entropy coding are specified in H.264/AVC. The simpler VLC method is supported in all profiles. In both cases, many syntax elements except for the quantized transform coefficients are coded using a regular, infinite-extent variable-length codeword set. In this study, the same set of Exp-Golomb codewords is used for each syntax element, but the mapping of codewords to decoded values is changed depending on the statistics associated with each element.

In the simpler entropy coding method, scans of quantized coefficients are coded using a Context-Adaptive Variable-Length Coding (CAVLC) scheme. In this method, one of the number of VLC tables is selected for each symbol, depending on the contextual information. This includes statistics from previously coded neighbouring blocks, as well as statistics from previously coded coefficient values within the current scan.

For improved compression, a more complex Context-Adaptive Binary Arithmetic Coding (CABAC) method can be used [27]. When CABAC is in use, scans of transform coefficients and other syntax elements of the macroblock level and below (such as reference picture indices and motion vector values) are encoded differently. In this method, each symbol is binarized, that is, converted to a binary code, and then the value of each bin (bit of the binary code) is arithmetically coded. To adapt the coding to nonstationary symbol statistics, context modeling is used to select one of several probability models for each bin, based on the statistics of previously coded symbols. The use of arithmetic coding allows for a noninteger number of bits to be used to code each symbol, which is highly beneficial in the case of very skewed probability distributions. Probability models are updated following the coding of each bit, and this high degree of adaptivity improves the compression capability even more than the ability to use a fractional number of bits for each symbol. Due to its bit-serial nature, the CABAC method entails greater computational complexity than the VLC-based method, particularly for highly parallel architectures.

10.4.1.7 *Frame/Field Adaptive Coding*

H.264/AVC includes tools for efficiently handling the special properties of interlaced video, since the two fields that compose an interlaced frame are captured at different instances of time. In areas of high motion, this can lead to less statistical correlation between adjacent rows within the frame, and greater correlation within individual fields, making field coding (where lines from only a single field compose a macroblock) a more efficient option. In addition to regular frame coding in which lines from both fields are included in each macroblock, H.264/AVC provides two options for special handling of interlaced video field coding and macroblock-adaptive field/frame coding (MB-AFF).

Field coding provides the option of coding pictures containing lines from only a single video field (i.e., a picture composed of only top field lines or only bottom field lines). Depending on the characteristics of the frame, an encoder can choose to code each input picture as two field frames, or as a complete frame. In the case of field coding, an alternate zig-zag coefficient scan pattern is used, and individual reference fields are selected for motion compensated prediction. Thus, frame/field adaptivity can occur at the picture level. Field coding is most effective when there is significant motion throughout the video scene for interlaced input video, as in the case of camera panning.

However, in MB-AFF coding, the adaptivity between frame and field coding occurs at a level known as the macroblock pair level. A macroblock pair consists of a region of the frame that has a height of 32 luma samples and width of 16 luma samples and contains two macroblocks. This coding method is most useful when there is significant motion in some parts of an interlaced video frame and little or no motion in other parts. For regions with little or no motion, frame coding is typically used for macroblock pairs. In this mode, each macroblock consists of 16 consecutive luma rows from the frame, thus lines from both fields are mixed in the same macroblock. In moving regions, field macroblock pair coding can be used to separate each macroblock pair into two macroblocks that each include rows from only a single field. Frame and field macroblock pairs are illustrated in Fig. 10.19. The unshaded rows compose the first (or top) macroblock in each pair, and the shaded rows compose the second (or bottom) macroblock. Figure 10.19 (a) shows a frame macroblock pair, in which samples from both fields are combined in each coded macroblock. Figure 10.19 (b) shows a field macroblock pair, in which all of the samples

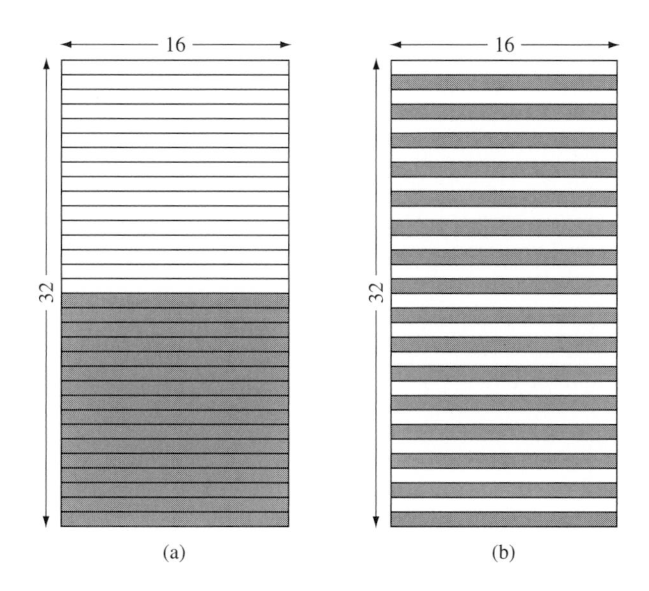

FIGURE 10.19

Illustration of macroblock pairs.

in each of the two macroblocks are derived from a single field. With MB-AFF coding, the internal macroblock coding remains largely the same as in ordinary frame coding or field coding. However, the spatial relationships that are used for motion vector prediction and other context determination become significantly more complicated in order to handle the low-level switching between field and frame based operation.

10.4.2 Profiles

Similar to other standards, H.264/AVC defines a set of "profiles" that each support only a subset of the entire syntax of the standard and are designed to target specific application areas. The Baseline Profile targets applications not requiring interlace support and requiring moderate computational complexity, such as videoconferencing and mobile. This profile also provides robustness for use on unreliable channels, where some of the video data may be lost or corrupted. The High Profile targets applications such as broadcast of standard-definition (SD) and high-definition (HD) video on more reliable channels, and storage on optical media. This profile includes more advanced coding tools, such as CABAC entropy coding and B slices, that can improve compression capability over that provided by the Baseline Profile, at the cost of higher complexity. The Main Profile contains a subset of the features of the High Profile—it is now primarily only of historical relevance after the development of the High Profile. The Extended Profile is intended for use in streaming and use in error prone transmission environments, such as wireless networks—although it has not been widely adopted.

A number of profiles have been defined for higher quality, more "professional", usage applications. The High 10, High 4:2:2, and High 4:4:4 Predictive profiles fall in this category. The High 10 Profile permits up to 10-bit sample precision; the High 4:2:2 Profile further permits both 4:2:0 and 4:2:2 chroma samplings; and the High 4:4:4 Predictive Profile also supports up to 14-bit samples and 4:4:4 chroma sampling. For production and contribution applications such as professional high-definition video acquisition and editing, the interpicture prediction feature is less necessary, and to target these applications the corresponding High 10 Intra, High 4:2:2 Intra, High 4:4:4 Intra, and CAVLC 4:4:4 Intra profiles have been specified. (The CAVLC Intra Profile is similar to the High 4:4:4 Intra Profile, but omits support of CABAC entropy coding to reduce computational complexity.) Since there is no interpicture prediction loop in the Intra-only profiles, and since these profiles are intended for applications that typically avoid significant artifacts, the deblocking filter is not normatively required for their decoding.

Three SVC profiles have also been specified, which are the Scalable Baseline, Scalable High, and Scalable High Intra profiles.[9]

The key features that are supported by some of these profiles[10] of H.264/AVC are shown in Table 10.2.

[9]The SVC extensions are a major addition to the standard, but are not discussed in depth in this study to limit the scope of this treatment for reasons of brevity.

[10]H.264/AVC currently has 14 profiles altogether, including main, professional, specialized, and SVC profiles. In addition, multiview video coding (MVC) extensions are being finalized.

TABLE 10.2 H.264/AVC profiles.

Supported functionalities	Profile name				
	Baseline	High	Extended	High 10 Intra	High 4:2:2 Intra
I slices	X	X	X	X	X
P slices	X	X	X		
B slices		X	X		
SI and SP slices			X		
In-loop deblocking filter	X	X	X		
CAVLC entropy decoding	X	X	X	X	X
CABAC entropy decoding		X		X	X
Weighted prediction		X	X		
Field pictures		X	X	X	X
MB-AFF		X	X	X	X
Multiple slice groups (Flexible macroblock ordering)	X		X		
Arbitrary slice order (ASO)	X		X		
Redundant pictures (RP)	X		X		
Data partitioning (DP)			X		
Quantization scaling matrices		X		X	X
8 × 8 transform		X		X	X
8 × 8 Intra prediction		X		X	X
10-bit samples				X	X
4:2:2 chroma sampling					X

10.5 MPEG-4 COMPRESSION PERFORMANCE

In this section, we first present experimental results that compare the coding performance of frame-based coding to object-based coding using an MPEG-4 Part 2 codec on some progressive-scan video content. Next, the compression capability of H.264/AVC is compared with that of the previous standards.[11]

10.5.1 MPEG-4 Part 2

Although MPEG-4 Part 2 yields improved compression capability over the previous coding standards, (e.g., 20–30% bit rate savings over MPEG-2 [28]), its main advantage

[11] For space reasons, for each comparison we provide an individual example rather than presenting test results obtained from a large body of experimental data. A proper evaluation of typical or average performance for a given application would require an expanded analysis.

is its object-based representation, which enables many desired functionalities and can yield substantial savings in bit rate for some low-complexity video sequences. In this study, we present an example that illustrates such a compression capability advantage.

The simulations were performed by encoding the color sequence *Bream* at CIF resolution (352 × 288 luma samples/frame) at 10 frames per second (fps) using a constant quantization parameter of 10. The sequence shows a moderate motion scene with a fish swimming and changing directions. We used the MPEG-4 Part 2 reference codec for encoding. The video sequence was coded 1) in frame-based mode and 2) in object-based mode at 10 fps. In the frame-based mode, the codec achieved a 56:1 compression ratio with relatively high reconstruction quality (34.4 dB). If the quantizer step size were larger, it would be possible to achieve up to a 200:1 compression ratio for this sequence, while still keeping the reconstruction quality above 30 dB. In the object-based mode, where the background and foreground (fish) objects are encoded separately, a compression ratio of 80:1 is obtained. Since the background object did not vary with time, the number of bits spent for its representation was very small. In this study, it is also possible to employ sprite coding by encoding the background as a static sprite.

The PSNR versus rate performance of the frame-based and object-based coders for the video sequence Bream is presented in Fig. 10.20. As shown in the figure, for this sequence, the PSNR versus bit rate tradeoffs of object-based coding are better than those of frame-based coding. This is due mainly to the constant background, which is coded only once in the object-based coding case. However, for scenes with complex and fast varying shapes, since a considerable amount of bits would be spent for shape coding,

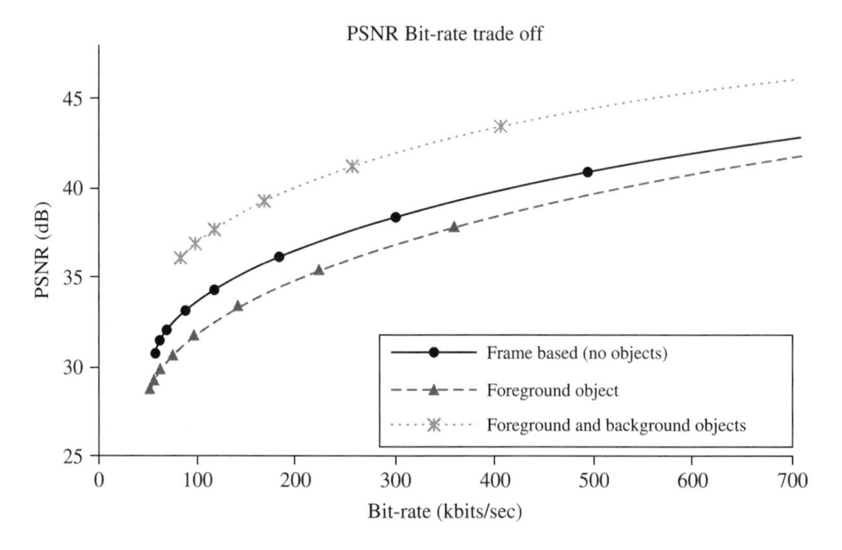

FIGURE 10.20

PSNR performance for the Bream video sequence using different profiles of the MPEG-4 Part 2 video coder.

frame-based coding would achieve better compression levels, but at the cost of a limited object-based access capability.

10.5.2 MPEG-4 Part 10: H.264/AVC

Next, the coding performance of H.264/AVC is compared with that of MPEG-2, H.263, and MPEG-4 Part 2. The results were generated using encoders for each standard that were similarly optimized for rate-distortion performance using Lagrangian coder control. The use of the same efficient rate-distortion optimization method, which is described in [28], allows for a fair comparison of encoders that conform to the corresponding standards. In this study, we provide one set of results comparing the standards in two key application areas low-latency video communications and entertainment-quality broadband video.

In the video communications test, we compare the rate-distortion performance of an H.263 Baseline Profile encoder, which was the most widely deployed conformance point for such applications prior to H.264/AVC, with an MPEG-4 SP encoder and an H.264/AVC Baseline Profile encoder. The rate-distortion curves generated by encoding the color sequence Foreman at CIF resolution (352 × 288 luma samples/frame) at a rate of 15 fps are shown in Fig. 10.21. As shown in the figure, the H.264/AVC encoder provides significant improvements in compression capability. More specifically,

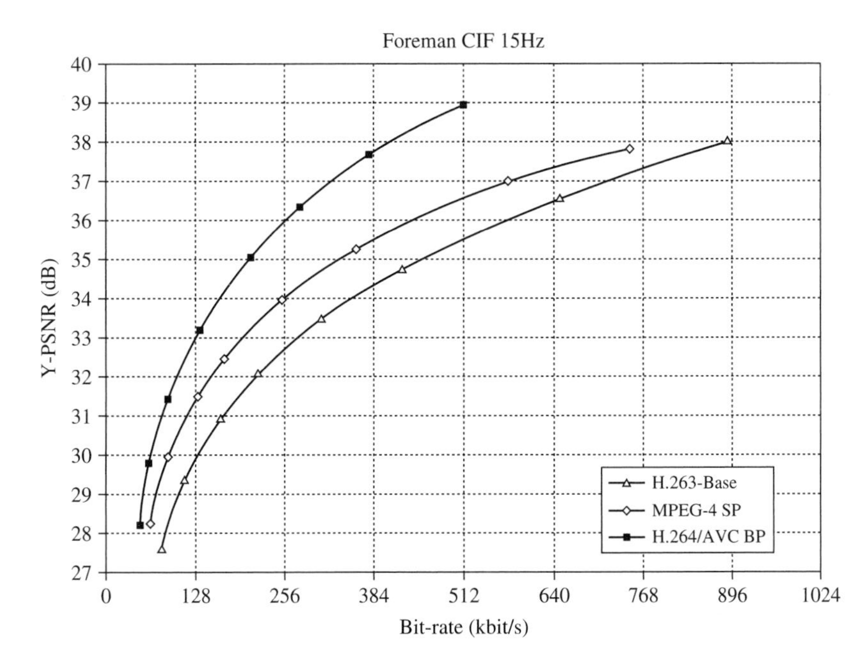

FIGURE 10.21

Comparison of H.264/AVC Baseline Profile and the H.263 Baseline Profile using the sequence Foreman.

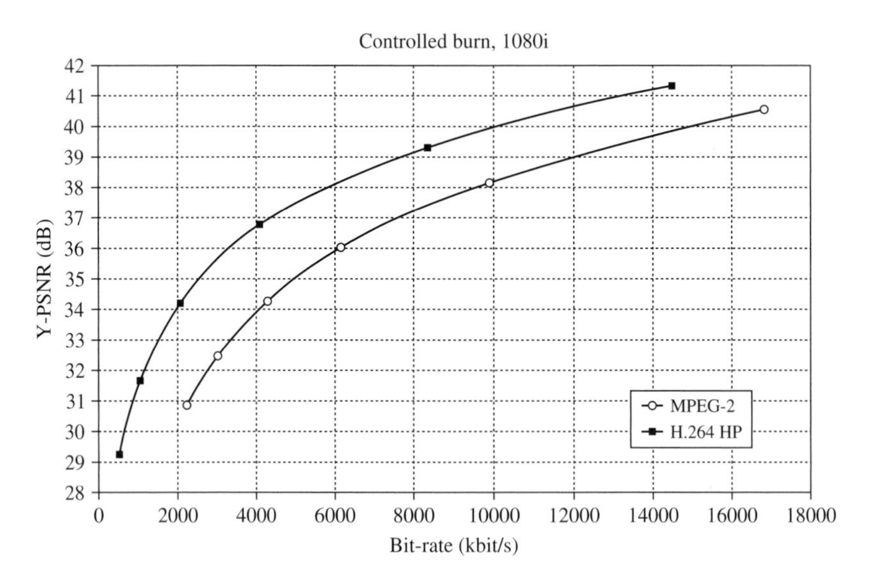

FIGURE 10.22

Rate distortion curves for the interlaced HD sequence Controlled Burn (1920×1080).

H.264/AVC coding yields bit rate savings of approximately 55% over H.263 Baseline profile and approximately 35% over MPEG-4 Part 2 SP.

A second comparison addresses broadband entertainment-quality applications, where higher resolution content is encoded and larger amounts of latency are tolerated. In this comparison, the H.264/AVC High Profile is compared with the widely implemented MPEG-2 Main Profile. Rate distortion curves for the interlaced HD sequence Controlled Burn (1920×1080) are given in Fig. 10.22. In this plot, we can see that H.264/AVC can yield similar levels of objective quality at approximately half the bit rate for this sequence. In addition to these objective rate-distortion results, extensive MPEG subjective testing of H.264/AVC has confirmed that similar subjective quality to that of MPEG-2 encoders can be achieved with H.264/AVC encoders at approximately half the bit rate used by the same MPEG-2 encoders [29].

10.6 MPEG-4 VIDEO APPLICATIONS

MPEG-4 Part 2 has been commercialized in a variety of software and hardware products. By far, the most popular implementations of MPEG-4 Part 2 use just its SP, and most of these implementations enable video playback in mobile and embedded consumer devices (e.g., cellular phones, PDAs, digital cameras, digital camcorders, and personal video recorders). The MPEG-4 Part 2 ASP which permits significantly better video quality than that of SP, has also been used in some professional video markets, particularly in surveillance cameras and higher quality video coding systems. The existence of two reference software implementations published by the MPEG standards committee, as well as the availability of some open source software for MPEG-4 Part 2 [30], has facilitated wider adoption of the standard. However, the ascendance of interest in the adoption of the more recent H.264/AVC standard has recently had a significant dampening effect on the further widespread deployment of MPEG-4 Part 2.

MPEG-4 Part 10 (H.264/AVC) has rapidly gained support from a wide variety of industry groups worldwide. This was mainly because of its excellent compression capability, and the availability of the JVT reference software implementation that helped industry deployment and conformance testing efforts. In the video conferencing industry, for example, where video codecs are generally implemented in software, companies such as UB Video (now part of Cisco) developed as early as 2002 H.264/AVC. Baseline software codecs that are now being used in most of the popular video conferencing products. Moreover, even before the first version of the H.264/AVC specification was completed, UB Video and other companies such as VideoLocus (now part of Magnum Semiconductor) demonstrated software/hardware products that were deployed in the broadcast market. In the past few years, companies such as Adobe Systems, Apple Computer, Sony, Polycom, Tandberg, Ahead, Nero Digital, MainConcept, VideoSoft, KDDI, and others have deployed H.264/AVC software products that address a wide market space, from real-time mobile and video conferencing applications to broadcast-quality video coding applications. In the broadcast domain, Scientific Atlanta (now part of Cisco) and Modulus Video (now part of Motorola) were among the first companies to productize H.264/AVC real-time hardware encoding of HD video. Other companies, such as Tandberg, Envivio and Harmonic, have deployed hardware H.264/AVC encoders and decoders for broadcast and streaming of SD/HD video. Additionally, real-time HDTV capable decoding, encoding, and transcoding Application Specific Integrated Circuits (ASICs) have been produced by companies such as Broadcom, Ambarella, Horizon Semiconductors, Conexant, Micronas, NXP, Sigma Designs, STMicroelectronics, Texas Instruments, and Magnum Semiconductor. The patent licensing situation surrounding the H.264/AVC standard has also developed in a manner that is generally considered favorable to widespread deployment.

10.7 CONCLUSIONS AND OUTLOOK

In this chapter, we have presented a detailed technical description, and a brief discussion of video applications of MPEG-4 (Parts 2 & 10).

MPEG-4 Part 2 has achieved widespread deployment in mobile and PC-based video applications, although its most ambitious object-coding features have not yet been widely adopted in the digital video market. The object coding aspects have become more of an academic research topic than a practical solution. Today, the MPEG-2 Part 2 SP is a widely supported feature in a variety of portable media players, advanced mobile phones, digital cameras, camcorders, and personal computing scenarios. The Advanced Simple Profile of MPEG-4 Part 2 has been widely used in PC-based internet video and, to some extent, in optical storage media players. This profile is often referred to as the DivX format due to its use in software known by that name and due to a conformance logo program operated by the company DivX Inc. [30].

Since its more recent development, MPEG-4 Part 10 (H.264/AVC)—which has demonstrated very high compression performance—has already achieved very wide adoption in essentially all digital video applications, as evidenced by a fast-growing number of H.264/AVC products. H.264/AVC is even beginning to significantly displace MPEG-2 in applications such as broadcast and optical disc storage media, which are applications that have historically been practically synonymous with MPEG-2 video. The transition to H.264/AVC in such applications is particularly evident for high-definition video content. Much of the market momentum of H.264/AVC is due to its excellent compression capability, which is a clear advancement over the prior state of the art of standardized video formats. Transcoding and splicing systems have begun to be deployed to facilitate the transition from MPEG-2 to H.264/AVC and to buy enough time to realize a good return on MPEG-2–related investments while also benefiting from H.264/AVC's strong compression capability advantage. We expect H.264/AVC to dominate the digital video market by the end of this decade or shortly thereafter. The recent adoption of FRExt, SVC, and MVC extensions to H.264/AVC [31][32] will extend the reach of this powerful new design into additional existing and new application domains.

ACKNOWLEDGMENT

We would like thank Adriana Dumitraş of Apple Computer for her contributions to the previous editions of this chapter.

REFERENCES

[1] ISO/IEC. Information technology—coding of moving pictures and associated audio for digital storage media at up to about 1.5 mbits/s: Video. 11172–2, 1993.

[2] ISO/IEC. Information technology—generic coding of moving pictures and associated audio information: Video. 13818–2, 1995.

[3] ITU-T. Video coding for low bit rate communication. Recommendation H.263, 1996.

[4] ITU-T. Video coding for low bit rate communication. Recommendation H.263 Version 2, 1998.

[5] ISO/IEC 14496-1. Information technology—coding of audio-visual objects—Part 1: Systems. 1999.

[6] ISO/IEC 14496-2. Information technology—coding of audio-visual objects—Part 2: Visual. 2001.

[7] ISO/IEC 14496–10. Information technology—coding of audio-visual objects—Part 10: Advanced video coding; also ITU–T Recommendation H.264: Advanced video coding for generic audiovisual services. 2003.

[8] ISO/IEC 14496-3. Information technology—coding of audio-visual objects—Part 3: Audio. 2001.

[9] R. Koenen. Overview of the MPEG-4 standard. In *ISO/IEC JTC1/SC29/WG11 N4668*, March 2002.

[10] MPEG-4 Video Group. MPEG–4 video verification model version 8.0. ISO/IEC JTC1/SC29/WG11 N3093, December 1999.

[11] MPEG-4 SNHC Group. SNHC verification model 9.0. ISO/IEC JTC1/SC29/WG11 MPEG98/ M4116, October 1998.

[12] R. M. Haralick and L. H. Shapiro. Image segmentation techniques. *Comput. Vis. Graph. Image Process.*, 29:100–132, 1985.

[13] A. K. Katsaggelos, L. P. Kondi, F. W. Meier, J. Ostermann, and G. M. Schuster. MPEG–4 and rate–distortion–based shape–coding techniques. *Proc. IEEE*, 86:1029–1051, 1998.

[14] M. Lee, W. Chen, C. B. Lin, C. Gu, T. Markoc, S. Zabinsky, and R. Szeliski. A layered video object coding system using sprite and affine motion model. *IEEE Trans. Circuits Syst. Video Technol.*, 7:130–146, 1997.

[15] A. M. Tekalp, P. V. Beek, C. Toklu, and B. Gunsel. Two–dimensional mesh–based visual–object representation for interactive synthetic/natural video. *Proc. IEEE*, 86:1126–1154, 1998.

[16] P. Kalra, A. Mangili, T. N. Magnenat, and D. Thalmann. Simulation of facial muscle actions based on rational free form deformations. *Proc. Eurographics*, 59–69, 1992.

[17] P. Doenges, T. Capin, F. Lavagetto, J. Ostermann, I. S. Pandzic, and E. Petajan. MPEG–4: Audio/video and synthetic graphics/audio for real–time, interactive media delivery. *Image Commun.*, 5(4):433–463, 1997.

[18] B. G. Haskell, P. G. Howard, Y. A. Lecun, A. Puri, J. Ostermann, M. R. Civanlar, L. Rabiner, L. Bottou, and P. Haffner. Image and video coding – emerging standards and beyond. *IEEE Trans. Circuits Syst. Video Technol.*, 8(7):814–837, 1998.

[19] S. A. Martucci, I. Sodagar, T. Chiang, and Y. Zhang. A zerotree wavelet video coder. *IEEE Trans. Circuits Syst. Video Technol.*, 7:109–118, 1997.

[20] W. Li. Overview of fine granularity scalability in MPEG-4 video standard. *IEEE Trans. Circuits Syst. Video Technol.*, 11(3):301–317, 2001.

[21] R. Talluri. Error resilient video coding in the MPEG-4 standard. *IEEE Commun. Mag.*, 26(6):112–119, 1998.

[22] S. Wenger. H.264/AVC over IP. *IEEE Trans. Circuits Syst. Video Technol.*, 13(7):645–656, 2003.

[23] T. Stockhammer, M. M. Hannuksela, and T. Wiegand. H.264/AVC in wireless environments. *IEEE Trans. Circuits Syst. Video Technol.*, 13(7):657–673, 2003.

[24] M. Flierl and B. Girod. Generalized B pictures and the draft JVT/H.264 video compression standard. *IEEE Trans. Circuits Syst. Video Technol.*, 13(7):657–673, 2003.

[25] P. List, A. Joch, J. Lainema, G. Bjontegaard, and M. Karczewicz. Adaptive deblocking filter. *IEEE Trans. Circuits Syst. Video Technol.*, 13(7):614–619, 2003.

[26] H. S. Malvar, A. Hallapuro, M. Karczewicz, and L. Kerofsky, Low–complexity transform and quantization in H.264/AVC. *IEEE Trans. Circuits Syst. Video Technol.*, 13(7):598–603, 2003.

[27] D. Marpe, H. Schwarz, and T. Wiegand. Context–adaptive binary arithmetic coding in the H.264/AVC video compression standard. *IEEE Trans. Circuits Syst. Video Technol.*, 13(7):620–636, 2003.

[28] T. Wiegand, H. Schwarz, A. Joch, F. Kossentini, and G. J. Sullivan. Rate–constrained coder control and comparison of video coding standards. *IEEE Trans. Circuits Syst. Video Technol.*, 13(7):688–703, 2003.

[29] ISO/IEC JTC 1/SC29/WG 11. Report of the formal verification tests on AVC, document N6231. December 2003.

[30] DIVX, "Open source MPEG–4 software," info available at http://www.divx.com/, 2004.

[31] H. Schwarz, D. Marpe, and T. Wiegand. Overview of the scalable video coding extension of the H.264/AVC standard. *IEEE Trans. Circuits Syst. Video Technol.*, 17(9):1103–1120, 2007.

[32] G. J. Sullivan, H. Yu, S. Sekiguchi, H. Sun, T. Wedi, S. Wittmann, Y.L. Lee, A. Segall, and T. Suzuki. New standardized extensions of MPEG4-AVC/H.264 for professional-quality video applications. *Int. Conf. Image Process.*, 1:13–16, 2007.

CHAPTER

11 Interframe Subband/Wavelet Scalable Video Coding

John W. Woods, Peisong Chen, Yongjun Wu, and Shih-Ta Hsiang

Center for Image Processing Research,
Department of Electrical, Computer, and Systems Engineering,
Rensselaer Polytechnic Institute, Troy, New York, USA

Abstract

A new class of video coders using 3D subband/wavelet transforms along the motion trajectory has emerged in recent years. These nonhybrid or "stateless" video coders can be highly scalable in bit rate, resolution, and frame rate without losing coding efficiency. This class of interframe subband/wavelet coders requires a highly accurate local motion field though. But this motion field can then be used for preprocessing at the coder as well as postprocessing at the decoder. In this chapter, we try to illustrate some of their important issues by using the motion-compensated video coder MC-EZBC as a concrete example. We compare results of scalable MC-EZBC to those of two other 3D transform coders and to an early version of the MPEG scalable video coder.

11.1 INTRODUCTION

MC-EZBC [1] is one of a family of motion-compensated (MC) subband/wavelet coders that exploit temporal correlation but are fully embedded in quality/bit rate, spatial resolution, and frame rate. The name EZBC stands for embedded zero-block coder. Figure 11.1 shows the basic structure of the MC-EZBC coder. This chapter will use this coder as an example of the class of fully embedded subband/wavelet transform (SWT)[1] coders that use an MC temporal filter (MCTF).

[1]More commonly called discrete wavelet transform (DWT) or just subband transform.

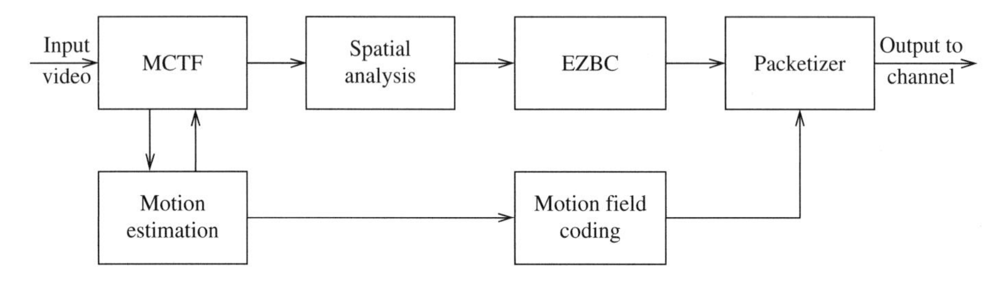

FIGURE 11.1

Basic structure of MC-EZBC video coder.

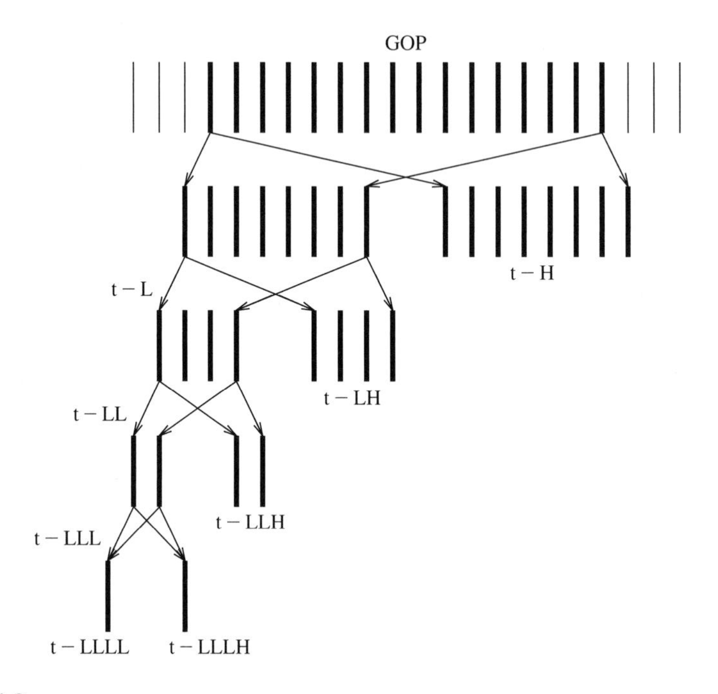

FIGURE 11.2

Octave-based five-band temporal decomposition [1] using Haar filters.

In Fig. 11.1, the incoming data is operated on first by the MCTF, followed by spatial SWT, and then the EZBC entropy coder. [2] By using the MCTF, this system does not suffer the drift problem exhibited by hybrid coders that have feedback loops. A Haar 2-tap MCTF structure is shown in Fig. 11.2 for a group of pictures (GOP) of size 16, yielding 4 + 1 temporal levels or frame rates. Since the complete MC temporal transform

[2]Here, we have separated out the entropy coder from the spatial SWT, but normally the EZBC coder includes the spatial transform as an integral part.

comes before the spatial transform, this structure has been labeled "$t + 2D$." Order matters here due to the nonlinear effects of the motion estimation/compensation on the MCTF.

Here for simplicity, we have used the 2-tap Haar filters in the MCTF, but longer filters, such as the LeGall and Tabatabai 5/3 can offer a modest boost in performance [2–4]. We discuss this extension later in Section 11.5 of this chapter. In [1], half-pixel accurate MCTF with perfect reconstruction was realized and significant coding gain was observed. The use of subpixel accurate motion compensation has also achieved substantial coding gain in MPEG2 [5], and H.264/AVC [6] even uses 1/4 pixel accurate motion compensation. Subpixel accurate motion compensation is most useful for low-resolution video, where the spatial sample rate is near to or below the Nyquist rate, making the power spectrum flatter.

Both MPEG-2 and H.264/AVC [6] are hybrid coders and the reconstructed frames are used as reference frames for motion-compensated prediction (MCP), the so-called temporal differential pulse-code modulation (DPCM). We are free to use any interpolation in the reference frames in such coders since the warped reference frames are only used for MCP and the residual is coded before display. Since they operate in a closed loop, any subpixel accurate interpolation can be duplicated at the decoder. Conversely, MCTF is open loop, and any interpolated values used at the encoder must be reinterpolated at the decoder. Hsiang et al. [7, 8] designed an invertible MCTF with half-pixel accuracy by incorporating the spatial interpolation as part of the temporal subband/wavelet filtering. More recently, lifting implementations of subband/wavelet filters have been used for MC temporal subband decomposition [9–11] to provide invertibility for arbitrary interpolation schemes. In all these approaches, the chosen spatial interpolation is incorporated into the perfect reconstruction subband/wavelet decomposition, effectively making the MC temporal filter somewhat spatial also. The use of a lifting implementation makes this incorporation easy, even for any kind of motion field. However, there is no reason to assume that the resulting motion-compensated SWT should be optimal in any sense [12], even though it is perfectly reconstructible in the absence of quantization errors.

In the sequel, we first cover motion estimation and compensation for MCTF, including detection of covered/uncovered pixels and the use of chroma for motion estimation in Section 11.2. This is followed in Section 11.3 by a discussion on improvements to MC temporal filtering, including the lifting implementation for invertibility and directional IBLOCKs for intraframe spatial coding. Section 11.4 then presents the lossy entropy coder EZBC and relates it to JPEG 2000. Section 11.5 presents the extension to longer filters, including an adaptive 5/3 filter. Objective and visual comparisons are presented in Section 11.6. The effect of multiple adaptations is shown in Section 11.7. Section 11.8 then shows some visual results, followed by a brief section on related coders, and then follows the conclusions.

11.2 MOTION ESTIMATION AND COMPENSATION FOR MCTF

The original frames in one GOP serve as the input to the Haar MCTF. The size of the GOP can vary but usually is in the range 8–64, the latter being approximately 1.1 s at

60 fps. A large GOP is only efficient if the motion is largely coherent over the corresponding time interval. On the other hand, the algorithm-inherent transmission delay goes up with GOP size. Next, we provide details of the system blocks in Fig. 11.1, starting with the MCTF and motion estimation.

We can see from Fig. 11.2 that motion vectors are only needed for successive pairs of input frames (for the Haar filter). In fact, the total number of motion vector estimations, totaled down the temporal levels, equals the number of frames [13]. Thus, the set of motion vectors V'_{xx} in Fig. 11.3 is only used for the prediction of motion vectors in the next lower temporal level [2], that is, we estimate the additional motion vectors V'_{00} but do not code them. Only $V'_{00} + V_{01}$ will be used as the starting point for the estimation of V_{10}, which then can be estimated in a reduced search range. Namely, if the original hierarchical variable-size block matching (HVSBM) search range is $\beta \times \beta$ pixels, we start a search from $V'_{00} + V_{01}$ in a reduced range of $\beta/2 \times \beta/2$ pixels.

Motion estimation is done through HVSBM, with block sizes varying from 64×64 down to 4×4 for each pair of frames as shown in Fig. 11.4. Block matching is a good choice for practical applications and the range of block sizes can achieve a fairly accurate motion field portrayal. This hierarchical aspect offers a computational savings and tends to reduce "noise" in the resulting motion vectors.

When deciding whether to split a motion block or not, it is appropriate to augment the spatial error, often chosen as mean absolute difference (MAD) plus a weighted term estimating the motion-vector coding rate,

$$MAD + \lambda_{\text{mv}} R_{\text{mv}},$$

and then choose whether to split or not based on this criteria. In Haar MC-EZBC, this optimization is done for the highest quality or bit rate. The resulting motion vectors may be approximated and combined to produce a layered and scalable motion vector stream, as will be discussed in Section 11.3.2. Also, this optimization can be combined with the

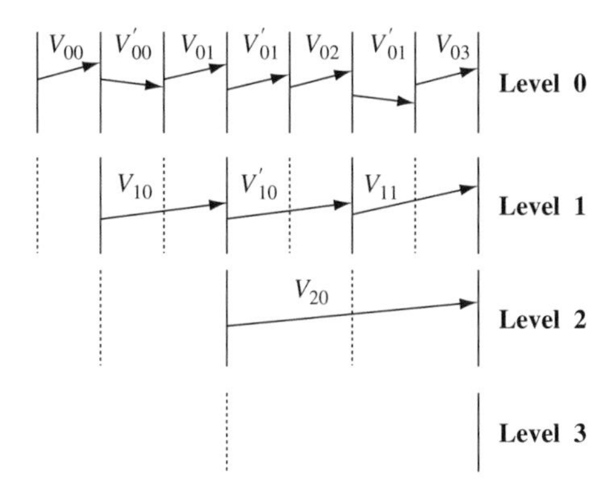

FIGURE 11.3

Motion estimation in a GOP with size 8 frames. While the V_{xx} are estimated and coded, the V'_{xx} may be estimated to provide an initial search point at the next lower temporal level.

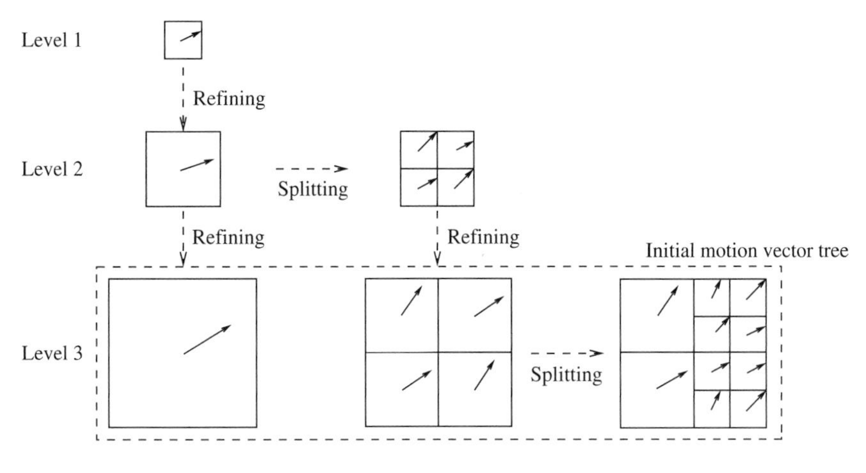

FIGURE 11.4

A three-level HVSBM showing three subband levels (© 2004 IEEE).

motion vector mode determination [4, 14] in an adaptive 5/3 MCTF as will be treated in Section 11.5 of this chapter.

11.2.1 Connected and Unconnected Blocks

When we do motion estimation, we find a motion vector for each block in the predicted frame. There are good matches and bad matches based on their displaced frame differences (DFD). For the *connected* pixels, that is, those with good matches, we perform the temporal filtering. For the remaining or *unconnected* blocks, we seek to place them in the output temporal t-L and t-H frames as appropriate. For a t-L frame, one can just put in the unconnected block directly. For a t-H frame, it is better to predict the block first and then insert into the t-H frame, the corresponding prediction error.

So, the question now is how to form the prediction for unconnected blocks. As illustrated in Fig. 11.5 taken from [15], in the predicted frame B, two blocks may choose the same reference block, that is, the block that will be warped to make the temporal high block in frame B. Here, frame A will become a temporal low frame t-L, and frame B will be replaced with the temporal high block t-H. One of the reasons for this multiconnection is the occlusion problem [16 and see Chapter 3 of this volume]. Using frame A as the reference frame, then those pixels in frame B that are revealed and not present in frame A are the *uncovered* pixels. For such pixels, frame B's immediately next frame most probably would contain their matched regions.[3]

[3]We ignore the possibility of an object being uncovered for only one frame. Usually, the frame rate is high enough for this assumption to be sound. Even when it happens, mostly such an occurrence can be suppressed in the coded data.

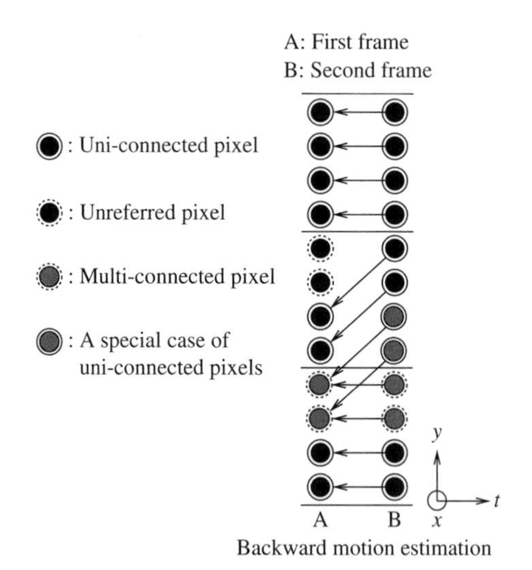

FIGURE 11.5

Allowed connections between blocks of pixels in pairs of frames [13] (© 2004 IEEE).

In MC-EZBC using Haar filters, we classify unconnected blocks into PBLOCKs and REVERSE blocks, depending whether their best reference choice is forward or behind. Some blocks do not benefit from prediction from either direction and these are classified as IBLOCKS, where some kind of spatial or intraframe coding is used. This was done using Chen's algorithm in [17] to try to find the best matches for the temporal filtering and not just what is determined by the scan order as in the past [15, 19]. This has been referred to as *bidirectional* motion compensation [18]. It has been performed for either choice of first *A* or second *B* frame for reference or t-H frame, but when the second frame is chosen, then processing within a GOP must proceed from last frame to first because of causality [15]. Either way, bidirectional MCTF reduces artifacts considerably in the t-L band and results in a more accurate motion field [15].

To show the importance of the bidirectional capability in MCTF, we present two frames below in Fig. 11.6(a) and (b). On the left in Fig. 11.6(a), we see the temporal low frame t-L4, four levels down at 1/16th the original frame rate for the unidirectional MCTF on the *Flower Garden* test clip. On the right in Fig. 11.6(b), we see the great reduction in artifacts achieved with the bidirectional MCTF.

A large number of unconnected pixels can be used to trigger an *adaptive GOP size*, meaning that the GOP terminates with this frame. An alternative is to do another level of temporal decomposition but at a reduced spatial resolution [17]. There is a substantial variation called unconstrained MCTF (UMCTF) [19], wherein the update step is completely eliminated, that is, leaving the temporal low frame unchanged from the next higher level. This method can be used to reduce the delay of MCTF coders but generally suffers a penalty in peak signal-to-noise ratio (PSNR) [20], especially at high qualities and bit rates. It makes MCTF into a kind of MPEG hierarchical B-frame decomposition.

FIGURE 11.6

Haar MCTF t-L4 output frames, (a) unidirectional, (b) bidirectional.

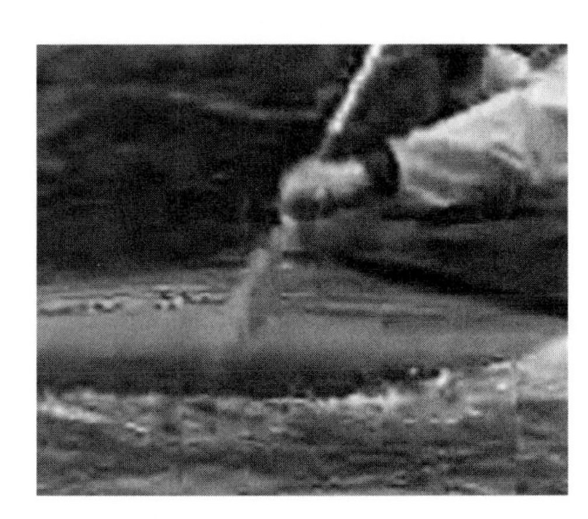

FIGURE 11.7

An example of HVSBM (left) versus color HVSBM (right) on *Conoa*.

11.2.2 Using Chroma for Motion Estimation

Often using chroma data U and V in addition to the luma data Y sometimes gives a more stable motion field. We call this *color HVSBM* [21] and have used 4:2:0 subsampled U and V frames. The computation time for color HVSBM is about 1.5 times that of HVSBM running on the luminance data alone. Each dimension of U and V is half that of Y in the chosen 4:2:0 color space and we need subpixel accuracy motion vectors, but we keep the full accuracy for the U and V motion vectors and saturate this accuracy at one-eighth pixel. An example of the improvement possible is shown in Fig. 11.7 below for the test clip *Conoa*.

After we form the full motion vector quadtree with bidirectional color HVSBM, we prune it back in the sense of MV rate versus MAD error in a Lagrangian optimization using a fixed λ. We also introduced an elementary type of scalable motion vector coder [15] based on context-based adaptive binary arithmetic coder (CABAC) [22] with respect to temporal, SNR, and resolution as described in the following.

11.2.3 Improving the Haar MCTF

The MCTF plays an essential role in motion-compensated 3D SWT coding. It will influence the coding efficiency and temporal scalability. Since MCTF is a kind of SWT, we can implement MCTF in the conventional (transversal) way or by using the lifting scheme [23]. The low-pass and high-pass filtering in the conventional implementation can be looked at as a parallel computation, but the low-pass and high-pass filtering in the lifting implementation of this transform is a serial computation. For integer-pixel accurate MCTF, these two schemes are equivalent, but for subpixel accurate MCTF, the lifting scheme can provide some nice properties. In the following, we compare these two schemes in subpixel accurate MCTF. Before that, we have to define connected pixels in the case of subpixel motion vectors. When a motion displacement $\{d_m, d_n\}$ from the second frame B points to a subpixel position in the first frame A, we can say $B\{m,n\}$ is connected to $A\{m-\overline{d}_m, n-\overline{d}_n\}$, where $\overline{d}_m$ and $\overline{d}_n$ are defined as follows. If (d_m, d_n) points to a subpixel position, $\{\overline{d}_m, \overline{d}_n\} = [\{d_m\}, \{d_n\}]$, where $[\cdot]$ denotes the nearest integer function (nint). All the pixels in uncovered blocks will be left for later.

11.2.3.1 Noninvertible Approach

This approach was proposed by Ohm [18] and extended by Choi and Woods [13]. In this approach, the subband analysis pair for connected pixels is

$$H(m,n) = \left[B(m,n) - \tilde{A}(m-d_m, n-d_n)\right]/\sqrt{2} \tag{11.1}$$

$$L\left(m-\overline{d}_m, n-\overline{d}_n\right) = \left[\tilde{B}\left(m-\overline{d}_m+d_m, n-\overline{d}_n+d_n\right) + A\left(m-\overline{d}_m, n-\overline{d}_n\right)\right]/\sqrt{2}. \tag{11.2}$$

The high-pass coefficient comes from the filtering of $B\{m,n\}$ and the interpolated reference pixel $\tilde{A}\{m-d_m, n-d_n\}$. Since we want the low-pass coefficients to be at integer pixel positions, we do the low-pass filtering between $\tilde{A}\{m-d_m, n-d_n\}$'s closest integer pixel $A\{m-\overline{d}_m, n-\overline{d}_n\}$ and the interpolated pixel $\tilde{B}\{m-\overline{d}_m+d_m, n-\overline{d}_n+d_n\}$ as illustrated in Fig. 11.8 below. Here, we use the inverse (reverse) of $B\{m,n\}$'s backward motion vector as $A\{m-\overline{d}_m, n-\overline{d}_n\}$'s forward motion vector. In general, this scheme is not invertible [24]. Since we need to use H and L to reconstruct A, but H only contains the information of interpolated pixels in A, if this interpolation itself is not invertible, we cannot reconstruct frame A exactly.

Unconnected pixels in B are processed like (11.1), and unconnected (unreferred) pixels in A are processed as

$$L(m,n) = \sqrt{2}A(m,n) \tag{11.3}$$

This open loop prediction is not good, but fortunately there are not a lot of unconnected pixels present, at least at the first few stages of temporal decomposition.

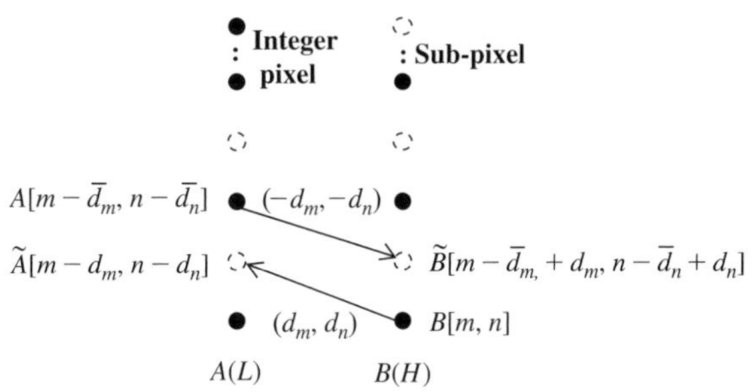

: Integer pixel

: Sub-pixel

$A[m-\bar{d}_m, n-\bar{d}_n]$ $(-d_m, -d_n)$

$\tilde{A}[m-d_m, n-d_n]$ $\tilde{B}[m-\bar{d}_m + d_m, n-\bar{d}_n + d_n]$

(d_m, d_n) $B[m, n]$

$A(L)$ $B(H)$

FIGURE 11.8

Subpixel accurate MCTF (Choi and Woods' scheme [16]) (© 2004 IEEE).

11.2.3.2 Invertible Half-Pixel Accurate MCTF

Hsiang et al. [7, 8] designed an invertible half-pixel accurate MCTF. In this approach, a composite block is constructed by merging a pair of linked motion blocks when the motion vector points to a half-pixel position. Then the composite block is decomposed by a two-channel subband analysis filter bank and the low-pass output and the high-pass output are put into the temporal low subband and the temporal high subband, respectively. So, perfect reconstruction can be realized. Effectively, the desired spatial interpolation has been incorporated into the subband/wavelet temporal filter.

11.2.3.3 Lifting Implementation

The so-called lifting scheme is a new realization for SWT [23]. This implementation has been introduced into subpixel MCTF to realize perfect reconstruction [9–11], as mentioned above. In the following, we first take a look at this implementation for connected pixels. If motion vectors have subpixel accuracy, the lifting scheme gets $H\{m,n\}$ in the same way as the conventional method [13],

$$H(m,n) = \left[B(m,n) - \tilde{A}(m-d_m, n-d_n)\right]/\sqrt{2} \tag{11.4}$$

For the low-pass filtering, there are two ways to get the forward motion vectors. One way in [11] used forward motion estimation, thereby requiring two motion fields to be transmitted. The other way, following a "homogeneous motion assumption" [18] is used here [9, 15],

$$L\left(m-\bar{d}_m, n-\bar{d}_n\right) = \tilde{H}\left(m-\bar{d}_m + d_m, n-\bar{d}_n + d_n\right) + \sqrt{2}A\left(m-\bar{d}_m, n-\bar{d}_n\right). \tag{11.5}$$

At the decoder, by using L and H, we can do the same interpolation on H and reconstruct A exactly if there is no quantization error,

$$A\left(m-\bar{d}_m, n-\bar{d}_n\right) = \left[L\left(m-\bar{d}_m, n-\bar{d}_n\right) - \tilde{H}\left(m-\bar{d}_m + d_m, n-\bar{d}_n + d_n\right)\right]/\sqrt{2}. \tag{11.6}$$

After A is available, we can do the same interpolation on A as the encoder and reconstruct B exactly as

$$B(m,n) = \sqrt{2}H(m,n) + \tilde{A}(m - d_m, n - d_n). \tag{11.7}$$

So, no matter how we interpolate those subpixels, if we interpolate pixels in the same way at the encoder and the decoder, we can realize perfect reconstruction. This should come as no surprise by now, since we have put the desired spatial interpolation into the guaranteed reconstructible lifting filters. In (11.6), we see L and H are still necessary for the reconstruction of A, and H still only contains the information of interpolated pixels in A. But this interpolated information is also available in L. So it is canceled out in (11.6). Thus, the interpolation algorithm has no influence on the perfect reconstruction. Of course, there is still the question of which interpolation is best to use.

Unconnected pixels in B are processed like (11.4), and unconnected (unreferred) pixels in A are processed as in (11.3).

11.2.3.4 Subpixel Interpolation

We saw that interpolated temporal high-pass band $\tilde{H}$ was used to update A. Then we have a question: can we make $L\{m - \overline{d}_m, n - \overline{d}_n\}$ the same as (11.2)? To answer this question, we use A and B to represent L

$$\begin{aligned} L\left(m-\overline{d}_m, n-\overline{d}_n\right) &= \tilde{H}\left(m-\overline{d}_m+d_m, n-\overline{d}_n+d_n\right)+\sqrt{2}A\left(m-\overline{d}_m, n-\overline{d}_n\right) \\ &= \left[\tilde{B}\left(m-\overline{d}_m+d_m, n-\overline{d}_n+d_n\right)-\tilde{\tilde{A}}\left(m-\overline{d}_m, n-\overline{d}_n\right)\right]/\sqrt{2} \\ &\quad +\sqrt{2}A\left(m-\overline{d}_m, n-\overline{d}_n\right) \\ &= \frac{1}{\sqrt{2}}\left[2A\left(m-\overline{d}_m, n-\overline{d}_n\right)-\tilde{\tilde{A}}\left(m-\overline{d}_m, n-\overline{d}_n\right)\right] \\ &\quad +\frac{1}{\sqrt{2}}\tilde{B}\left(m-\overline{d}_m+d_m, n-\overline{d}_n+d_n\right), \end{aligned} \tag{11.8}$$

where$\tilde{\tilde{A}}\{m - \overline{d}_m, n - \overline{d}_n\}$ results from two interpolations: the first interpolation is using integer pixels in A to obtain interpolated subpixels, which as seen in (11.4)) are stored in the integer positions of H; the second interpolation happens in (11.5)) when we use integer pixels in H (subpixels of A in effect) to interpolate subpixels in H (integer pixels of A in effect). If we can achieve

$$A\left(m-\overline{d}_m, n-\overline{d}_n\right) = \tilde{\tilde{A}}\left(m-\overline{d}_m, n-\overline{d}_n\right),$$

then

$$L\left(m-\overline{d}_m, n-\overline{d}_n\right) = \frac{1}{\sqrt{2}}\left[A\left(m-\overline{d}_m, n-\overline{d}_n\right)+\tilde{B}\left(m-\overline{d}_m+d_m, n-\overline{d}_n+d_n\right)\right], \tag{11.9}$$

which is the same as (11.2). Since the pixels in $\tilde{\tilde{A}}\{m - \overline{d}_m, n - \overline{d}_n\}$ undergo an interpolation from integer pixels to subpixels and then from subpixels back to integer pixels, the necessary condition is *sinc* function interpolation and a constant motion vector within the support region of the interpolation filter. For different s at 1/4, 1/2, and 3/4 we can get three different interpolation filters

$$f(n+s) = \sum_{m} f(m) \frac{\sin \pi(n+s-m)}{\pi(n+s-m)}, \quad 0 < s < 1. \tag{11.10}$$

The finite impulse response (FIR) interpolation can be designed as separable and using Hamming window. For relevant values of the shift s, our filter has the following coefficients:

$$s = \frac{1}{4} : [-0.0110 \quad 0.0452 \quad -0.1437 \quad 0.8950 \quad 0.2777 \quad -0.0812 \quad 0.0233 \quad -0.0053]$$

$$s = \frac{1}{2} : [-0.0053 \quad 0.0233 \quad -0.0812 \quad 0.2777 \quad 0.8950 \quad -0.1437 \quad 0.0452 \quad -0.0110]$$

$$s = \frac{3}{4} : [-0.0105 \quad 0.0465 \quad -0.1525 \quad 0.6165 \quad 0.6165 \quad -0.1525 \quad 0.0465 \quad -0.0105].$$

With these filters, we can interpolate 1/4, 1/2, and 3/4 pixels. We first interpolate subpixel at integer columns and integer rows, where only 1D filtering of integer pixels is used. Then at each subpixel column, we use those already generated subpixels to interpolate the remaining subpixels (see Fig. 11.9). We can also implement by each subpixel row.

11.2.3.5 Comparison of Different Interpolation Filters

We use bilinear filter and our designed 8-tap FIR filters for 1/2 pixel accurate MC-EZBC. From the coding results on the *Mobile* test sequence as seen in Fig. 11.10, we can see that the 8-tap FIR filter gives better PSNR than bilinear filter.

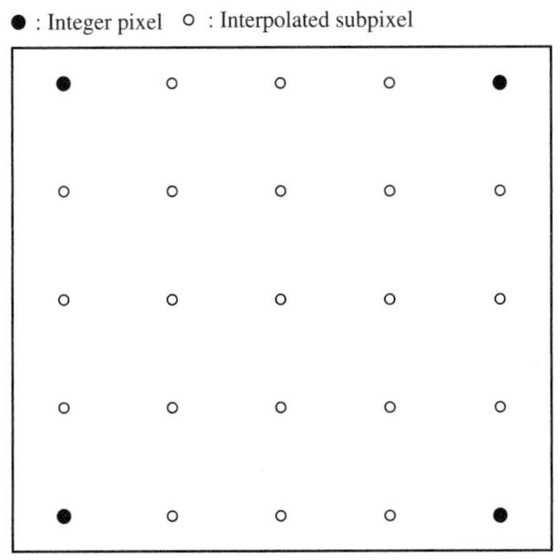

● : Integer pixel ○ : Interpolated subpixel

FIGURE 11.9

An illustration of subpixel interpolation in 1/4 pixel accuracy case.

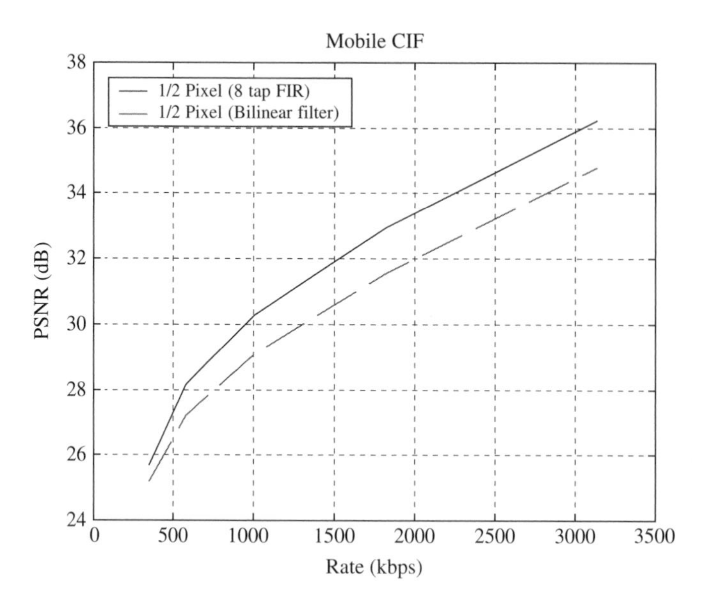

FIGURE 11.10

Coding performance with 8-tap FIR and bilinear filters.

11.2.3.6 Comparison of Integer-Pixel, 1/2 Pixel, and 1/4 Pixel Accuracy

Figure 11.11 shows operational rate-distortion curves of MC-EZBC of *Mobile* with different motion accuracies. The eighth pixel accurate case is omitted but it would lie just slightly above the 1/4 pixel accurate case. We can see that the MCTF coding can benefit from quite an accurate motion compensation, with a gain of 4 dB occurring at 2.5 Mbps for this CIF test sequence at 30 fps.

11.3 NEW HAAR MCTF

We have already seen that because of the covered and uncovered pixels, for the temporal high-frequency frame, it is advantageous to use frames on either side as an additional references, and we called this bidirectional MCTF. Now, when Ohm introduced the idea of bidirectional MCTF in [18], the temporal high subband was in the position of the first frame. However, he did not include a test to find the best matched blocks for the temporal filtering as done in [15]. Here, we return to Ohm's original order for the temporal low and high frames for bidirectional MCTF while retaining the detection test for connected/unconnected of [15], thus avoiding the need to process each GOP in reverse order. An illustration of this new MCTF diagram is shown as Fig. 11.12 below.

We classify the motion blocks into three categories or modes in this new Haar MC-EZBC. With reference to Fig. 11.13 below, they are DEFAULT for connected blocks,

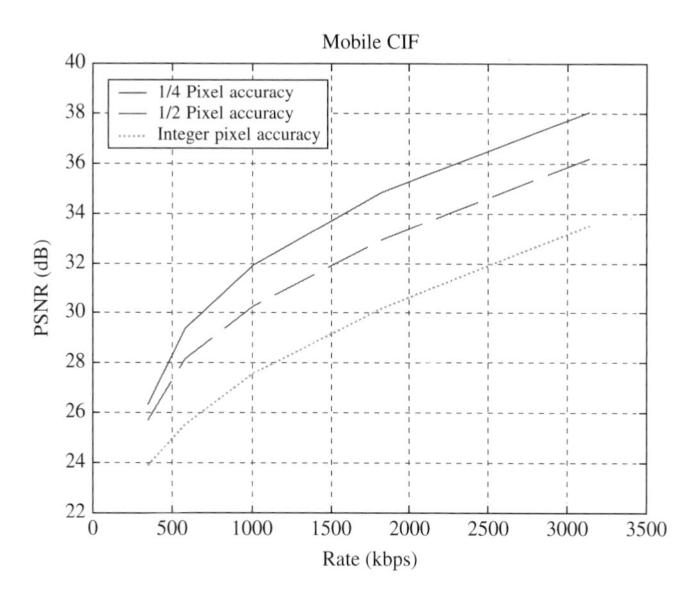

FIGURE 11.11

PSNR variation versus subpixel accuracy.

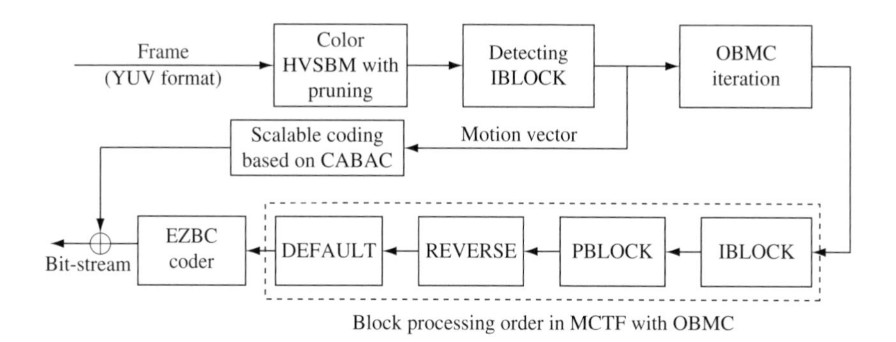

FIGURE 11.12

A detailed diagram of new Haar MC-EZBC coder (© 2007 IEEE).

where we do the lifting predict step in A^t and the update step in B^t. REVERSE blocks with motion vector between A^t and B^{t-1}, where we do MCP from a reference block in B^{t-1}. The PBLOCK has a motion vector between A^tandB^t, where we do a MCP step.

We look for IBLOCK candidates from the set of PBLOCKs and REVERSE blocks, and performs an adaptive directional spatial prediction, called *directional* IBLOCKs [25] similarly to H.264/AVC [6]. An interesting alternative approach is presented in [14],

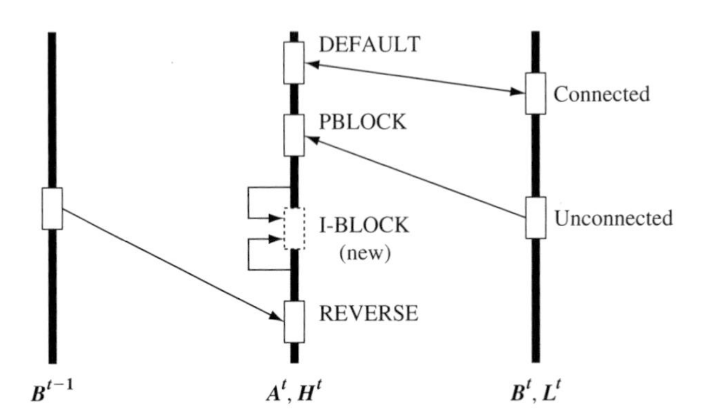

FIGURE 11.13

Bidirectional MCTF with temporal high frame first.

where the authors use Lagrange optimization to jointly determine motion vector block size and the mode classification. They only use VSBM with no resolution hierarchy for motion estimation though.

In the next section, we go on to describe the blocks in Fig. 11.12 in some detail.

11.3.1 Overlapped Block Motion Compensation

In MC-EZBC, after IBLOCK detection [15], we apply overlapped block motion compensation (OBMC) to the variable size blocks as shown in Fig. 11.3. In our OBMC framework, we view data received by the decoder prior to the decoding of frame k as a source of information about the true prediction scheme finally used. For simplicity, we limit the relevant information for each block to be the two horizontal, two vertical nearest neighbors and itself [26], and we assume stationarity of the image and block motion field.

We use a modified 2D bilinear window, whose performance is only a little worse than the iterated optimal window design presented by Orchard and Sullivan in [27]. Each block size has its corresponding weighting window. Since a large block probably has a different motion vector from its small neighbor, a *shrinking* scheme is introduced between the different sized blocks to reduce smoothing at a motion discontinuity. We do OBMC for all the types of blocks classified above, but we use a different weighting window for IBLOCKs to emphasize the continuity between motion compensation and spatial interpolation/prediction.

To further optimize OBMC, we use an iterative procedure and minimize the MAD distortion. Since bidirectional color HVSBM runs on both luma and chroma data, it follows naturally that the OBMC iterations should be applied to YUV simultaneously. We know U and V are subsampled frame data after the color space transform from RGB data, so the weighting windows used for U and V are also subsampled versions of those used for Y [28].

11.3.1.1 Incorporation into MCTF

As shown in Fig. 11.11, we do OBMC in a lifting implementation for DEFAULT blocks, that is, with the prediction and update steps in normal order to reduce the noise in the area of good motion. The specific equations for OBMC in our lifting implementation are as follows [28],

$$H(m,n) = \frac{1}{\sqrt{2}}B(m,n) - \frac{1}{\sqrt{2}}\sum_k h_k(m,nt)\tilde{A}\left(m - d_{mk}, n - d_{nk}\right), \tag{11.11}$$

$$L\left(m - \overline{d}_m, n - \overline{d}_n\right) = \tilde{H}\left(m - \overline{d}_m + d_m, n - \overline{d}_n + d_n\right) + \sqrt{2}A\left(m - \overline{d}_m, n - \overline{d}_n\right). \tag{11.12}$$

The OBMC approach treats the displacement vector (d_m, d_n) as random, meaning that pixel $B\{m,n\}$in frame B has motion vector (d_{mk}, d_{nk}) with probability $h_k\{m,n\}$ as determined by its corresponding probability weighting window. So, the prediction is the weighted average of the self and nearest neighbor predictions. As above, in these equations, $(\overline{d}_m, \overline{d}_n)$ is the nearest integer to (d_m, d_n). Although the form of the low temporal frame (11.12) seems to be the same as that without OBMC, actually OBMC affects both high and low temporal frames. To see this, note that in this lifting implementation, the H value in (11.11) is shifted and interpolated and then becomes the $\tilde{H}$ value in (11.12). Also, importantly, the low temporal frames from OBMC are visually preferred and more suitable for further stages of MCTF (i.e., much more nearly block-artifact free).

For PBLOCK, there is only a prediction step (11.1) from a future frame, and for REVERSE block, there is only a prediction step (11.1) from a previous frame. The IBLOCK is predicted with the weighted average of spatial prediction/interpolation from spatial neighbors in the corresponding frame.

11.3.2 Scalable Motion Vector Coding

After careful arrangement of the bit stream for motion vectors, namely grouping the motion vectors in each temporal layer with the frame data in that level, temporal scalability of motion vectors follows naturally. But the bit stream for motion vectors was still not scalable with respect to SNR and resolution. Here, we describe a scalable motion vector coder based on CABAC [22]. A layered structure for motion vector coding and alphabet general partition (AGP) of motion vector symbols are used for SNR and resolution scalability of the motion vector bit stream [29, 30]. With these two features and the careful arrangement of the motion vector bit stream in MC-EZBC, we can have a layered form of temporal, SNR, and resolution scalability for motion vectors, and we can improve significantly both visual and objective results for low bit rates and low resolution with slight PSNR loss (approximately 0.05 dB) and unnoticeable visual loss at high bit rates. First, results on scalable motion vector coding for MC-EZBC were obtained in [31].

11.3.2.1 Scan/spatial Prediction for Motion Vectors

In Fig. 11.14, we show the motion vector mv in the current block being predicted from its three nearest previously processed spatial neighbors mv1, mv2, and mv3 similar to

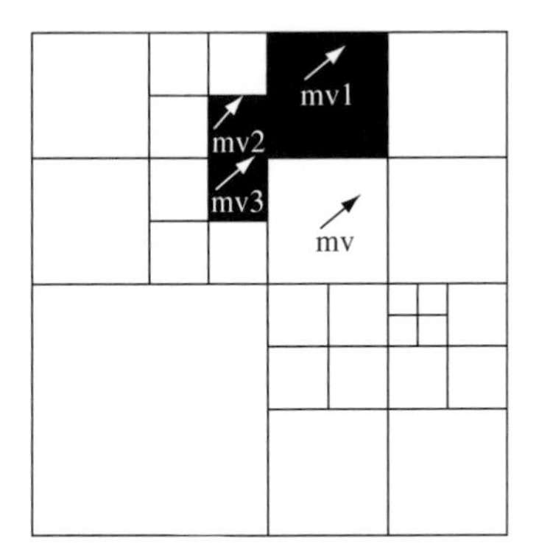

FIGURE 11.14

Spatial prediction from three direct neighbors of target mv block.

the approach in [2]. This spatial prediction scheme can predict the motion vector quite efficiently. The motion vectors in DEFAULT and PBLOCK are between current and next frames (defined as *normal* motion vectors) and those in REVERSE blocks are between current and previous frames (defined as *reverse* motion vectors).

We found experimentally that the characteristics of normal motion (forward) vectors and reverse ones are quite different. So, we predict and code the two sets of motion vectors separately to improve the prediction. Then, there are two loops for motion vector coding. The first loop is for normal motion vector coding, and the second loop is for reverse motion vector coding.

For the motion vector mv in a target block, if there is at least one motion vector in mv1, mv2, and mv3 with the same type (normal or reverse) as that in this block, we predict from the spatial neighbors. If there is no motion vector with the same type in these three neighbors, we predict mv in the target block from the previous motion vector with the same type in quadtree scan order. We consistently use this combined spatial and scan-order prediction, and the prediction residual is then coded by CABAC.

11.3.2.2 Alphabet General Partition of Motion Vector Symbols

In MC-EZBC, motion estimation is typically done at 1/8 pixel accuracy. Although they can reduce mean squared error (MSE) after motion compensation, the quarter and eighth pixel bits of motion vectors are quite random due to the camera and quantizer noises. We can therefore model the motion vectors as follows (for simplicity we use one dimensional notation),

$$r_k = s_k + n_k.$$

Here, r_k is the estimated kth motion vector, s_k is the true kth motion vector, and n^k is the noise vector due to the inaccuracies in the frame data. All the components are 1/8 pixel accuracy. Since the noise in frame data is quite small, we believe it can contaminate only the quarter pixel and eighth pixel bits of the estimated motion vector. Thus, we divide the estimated motion vector r_k into two parts,

$$r_k = r_{k1} + r_{k2},$$

where $r_{k1}$is the *major symbol* at 1/2 pixel accuracy, and $r_{k2}$is the *subsymbol* for quarter and eighth pixel accuracy. For example, if the estimated motion vector $r_k = -1.625$, then $r_{k1} = -1.5$ and $r_{k2} = -0.125$. We predict the major symbol with the above scheme and code the prediction residual with CABAC but code the subsymbol directly as a binary number.

Actually, we do not need to code the sign for subsymbols because we can find the sign from the major symbol. The only exception is for those motion vectors in the range $[-0.375, +0.375]$, for which the major symbol $r_{k1} = 0$. Then we don't know the sign for subsymbol r_{k2}, so we need one additional bit to indicate whether the sub-symbol is positive or negative in this case.

At high bit rates and full resolution, we transmit all the three parts of the motion vector bit stream. At the decoder, we use the lossless motion vectors to reconstruct the frames. But at low bit rates, we can discard the subsymbol and sign parts at the encoder and use more bits for coding the temporal subband frames. Although the motion vectors are lossy coded, this can be compensated by the increased accuracy of the temporal subband frame data, and thus, the total performance can be improved. Also at low resolution, since in MC-EZBC the motion vectors will be scaled down, obviously we do not need the same accuracy for motion vectors as in the full resolution case. So the subsymbol and sign parts can also be discarded, making more room for the SWT coefficients data.

11.3.2.3 Layered Structure of Motion Vector Coding

After one spatial resolution level down, the motion vector block size is halved, that is, $16 \times 16, 8 \times 8$, and 4×4 blocks become $8 \times 8, 4 \times 4$ and 2×2 blocks, respectively. Thus, after one or two spatial levels down, we do not need the same number of motion vectors as at full resolution. Moreover, the motion vectors are also scaled down by a factor of 2 after each spatial level downwards since if two adjacent motion vectors differ by less than two pixels at full resolution, then at half resolution they will differ by less than one pixel.

Assume four grouped blocks (children) in the quadtree structure have similar motion vectors. We can replace the four motion vectors by one representative at one spatial level down. Currently, we just choose the first motion vector in quadtree scan order as the representative for the four motion vectors among the children.

As mentioned above, the normal and the reverse motion vectors have different characteristics. So, we reserve up to two representatives for the four children, one for normal motion vectors and another for reverse motion vectors in the four child blocks.

We use a subsample selection scheme from bottom-up to form the layered structure for motion vector coding. In each layer, we use the scan/spatial prediction as stated above.

We also continue using the context model from layer to layer. Since we simply choose the first normal and reverse motion vector in quadtree scan as the representative, the coded number of motion vectors is the same as in nonlayered coding. Then, we code the prediction residuals in larger blocks in the base layer of the motion vectors, with successively smaller blocks as the enhancement layers. When coding enhancement layers, one should use all the information up to that layer, that is, updated motion field and updated context models from previous layers.

This coarse layer structure for motion vector coding shows good results [29, 30]. As an alternative, the local gradient of the compensated frame data has been used with success by Secker and Taubman [32]. If an image frame area is very smooth, then the difference among the motion vectors will not affect the reconstructed frame's distortion very much. However, if the area is full of texture, then a little difference among the motion vectors may cause a large distortion. In [32], a local average of the energy in the gradient is used to modulate the portion of the total bit assignment going to the motion vectors, determined over a triangular mesh, which are also coded with a lossless SWT coder.

11.4 EZBC CODER

After MCTF analysis with OBMC, a set of high temporal frames and one low temporal frame per GOP, as shown in Fig. 11.2, will be spatially analyzed by 2D SWT and then input to the EZBC coder. The entropy coder EZBC then codes the spatially transformed frame data in a scalable manner with respect to SNR, frame rate, and spatial resolution by interleaving the bit streams of the spatial and temporal subbands. Please refer to [1] for the details about the EZBC coder for fine scalable video compression. Unlike JPEG 2000 and EBCOT [33] on which it was based, there is no rate-distortion optimized bit assignment over code blocks, as the entire subbands are coded together. This results in a complexity advantage for EZBC but may require more high-speed storage to accomplish.

EZBC [34] is an embedded subband/wavelet image coding algorithm that jointly exploits context modeling and set partitioning for entropy coding of SWT coefficients. The extensive experimental results in [35] show that the EZBC coder significantly outperforms the well-known set-partitioning coders SPIHT and SPECK [36], and also slightly improves upon JPEG 2000 and EBCOT [33] in PSNR performance in the coding of natural image data. Such performance gains are achieved without rate-distortion optimized bit allocation among codeblocks (there are no codeblocks!) and multipass scanning/coding of individual bit planes as employed by JPEG 2000 and EBCOT, and are generally considered computationally expensive operations. In addition, the results in [35] further show that the EZBC coder based on the set partitioning approach can represent the bit-plane samples in groups and thus has much fewer binary symbols to encode in comparison to a conventional pixel-wise bit-plane coder such as JPEG2000, leading to another saving in computational cost.

11.4.1 Coding Process

The EZBC coding process begins with establishment of quadtree representations for the individual subbands and frames. The bottom quadtree level, or pixel level, consists of the magnitude of each subband coefficient. Each quadtree node of the next higher level is then set to the maximum value of its four corresponding nodes at the current level. By recursively grouping each 2×2 vector this way, the number of pixels covered by one quadtree node exponentially grows with its level number. In the end, the top quadtree node just corresponds to the maximum magnitude of all the coefficients from the same subband.

Similarly to conventional embedded image coders, we progressively encode the bit planes of subband coefficients from the most significant bit (MSB) toward the least significant bit. Such a bit-plane coding method is equivalent to successive approximation quantization with threshold 2^n for coefficient bit plane n. Whenever a node tests *significant* (to be defined later) against the current threshold, it is split into four *descendent* nodes. This testing and splitting procedure is recursively performed until the bottom (pixel) level. Once a pixel first tests significant, its sign bit is coded. Each bit-plane coding pass is finished with a bit-plane refinement subpass which further refines the significant subband coefficients from the previous bit-plane passes. In this way, we can send data in the order of their importance.

Like SPIHT and other hierarchical bit-plane coders, EZBC uses lists for tracking the set-partitioning information. Two arrays of lists, LIN and LSP, are defined for each subband. A node is added to the end of LIN once it tests significant. LSP contains a full list of all significant pixels. Whereas the lists in EZBC are separately maintained for nodes from different subbands and quadtree levels, former algorithms such as SPIHT typically have each list used to group SWT coefficients from all subband regions. This modification is necessary to provide a resolution scalable code stream. The context statistics can also be more effectively exploited in this way, as to be detailed in the next subsection.

The complete coding algorithm is outlined as follows.

11.4.1.1 Definitions

- $c(i,j,t)$: integer part of subband/wavelet coefficients at position (i,j,t) after proper scaling.
- $\mathrm{QT}_k[l](i,j,t)$: quadtree representation of SWT coefficients for subband k, at quadtree level l, and spatiotemporal position (i,j,t)

$$\mathrm{QT_k}[0](i,j,t) \stackrel{\Delta}{=} |c_k(i,j,t)|$$

$$\begin{aligned}\mathrm{QT_k}[l](i,j,t) \stackrel{\Delta}{=} \max\{&\mathrm{QT_k}[l-1](2i,2j,t), \mathrm{QT_k}[l-1](2i,2j+1,t),\\ &\mathrm{QT_k}[l-1](2i+1,2j,t), \mathrm{QT_k}[l-1](2i+1,2j+1,t)\}\end{aligned}$$

- $m(i,j,t)$: MSB of quadtree node (i,j,t).

- D_k: depth of the quadtree of subband k
- QTR_k: collection of all quadtree roots of subband k
- D_{max}: $\max_k\{D_k\}$
- K: total number of subbands
- n: index of the current bit-plane pass, corresponding to quantization threshold 2^n
- $S_n(i,j,t)$: significance of quadtree node (i,j,t) against threshold 2^n,

$$S_n(i,j,t) \equiv \begin{Bmatrix} 1, & n \le m(i,j,t), \\ 0, & \text{otherwise.} \end{Bmatrix}$$

  Node (or pixel or coeff.) is significant if $S_n(i,j,t) = 1$.
- $\text{LIN}_k[l]$: list of insignificant nodes from quadtree level l of subband k
- LSP_k: list of significant pixels from subband k
- $\text{CodeLIN}(k,l)$: function for processing insignificant nodes in $\text{LIN}_k[l]$
- $\text{CodeLSP}(k)$: function for coefficient refinement
- $\text{CodeDescendentNodes}(k,l,i,j,t)$: function for coding significance of all descendent nodes of $\text{QT}_k[l](i,j,t)$, which just tested significant against the current threshold.

11.4.1.2 Coding Steps

Initialization

$$\text{LIN}_k[l] = \begin{Bmatrix} \text{QTR}_k, & l = D_k \\ \phi, & \text{otherwise,} \end{Bmatrix}$$

$$\text{LSP}_k = \phi,$$

$$n = \left(\log_2\left\{\max_{(k,i,j,t)}\left[|c_k(i,j,t)|\right]\right\}\right).$$

step 1 : for $l = 0 : D_{max}$

for $k = 0 : K - 1$

codeLIN(k,l)

end

end

for $k = 0 : K - 1$

codeLSP(k)

end

$n \to n - 1$ and return to step 1, stopping when target bit rate is reached.

11.4.1.3 Pseudo-Code

```
CodeLIN(k, l) {
for each (i, j, t) in LIN_k[l]
        code S_n(i, j, t)
        if (S_n(i, j, t) = 0, (i, j, t) remains in LIN_k[l]
        else
            if (l = 0), then output sign bit of c_k(i, j, t) and add (i, j, t, l) to LSP_k
            else CodeDescendentNodes (i, j, t, l)
    }
CodeLSP {
    for each pixel (i, j, t) in LSP_k, code bit n of |c_k(i, j, t)|
    }
CodeDescendentNodes (k, l, i, j, t) {
    for each node (x, y, t) in {(2i, 2j, t), (2i, 2j + 1, t), (2i + 1, 2j, t),
                                                (2i + 1, 2j + 1, t)} of
    quadtree level l − 1 and subband k
        output S_n(x, y, t)
        if (S_n(x, y, t) = 0), add (x, y, t, l − 1) to LIN_k[l − 1]
        else
            if (l = 1) output the sign bit of c_k(x, y, t) and add (x, y, t) to LSP_k
            else CodeDescendentNodes (k, l − 1, x, y, t)
}
```

11.4.2 Context Modeling

In contrast to the conventional pixel-wise bit-plane coders, EZBC deals with nodes from individual quadtree levels. Four types of binary symbols are coded in EZBC: (1) significance of nodes from LIS (in routine CodeLIN); (2) significance of descendants (in routine CodeDescendentNodes); (3) sign of significant coefficients; and (4) coefficient refinement. Unlike most other set-partition coders, the lists in EZBC are separately maintained for individual subbands and quadtree levels. Therefore, the separate context models are allowed to be built up for the nodes from different subbands and quadtree levels. In this way, distinctive statistical characteristics of quadtree nodes from different orientations, subsampling factors, and amplitude distributions are not be mixed. However, in the actual 3D-EZBC implementation, the same set of context models is shared among quadtree level 2 and higher to reduce the actual model cost.

To code the significance of the quadtree nodes, we include eight neighbor nodes of the *same* quadtree level in the spatial context, as illustrated in Fig. 11.15. This spatial context has been widely adopted for coding of the significance of SWT samples/coefficients. However, the information across scale is given by the node of the parent subband at the *next lower* quadtree level, as shown in the same figure. This choice is based on the fact that at the same quadtree level, the dimension of the region a node corresponds to in the input image is halved in the parent subband as a result of subsampling at the transform stage.

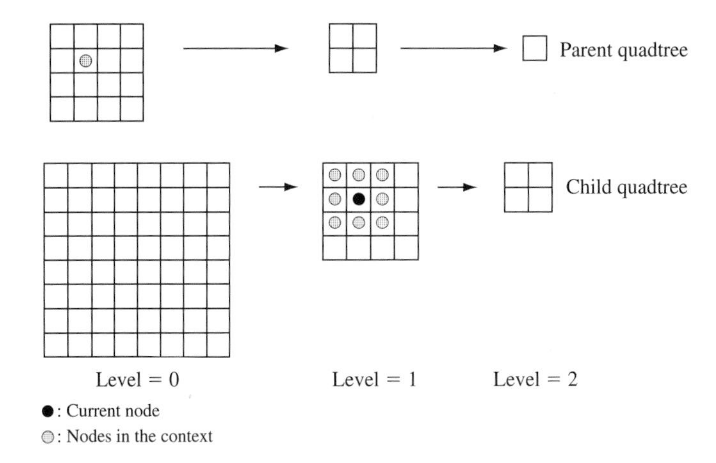

FIGURE 11.15

Illustration of the context models of a quadtree node.

The model selection of the arithmetic coding is just based on a 9-bit string with each bit indicating the significant status of nodes in this context. To lower model cost, instead of including all context states (2^9 totally), we adopted a similar method to EBCOT [33] for context reduction. The sign coding scheme of EBCOT is also used in this algorithm.

We should also mention that a tree-structured enhancement to the EZBC context model has been published in [37].

11.4.3 Scalability

The layout of the bit stream for a GOP is shown in Fig. 11.16(a). The 3D SWT coefficients are coded from the MSB. The quantizer step size is halved after each bit coding pass. The quality (or the quantizer step size) of the video can be controlled by just dropping the corresponding sub-bit-streams for the remaining bits of subband coefficients. The bit stream is therefore scalable in SNR. Since the bit stream itself is embedded, it can be truncated at any point to meet desired video quality or rate constraint. Hence, it is also scalable in coding rate.

In MC-EZBC, context models and lists are separately established for individual subbands. Any subband can be separately decoded from its corresponding sub-bit-stream once its *dependent subband* has been processed up to the current bit coding pass. This restriction in the decoding order is due to the fact that the inter-subband models are used for context-based arithmetic coding. The resolution and frame-rate scalabilities can be achieved by discarding the sub-bit-streams for irrelevant spatial/temporal subbands. For example, the video can be reconstructed at half resolution and at half frame rate by throwing away sub-bit-streams for s-LH, s-HL, s-HH, and t-H subbands. For ease of extraction of sub-bit-streams without transcoding, the sub-bit-streams for individual subbands are packetized as illustrated in Fig. 11.16(b). The actual decoding/transmission

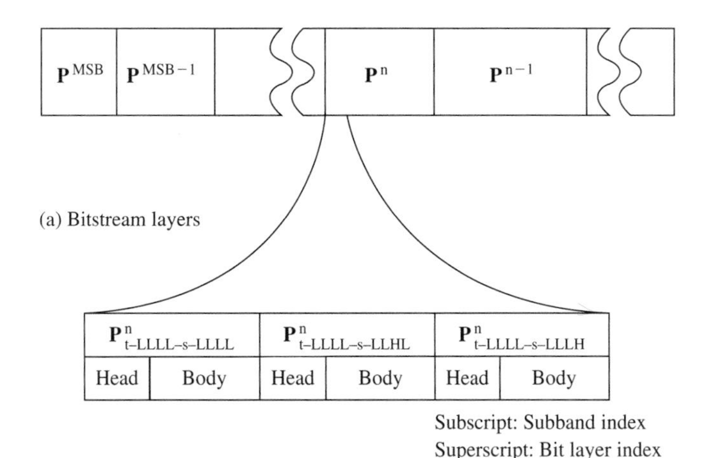

FIGURE 11.16

Coded bit stream structure for a GOP in EZBC.

order for individual subbands is not specified. We may, for instance, transmit packets for all spatial low-frequency subbands first.

11.4.4 Packetization

As the last step of MC-EZBC, packetization realizes quality control. In our coding system with GOP structure, we try to realize constant quality both inside a GOP and across the many GOPs of the video clip. Since we use a near orthonormal transformation both in the spatial and temporal domains, corresponding bit planes of different subbands have the same importance. In bit-plane scanning, we proposed to interleave spatial subbands of all temporal subband frames of a GOP and interleave their coding passes further [17] to realize constant quality inside a GOP.

To realize constant quality across GOPs, we proposed a two-step coding scheme for the long-term constant quality problem [15, 17]. In this scheme, we first encode the entire sequence and stop the bit-plane encoding at some bit plane, which will ensure the needed quality. Effectively, we have just created an archive. Our strategy of bit allocation over GOPs is to make the bit-plane coding of all the GOPs stop at the same bit plane and realize approximately constant quality. While this type of constant quality does not optimize the total average performance, it comes close and may well provide the visually better performance.

11.4.5 Frequency Roll-Off

The extracted low-resolution video from a motion-compensated 3D subband/wavelet scalable video coder (SVC) is unnecessarily sharp and sometimes contains significant aliasing, compared to that obtained by the MPEG4 low-pass filter. Frequency roll-off is a content adaptive method for aliasing reduction in subband/wavelet scalable image and video coding. We try to make the low-resolution frame (LL subband) visually and energy-wise similar to that of the MPEG4 decimation filter through frequency roll-off. Scaling of the subbands is introduced to make the variances of the subbands comparable in these two cases. Thanks to the embedded properties of the EZBC coder, we can achieve the needed scaling of energies in each subband by sub-bit-plane shift in the extractor and value (coefficient) scaling in the decoder. Two methods are presented in [38], which offer substantial PSNR gain for lower spatial resolution, as well as substantial reduction in visible aliasing, with little or no reduction in full-resolution PSNR or visual quality.

Referring to Fig. 11.17 below, we can expect that the MPEG4 low-pass separable spatial filter almost completely avoids aliasing in its corresponding low-pass video frame at the expense of a softer image, which by the way should be easier to code. To provide a similar image softness in the image frames created by CDF 9/7 will require us to further low-pass filter the resulting LL subbands. This in turn can be accomplished by further SWT decomposition of this subband and then scaling down the resulting components. If this scaling is done with powers of two, then the scaling can be hidden in the ensuing EZBC bit-plane-embedded coding.

Figure. 11.18 below shows a coding result for the 4CIF test clip City as coded by MC-EZBC. The displayed images in Fig. 11.18 are without frequency roll-off (left) and with frequency roll-off (right). Both are one level down in resolution CIF, at half-frame-rate (15 fps), and coded at 512 kbps. Details of the method along with more experimental results are contained in [38].

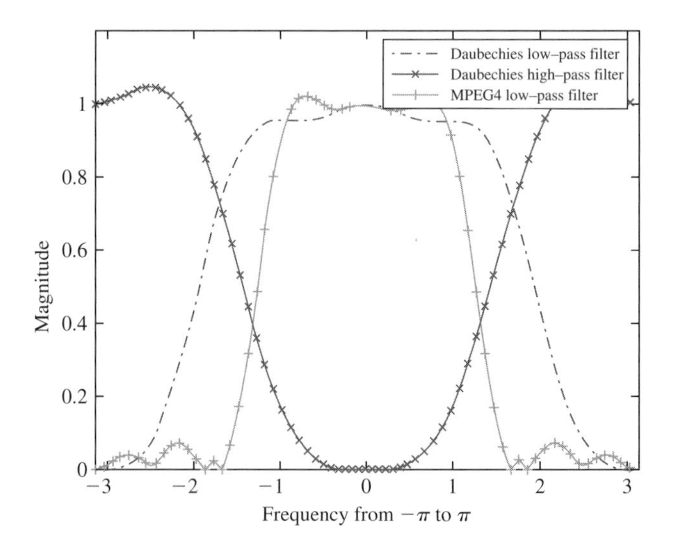

FIGURE 11.17

Frequency responses of MPEG4 and CDF 9/7 filters (© 2008 IEEE).

FIGURE 11.18

Illustration of frequency roll-off: (a) without roll-off, (b) with roll-off.

11.5 EXTENSION TO LEGALL AND TABATABAI 5/3 FILTERING

In the above development, we explicitly considered only the use of the 2-tap Haar filters for the MCTF. Actually in the presence of strong motion with good motion estimates, it can be advantageous to use longer filters. Here, we consider the use of the LeGall and Tabatabai (LGT) 5/3 filters for this purpose. Early results were obtained in [2], where the need for a sliding window approach was realized on account of the overlap of the transform windows for these longer filters. However, PSNR improvement could only be obtained at the higher bit rates because of a higher motion vector load of these longer filters.

To allow for cases where the motion information is not so good, we consider an adaptive 5/3 MCTF that can also perform Haar filtering when appropriate, as shown in Fig. 11.18. This adaptive 5/3 MCTF was first presented in [3, 4]. As seen earlier in (11.4) and (11.5), the Haar MCTF only requires one motion vector for warping motion blocks in reference frames A predict blocks in target frames B, but the 5/3 MCTF involves prediction from two reference frames, both just before and just after target frames, the three frames being denoted as A_{t-1}, B_t, and A $_{t+1}$ in the figure. Since two motion vectors are involved, the overhead of transmitting these two vectors could be a problem. Hence, the ability of the adaptive MCTF to perform Haar filtering when appropriate, for example, object occlusion or scene change (refer frame B_{t+2} in Fig. 11.19). Thus, (11.4) in Section 11.2.3 is generalized in this way to permit two or one prediction as required. Note that the adaptive 5/3 MCTF framework also subsumes the bidirectional Haar MCTF as presented in Section 11.6.

We complete the 5/3 MCTF with the generalization of update (5) as shown in the bottom part of Fig. 11.19 above, with the required warping of H being denoted as inverse motion compensation (IMC). Again the homogeneous motion assumption can be used

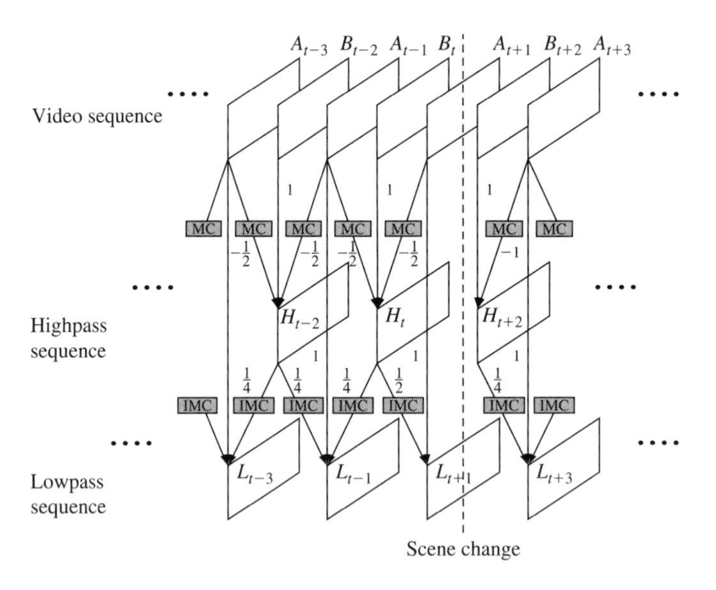

FIGURE 11.19

Illustration of Adaptive 5/3 MCTF [39].

TABLE 11.1 Block modes in adaptive 5/3 MCTF [39].

Modes	Block Mode	Reference Frame		Update Step	Motion Vectors
		Left	Right		
1	biconnected	Y	Y	Y	2
2	left-connected	Y	N	Y	1
3	right-connected	N	Y	Y	1
4	bipredicted	Y	Y	N	2
5	left-predicted	Y	N	N	1
6	right-predicted	N	Y	N	1
7	parallel	Y	Y	Y	1
8	spatial-direct	Y	Y	Y	0

here, thus avoiding the need for a third motion estimation. At the decoder, the inverse 5/3 adaptive MCTF can be easily accomplished by simply undoing the above operations, starting with H, similarly to (11.6) and (11.7) in the Haar case.

In [3, 4], the authors also introduced a set of block modes for the adaptive 5/3 MCTF as shown in Table 11.1 above.

Again, rate distortion (prediction error norm) is used to find the best mode choice for a given block, which is then coded and transmitted as a small amount of overhead. With reference to this table, we see that modes 1 and 7 are LGT 5/3 modes, while modes 2 and 3 correspond to Haar filtering, and modes 4–6 handle various types of unconnected pixels, such as PBLOCKs earlier. Mode 8 is essentially what we have called IBLOCK earlier. For more detailed information, please consult [39]. One should note that the adaptive 5/3 MCTF is able to achieve modes similar to the bidirectional Haar MCTF but without the need to use MCP that leads to increased coding error.

The most recent version of our coder *enhanced* MC-EZBC [39] is a combination of the tools presented earlier for the Haar case, that is, OBMC, HVSBM, scalable motion coding, and frequency roll-off integrated into the adaptive 5/3 MCTF from [3, 4]. In the next section, we present results from enhanced MC-EZBC in comparison with some other scalable coders from the literature and an early version of the new MPEG SVC mentioned earlier.

11.6 OBJECTIVE AND VISUAL COMPARISONS

Since MCTF coders, and in fact all scalable coders, produce the low-resolution and low-frame-rate video, they in effect create their own *references*[4]. Since each scalable coder is expected to generate different references, there is the problem of doing an objective (PSNR say) comparison. This is a new problem that arises in comparing scalable coders. It has not arisen in the past since the references for scalable hybrid coders were always generated in the same way, that is, spatially standard downsampling filters, and temporally just frame skipping.

The new SVC from MPEG is an extension of H.264/AVC that makes use of hierarchical coding and constructs its low-resolution frames using a standard MPEG 4 low-pass filter and constructs its low-frame-rate versions by frame skipping. Here, we make a comparison with an early version of the SVC coder (JSVM 1.0) and the enhanced version of MC-EZBC above. In this comparison, there were five so-called testing points at various bit rates, resolutions, and frame rates. In Fig. 11.20 from [39], we show Y-PSNRs for the well-known *Bus* sequence at testing points 1: 64 kbps 15 fps QCIF, 2: 96 kbps 15 fps QCIF, 3: 192 kpbs 15 fps QCIF, 4: 384 kbps 15 fps CIF, 5: 512 kbps 30 fps CIF.

Another comparison from [39] in Fig. 11.21 below plots Y-PSNR of enhanced MC-EZBC versus two other coders in the literature. The coder EBCOT-SMV is taken from [32] and used three levels of temporal analysis. Performing better in terms of luma PSNR were MC-EZBC (Haar version) with both 3- and 4-level temporal analysis. Top performing were enhanced MC-EZBC and the codec of [14] referred to here as MSRA-codec.

11.6.1 Some Visual Results

Here, we show some visual results for MC-EZBC coding of the *Foreman* clip in CIF resolution at frame rates of 30 fps and 15 fps, with and without OBMC and IBLOCKs.

[4]The reference for this purpose is the ideal frame to be used for calculating MSE and PSNR. It must be created (by the scalable coder). It is a different meaning from that used when discussing motion compensation.

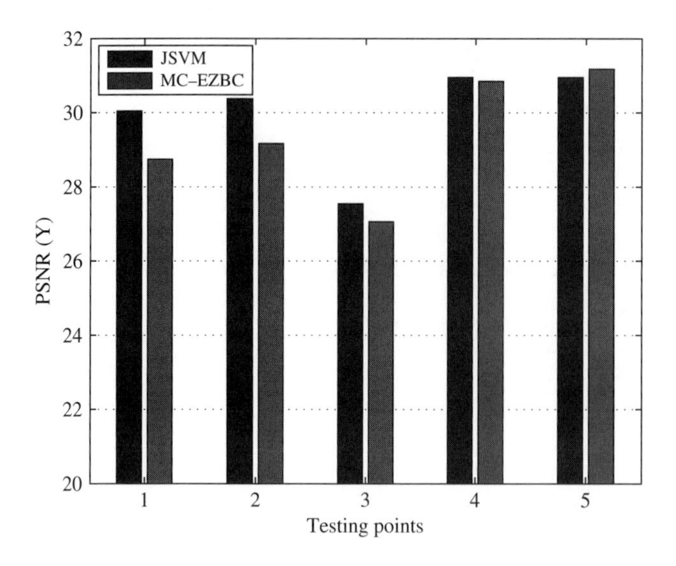

FIGURE 11.20

Enhanced MC-EZBC versus SVC JSVM 1.0 at various test points for CIF sequence *Bus* [41].

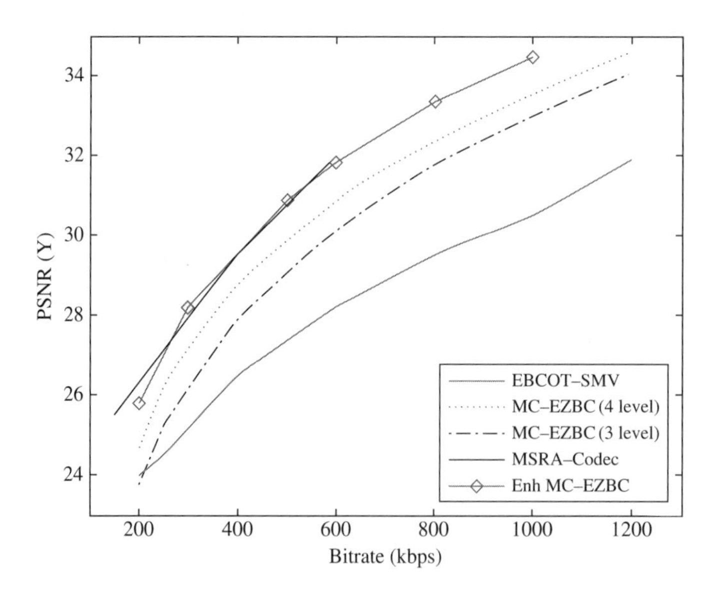

FIGURE 11.21

Y-PSNR comparisons for test clip *Bus* at full frame rate (30 fps) and full (CIF) resolution (© 2007 IEEE).

FIGURE 11.22

(a) Frame 141 in *Foreman* with OBMC, IBLOCK, and color HVSBM at bit rate 512 kbps at 30 fps. (b) Frame 141 in *Foreman* without OBMC or IBLOCK, but with color HVSBM at bit rate 512 kbps at 30 fps.

FIGURE 11.23

(a) Frame 43 in *Foreman* with OBMC, IBLOCK, and color HVSBM (corresponding to original frame 86) at rate 256 kbps at 15 fps, half the original frame rate. (b) Frame 43 in *Foreman* without OBMC and IBLOCK, but with color HVSBM (corresponding to original frame 86) at rate 256 kbps at 15 fps, half the original frame rate.

Visual improvement can be significant, as seen in the Fig. 11.22(a) and (b), which show comparisons for full frame rate, with OBMC (Fig. 11.22(a)) and without OBMC (Fig. 11.22(b)). We see a clear decrease in blocking artifacts. The results after one adaptation to half-frame-rate are shown in Fig. 11.23(a) and (b).

11.7 MULTIPLE ADAPTATIONS

So far we have presented scalable coding as a method for compressing the video once at the source and placing it onto a video server, from which various qualities, bit rates, spatial resolutions, and frame rates can be extracted. The individual spatiotemporal subbands are stored in embedded form and the context modeling is based on lower resolution data. A useful abstraction of the packetized data is the 3D data structure shown in Fig. 11.24 (borrowed with permission from [40]), where each packet is termed an *atom*. The three dimensions are called tiers and in our case would correspond to bit rate, resolution, and frame rate. So, this is the data structure present at the video server but in this figure quantized down to a small number of atoms for display purposes. Because of the context dependence of the arithmetic coding, there is a dependency here, with each atom depending on its lower resolution atoms. We can scan the cube in various directions, always in increasing order in each dimension to satisfy the context dependencies, to perform the required scalabilities, as also indicated with a different notation in Section 11.4.3. This is the so-called "pull function" of scalable coders, which creates the actual (one-dimensional) bit stream to be sent out over the channel or network link.

In a network application though, we may have to *adapt* the data further inside the network by adapt we mean reduce one or more of the dimensions: bit rate, spatial resolution, and frame rate. Adaptation may be required multiple times inside a network due to the use of multicast and broadcast to serve a number of users, with their heterogeneous set of displays and compute capability. Another motivation is the need to respond to congestion and packet loss inside the network. The real practical case would combine these

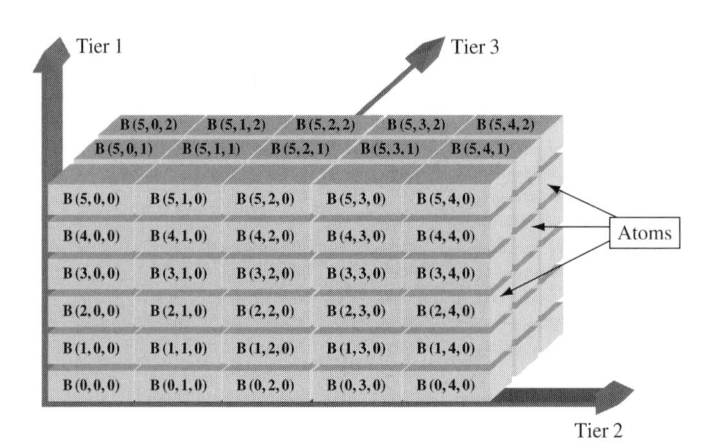

FIGURE 11.24

3D scalable data structure [30].

two. For a nonscalable coder, one would have to adapt the data by using transcoding, with its attendant computational load and quality loss. A transcoding solution would also require the network node processors to understand the coding algorithm used. In contrast, a scalable coder merely sheds unneeded atoms, making data adaptation much easier. To do this, some additional small amount of information has to be packetized and transmitted with the coded data to indicate the locations in the bit stream, where the various spatiotemporal subbands are located (see Fig. 11.16(b)). Such adaptations would not be performed in the network itself but on a content delivery network (CDN) superimposed on the underlying network.

Below we give some illustrative results using MC-EZBC with a $5 \times 6 \times 7$ scalability structure, meaning five temporal or frame-rate layers, six spatial resolution layers, and seven bit rates or quality layers, all rate controlled on a GOP basis. Each GOP will then have a header that contains information as to how many temporal and spatial layers are included. Further, each spatiotemporal layer is proceeded by a length field. When an adaptation is made on a GOP, not only should appropriate parts of the bit stream be removed but also the information in the header and length fields must be updated appropriately to enable correct decoding. Since we downsize the description file to seven quality layers to save on bits for the resource description file, at the following adaptations we cannot access the bit streams at the fractional bit-plane level. Figure 11.25(a) shows the PSNR performance after a second adaptation (transcoding) for the *Mobile* clip. We see that the dashed curve (a second adaptation) is nearly coincident with the solid curve (first adaptation). The data in these curves is at the indicated rate points, which are $7 + 1$ in number. Figure 11.25(b) shows a similar performance for the *Coastguard* clip, where slightly more loss is evident but mostly under 0.5 dB. For any fully embedded coder, the loss at the indicated rate points is solely due to the overhead of carrying the additional header information needed for the multiple adaptations. Visually, there is no apparent degradation of either clip after two adaptations. Further adaptations should suffer no further loss since the header information is the same. In fact, further adaptations would truncate the resource description file, deleting unneeded parts, and of course, reducing the slight overhead it causes.

11.8 RELATED CODERS

The MC-EZBC coder performs the temporal transform first followed by the spatial transform and is often denoted "$t + 2D$." Another class of interframe SWT coders, denoted "$2D + t$", performs the spatial transform first, followed by motion compensation of the subbands, often involving a complete to overcomplete SWT. Good examples that form this category are [14, 41, 42]. A summary presentation of these and other interframe/wavelet coders is contained in the special issue [43]. Interested readers can also see the review article [42].

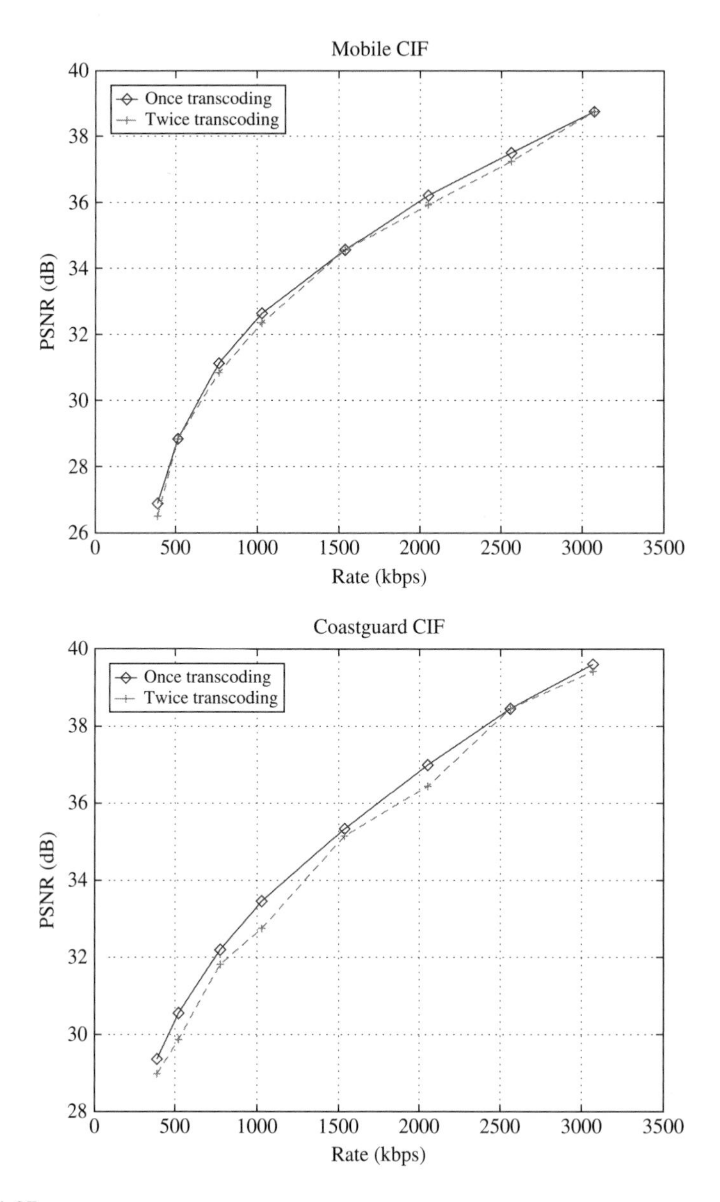

FIGURE 11.25

(a) PSNR performance in multiple transcoding *Mobile*. (b) PSNR performance in multiple transcoding *Coastguard*.

11.9 CONCLUSIONS

In this chapter, we have considered highly SVCs producing embedded bit streams that can be adapted to produce reduced bit rates, resolutions, and frame rates. Of course, the trick is to do this with very little loss in performance over that of a state-of-art coder at each comparison point. The classical hybrid coders are not capable of doing this well due to the coder state implied by their internal temporal DPCM loop. The recent class of MCTF coders has an inherent advantage in this respect and compare very well to the best nonscalable coders in compression performance. Many important design issues were illustrated in the context of the MC-EZBC video coder.

Finally, we should mention a possibility of using the MCTF in such coders to preprocess the data prior to encoding, as well as postprocess the data at decoding. Due to the high accuracy of the MCTF motion compensation, such filtering can be very beneficial to reduce sensor and other noise in the input data, as well as being useful for postprocessing at the receiver.

REFERENCES

[1] S.-T. Hsiang and J. W. Woods. Embedded video coding using invertible motion compensated 3-D subband/wavelet filter bank. *Signal Process. Image Commun.*, 16:705–724, 2001.

[2] A. Golwelkar and J. W. Woods. *Improved Motion Vector Coding for the Sliding Window (SW-)EZBC Video Coder*, JTC1/SC29/WG11, MPEG2003/M10415, 2003, Hawaii, USA. See also A. Golwelkar and J. W. Woods. Motion-compensated Temporal Filtering and Motion Vector Coding using Biorthogonal Filters. *IEEE Trans. Circ. Sys. Video Technol.*, 17(4):417–428, 2007.

[3] M. Wien, T. Rusert, and K. Hanke. IENT/RWTH proposal for scalable video coding technology. JTC1/SC29/WG11/MPEG2004/M10569/S16, 2004, Munich, Germany.

[4] T. Rusert, K. Hanke, and M. Wien. Optimization for locally adaptive MCTF based on 5/3 filtering. *Proc. Picture Coding Symposium* (PCS), volume 6, 2004, San Francisco, CA.

[5] B. Haskell, A. Netravali, and A. Puri. *Digital Video: An Introduction to MPEG-2.* Kluwer Academic Publisher, New York, 1996.

[6] G. J. Sullivan and T. Wiegand. Video compression – from concepts to the H.264/AVC standard. *Proc. of the IEEE,* 93(1):18–31, 2005. See also Chapter 10 of this volume.

[7] S.-T. Hsiang and J. W. Woods. Invertible three-dimensional analysis/synthesis system for video coding with half-pixel-accurate motion compensation. *Proc. Visual Comm. and Image Process. (VCIP)*, SPIE 3653:537–546, 1999.

[8] S.-T. Hsiang, J. W. Woods, and J.-R. Ohm. Invertible temporal subband/wavelet filter banks with half-pixel-accurate motion compensation. *IEEE Trans. Image Process.*, 13:1018–1128, 2004.

[9] B. Pesquet-Popescu and V. Bottreau. Three-dimensional lifting schemes for motion compensated video compression. *Proc. Int. Conf. Accoust, Speech, Signal Process. (ICASSP)*, IEEE, 3:1793–1796, 2001.

[10] L. Luo, J. Li, S. Li, Z. Zhuang, and Y.-Q. Zhang. Motion compensated lifting wavelet and its application in video coding. *Proc. Intl. Conf. Multimedia Expo (ICME)*, IEEE, Tokyo, Japan, 18(2):214–222, 2001.

[11] A. Secker and D. Taubman. Motion-compensated Highly Scalable Video Compression Using An Adaptive 3-D Wavelet Transform Based on Lifting. *Proc. Int. Conf. Image Process.* (*ICIP*), IEEE, 2:1029–1032, 2001.

[12] J. Konrad. Transversal versus lifting approach to motion-compensated temporal discrete wavelet transform of image sequences: equivalence and tradeoffs. *Proc. VCIP*, SPIE, 17(7):907–911, 2004, San Jose, CA.

[13] S.-J. Choi and J. W. Woods. Motion-compensated 3D subband coding of video. *IEEE Trans. on Image Process.*, 8:155–167, 1999.

[14] R. Xiong, J. Zu, F. Wu, S. Li, and Y. Zhang. Layered motion estimation and coding for fully scalable 3D wavelet video coding. *Proc. ICIP*, 2271–2274, 2004, Singapore. See also R. Xiong, F. Wu, S. Li, Z. Xiong, and Y.-Q. Zhang. Exploiting temporal correlation with adaptive block-size motion alignment for 3D wavelet coding. *Proc. VCIP*, SPIE, 2004, San Jose, CA.

[15] P. Chen and J. W. Woods. Bi-directional MC-EZBC with lifting implementation. *IEEE Trans. Circ. Sys. Video Technol.*, 14:1183–1194, 2004.

[16] J. W. Woods. *Multidimensional Signal, Image, and Video Processing and Coding*, Academic Press – Elsevier, Burlington, MA, 2006.

[17] P. Chen and J. W. Woods. *Comparison of MC-EZBC and H.26L TML 8 on Digital Cinema Test Sequences*, ISO/IEC JTC1/SC29/WG11, MPEG2002/8130, Cheju Island, Korea, 2002.

[18] J.-R. Ohm. Three dimensional subband coding with motion compensation. *IEEE Trans. Image Process*, 3:559–571, 1994.

[19] D. S. Turaga and M. v. d. Schaar. *Unconstrained Motion Compensated Temporal Filtering*, ISO/IEC JTC1/SC29/WG11, MPEG2002/M8520, Klagenfurt, Austria, 3:81–84, 2002.

[20] D. Taubman. Successive refinement of video: fundamental issues, past efforts and new directions. *Proc. VCIP*, SPIE, volume 5150, Lugano, IT, 2003.

[21] Y. Wu and J. W. Woods. *Recent Improvements in the MC-EZBC Video Coder*, ISO/IEC/JTC1/SC29/WG11, MPEG2003/M10158, Hawaii, USA, 2003.

[22] D. Marpe, H. Schwarz, and T. Wiegand. Context-based adaptive binary arithmetic coding in the H.264/AVC video compression standard. *IEEE Trans. Circ. Sys. Video Technol.*, 13:620–636, 2003.

[23] W. Sweldens. The lifting scheme: A new philosophy in biorthogonal wavelet constructions. *Wavelet Appl. Signal Image Process. III*, SPIE, 2569:68–79, 1995.

[24] S.-J. Choi. *Three-dimensional Subband/Wavelet Video of Video with Motion Compensation*, PhD Thesis, Department of ECSE, Rensselaer Polytechnic Institute, Troy, NY, 1996.

[25] Y. Wu and J. W. Woods. Directional spatial IBLOCK for the MC-EZBC video coder. *Proc. ICASSP*, IEEE, Montreal, CA, 2004.

[26] Y. Wang, J. Ostermann, and Y. Zhang. *Video processing and communications,* Prentice Hall, Englewood Cliffs, NJ, 296–300, 2002.

[27] M. T. Orchard and G. J. Sullivan. Overlapped block motion compensation: an estimation-theoretical approach. *IEEE Trans. Image Process.*, 3:693–699, 1994.

[28] Y. Wu, R. A. Cohen, and J. W. Woods. *An Overlapped Block Motion Estimation for MC-EZBC*, ISO/IEC/JTC1/SC29/WG11, MPEG2003/M10158, Brisbane, AU, 2003.

[29] Y. Wu, A. Golwelkar, and J. W. Woods. *MC-EZBC video proposal from Rensselaer Polytechnic Institute*, ISO/IEC JTC1/SC29/WG11, MPEG04/M10569/S15, Munich, 2004.

[30] Y. Wu and J. W. Woods. Scalable motion vector coding based on CABAC for MC-EZBC. *IEEE Trans. Circ. Sys. Video Technol.*, 17(6):790–795, 2007.

[31] S. S. Tsai, H.-M. Hang, and T. Chiang. *Motion Information Scalability for MC-EZBC: Response to Call for Evidence on Scalable Video Coding*, ISO/IEC JTC1/SC29/WG11, MPEG03/M9756, 2003.

[32] A. Secker and D. Taubman. Highly scalable video compression with scalable motion vector coding. *IEEE Trans. Image Process.*, 13:1029–1041, 2004.

[33] D. Taubman. *EBCOT (Embedded Block Coding with Optimized Truncation): A complete reference*, ISO/IEC JTC1/SC29/WG1, JPEG1998/N988, 1998.

[34] S.-T. Hsiang and J. W. Woods. Embedded image coding using zero-blocks of subband/wavelet coefficients and context modeling. *Proc. Intl. Sympos. Cir. Sys. (ISCAS)*, IEEE, 662–665, Geneva, 2000.

[35] S.-T. Hsiang. *Highly Scalable Subband/Wavelet Image and Video Coding*, PhD Thesis. Department of ECSE, Rensselaer Polytechnic Institute, Troy, NY, 2002.

[36] A. Islam and W. A. Pearlman. An embedded and efficient low-complexity hierarchical image coder. *Proc. VCIP*, SPIE, 3653:294–305, 1999.

[37] T. Zgaljic, M. Mrak, and E. Izquierdo. Omptimized compression strategy in wavelet-based video coding using improved context models. *Proc. ICIP*, III-401–III-404, San Antonio, TX, 2007.

[38] Y. Wu and J. W. Woods. Aliasing reduction via frequency roll-off for scalable image/video coding. *IEEE Trans. Circ. Sys. Video Technol.*, 18(1):48–51, 2008.

[39] Y. Wu, K. Hanke, T. Rusert, and J. W. Woods. Enhanced MC-EZBC scalable video coder. *IEEE Trans. Circ. Sys. Video Technol.*, in press.

[40] D. Mukherjee and A. Said. *Structured Content Independent Scalable Meta-formats (SCISM) for Media Type Agnostic Transcoding: Response to CfP on DIA/MPEG-21*, ISO/IEC JTC1/SC29/WG11, MPEG2002/M8689, 2002, Klagenfurt, AT. See also D. Mukherjee, A. Said, and S. Liu. A framework for fully format-independent adaptation of scalable bit streams. *IEEE Trans. Circ. Sys. Video Technol.*, 15(11):1280–1290, 2005.

[41] Y. Andreopoulos, A. Munteanu, J. Barbarien, M. v. d. Schaar, J. Cornelis, and P. Schelkens. In-band motion compensated temporal filtering. *Signal Process. Image Commun.*, 19:2004.

[42] N. Adami, A. Signoroni, and R. Leonardi. State-of-the-art and trends in scalable video compression with wavelet-based approaches. *IEEE Trans. Circ. Sys. Video Technol.*, 17(9):1238–1255, 2007.

[43] J. W. Woods and J.-R. Ohm, eds., Special issue on interframe subband/wavelet video coding. *Signal Process. Image Commun.*, 19:2004.

[44] T. Rusert, K. Hanke, and J. Ohm. Transition filtering and optimized quantization in interframe wavelet video coding. *Proc. VCIP*, SPIE, 5150:682–693, Lugano, Italy, 2003.

[45] L. Luo, F. Wu, S. Li, and Z. Zhuang. Advanced lifting-based Motion-Threading (MTh) technique for the 3D wavelet video coding. *Proc. VCIP 2003*, SPIE, volume 5150, Lugano, IT, 2003.

[46] R. Xiong, J. Xu, F. Wu, and S. Li. Barbell-lifting based 3-D wavelet coding scheme. *IEEE Trans. Circ. Sys. Video Technol.*, 17(9):1256–1269, 2007.

CHAPTER

Digital Video Transcoding

12

Shizhong Liu[1] and Alan C. Bovik[2]

[1] *Qualcomm Incorporated, San Diego, USA*
[2] *Dept. of Electrical and Computer Engineering, The University of Texas at Austin, USA*

12.1 INTRODUCTION

The rapid advent of digital multimedia technologies and the phenomenal growth of the internet have created numerous networked multimedia applications such as Video on Demand (VoD), Internet Protocol Television (IPTV), and video conferencing. For example, YouTube has become the most popular website for User Generated Video. People can go to YouTube website to search and view all kinds of video contents. The U-verse IPTV service provided by AT&T has also gained traction and, subsequently, its customer base has been growing exponentially. Meanwhile, with the wide deployment of 3G wireless cellular systems around the world, mobile TV, which sends digital TV signals directly to cellular phone or other portable mobile devices, has become a hot topic in the mobile communication industry. As a result, various mobile TV broadcast standards have been designed and developed, such as DVB-H, MediaFLO, and so forth.

Most digital video content is represented in some kind of compressed format to conserve the storage space and communication bandwidth. With more video compression techniques being developed (such as MPEG1/2/4, H.261/263/264, etc.), the diversity of digital video content available on the internet has grown explosively in the past few years. However, a given video receiving device may only be able to support a small subset of those existing compressed video formats. The unsupported video format has to be modified to the one supported by the device in order for the user to view the video on the device. In addition, the bandwidth of network connections used to deliver video bit streams from server to user device can vary significantly from several kilobits to gigabits per second. Even worse, the bandwidth of a given connection may undergo dynamic changes over time. The compressed video bit stream has to be manipulated dynamically to adapt to the network connections to ensure a continuous video-viewing experience.

Various scalable video coding schemes have been proposed to support video communication over heterogeneous networks. The main idea of scalable video coding is to encode the video as one base layer and a few enhancement layers so that lower bit rates, spatial resolutions and/or temporal resolutions could be obtained by simply truncating certain layers from the original bit stream to adapt to the communication channel bandwidth and/or user device capabilities. Several existing video coding standards such as MPEG2/4, H.263 support scalable video coding. More details on scalable video coding can be found in Chapters 10–11 and the references therein. One problem of scalable video coding is that it does not support interoperability between different video coding formats. Consequently, a device that can only decode a single video-coding format may not be able to access those videos coded in other formats. In addition, scalable video coding usually suffers from lower coding efficiency compared to single layer video-coding solutions. Since large numbers of video streams coded without scalability are available, such as DVDs, alternative technologies are needed to support universal access to those videos.

Video transcoding, in which video is converted from one compressed format to another compressed format for adaptation of channel bandwidth or receiver or both, is another technique proposed to adaptively deliver video streams across heterogeneous networks. In general, video transcoding can be used for bit rate reduction, spatial resolution, and/or temporal resolution reduction as well as for video-coding format conversion. Quite a bit of research has been done on video transcoding due to its wide range of applications. Early research work on video transcoding mainly focused on bit rate reduction to adapt the compressed video to available channel bandwidth [1–4]. Owing to the coexistence of multiple video-coding standards as well as the emergence of new classes of devices with limited display and processing power, heterogeneous video transcoding—where the video-coding format, spatial resolution, and/or temporal resolution might all change—has also been investigated recently [5–8]. Lately, error-resilient video transcoding has gained a significant amount of attention with the introduction of video applications over wireless networks [10, 11]. Figure 12.1 shows an example MediaFLO system architecture designed and developed by Qualcomm Inc. for mobile TV service, where video-transcoding techniques are used to transcode the video formats

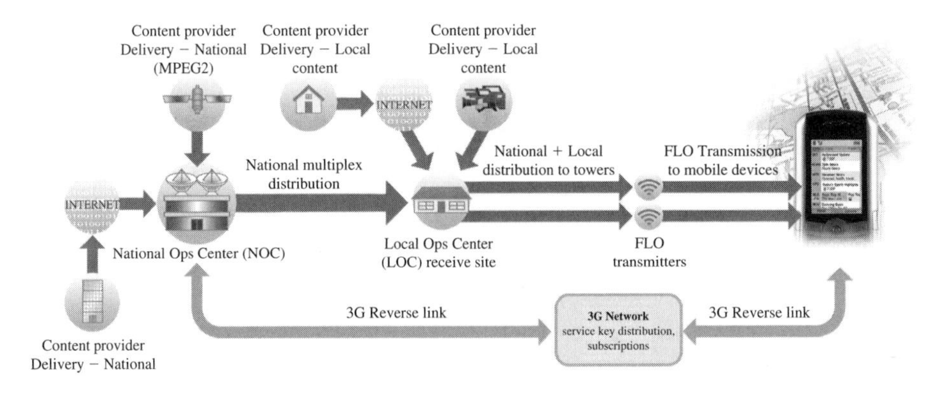

FIGURE 12.1

Illustration of MediaFLO system architecture (Adapted from Qualcomm MediaFLO whitepaper: http://www.qualcomm.com/common/documents/brochures/tech_overview.pdf).

from content providers to the H.264 format that is supported by the MediaFLO system. In the system, the National Ops Center (NOC) and Local Ops Center (LOC) receive digital video content from a content provider, typically through a C-band satellite in MPEG-2 format (704 or 720 × 480 or 576 pixels), utilizing off-the-shelf infrastructure equipment. Since it uses the H.264 video-compression standard as video format and targets the QVGA (320 × 240) resolution supported by most mobile devices, the MediaFLO system must transcode the incoming high resolution MPEG2 video into the H.264 format with QVGA resolution before transmitting over the MediaFLO wireless links.

In all video-transcoding applications, it is possible to employ a cascaded pixel-domain approach that decodes the incoming video stream, performs certain processing (if any), and reencodes the processed signal subject to any new constraints. However, this straightforward approach has a rather high computational complexity, and the quest for more efficient techniques is the major driving force behind most video-transcoding research activities, where the information extracted from the incoming video bit stream is utilized to help reduce the computational complexity of video transcoding. Certainly, any gains in efficiency should have negligible impact on the visual quality of the transcoded video. Since most current video-coding standards are block-based, video-coding schemes that employ the discrete cosine transform (DCT) and motion estimation and compensation (more details can be found in Chapters 8–10), we will concentrate on block-based, video-transcoding techniques in this chapter. Considering that the chrominance components can usually be handled similarly as the luminance components, we will only discuss the processing techniques done on the luminance components of the video.

The rest of this chapter is organized as follows. In Section 12.2, video-transcoding techniques for bit rate reduction and corresponding architectures are reviewed. Section 12.3 reviews the progress made in heterogeneous video transcoding, including spatial resolution reduction, motion vector (MV) reestimation after spatial and/or temporal resolution reduction, as well as macro-block coding-type decision. In Section 12.4, we discuss bit rate control techniques in video transcoding. Section 12.5 briefly discusses recent research activities on error-resilient video transcoding for wireless video. Finally, we conclude the Chapter in Section 12.6.

12.2 VIDEO TRANSCODING FOR BIT RATE REDUCTION

Video transcoding for bit rate reduction has been studied intensively due to its wide range of applications including television broadcast and video streaming over the internet. The most straightforward way to achieve this is to decode the incoming video bit stream and fully reencode the reconstructed video at the new bit rate as illustrated in Fig. 12.2. Although this approach can achieve the best video-transcoding results, its high computational complexity prevents it from being a practical solution. It is the video transcoder's aim to significantly reduce the complexity, while still maintaining acceptable quality, by exploiting the information contained in the incoming video bit stream such as MVs.

In the following, we first analyze the video transcoding of both intracoded frames and intercoded frames based on Fig. 12.2. In the analysis, we assume that the picture

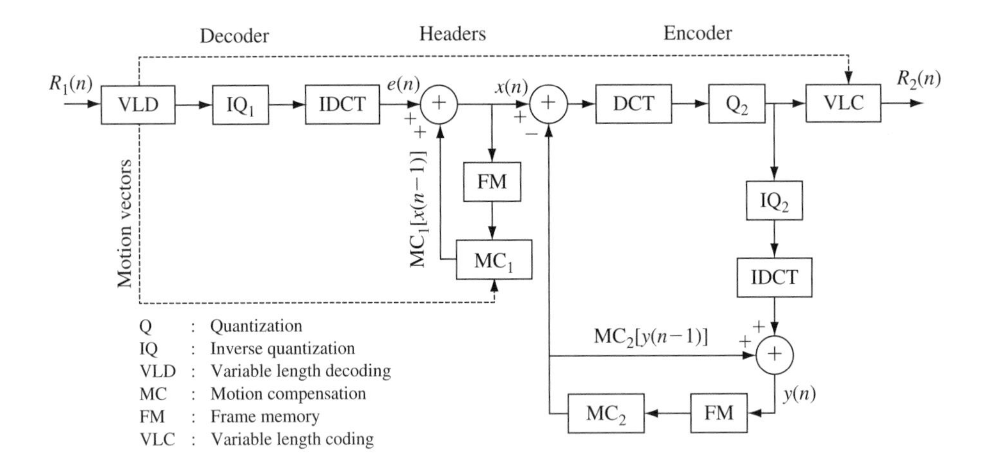

FIGURE 12.2

Diagram of the cascaded pixel domain video transcoder. $R_1(n)$ is the incoming video bit stream and $R_2(n)$ is the output stream with lower bit rate.

coding type does not change after transcoding. For instance, an intracoded frame will be transcoded into another intracoded frame. Then, we review and evaluate several fast video-transcoding architectures proposed over the past few years for bit rate reduction. Finally, we will briefly discuss the recent advances in DCT domain inverse motion compensation (IMC) used in DCT domain video-transcoding architectures.

12.2.1 Transcoding of Intracoded Frame

Intracoded pictures are transcoded by coarsely encoding the decoded picture denoted by $x(n)$. Thus, from Fig. 12.2, we have

$$R_2(n) = Q_2\{\text{DCT}[x(n)]\}, \tag{12.1}$$

where $x(n)$ is given by

$$x(n) = \text{IDCT}\left\{Q_1^{-1}[R_1(n)]\right\} \tag{12.2}$$

Substituting Eq. (12.2) to Eq. (12.1) gives the transcoding Eq. (12.3) for intracoded frames:

$$R_2(n) = Q_2\left[\text{DCT}\left(\text{IDCT}\left\{Q_1^{-1}[R_1(n)]\right\}\right)\right] = Q_2\left\{Q_1^{-1}[R_1(n)]\right\} \tag{12.3}$$

The transcoding distortion for intracoded frame consists of the following three components [3]:

1. Some nonzero AC coefficients of the input frame become zero after coarse requantization.

2. The quantization error for those nonzero coefficients.
3. The requantization error.

Although the former two are well-known causes of distortions, the third one is not obvious. In certain cases, requantization can lead to an additional error, which would not be introduced had the original DCT coefficients been quantized with the same coarser quantization step size. This is illustrated in Fig. 12.3, which shows how requantization can lead to higher distortion than that produced by quantizing the original DCT coefficients using the same quantization step size.

In Fig. 12.3, the reconstructed levels of the DCT coefficients X, Y, quantized with a quantizer step size of Q_1, are Q_1^X and Q_1^Y, respectively. If the coarser quantization step size Q_2 was used, then both would be reconstructed to the same level $Q_2^X = Q_2^Y$. However, if X, Y are first quantized with Q_1 and then quantized with Q_2, the reconstructed level of Y will be the same as that of direct coarser quantization with Q_2, that is, $Q_2^Y = Q_{12}^Y$, whereas in the case of X, the reconstruction value will be different from that of direct coarser quantization, that is, $Q_2^X \neq Q_{12}^X$. Therefore, the requantization error of Y is zero, whereas that of X is not.

Whenever the coarse quantization interval contains entirely the corresponding finer one, the direct coarse quantization and requantization distortions are equal. However, if the finer interval overlaps between two different coarser intervals, then the requantization distortion is larger in the case where the reconstruction value of the first quantization and the original coefficient fall into different coarser quantization intervals. Werner [12] analyzed the requantization problem theoretically and designed the quantizer Q_2 to reduce the overall requantization distortion by up to 1.3 dB, compared to the quantizer used in the MPEG2 reference coder TM5. A selective requantization method was proposed to reduce the requantization errors in video transcoding by avoiding critical ratios of

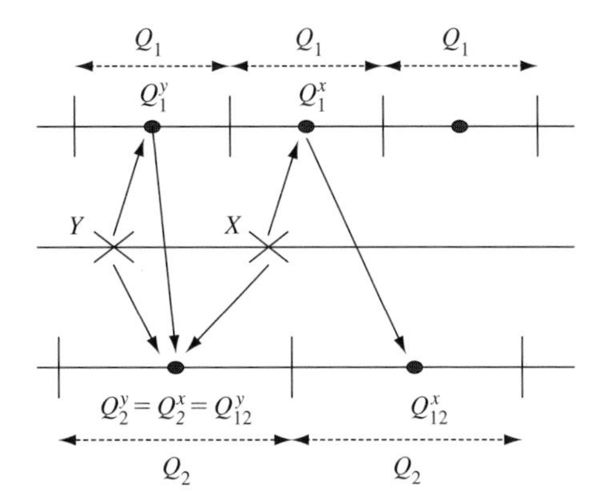

FIGURE 12.3

Requantization of DCT coefficients (Adapted from ref. [3]).

the two cascaded quantizations that either lead to larger transcoding errors or require a higher bit budget [13].

12.2.2 Transcoding of Intercoded Frame

In Fig. 12.2, the incoming intercoded frame is first reconstructed at the decoder by motion compensation (MC), that is,

$$x(n) = \mathrm{IDCT}\left\{Q_1^{-1}[R_1(n)]\right\} + \mathrm{MC}_1[x(n-1)]. \tag{12.4}$$

At the encoder side, we have

$$R_2(n) = Q_2\left(\mathrm{DCT}\left\{x(n) - \mathrm{MC}_2[y(n-1)]\right\}\right) \tag{12.5}$$

Substituting Eq. (12.4) into Eq. (12.5) gives the transcoding Eq. (12.6) for intercoded frames:

$$\begin{aligned} R_2(n) &= Q_2\left[\mathrm{DCT}\left(\mathrm{IDCT}\left\{Q_1^{-1}[R_1(n)]\right\} + \mathrm{MC}_1[x(n-1)] - \mathrm{MC}_2\left[y(n-1)\right]\right)\right] \\ &= Q_2\left(Q_1^{-1}[R_1(n)] + \mathrm{DCT}\left\{\mathrm{MC}_1[x(n-1)] - \mathrm{MC}_2\left[y(n-1)\right]\right\}\right). \end{aligned} \tag{12.6}$$

Note that Eq. (12.6) implies that, for transcoding an intercoded frame, the transcoding error of the previous anchor picture has to be added to the incoming DCT coefficients and then coarsely quantized. Otherwise, the transcoding error in the intercoded frames will be accumulated until the next intracoded frame is met. This error accumulation is known as *drift*. In MPEG bit streams, intracoded frames can terminate the *drift* within a group of picture structure. However, this refreshing by intracoded frame is not always available in other video streams. For instance, in H.263, the insertion of intracoded frames is not specified explicitly. Furthermore, in relatively complicated video sequences, the *drift* can be significant even with a small number of intercoded frames involved. In the following, various fast video-transcoding architectures will be discussed and the *drift* in each transcoder will be investigated.

12.2.3 Fast Video-Transcoding Architectures

Several different architectures for video transcoding have been proposed. A simple architecture uses open-loop transcoding as shown in Fig. 12.4, where the incoming bit-rate is down-scaled by truncating the high frequency DCT coefficients or performing a requantization process [1].

Since the transcoding is done in the coded domain, a very simple and fast transcoder is possible. However, since the transcoding error associated with the anchor picture is not added to the subsequent intercoded frames; the open-loop transcoding produces an increasing distortion caused by the *drift*, which can degrade the video quality dramatically.

Drift-free transcoding is made possible by using a decoder to decode the incoming video and the using an encoder to reencode the decoded signal at a lower bit rate as shown in Fig. 12.2. However, its high computational complexity makes it difficult to be used in

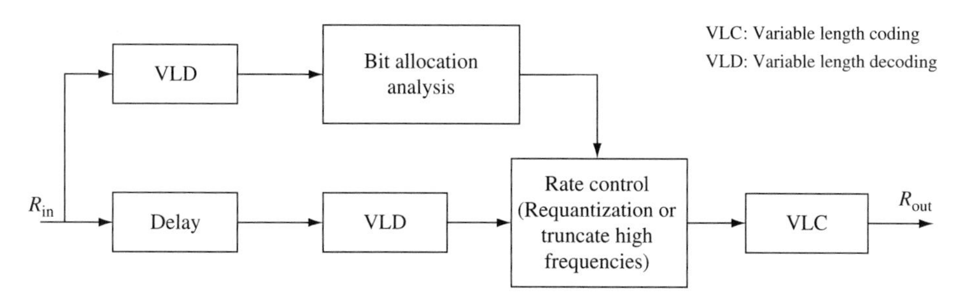

FIGURE 12.4

Fast pixel domain open-loop transcoder.

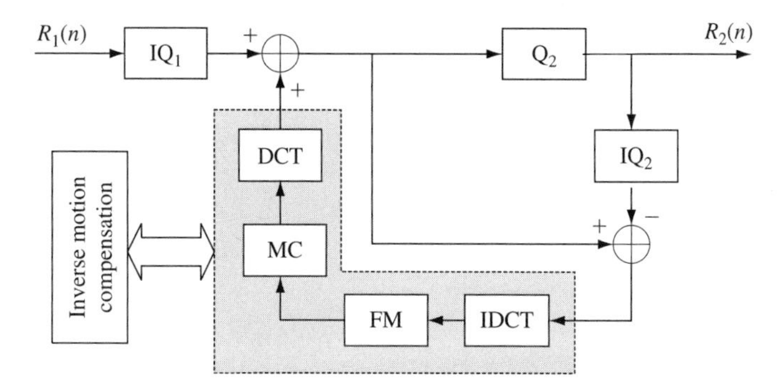

FIGURE 12.5

Fast pixel domain video transcoder.

real-time applications. Since an incoming video bit stream arriving at the transcoder already carries much useful information such as the picture type, MVs, quantization step-size, bit-allocation statistics, and so forth, it is possible to reduce the complexity of the video transcoder by reusing some of the available information. By reusing the MVs and macro-block coding mode decision information contained in the incoming video bit stream, a simplified architecture with less computational cost can be obtained [2–4]. By reusing MVs, that is, $MC_1 = MC_2 = MC$, Eq. (12.6) becomes

$$\begin{aligned} R_2(n) &= Q_2\left(Q_1^{-1}[R_1(n)] + \mathrm{DCT}\left\{MC_1[x(n-1)] - MC_2[y(n-1)]\right\}\right) \\ &= Q_2\left(Q_1^{-1}[R_1(n)] + \mathrm{DCT}\left\{MC[x(n-1) - y(n-1)]\right\}\right), \end{aligned} \tag{12.7}$$

which leads to a simplified architecture depicted in Fig. 12.5. In this architecture, the transcoding error associated with the reference picture is compensated back to the predictive picture so that the error drift problem is eliminated. This operation is called inverse motion compensation (IMC) as shown in Fig. 12.5.

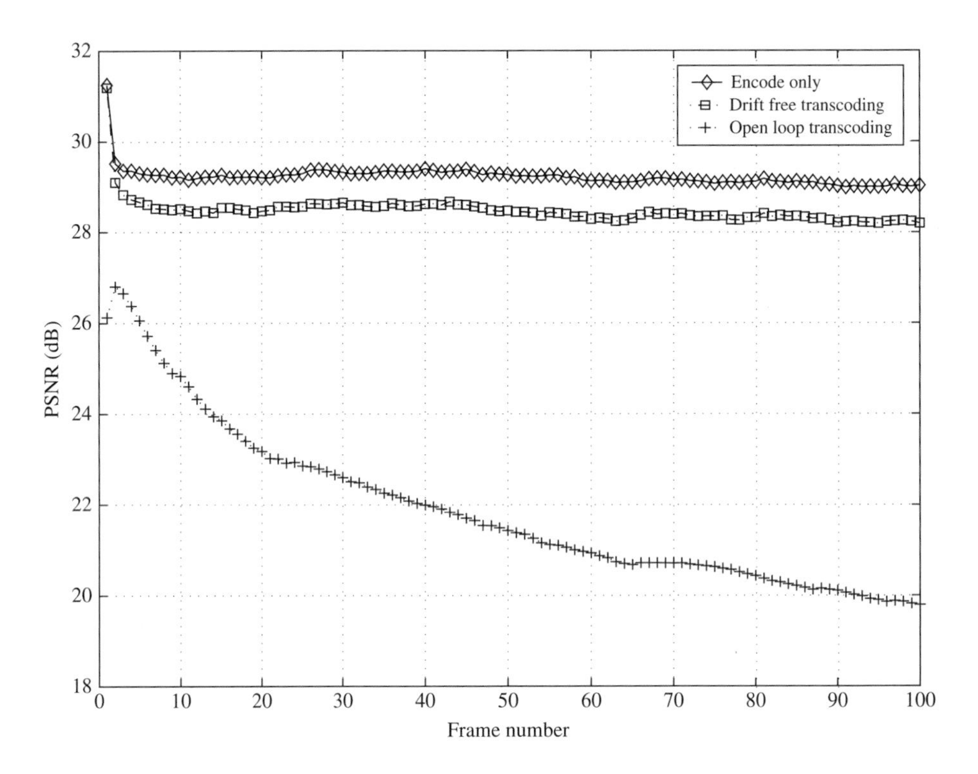

FIGURE 12.6

Performance of different video-transcoding architectures (mobile sequence, $Q_1 = 5, Q_2 = 10$).

This simplified architecture can save motion estimation, one frame memory, and one inverse DCT (IDCT) operation compared to that shown in Fig. 12.2. The performance of different video-transcoding architectures is depicted in Fig. 12.6 in which the *mobile* sequence was encoded using H.263 with the fixed quantization step size of $Q_1 = 5$. Figure 12.6 shows the peak signal-to-noise ratio (PNSR) value of each frame, obtained by the following three methods, respectively.

- Encode the original mobile sequence using H.263 with fixed quantization step size of $Q_1 = 10$.
- Trancoded the bit stream with $Q_1 = 5$ into another one with $Q_2 = 10$, using the open-loop transcoder shown in Fig. 12.4.
- Transcode the bit stream with $Q_1 = 5$ into another one with $Q_2 = 10$, using the drift-free fast pixel domain transcoder shown in Fig. 12.5.

Clearly, the transcoding errors are accumulated in the open-loop video transcoder, causing a serious video quality drift problem with more than 6 dB PSNR degradation in the last frame. By contrast, the difference between that obtained by method 1 and that by method 3 is about 0.7 dB for all frames.

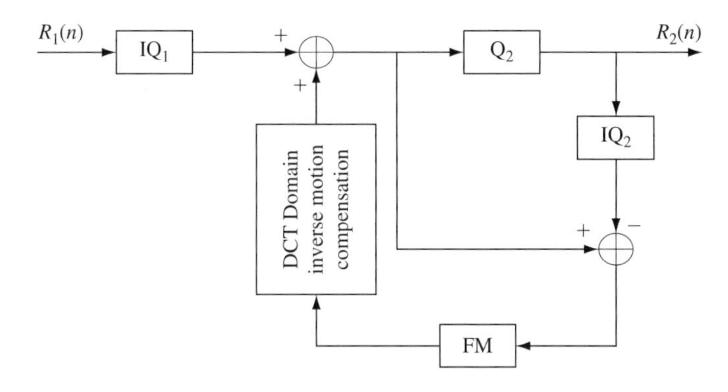

FIGURE 12.7

Fast DCT domain video transcoder.

In the fast pixel domain transcoder shown in Fig. 12.5, the coded quantization errors of the second stage quantization are decoded into the pixel domain through inverse DCT transformation and then stored in the frame memory as pixel values. The MC operation is performed in the pixel domain and then transformed into the DCT domain for the prediction operation. The whole process required one IDCT transform, one DCT transform, and one MC. By keeping everything in the DCT domain, the computation for IDCT/DCT can be avoided, which leads to the DCT domain-transcoding architecture shown in Fig. 12.7. In addition, for most video sequences, the DCT block is quite sparse; hence, the data volume to be processed can be significantly reduced while achieving the same functionality. The efficiency of this transcoder mainly depends on the computational complexity of DCT domain IMC. In the next subsection, we will discuss the problem of DCT domain IMC and briefly review several fast algorithms proposed in the last few years.

12.2.4 DCT Domain IMC

IMC is a necessary step in most video transcoding applications in order to convert the intercoded frames to intracoded frames or to compensate the transcoding errors in the anchor frames back to the prediction error frames for *drift-free* video transcoding. Since the data is organized block-by-block in the DCT domain, IMC in the DCT domain is more complex than its counterpart in the spatial domain and hence becomes the bottleneck of DCT domain video-transcoding algorithms.

The general setup of DCT domain IMC is illustrated in Fig. 12.8, where $\hat{x}$ is the current block of interest, x_1, x_2, x_3, and x_4 are the reference blocks from which $\hat{x}$ is derived. In the spatial domain, $\hat{x}$ can be expressed as a superposition of the appropriate windowed and shifted versions of x_1, x_2, x_3, and x_4 as follows:

$$\hat{x} = \sum_{i=1}^{4} q_{i1} x_i q_{i2}, \tag{12.8}$$

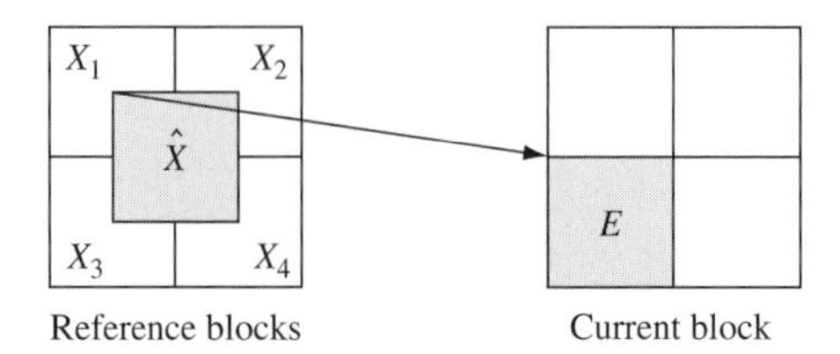

FIGURE 12.8

Illustration of DCT domain inverse motion compensation.

where $q_{ij} i = 1,\ldots,4,\ j = 1,2$, are sparse 8×8 matrices of zeros and ones that perform windowing and shifting operations. For example, for $i = 1$,

$$q_{11} = \begin{pmatrix} 0 & I_h \\ 0 & 0 \end{pmatrix},\ q_{12} = \begin{pmatrix} 0 & 0 \\ I_w & 0 \end{pmatrix}, \tag{12.9}$$

where I_h and I_w are identity matrices of dimension $h \times h$ and $w \times w$, respectively. The values h and w are determined by the MV corresponding to $\hat{x}$. By using the linear and distributive properties of the DCT, we can write the DCT domain IMC as

$$\hat{X} = \sum_{i=1}^{4} Q_{i1} X_i Q_{i2} \tag{12.10}$$

where $\hat{X}, X_i, Q_{i1}$, and Q_{i2} are the DCT's of $\hat{x}, x_i, q_{i1}$, and q_{i2}, respectively. Note that Q_{i1} and Q_{i2} are constant and hence can be precomputed and stored in memory [14]. Obviously, the objective of DCT domain IMC is to compute the DCT coefficients of directly from the DCT's of $x_i, i = 1,\ldots 4$.

Brute-force computation of Eq. (12.10) in the case where the reference block $\hat{x}$ is not aligned in any direction with the block structure requires eight floating-point matrix multiplications and three matrix additions. Several algorithms have been proposed to reduce the computational complexity of the DCT domain IMC. Merhav et al. [15] proposed to factorize the constant matrices Q_{ij} into a series of relatively sparse matrices instead of fully precomputing them. As a result, some of the matrix multiplications in Eq. (12.10) can be replaced by simple addition and permutation operations such that computational complexity can be reduced. The authors [3] approximated the elements of Q_{ij} by binary numbers so that all multiplications can be implemented by basic integer operations such as *shift* and *add*. In general, MC is done on a macro-block basis, meaning that all blocks in the same macro-block have the same MV. Based on this observation, Song et al. [16] presented a fast algorithm for DCT IMC by exploiting the shared information among the blocks within the same macro-block instead of constructing the DCT domain values of each target block independently like the previous two methods. However, this method does not apply to the case where one macro-block has multiple MVs. For instance, MPEG4 and H.264 support four MVs per macro-block. Although all three methods above adopted the two-dimensional (2D) procedure shown in Fig. 12.8, Acharya et al. [17] developed a separable implementation diagram in which the 2D problem was decomposed into two separate 1D problems. They showed that the decomposition is more efficient than computing the combined operation.

In ref. [18], the authors approached the problem of Eq. (12.10) from a different angle by analyzing the statistical properties of nature image/video data. The proposed algorithm first estimates the local bandwidth of the target block to be reconstructed from the reference blocks by modeling a natural image as a 2D separable Markov random field. The algorithm can reduce the processing time by avoiding the computations of those DCT coefficients outside the estimated local bandwidth. Note that the DCT coefficients inside the estimated local bandwidth can be computed by employing other fast algorithms such as the ones discussed earlier. No significant distortion is introduced by the algorithm as shown in the experimental results [18].

12.3 HETEROGENEOUS VIDEO TRANSCODING

Heterogeneous video transcoding has become increasingly important for supporting universal multimedia access due to the fact that multiple video-coding standards coexist and a wide range of new devices with different characteristics have been developed for networked multimedia applications. In heterogeneous video-transcoding, the incoming video bit stream may be subject to coding-format conversion, spatial and/or temporal resolution reduction in order to satisfy the constraints imposed by the video communication channel and/or receiving devices. As such, the decoder and encoder loops in the heterogeneous video transcoder could not be combined to simplify the transcoder, as in the fast video-transcoding architectures described in Section 2. Instead of simply cascading the video decoder and encoder, various efficient video-transcoding techniques have been proposed in the literature by taking advantage of the information extracted from the incoming video bit stream (e.g., MVs, macro-block-coding types as well as bit allocation statistics) to help reduce the computational complexity of the encoder. For instance, the MVs extracted from the input video can be reused or employed to derive new MVs in the video encoder such that full scale–motion estimation, which comprises more than 60% of the encoding complexity, can be avoided [5].

Figure 12.9 shows a typical video-transcoding architecture in which an input MPEG2 video is transcoded to a MPEG4 video bit stream for wireless video applications. It first decodes the incoming MPEG2 bit stream to the pixel domain by performing variable length decoding, inverse quantization, IDCT, and MC; then down-scales the decoded video by a factor of two in both the horizontal and vertical directions in the pixel domain; and finally reencodes the down-scaled video into an outgoing MPEG4 video bit stream. Note that the MC is performed using the original MVs. The MVs extracted from the incoming bit stream are used to speed up the motion estimation process in the MPEG4 encoder.

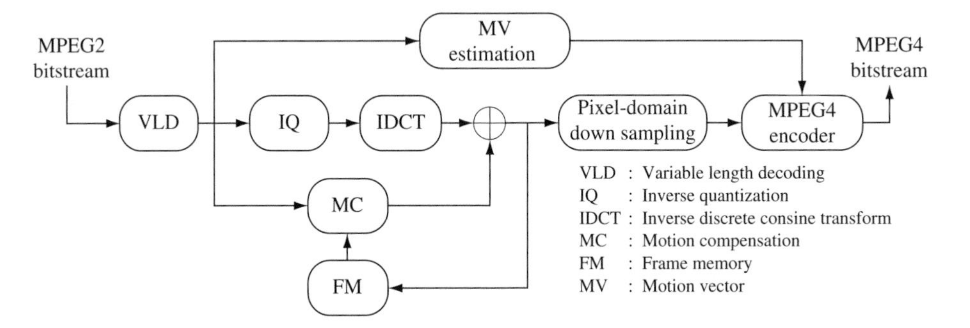

FIGURE 12.9

Pixel domain video-transcoding architecture to transcode MPEG2 videos to MPEG4 bit streams with spatial and/or temporal resolution reduction.

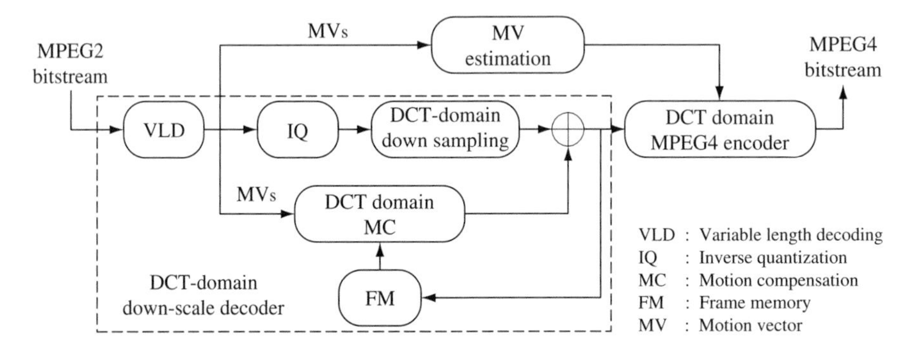

FIGURE 12.10

DCT domain video transcoder to transcode MPEG2 videos to MPEG4 bit streams with spatial and/or temporal resolution reduction.

If both the input and output video-coding formats employ the DCT transform, then DCT domain video-transcoding architectures, where all operations are performed in the DCT domain to avoid DCT/IDCT process, can be used. Significant effort has been applied toward developing efficient DCT domain video-transcoding architectures and algorithms, such as DCT domain spatial resolution reduction and DCT domain MC. For example, a fast and memory-efficient DCT domain video-transcoding architecture is illustrated in Fig. 12.10, in which the input video is directly decoded to a lower resolution video by a so-called DCT domain down-scale video decoder [19]. In Fig. 12.10, the DCT domain MC is basically the same problem as the DCT domain IMC discussed in Section 2. Thus, fast algorithms for DCT domain IMC also apply to DCT domain MC. In comparison with the pixel domain approach shown in Fig. 12.9, this transcoder can save more than 50% of required memory, since there is no buffer needed for the original high resolution video frame. In addition, the computational cost is reduced by more than

70%. However, the video quality achieved by both approaches is hardly distinguishable at target bit rates of 384 kb/s and 256 kb/s according to the experimental results reported in ref. [19].

In the following, we will review some recent progress made in heterogeneous video-transcoding techniques, which include MV estimation, video down-sampling as well as macro-block coding type decision in the encoder. In the following discussion, we will assume the input video is down-sampled by 2 in both the horizontal and vertical resolutions. The techniques are extended to arbitrary scaling factors in refs. [20, 21].

12.3.1 MV Estimation for Spatial Resolution Reduction

To avoid full-scale motion estimation in the encoder, the MVs extracted from the incoming video bit stream can be employed to derive new MV(s) for the down-sampled outgoing video sequence. Because of the down-sampling in spatial resolution, one macro-block in the outgoing video corresponds to four macro-blocks in the original video sequence. Each intercoded macro-block in the original video is assumed to have one MV. Thus, there will be four MVs associated with each 16×16 macro-block in the lower spatial resolution picture as shown in Fig. 12.11. If the outgoing video bit stream supports four 8×8 MVs in each 16×16 macro-block (such as MPEG4), then the input MVs of the original video can be employed directly for each 8×8 block in the lower spatial resolution video with appropriate scaling by two. However, it is sometimes inefficient to use four MVs in each macro-block since more bits must be used to code four MVs. In the case where each 16×16 macro-block has one MV in the outgoing video, various techniques have been investigated to derive the output MV from the four input MVs. Three methods of deriving a new MV from the four input MVs are studied in ref. [5] including the median value, the majority, and the mean value of the four input MVs. They showed that the median value gives the best results. The median vector is

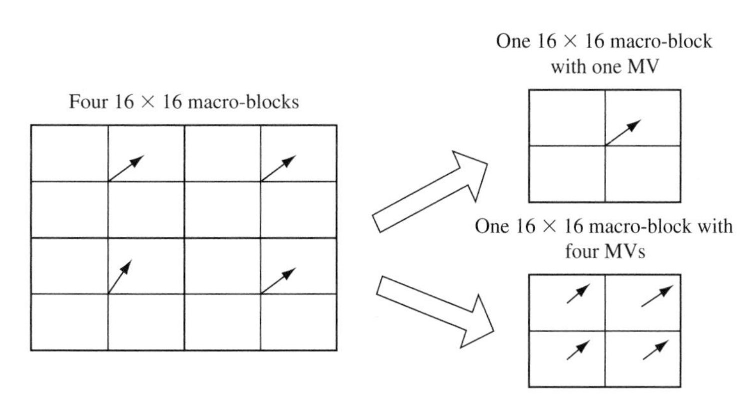

FIGURE 12.11

Illustration of motion vector (MV) mapping.

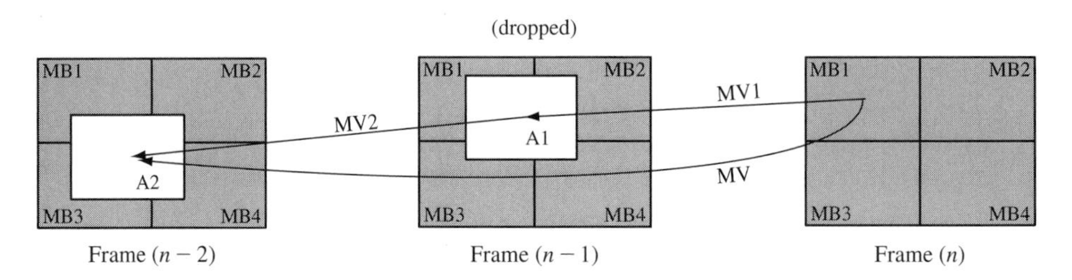

FIGURE 12.12

Motion vector (MV) reestimation. A new MV from frame n to frame $n-2$ will be derived from the incoming MVs, MV1 and MV2 since the frame $n-1$ is dropped.

defined as one of the four input MVs that has the least Euclidean distance from all, that is,

$$\bar{v} = \arg\min_{v_k \in V} \left\{ \sum_{\substack{j=1 \\ j \neq k}}^{4} \|v_k - v_j\| \right\}, \tag{12.11}$$

where $V = (v_1, v_2, v_3, v_4)$ and $v_j, j = 1, \ldots, 4$ are the incoming MVs and $\bar{v}$ is the candidate MV for the down-scaled video. Note that the magnitude of the estimated vector v should be down-scaled by a factor of two and those intracoded or skipped macro-blocks in the incoming bit stream are usually viewed as predicted macro-blocks with zero-valued MVs. Instead of simply averaging the four input MVs, it has been proposed [22] to estimate the new MV by using the weighted average of the four incoming MVs, where the weights correspond to the block activity. The proposed method produces higher video quality than the simple averaging method. However, the comparison between the median value method and the proposed method was not given in ref. [22].

To further improve the output video quality, it has also been suggested to refine, within a small searching area, the MV derived from the four input vectors. The experimental results from ref. [5] show that a refinement within 0.5 pixel range delivers satisfactory results. Nevertheless, the authors in ref. [23] present an adaptive MV-refinement algorithm called variable step-size search, where the search area is adapted to the MV magnitude itself. The algorithm achieves better video quality at the expense of higher computational cost.

12.3.2 MV Estimation for Temporal Resolution Reduction

In order to achieve higher bit rate reduction or to adapt to the processing power of the receiving device, the video transcoder may have to reduce the temporal resolution of the incoming video by dropping some of the encoded frames. For instance, a mobile terminal may only be capable of decoding and displaying 10 frames per second, whereas most DVD videos are encoded at 30 frames per second.

As discussed earlier, MVs from the original video bit stream are usually reused or employed to derive the new MVs for the outgoing bit stream in bit rate reduction and spatial resolution reduction video transcoders. For temporal resolution reduction, the problem is similar in that it is necessary to derive the MVs from the current frame to the previous nonskipped frame by using the MVs extracted from the skipped video frame(s). Figure 12.12 illustrates a situation where a frame is dropped. One way to obtain MV without performing full-scale motion estimation is to use the summation of MV1 and MV2. However, since the corresponding area of MV2 in frame $n-1$ is usually not aligned with macro-block boundary, MV2 is not available from the incoming bit stream. So the problem is boiled down to estimate MV2 from its four neighboring MVs obtained from the incoming bit stream. One method is to use bilinear interpolation from the four neighboring MVs. Another method is the forward dominant vector selection (FDVS) algorithm proposed in ref. [24]. It selects the MV of the macro-block that has the largest overlapping portion with the area pointed at by the incoming MV MV1 in frame $n-1$. For instance, in Fig. 12.12, the FDVS would select the MV of MB1 as MV2 since the macro-block, MB1, has the largest overlapping area with the area A1. The FDVS usually yields much better video performance than the bilinear interpolation scheme [24]. However, when the overlapping areas among the four macro-blocks are very close, the MV selected by FDVS may not be optimal. Based on this observation, the authors in ref. [23] proposed a so-called activity-dominant vector selection (ADVS) algorithm by utilizing the activity of each macro-block to determine which MV should be selected. The activity information of a macro-block can be represented by the number of nonzero DCT coefficients. From the simulation results reported in ref. [23], the ADVS algorithm is superior to the FDVS method, especially in high motion video sequences.

12.3.3 Spatial Resolution Reduction

In pixel domain video transcoding with spatial resolution reduction (Fig. 12.9), two methods can be employed to perform spatial resolution reduction. The first one is pixel-averaging method in which every 2×2 pixel block is represented by a single pixel of their average value. The second is the subsampling method that applies low-pass filtering to the original image, then down-samples the low-pass filtered image by dropping every alternative pixel in both horizontal and vertical directions [25]. The first method is simpler to implement, whereas the second usually gives better performance in video quality if a proper low-pass filter is employed. With the popularity of DCT domain video-transcoding techniques, DCT domain algorithms for spatial resolution reduction have also been studied. By utilizing the linear, distributive, and unity properties of the DCT transform, it is always possible to find a counterpart in the DCT domain for any given pixel domain linear operation [15, 26, 27]. Since most of the signal energy is concentrated at the lower frequency band in the DCT domain, the authors in ref. [19] proposed to perform spatial resolution reduction by only decoding the top-left 4×4 DCT coefficients of each 8×8 block in the incoming bit stream. Then every four 4×4 DCT blocks are transformed into one 8×8 DCT block in the DCT domain as shown in Fig. 12.13.

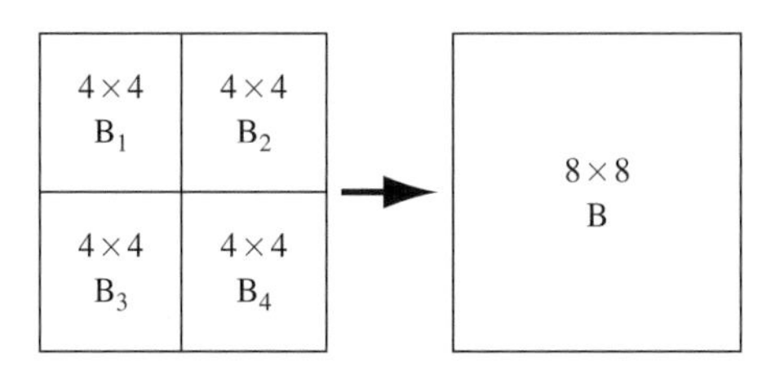

FIGURE 12.13

Convert four 4 × 4 sub-blocks to one 8 × 8 block.

To show this, let T and T_4 denote the 8×8 and 4×4 DCT operator matrices, respectively. Then we have

$$\begin{aligned} B = TbT^t &= [T_L \quad T_R] \begin{bmatrix} b_1 & b_2 \\ b_3 & b_4 \end{bmatrix} \begin{bmatrix} T_L^t \\ T_R^t \end{bmatrix} \\ &= \frac{1}{2} \begin{bmatrix} T_L & T_R \end{bmatrix} \begin{bmatrix} T_4^t B_1 T_4 & T_4^t B_2 T_4 \\ T_4^t B_3 T_4 & T_4^t B_4 T_4 \end{bmatrix} \begin{bmatrix} T_L^t \\ T_R^t \end{bmatrix} \\ &= \frac{1}{2}(T_L T_4^t)[B_1(T_L T_4^t)^t + B_2(T_R T_4^t)^t] + \frac{1}{2}(T_R T_4^t)[B_3(T_L T_4^t)^t + B_4(T_R T_4^t)^t] \end{aligned} \tag{12.12}$$

where b, b_1, b_2, b_3, b_4 are the pixel-domain representations of the DCT blocks B, B_1, B_2, B_3, B_4 (shown in Fig. 12.13), respectively; T_L, T_R are 8×4 matrices denoting the first and last four columns of the eight-point DCT kernel T, respectively, and t denotes the matrix transpose. Let $C = T_L T_4^t + T_R T_4^t$ and $D = T_L T_4^t - T_R T_4^t$, then we have $T_L T_4^t = \frac{C+D}{2}$ and $T_R T_4^t = \frac{C-D}{2}$. Hence, Eq. (12.12) can be rewritten as

$$\begin{aligned} B &= \frac{1}{8}\left\{[C(B_1 + B_3) + D(B_1 - B_3)](C + D)^t + [C(B_2 + B_4) + D(B_2 - B_4)](C - D)^t\right\} \\ &= \frac{1}{8}\left[X(C + D)^t + Y(C - D)^t\right] \\ &= \frac{1}{8}\left[(X + Y)C^t + (X - Y)D^t\right] \end{aligned} \tag{12.13}$$

with

$$X = C(B_1 + B_3) + D(B_1 - B_3) \tag{12.14}$$

$$X = C(B_2 + B_4) + D(B_2 - B_4) \tag{12.15}$$

It can be shown that more than 50% of the elements in the matrices C and D are zeros, hence Eqs. (14) and (15) can be computed very efficiently [31].

12.3.4 Macro-block-Coding Type Decision

In video transcoding with spatial resolution reduction, every four macro-blocks in the incoming bit stream will collapse to one macro-block in the outgoing bit stream as shown in Fig. 12.11. Since the four incoming macro-blocks may have different coding types, a new macro-block-coding type has to be designated to the output macro-block in the output bit stream. One way is to derive a new macro-block-coding type by comparing the coding complexity of different coding types as in a standalone video encoder. However, the cost of carrying out a new macro-block coding type decision is rather high [5, 32]. Another method is to select a macro-block-coding type from the four incoming macro-block-coding types. For instance, we can use the majority of the input macro-block-coding types. In other words, if three of the four input macro-blocks are intercoded, then the output macro-block will be intercoded. The intracoded input macro-block will be considered as the intercoded one but with zero-valued MVs. Experimental results reported in ref. [5] show that the video quality degradation in the second method is less than 0.2 dB, compared to the one achieved by the first method.

12.4 BIT RATE CONTROL IN VIDEO TRANSCODING

One important application of video transcoding is to adapt the bit rate of precoded video bit streams to various network connections. In video transcoding, the performance of bit rate adaptation heavily depends on the bit rate control technique employed in the video transcoder. Therefore, bit rate control is one of the key components in a video transcoder. In the following, we discuss the difference between the problem of bit rate control in video transcoding and that in video encoding.

In general, bit rate control is a budget constrained bit allocation problem. A stand-alone video encoder usually relies on a preanalyzed rate-distortion (R-D) model to solve the bit allocation problem. The RD model determines the relationship between the number of produced bits and the quantization step size. Various R-D models have been proposed in the literature based on information theory and the characteristics of the human vision system [33–37], such as the TMN8, TMN10 models developed in the H.263 standard. Since the R-D model is usually derived from extensive experimental data, it may not match the R-D characteristics of any particular video sequence. The mismatch may produce large bit rate control error in video encoding. To reduce the mismatch, most bit rate control techniques update the R-D model by using the actual bit usage information in the previous frame or macro-black and then apply the updated R-D model to the current frame or macro-block. However, if the statistical properties of the current frame or macro-block are significantly different from the previous one, the mismatch problem still exists. For example, during a scene change, the current frame may be completely different from the previous one. Thus, the R-D model based on the statistics of the previous frame may not fit the current frame at all.

Although the bit rate control techniques used in video encoding can be directly employed in video transcoding, the bit allocation and rate control in video transcoding is different from that in video encoding. The input of video transcoding is a preencoded video bit stream that contains much information such as bit usage of each macro-block, macro-block-coding types, and MVs. The bit usage information extracted from the input bit stream can be reused by the video transcoder to build a R-D model that matches the actual statistics of the video frame or macro-block being encoded so that more accurate bit allocation and rate control can be achieved [38–40]. In ref. [38], the authors proposed an accurate bit allocation and rate control algorithm for video transcoding by exploiting the approximate relationship between the number of VLC code words and the number of produced bits for encoding those VLC code words. Since both the number of VLC code words and the number of produced bits are already available in the incoming bit stream, an accurate bit allocation model can be derived from the information contained in the incoming bit stream. The proposed rate-control scheme can adaptively allocates bits to each macro-block in a frame and make the total bits of a frame meet the target bits budget. Compared to TMN-8, the proposed technique can allocate the bit budget more efficiently to the macro-blocks in a frame to produce better video quality.

12.5 ERROR-RESILIENT VIDEO TRANSCODING

With the emergence of various advanced wireless networks such as 3G cellular systems, wireless video has become increasingly popular and is attracting great interest [41–43]. Transmitting coded video stream over wireless channels is even more challenging than over wired channels since wireless channels usually have lower bandwidth and higher bit error rates than wired channels.

Error-resilient video transcoding increases the error resilience of a video stream so as to produce acceptable video quality at the receiver side. For example, the authors in ref. [11] proposed an error-resilient transcoding technique that is built on three fundamental blocks, which are as follows:

1. Increase the spatial resilience by reducing the number of macro-blocks per slice and enhance the temporal resilience by increasing the proportion of intracoded macro-blocks in each frame. The amount of resilience is tailored to the content of the video and the prevailing error conditions, as characterized by bit error rate.
2. Derive an analytical model that characterizes how the bit errors propagate in a video that is compressed using motion-compensated encoding and subjected to bit errors.
3. Compute the optimal bit allocation between spatial resilience, temporal resilience, and video signal itself so that an optimal resilience will be injected into the video stream.

In ref. [10], the authors presented a novel fully comprehensive mobile video communication system that exploits the useful rate management features of the video transcoders and combines them with error resilience for transmissions of coded video streams over general packet radio service (GPRS) mobile access networks. The error resilient–video transcoding operation takes place in a video proxy, which not only performs bit rate adaptation but also increases error resilience. In the proposed system, two resilience schemes are employed, which are adaptive intra refresh (AIR) and feedback control signaling (FCS) methods. These two schemes can work independently or combined. The system adjusts the output bit rate from the proxy by monitoring the current channel conditions such as bit error rate, delay, and so forth. Meanwhile, the amount of resilience added to the video data can also be controlled by monitoring the proxy output rate and the change in the error conditions of the network connection. AIR is mainly used to stop the error propagation across different frames. Experiments showed that the combined AIR–FCS method gave superior transcoding performances over error-prone GPRS channels relative to nonresilient video streams.

12.6 CONCLUDING REMARKS

Besides the transcoding techniques discussed in this chapter, other video-transcoding techniques have also been proposed and developed for various applications. These include content-based transcoding [44, 45], joint transcoding of multiple video streams [46], transcoding for fast forward and reverse playback [47], and other more exotic schemes.

In summary, most video-transcoding research activities have been focusing on the following two issues:

1. Reduce the computational cost of video transcoding for real-time operations. Due to the high data volume and computational complexity of video decoding and encoding, efficient algorithms are needed for real-time video transcoding.
2. Develop efficient bit rate control algorithms to achieve better video quality by exploiting the information extracted from the input bit stream and the properties of the HVS.

While a large number of video-transcoding techniques have been developed, the two problems above are still not fully solved, especially the second issue. For any given target bit rate, various transcoding techniques can be employed to achieve the target bit rate. For example, we can choose to increase the quantization step size, or reduce the spatial/temporal resolution, drop the chromatic components, and soforth. Yet, there is a lack of a unified video-transcoding strategy that can automatically determine which transcoding technique to be used to achieve the best visual quality to the end users. This requires developing utility functions that can gauge a user's satisfaction of a coded video [48, 49]. Although great progress has been made in video quality assessment recently (See Chapter 14), further study is still needed toward developing algorithms that measure and compare transcoded video quality across spatiotemporal scales in a perceptually consistent way.

REFERENCES

[1] H. Sun, W. Kwok, and J. W. Zdepski. Architectures for MPEG compressed bitstream scaling. *IEEE Trans. Circuits Syst. Video Technol.*, 6:191–199, 1996.

[2] Kou-Sou Kan and Kuo-Chin Fan. Video transcoding architecture with minimum buffer requirement for compressed MPEG-2 bitstream. *Signal Process.*, 67:223–235, 1998.

[3] A. Pedro, A. Assuncao, and M. Ghanbari. A frequency-domain video transcoder for dynamic bit-rate reduction of MPEG-2 bit streams. *IEEE Trans. Circuits Syst. Video Technol.*, 8:953–967, 1998.

[4] G. Keesman, R. Hellinghuizen, F. Hoeksema, and G. Heideman. Transcoding of MPEG bitstreams. *Signal Process. Image Commun.*, 8:481–500, 1996.

[5] T. Shanableh and M. Ghanbari. Heterogeneous video transcoding to lower spatio-temporal resolutions and different encoding formats. *IEEE Trans. Multimedia*, 2(2):101–110, 2000.

[6] Yap-Peng Tan and Haiwei Sun. Fast motion re-estimation for arbitrary downsizing video transcoding using H.264/AVC standard. *IEEE Trans. Consum. Electron.*, 50(3):887–894, 2006.

[7] G. Shen, Y. He, W. Cao, and S. Li. MPEG-2 to WMV Transcoder With Adaptive Error Compensation and Dynamic Switches. *IEEE Trans. Circuits Syst. Video Technol.*, 16:1460–1476, 2007.

[8] H. Kato, Y. Takishima, and Y. Nakajima. A Fast DV to MPEG-4 Transcoder Integrated With Resolution Conversion and Quantization. *IEEE Trans. Circuits Syst. Video Technol.*, 17:111–119, 2007.

[9] J. Nakajima, H. Tsuji, Y. Yashima, and N. Kobayashi. Motion vector re-estimation for fast video transcoding from MPEG-2 to MPEG-4, MPEG-4. *2001 Proc. Workshop Exhib.*, 18–20:87–90, 2001.

[10] S. Dogan, A. Cellatoglu, M. Uyguroglu, A. H. Sadka, and A. M. Kondoz. Error-resilient video transcoding for robust internetwork communications using GPRS. *IEEE Trans. Circuits Syst. Video Technol.*, 12(6):453–464, 2002.

[11] G. de los Reyes, A. R. Reibman, S.-F. Chang, and J. C.-I. Chuang. Error-Resilient Transcoding for Video over Wireless Channels. *IEEE J. Sel. Areas Commun.*, 18:1063–1074, 2000.

[12] O. Werner. Requantization for Transcoding of MPEG-2 intraframe. *IEEE Trans. Image Process.*, 8:179–191, 1999.

[13] H. Sorial, W. E. Lynch, and A. Vincent. Selective Requantization for Transcoding of MPEG compressed video. *Multimedia and Exp, ICME 2000. IEEE Int. Conf.*, 1:217–220, 2000.

[14] S. F. Chang and D. G. Messerschmitt. Manipulation and compositing of MC-DCT compressed video. *IEEE J. Sel. Areas Commun.*, 13:1–11, 1995.

[15] N. Merhav and V. Bhaskaran. Fast Algorithm for DCT-domain Image Down-Sampling and for Inverse Motion Compensation. *IEEE Trans. Circuits Syst. Video Tech.*, 7:468–476, 1997.

[16] J. Song and B.-L. Yeo. A Fast Algorithm for DCT-domain Inverse Motion Compensation based on shared information in a macroblock. *IEEE Trans. Circuits Syst. Video Tech.*, 10:767–775, 2000.

[17] S. Acharya and B. Smith. Compressed domain transcoding of MPEG. *Proceedings of the Int. Conf. Multimedia Comput. Syst. (ICMCS)*, Austin, TX, June 1998.

[18] S. Liu and A. C. Bovik. Local bandwidth constrained fast inverse motion compensation for DCT-domain video transcoding. *IEEE Trans. Circuits Syst. Video Technol.*, 12(5):309–319, 2002.

[19] S. Liu and A. C. Bovik. A fast and memory efficient video transcoder for low bit rate wireless communications. *Acoustics, Speech, and Signal Process., 2002. Proc. (ICASSP '02). IEEE Int. Conf.*, 2:1969–1972, 2002.

[20] Y.-P. Tan, H. Sun and Y. Q. Liang. On the methods and applications of arbitrarily downsizing video transcoding. *Multimedia and Exp, 2002. ICME '02. Proceedings. 2002 IEEE Int. Conf.*, 1:609–612, 2002.

[21] Y. Q. Liang, L.-P. Chau, and Y.-P. Tan. Arbitrary downsizing video transcoding using fast motion vector reestimation. *Signal Process. Lett., IEEE*, 9(11):352–355, 2002.

[22] B. Shen, I. K. Sethi, and B. Vasudev. Adaptive motion-vector resampling for compressed video downscaling. *Circuits Syst. Video Technol., IEEE Trans.*, 9(6):929–936, 1999.

[23] M.-J. Chen, M.-C. Chu, and C.-W. Pan. Efficient Motion_Estimation Algorithm for Reduced Frame-Rate Video Transcoder. *Circuits Syst. Video Technol., IEEE Trans.*, 12(4):269–275, 2002.

[24] J. Youn and M. Sun. Motion Vector refinement for high-performance transcoding. *IEEE Trans. Multimedia*, 1:30–40, 1999.

[25] Video codec Test Model, TM5, Jan. 31, 1995. ITU Telecommunication Standardization Sector LBC-95, Study Group 15, Working Party 15/1.

[26] W. Kou and T. Fjalbrant. A direct computation of DCT coefficients for a signal block taken from two adjacent blocks. *IEEE Trans. Signal Process.*, 39:1692–1695, 1991.

[27] J. B. Lee and B. G. Lee. Transform domain filtering based on pipelining structure. *IEEE Trans. Signal Process.*, 40:2061–2064, 1992.

[28] Zhijun Lei and N. D. Georganas. H.263 video transcoding for spatial resolution downscaling. *Information Technology: Coding and Computing, 2002. Proc. Int. Conf.*, 425–430, 2002.

[29] Peng Yin, Min Wu, and Bede Liu. Video transcoding by reducing spatial resolution," Image Process., 2000. *Proc. 2000 Int. Conf.*, 1:972–975, 2000.

[30] Nam-Hyeong Kim, Yoon Kim, Goo-Rak Kwon, and Sung-Jea Ko. A fast DCT domain downsampling technique for video transcoder. *Consumer Electronics, 2003. ICCE. 2003 IEEE Int. Conf.*, 36–37, 2003.

[31] R. Dugad and N. Ahuja. A fast scheme for image size change in compressed domain. *IEEE Trans. Circuits Syst. Video Tech.*, 11:461–474, 2001.

[32] P. Yin, A. Vetro, B. Liu, and H. Sun. Drift compensation for reduced spatial resolution transcoding. *IEEE Trans. Circuits Syst. Video Technol.*, 12:1009–1020, 2002.

[33] Y. Yang and Sheila S. Hemami. Generalized rate-distortion optimization for motion-compensated video coders. *IEEE Trans. CSVT*, 10(6):942–945, 2000.

[34] H.-M. Hang and J.-J. Chen. Source model for transform video coder and its applications – Part I: Fundamental theory. *IEEE Trans. CSVT*, 7(2):287–298, 1997.

[35] Z. He, Y. K. Kim, and S.K Mitra. Low-delay rate control for DCT video coding via ρ-domain source modeling. *IEEE Trans. CSVT*, 11:928–940, 2001.

[36] T. Chiang and Y.-Q. Zhang. A new rate control scheme using quadratic rate distortion model. *IEEE Trans. CSVT*, 7(1):246–250, 1997.

[37] A. Puri and R. Aravind. Motion-compensated video coding with adaptive perceptual quantization. *IEEE Trans. On CSVT*, 1(4):351–361, 1991.

[38] Z. Lei and N. D. Georganas. Accurate bit allocation and rate control for DCT domain video transcoding. *Electrical and Computer Engineering, 2002. IEEE CCECE 2002. Can. Conf.*, 2:968–973, 2002.

[39] S. Liu and C.-C. J. Kuo. Joint temporal-spatial rate control for adaptive video transcoding. *Multimedia and Expo., 2003. ICME '03. Proc. 2003 Int. Conf.*, 2:225–228, 2003.

[40] Y. Sun, X.-H. Wei, and I. Ahmad. Low delay rate-control in video transcoding. *Proc. 2003 Int. Symp. Circuits Syst., 2003. ISCAS '03.*, 2:660–663, 2003.

[41] J. Cai and C. W. Chen. A high-performance and low-complexity video transcoding scheme for video streaming over wireless links. *Wireless Commun. Networking Conf., 2002. WCNC2002. 2002 IEEE*, 2:913–917, 2002.

[42] O. Iwasaki, T. Uenoyama, A. Ando, T. Nishitoba, T. Yukitake, and M. EtoH. Video transcoding technology for wireless communication systems. *Vehicular Technol. Conf. Proceed., 2000. VTC 2000-Spring Tokyo. 2000 IEEE 51st*, 2:1577–1580, 2000.

[43] R. Asorey-Cacheda and F. G. Gonzalez-Castano. Real-time transcoding and video distribution in IEEE 802.11b multicast networks. *IEEE Int. Symp. Proc. Computers Commun., 2003. (ISCC 2003)*, 693–698, 2003.

[44] A. Cavallaro, O. Steiger, and T. Ebrahimi. Semantic segmentation and description for video transcoding. *Multimedia and Expo., 2003. ICME '03. Proc. 2003 Int. Conf.*, 3:597–600, 2003.

[45] Y. Q. Liang and Y.-P. Tan. A new content-based hybrid video transcoding method. *Image Process. 2001. Proceed. 2001 Int. Conf.*, 1:429–432, 2001.

[46] H. Sorial, W. E. Lynch, and A. Vincent. Joint transcoding of multiple MPEG video bitstreams. In *Proc. IEEE Int. Symp. Circuits Syst.*, Orlando, FL, May 1999.

[47] Y. P. Tan, Y. Q. Liang, and J. Yu. Video transcoding for fast forward/reverse video playback. In *Proc. IEEE Int. Conf. Image Process.*, Vol. 1, pp. 713–716. Rochester, NY, 2002.

[48] Z. Yu and H. R. Wu. Human visual system based objective digital video quality metrics. *Signal Processing Proceedings, 2000. WCCC-ICSP 2000. 5th Int. Conf.*, 2:1088–1095, 2000.

[49] Zhou Wang, Ligang Lu, and A. C. Bovik. Video quality assessment using structural distortion measurement. *Image Processing. 2002. Proc. 2002 Int. Conf.*, 3:65–68, 2002.

CHAPTER

Embedded Video Codecs 13

Minhua Zhou and Raj Talluri

Texas Instruments, USA

13.1 INTRODUCTION

Digital video capture and playback has become a standard feature in many handheld consumer electronic products. Figure 13.1 illustrates some of the popular, handheld, portable digital multimedia products such as digital still cameras (DSCs), 3G cellular phones, digital camcorders, and personal digital assistants (PDAs). This feature allows the user to capture short video clips in a popular file format such as Apple QuickTime, Microsoft ASF, or MPEG mp4. These captured videos can then be instantly viewed, can be shared via e-mail, or used for later viewing on desktops or lap-tops. Cellular phones, traditionally designed for voice and data transmission, are becoming increasingly popular and are now providing options to capture still images and video clips and, thus, enabling applications such as video messaging, video-phoning, video conferencing, or mobile video broadcasting. The 3G cellular phone, for example, allows users to capture the exciting moments as video messages, share them with friends or loved ones through 3G MMS (multimedia message service). It is also possible to download and watch televised news, movies, or the satellite or terrestrial live television broadcasting on some mobile phones. Video camcorders that have traditionally used analog format to record to magnetic tape are now being replaced by digital format and can record to tape, optical disk, removable memory cards, and, recently, even hard-disk drives. Inexpensive removable memory cards and advanced video-compression technology make it possible to record a huge amount video in digital format, which is easy for editing, encryption, storage, and information sharing.

The tremendous amount of memory and bandwidth requirements of raw video necessitates the need for efficient compression techniques making video-compression technology an integral part of these consumer products. To enable interoperability, most products use standards-based video-compression techniques. The ISO MPEG-1 [1], MPEG-2 [2], MPEG-4 [3], and the ITU-T H.261 [4], H.263 [5], and H.264 [6] are some of the popular international video-compression standards. The essential underlying technology in each of these video-compression standards is very similar [i.e., motion compensation, transform, quantization (Q), entropy coding, and deblocking filter (post

FIGURE 13.1

Embedded video codec application in handheld and portable multimedia products.

or in-loop)] [34]. The standards differ in the applications they address. Each standard is tuned to perform optimally for a particular application in terms of bit rates and computational requirements: MPEG-1 for CD-ROM, MPEG-2 for digital video broadcasting (DVB), and digital video device (DVD), H.263 for videophones, H.261 for videoconferencing, and MPEG-4 for wireless and internet. H.264, the most recent video-compression standard jointly developed by MPEG and ITU-T, provides up to 50% more bit-rate reduction at the same quality of other existing standards. It will most likely be used in a wide variety of applications as a result of this significant advancement in compression technology.

In the handheld products discussed earlier, the video functionality is provided along with other multimedia features that include speech, audio, and graphics. The consumer price points that these products need to meet call for a high level of integration in the silicon solutions used to power these products. These silicon solutions, thus, need to be able to support these multimedia functionalities in a highly integrated and cost-effective manner. In addition, most of these products are battery operated and need to be compact. The long battery life demands and the portable nature of these products require the engines that power these products be extremely energy efficient. In addition, with the ever increasing image and video resolutions, and the complexity of the compression technologies, these silicon solutions need to be extremely high performance.

In the past few years, the rapid progress made in the very large-scale integration technology has dramatically changed the embedded world. Today's high-performance, programmable signal processing chips can comfortably satisfy the high computational and memory bandwidth requirement of video processing while continuing to satisfy the low-power requirements and stringent cost pressures. Although it was hard to implement a cost-effective embedded video codec on programmable processors 5 years ago, it is fairly common nowadays to realize high-quality video capture and playback along with other multimedia functionality on a low-cost digital signal processor (DSP) [35]. The programmable nature of embedded solution makes it economical to support multiple audio/visual formats or add upgrades and new applications with rapid time-to-market.

However, implementing a good-quality embedded video codec on these, highly integrated, low-cost, high-performance, low-power, programmable processors requires many algorithmic and engineering tradeoffs. This chapter describes some of these tradeoffs.

The rest of chapter is organized as follows: Section 13.2 presents an overview of the popular block-based video-coding technique. Section 13.3 addresses the embedded video codec design requirements and constraints. An example for embedded video codec development flow is presented in Section 13.4, and new trends for embedded video codecs are discussed in Section 13.5 followed by a summary in Section 13.6.

13.2 BLOCK-BASED VIDEO CODING

The essential technology in the MPEG (MPEG1, MPEG2, and MPEG4) and ITU-T (H.261, H.263, H.264) video coding standards is a block-based, predictive, differential video-coding scheme. As mentioned in Section 13.1, the fundamental coding techniques involved are motion-compensated prediction, transform ($4 \times 4/8 \times 8$ integer transform for H.264, and 8×8 discrete cosine transform [DCT] for the rest of the standards), Q, entropy coding, and loop-filtering (for H.264 only). In the block-based video coding [21], the sequence of pictures to be encoded is divided into several group of pictures (GOPs). A GOP normally contains an intracoded picture (I-picture) followed by several intercoded pictures [P-pictures (predictive pictures) and B-pictures (bidirectional pictures)]. An I-picture is independently encoded without any relation to the previous pictures (hence the name intracoded), whereas in an intercoded P-picture or B-picture, the current picture is predicted from other reference pictures, and the difference between the current picture and the predicted picture is encoded. A P-picture can only reference the reference pictures in the past, whereas a B-picture can predict from reference pictures both in the past and in the future.

To encode a picture, it is first decomposed into a set of macroblocks. A macroblock contains a 16×16 luminance area and the corresponding chrominance areas of two chrominance components (for 4:2:0 chroma-format, a macroblock contains a 16×16

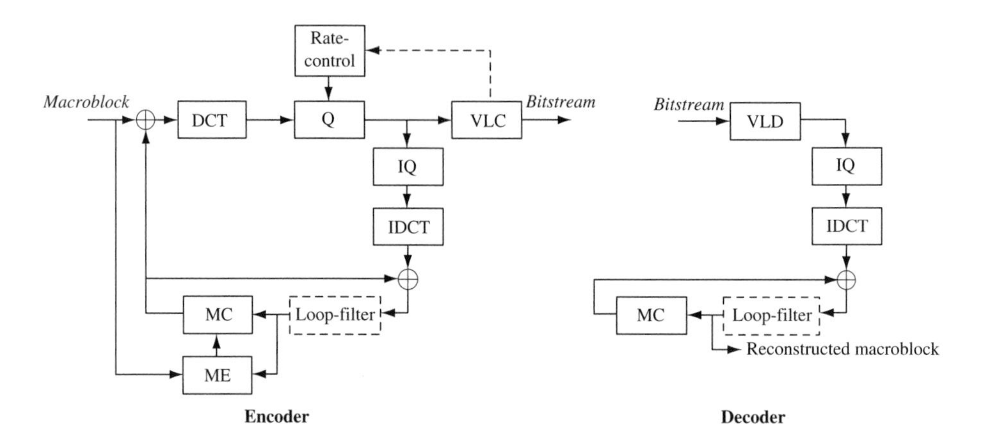

FIGURE 13.2

Block diagram of MPEG/ITU-T video encoder/decoder.

luminance block and an 8 × 8 chrominance block from each of the chrominance components). As shown in Fig. 13.2, on the encoder side, motion vectors are estimated for each intercoded macroblock then the motion-compensated prediction is used to reduce the temporal redundancy. The prediction errors are further compressed by using a transform technique, such as DCT, to remove the spatial correlation. The transform coefficients are then quantized in an irreversible manner that discards the less important information. The motion vectors associated with intercoded macroblocks are differentially predicted. Finally, the differential vectors are combined with the quantized transform information, and are encoded by using an entropy-coding technique such as variable-length coding (VLC). In the case of H.264, an in-loop deblocking filter (loop-filter) is also applied on the reconstructed macroblock to reduce the blocking-coding artifacts [36].

On the decoder side, all the necessary information for picture reconstruction, including motion vectors and transform coefficients, are retrieved from encoded video bitstream by using an entropy-decoding technique such as variable-length decoding (VLD). Inverse *Q* and inverse transform are then used to reconstruct the residual macroblock. A similar motion-compensation technique as on the encoder side is also performed in the decoder to obtain the motion-compensated macroblock. The final reconstructed macroblock is formed by adding the residual and motion-compensated macroblock together and clipping all the pixels to (0:255) range. In the case of H.264, analogous to the encoder side, the loop-filter is applied to the reconstructed macroblock at the decoder also.

This macroblock based–coding technique requires only a small amount of fast data memory, thus makes the video-coding implementation friendly to programmable multimedia processors. In these processors, the high-speed internal memory size is usually very limited due to the low-cost and low-power requirements. In addition, the basic algorithms involved video coding such as motion compensation, transform and *Q* offer a high level of parallelism, which enables the effective real-time implementation on multimedia chips.

13.3 EMBEDDED VIDEO CODEC REQUIREMENTS AND CONSTRAINTS

Low cost, low-power consumption, and high video quality are key requirements in designing a programmable multimedia engine used for handheld applications. To keep the chip cost compellingly low, the on-chip memory size of these multimedia chips needs to be small. Typically, the on-chip program and data memory sizes are in the order of 1000 bytes (e.g., 64 KB, 128 KB, or at best 192 KB). Because of low-power consumption requirements, the clock rate of processors cannot be too high, which restricts the processing power (usually measured in millions instruction per second [MIPS]) and memory bandwidth of these chips. Sophisticated power management logic is also incorporated into these chips to manage power consumption at the lowest possible level. Despite those cost and power consumption constraints, video quality still needs to be high enough to meet the customer's expectation.

In developing an embedded video codec, because of the limited memory and processing resources of the chip, particular attention should be paid to code size, code efficiency, and memory allocation. Code should be written in a way that the entire codec can fit into the on-chip program memory. For devices with on-chip Instruction cache, the cache miss rate should be minimized. There is usually a trade-off between the code size and code efficiency. For example, for an iteration loop with several conditional cases, there are two different ways to write the code. One possible way is to keep a single execution loop that contains all the conditional cases. In this case, the code size is small but the code is less efficient because the in-loop conditional checks consume extra cycles. The other way is to use multiple loops with each loop dealing with a single case of processing. Here the code size is larger but the code executes faster because the conditional checks are now moved outside the loop. Other similar trade-offs should be made between the code size and code efficiency based on the on-chip program memory or cache size.

Memory allocation is another key factor that has significant impact on code efficiency. As the on-chip data memory or data cache is extremely limited, only small amount of data can be buffered on chip. One common strategy is to allocate the most frequently accessed data to on-chip data memory, whereas leaving the less frequently accessed data on off-chip memory. The on-chip data include the static tables used for entropy encoding and decoding, motion vectors, transform coefficients, reference blocks used for motion compensation, and reconstructed macroblocks. Off-chip memory is used to store less frequently used data such as decoded pictures. To minimize intermediate memory requirements, a memory-overlapping technique is often applied to reuse memory space that is freed up by the preceding function blocks. To minimize the time and processor resources spent on data transfers from external memory to processor internal memory, transfers are usually accomplished by an on-chip direct memory access (DMA) controller. Attempts to directly access the off-chip data through general-purpose memory interface will significantly decrease the codec performance because such types of memory access are usually much slower.

For handheld multimedia applications, a cost-effective video codec should keep the program size, on-chip data size, and computational complexity in a reasonable range. To do this, certain compromises between the coding quality and complexity must be made in the embedded video codec design.

Developing a good embedded video codec is about finding a good trade-off between the coding quality and computational complexity. The best possible video-coding quality is usually achievable only for nonreal-time applications such as off-line video content productions for DVB and DVD production. In these applications, the cost of the video encoder is not a critical issue. In these nonreal-time applications, the video content can be encoded using sophisticated and computationally intensive video-encoding algorithms. The encoder can even use brute force, exhaustive search methods to make more optimal decisions on things such as mode selection and motion estimation. However, for the embedded video encoders running on handheld devices because the memory and MIPS are quite limited, only lightweight algorithms can be used to achieve the target speed at the cost of certain quality loss. An embedded video-encoder design is, thus, about achieving optimal video quality for a given memory/MIPS budget and encoding speed requirement.

Some of the most challenging parts of encoder implementation are motion estimation, mode decision, rate control, and rate-distortion (R-D) optimization. These tools have a significant impact on coding quality and speed. Motion estimation generates the motion vectors for the current macroblock so as to reduce the temporal redundancy with the motion-compensated prediction. The mode decision stage is responsible for choosing the best mode for the current macroblock from the all the possible modes supported by the standard. Rate control ensures a video sequence is encoded at the target bit rate. R-D optimization optimizes the video compression by jointly considering the quality degradation and number of bits used. It can be combined with motion estimation, mode decision, and rate control to provide significant quality enhancement in video coding.

Of the encoding tools, motion estimation is usually the most time-consuming part of the encoder in terms of memory access and computation. The complexity of motion estimation depends on the extent of the search range and the number of vectors to be estimated for each macroblock. Search range also determines the size of reference blocks that need to be loaded from off-chip memory to on-chip internal data memory of the processor. In embedded video codecs, it is not realistic to use a full-search, block-matching algorithm to estimate motion vectors for prediction because of the extensive memory and computational requirements of this algorithm. Instead, a fast, lightweight, motion estimation algorithm is preferable to achieve a good trade-off between encoding speed and video quality.

For sequences with small motion or small picture resolutions, the N-step (e.g., $N = 3, 4$) search motion estimation algorithm [11, 12] or diamond search algorithm [13] is widely used. Typical search ranges of $(-16, 15)$ horizontally and vertically are used. The N-step search reduces computational complexity by a factor about 30–40 but still provides a better coding efficiency than that of a full search. However, the reference block size used in N-step search is still a function of search range, and N-step search is also not suited for sequences with large motion or large picture sizes.

Other categories of fast motion estimation algorithms that overcome the shortcomings of the N-step search algorithm are described in refs. [14, 15]. These algorithms use two steps to estimate the motion vectors. First, an initial vector is determined from candidate vectors of its neighboring temporal and spatial blocks, the candidate with the smallest prediction error is chosen as the initial vector. In the second step, a vector refinement with a small search range is carried out around the initial vector to obtain the final vector for the block. This type of fast motion estimation algorithm [15] can provide a factor about 100:1 speed up, usually with a quality loss of about 0.3–0.6 dB. Since the algorithm uses a fixed number of candidate vectors and refinement search range is of fixed size, the memory requirements and the computational complexity of the algorithm are constant. In addition, due to its self-adjustable nature to motion transitions, the algorithm adapts to the motion in the video sequence and, hence, is equally efficient for sequences with slow and fast motion, which makes the algorithm work well for a wide range of sequences.

In developing an embedded video encoder, the mode-decision process needs to be simplified as well. Instead of supporting all the features allowed by the video standard, the encoder may drop some less critical features up front. Doing so will sacrifice some video quality but can dramatically reduce the coding complexity. To minimize the quality loss for a given complexity, careful investigation should be conducted to test the quality impact and estimate the complexity of individual modes so that the least important modes can be dropped in the encoder design. For example, in H.264 baseline profile [6] up to 16 motion vectors per macroblock and 16 reference frames can be used. Searching through all these vectors and then selecting a motion mode with least prediction error would require huge amount of processing power. To avoid this, one may first consider dropping all the vectors of block size below 8×8 and limiting the maximum number of reference frames to two because study [16] shows that the quality impact of such a simplification is marginal but complexity reduction is substantial. Another consideration is to identify the less useful features on the basis of the characteristics of the applications. Dropping those features may not impact the coding efficiency but can ease the computational burden quite a bit. For example, one may skip H.264 features like the slice group map, reference frame reordering and reference frame-marking process from the encoder side in the mobile DVB with H.264 [6].

Rate control is another tool that impacts video quality significantly. Rate-control algorithms are needed to do frame-level bit-allocation according to the target bit rate and frame rate and to compute the Q-scale for macroblocks. The macroblock-level Q-scale computation is expensive because it can require 32-bit multiply and integer division. For constant bit rate applications, additional control needs to be added to the algorithm to guarantee that there is no buffer underflow or overflow at anytime [2, 3]. Special rate-control strategies are usually applied to enhance the coding quality at certain special video situations such as scene cuts, fades, and dissolves. For a wireless video application in which there is normally only one I-frame at beginning in the entire transmission session, the rate-control algorithm must provide intramacroblock refreshment to ensure error recovery. More critically, for two-way video applications like video-phony, the rate control may need to support the functionality such as fast I-frame update (if there are burst errors in wireless transmission [20]) or run-time modification of bit rate, frame rate, and picture size (fading channel [21]).

Because of the limited MIPS in the embedded systems, some computationally intensive features like rate-control strategy dealing with scene changes or local complexity measurement [22] to utilize the properties of human visual systems improving visual quality must be dropped from the rate-control algorithm. Instead of macroblock-based *Q*-scale computation, sometimes the *Q*-scale adjustment may be performed at slice level or even at picture level to save on computation. Nevertheless, the rate control of an embedded video encoder still needs to hit the target bit rate with satisfactory accuracy and support some advanced features such as intrarefreshment and run-time modification of bit rate, frame rate, and picture size for wireless video applications.

R-D optimization [23] has proven to be the most efficient encoding optimization tool that significantly improves the video quality [16]. However, it requires multiple passes of encoding to calculate the R-D cost of different modes. Because of its extremely high complexity, it is very hard to apply the R-D optimization to an embedded video encoder. However, some lightweight R-D optimization concepts, such as joint motion text coding, might be implemented in an embedded system to achieve some additional quality improvements.

Unlike the encoder implementation, which uses algorithms such as motion estimation, mode decision, rate control, and R-D optimization that need careful tradeoffs in design, the decoder implementation is more or less fixed. A decoder compliant to a certain profile of standard must support all the features that the profile specifies. A standard compliant decoder has no freedom to drop any normative feature from implementation. However, for error-prone application like wireless video, error detection and error concealment must be implemented on the decoder side. There is no standardized way to do error detection and concealment; the quality and complexity tradeoff should be made in the design.

Error detection [25] is to ensure no decoder crash in any circumstances, even if nonvideo (e.g., audio/speech) bitstreams are fed into the video decoder. Errors in the bitstream can be detected by checking events such as decoded syntactic element has illegal value, the motion vectors are outside the predefined range, and the number of transform coefficients decoded exceeds the block size [26]. The more conditional checks are put into the decoder, the more robust the decoder will be. However, the conditional checks are expensive instructions on programmable devices; thus, too many checks will surely decrease the decoder speed. A good tradeoff is to implement only the essential conditional error checks into the decoder to minimize the overhead for error detection but still satisfy the error-concealment quality requirements. Such tradeoff requires careful study and extensive testing of the decoder.

The video-coding standards provide error-resilience tools to ensure error recovery once the error is detected. For example, in MPEG-4 [24] error-resilience tools like resynchronization marks, data partitioning (DP), and reversible variable-length coding (RVLC) are specified. Those tools allow decoder to resume decoding at the next resynchronization point and perform decent error concealment for impaired slices by using the preserved motion vectors (DP) or recover partial text data by decoding the packets backwards (RVLC).

Error concealment is employed to repair the lost or corrupted macroblocks of a decoded picture after the errors are detected. In the past decade, several error-concealment techniques have been investigated and developed. The error concealment can be done in the frequency domain [27], in spatial and temporal domain [29–31], or by using the recursive block matching on the decoder side [28].

However, as a result of the high complexity of those algorithms, most of them have not found practical use in the handheld video products. Instead, the most common way to do error concealment is using the memory copy, which replaces the erroneous macroblocks of the current decoded picture with the colocated macroblocks of previous decoded pictures. If there are more MIPS left for the decoder, more advanced error-concealment algorithms can be used. For example, instead of simple memory copy, an erroneous macroblock can be concealed by motion compensation using the motion vectors of its neighboring macroblocks. More sophistically, scene cut detection can be applied on the decoder side to enable adaptive spatial and temporal error-concealment for better error-concealment quality. Just like error detection, the designer has to decide what kind of error-concealment algorithm to apply, by jointly considering the computational and memory requirements of the algorithm and the bit-error rate of the application.

13.4 EMBEDDED VIDEO CODEC DESIGN FLOW

Typically, the following steps are involved in a good embedded video codec development:

- Understanding the chip architecture
- Understanding the codec algorithms
- Modularity and application programming interface (API) definition
- Reference software development in Golden C
- Platform specific development and porting
- Kernel optimization and integration
- Concurrent processing
- Overall optimization
- Stress and conformance testing.

13.4.1 Understanding the Chip Architecture

To achieve the maximum performance, it is very important to first understand the chip architecture. The main factors that should be taken into consideration are processor architecture, memory map, memory access types, instruction set, and any coprocessors supporting the main processor. The chip architecture determines the functional partitioning, memory allocation, and data flow of the codec.

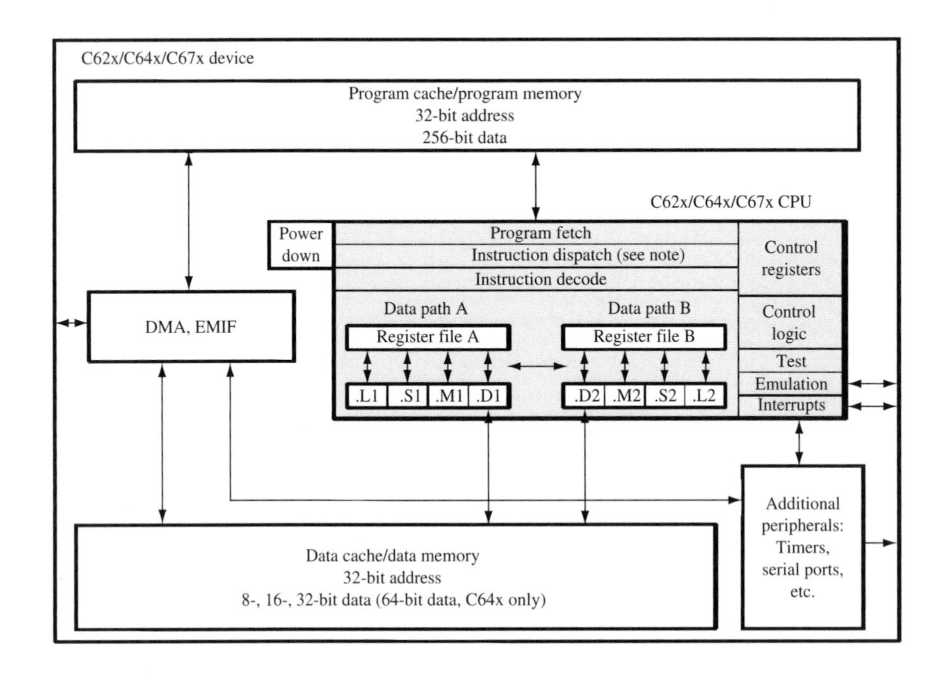

FIGURE 13.3

Block diagram of TMS320C6X.

Texas Instruments TMS320C6X [9] and TMS320C5X [10] DSPs are two typical examples of embedded multimedia processors that are finding many applications in consumer electronic equipment. As shown in Fig. 13.3, The TMS320C6X devices are based on VelociTI™, an advanced very long instruction word (VLIW) architecture. The C6000™ DSP core has eight highly orthogonal functional units, including two multipliers and six arithmetic units, and 32 general-purpose registers of 32-bit length, providing the compiler and assembly optimizer with many execution resources. Eight 32-bit RISC-like instructions are fetched by the computer processing unit (CPU) each cycle. VelociTI's instruction packing features allow these eight instructions to be executed in parallel, in serial, or in parallel/serial combinations. All instructions operate on registers and can execute conditionally. Its memory maps consist of internal memory, internal peripherals, and external memory. The internal program memory can be mapped into the CPU address space or operate as a cache. The internal and external data memory can be accessed by the CPU or DMA, reading or writing in 8-bit bytes, 16-bit half-words or 32-bit words.

The TMS320C54X DSP architecture is different from that of TM320C6X. C54X was primarily designed for speech applications. It includes hardware acceleration for common speech functions such as Viterbi accelerator and single-cycle instructions for FIR filter. Unlike TMS320C6X in which the architecture is highly pipelined, C54X architecture mainly executes instructions in sequential manner. The C54X instruction sets enable

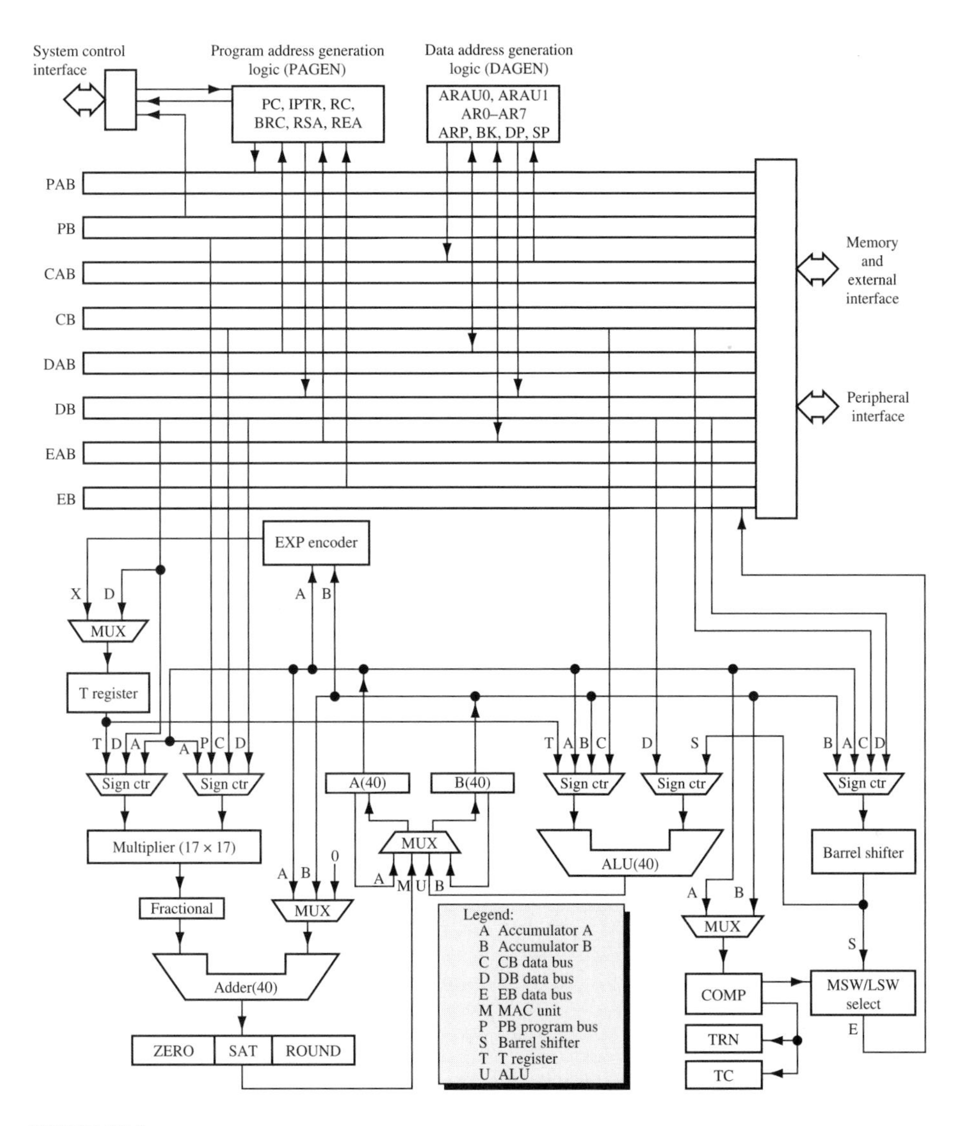

FIGURE 13.4

Block diagram of TMS320C54X.

extremely small code size for DSP functions, allowing the DSPs to take maximum advantage of on-chip RAM. The internal memory can be accessed by CPU in 16-bit or in 32-bit but not in 8-bit byte. There are no conditional instructions on the C54X. As shown in Fig. 13.4, the other key features of the architecture include the following:

- A 40-bit adder and two 40-bit accumulators support crucial parallel instructions that execute in only one instruction cycle.

- A second 40-bit adder available at output of the multiplier allows nonpipelined multiply accumulator (MAC) operation as well as dual addition and multiplication in parallel.
- Single-cycle normalization and exponential encoding support floating-point arithmetic that is useful in voice coding.
- A 17×17 multiplier allows 16-bit signed or unsigned multiplication, with rounding and saturation control—all in one instruction cycle.
- Eight auxiliary registers and a software stack enable an advanced fixed-point DSP C-compiler.

Different chip architectures lead to different programming styles. To achieve the desired performance, the code should be developed in a way that fits the architecture of target device. On TM320C6X, where the CPU can execute up to eight instructions per cycle, the code should be written in a parallel fashion with least amount of memory dependency, so that as many instructions as possible can be scheduled in parallel by the compiler. For iteration loops which contain small amount of instructions inside, it is necessary to unroll the loop. The loop unrolling expands the small loops so that enough number of instructions available to execute in parallel, thus, it utilizes full resources of the C6X architecture.

Since the TM320C54X CPU normally executes only one instruction per cycle, the code can be written in sequential fashion. But attention should be paid to the conditional operations and memory accesses. Because there is no conditional instruction on C54X, the local branches are expensive thus should be minimized. Best effort should be made to avoid having local branches in inner iteration loops. Moreover, video coding mostly uses 8-bit byte per pixel storage format–for example, in motion compensation, the reference blocks are stored in 8-bit per pixel. In this case, two reference pixels (2 bytes) have to be combined and loaded into C54X accumulator as one 16-bit word, then unpacked as 2 bytes at run-time for processing. The codec designer should carefully consider such restrictions of the chip architecture when developing the code to get the most performance out of the chip.

13.4.2 Understanding the Codec Algorithms

Developing an efficient embedded video codec requires not only the knowledge of the chip architecture but also deep understanding of the codec algorithms. Although, the major coding tools used in the video standards are essentially very similar at high-level, they could be significantly different at a more detailed level. Inside each a video-coding algorithm, it is important to understand the dynamic range and semantics of each syntactic element to define the appropriate data type for the element and implement appropriate error detection on the decoder side. For each of the coding tools such as transform, Q, entropy coding, motion compensation, one needs to understand not only the algorithmic detail but also the implementation details such as input/output data accuracy, memory requirement, memory dependency, and any potential parallelism in implementation.

Different chip architectures may lead to totally different implementation for a same coding tool. For example, the most efficient realization of 8×8 DCT transform on TM320C6X is to implement it as fast "butterfly" structure [33], whereas on TM320C54X the fastest execution of the same DCT transform is to implement it as "direct matrix multiply." This is due to the fact that on C6X there are only two multiply and two memory access units among the eight parallel function units, and hence these become the bottleneck for "direct matrix multiply" type implementation of DCT. However, a butterfly structure reduces the number of multiplication and memory accesses, and hence makes it fully suitable for implementation on C6X. On TMS320C54X, there are instructions that can execute two memory accesses, one multiplication, and one addition in a single cycle. These accelerated instructions make the DCT implementation with direct matrix multiply most efficient on C54X.

13.4.3 Modularity and APIs Definitions

Another key concern in developing embedded video codecs is to make the implementation reusable and portable across multiple platforms. One method to accomplish this is to make the codec modular and structured. To enable this, we need to develop APIs for the various components of the software. Figure 13.5 shows two kinds of APIs–high-level APIs and low-level kernels that are defined in an example video codec implementation.

Different applications may have different requirements in the actual instantiation of the video codec based on the use scenario. For example, a video encoder might be required to support run-time modification of bit rate, frame rate, and picture size in the rate-control algorithm. It may also be required to support fast I-picture update whenever it is requested by decoder or to support the conditional intramacroblock refreshment for error-prone applications. Similarly, a decoder might be required to return the decoded vector field, which could be used for implementing different error-concealment or video-stabilization strategies based on the application. In most applications, a video codec is normally integrated with audio/speech codec to realize audio/visual applications. Hence,

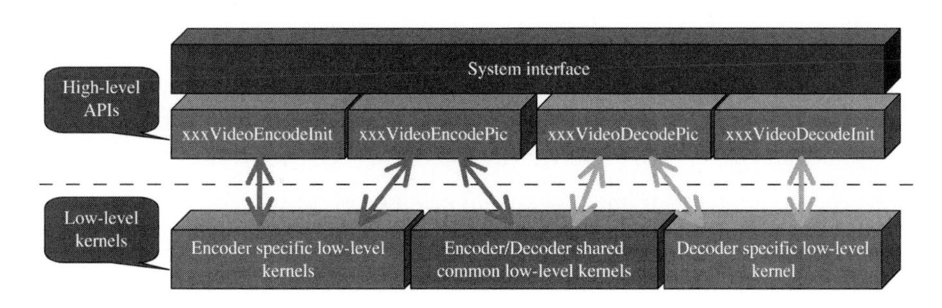

FIGURE 13.5

The relationship between the high-level APIs and low-level kernels in embedded video codec.

it is important to make the video codec modular, so that video system interface can be tailored to integrate easily with the audio encoding and playback. The high-level APIs should be defined to enable such required functionalities easily. We can broadly classify the low-level kernels, into three categories–encoder specific, decoder specific, and encoder/decoder-shared common kernels. It is important to identify the common set of the encoder and decoder functional blocks to minimize the code size, enable reuse, and reduce the development efforts.

We describe below an example for embedded video codec implementation and the associated high-level APIs and low-level kernels. There are four high-level APIs in this example.

1. *VideoEncodeInit(EncodeInitInputPars, EncodeInitOutputPars)* Video-encoder initialization, including encoder memory allocation, sequence-level and picture-level coding parameters initialization, and coprocessor (if any) parameters initialization.

2. *VideoEncodePic(EncodePicInputPars, EncodePicOutputPars)* VideoEncodePic encodes one picture. The input parameters may include the initial quantization scale, target number of bits for the picture, memory location of reference pictures, original picture and reconstructed picture, and memory location of output bitstream. The output might include coding status, number of bits generated by picture, bitstream, and final quantization scale of the picture.

3. *VideoDecodeInit(DecodeInitInputPars, DecodeInitOutputPars)* Video-decoder initialization includes decoder memory location and coprocessor (if any) parameters initialization.

4. *VideoDecodePic(DecodePicInputPars, DecodePicOutputPars)* VideoDecodePic decodes one picture. The input parameters may include memory location of reference pictures and reconstructed picture, memory location of input bitstream, and output picture format. The output may include the decoding status, number of bits used by the decoded picture, motion vector field, and so forth.

The high-level APIs call the low-level kernels to realize the encoding/decoding functionality. The encoder-specific, low-level kernels include the following:

1. Encoder memory allocation: centralized memory allocation on the encoder side.

2. Encoder high-level parameters initialization: initialization of parameters such as picture size, target bit rate, and so forth.

3. Encoder coprocessor initialization: initialization of the coprocessors.

4. Rate control: bit-allocation to each macroblock according to bitstream buffer status and macroblock content.

5. Motion estimation: estimation of the motion vectors for a macroblock.

6. Mode decision: selecting the best-coding mode (e.g., intra/inter, one or four motion vector mode) for the current macroblock.

7. Forward transform: For example, 8 $\times$ 8 DCT transform.
8. Forward Q of transform coefficients.
9. Sequence-level header encoding.
10. Picture-level header encoding.
11. Slice-level header encoding.
12. Macroblock overhead encoding: encoding macroblock and block type such as Intra, Inter, codec, skipped, and so forth.
13. Motion vector encoding.
14. Entropy encoding of transform coefficients.
15. Intra (AC/DC) prediction.
16. Encoder utility routines for bitstream handling.

The decoder-specific, low-level kernels are defined as below (error detection is normally implemented in kernels 1–6):

1. Sequence-level header decoding
2. Picture-level header decoding
3. Slice-level header decoding
4. Macroblock overhead decoding: decode macroblock type (MTYPE), coded block pattern (CBP), and so forth
5. Motion vector decoding
6. Entropy decoding of transform coefficients
7. Inverse intra (AC/DC) prediction
8. Decoder utility routines for bitstream handling
9. Error concealment: to conceal corrupted or lost macroblocks by using certain error-concealment scheme such as the memory copy.

The common low-level kernels shared by encoder and decoder are as follows:

1. Inverse quantization of transform coefficients
2. Inverse transform
3. Motion compensation
4. Macroblocks reconstruction: adding prediction blocks and inverse transform output together and clipping results to [0:255] range
5. Loop-filter (for H.264 only)

6. Boundary macroblock padding: prepad the reference picture boundaries for unrestricted motion compensation (for MPEG4, H.263, and H.264 only)
7. Data reorganization (e.g., YUYV interleaved to YUV separated).

To share the common low-level kernels, the encoder and decoder must have common picture storage format.

13.4.4 Reference Codec Software Development in Golden C

One of the key recommended steps in developing an embedded video codec is to first develop a platform independent, video codec implementation in what is called as *Golden C*. The C programming language is usually the language of choice for embedded video codec development. The goal of this implementation is to be platform independent but serve as an efficient reference for the final platform-specific implementation. One of the key uses of this is that we can always refer back to this implementation to ensure compliance of the final implementation to the standard. Other advantages include the ability to quickly be able to target this Golden C implementation to various platforms and various applications as needed.

For most of the popular video-coding standards, the trend nowadays is for the standards bodies to provide a public domain reference software implementation of the standard. However, this reference code usually meant to be instructional about the standard and is not written to be an efficient implementation of the standard, particularly from an embedded system implementation point of view. Therefore, in creating an efficient and optimal implementation of a standard, it is recommended to implement the entire codec in Golden C from scratch, taking the standards-based public domain codec as the functional reference code.

If both the encoder and decoder need to be developed, it is recommended to first start with the decoder implementation. For the decoder, there is a conformance bitstream test suite available for exercising all the features required by standard. The decoding of all the conformance bitstreams will guarantee a standard-compliant implementation of decoder. Once the compliant decoder is in place, the encoder can then be built up by reusing the common low-level kernels by developing the encoder counterparts of the decoder-specific kernels, and finally adding the encoder-specific motion estimation, mode decision, and rate-control algorithms. In this manner, most parts of code are similar for both the encoder and decoder, making the debugging of the codec much easier.

Some of the recommended programming rules for the Golden C implementation are as follows:

1. No usage of dynamic memory allocation such as "malloc()." All the memory allocation should the centralized and performed in the codec initialization API. Certain memory overlapping may be used to minimize the "on-chip" memory requirement. Carefully check memory leak and memory-initialization issues by using tools such as "purify." Memory leak can easily cause failures when the code is ported on the embedded platforms which are highly memory constrained.

2. Picture buffers should always be located in the on "off-chip" memory external memory. No DMA to the picture buffer should be assumed. Dedicated block writing/reading routines should be implemented to perform data transfers between the picture buffer and "on-chip" data buffer. The typical operations are writing or reading bitstream to and from off-chip memory, loading the original macroblocks and reference blocks from the picture buffer, and writing out reconstructed macroblocks to the picture buffer. The dedicated data-transfer routines will be replaced by the DMA routines when the code is ported on hardware board.
3. No global variables should be used. Even for the static tables such as entropy encoding and decoding tables, they should be made reallocable and passed down to functions as pointers. Using global variables makes code migration from one codec to other difficult.
4. No big arrays (larger than 128 bytes) should be opened inside the local functions. Big local data arrays could cause stack overflow problem.
5. Use the fixed-point implementation only. Minimize the usage of divisions, covert divisions as multiplication, and right-shift operations whenever it is possible.
6. Run-time support functions such "printf" should be called by using defined macros so that they can be easily removed from code at compiling time if needed.

13.4.5 Platform-Specific Development and Porting

Starting from the Golden C implementation, the platform-specific implementation is then developed by taking the hardware architecture of target platform into account. The platform-specific implementation is designed to be able to compile and run on the target device and to produce the bit-exact results as the Golden C implementation. Certain modifications need to be introduced to the Golden C code to form the platform-specific implementation. The major changes are as follows:

1. Memory map: The codec initialization API to do memory allocation for the on-chip and off-chip data buffers need to be modified according to the actual memory map of the chip.
2. Memory addressing: Some platforms such as TM320C5x may not support 8-bit byte memory accesses. Hence, the related routines should be rewritten to avoid unsupported memory access type.
3. Data transfer between on-chip and off-chip memory: Data transfer routines that are used for data transfer between on-chip and off-chip memory should be replaced by the actual DMA utility routines.
4. Run-time support library: Functions such as "printf" should be removed because of the large code-size of these functions.

5. System support routines: The hardware system initialization routines and other system support routines should be integrated into the project when compiling the code on the hardware device.

The platform-specific C-code should compile and run on both the device simulation platforms like PC/UNIX and the target device hardware system. It is suggested to keep Golden-C part of code for item 1, 3, and 4, and to turn on item 5 only when code is complied for hardware device. Compiler directives can be used to switch between the Golden C implementation and the platform-specific implementation for related routines. Keeping the platform-specific software executable on PC/UNIX simulators makes it easy to introduce future changes to the code since these changes can be first debugged using the more advanced development tools on PC/UNIX device simulators and then later ported on the device hardware platform. Moreover, having a bit-exact platform-specific C reference code running on PC/UNIX is extremely important for efficiently resolving any quality or conformance issues discovered after the codec is optimized on the target platform.

13.4.6 Kernel Optimization and Integration

At this point in the flow, the platform-specific implementation running on the device is still not optimized in terms of performance. It mainly serves as a functional reference for next stage of kernel-level optimization on the target device. The final optimized version of codec is normally a hybrid version, with high-level control logic in C and low-level kernels in assembly or other coprocessor languages. All the encoder/decoder-specific and common low-level kernels are candidates to be optimized. The kernel optimization is usually the most time-consuming stage in the embedded video codec development.

The chip architecture determines the specific kernels to be optimized. For devices such as TMS320C6X, in which a highly efficient C-compiler is available, C-level kernel optimization with loop unrolling and intrinsic operators can already provide about 90% of performance optimization for highly pipelined kernels such as sum of absolute differences computation in motion estimation and block interpolation in motion compensation. Only highly interdependent kernels such as entropy coding, need to be carefully scheduled and optimized in linear or hand assembly. On the platforms like TMS320C54X, however, the C-compiler can achieve only about 50% of performance optimization on the well-designed C-code. Thus, assembly optimization for all the kernels is the only method to achieve the best possible performance for the codec.

It is more challenging to carry out kernel optimization on platforms with multiple processors. Nowadays, the multimedia chips that support video functionality often times consist of several processors, with hard-wired blocks for some of the computationally intensive function.

Memory access can also consume significant amount of cycles. Sometimes, the memory access is very expensive if the clock rate is high. Therefore, it is important to minimize the number of memory accesses. For a device like TM320C6X with single cycle, 32-bit, 16-bit, and 8-bit memory access, it would be more efficient to modify the kernels to combine 16-bit or 8-bit memory access into 32-bit memory access. Sometimes, it is advantageous to restructure the kernels to reduce the number memory accesses. For example, combining entropy decoding and inverse *Q* kernel into one on the decoder side can cut the memory access cycles by half of the number that would be needed if those two kernels were implemented separately. Avoiding or minimizing the memory bank conflicts in these memory accesses is also important for the performance.

Another optimization step is to minimize the DMA-transfer overhead. For some platforms, there is significant overhead caused by issuing DMA requests. Hence, the number of DMA requests per macroblock should be kept as low as possible. If there is enough on-chip data memory on the device, the common strategy is to combine data of several macroblocks into one group, and use DMA engine to move the data piece and piece instead of macroblock by macroblock to minimize the number of DMA transfers.

Once the kernel is optimized, it is strongly recommended to integrate the kernels into the codec one by one. Because there is a standalone C-reference function available for each kernel, this methodology is advantageous for debugging the optimized kernels and localizing the bugs if there are problems in the integration phase.

13.4.7 Concurrent Processing

Once the kernel optimization and integration is done, the next performance optimization step is concurrent processing. If there are multiple processors on the chip that can run independently, then some of the tasks can be run concurrently to further improve the codec performance.

Figure 13.6 illustrates one case of concurrent processing on the video-decoder side. In this case, there are two processing units on the chip, one is the CPU (e.g., TMS320C6X core) and the other is DMA. DMA can run without intervention of CPU, once it is set up and kicked off by the CPU. As shown in Fig. 13.6, after the motion vectors are decoded for the current macroblock, the location of reference blocks in SDRAM is known. The CPU can setup the DMA and issue the DMA request to load the reference blocks for the motion compensation. While the CPU is performing the VLD, inverse *Q* and IDCT, the DMA is loading the reference blocks from SDRAM to on-chip data memory in the background

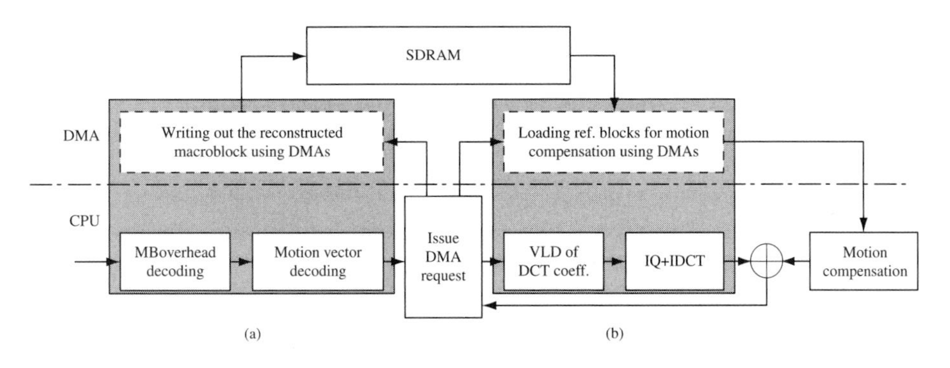

FIGURE 13.6

Example of concurrent processing between DMA and CPU for video decoding.

(see block b of Fig. 13.6). After the motion compensation and the reconstruction of the current macroblock, another DMA request can be issued to write the reconstructed macroblock from on-chip memory to SDRAM. Here the CPU is doing the macroblock overhead and vector decoding for the next macroblock, while the DMA is writing out the current reconstructed macroblock (see block a of Fig. 13.6). Hence, with the concurrent processing between the CPU and DMA, the decoder performance is improved because the data transfer between on-chip and off-chip memory is hidden behind the processor tasks.

The more processing units a chip has, the more complicated the concurrent processing will be. If a device has several processing units (e.g., on TMS320DM270 [8] there are five processing units, which are as follows: C54X, iMX, VLCD/QIQ, DMA, and ARM7), the concurrent scheduling becomes complex because it needs to resolve the resource conflicts between tasks running on the different processing units. The concurrent scheduling normally involves the following steps:

1. Benchmarking the entire codec; getting accurate cycle count measurement for all the kernels
2. Partitioning the encoding and decoding processes into tasks–one task may contain several steps of processing (kernels)
3. Drawing the dependency (processing unit and memory buffer usage, timing dependency) diagram of tasks and repeating step 2 to resolve the resource conflicts. Computing the cycle counts for each task according to kernel benchmark
4. Drawing the concurrent scheduling diagram according to the task cycle count, timing, and resource usage
5. Modifying the code according to the scheduling diagram to realize the concurrent processing.

13.4.8 Overall Optimization

After the concurrent processing is done, the final tuning of the code can further increase the codec performance. The overall optimization is mainly for the embedded video codec running on a device with multiple processors. The main idea is to measure the loading of each processor, identify which processor is the bottleneck, and then consciously move some kernels from one processor to another to balance the loading among different processors. If a kernel needs to be moved to a different processor for the purpose of balancing processor loading, optimization for the particular kernel will need to be redone according to the architecture of the new processing unit, and the concurrent scheduling might need to be modified as well.

A typical example of such a multiple processor chip is TMS320DM270. The TMS320DM270 programmable DSP-based solution from Texas Instruments is a highly integrated video and imaging engine offering excellent performance, leading edge process technology, and flexibility for next generation portable media products. The DM270 (see Fig. 13.7), which contains the TMS320C54XTM DSP, the ARM7TDMI®RISC

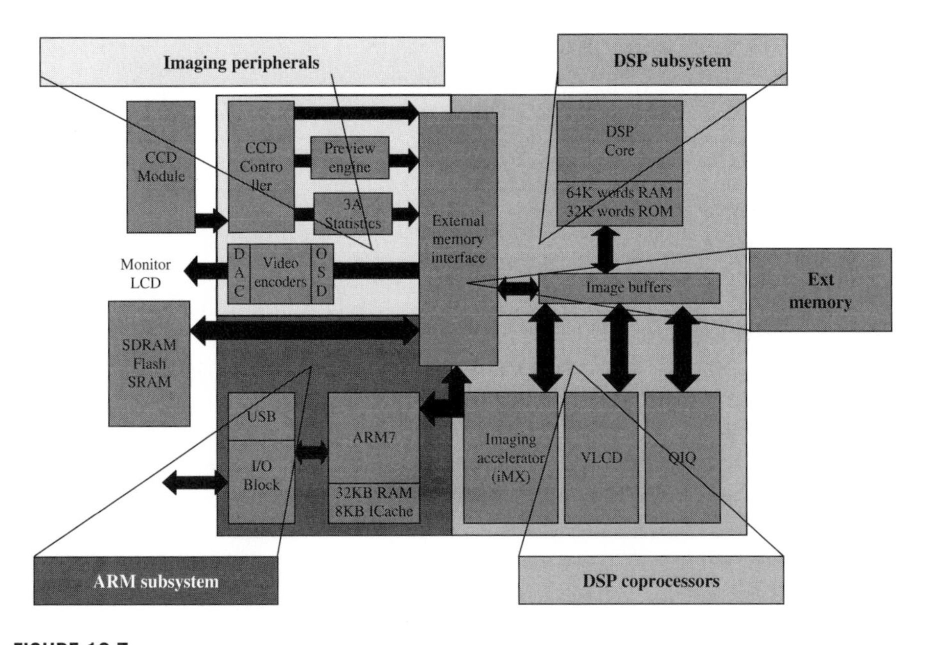

FIGURE 13.7

Diagram of TMS320DM270.

processor, imaging peripherals plus video and imaging coprocessors, is a highly integrated, programmable platform for the DSC and other multimedia applications including portable multimedia jukeboxes, camera phones, DVD players, televisions, and digital video recorders. DM270's programmability comes from a DSP-based imaging coprocessor that enables manufacturers to implement their own proprietary image processing algorithms in software. The interface is flexible enough to support various types of CCD and CMOS sensors, signal-conditioning circuits, power management, SDRAM, shutter, iris, and auto-focus motor controls.

The programmable DM270 supports a variety of video, imaging, audio- and voice-compression standards, including, but not limited to, JPEG, motion-JPEG, MPEG-1, MPEG-2, MPEG-4, H.263, H.264, DivX, Windows Media Video (WMV), as well as audio standards such as MP3, Advanced Audio Coding (AAC), and Windows Media Audio (WMA). Supported voice standards include G.711, G.723.1, and G.726. The DM270 system-on-a-chip (SoC) has the ability to run various operating systems, including Nucleus, Linux, ulTRON, and VxWorks.

The DM270 architecture represents the cutting-edge trend for the programmable multimedia chip design, with the most critical parts of the imaging/video function blocks in hardware and the less critical parts in software. This architecture provides customers with both the high performance and flexibility. The DM270 DSP-subsystem, on which imaging/video function is realized, is made of the C54X DSP, IMX accelerator, VLCD/QIQ coprocessor, DMA controller plus the dedicated on-chip memory buffers. The VLCD/QIQ coprocessor supports VLC/VLD/Q/IQ of DCT-coefficients for MPEG1, MPEG2, MPEG4, H.261, and H.263. The imaging-processing accelerator, iMX, is designed for speedup of high-pipelined image/video processing functions, such as motion estimation, motion compensation, and DCT transform.

For the MPEG4 Simple Profile [3] decoder implementation on DM270, for example, kernels are running on different processors. The iMX is used for all the parallel operations like motion compensation, inverse DCT transform, macroblock reconstruction, data reorganization, and so forth. The VLCD/QIQ coprocessor performs the MPEG4 VLD and inverse Q, while the C54X DSP is responsible for the VOL, VOP, Slice and macroblock header decoding, and other high-level control code.

In implementing an MPEG-4 codec on DM270, no assembly optimization will be needed for Q, inverse Q, VLC/VLD of the DCT-coefficients. Simple register setup for the QIQ/VLCD engine will do the job as there are dedicated coprocessors for these functions. While for kernels like motion estimation, motion compensation with high-level of parallelism, iMX coprocessor should be the used. The rest of the kernels are optimization on C54X in assembly language.

With the overall optimization done, the DM270 is able to provide simultaneous playback of 30 fps MPEG-4 simple profile at VGA (640×480) resolution simultaneously with AAC-LC audio decoder.

13.4.9 Stress and Conformance Testing

After the codec optimization is done, extensive testing should be carried out before the code is released in market. Code testing under unusual conditions may uncover some deep system-level bugs rarely seen under normal circumstances. The code testing may prove to be a very time-consuming process for embedded video codec development.

For an embedded video encoder, the testing is to make sure that the encoder generates the standard compliant bitstreams with market acceptable quality for target applications. It would be helpful if the encoder can be tested with field trial sequences to verify the coding quality, effectiveness of the designed encoder functionality and the robustness of the code. The encoder also needs to run stressing tests to make sure that it does not crash after a long time of continuous encoding. The encoder also needs to be tested with video sequences of varying levels of scene complexity from low motion to high motion and low spatial detail to high spatial detail. The standards bodies also usually provide a set of standard video sequences to test the video codec. It is also important to test the codec under varying lighting conditions.

The testing of an embedded video decoder is more complicated. For the functional tests, the conformance bitstream test suite can be used to verify whether the decoder output on the device is identical to the anchor sequences corresponding to these bitstreams. For robustness tests, the decoder can be tested with long bitstreams. However, for the error protection part of tests, significant amount of time will have to be spent. The decoder is required to be functional under all circumstances. The decoder is not supposed to crash even if the decoder is fed with "garbage" bitstreams. A common method to develop the robust error-protection scheme is to take a compliant bitstream, which excises all major coding features, corrupt the data byte by byte so that error type and position are known in advance, and input the corrupted data into the decoder to see if it crashes. If it does crash, knowing the error type and position is a great help to find the crash reason and to fix the problem.

Doing more up-front C-reference encoder/decoder testing before porting the code on the platform may prove to be an efficient way of reducing the overall codec testing time. Compared to the platform codec testing, which is expensive and time-consuming, C-code has advantages of fast running speed, easy-upgrade and friendly debugging, which enables testing the codec quality and conformance with a large amount of testing vectors in a relatively short amount of time. Moreover, PC or UNIX platforms are normally well-tested and stable, which allows developers to focus on debugging software when codec quality or conformance issues arise; this avoids confusion about whether the issues are caused by bugs in hardware, or in software, or in both, as developers always face on embedded platforms that oftentimes are not stable enough especially in sample chip phase. Having a bug free and quality proven C-reference code upfront will help avoid most of quality and conformance issues in the platform-optimized code, as long as the bit-exactness to the C-reference code is strictly maintained during the codec optimization. This essentially shortens the time for the platform codec testing and improves the overall codec testing process.

13.5 NEW TRENDS

The image and video resolutions on handheld devices are increasing rapidly. Nowadays, many portable digital multimedia products can already provide video capture and playback feature at 720@30p. As the next requirement in the portable video space the video resolution will soon move to 1080@30p. But this is not the end of story, industry is interested in 1080@50/60p and 1080@120p video on portable devices in the foreseeable future; 4Kx2K@24/30/60p video [37], now defined for studio video applications, could be reality on portable customer products in long run.

While the video resolution moving to high definition (HD), the number of required video codec formats is also increasing. Particularly in the cellular phone space, it is required to support not only the MPEG and ITU-T video standards such as MPEG-4, H.264, H.263, but also several proprietary codecs such as Microsoft VC1, Real video, ON-2, Sorenson Spark H.263, and DivX.

The ever growing requirements in video resolutions and number of codec formats pose new challenges for chip design and video codec development to come up with low-cost, low-power consumption, and high-quality embedded video solutions. To support the HD video resolution, a significant increase in processing power, memory bandwidth, and memory size is inevitable, which imposes burden on both power consumption and chip cost. The requirement of multiple codec support puts further pressure on chip area because of additional logic required.

Although, we can still rely on the advance in the silicon process technologies to reduce chip cost and lower the power consumption, process technology alone will not be able to address all the challenges that we are facing. Trend for improving processor performance to meet the HD video requirements, and in the meantime keeping chip cost and power consumption in check is to move away from single-processor architectures to multicore processors or multiprocessor architectures with more dedicated hardware accelerations for video processing. An increasing number of video functional blocks (e.g., motion estimation, motion compensation, entropy coding, transform and Q, loop-filters, and so forth) are now hardwired and off-loaded to the hardware accelerators, leaving only essential operations remain on the programmable processor for customers to create differentiations in rate control, mode decision, and error concealments, and to support future unknown video standards. The multiprocessor architecture has the ability of achieving a same throughput at a low clock-speed by processing multiple macroblocks in parallel. The low clock-speed results in a lower supply voltage for the chip, and thus enables the HD video capture with low-power consumption.

The growing number of hardware accelerators on the multiprocessor architecture has significantly changed the landscape of the embedded video codec development. Since almost all the critical video functions are hardwired, codec performance is no longer dependent on the kernel optimization, instead, it now heavily relies on how the concurrent schedule is deigned to run multiple processors (hardware accelerators) in parallel. Therefore, the concurrent scheduler design becomes the key for exploring the full chip potential, and a challenging task as well. It is highly recommended not to use hard-coded scheduler. Designing a hard-coded scheduler for best codec performance is only possible if the number of processors on the chip is very limited (e.g., less than three). When the number of processors goes higher, it becomes difficult to find out a hard-coded scheduler for optimal performance. Moreover, the readability of the scheduler code deteriorates fast, which makes it very time-consuming for modifying the concurrent processing if codec requirements change. Therefore, it is suggested to employ a programmable concurrent scheduler, which is lightweight but has high code readability and flexibility for rescheduling the concurrent processing to accommodate new features or requirements.

The full-scale dedicated video hardware accelerations on multiprocessor architectures also make the chip architecture validation much more complicated. In the past, for the full programmable single processor architecture with general-purpose instruction sets, the codec development and architecture validation were separate steps: The architecture validation validated functionality and performance of the specified instruction sets, whereas codec developers managed to map the codec on the platform by using the instruction sets, and validated the codec conformance and performance afterwards. Today's multiprocessor architectures require full validation of video codecs before architecture freeze. That is, not only the functionality of each video hardware accelerator needs to be validated, but also the performance and conformance of each supported video codec are verified by running hardware accelerators concurrently. This essentially combines the chip architecture validation with the codec development into a single step. The common practice is to develop the video codecs on a chip simulator before the silicon. On the simulator, the underlying hardware accelerators are simulated by using C-models, the chip architecture is validated in terms of codec functionality and performance by running the codecs on the simulator with testing vectors; when silicon comes out, the video codecs are quickly made available by simply replacing the C-models of hardware accelerators with actual hardware setups. Therefore, the embedded video codec development is becoming an integral part of chip design process and plays an important role for chip architecture validation.

Having a bit-exact C-reference code upfront becomes more important than ever because of HD video requirement and high degree of hardwired video acceleration mentioned earlier. The HD video requirement puts the cycle budget for video capture and playback on the HD video chips in the matter of 1000 cycles per macroblock, which leaves no room for software workaround methods to correct hardware design bugs from the performance point of view. Therefore, it is highly recommended to use the C-reference code to test the functionality, quality, and performance of the video codecs before architecture freeze, to identify any software/hardware interfacing issues, missing features of the hardware accelerators, codec quality, and performance issues upfront; failing to do so could jeopardize the product development for the simple fact that correcting hardware design errors after silicon is out is very expensive and time-consuming.

Moving forward, the low-cost, low-power consumption should be the key requirements for the new video technology development [38, 39]. The embedded video market has been growing fast, the annual shipment already reach hundreds of millions of units, but this market is still relatively new. In fact, none of existing video standards have seriously considered the special requirements and constraints of portable video applications during the standardization because this market did not exist in large scale, when standards were being specified. Taking H.264 as example, lots of advanced video-quality enhancement tools are not used on the portable devices because of memory bandwidth and computational complexity constraints. As video goes to HD resolution in portable video space, memory bandwidth and computational complexity constraints will be an even bigger limiting factor for video quality. Therefore, taking those constraints into consideration would help develop new coding tools that provide better tradeoffs on portable devices and thus lead to quality improvements for embedded video applications. To lower the power consumption in a most cost-effective way, it is strongly desired to increase algorithm-level parallelism in the future codec algorithm design [40] for parallel processing at algorithm-level does not require memory replication as slice-level or picture-level parallelism does. As the video requirement increases in terms of frame rate and picture resolution, the power consumption spent on the on-chip/off-chip data transfer becomes more significant; the in-loop memory compression technique [41] is another promising tool, which can lower the codec power consumption and extend battery life for the batter-driven devices.

13.6 SUMMARY

Digital video capture and playback has become a standard feature for many handheld consumer electronic products such as DSCs, camera phones, camcorders, and PDAs. Video compression is used in these products for accommodating the limited storage and transmission requirements. Video-compression standards (MPEG/ITU-T) are popular as they guarantee interoperability among the products. The block-based nature of these standards makes them implementation friendly to embedded video solutions on programmable multimedia chips. Programmable nature of embedded solution provides a cost-effective way to support multiple audio/visual formats or add upgrades and new applications with rapid time-to-market. Due to the limited memory size and processing power of programmable multimedia chips used in consumer appliances, the embedded video codec design needs to make good tradeoff between quality and complexity. Particularly, efforts should be made to develop the lightweight algorithms for motion estimation, mode decision, rate-control, error detection, and error concealment. Understanding of both the chip architecture and codec algorithm is the key for designing the most efficient embedded video codec. Characteristics of chip architecture and codec algorithm determine the code partitioning, data flow and memory allocation of the codec. A good tradeoff between code size and code efficiency, minimization of memory access, memory bank conflicts, cache miss rate and DMA transfer overhead, and balanced processor loading in concurrent mode are some of the key factors that contribute to an efficient implementation of embedded video codec. As the video goes to HD on portable devices, multiprocessor architectures with more dedicated hardware accelerations for video processing are becoming more popular. More upfront testing with C-reference code, a combined architecture validation and embedded video codec development, and concurrent processing with a programmable scheduler are the key for developing low-cost, low-power consumption, and high quality HD video solutions for embedded video applications.

REFERENCES

[1] ISO/IEC 11172-2 (MPEG1), Information technology - Coding of moving pictures and associated audio for digital storage media at up to about 1.5 Mbit/s - Part 2 Video, 1993.

[2] ISO/IEC 13818-2, ITU-T Rec. H.262: 1995, Information technology - Generic coding of moving pictures and associated audio information - Part2: Video.

[3] ISO/IEC 14496-2, November 1998, International Organization for Standardization, Final Draft of Internal Standard: Information technology Coding of Audio-Visual objects: Visual.

[4] International Telecommunications Union Telecommunications Standardization Sector, Recommendation H.261: video codec for audiovisual services at px64 kbits, Helsinki, 1993.

[5] International Telecommunications Union Telecommunications Standardization Sector, Recommendation H.263: video coding for low bitrate communication, Geneva, 1996.

[6] ISO/IEC 13818-10, ITU-T Rec. H.264: 2003, Information technology - Coding of Audio-Visual objects: Part 10: Advanced Video Coding.

[7] M. Zhou and R. Talluri. DSP-Based Real-Time Video Decoding, *IEEE Digest of ICCE'99 (International Conference on Consumer Electronics '99)*, pp. 296–297, Los Angeles, CA, 1999.

[8] TMS320DM270 product fact sheet, from Texas Instruments, http://focus.ti.com/pdfs/vf/vidimg/dm270fs.pdf.

[9] TMS320C6X User Guides, from Texas Instruments, http://dspvillage.ti.com/docs/catalog/resources/techdocs.jhtml?familyId=132&navSection=user_guides.

[10] TMS320C5X User Guides, from Texas Instruments, http://dspvillage.ti.com/docs/catalog/resources/techdocs.jhtml?familyId=114&navSection=user_guides.

[11] R. Li, B. Zeng, and M. L. Liou. A new three-step search algorithm for block motion estimation. *IEEE Trans. Circuits Syst. Video Technol.*, 4(4):438–442, 1994.

[12] L. M. Po and W. C. Ma. A novel four-step search algorithm for fast block motion estimation. *IEEE Trans. Circuits Syst. Video Technol.*, 6(3):313–317, 1996.

[13] S. Zhu and K.-K. Ma. A new diamond search algorithm for fast block matching motion estimation. *Int. Conf. Inform. Commun. Signal Proc.* (ICICS '97), Singapore, pp. 292–296, 1997.

[14] L. K. Liu and E. Feig. A block-based gradient descent search algorithm for block motion estimation in video coding. *IEEE Trans. Circuits Syst. Video Technol.*, 6(4):419–423, 1996.

[15] M. Zhou. A fast motion estimation algorithm for MPEG-2 video encoding. *SPIE Proc. VCIP'99 (Visual Communication and Image Processing '99)*, 1487–1495, San Jose, CA, 1999.

[16] M. Zhou. Simplification of H.26l baseline coding tools. JVT-B030, 2nd JVT meeting, Geneva, 2002

[17] H. S. Kong, A. Vetro, and H. Sun. Combined Rate Control and Mode Decision Optimization for MPEG-2 Transcoding with Spatial Resolution Reduction. *IEEE Int. Conf. Image Process. (ICIP)*, 1:161–164, 2003.

[18] D. Turaga and T. Chen. Classification based mode decisions for video over network. *IEEE Trans. Multimedia* 3(1):41–52, 2001.

[19] M. Hamdi, J. W. Roberts and P. Rolin. Rate Control for VBR Video Coders in Broadband Networks. *IEEE JSAC - J. Sel. Areas Commun.*, 15(6):1040–1051, 1997.

[20] C. Y. Hsu, A. Ortega, and M. Khansari. Rate Control for Robust Video Transmission over Burst-Error Wireless Channels. *IEEE J. Sel. Areas Commun.*, 17(5):756–773, 1999.

[21] J. Razavilar, K. J. Ray Liu, and S. I. Marcus. Jointly optimized bit-rate/delay control policy for wireless packet networks with fading channels. *IEEE Trans. Commun.*, (3)484–494, 2002.

[22] ISO/IEC JTC1/SC2/WG11/N0400, April 1993, Test Model 5, Draft Revision 2.

[23] G. J. Sullivan and T. Wiegand. Rate-Distortion Optimization for Video Compression. *IEEE Signal Process. Magazine*, 15(6):74–90, 1998.

[24] Raj Talluri. Error-Resilient Video Coding in the ISO MPEG-4 Standard. *IEEE Commun. Magazine*, 36(6):112–119, 1998.

[25] K. Ekram, G. Hiroshi, S. Lehmann, and G. Mohammed. Error Detection and Correction in H.263 coded video over wireless network. The 12th International Packetvideo Workshop (PV 2002), 2002.

[26] K. Bhattacharyya and H. S. Jamadagni. DCT coefficient-based error detection technique for compressed video stream. *IEEE Int. Conf. Multimedia Expo.*, 3:1483–1486, 2000.

[27] Y. S. Lee, K.-K. Ong, and C.-Y. Lee. Error-resilient image coding (ERIC) with smart-IDCT error concealment technique for wireless multimedia transmission. *IEEE Trans. Circuits Syst. Video Technol.*, 13(2):176–181, 2003.

[28] M.-J. Chen, C.-S. Chen, and M.-C. Chi. Recursive block-matching principle for error concealment algorithm. *Proc. 2003 Int. Symp. Circuits Syst. (ISCAS2003)*, 2:528–531, 2003.

[29] W. Zeng. Spatial-temporal error concealment with side information for standard video codecs. *Proc. Int. Conf. Multimedia Expo. (ICME)*, 2:113–116, 2003.

[30] S. Belfiore, M. Grangetto, E. Magli, and G. Olmo. Spatial-temporal video error concealment with perceptually optimized mode selection. *Proc. IEEE Int. Conf. Acoust. Speech Signal Process. (ICASSP)*, 5:748–751, 2003.

[31] W. Y. Kung, C. S. Kim, and C.-C. J. Kao. A spatial-domain error concealment method with edge recovery and selective directional interpolation. *Proc. IEEE Int. Conf. Acoust. Speech Signal Process. (ICASSP)*, 5:700–703, 2003.

[32] T. P. Chen and T. Chen. Second-generation error concealment for video transport over error prone channels. *IEEE ICIP 2002*, paper MA-L1, Rochester New York, 2002.

[33] E. Feig and S. T. Winograd. Fast Algorithms for Discrete Cosine Transform. *IEEE Trans. Signal Proc.*, 40:2174–2193, 1992.

[34] A. N. Netravali and B. G. Haskell. Digital Pictures—Representation, Compression, and Standards, 2nd ed. Plenum Press, New York, 1995.

[35] D. Talla, C. Y. Hung, R. Talluri, F. Brill, D. Smith, D. Brier, B. Xiong, and D. Huynh. Anatomy of a portable digital media processor. *IEEE Micro*, 24(2):32–39, 2004.

[36] I. E. G. Richardson. H.264 and MPEG-4 Video Compression, John Wiley & Sons, New York, 2003.

[37] ISO/IEC/JTC1/SC29/WG11/N9435, "Text of ISO/IEC 14496-2:2004/PDAM 5 Simple studio profile levels 5 and 6", Shenzhen, China, 2007.

[38] Texas Instruments, Inc., "Requirements for next generation video coding standards," Document COM 16 – D 267 – E, ITU-T SG16 contribution, Geneva, 2006.

[39] Texas Instruments, Nokia, Polycom, Samsung AIT, Tandberg, "Desired features in future video coding standards," Document COM 16 – C 215 – E, ITU-T SG 16 contribution, Geneva, 2007.

[40] Texas Instruments Inc., "Parallel CABAC," Document COM 16 – C 334 – E, ITU-T SG 16 contribution, Geneva, 2008.

[41] M. Budagavi and M. Zhou. Video coding using compressed reference frames. *Proc. IEEE Int. Conf. Acoust. Speech Signal Process.*, pp. 1165–1168, Las Vegas, 2008.

CHAPTER

Video Quality Assessment

14

Kalpana Seshadrinathan and Alan C. Bovik

Laboratory for Image and Video Engineering (LIVE), Department of Electrical and Computer Engineering, The University of Texas at Austin

14.1 INTRODUCTION

The arrival of the personal computer and the Internet has ushered in a digital video revolution in an astonishingly short span of less than 30 years. There has been an explosion of video and multimedia applications such as biomedical imaging, video on demand, video teleconferencing, video telephony, digital cinema, high-definition television, video transmission over IP networks (IPTV, Youtube, streaming video), video transmission over wireless networks (inflight entertainment, mobile TV), and consumer electronics (video camcorders, displays). In a vast majority of applications, the ultimate end user of these videos is a human observer. Therefore, it is of interest to evaluate the perceptual degradations introduced by acquisition, communication, and processing of videos. Video quality assessment (VQA) is the term used for techniques that attempt to quantify the quality of a video signal as seen by a human observer. VQA algorithms play a key role in almost every aspect of video processing. Applications of quality assessment can broadly be categorized as follows:

- Quality Monitoring: Video quality is affected by numerous factors in a video communication system such as compression, noise, errors, congestion, and latency in networks. Quality monitoring of video can allow service providers to meet their quality of service (QoS) requirements by changing resource allocation strategies.
- Performance Evaluation: Systematic evaluation of both hardware and software video processing systems that target human users are greatly facilitated by reliable means of VQA. Perceptual quality of images, and video generated by video devices (cameras, camcorders, displays, scanners, printers) and video processing algorithms (implemented in graphics cards, set-top boxes, encoders, super resolution, and enhanced displays) can be automatically measured for competitive evaluation using VQA systems.
- Optimization of video processing systems: Several video processing systems are designed by either specifying a maximum distortion of the video signal or

alternately, minimizing the video distortion for a specified system configuration. Example applications include bit-rate allocation and rate-distortion design of image and video communication systems. Significant reduction in resource requirements can be achieved using perceptual VQA techniques that closely match visual perception in video system design.

The only reliable means of assessing the quality of a video as seen by a particular human observer is to ask the human subject for their opinion of the visual quality of the video on a numerical or qualitative scale. This process is known as subjective VQA. Subjective studies are, however, cumbersome and expensive to do. To account for human variability in assessing quality and to have some statistical confidence in the score assigned by the subject, a large number of subjects may be required to view the same image. Also, in such an experiment, the assessment is dependent on several factors such as the display device, viewing distance, content of the image, whether or not the subject is a trained observer who is familiar with processing of images etc. Further, it is impossible to subjectively assess the quality of each and every video of interest. Subjective studies are valuable in providing a benchmark to evaluate the performance of objective or automatic methods of VQA, but are certainly not practical for most of the applications mentioned above.

In this chapter, we discuss objective methods of VQA that attempt to automatically predict the quality that an average human observer would assign to a given video. We begin by reviewing some of the essential concepts involved in VQA. First, there are three loosely agreed upon categories of VQA algorithms—Full Reference VQA; Reduced Reference VQA; and No Reference VQA, also known as blind VQA.

Full reference VQA algorithms operate on distorted videos while having a pristine, ideal "reference" video available for comparison. The goal of full reference VQA is to evaluate the fidelity of the test or distorted video with respect to the reference video. The vast majority of VQA algorithms fall into this category, because of the relative simplicity of making quality judgments relative to a standard.

Reduced reference algorithms operate without the use of a pristine reference, but do make use of additional (side) information along with the distorted video. Reduced reference algorithms may use features such as localized spatiotemporal activity information or edge locations extracted from an original reference; embedded marker bits in the video stream as side information to estimate the distortion of the channel and so on [1, 2]. Other algorithms use knowledge that has been independently derived regarding the distortion process (such as foreknowledge of the nature of the distortion introduced by a compression algorithm, e.g., blocking, blurring, or ringing) to assist in the QA process [1, 3], and possibly to correct the distortions [2, 4–7]. Sometimes algorithms of this latter type are referred to as "blind," but in our view, these should be categorized separately or as reduced reference algorithms.

No reference or blind algorithms attempt to assess image/video quality without using any other information than the distorted signal (using our interpretation of blind techniques as not assuming foreknowledge of the distortion process). This process has proved daunting, and there is very little substantive work on this topic. However, human beings can perform the task almost instantaneously, which suggests that there is hope in this direction—but in the long term. Currently, it is our view that much yet remains to be learned regarding full reference and reduced reference VQA, and especially regarding human visual perception of quality, before generic blind algorithms become feasible.

The mean squared error (MSE), or equivalently the peak signal-to-noise ratio (PSNR), computed between the reference and test videos is often used as a measure of quality because of its simplicity and mathematical convenience. However, it is well known that the MSE correlates very poorly with visual quality. The reader is directed to Chapter 21 of the companion volume, *The Essential Guide to Image Processing*, for a discussion on the failure of the MSE as a reliable perceptual quality predictor.

Most VQA systems include a preprocessing or calibration stage at the front end. This stage accounts for registration (spatial and temporal alignment of reference and test video patches), modeling of display device (gamma correction, eccentricity), viewing distance calibration, determination of the valid region of the video (accounting for nontransmitted pixels and lines in digital video), gain and offset calibration and so on [8].

In this chapter, we will discuss three classes of VQA algorithms that measure video quality using these calibrated videos. A substantial body of work in the literature has focused on using models of the human visual system (HVS) in the design of VQA algorithms, which we discuss in Section 14.2. There has been a shift toward VQA techniques that attempt to characterize features that the human eye associates with loss of quality; for example, blur, blocking artifacts, fidelity of edge, and texture information, color information, contrast and luminance of patches, and so on. We review such approaches to VQA in Section 14.3. In Section 14.4, we review recent work that focuses on using models of motion and temporal distortions in video in the VQA process to better match visual perception. The design of most VQA systems incorporates some perceptual modeling based on the properties of human vision since they are intended to mimic human behavior, and the VQA systems discussed in Sections 14.3 and 14.4 are no exception. The distinction that we draw between the so-called "HVS based" methods and other methods is the fact that HVS-based methods utilize computational models of various stages of processing that occurs in human vision in constructing a VQA system. The methods discussed in Sections 14.3 and 14.4 primarily utilize features and statistics computed from the reference and test videos to construct the quality metric, and properties of visual perception are used in the design often as weighting and masking models.

14.2 HVS MODELING BASED METHODS

The HVS derives information about the environment from light that is either emitted, transmitted or reflected from different objects in the environment [9, 10]. Inferences about the shape, size, location, color, motion, and depth of various objects in the environment are inferred using the two 2D images formed by the two eyes (optical systems of vision). Vision is an extraordinarily difficult problem, and considerable resources in the human (and other animals') brain is devoted to the visual processing tasks performed by humans (and other biological organisms). See [9–11] for a more detailed discussion of biological vision systems and the visual processing tasks performed by these systems.

In as much as the goal of VQA is to as nearly match the subjective assessment behavior of the HVS as possible, the last 30 years of research on VQA has largely focused on metrics based on models of the HVS. The premise behind such HVS-based metrics is to process the visual data by simulating the visual pathway of the eye-brain system. The reference and distorted videos are passed through computational models of various stages of processing that occur in the HVS, and visual quality is defined as an error measure between these signals computed at the output of the visual perception model. Most HVS-based methods, model the properties of the optical systems (eyes), and early stages of visual processing that occur in the human brain. HVS-based VQA algorithms have mostly been extensions of HVS-based image quality assessment (IQA) techniques that have been discussed in detail in Chapter 21 of the companion volume, *The Essential Guide to Image Processing.* HVS-based IQA methods use elaborate models of spatial processing that occurs in the HVS. HVS-based VQA techniques include models of temporal processing that occurs in the HVS to extend still image methods to video. We will primarily focus on the temporal processing mechanisms in HVS-based VQA techniques in this chapter and direct the reader to the companion volume for an in-depth discussion of the spatial processing mechanisms.

As depicted in Fig. 14.1, HVS-based IQA systems typically begin by preprocessing the signal to correct for nonlinearities, since lightness perception is a nonlinear function of luminance. A filterbank in the "Linear Transform" stage decomposes reference, and distorted (test) signals into multiple spatial frequency- and orientation-tuned channels in an attempt to model similar processing by neurons in the early stages of the visual pathway [10]. The contrast sensitivity, luminance, and contrast masking features of the HVS are then modeled to account for perceptual error visibility as a function of luminance and contrast of the image patches. A space-varying threshold map is created for each channel describing local spatiospectral error sensitivity, and is used to normalize the differences between reference and test images. In the final stage, the normalized errors for all channels are pooled through a suitable metric such as a weighted MSE, to generate a space-varying quality map. This approach to IQA is intuitive and has met with considerable success.

A block diagram of a generic HVS-based VQA system is illustrated in Fig. 14.2. This system is identical to the generic HVS-based IQA system in Fig. 14.1, but for the inclusion of a block labeled "temporal filtering." In addition to the spatial filtering stage in IQA systems depicted in the "Linear Transform" block, VQA systems utilize a temporal filtering stage in cascade, which is equivalent to filtering the videos using a spatiotemporal filterbank that is separable along the spatial and temporal dimensions. The "temporal

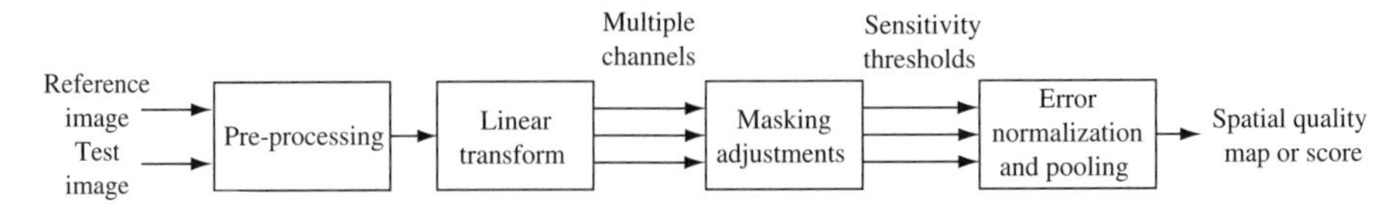

FIGURE 14.1

Block diagram of HVS-based IQA system.

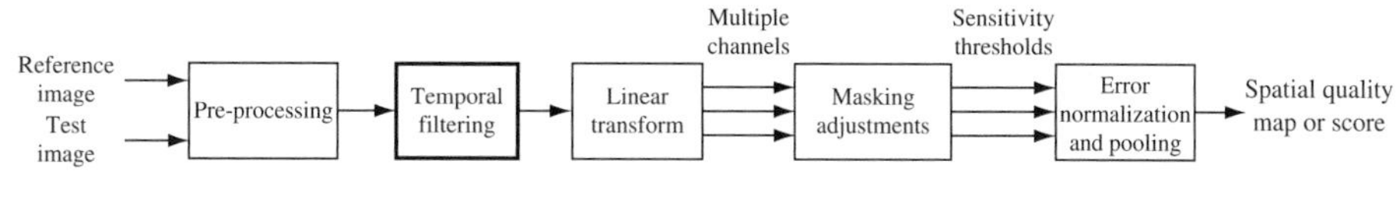

FIGURE 14.2

Block diagram of HVS-based VQA system.

filtering" block typically models two kinds of temporal mechanisms that exist in the early stages of processing in the visual cortex, that are often modeled using linear lowpass and bandpass filters applied along the temporal dimension of the videos. One of the first HVS-based metrics for video signals was known as the moving pictures quality metric (MPQM) [12]. This model utilized a Gabor filterbank in the spatial frequency domain, and the temporal mechanism consisted of one bandpass and one lowpass filter. Additionally, measurement of the contrast sensitivity function as a nonseparable function of spatial and temporal frequencies was performed using psychophysical experiments. This metric was improved upon further by Winkler et al. using more recent models for the temporal mechanisms in the HVS to develop the Perceptual Distortion Metric (PDM) [13]. This metric used two infinite impulse response (IIR) filters to model the lowpass and bandpass mechanisms that were developed by Fredericksen et al. [14]. Additionally, this model chose to use the steerable pyramid for the spatial decomposition, instead of the Gabor filterbank. The Lubin model for video includes two temporal filters, whose impulse response is differ only slightly from the ones used in PDM [15]. A computationally efficient metric using the Discrete Cosine Transform (DCT) in the linear transform stage was proposed by Watson, known as the digital video quality (DVQ) metric [16]. This model utilizes a simple single channel temporal mechanism, which was implemented as an IIR filter. A scalable wavelet-based video distortion metric was proposed in [17]. This metric uses a single channel finite impulse response (FIR) filter to model the temporal mechanisms in the HVS, and an orthonormal wavelet decomposition spatially using the Haar wavelet. This metric is designed to be scalable and can operate in either full reference or reduced reference mode.

Several HVS-based VQA systems have been implemented in commercial products that are used for video quality monitoring in applications such as broadcast television, IPTV, and so on. The Lubin model forms the basis for the Sarnoff JNDMetrix technology from Sarnoff Corporation, which won an Emmy award for Outstanding Achievement in Technological Advancement [18]. JNDMetrix was used by Tektronix Corporation in their Picture Quality Analyzer (PQA) 200 system [19] and forms the basis for the latest generation PQA 500 system. Winkler cofounded Genista Corporation that delivered a VQA system based on the MPQM model, which was later purchased by Symmetricom and incorporated into their V-Factor Quality of Experience platform that delivers a perceptual video quality solution for cable operators. The Video Quality Analyzer (VQA) from AccepTV is based on a HVS model, that utilizes contrast sensitivity and masking models from Daly's Visible Differences Predictor, an IQA model described in Chapter 21 of the companion volume [20, 21].

14.3 FEATURE BASED METHODS

Feature based methods for VQA utilize statistics and features computed from the reference and test videos to predict the visual quality of the test video. Feature based approaches form the backbone of several no reference VQA systems, and the reader is directed to [3] and the references therein for a discussion of no reference IQA and VQA systems. We focus on feature based full reference VQA systems in this section.

Very recently, a performance evaluation contest was completed by ITU-T for standardization of VQA systems for use in multimedia applications, and four full reference VQA systems were approved and standardized as a formal recommendation [22, 23]. This includes a full reference video quality model developed by NTT Service Integration Laboratories for IP-based video delivery services [22]. The NTT model is a feature-based system that estimates perceptual quality degradation due to blocking, blurring, and packet losses in network transmission of video [24]. Another model approved for standardization is the Perceptual Evaluation of Video Quality (PEVQ) model from Opticom [25–27]. PEVQ is based on an earlier model known as the Perceptual Video Quality Measure (PVQM) developed by Swisscom/KPN Research, Netherlands [28]. PVQM measures three different quantities from the reference and distorted videos to compute video quality—an edginess indicator, a temporal indicator, and a chrominance indicator. The edginess indicator compares the edginess of the luminance signal between the reference and distorted videos. The local edginess of the reference and test videos are computed using approximations to the local gradients of the luminance of the videos. The temporal indicator describes the amount of movement or change in the reference video sequence, which is used to decrease sensitivity to details in fast moving pictures to match visual perception of videos. The temporal indicator is defined as a normalized cross-correlation computed between adjacent frames of the reference video. The chrominance indicator describes the perceived difference in chrominance or color information between the reference and distorted videos. The chrominance indicator accounts for perceptual effects such as higher sensitivity to chrominance errors in smooth areas as compared to edges and reduced sensitivity to chrominance errors in areas with saturated colors and bright luminance. Finally, a mapping from these indicators to the overall perceived video quality is defined, that predicts the visual quality of the test video.

Another prominent VQA system was developed at the National Telecommunications and Information Administration (NTIA) and is known as the Video Quality Metric (VQM) or the NTIA General Model [8]. VQM and its associated calibration techniques have been adopted as a North American standard by the American National Standards Institute (ANSI) in 2003. The International Telecommunication Union (ITU) has also included VQM as a normative method for digital cable television systems [29, 30]. Due to the popularity and success of VQM, we describe it in detail in Section 14.3.1.

14.3.1 VQM

The first stage of processing in VQM is the calibration of the reference and test video sequences, which includes spatial and temporal alignment, extraction of valid regions from the videos, gain and offset correction. The calibration stage of VQM is included as a normative method in an ITU recommendation, and further details can be found in [8, 29]. This stage is followed by feature extraction, spatiotemporally local quality parameter computation, and pooling of local quality indices into a single quality score for the entire video sequence discussed in the following sections.

14.3.1.1 Feature Computation

A quality feature in VQM is defined as a quantity of information extracted from spatio-temporal subregions of a video stream. The feature streams are a function of space and time and may be extracted from the reference and/or test video sequence. VQM contains seven parameters of which four are based on features extracted from spatial gradients of the luminance component of the video, two parameters are based on features extracted from the chrominance component, and one parameter is based on the product of features that measure contrast and temporal information, both of which are extracted from the luminance component of the video.

The feature computation stage in VQM first applies a filter to the video stream to enhance some property of perceived video quality (edges, color). After this filtering, the video streams are divided into abutting spatiotemporal regions containing a window of pixels spanning a specified number of rows, columns, and frames. Features are extracted from corresponding spatiotemporal subregions of the reference and test video sequence using two mathematical functions—the mean which estimates the average feature value or the standard deviation, which measures the spread of the feature value. Finally, a threshold is applied to the extracted features to prevent them from measuring impairments that are imperceptible. The different filters that are used in VQM to compute perceived video quality are described here. The reader is referred to [8] for further details on the steps in feature computation and the thresholding procedure.

Spatial gradients along the horizontal and vertical directions of the video are computed by the application of a linear filter (termed "spatial information filter" in VQM) applied separably on the luminance or Y component of the video. This filter spans 13 pixels and is developed to measure perceptually significant edge impairments. The resulting horizontal and vertical filtered videos denoted using I_h and I_v are used to measure four parameters of VQM. One captures decrease of spatial information or blurring in the video sequence and is computed using the Euclidean distance between I_h and I_v. Two parameters detect a shift of edges from horizontal and vertical orientation to diagonal orientation, or a shift of edges from diagonal orientation to horizontal or vertical orientation (this may occur due to blocking impairments) using I_h and I_v. Finally, the fourth parameter is the only quality improvement parameter in VQM and captures any improvements in quality due to edge sharpening or enhancement. This parameter is also computed using the filtered videos, I_h and I_v.

Two parameters in VQM operate on the chrominance or color components of the video, that is, the C_R and C_B components in YC_RC_B video. One parameter detects changes in the spread of the distribution of the color samples, whereas the other detects severe localized color impairments often produced by errors in digital transmission of video.

The mean of the C_R and C_B components in local spatiotemporal windows is used to compute both these features, however with different pooling techniques.

Perceptibility of spatial impairments in video can be reduced in the presence of large motion in the video, and similarly, perceptibility of temporal impairments can be influenced by the amount of spatial detail. The final parameter in VQM is the product of a contrast feature that measures the amount of spatial detail in the video, and a temporal information feature to account for these perceptual effects. The contrast feature is computed as the standard deviation of local spatiotemporal windows of pixels in the luminance or Y component of the video. To measure temporal information, an absolute temporal information (ATI) filter is applied on the video. The ATI is a simple filter that computes a pixel-by-pixel absolute difference between adjacent frames of the video. The temporal information feature is the standard deviation computed over local windows of the output of the ATI filter.

14.3.1.2 *Quality Parameters*

The seven different quality features computed by VQM in the previous section quantify some perceptual aspect of a video stream. Quality parameters computed by VQM compare with corresponding features from the reference and distorted videos to obtain a measure of video distortion. In the quality parameter estimation step, the feature values for each local region of the reference and distorted video, denoted as f_R and f_D, are compared using one of three different comparison functions that emulate the perception of impairments. The three comparison functions in VQM have been developed to account for visual masking properties of videos; see Chapter 21 in the companion volume for a discussion of contrast masking.

The three comparison functions used are a Euclidean distance computed between the features, a ratio comparison function or a log comparison function:

$$p = \sqrt{(f_R - f_D)^2} \tag{14.1}$$

$$p = \frac{f_D - f_R}{f_R} \tag{14.2}$$

$$p = \log_{10}\left(\frac{f_D}{f_R}\right) \tag{14.3}$$

14.3.1.3 *Pooling*

The quality parameters computed by VQM so far form 3D arrays spanning the temporal axis and the two spatial axes (horizontal and vertical). The pooling stage in VQM computes a single quality score for the entire video sequence by combining these local quality parameters appropriately. In the first stage, a spatial pooling is performed wherein the space-varying quality parameters computed at a single time instant (e.g., a frame) of the video sequence are pooled to obtain a quality index for that instant of time. The spatial pooling stage utilizes the mean, standard deviation or some form of worst case processing of the quality parameter values, that is, rank sorting with percent threshold selection, which is also known as quantile computation in statistics [31]. In the spatial pooling stage, quantile pooling in VQM always uses some form of worst case processing such as taking the average of the worst 5% of the distortions observed at that point of time, which we refer to as worst-case quantiles. This is because viewer attention is drawn to localized errors in the video sequence and becomes the predominant factor in subjective judgment of quality.

At the end of the spatial pooling stage, a time history of parameter values is obtained that is combined into a single quality estimate using a temporal pooling function. The temporal pooling functions used in VQM are the mean, standard deviation, worst-case quantiles, and best-case quantiles (to detect distortions that are nearly always present in the video) of the time history of quality parameters. Each of the seven quality parameters in VQM is pooled spatially and temporally using some combination of these pooling strategies, and some parameters are also subjected to a nonlinear mapping and thresholding (clipping) after pooling [8].

The overall VQM index of the distorted video sequence is computed as a linear combination of the seven parameters.

14.4 MOTION MODELING BASED METHODS

HVS-based VQA systems discussed in Section 14.2, and feature based systems discussed in Section 14.3 use elaborate mechanisms to capture spatial distortions in video and model various aspects of spatial vision. Spatial distortions refer to artifacts that occur within video frames and do not arise from temporal processes. These include blocking artifacts from DCT coefficient quantization in MPEG and H.263/264; ringing distortions from finite-length basis functions and quantization in video encoders; mosaic patterns from mismatches between adjacent compressed blocks due to coarse quantization; and false contouring from quantization in smooth image regions [32]. Spatial aspects of human vision that are modeled in the VQA techniques discussed so far include spatial contrast masking, color perception, response to spatial frequencies and orientations, and so on.

Although videos do suffer from spatial distortions, videos also suffer from significant temporal distortions. By temporal distortions, we refer to artifacts that predominantly arise from motion in the video sequence and alter the movement trajectories of various pixels of the video. Temporal artifacts in video include ghosting—trailing ghosts behind fast-moving objects from temporal noise filtering; motion blocking in block motion compensated compression algorithms; motion compensation mismatches when multiple motions occur within macro blocks; mosquito effect, or brightness fluctuations near moving edges; jerkiness from temporal aliasing in fast-moving videos or transmission delays in networked video; and smearing from noninstantaneous acquisition or exposure time [32]. Further, the human eye is very sensitive to motion and can accurately judge the velocity and direction of motion of objects in a scene. This does not seem surprising in view of the fact that the ability to detect motion accurately is crucial to survival and performance of tasks such as navigating through the environment, avoiding danger, and so on. Motion attracts visual attention and affects spatiotemporal aspects of human vision (such as reduced sensitivity to fine spatial detail in fast moving videos). The

VQA algorithms discussed so far do not do an adequate job in capturing these temporal distortions or modeling the important role of motion in human vision, and hence, visual perception of videos.

HVS-based VQA systems suffer from inaccurate modeling of the temporal mechanisms in the HVS. All VQA systems discussed in Section 14.2 use separable spatial and temporal linear filters, and either one or two temporal channels that model the temporal tuning of neurons in the front end of the human eye-brain system. The front end that is modeled includes the retina, lateral geniculate nucleus (LGN) and Area V1 of the visual cortex [10]. However, it is well known that Area MT/V5 of the extrastriate cortex plays an important role in motion perception. Most of the neurons in the front end of the HVS respond to specific spatial frequencies and orientations, and respond best to a stimulus moving in a particular direction. This represents basic spatial processing mechanisms and the first stage of motion processing in the HVS. Neurons in this area are reasonably well-modeled by separable, linear, spatial, and temporal filters that are used in the linear decomposition stage of HVS-based systems. However, visual data from Area V1 are transported along the ventral stream to Area MT/V5, which represents a second stage of motion processing in the HVS. Area MT/V5 integrates local motion information computed in V1 of oriented components in the video into global percepts of motion of complex patterns that typically occur in video sequences [33]. Area MT/V5 also plays a role in the guidance of some eye movements, segmentation, and structure computation in 3D space [34], properties of human vision that play an important role in visual perception of videos. The response of neurons in Area MT have been studied, and models of motion sensing in the human eye have been proposed, but none of the existing HVS-based systems have incorporated these models in VQA.

Feature based methods discussed in Section 14.3 also predominantly capture spatial distortions in the video, and fail to do an adequate job in capturing temporal distortions in video. The only temporal component of PVQM computes local normalized cross-correlation computed between adjacent frames of the video, whereas VQM uses absolute pixel-by-pixel differences computed between adjacent frames of the video sequences. Such minimal temporal feature computation is unlikely to adequately match visual perception, since it does not capture the visual perception of motion and motion related distortions experienced by a human observer who watches the video. Motion models have been used to advantage in the design of pooling strategies, that is, combining local spatiotemporal quality indices of a video into a single quality score for the entire video [35, 36]. In this study, the structural similarity (SSIM) index that was developed for still images (described in Chapter 21 of the companion volume) is applied frame-by-frame on the video sequence, and motion vectors computed from the reference video are used to pool local SSIM indices into a single video quality score. However, since the quality model used is the SSIM index operating frame-by-frame, it does not adequately capture temporal distortions in the video or the visual perception of motion experienced by human observers.

Accurate representation of motion in video sequences, as well as of temporal distortions, have great potential to advance video quality prediction. Recently, we have developed a framework for evaluating spatial and temporal (and spatiotemporal) aspects of distortions in video [37, 38], based on which an algorithm known as the MOtion based Video Integrity Evaluation or MOVIE index was defined. In this framework, video quality is evaluated not only in space and time, but also in space-time, by evaluating motion quality along computed motion trajectories. It is our view that using motion models in VQA represents a significant step forward in reaching the ultimate goal of VQA—matching human perception of videos.

In our motion-based framework for VQA, separate components for spatial and temporal quality are defined [37, 39]. First, the reference and test videos are decomposed into spatiotemporal bandpass channels using a Gabor filter family. Spatial quality measurement is accomplished by computing an error index between the bandpass reference and distorted Gabor channels, using models of the contrast masking property of visual perception. Temporal quality is measured using motion information computed from the reference video. Motion of the reference video is estimated in the form of optical flow fields using our own multiscale extension of the Fleet and Jepson phase-based optical flow estimation technique [37, 40]. Finally, the spatial and temporal quality scores are pooled to produce an overall video integrity score.

In the MOVIE index, both the reference and distorted videos are filtered spatiotemporally using a family of bandpass Gabor filters. This results in a sequence of Gabor filtered reference and test videos, denoted X_k and Y_k respectively, where k indexes over the Gabor filters. Gabor filters are particularly attractive for VQA since the receptive fields of neurons in the visual cortex are well modeled by Gabor filters [41]. Additionally, Gabor filters have been used in several motion estimation algorithms for video [40, 42], and models of human visual motion sensing [42–45]. MOVIE uses multiple scales of Gabor filters with constant octave bandwidth that are separable in space and time. The filter design is very similar to the filters used in the Fleet and Jepson algorithm [40]; however, MOVIE uses multiple scales of filters with narrower bandwidth as described in [37]. Figure 14.3 shows one scale of the Gabor filterbank used in MOVIE. The center frequencies of all Gabor filters at one scale lie on the surface of a sphere in frequency. Each ball in Fig. 14.3 represents a Gabor filter since the filters have equal spread along both spatial frequency coordinates and temporal frequency. The Gabor filters are designed to intersect each other at one standard deviation of the Gabor frequency response, and the entire filterbank samples the spatio-temporal frequencies present in the videos densely. The frequency decomposition achieved by the Gabor filterbank used in MOVIE is far more elaborate spatiotemporally than other existing VQA systems (e.g., HVS-based methods described in Section 14.2). This is because the Gabor filter family used in MOVIE is also used to compute motion information from the reference video sequence, which is quite significantly in the context of VQA, a visual task that is also performed by human beings using the sequence of time-varying images (or video) captured by the eye and optical system.

The outputs of the Gabor filters, namely the bandpass filtered reference and test videos, are first used to measure spatial quality degradations in the video. MOVIE computes spatiotemporally local quality indices by considering vectors of coefficients extracted from corresponding locations in the Gabor filtered reference and distorted videos using a window function. Then, a spatial error index at each pixel and Gabor subband is

FIGURE 14.3

Geometry of the Gabor filterbank in frequency. Each ball represents the iso-surface contour of one Gabor filter, and the figure shows all Gabor filters at one scale. The horizontal axis in the 3D view depicts horizontal and vertical spatial frequency coordinates, and the vertical axis denotes temporal frequency.

computed, as the MSE between vectors of coefficients normalized by a masking function. The masking model is intended to account for the contrast masking property of human vision and is closely related to multiscale structural similarity and informational theoretic metrics for IQA [46]. However, unlike these IQA methods, Spatial MOVIE uses a mutual masking function in the subband domain, wherein the mask is computed using both the reference and distorted videos. Such a mutual masking function is better able to account for visual perception of artifacts in videos [37].

The Spatial MOVIE index primarily captures spatial distortions in the video such as blur, ringing, and so on described earlier. However, since the Spatial MOVIE index is computed using the outputs of a spatiotemporal Gabor filter family, it also responds in a limited fashion to temporal distortions in the video. Spatial MOVIE is conceptually similar to the VQA systems described in Sections 14.2 and 14.3 in that it uses a linear, spatiotemporal decomposition and computes perceptual error indices on these bandpass channels, while accounting for visual masking effects. After spatial quality computation, the Temporal MOVIE index captures temporal distortions by tracking video quality along the motion trajectories of the reference video and as we will see, the Temporal MOVIE index integrates concepts from motion perception and can match human perception of videos.

To compute temporal quality, as a first step, optical flow fields are computed from the Gabor filtered reference videos X_k using a multiscale extension of the Fleet and Jepson algorithm [37]. MOVIE assumes that the reference and test videos can be described by local image patches undergoing translational motion. This assumption is commonly used in video encoders (see Chapters 14.8–14.10 in this book) and is reasonable since translation represents a first order approximation to more complex motions. The spectrum of an infinitely translating image patch is an oriented plane in the frequency domain, whose orientation is solely determined by the velocity and direction of motion [47]. Figure 14.4 illustrates the spectrum of a video resulting from infinite translation of an image patch. Figure 14.4(a) shows the spectrum of a static video, that is, a video that consists of a single image repeated across frames. Observe that the spectrum of such a video lies along

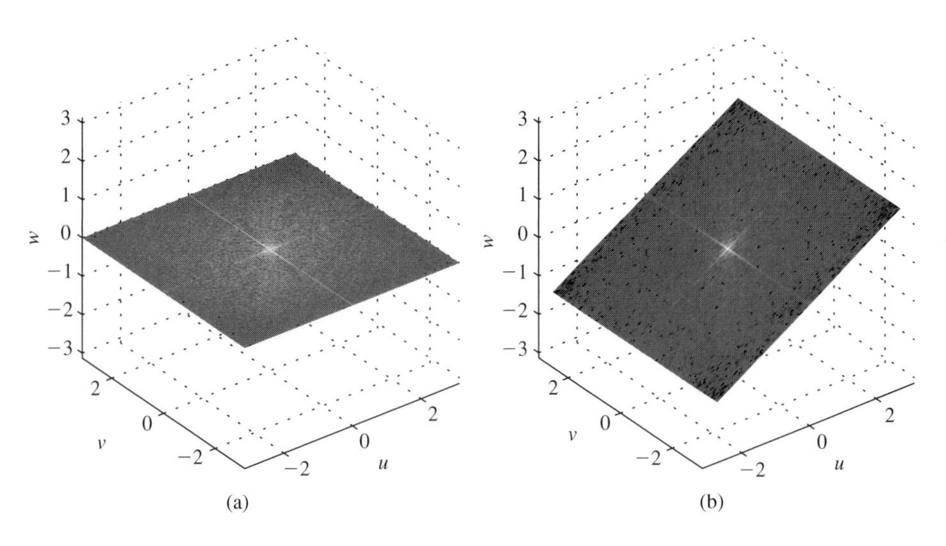

FIGURE 14.4

(a) Spectrum of a static video consisting of an image patch repeated across frames. (b) Spectrum of a translating image patch. u denotes the horizontal spatial frequency axis, v denotes vertical spatial frequency, and w denotes temporal frequency.

the plane of zero temporal frequency, since the video is static and does not contain any temporal frequencies. When this image patch translates at a certain velocity, the spectrum of the video is sheared to an orientation determined by the velocity of translation and is shown in Fig. 14.4(b).

The Gabor filter family used in MOVIE responds to such oriented components in frequency and a subset of Gabor filters that lie along the orientation of the spectral plane are activated by the video. This is illustrated in Fig. 14.5 for the static and oriented video spectra in Fig. 14.4. In the case of the static video, Gabor filters tuned to zero temporal frequency as shown in Fig. 14.5(a) produce large outputs. In the case of the translating patch, Gabor filters shown in Fig. 14.5(b) that lie along the same orientation as the spectrum of the translating patch produce large outputs.

To compute temporal quality, MOVIE computes reference motion-tuned responses from both the reference and distorted videos. This is conceptually similar to motion compensated video filtering mechanisms, which is why MOVIE can be said to track video quality along the motion trajectories of the video. Motion-tuned responses in MOVIE are constructed using a weighted combination of the energies of the Gabor outputs X_k and Y_k, where the weights are determined based on the local motion of the reference video. Weights are computed for the Gabor filters to create an orientation tuned response in spatiotemporal frequency that matches the orientation of the spectral plane of the reference video, while also accounting for spatiotemporal masking effects. Note that the motion-tuned responses are tuned to the motion of the reference video, and any mismatches in the motion trajectories between the reference and distorted videos

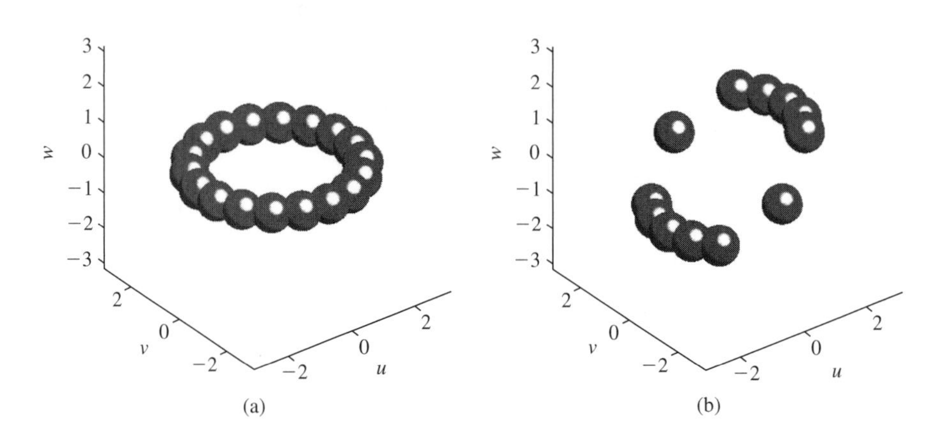

FIGURE 14.5

(a) Subset of Gabor filters activated by the static video spectrum in Fig. 14.4(a). (b) Subset of Gabor filters activated by the oriented spectrum in Fig. 14.4(b). u denotes the horizontal spatial frequency axis, v denotes vertical spatial frequency, and w denotes temporal frequency.

(such as motion compensation mismatches, mosquito noise, jerkiness, and so on) will be captured by this mechanism. The Temporal MOVIE index is then defined as the MSE between these motion-tuned responses. The temporal quality computation in MOVIE is very closely related to computational models that have been proposed for motion processing in the HVS [37]. In particular, the temporal quality computation in MOVIE shares commonalities with computational models of neurons in Area MT [45] and can match visual perception of motion, which is a critical component of any subjective video judgment.

The spatial and temporal quality computations in MOVIE result in maps that show the local spatial and temporal quality of the video at each pixel of the video sequence. MOVIE maps are shown in Fig. 14.6 for one frame of an example video. Note that the reference video, test video, and MOVIE maps are displayed as images corresponding to one frame of the video for illustrative purposes. The computation of the MOVIE index is spatiotemporal and uses a neighborhood of frames that is centered on the frames shown in Fig. 14.6. Figures 14.6(a) and 14.6(b) show corresponding frames from the reference and distorted videos. Figure 14.6(c) shows the Spatial MOVIE index computed at each pixel of the frame in Fig. 14.6(b). Observe that the Spatial MOVIE index captures spatial distortions in the video, that is, the blur and loss of sharpness throughout the video such as the hair of the woman and the plant in the background. Figure 14.6(d) shows the Temporal MOVIE index computed at each pixel of the frame in Fig. 14.6(b). The Temporal MOVIE index captures temporal distortions in the video, that is, motion compensation mismatches around the edges of the harp, the strings and the hand of the woman that result in flickering artifacts in the test video.

Finally, these local Spatial and Temporal MOVIE indices are pooled to obtain a single prediction of the visual quality of the entire video, which is termed the MOVIE index.

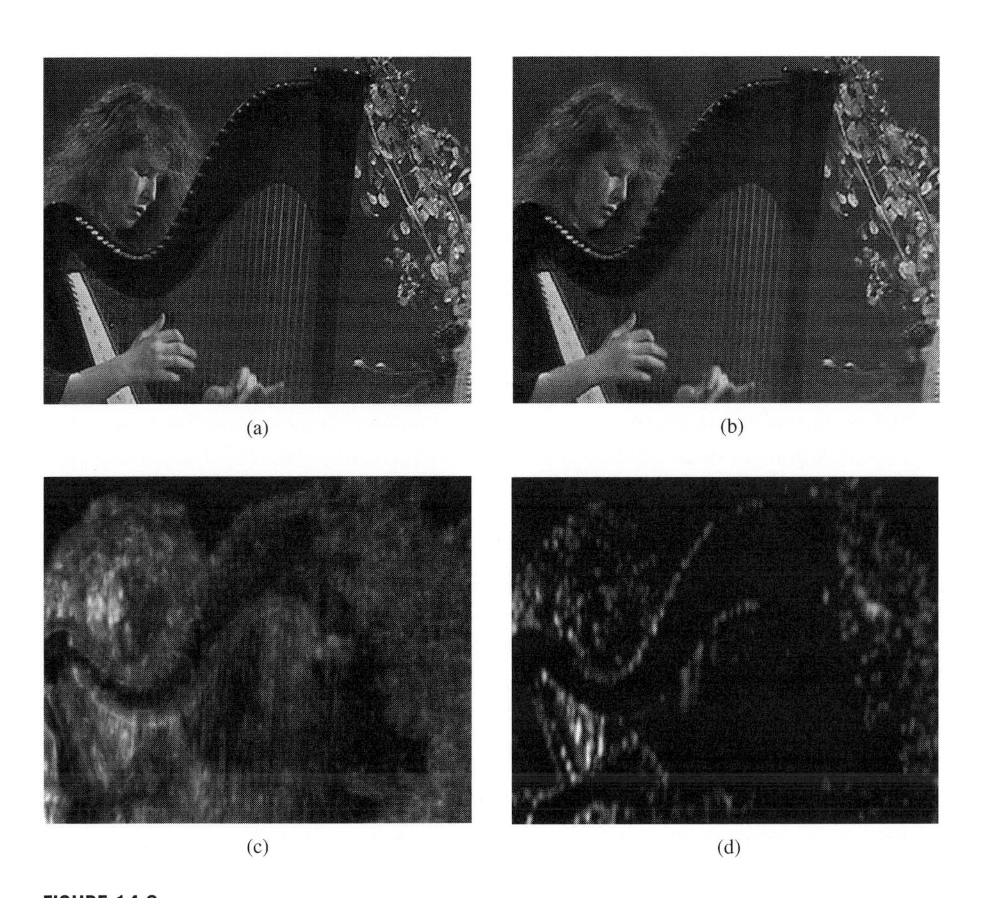

FIGURE 14.6

(a) Frame from reference video. (b) Corresponding frame from distorted video. (c) Spatial MOVIE map. (d) Temporal MOVIE map. Bright regions in the Spatial and Temporal MOVIE maps correspond to regions of poor quality.

14.5 PERFORMANCE

The goal of objective VQA algorithms is to match subjective or human perception of videos as closely as possible. Performance evaluation of objective VQA systems is hence achieved by correlating VQA model predictions of visual quality against quality scores provided by human subjects. Human opinion of visual quality is obtained through subjective studies, which are quite challenging and difficult to conduct. One of the prominent, commonly available subjective studies in the field of full reference VQA to date was conducted by the Video Quality Experts Group (VQEG) in 2000 known as the VQEG FR-TV Phase I database.

The performance of many VQA systems has been evaluated on the VQEG FR-TV Phase I database. Indicators of the performance of an objective VQA system in matching visual perception of videos include metrics such as Spearman rank order correlation coefficient (SROCC), linear correlation coefficient (LCC), and root mean square error (RMSE) etc. computed between VQA model predictions of visual quality, and human mean opinion scores (MOS) obtained from subjective studies of video quality [48].

In this study, we present the performance of some of the VQA systems described in this chapter in terms of two metrics—the SROCC and LCC after nonlinear regression. In computing the LCC, to account for any nonlinearities in objective model prediction with respect to MOS, model predictions are fitted to MOS scores using a nonlinear logistic function [48]. Note that the SROCC does not require this regression, since the SROCC is independent of the form of the relation between the subjective and objective scores. The study conducted by the VQEG tested the performance of ten proponent models including PSNR; HVS-based VQA systems—DVQ from the National Aeronautics and Space Administration (NASA), a model from Tektronix (based on the Sarnoff JND Vision Model); and feature based methods—models from NTIA (a precursor to VQM) and KPN/Swisscom (PVQM) [48]. We report the results of this study, along with the reported performance of the motion weighting based SSIM index for video [35] and the MOVIE index [37] in Table 14.1.

Note that the reported performance of the VQEG proponents is from [48], where the proponents did not have access to the VQEG database. Some of these algorithms have been modified, since the publication of the study in 2000 [48]. However, the results in Table 14.1 provide some insight into the relative performance of various VQA systems. PSNR provides a baseline for the performance of VQA systems, and it is clear that PSNR does not correlate very well with visual quality. The MOVIE index correlates quite well with visual quality, which powerfully illustrates the importance of incorporating models of motion perception and temporal distortions in video.

Some discussion of video quality databases is relevant. The VQEG FR-TV Phase I Database on which we have stated performance results has been criticized in several regards. First, the database contains a limited set of distortion types (mostly arising from compression) that occur in television systems, and does not include distortions created by modern compression algorithms such as H.264. Second, the database utilizes only

TABLE 14.1 Comparison of the performance of VQA algorithms.

Prediction Model	SROCC	LCC
Peak Signal - Noise Ratio	0.786	0.779
Proponent P2 (Sarnoff)	0.792	0.805
Proponent P5 (PDM from EPFL)	0.784	0.777
Proponent P7 (DVQ from NASA)	0.786	0.770
Proponent P8 (Swisscom)	0.803	0.827
Proponent P9 (Precursor to VQM from NTIA)	0.775	0.782
SSIM (weighting)	0.812	0.849
MOVIE	0.833	0.821

interlaced videos, which presents the VQA algorithm designer with having to incorporate deinterlacing into an algorithm to test it on the VQEG database. De-interlacing is subject to distortions, which confuses the testing issue. Moreover, interlaced video is not relevant to all applications, a notable exception being multimedia applications. Third, the distortions in the VQEG database are heavily clustered into two groups of very good quality, and low quality videos. Within each cluster the perceptual clustering is tight, making it difficult for any algorithm to perform with particular distinction. Very simply, separating videos into high and low quality on the VQEG Database is easy, but separating videos within each class is difficult for both humans and algorithms. As such, the leading algorithms in the past have struggled to outperform the MSE/PSNR for nearly a decade, and in the early tests all algorithms were statistically indistinguishable from PSNR on the VQEG Phase I Database.

Realizing this, the VQEG has been conducting further studies with better effect, but the database of videos and the study details are not made public. Toward this end, we have been conducting a large-scale VQA study, which will include a broader diversity of video distortions with realistic perceptual separations. We have conducted a large-scale human study on the database and are in the process of comparing many different competitive VQA algorithms on the database. This database may be viewed as a companion to the widely used LIVE Image Quality Database [49], which has become a de facto standard for IQA and has been downloaded by more than 500 organizations. The upcoming LIVE Video Quality Database, which unfortunately will not be ready as this volume goes to press, will be available in the future at the same web site [49]. Stay tuned!

14.6 CONCLUSIONS

The topics of image and VQA remain very active ones, and there remains a lot of work to be done to bring them to fruition in the broad diversity of applications they will affect in the future. First, current algorithms are largely content-independent, meaning that they operate the same regardless of whether there is an attractive person or automobile in the video. It may be hypothesized that content distortion may play a large role in perceptual video quality. Second, the gaze of the viewer is becoming an increasingly important consideration in all aspects of video presentation as display sizes increase. The perspicacious moviegoer may have noticed, for example, that IMAX movies often have better quality near the center of the screen than near the periphery; human gaze is naturally drawn toward the center of the field, unless attracted by motion, interesting objects, color, or other video features. Finding these attractors and accounting for them in video quality is an active area of research as well [50]. Of course, the door has barely been opened on more difficult problems, such as blind VQA, which will require better models of the source (statistical video modeling), or the channel (statistical distortion modeling), and the receiver (better perceptual modeling). We look forward to these developments going forward.

REFERENCES

[1] Z. Wang and A. C. Bovik. *Modern Image Quality Assessment.* Morgan and Claypool Publishing Co., New York, 2006.

[2] J. Y. L. V. M. Liu and K. G. Wang. Objective image quality measure for block-based DCT coding. *IEEE Trans. Consum. Electron.*, 43:511–516, 1997.

[3] H. R. Sheikh, A. C. Bovik, and L. Cormack. No-reference quality assessment using natural scene statistics: JPEG2000. *Image Process. IEEE Trans.*, 14(11):1918–1927, 2005.

[4] Z. Wang, A. C. Bovik, and B. L. Evans. Blind measurement of blocking artifacts in images. *IEEE Int. Conf. Image Process.*, 3:981–984, 2000.

[5] K. T. Tan and M. Ghanbari. Frequency domain measurement of blockiness in MPEG-2 coded video. *IEEE Int. Conf. Image Process.*, 3:977–980, 2000.

[6] A. C. Bovik and S. Liu. DCT-domain blind measurement of blocking artifacts in DCT-coded images. *IEEE Int. Conf. Acoust. Speech Signal Process.*, 3:1725–1728, 2001.

[7] S. Liu and A. C. Bovik. Efficient DCT-domain blind measurement and reduction of blocking artifacts. *IEEE Trans. Image Process.*, 12:1139–1149, 2002.

[8] M. H. Pinson and S. Wolf. A new standardized method for objectively measuring video quality. *IEEE Trans. Broadcast.*, 50(3):312–322, 2004.

[9] W. S. Geisler and M. S. Banks. Visual performance. In M. Bass, editor, *Handbook of Optics*, McGraw-Hill, New York, 1995.

[10] B. A. Wandell. *Foundations of Vision.* Sinauer Associates Inc., Sunderland, MA, 1995.

[11] L. K. Cormack. Computational models of early human vision. In A. C. Bovik, editor, *The handbook of image and video processing*, 325–346. Elsevier, New York, 2005.

[12] C. J. van den Branden Lambrecht and O. Verscheure. Perceptual quality measure using a spatiotemporal model of the human visual system. In *Proc. SPIE*, Vol. 2668(1), 450–461, SPIE, San Jose, CA, March 1996.

[13] S. Winkler. Perceptual distortion metric for digital color video. *Proc. SPIE*, 3644(1):175–184, 1999.

[14] R. E. Fredericksen and R. F. Hess. Temporal detection in human vision: dependence on stimulus energy. *J. Opt. Soc. Am. A (Opt. Image Sci. Vis.)*, 14(10):2557–2569, 1997.

[15] J. Lubin. The use of psychophysical data and models in the analysis of display system performance. In A. B. Watson, editor, *Digital Images and Human Vision*, 163–178. The MIT Press, Cambridge, MA, 1993.

[16] A. B. Watson, J. Hu, and J. F. McGowan III. Digital video quality metric based on human vision. *J. Electron. Imaging*, 10(1):20–29, 2001.

[17] M. Masry, S. S. Hemami, and Y. Sermadevi. A scalable wavelet-based video distortion metric and applications. *Circuits Syst. Video Technol. IEEE Trans.*, 16(2):260–273, 2006.

[18] J. Lubin. A human vision system model for objective picture quality measurements. *IEEE Int. Broadcast. Conv.*, 498–503, 1997.

[19] Tektronix. [Online] Available: http://www.tek.com/products/video_test/pqa500/

[20] AccepTV. [Online] Available: http://www.acceptv.com/

[21] M. Carnec, P. Le Callet, and D. Barba. Objective quality assessment of color images based on a generic perceptual reduced reference. *Image Communication*, 23(4):239–256, 2008.

[22] ITU-T Rec. J. 247. *Objective perceptual multimedia video quality measurement in the presence of a full reference.* International Telecommunications Union Std. 2008.

[23] Final report from the video quality experts group on the validation of objective models of multimedia quality assessment, Phase I. 2008. Available: http://www.its.bldrdoc.gov/vqeg/projects/multimedia/

[24] NTT. NTT News Release. 2008. Available: http://www.ntt.co.jp/news/news08e/0808/080825a.html

[25] Opticom. [Online] Available: http://www.opticom.de/technology/pevq_video-quality-testing.html

[26] M. Malkowski and D. Claben. Performance of video telephony services in UMTS using live measurements and network emulation. *Wirel. Pers. Commun.*, 1:19–32, 2008.

[27] M. Barkowsky, J. Bialkowski, R. Bitto, and A. Kaup. Temporal registration using 3D phase correlation and a maximum likelihood approach in the perceptual evaluation of video quality. *IEEE Workshop Multimed. Signal Process.*, 195–198, 1–3 October 2007.

[28] A. P. Hekstra, J. G. Beerends, D. Ledermann, F. E. de Caluwe, S. Kohler, R. H. Koenen, S. Rihs, M. Ehrsam, and D. Schlauss. PVQM - A perceptual video quality measure. *Signal Proc. Image Commun.*, 17:781–798, 2002.

[29] ITU-T Rec. J. 144. *Objective perceptual video quality measurement techniques for digital cable television in the presence of a full reference.* International Telecommunications Union Std. 2004.

[30] The Video Quality Experts Group. *Final VQEG report on the validation of objective models of video quality assessment.* 2003. Available: http://www.its.bldrdoc.gov/vqeg/projects/frtv_phaseII

[31] P. J. Huber. *Robust Statistics.* John Wiley & Sons, Inc., New York, 1981.

[32] M. Yuen and H. R. Wu. A survey of hybrid MC/DPCM/DCT video coding distortions. *Signal Process.*, 70(3):247–278, 1998.

[33] J. A. Movshon and W. T. Newsome. Visual response properties of striate cortical neurons projecting to Area MT in macaque monkeys. *J. Neurosci.*, 16(23):7733–7741, 1996.

[34] R. T. Born and D. C. Bradley. Structure and function of visual area MT. *Annu. Rev. Neurosci.*, 28:157–189, 2005.

[35] Z. Wang, L. Lu, and A. C. Bovik. Video quality assessment based on structural distortion measurement. *Signal Process. Image Commun.*, 19(2):121–132, 2004.

[36] Z. Wang and Q. Li. Video quality assessment using a statistical model of human visual speed perception. *J. Opt. Soc. Am. A Opt. Image Sci. Vis.*, 24(12):B61–B69, 2007.

[37] K. Seshadrinathan and A. C. Bovik. Spatio-temporal quality assessment of natural videos. *IEEE Trans. Image Process.*, submitted for publication.

[38] ——— A structural similarity metric for video based on motion models. *IEEE Int. Conf. Acoust., Speech Signal Process.*, 1:869–872, 2007.

[39] K. Seshadrinathan. Video quality assessment based on motion models. *Ph.D. dissertation*, University of Texas at Austin, 2008.

[40] D. J. Fleet and A. D. Jepson. Computation of component image velocity from local phase information. *Int. J. Comput. Vis.*, 5(1):77–104, 1990.

[41] J. G. Daugman. Uncertainty relation for resolution in space, spatial frequency, and orientation optimized by two-dimensional visual cortical filters. *J. Opt. Soc. Am. A (Opt. Image Sci.)*, 2(7): 1160–1169, 1985.

[42] D. J. Heeger. Optical flow using spatiotemporal filters. *Int. J. Comput. Vis.*, 1(4):279–302, 1987.

[43] E. H. Adelson and J. R. Bergen. Spatiotemporal energy models for the perception of motion. *J. Opt. Soc. Am. A*, 2(2):284–299, 1985.

[44] N. J. Priebe, S. G. Lisberger, and J. A. Movshon. Tuning for spatiotemporal frequency and speed in directionally selective neurons of macaque striate cortex. *J. Neurosci.*, 26(11):2941–2950, 2006.

[45] E. P. Simoncelli and D. J. Heeger. A model of neuronal responses in visual area MT. *Vis. Res.*, 38(5):743–761, 1998.

[46] K. Seshadrinathan and A. C. Bovik. Unifying analysis of full reference image quality assessment. *IEEE Int. Conf. Image Process.*, 2008.

[47] A. B. Watson and J. Ahumada, A. J., Model of human visual-motion sensing. *J. Opt. Soc. Am. A (Opt. Image Sci.)*, 2(2):322–342, 1985.

[48] The Video Quality Experts Group. *Final report from the video quality experts group on the validation of objective quality metrics for video quality assessment.* 2000. Available: http://www.its.bldrdoc.gov/vqeg/projects/frtv_phaseI

[49] LIVE image quality assessment database. 2003. Available: http://live.ece.utexas.edu/research/quality/subjective.htm

[50] A. Moorthy and A. C. Bovik. Visual importance pooling for image quality assessment. *IEEE J. Sel. Top. Signal Process. Special Issue Visual Media Quality Assessment*, 3(2):193–201, 2008.

CHAPTER

A Unified Framework for Video Indexing, Summarization, Browsing, and Retrieval

15

Ziyou Xiong[1], Regunathan Radhakrishnan[2], Yong Rui[3], Ajay Divakaran[4], Tsuhan Chen[5], and Thomas S. Huang[6]

[1] *United Technologies Research Center, East Hartford, Connecticut, USA*
[2] *Dolby Laboratories, San Fransecso, California, USA*
[3] *Microsoft Research, Redmond, Washington, USA*
[4] *Sarnoff Corporation, Princeton, New Jersey, USA*
[5] *Department of Electrical and Computer Engineering, Carnegie Mellon University, Pittsburgh, Pennsylvania, USA*
[6] *Department of Electrical and Computer Engineering, University of Illinois at Urbana-Champaign, Urbana, Illinois, USA*

15.1 INTRODUCTION

The amount of digital content, in the form of images and video, has been increasing exponentially in recent years. With increasing computing power and electronic storage capacity, the potential for large digital image/video libraries is growing rapidly. In particular, the World Wide Web has seen an increased use of digital images and video, which form the base of many entertainment, educational, and commercial applications. As a result, it has become more challenging for a user to search for the relevant information among a large amount of digital images or video. Image and video libraries therefore need to provide easy informational access, and the retrieval information must be easy to locate, manage, and display.

As the size of accessible image and video collections grows to thousands of hours, potential viewers will need abstractions and technology that help them browse effectively and efficiently. Text-based search algorithms offer some assistance in finding specific images or segments among large video collections. In most cases, however, these systems output many irrelevant images or video segments to insure retrieval of pertinent information. Intelligent indexing systems are essential for optimal retrieval of image and video data.

15.1.1 Content Categories

Video content can be accessed by using either a top-down approach or a bottom-up approach [1–4]. The top-down approach, that is, video browsing, is useful when we need to get an essence of the content. The bottom-up approach, that is, video retrieval, is useful when we know exactly what we are looking for in the content, as shown in Fig. 15.1. In video summarization, what essence the summary should capture depends on whether the content is scripted or not. Because scripted content, such as news, drama, and movie, is carefully structured as a sequence of semantic units, one can get its essence by enabling a traversal through representative items from these semantic units. Hence, table of contents (ToC)-based video browsing caters to summarization of scripted content. For instance, a news video composed of a sequence of stories can be summarized/browsed using a key-frame representation for each of the shots in a story. However, summarization of unscripted content, such as surveillance and sports), requires a "highlights" extraction framework that only captures remarkable events that constitute the summary.

Considerable progress has been made in multimodal analysis, video representation, summarization, browsing, and retrieval, which are the five fundamental bases for accessing video content. The first three bases focus on metadata generation and organization while the last two focus on metadata consumption. Multimodal analysis deals with the signal processing part of the video system, including shot boundary detection,

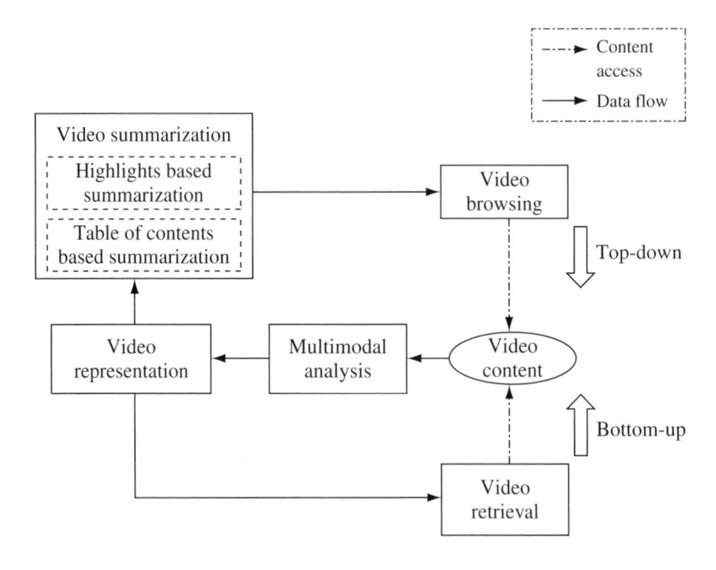

FIGURE 15.1

Relation between five research areas.

key-frame extraction, key object detection, audio analysis, closed caption analysis, etc. Video representation is concerned with the structure of the video. Again, it is useful to have different representations for scripted and unscripted content. An example of a video representation for scripted content is the tree-structured key-frame hierarchy [3, 5]. Built on top of the video representation, video summarization, either based on ToC generation or highlights extraction, deals with how to use the representation structure to provide the viewers top-down access using the summary for video browsing. Finally, video retrieval is concerned with retrieving specific video objects. The relationship between these five bases is illustrated in Fig. 15.1.

As seen in Fig. 15.1, video browsing and retrieval directly support users' access to the video content. For accessing a temporal medium, such as a video clip, summarization, browsing, and retrieval are equally important. As mentioned earlier, browsing enabled through summarization helps a user to quickly grasp the global picture of the data, whereas retrieval helps a user to find the results of a specific query.

An analogy explains this argument. How does a reader efficiently access the content of a 1000-page book? Without reading the whole book, he can first go to the book's ToC to find which chapters or sections suit his needs. If he has specific questions (queries) in mind, such as finding a term or a key word, he can go to the Index at the end of the book and find the corresponding book sections addressing that question. On the other hand, how does a reader efficiently access the content of a 100-page magazine? Without reading the whole magazine, he can either directly go to the featured articles listed on the front page or use the ToC to find which article suits his needs. In short, the book's ToC helps a reader browse and the book's index helps a reader retrieve. Similarly, the magazine's featured articles also help the reader browse through the highlights. All these three aspects are equally important in helping users access the content of the book or the magazine. For today's video content, techniques are urgently needed for automatically (or semi-automatically) constructing video ToC, video Highlights, and video Indices to facilitate summarization, browsing, and retrieval.

A great degree of power and flexibility can be achieved by simultaneously designing the video access components (ToC, Highlights, and Index) using a unified framework. For a long and continuous stream of data, such as video, a back-and-forth mechanism between summarization and retrieval is crucial.

15.1.2 Storage and Compression

In analysis of digital content, compression schemes offer increased storage capacity by utilizing statistical characteristics of images and video. Images and video are compressed and stored as discrete cosine transform (DCT) coefficients and motion vectors. One drawback to these compression schemes is loss in quality. Bit streams created by lossy compression schemes, however, typically preserve some statistical information of the original video in an explicit manner. For example, the DCT coefficients preserve colors, texture, and other spatial domain characteristics, and motion vectors preserve object motion, camera pan and zoom, and other temporal characteristics. Lossless schemes, such as Run Length Encoding (RLE) and Huffman coding, do not sacrifice quality but provide lower compression ratios. Furthermore, bit streams created by lossless algorithms do not explicitly contain any statistical information of the original video. Many algorithms provide compression as high as 100 to 1 and often use DCT and motion compensation for compression. The parameters of the DCT may be used for video segmentation while the motion compensation statistics may be used as a form of optical flow, as discussed in Section 15.2.

15.1.3 Terminology

Before we go into the details of the discussion, it will be beneficial to first introduce some important terms used in the digital video research field.

- Scripted content is a video that is carefully produced according to a script or plan that is later edited, compiled, and distributed for consumption. News videos, dramas, and movies are examples of scripted content. Video content that is not scripted is then referred to as unscripted content. In unscripted content, such as surveillance video, the events happen spontaneously. One can think of varying degrees of "scriptedness" and "unscriptedness" from movie content to surveillance content.
- Video shot is a consecutive sequence of frames recorded from a single camera. It is the building block of video streams.
- Key frame is the frame that represents the salient visual content of a shot. Depending on the complexity of the content of the shot, one or more key frames can be extracted.
- Video scene is defined as a collection of semantically related and temporally adjacent shots depicting and conveying a high-level concept or story. Although shots are marked by physical boundaries, scenes are marked by semantic boundaries.
- Video group is an intermediate entity between the physical shots and semantic scenes and serves as the bridge between the two. Examples of groups are temporally adjacent shots [5] or visually similar shots [3].
- Play and break is the first level of semantic segmentation in sports video and surveillance video. In sports video (e.g., soccer, baseball, golf), a game is in play when the ball is in the field and the game is going on; break, or out of play, is the complement set, that is, whenever the ball has completely crossed the goal line or touch line, whether on the ground or in the air or the game has been halted by the referee [6]. In surveillance video, a play is a period in which there is some activity in the scene.
- Audio marker is a contiguous sequence of audio frames representing a key audio class that is indicative of the events of interest in the video. An example of an

audio marker for sports video can be the audience reaction sound (cheering and applause) or commentator's excited speech.

- Video marker is a contiguous sequence of video frames containing a key video object that is indicative of the events of interest in the video. An example of a video marker for baseball videos is the video segment containing the squatting catcher at the beginning of every pitch.
- Highlight candidate is a video segment that is likely to be remarkable and can be identified using the video and audio markers.
- Highlight group is a cluster of highlight candidates.

In summary, scripted video data can be structured into a hierarchy consisting of five levels: video, scene, group, shot, and key frame, which increase in granularity from top to bottom [4] (see Fig. 15.2). Similarly, the unscripted video data can be structured into a hierarchy of four levels: play/break, audiovisual markers, highlight candidates, highlight groups, which increase in semantic level from bottom to top (see Fig. 15.3).

The goals of this chapter are to develop novel techniques for constructing the video ToC, video Highlights, and video Index as well as how to integrate them into a unified framework for video summarization, video browsing, and video retrieval. The rest of the chapter is organized as follows. We review image and video features, video analysis, representation, browsing, and retrieval in Sections 15.2–15.6, respectively. In Section 15.7, we describe in detail a unified framework for video summarization, video browsing, and video retrieval. Conclusions and future research directions are summarized in Section 15.8.

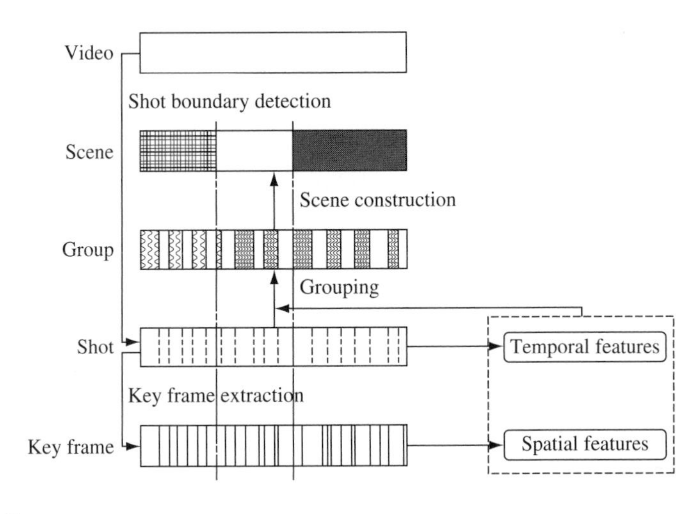

FIGURE 15.2

A hierarchical video representation for scripted content.

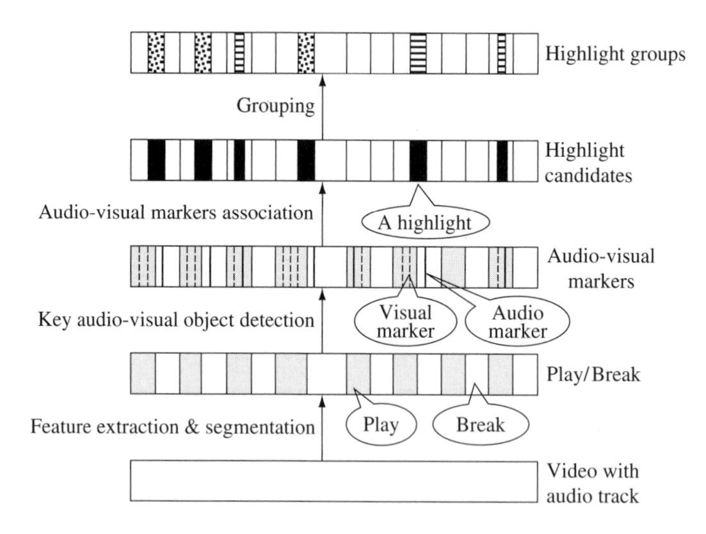

FIGURE 15.3

A hierarchical video representation for unscripted content.

15.2 IMAGE AND VIDEO FEATURES

A feature is defined as a descriptive parameter that is extracted from an image or video stream. Features may be used to interpret visual content or as a measure for similarity in image and video databases. In this chapter, features are described in the following categories:

- Statistical features are extracted from an image or video sequence without regard to content. These include parameters derived from such algorithms as image difference and camera motion.
- Compressed-domain features are extracted from a compressed image or video stream without regard to content.
- Content-Based Features are features that are derived for the purpose of describing the actual content in an image or video stream.

In the sections that follow, we describe examples of each feature and potential applications in image and video databases.

15.2.1 Statistical Features

Certain features may be extracted directly from image pixels without regard to the content. These features include such analytical features as scene changes, motion flow, and video structure in the image domain and sound discrimination in the audio domain. In this section, we describe techniques for image difference and motion analysis as statistical features.

15.2.1.1 Image Difference

A difference measure between images serves as a feature to measure similarity. We describe two fundamental methods for image difference: absolute difference and histogram difference. The absolute difference requires less computation, but is generally more susceptible to noise and other imaging artifacts, as described below.

15.2.1.1.1 Absolute Difference

The image difference of two images is defined as the sum of the absolute difference at each pixel. The first image I_t is analyzed with a second image, I_{t-T}, at a temporal distance T. The difference value is defined as

$$D(t) = \sum_{i=0}^{M} \left| I_{(t-T)}(i) - I_t(i) \right|,$$

where M is the resolution or number of pixels in the image. This method for image difference is noisy and extremely sensitive to camera motion and image degradation. When applied to subregions of the image, $D(t)$ is less noisy and may be used as a more reliable parameter for image difference.

$$D_s(t) = \sum_{j=S}^{\frac{H}{n}} \sum_{i=S}^{\frac{W}{n}} \left| I_{(t-T)}(i,j) - I_t(i,j) \right|$$

$D_s(t)$ is the sum of the absolute difference in a subregion of the image, where S represents the starting position for a particular region and n represents the number of subregions.

We may also apply some form of filtering to eliminate excess noise in the image and subsequent difference. For example, the image on the right in Fig. 15.4 represents the output of a Gaussian filter on the original image on the left.

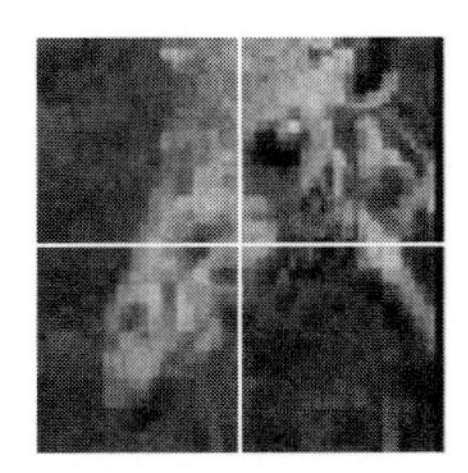
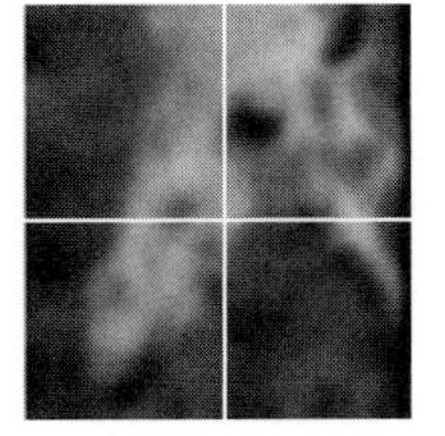

FIGURE 15.4

Left: original; right: filtered.

15.2.1.1.2 Histogram Difference

A histogram difference is less sensitive to subtle motion and is an effective measure for detecting similarity in images. By detecting significant changes in the weighted color histogram of two images, we form a more robust measure for image correspondence. The histogram difference may also be used in subregions to limit distortion due to noise and motion.

$$D_H(t) = \sum_{v=0}^{N} \left| H_{(t-1)}(v) - H_t(v) \right|$$

The difference value, $D_H(t)$, will rise during scene changes, image noise, and camera or object motion. In the equation below, N represents the number of bins in the histogram, typically 256. Two adjacent images may be processed, although this algorithm is less sensitive to error when images are separated by a spacing interval, Di. Di is typically on the order of 5–10 frames for video encoded at standard 30 fps. An empirical threshold may be set to detect values of $D_H(t)$ that correspond to scene changes. For inputs from multiple categories of video, an adaptive threshold for $D_H(t)$ should be used.

$$D_{H-R}(t) = \sum_{v=0}^{N} \left| H_{R(t-1)}(v) - H_{Rt}(v) \right|$$

$$D_{H-G}(t) = \sum_{v=0}^{N} \left| H_{G(t-1)}(v) - H_{Gt}(v) \right|$$

$$D_{H-B}(t) = \sum_{v=0}^{N} \left| H_{B(t-1)}(v) - H_{Bt}(v) \right|$$

$$D_{H-\mathrm{RGB}}(t) = \frac{\sum (D_{H-R}(t) + D_{H-G}(t) + D_{H-B}(t))}{3}$$

If the histogram is actually three separate sets for RGB, the difference may simply be summed. An alternative to summing the separate histograms is to convert the RGB histograms to a single color band, such as Munsell or LUV color.

15.2.1.2 Video Segmentation

An important application of image difference in video is the separation of visual scenes. A simple image difference represents one of the more common methods for detection of scene changes. The difference measures, $D(t)$ and $D_H(t)$, may be used to determine the occurrence of a scene change. By monitoring the difference of two images over some time interval, a threshold may be set to detect significant differences or changes in scenery. This method provides a useful tool for detecting scene cuts but is susceptible to errors during transitions. A block-based approach may be used to reduce errors in difference calculations. This method is still subject to errors when subtle object or camera motion occurs.

The most fundamental scene change is the video cut. For most cuts, the difference between image frames is so distinct that accurate detection is not difficult. Cuts between similar scenes, however, may be missed when using only static properties such as image difference. Several research groups have developed working techniques for detecting scene changes through variations in image and histogram differencing.

A histogram difference is less sensitive to subtle motion and is an effective measure for detecting scene cuts and gradual transitions. By detecting significant changes in the weighted color histogram of each successive frame, video sequences can be separated into scenes. This technique is simple and yet robust enough to maintain high levels of accuracy. An illustration of histogram-based segmentation is shown in Fig. 15.5.

15.2.1.2.1 Scene Change Categories

There are a variety of complex scene changes used in video production, but the basic premise is a change in visual content. The video cut, as well as other scene change procedures, is discussed below.

Fast cut—A sequence of video cuts, each very short in duration, represents a fast cut. This technique heightens the sense of action or excitement. To detect a fast cut, we may look for a sequence of scene changes that are in close proximity.

Distance cut—A distance cut occurs when the camera cuts from one perspective of a scene to another some distance away. This shift in distance usually appears as a cut from a wide shot to a close-up shot or vice versa.

Intercutting—When scenes change back and forth from one subject to another, we say the subjects are intercut. This concept is similar to the distance cut, but the images are separate and not inclusive of the same scenes. Intercutting is used to show a thought process between two or more subjects.

Dissolves and fades—Dynamic imaging effects are often used to change from one scene to another. A common effect in all types of video is the fade. A fade occurs when a scene changes over time from its original color scheme to a black background. This procedure is commonly used as a transition from one topic to another. Another dynamic effect is the dissolve. Similar to the fade, this effect occurs when a scene changes over time and morphs into a separate scene. This transition is less intrusive and is used when subtle change is needed.

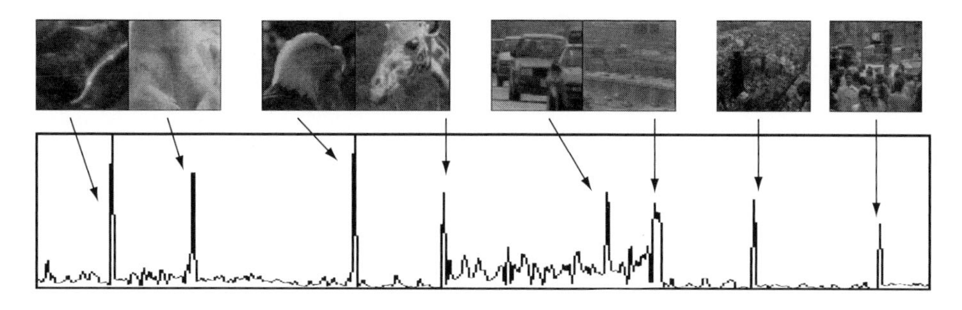

FIGURE 15.5

Histogram difference, $D_{H-\mathrm{RGB}}(t)$, for scene segmentation.

Wipes and blends—These effects are most often used in news video. The actual format of each may change from one show to the next. A wipe usually consists of the last frame of a scene being folded like a page in a book. A blend may be shown as pieces of two separate scenes combined in some artistic manner. Like the fade and dissolve, wipes and blends are usually used to transition to a separate topic. In feature films, a wipe is often used to convey a change in time or location.

15.2.1.2.2 Alternative Segmentation Technology

An alternative form of scene segmentation involves the use of traditional edge detection characteristics. Edges in images are useful information about the changes in background and object distribution between scenes. An effective algorithm for detecting cuts and gradual transitions was developed at Cornell University using edge detection technology [7].

An analysis of the global motion of a video sequence may also be used to detect changes in scenery. For example, when the error in optical flow is high, this is usually attributed to its inability to track a majority of the motion vectors from one frame to the next. Such errors can be used to identify scene changes. A motion-controlled temporal filter may also be used to detect dissolves and fades, as well as separate video sequences that contain long pans. The use of motion as a statistical feature is discussed in the following section. The methods for scene segmentation described in this section may be used individually or combined for more robust segmentation.

15.2.1.3 Motion Analysis

Motion characteristics represent an important feature in video indexing. One aspect is based on interpreting camera motion [8, 9]. Many video scenes have dynamic camera effects but offer little in the description of a particular segment. Static scenes, such as interviews and still poses, contain essentially identical video frames. Knowing the precise location of camera motion can also provide a method for video parsing. Rather than simply parse a video by scenes, one may also parse a video according to the type of motion. An important kind of video characterization is defined not just by the motion of the camera but also by motion or action of the objects being viewed.

An analysis of optical flow can be used to detect camera and object motion. Most algorithms for computing optical flow require extensive computation, and more often, researchers are exploring methods to extract optical flow from video compressed with some form of motion compensation. Section 15.3 describes the benefits of using compressed video for optical flow and other image features.

Statistics from optical flow may also be used to detect scene changes. Optical flow is computed from one frame to the next. When the motion vectors for a frame are randomly distributed without coherency, this may suggest the presence of a scene change. In this sense, the quality of the camera motion estimate is used to segment video. Video segmentation algorithms often yield false scene changes in the presence of extreme camera or object motion. An analysis of optical flow quality may also be used to avoid false detection of scene changes.

Optical flow fields may be interpreted in many ways to estimate the characteristics of motion in video. Two such interpretations are the camera motion and object motion.

15.2.1.3.1 Camera Motion

An affine model is used to approximate the flow patterns consistent with all types of camera motion.

$$u(xi, yi) = axi + byi + c$$

$$v(xi, yi) = dxi + eyi + f$$

Affine parameters a, b, c, d, e, and f are calculated by minimizing the least squares error of the motion vectors.

$$\begin{bmatrix} \sum x2 & \sum xy & \sum x & 0 & 0 & 0 \\ \sum xy & \sum x & \sum y & 0 & 0 & 0 \\ \sum x & \sum y & \sum N & 0 & 0 & 0 \\ 0 & 0 & 0 & \sum x2 & \sum xy & \sum x \\ 0 & 0 & 0 & \sum xy & \sum x2 & \sum y \\ 0 & 0 & 0 & \sum x & \sum y & \sum N \end{bmatrix} \begin{bmatrix} a \\ b \\ c \\ d \\ e \\ f \end{bmatrix} = \begin{bmatrix} \sum ux \\ \sum uy \\ \sum u \\ \sum vx \\ \sum vy \\ \sum v \end{bmatrix}$$

We also compute average flow $\bar{v}$ and $\bar{u}$. Where $\bar{v}$ and $\bar{u}$,

$$\bar{u} = \sum_{i-0}^{N} axi + byi + c$$

$$\bar{v} = \sum_{i-0}^{N} dxi + eyi + f.$$

Using the affine flow parameters and average flow, we classify the flow pattern. To determine if a pattern is a zoom, we first check if there is the convergence or divergence point $(x0, y0)$, where $u(xi, yi) = 0$ and $v(xi, yi) = 0$. To solve for $(x0, y0)$, the following relation must be true:

$$\begin{vmatrix} a & b \\ d & e \end{vmatrix} = 0$$

If the above relation is true, and $(x0, y0)$ is located inside the image, then it must represent the focus of expansion. If $\bar{v}$ and $\bar{u}$ are large, then this is the focus of the flow and camera is zooming. If $(x0, y0)$ is outside the image, and or are large, then the camera is panning in the direction of the dominant vector.

If the above determinant is approximately 0, then $(x0, y0)$ does not exist and the camera is panning or static. If $\bar{v}$ or $\bar{u}$ are large, the motion is panning in the direction of the dominant vector. Otherwise, there is no significant motion and the flow is static. We may eliminate fragmented motion by averaging the results in a W frame window over time. Examples of the camera motion analysis results are shown in Fig. 15.6.

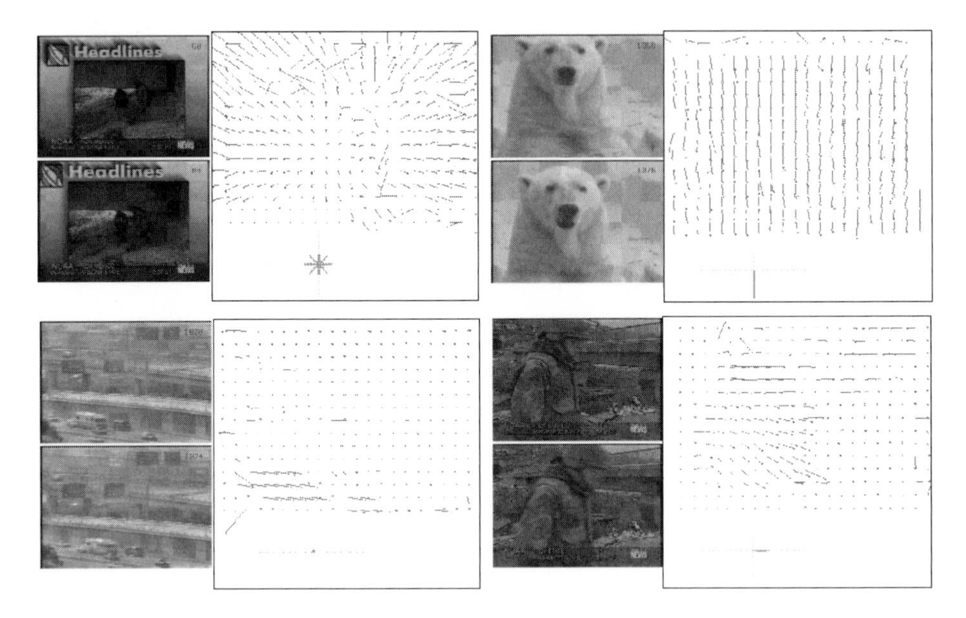

FIGURE 15.6

Optical flow fields for a pan (top right), zoom (top left), and object motion.

15.2.1.3.2 Object Motion

Object motion typically exhibits flow fields in specific regions of an image, whereas camera motion is characterized by flow throughout the entire image. The global distribution of motion vectors distinguishes between object and camera motion. The flow field is partitioned into a grid as shown in Fig. 15.7. If the average velocity for the vectors in a particular grid is high (typically >2.5 pixels), then that grid is designated as containing motion. When the number of connected motion grids, G_m,

$$G_m(i) = \begin{cases} 0 & (G_m(i-1) = 0, G_m(i+1) = 0, \ldots M) \\ 1 & \text{otherwise} \end{cases}$$

is high (typically $G_m > 7$), the flow is some form of camera motion. $G_m(i)$ represents the status of motion grid at position i and M represents the number of neighbors. A motion grid should consist of at least a 4×4 array of motion vectors. If G_m is not high, but greater than some small value (typically two grids), the motion is isolated in a small region of the image and the flow is probably caused by object motion. This result is averaged over a frame window of width W_A, just as with camera motion, but the number of object motion regions needed is typically on the order of 60%. This is 12 object motion frames for a typical W_A of 20 frames. Examples of the object motion analysis results are shown in Fig. 15.7.

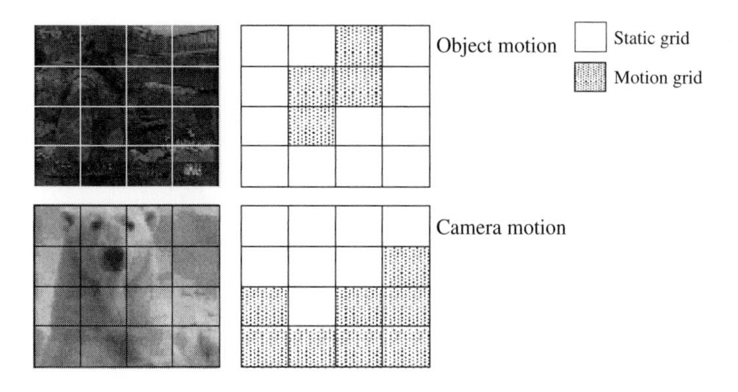

FIGURE 15.7

Camera and object motion detection.

15.2.1.4 *Alternative Statistical Features*

15.2.1.4.1 Texture

Analysis of image texture is useful in the discrimination of low-interest video from video containing complex features. A low-interest image may also contain uniform texture, as well as uniform color or low contrast. Perceptual features for individual video frames were computed using common textual features such as, coarseness, contrast, directionality, and regularity.

15.2.1.4.2 Shape and Position

The shape and appearance of objects may also be used as a feature for image correspondence. Color and texture properties will often change from one image to the next, making image difference and texture features less useful. An example of this is shown in Fig. 15.8, where the feature of interest is an anchorperson, but the color, texture, and position of the subjects are different for each image. Commercial systems for shape-based image correspondence are discussed in Section 15.7.

15.2.1.4.3 Audio Features

In addition to image features, certain audio features may be extracted from video to assist in the retrieval task. Loud sounds, silence, and single frequency sound markers may be detected analytically without actual knowledge of the audio content. Loud sounds imply a heightened state of emotion in video, and are easily detected by measuring a number of audio attributes, such as signal amplitude or power. Silent video may signify an area of less importance and can also be detected with straightforward analytical estimates. A video producer will often use single frequency sound markers, typically a 1000 Hz. tone, to mark a particular point in the beginning of a video. This tone may be detected to determine the exact point in which a video will start.

FIGURE 15.8

Images with similar shapes (human face and torso).

15.2.1.5 *Hierarchical Video Structure*

Most video is produced with a particular format and structure. This structure may be taken into consideration when analyzing particular video content. News segments are typically 30 min in duration and follow a rigid pattern from day to day. Commercials are also of fixed duration, making detection less difficult.

Another key element in video is the use of the black frame. In most broadcast video, a black frame is shown between a transition of two segments. In news broadcast, this usually occurs between a story and a commercial. By detecting the location of black frames in video, a hierarchical structure may be created to determine transitions between segments. A black frame or any single intensity image may be detected by summing the total number of pixels in a particular color space, P_s.

$$P_s(t) = \sum_{i=0}^{M} \begin{cases} 0 & (i > I_{\text{high}} || i < I_{\text{low}}) \\ 1 & \text{otherwise} \end{cases}$$

In the detection of the black frame, I_{high}, the maximum allowable pixel intensity is on the order of 20% of the maximum color resolution (51 for a 256-bit image), and I_{low}, the minimum allowable pixel intensity, is 0. The separation of segments in video is crucial in retrieval systems, where a user will most likely request a small segment of interest and not an entire full-length video. There are a number of ways to detect this feature in video, the simplest being to detect a high number of pixels in an image that are within a given tolerance of being a black pixel.

15.2.2 Compressed-Domain Features

In typical applications of multimedia databases, the materials, especially the images and video, are often in a compressed format. Given large amounts of compressed materials (e.g., Moving Picture Experts Group [MPEG]), how do we index and retrieve the content rapidly? To deal with these materials, a straightforward approach is to decompress all the data and utilize the same features as mentioned in previous section. Doing so, however, has some disadvantages. First, the decompression implies extra computation. Second, the process of decompression and recompression, often referred to as "recoding," results in further loss of image quality. Finally, since the size of decompressed data is much larger than the compressed form, most hardware and CPU cycles are needed to process and store the data. The solution to these problems is to extract features directly from the compressed data. We call these the compressed-domain features, and these features can be useful for indexing and retrieval [8, 10, 11]. The question is how to explore unique information available in the compressed domain. We start by introducing a number of commonly used compressed-domain features:

The motion vectors that are available in all video data compressed using standards such as H.261/H.263 and MPEG-1/2 are very useful. Analysis of motion vectors can be used to detect scene changes and other special effects such as dissolve, fade in, and fade out. For example, if the motion vectors for a frame are randomly distributed without coherency, it may suggest the presence of a scene change. Segmentation of a field of motion vectors into regions of similar vectors can be used to detect moving objects and track their positions. They can also be used to derive camera motions such as zoom and pan [8, 9]. Essentially, since motion vectors represent a low-resolution optical flow in the video, they can be used to extract all information that can be extracted using the optical flow method.

The percentage of each type of block in a picture is also a good indicator of scene changes, too. For a P-frame, a large percentage of intrablocks implies a lot of new information for the current frame that cannot be predicted from the previous frame. Therefore, such a P-frame indicates the beginning of a new scene right after a scene change. For a B-frame, the ratio between the number of forward-predicted blocks and the number of backward-predicted blocks can be used to conclude whether the scene change happens before this B-frame or after this B-frame. If the number of forward-predicted blocks is larger than the number of backward-predicted blocks, that is, there is more correlation between the previous frame and the current B-frame than there is between the current B-frame and the following frames, then one can conclude that the scene change happens after the B-frame. If the number of forward-predicted blocks is smaller than the number of backward-predicted blocks, then one can conclude that the scene change happens before the B-frame.

The DCT provides a decomposition of the original image in the frequency domain. Therefore, DCT coefficients form a natural representation of texture in the original image. In addition to texture analysis, DCT coefficients can also be used to match images and to detect scene changes. If only the DC components are collected, we have a low-resolution representation of the original image, averaged over 8×8 blocks. This is very helpful because it means much less data to manipulate, and it is found that for some applications, DC components already contain sufficient information. For color analysis, usually only the DC components are used to estimate the color histogram. For scene change detection, usually only the DC components are used to compare the content in two consecutive frames.

Not only can information be extracted from the compressed data for indexing and retrieval, the parameters in the compression process that are not explicitly specified in the bit stream can be very useful as well. One example is the bit rate, that is, the number of bits used for each picture. For intracoded video (i.e., no motion compensation), the number of bits per picture should remain roughly constant for a scene segment and should change when the scene changes. For example, a scene with simple color variation and texture requires fewer bits per picture compared to a scene that has detailed texture. For intercoding, the number of bits per picture is proportional to the action between the current picture and the previous picture. Therefore, if the number of bits for a certain picture is high, we can often conclude that there is a scene cut.

The compressed-domain approach does not solve all problems, though. To identify useful features from compressed data is typically difficult because each compression technique poses additional constrains, for example, nonlinear processing, rigid data structure syntax, and resolution reduction.

Another issue is that compressed-domain features depend on the underlying compression standard. For different compression standards, different feature extraction algorithms have to be developed. Ultimately, we would like to have new compression standards with maximal content accessibility. MPEG-4 and MPEG-7 already have considered this aspect. In particular, MPEG-7 is a standard that goes beyond the domain of "compression" and seeks efficient "representation" of image and video content. In conclusion, the compressed-domain approach provides significant advantages but also brings new challenges.

15.2.3 Content-Based Features

Section 15.2.1 described a number of image and video features that can be extracted using well-known techniques in image processing. Section 15.2.2 described how many of these features are computed or approximated using encoded parameters in image and video compression. Although in both cases there is considerable understanding of the structure of the video, the features in no way estimate the actual image or video content.

In this section, we describe several methods to approximate the actual content of an image or video. The desired result has less to do with analytical features such as color, or texture, and more with the actual objects within the image or video.

15.2.3.1 Object Detection

Identifying significant objects that appear in the video frames is one of the key components for video characterization. Several working systems have generated reasonable results for the detection of a particular object, such as human faces, text, or automobile. These limited domain systems have much greater accuracy than do broad domain systems that attempt to identify any object in the image.

15.2.3.1.1 Human Subjects

The "talking head" image is common in interviews and news clips and illustrates a clear example of video production focusing on an individual of interest. A human interacting within an environment is also a common theme in video. The detection of a human subject is particularly important in the analysis of news footage. An anchorperson will often appear at the start and end of a news broadcast, which is useful for detecting segment boundaries. In sports, anchorpersons will often appear between plays or commercials.

The detection of humans in video is possible using a number of algorithms. Fig. 15.9 shows examples of faces detected using the Neural Network Arbitration method [12]. Most techniques are dependent of scale and rely heavily on lighting conditions, limited occlusion, and limited facial rotation.

15.2.3.1.2 Captions and Graphics

Text and graphics are used in a variety of ways to convey content to the viewer. They are most commonly used in broadcast news, where information must be absorbed in a short time. Examples of text and graphics in video are discussed below.

Text in video provides significant information as to the content of a scene. For example, statistical numbers and titles are not usually spoken but are included in captions for viewer inspection. Moreover, this information does not always appear in closed captions so detection in the image is crucial for identifying potentially important regions.

In news video, captions of the broadcasting company are often shown at low opacity as a watermark in a corner without obstructing the actual video. A ticker tape is widely used in broadcast news to display information such as the weather, sports scores, or the stock market. In some broadcast news, graphics such as weather forecasts are displayed in a ticker-tape format with the news logo in the lower right corner at full opacity. Captions that appear in the lower third portion of a frame are almost always used to describe a location, person of interest, title, or event in news video. In Fig. 15.9, the anchorperson's location is listed.

FIGURE 15.9

Recognition of captions and faces [34].

Captions are used less frequently in video domains other than broadcast news. In sports, a score or some information about an ensuing play is often shown in a corner or border at low opacity. Captions are sometimes used in documentaries to describe a location, person of interest, title, or event. Almost all commercials use some form of captions to describe a product or institution, because their time is limited to only a few minutes.

For feature films, a producer may use text at the beginning or end of a film for deliberate viewer comprehension, such as character listings or credits. A producer may also start a film with an introduction to the story being told. Throughout a film, captions may be used to convey a change in time or location, which would otherwise be difficult and time consuming for a video producer to create. A producer will seldom use fortuitous text in the actual video unless the wording is noticeable and easy to read in a short time. A typical text region can be characterized as a horizontal rectangular structure of clustered sharp edges because characters usually form regions of high contrast against the background. By detecting these properties, we can extract potentially important regions from video frames that contain textual information. Most captions are high-contrast text such as the black and white chyron commonly found in news video. Consistent detection of the same text region over a period of time is probable since text regions remain at an exact position for many video frames. This may also reduce the number of false detections that occur when text regions move or fade in and out between scenes.

A typical text region can be characterized as a horizontal rectangular structure of clustered sharp edges because characters usually form regions of high contrast against the background. By detecting these properties, we can extract regions from video frames that contain textual information. Figure 15.10 illustrates the process of text detection, primarily, regions of horizontal titles and captions. We first apply a global horizontal differential filter, F_{HD}, to the image:

$$F_{HD} = \left[-\frac{1}{2} \ \ 1 \ \ \frac{1}{2} \right].$$

An appropriate binary threshold should be set for extraction of vertical edge features. A smoothing filter, F_S, is then used to eliminate extraneous fragments and to connect character sections that may have been detached:

$$F_S = \left[\frac{1}{3} \ \ \frac{1}{3} \ \ \frac{1}{3} \right].$$

Individual regions must be identified through cluster detection. A bounding box, B_B, should be computed for selection of text regions. We now select clusters with bounding regions that satisfy constraints in cluster size, C_S, cluster fill factor, C_{FF}, and horizontal—vertical aspect ratio:

$$C_s(n) = \sum_{i=0}^{P} C_n(i)$$

$$C_{FF}(n) = \frac{C_S(n)}{B_{B\ \text{area}}(n)}.$$

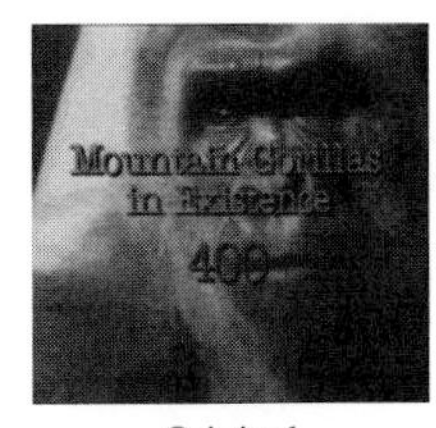
Original

HDF filter

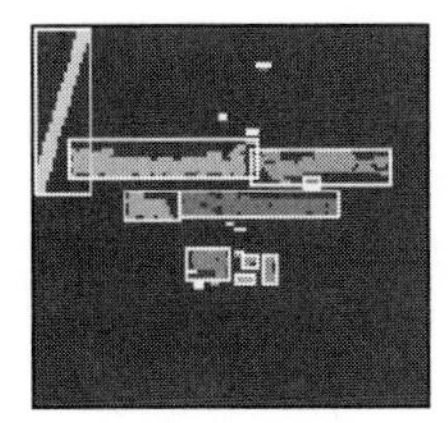
Clustering

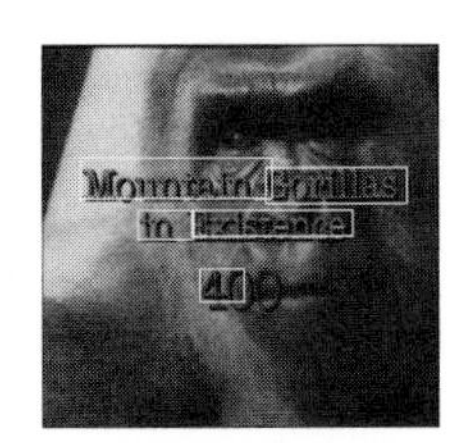
Region extraction

FIGURE 15.10

Text detection in video.

A sample set of parameters for the font size in Fig. 15.10 is listed below:

Cluster size > 70 pixels

Cluster fill factor > 0.45

Horizontal—vertical aspect ratio > 0.75

Maximum cluster height = 50 pixels

Minimum cluster height = 10 pixels

Maximum cluster width = 150 pixels

Minimum cluster width = 15 pixels

A cluster's bounding region must have a small vertical-to-horizontal aspect ratio as well as satisfying various limits in height and width. The fill factor of the region should be high to insure dense clusters. The cluster size should also be relatively large to avoid small fragments. Other controlling parameters are listed below.

Finally, we examine the intensity histogram of each region to test for high contrast. This is because certain textures and shapes appear similar to text but exhibit low contrast when examined in a bounded region.

For some fonts, a generic optical character recognition (OCR) package may accurately recognize video captions. For most OCR systems, the input is an individual character. This presents a problem in digital video since most of the characters experience some degradation during recording, digitization, and compression. For a simple font, we can search for blank spaces between characters and assume a fixed width for each letter [13].

A graphic is usually a recognizable symbol, which may contain text. Graphic illustrations or symbolic logos are used to represent many institutions, locations, and organizations. They are used extensively in news video, where it is important to describe the subject matter as efficiently as possible. A logo representing the subject is often placed in a corner next to an anchorperson during dialogue. Detection of graphics is a useful method for finding changes in semantic content. In this sense, its appearance may serve as a scene break. Recognition of corner regions for graphics detection may be possible through an extension of the scene change technology. Histogram difference analysis, $D_{Hs}(t)$, of isolated image regions instead of the entire image can provide a simple method for detecting corner graphics. An example of a graphics logo detected with $D_{Hs}(t)$ is shown in Fig. 15.11. In this example, a change is detected in the upper corner, although no scene change is detected.

$$D_{Hs}(t) = \sum_{j=0}^{\frac{H}{2}} \sum_{i=\frac{W}{2}}^{W} \left| H_{(t-T)}(i,j) - H_t(i,j) \right|$$

15.2.3.1.3 Articulated Objects

A particular object is usually the emphasis of a query in image and video retrieval. Recognition of articulated objects poses a great challenge and represents a significant step in content-based feature extraction. Many working systems have demonstrated accurate recognition of animal objects, segmented objects, and rigid objects such as planes or automobiles.

The recognition of a single object is only one potential use of image-based recognition systems. Discrimination of synthetic and natural backgrounds or an animated or mechanical motion would yield a significant improvement content-based feature extraction.

15.2.3.2 Audio and Language

An important element in video indexing creation is the audio track. Audio is an enormous source for describing video content. Words specific to the actual content or "key words" can be extracted using a number of language processing techniques [14, 15]. Key words may be used to reduce indexing and provide abstraction for video sequences. There are many possibilities for language processing in video, but the audio track must first exist as an ASCII document or speech recognition is necessary.

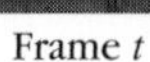

Frame t Frame $t+T$

FIGURE 15.11

Graphics detection through subregion histogram differencing.

Audio segmentation is needed to distinguish spoken words from music, noise, and silence. Further analysis through speech recognition is necessary to align and translate these words into text. Audio selection is made on a frame by frame basis, so it is important to achieve the highest possible accuracy. At a sampling rate of 8 kHz, one frame corresponds to 267 samples of audio. Techniques in language understanding are used for selecting the most significant words and phrases.

To understand the audio track, we must isolate each individual word. To transcribe the content of the video material, we recognize spoken words using a speech recognition system. Speaker independent recognition systems have made great strides as of late and offer promise for application in video indexing [16]. Speech recognition works best when closed-captioned data is available. Captions usually occur in broadcast material, such as sitcoms, sports, and news. Documentaries and movies may not necessarily contain captions. Closed captions have become more common in video material throughout the United States since 1985 and most televisions provide standard caption display.

15.2.3.3 Rule-Based Features

The features described in previous sections may be used with rules that describe a particular type of video scene to create an additional set of content-based features [17]. By using examples from video production standards, we can identify a small set of heuristic rules. In some cases, these rules involve the integration of image processing features with audio and language features. Below is a description of three rule-based features suitable for most types of video.

15.2.3.3.1 Introduction Scenes

The scenes prior to the introduction of a person usually describe their accomplishments and often precede scenes with large views of the person's face. A person's name is generally spoken and then followed by supportive material. Afterward, the person's actual face is shown. If a scene contains a proper name, and a large human face is detected in the scenes that follow, we call this an introduction scene. Characterization of this type is useful when searching for a particular human subject because identification is more reliable than using the image or audio features separately. Introduction scenes must meet the following criteria:

$$\text{Scene}_{\text{Introduction}}(i) = \begin{cases} 1, & (\text{Face}_i = \text{TRUE}\,\&\&\,\text{WORD}_i = \text{PROPER\_NAME}) \\ 0, & (\text{otherwise}) \end{cases}.$$

15.2.3.3.2 Adjacent Similar Scenes

The color histogram difference measure gives us a simple routine for detecting similarity between scenes. Scenes between successive shots of a human face usually imply illustration of the subject. For example, a video producer will often interleave shots of research between shots of a scientist. Images that appear between two similar scenes that are less than T_{SS} seconds apart are characterized as an adjacent similar scene. Scene(i) is an adjacent similar scene if it meets the following criteria:

$$\text{Scene}_{\text{Adjacent similar}}(i) = \begin{cases} 1, (\text{Scene}(i-T) = \text{Scene}(i+T)) \\ \text{AND} \\ (|\text{Scene}_{\text{Start}}(i-T) - \text{Scene}_{\text{Start}}(i+T)| < T_{SS}) \\ 0, (\text{otherwise}) \end{cases},$$

where T_{SS} is on the order of 10 s or less.

15.2.3.3.3 Short Successive Scenes

Short successive shots often introduce an important topic. By measuring the duration of each scene, S_D, we can detect these regions and identify short successive sequences. A set of scenes is short successive if a group of five or more scenes meet the following criteria:

$$\text{Scene}_{\text{Short successive}}(i) = \begin{cases} 1, \begin{pmatrix} \text{Scene}_{\text{Duration}}(i-T) < S_D \,\&\&\, \text{Scene}_{\text{Duration}}(i+T) < S_D \,\&\& \\ \text{Scene}_{\text{Duration}}(i+2T) < S_D \,\&\&\, \text{Scene}_{\text{Duration}}(i+3T) < S_D\mathbf{K} \end{pmatrix}, \\ 0, (\text{otherwise}) \end{cases}$$

where S_D for each scene is on the order of 3 s or less.

15.2.3.4 Embedded Video Features

A final solution for content-based feature extraction is the use of known procedures for creating video. Video production manuals provide insight into the procedures used during video editing and creation. There are many documents that describe the editing and production procedures for creating video segments, but one of the most recent is published by Pryluck [18].

One of the most common elements in video production is the ability to convey climax or suspense. Producers use a variety of different effects, ranging from camera positioning, lighting, and special effects to convey this mood to an audience. Detection of procedures such as these is beyond the realm of present image and language understanding technology. However, many of the important features described in Sections 15.2–15.4 were derived from research in the video production industry.

Structural information as to the content of a video is a useful tool for indexing video. For example, the type of video being used (documentaries, news footage, movies, and sports) and its duration may offer suggestions to assist in object recognition. In news footage, the anchorperson will generally appear in the same pose and background at different times. The exact locations of the anchorperson can then be used to delineate story breaks. In documentaries, a person of expertise will appear at various points throughout the story when topical changes take place. There are also many visual effects introduced during video editing and creation that may provide information for video content. For example, in documentaries, the scenes prior to the introduction of a person usually describe their accomplishments and often precede scenes with large views of the person's face.

A producer will often create production notes that describe in detail action and scenery of a video, scene by scene. If a particular feature is needed for an application in image or video databases, the description may have already been documented during video production.

Another source of descriptive information may also be embedded in the video stream in the form of timecode and geospatial (GPS/GIS) data. These features are useful in indexing precise segments in video or a particular location in spatial coordinates. Aeronautic and automobile surveillance video will often contain GPS data that may be used as a source for indexing.

15.3 VIDEO ANALYSIS

As can be seen from Fig. 15.1, multimodal analysis is the basis for later video processing. It includes shot boundary detection and key-frame extraction for scripted content. For unscripted content, it includes play/break segmentation, audio marker detection, and visual marker detection.

15.3.1 Shot Boundary Detection

It is not efficient (sometimes not even possible) to process a video clip as a whole. It is beneficial to first decompose the video clip into shots and do the signal processing at the shot level. In general, automatic shot boundary detection techniques can be classified into five categories: pixel-based, statistics-based, transform-based, feature-based, and histogram-based. Pixel-based approaches use pixel-wise intensity difference to mark shot boundaries [1, 19]. However, they are highly sensitive to noise. To overcome this problem, Kasturi and Jain propose to use intensity statistics (mean and standard deviation) as shot boundary detection measures [19]. To achieve faster processing, Arman, Hsu, and Chiu propose to use the compressed DCT coefficients (e.g., MPEG data) as the boundary measure [20]. Other transform-based shot boundary detection approaches make use of motion vectors, which are already embedded in the MPEG stream [21, 22]. Zabih et al. address the problem from another angle. Edge features are first extracted from each frame. Shot boundaries are then detected by finding sudden edge changes. So far, the histogram-based approach is the most popular. Instead of using pixel intensities directly, the histogram-based approach uses histograms of the pixel intensities as the measure. Several researchers claim that it achieves a good trade-off between accuracy and speed [1]. Representatives of this approach are Refs. [1, 23–26]. More recent work has been based on clustering and postfiltering [22], which achieves fairly high accuracy without producing many false positives. Two comprehensive comparisons of shot boundary detection techniques are presented in [27, 28].

15.3.2 Key-Frame Extraction

After the shot boundaries are detected, corresponding key frames can then be extracted. Simple approaches may just extract the first and last frames of each shot as the key frames

[21]. More sophisticated key-frame extraction techniques are based on visual content complexity indicators [29], shot activity indicators [23], and shot motion indicators [24, 25]. The following three analysis steps mainly cater to the analysis of unscripted content.

15.3.3 Play/Break Segmentation

Since unscripted content has short periods of activity (plays) between periods of inactivity (breaks), it is useful to first segment the whole content into these units. This helps in reducing the amount of content to be analyzed for subsequent processing that looks for highlight segments within plays. Play/break segmentation for sports, both in an unsupervised and supervised manner using low-level features, has been reported in [26]. Play/break segmentation in surveillance has been reported using adaptive background subtraction techniques that identify periods of object activity from the whole content [30].

15.3.4 Audio Marker Detection

Audio markers are key audio classes that are indicative of the events of interest in unscripted content. In our previous work on sports, audience reaction and commentator's excited speech are classes that have been shown to be useful markers [27, 28]. Nepal et al. [31] detect basketball "goal" based on crowd cheers from the audio signal using energy thresholds. Another example of an audio marker, consisting of key words such as "touchdown" or "fumble," has been reported in [32].

15.3.5 Video Marker Detection

Visual markers are key video objects that are indicative of the events of interest in unscripted content. Some examples of useful and detectable visual markers are the squatting baseball catcher pose for baseball, the goal post for soccer, etc. Kawashima et al. [33] detect bat-swings as visual markers using visual features. Gong et al. [34] detect and track visual markers such as the soccer court, the ball, the players, and the motion patterns.

15.4 VIDEO REPRESENTATION

Considering that each video frame is a 2D object and the temporal axis makes up the third dimension, a video stream spans a 3D space. Video representation is the mapping from the 3D space to the 2D view screen. Different mapping functions characterize different video representation techniques.

15.4.1 Video Representation for Scripted Content

Using an analysis framework that can detect shots, key frames, and scenes, it is possible to come up with the following representations for scripted content.

15.4.1.1 Representation Based on Sequential Key Frames

After obtaining shots and key frames, an obvious and simple video representation is to sequentially lay out the key frames of the video, from top to bottom and from left to right.

This simple technique works well when there are few key frames. When the video clip is long, this technique does not scale, since it does not capture the embedded information within the video clip, except for time.

15.4.1.2 Representation Based on Groups

To obtain a more meaningful video representation when the video is long, related shots are merged into groups [3, 5]. In [5], Zhang et al. divided the entire video stream into multiple video segments, each of which contains an equal number of consecutive shots. Each segment is further divided into subsegments, thus constructing a tree-structured video representation. In [3], Zhong et al. proposed a cluster-based video hierarchy, in which the shots are clustered based on their visual content. This method again constructs a tree-structured video representation.

15.4.1.3 Representation Based on Scenes

To provide the user with better access to the video, the construction of a video representation at the semantic level is needed [2, 4]. It is not uncommon for a modern movie to contain a few thousand shots and key frames. This is evidenced in [35]—there are 300 shots in a 15-min video segment of the movie *Terminator 2—Judgment Day* and the movie lasts 139 min. Because of the large number of key frames, a simple 1D sequential presentation of key frames for the underlying video (or even a tree-structured layout at the group level) is almost meaningless. More importantly, people watch the video by its semantic scenes rather than the physical shots or key frames. Although shot is the building block of a video, it is scene that conveys the semantic meaning of the video to the viewers. The discontinuity of shots is overwhelmed by the continuity of a scene [2]. Video ToC construction at the scene level is thus of fundamental importance to video browsing and retrieval. In [2], a scene transition graph (STG) of video representation is proposed and constructed. The video sequence is first segmented into shots. Shots are then clustered by using time-constrained clustering. The STG is then constructed based on the time flow of the clusters.

15.4.1.4 Representation Based on Video Mosaics

Instead of representing the video structure based on the video-scene-group-shot-frame hierarchy as discussed above, this approach takes a different perspective [36]. The mixed information within a shot is decomposed into three components:

- Extended spatial information captures the appearance of the entire background imaged in the shot and is represented in the form of a few mosaic images.
- Extended temporal information captures the motion of independently moving objects in the form of their trajectories.
- Geometric information captures the geometric transformations that are induced by the motion of the camera.

15.4.2 Video Representation for Unscripted Content

Highlights extraction from unscripted content requires a different representation from the one that supports browsing of scripted content. This is because shot detection is known to be unreliable for unscripted content. For example, in soccer video, visual features are so similar over a long period of time that almost all the frames within it may be grouped as a single shot. However, there might be multiple semantic units within the same period such as attacks on the goal, counter attacks in the midfield, etc. Furthermore, the representation of unscripted content should emphasize detection of remarkable events to support highlights extraction while the representation for scripted content does not fully support the notion of an event being remarkable compared to others. For unscripted content, using an analysis framework that can detect plays and specific audio and visual markers, it is possible to come up with the following representations.

15.4.2.1 Representation Based on Play/Break Segmentation

As mentioned earlier, play/break segmentation using low-level features gives a segmentation of the content at the lowest semantic level. By representing a key frame from each of the detected play segments, one can enable the end user to select just the play segments.

15.4.2.2 Representation Based on Audiovisual Markers

The detection of audiovisual markers enables a representation that is at a higher semantic level than play/break representation is. Since the detected markers are indicative of the events of interest, the user can use either or both of them to browse the content based on this representation.

15.4.2.3 Representation Based on Highlight Candidates

Association of an audio marker with a video marker enables detection of highlight candidates that are at a higher semantic level. Such a fusion of complementary cues from audio and video helps eliminate false alarms in either of the marker detectors. Segments in the vicinity of a video marker and an associated audio marker give access to the highlight candidates for the end user. For instance, if the baseball catcher pose (visual marker) is associated with an audience reaction segment (audio marker) that follows it closely, the corresponding segment is highly likely to be remarkable or interesting.

15.4.2.4 Representation Based on Highlight Groups

Grouping of highlight candidates would give a finer resolution representation of the highlight candidates. For example, golf swings and putts share the same audio markers (audience applause and cheering) and visual markers (golfers bending to hit the ball). A representation based on highlight groups, supports the task of retrieving finer events such as "golf swings only" or "golf putts only."

15.5 VIDEO BROWSING

These two functionalities are the ultimate goals of a video access system, and they are closely related to (and built on top of) video representations. The representation techniques for scripted content discussed above are suitable for browsing through ToC-based summarization while the last can be used in video retrieval. On the other hand, the representation techniques for unscripted content are suitable for browsing through highlights-based summarization.

15.5.1 Video Browsing Using ToC-Based Summary

For representation based on sequential key frames, browsing is obviously sequential browsing, scanning from the top-left key frame to the bottom-right key frame. For representation based on groups, a hierarchical browsing is supported [3, 5]. At the coarse level, only the main themes are displayed. Once the user determines which theme he is interested in, he can then go to the finer level of the theme. This refinement process can go on until the leaf level. For the STG representation, a major characteristic is its indication of time flow embedded within the representation. By following the time flow, the viewer can browse through the video clip.

15.5.2 Video Browsing Using Highlights-Based Summary

For representation based on play/break segmentation, browsing is also sequential, enabling a scan of all the play segments from the beginning of the video to the end. Representation based on audiovisual markers supports queries such as find me video segments that contain the soccer goal post in the left-half field, find me video segments that have the audience applause sound, or find me video segments that contain the squatting baseball catcher. Representation based on highlight candidates supports queries such as find me video segments where a golfer has a good hit or find me video segments where there is a soccer goal attempt. Note that "a golfer has a good hit" is represented by the detection of the golfer hitting the ball followed by the detection of applause from the audience. Similarly, that there is a soccer goal attempt is represented by the detection of the soccer goal post followed by the detection of long and loud audience cheering. Representation based on highlight groups supports more detailed queries than the previous representation. These queries include find me video segments where a golfer has a good swing, find me video segments where a golfer has a good putt, or find me video segments where there is a good soccer corner kick, etc.

15.6 VIDEO RETRIEVAL

As discussed in Section 15.1, the ToC, Highlights, and Index are all equally important for accessing the video content. Unlike the other video representations, the mosaic representation is especially suitable for video retrieval. Three components, moving objects, backgrounds, and camera motions, are perfect candidates for a video index. After constructing such a video index, queries such as find me a car moving like this, find me a conference room having that environment, etc. can be effectively supported.

In Section 15.2, we described analytical- and content-based features, which can be extracted from image and video segments. In this section, we describe techniques for establishing correspondence between these features.

15.6.1 Feature-Based Retrieval (Statistical and Compressed)

Correspondence between analytical features is established with the difference measures described in Section 15.2. This is straightforward for image matching features, where a match is based on the minimum absolute difference, $D(t)$, or histogram difference, $D_H(t)$. In the case of features based on motion, texture, and shape, the difference is based on the Euclidean distance between the parameters of the perspective feature. The difference measures may be applied to the entire image or a subregion of the image for better correspondence between objects in the image. Regardless of the granularity in applying difference measures, a key problem with color image matching is that similar colors do not necessarily provide similar content.

Image correspondence is important for identifying scenes that appear often in a video segment. The color histogram is not only useful for detecting scene changes but serves as an adequate method for image correspondence. A histogram from the first video frame of each scene is stored and compared with that of video frames in subsequent scenes. An analysis of the entire image requires less computation than subregion differencing, but the image match is less robust to foreground objects. Global image matching is particularly useful with images of uniform color and texture.

In news footage, an icon or logo is often used to symbolize the subject of the video. This icon is usually placed in the upper-quarter of the image. Although the background of the image remains the same, changes in this icon represent changes in content. By applying histogram differencing to a small region in the image, we can detect changes in news icons. Processing of subregions requires more computation, but the resulting image match is usually more robust. Objects that appear away from the background are usually easier to match with subregion differencing. Subregion differencing is also more affective with images of complex color and texture.

15.6.2 Content-Based Retrieval

The main problem in image in video indexing is that users query content and most systems only match statistical features such as color and texture. Sometimes two images have essentially identical content but almost no similarities in color, shape, texture, or motion. In this case, the motion of the players is similar, but the angle of camera will yield two separate forms of object motion from the original video sequence.

Content matching attempts to correlate actual objects with a given query. The user is not limited to selections based on similar color properties but rather a collection based on content. In this form of matching, the query may be an image or text. The content features, such as caption and face detection, correspond to textual descriptions so a query need not be an image.

A number of content-based image and video systems are applicable to the features described in this chapter. In Table 15.1, we list several potential query applications associated with content-based and statistical features.

Several working systems have demonstrated the potential of content-based matching for identifying specific objects and stories. Three of the more interesting systems are discussed below.

Name-It is a system for matching a human face to a name in news video [37]. It approximates the likelihood of a particular face belonging to a name in close proximity within the transcript. Integrated language and image understanding technology make the automation of this system possible.

Spot-It is a topological system that attempts to identify known characteristics in news video for indexing and classification [38]. It has reasonable success in identifying common video themes such as interviews, group discussions, and conference room meetings.

Pictorial Transcripts, a working system at AT&T Research Laboratories, has shown promising results in video summarization when closed captions are used with statistical visual attributes [39]. CNN video is digitized and displayed in an HTML environment with text for audio and a static image for every paragraph. More than 3000 h of processed video can be searched and browsed.

15.6.3 Relevance Feedback

Widely used in text retrieval, relevance feedback was first proposed by Rui et al. as an interactive tool in content-based image retrieval [40]. Since then, it has been proven to be a powerful tool and has become a major focus of research in this area. Relevance feedback often does not accumulate the knowledge the system learned. That is because the end-user's feedback is often unpredictable and inconsistent from user to user or even

TABLE 15.1 Potential query applications.

Query Type	Associated Feature
Pans or zooms in video	Camera motion
Action or moving objects	Object motion
Important scenes	Short sequences, adjacent similar scenes, introduction scenes
Human subjects	Face detection, introduction scenes, video text detection
Video captions	Video text detection
Subject location	GPS, video text detection
Image scenery	Color difference, texture
Name or description	Audio and language analysis
Simple objects	Color difference
Segment boundaries	Face detection, scene changes, black frames

query to query. If the user who gives the feedback is trustworthy and consistent, feedback can be accumulated and added to the knowledge of the system, as was suggested by Lee et al. [41].

15.6.4 Query-Concept Learner

In a query by example system, it is often hard to initialize the first query, because the user may not have a good example to begin with. Having got used to text retrieval engines such as Google, users may prefer to query the database by key word. Many systems with key word annotations can provide such kind of service. Chang et al. recently proposed the SVM Active Learning system [42] and MEGA system [43], which can be alternate solutions. SVM Active Learning and MEGA have similar ideas but with different tools. They both want to find a query-concept learner that learns query criteria through an intelligent sampling process. No example is needed as the initial query. Instead of browsing the database completely randomly, these two systems ask the user to provide some feedback and try to quickly capture the concept in the user's mind. The key to success is to maximally utilize the user's feedback and quickly reduce the size of the space that the user's concept lies in. Active learning is the answer.

Active learning is an interesting idea in the machine learning literature. Although in traditional machine learning research, the learner typically works as a passive recipient of the data, active learning enables the learner to use its own ability to respond to collect data and to influence the world it is trying to understand. A standard passive learner can be thought of as a student who sits and listens to a teacher, whereas an active learner is a student who asks the teacher questions, listens to the answers, and asks further questions based on the answer. Active learning has shown very promising results in reducing the number of samples required to finish a certain task.

In practice, the idea of active learning can be translated into a simple rule: if the system is allowed to propose samples and get feedback, always propose those samples that the system is most confused of or that can bring the greatest information gain. Following the rule, SVM Active Learning becomes very straightforward. In SVM, objects far away from the separating hyperplane are easy to classify. The most confused objects are those that are close to the boundary. Therefore, during the feedback loop, the system will always propose the images closest to the SVM boundary for the user to annotate.

15.6.5 Efficient Annotation through Active Learning

Key word annotation is a very expensive work, as it can only be done manually. It is natural to look for methods that can improve the annotation efficiency. Active learning turns out to be also suitable for this job. In [44], Zhang and Chen proposed a framework for active learning during the annotation. For each object in the database, they maintain a list of probabilities, each indicating the probability of this object having one of the attributes. During training, the learning algorithm samples objects in the database and presents them to the annotator to assign attributes to. For each sampled object, each probability is set to be one or zero depending on whether or not the corresponding attribute is assigned by the annotator. For objects that have not been annotated, the learning algorithm estimates their probabilities with biased kernel regression. Knowledge gain is then defined to determine, among the objects that have not been annotated, which one the system is the most uncertain of. The system then presents it as the next sample to the annotator to assign attributes to.

Naphade et al. proposed a very similar work in [45]. However, they used a support vector machine to learn the semantics. They have essentially the same method as Chang et al.'s SVM Active Learning [42] to choose new samples for the annotator to annotate.

15.6.6 Considerations in Multimedia Databases

The retrieval of an image or video segment is often limited in practical multimedia databases. There are many factors to consider when creating an image or video database, such as optimization for large databases, the type of query, the presentation of results, and the measure of success.

Retrieval efficiency is an important concern for image and video databases. Flat file systems are sufficient when the size of a collection is moderate. However, a more robust solution is necessary when image and video libraries grow to several thousand units of data. Researchers have developed tree structure optimization systems that greatly reduce the search space by clustering image characteristics into small subsets for later retrieval [46].

15.6.6.1 Queries: Image or Text

For most image and video retrieval systems, the query is an image. When the comparison is based on analytical features, the results can often be ambiguous, as shown in Section 15.5.1. Content-based features provide a more accurate match to the given query, but the results are based on image processing technology, which is only capable of recognizing a small number of objects.

Text queries eliminate ambiguity in the query and work only with content-based features. There is still a dependence on content-based feature extraction, but there is limited uncertainty in the query. This type of the query may also be used to match the title of the image or the transcript of the video.

15.6.6.2 Presentation of Results

The presentation of a query result is an important part of the image or visual query system. Presentations are usually visual and textural in layout. Textual presentations provide more specific information and are useful when presenting large collections of data. Visual, or iconic, presentations are more useful when the content of interest is easily recalled from imagery. This is quite often the case in stock footage video where there is no audio to describe the content. Section 15.7 describes current working systems for presentation of image and video results.

15.6.6.3 Testing and Evaluation

In image and video databases, accuracy is based on the relevance of the output set of images or video to a particular query. A user defines the level of quality; therefore, the evaluation of an image or video retrieval system cannot be based on traditional analytical measures. The accuracy of these systems is purely subjective, which requires that some human intervention takes place during evaluation.

User studies or some form of subjective rating is essential during the design and development of an image and video database systems. Researchers have successfully demonstrated the utility of user studies in testing image and video retrieval applications [47, 48]. A subject is generally shown a query and asked to rank the resulting image or video segments on a scale. For example, in a video database, the user might be asked to rate the quality of selection on a scale ranging from High Relevance to Low Relevance.

15.7 A UNIFIED FRAMEWORK FOR INDEXING, SUMMARIZATION, BROWSING, AND RETRIEVAL

The above two subsections described video browsing (using ToC generation and highlights extraction) and retrieval techniques separately. In this section, we integrate them into a unified framework to enable a user to go "back and forth" between browsing and retrieval. Going from the Index to the ToC or the Highlights, a user can get the context where the indexed entity is located. Going from the ToC or the Highlights to the Index, a user can pinpoint specific queries. Figure 15.12 illustrates the unified framework.

An essential part of the unified framework is composed of the weighted links. The links can be established between Index entities and scenes, groups, shots, and key frames

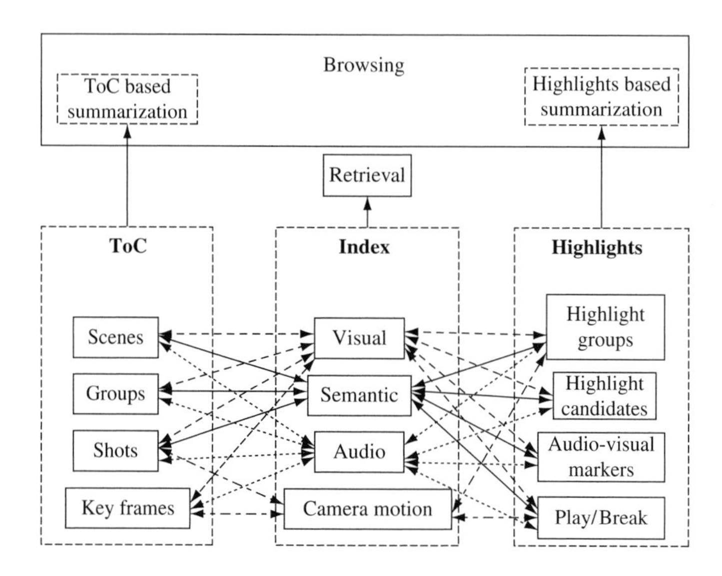

FIGURE 15.12

A unified framework.

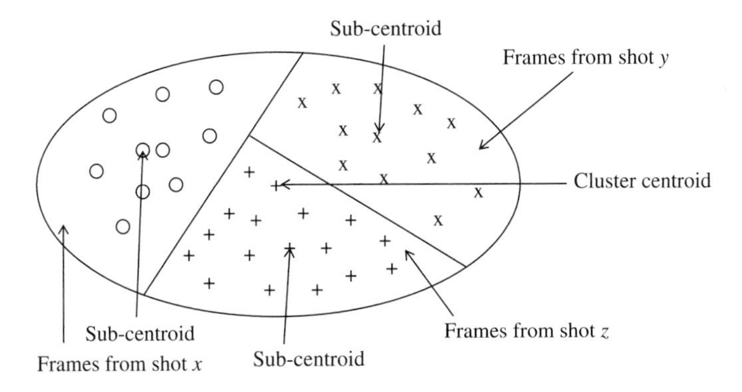

FIGURE 15.13

Subclusters.

in the ToC structure for scripted content and between Index entities and finer-resolution highlights, highlight candidates, audiovisual markers and plays/breaks.

For scripted content, as a first step, in this article, we focus our attention on the links between Index entities and shots. Shots are the building blocks of the ToC. Other links are generalizable from the shot link. To link shots and the visual Index, we propose the following techniques. As we mentioned before, a cluster may contain frames from multiple shots. The frames from a particular shot form a subcluster. This subcluster's centroid is denoted as "csub" and the centroid of the whole cluster is denoted as "c." This is illustrated in Fig. 15.13. Here, c is a representative of the whole cluster (and thus the visual Index) and csub is a representative of the frames from a given shot in this cluster.

15.8 CONCLUSIONS AND PROMISING RESEARCH DIRECTIONS

In this chapter, we

1. reviewed and discussed recent research progress in multimodal (audiovisual) analysis, representation, summarization, browsing, and retrieval;
2. introduced the video ToC, Highlights, and Index and presented techniques for constructing them; and
3. proposed a unified framework for video summarization, browsing, and retrieval and proposed techniques for establishing the link weights between the ToC, Highlights, and Index.

We should be aware that video is not just an audiovisual medium. It contains additional text information and is thus "true" multimedia. We need to further extend our investigation to the integration of closed-captioning into our algorithm to enhance the construction of ToCs, Highlights, Indexes, and link weights.

ACKNOWLEDGMENT

Part of this work (Rui and Huang) was supported in part by ARL Cooperative Agreement No. DAAL01-96-2-0003 and in part by a CSE Fellowship, College of Engineering, UIUC. The authors (Xiong, Radhakrishnan, and Divakaran) would like to thank Dr. Mike Jones of MERL for his help in visual marker detection. They also would like to thank Mr. Kohtaro Asai of Mitsubishi Electric Corporation (MELCO), Japan for his support and encouragement and Mr. Isao Otsuka of MELCO, for his valuable application-oriented comments and suggestions. The authors (Rui and Huang) would like to thank Sean X. Zhou, Atulya Velivelli, and Roy R. Wang for their contribution.

REFERENCES

[1] H. Zhang, A. Kankanhalli, and S. W. Smoliar. Automatic partitioning of full-motion video. *ACM Multimed. Syst.*, 1(1):1–12, 1993.

[2] R. M. Bolle, B.-L. Yeo, and M. M. Yeung. Video query: Beyond the keywords. *Technical Report*, IBM Research, October 17, 1996.

[3] D. Zhong, H. Zhang, and S.-F. Chang. Clustering methods for video browsing and annotation. *Technical Report*, Columbia University, New York, 1997.

[4] Y. Rui, T. S. Huang, and S. Mehrotra. Exploring video structures beyond the shots. *Proc. IEEE Conf. Multimed. Comput. Syst.*, 1998.

[5] H. Zhang, S. W. Smoliar, and J. J. Wu. Content-based video browsing tools. *Proc. IS&T/SPIE Conf. Multimed. Comput. Network.*, 1995.

[6] L. Xie, S.-F. Chang, A. Divakaran, and H. Sun. Structure analysis of soccer video with domain knowledge and hidden Markov models. *Pattern Recognit. Lett.*, 25(7):767–775, 2004.

[7] R. Zabih, J. Miller, and K. Mai. A feature-based algorithm for detecting and classifying scene breaks. *Proc. ACM Int. Conf. Multimed.*, San Francisco, CA, November 1995.

[8] A. Akutsu and Y. Tonomura. Video tomography: An efficient method for camerawork extraction and motion analysis. *Proc. ACM Multimed. '94*, 349–356, San Francisco, CA, October 1994.

[9] Y. T. Tse and R. L. Baker. Global zoom/pan estimation and compensation for video compression. *Proc. ICASSP*, 2725–2728, 1991.

[10] J. Meng and S.-F. Chang. Tools for compressed-domain video indexing and editing. *SPIE Conf. Storage Retr. Image Video Database*, San Jose, CA, February 1996.

[11] H. Wang and S. F. Chang. A highly efficient system for automatic face region detection in MPEG video sequences. *IEEE Trans. Circuits Syst. Video Technol., Special Issue on Multimed. Syst. Technol.*, 7(4):615–628, 1997.

[12] H. Rowley, S. Baluja, and T. Kanade. Neural network-based face detection. *IEEE Trans. Pattern Anal. Mach. Intell.*, 20(1):23–38, 1998.

[13] T. Sato, T. Kanade, E. Hughes, and M. Smith. Video OCR for digital news archives. *IEEE Workshop Content-Based Access Image Video Databases* (CAIVD'98), Bombay, India, January 1998.

[14] M. Mauldin. *Conceptual Information Retrieval: A Case Study in Adaptive Partial Parsing*. Kluwer Press, New York, 1991.

[15] TREC 93. *Proc. Second Text Retrieval Conf.*, D. Harmon, ed., sponsored by ARPA/SISTO, August 1993.

[16] A. Hauptmann and M. Smith. Text, speech, and vision for video segmentation. *AAAI Fall 1995 Symp. Comput. Models Integrat. Lang. Vision*, 1995.

[17] M. A. Smith and T. Kanade. Video skimming and characterization through the combination of image and language understanding techniques. *Comput. Vis. Pattern Recognit.*, San Juan, Puerto Rico, 1997.

[18] C. Pryluck, C. Teddlie, and R. Sands. Meaning in film/video: Order, time and ambiguity. *J. Broadcast.*, 26:685–695, 1982.

[19] R. Kasturi and R. Jain. Dynamic vision. In *Proceedings of Computer Vision: Principles*, R. Kasturi and R. Jain, eds. (IEEE Computer Society Press, Washington, DC, 1991).

[20] F. Arman, A. Hsu, and M.-Y. Chiu. Feature management for large video databases. *Proc. SPIE Storage Retr. Image Video Databases*, 1993.

[21] H. Zhang, C. Y. Low, S. W. Smoliar, and D. Zhong. Video parsing, retrieval and browsing: An integrated and content-based solution. *Proc. ACM Conf. Multimed.*, 1995.

[22] M. R. Naphade, R. Mehrotra, A. M. Ferman, T. S. Huang, and A. M. Tekalp. A high performance algorithm for shot boundary detection using multiple cues. *Proc. IEEE Int. Conf. Image Process.*, Chicago, IL, October 1998.

[23] P. O. Gresle and T. S. Huang. Gisting of video documents: A key frames selection algorithm using relative activity measure. *Proc. 2nd Int. Conf. Vis. Inform. Syst.*, 1997.

[24] W. Wolf. Key frame selection by motion analysis. *Proc. IEEE Int. Conf. Acoust., Speech, Signal Process.*, 1996.

[25] A. Divakaran, K. Peker, R. Radhakrishnan, Z. Xiong, and R. Cabasson. Video summarization using MPEG-7 motion activity and audio descriptors. In *Video Mining*, A. Rosenfeld, D. Doermann, and D. DeMenthon, eds. (Kluwer Academic Publishers, New York, 2003).

[26] L. Xie, S. Chang, A. Divakaran, and H. Sun. Structure analysis of soccer video with hidden markov models. *Proc. Int. Conf. Acoust., Speech Signal Process.*, (ICASSP-2002), Orlando, FL, May 2002.

[27] Z. Xiong, R. Radhakrishnan, and A. Divakaran. Effective and efficient sports highlights extraction using the minimum description length criterion in selecting GMM structures. Accepted into *Int. Conf. Multimed. Expo(ICME)*, June 2004.

[28] Y. Rui, A. Gupta, and A. Acero. Automatically extracting highlights for TV baseball programs. *Eighth ACM Int. Conf. Multimed.*, 105–115, 2000.

[29] Y. Zhuang, Y. Rui, T. S. Huang, and S. Mehrotra. Adaptive key frame extraction using unsupervised clustering. *Proc. IEEE Int. Conf. Image Process.*, 1998.

[30] T. C. Wren, A. Azarbayejani, and A. Pentland. Pfinder: Real-time tracking of the human body. *IEEE Trans. Pattern Anal. Mach. Intell.*, 19:780–785, 1997.

[31] S. Nepal, U. Srinivasan, and G. Reynolds. Automatic detection of 'goal' segments in basketball videos. *Proc. ACM Conf. Multimed.*, 2001.

[32] Y.-L. Chang, W. Zeng, I. Kamel, and R. Alonso. Integrated image and speech analysis for content-based video indexing. *Proc. IEEE Int. Conf. Multimed. Comput. Syst.*, June 1996.

[33] T. I. T. Kawashima, K. Tateyama, and Y. Aoki. Indexing of baseball telecast for content-based video retrieval. *Proc. IEEE Int. Conf. Image Process.*, 1998.

[34] Y. Gong, L. Sin, C. Chuan, H. Zhang, and M. Sakauchi. Automatic parsing of TV soccer programs. *IEEE Int. Conf. Multimed. Comput. Syst.*, 167–174, 1995.

[35] M. Yeung, B.-L. Yeo, and B. Liu. Extracting story units from long programs for video browsing and navigation. *Proc. IEEE Conf. Multimed. Comput. Syst.*, 1996.

[36] M. Irani and P. Anandan. Video indexing based on mosaic representations. *Proc. IEEE*, 86:905–921, 1998.

[37] S. Satoh, T. Kanade, and M. Smith. NAME-IT: Association of face and name in video. *Comput. Vis. Pattern Recognit.*, San Juan, Puerto Rico, June 1997.

[38] Y. Nakamura and T. Kanade. Semantic analysis for video contents extraction—spotting by association in news video. *Proc. Fifth ACM Int. Multimed. Conf.*, October 1997.

[39] B. Shahraray and D. Gibbon. Authoring of hypermedia documents of video programs. *Proc. Third ACM Conf. Multimed.*, 401–409, San Francisco, CA, November 1995.

[40] Y. Rui, T. S. Huang, M. Ortega, and S. Mehrotra. Relevance feedback: A power tool for interactive content-based image retrieval. *IEEE Trans. Circuits Syst. Video Technol.*, 8(5):644–655, 1998.

[41] C. S. Lee, W.-Y. Ma, and H. J. Zhang. Information embedding based on user's relevance feedback for image retrieval (invited paper). *SPIE Int. Conf. Multimed. Storage Arch., Syst. IV*, 19–22, Boston, MA, September 1999.

[42] S. Tong and E. Chang. Support vector machine active learning for image retrieval. *ACM Multimed.*, 9:107–108, 2001.

[43] E. Chang and B. Li. MEGA—The maximizing expected generalization algorithm for learning complex query concepts. UCSB Technical Report, August 2001.

[44] C. Zhang and T. Chen. An active learning framework for content based information retrieval. *IEEE Trans. Multimed., Special Issue on Multimed. Database*, 4(2):260–268, 2002.

[45] M. R. Naphade, C. Y. Lin, J. R. Smith, B. Tseng, and S. Basu. Learning to annotate video databases. *SPIE Conf. Storage Retr. Media Databases*, 2002.

[46] M. Flickner et al. Query by image content. *IEEE Comput.*, 23–32, 1995.

[47] M. G. Christel, D. B. Winkler, and C. R. Taylor. Improving access to a digital video library. *Human-Computer Interaction: INTERACT97, the 6th IFIP Conf. Hum.-Comput. Interact.*, Sydney, Australia, July 14–18, 1997.

[48] L. Ding et al. Previewing video data: Browsing key frames at high rates using a video slide show interface. *Proc. Int. Symp. Res. Dev. Pract. Digit. Libr.*, 151–158, Tsukuba Science City, Japan, November 1997.

CHAPTER

Video Communication Networks

16

Dan Schonfeld

Multimedia Communications Laboratory, Department of Electrical and Computer Engineering (M/C 154), 851 South Morgan Street—1020 SEO, University of Illinois, Chicago, Illinois, USA

Abstract

In this presentation, a broad overview of video communication networks is provided. Numerous video communication applications are currently being developed including digital television, video streaming, video-on-demand, and video conferencing. Efficient storage and communication require video data to be represented in compressed form. Various video compression standards have been developed by industrial organizations. However, among them the Motion Photographic Expert Group (MPEG)-2 compression standard still remains the most popular. It is currently the most powerful compression scheme for high-quality data representation and has been adopted by high-definition television (HDTV) and DVD. For simplicity, this presentation will focus primarily on the MPEG-2 video compression standard. The basic techniques used for video communication are illustrated over a variety of communication networks: Hybrid Fiber-Coax, Digital Subscriber Loop, Wireless, Fiber Optics, Integrated Services Digital Network, Asynchronous Transfer Mode (ATM), and Internet Protocol networks. The quality of video communications, especially over the Internet, is of critical concern for practical applications. Numerous protocols designed to allow users to enhance the quality-of-service of video transmission over communication networks have been proposed. Several of these protocols are described including MBONE, real-time transport protocol, real-time transport control protocol, real-time streaming protocol, session initiation protocol, resource reservation protocol, and DiffServ protocol.

16.1 INTRODUCTION

Paul Baran from the RAND Corporation first proposed the notion of a distributed communication network in 1964. His aim was to develop a decentralized communication system that could survive the impact of a nuclear attack. This proposal used a new approach to data communication based on packet switching.

Construction of a communication network based on packet switching was initiated by the Department of Defense through the Advanced Research Projects Agency (ARPA). This agency commissioned the ARPANET, later known as the Internet, in 1969. The ARPANET was initially an experimental communication network that consisted of four nodes: UCLA, UCSB, SRI, and the University of Utah.

Throughout the 1970s, various protocols had been adopted to facilitate services such as remote connection (telnet), file transfer (ftp), electronic mail, and news distribution. Initially, the ARPANET used the Network Control Protocol (NCP) for network and transport services. In 1983, the now ubiquitous Transport Control Protocol (TCP)/Internet Protocol (IP) protocol suite—TCP and IP stack developed in the early 1970s by Cerf and Khan for packet communication networks—had replaced NCP.

Evolution of the Internet was accelerated by the creation of the National Science Foundation Network (NSFNET) in 1986. In its infancy, the NSFNET used a backbone consisting of five supercomputer centers connected at 56 kbps. The NSFNET backbone, managed by National Science Foundation (NSF) and Merit Corp—a partnership formed by IBM and MCI—served in excess of 10 000 nodes in 1987.

To satisfy the increased demand on the Internet the NSFNET backbone was upgraded to T-1 (1.544 Mbps) in 1988. The Internet grew very rapidly to encompass over 100 000 nodes by 1989 connecting research universities and government organizations around the world. Management of the NSFNET backbone was delegated to Advanced Network and Services, Inc (ANS)—an independent nonprofit organization spun off from the partnership between Merit, IBM, and MCI.

Among the most important contributors to the proliferation of the Internet was the release of the World Wide Web (WWW) in 1991. Tim Berners-Lee proposed the WWW for the Corporation for Education and Research Networking (CERN)—the European center for nuclear research—in 1989. The web grew out of a need for physics researchers from around the world to collaborate using a large and dynamic collection of scientific documents.

The WWW provides a powerful framework for accessing linked documents throughout the Internet. The wealth of information available over the WWW has attracted the interest of commercial businesses and individual users alike. Its enormous popularity is enhanced by the graphical interfaces available for browsing multimedia information over the Internet.

The NSFNET backbone was upgraded to T-3 (44.736 Mbps) in 1991. Efforts to incorporate multimedia services were advanced with the introduction of the multicast backbone (MBONE) in 1992. The MBONE network intended to serve multicast real-time traffic over the Internet. It provided users with the capability to transmit audio and video multicast streams.

The enormous popularity of the WWW grew to over 10 million nodes by the mid-1990s. The NSF decommissioned the NSFNET and delegated commercial traffic to private backbones in 1995. The same year, the NSF has restructured its data networking architecture by providing the very-high-speed Backbone Network Service (vBNS) through a partnership with MCI Worldcom. In 1999, the vBNS was upgraded from OC-12 (622 Mbps) to OC-48 (2.5 Gpbs).

The NSF efforts to improve the communication network backbone were coupled with two related initiatives: Next Generation Internet (NGI) and Internet 2. In 1996, President Clinton introduced the NGI initiative in an effort to provide a faster and higher capacity Internet. This initiative was continued by the Large Scale Networking (LSN) coordinating group in an effort to advance networking technologies and services.

Internet 2 is an independent project coordinated by academic institutions whose goal is to accelerate the development of the Internet. This goal is addressed by deploying advanced network applications and technologies. Much of the effort of Internet 2 members has focused on the Abilene network. Abilene is a high-performance backbone network formed by partnership between Internet 2 and industry in 1999. Initially, Abilene provided communication at OC-48 (2.5 Gbps). Currently, the Abilene backbone has been upgraded to OC-192 (10 Mbps).

Improvements in communication networks' infrastructure are aimed at improving data communications and expanding applications. Efforts are underway to increase the communication bandwidth and support real-time services such as audio and video communications. The tremendous bandwidth required by video communications makes it among the most challenging of the applications envisioned in the next generation networks.

In the future, video communication networks will be used for a variety of applications including digital television, video streaming, video-on-demand (VoD), and video conferencing. An illustration of video communication services is depicted in Fig. 16.1. In this chapter, we will explore the current techniques used for video communications over data networks.

16.2 VIDEO COMPRESSION STANDARDS

16.2.1 Introduction

Video communications almost always relies on compressed video streams. Transmission of raw (uncompressed) video streams is impractical: excessive bandwidth is needed for both the communication channel and storage devices. Moreover, computer processing and memory limitations often impose serious constraints on transmission rates. Representation of video streams in compressed form is therefore required for efficient video communication systems.

Numerous video compression standards have been released by technical organizations and industrial corporations over the past couple of decades. The main organizations involved in adoption of video communication standards include the International

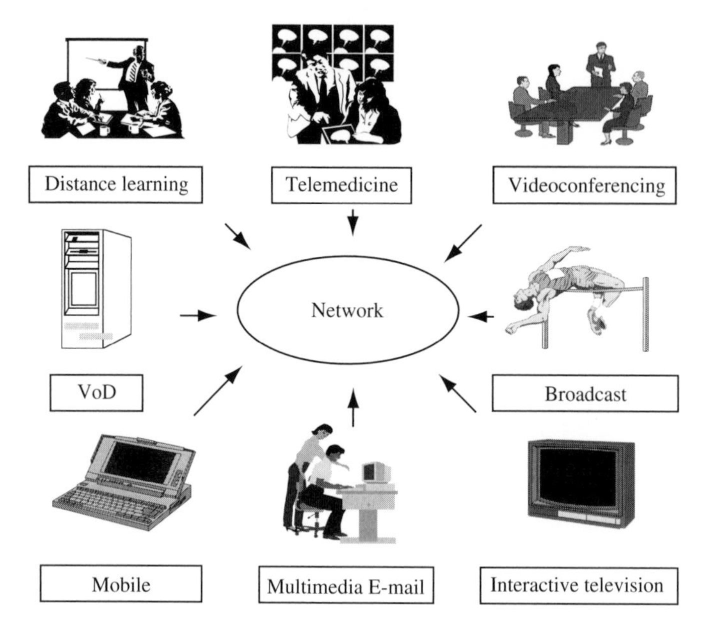

FIGURE 16.1

Video communication services.

Standards Organization (ISO) and International Telecommunications Union (ITU). Currently, the most widely used video compression method remains the MPEG-2 standard. It has been adopted for video communication applications such as HDTV and DVD. Our presentation will therefore be focused on the MPEG-2 standard.

16.2.2 Overview

Two main video compression standard families have emerged: MPEG and H.26X. MPEG standards have been developed by the ISO and are primarily aimed at motion picture storage and communications. H.26X compression schemes have been proposed by the ITU and focus on videoconferencing applications. The evolution of compression standards generated by MPEG and H.26X are very closely related. Many of the techniques adopted by MPEG's latest compression standard borrow from recent developments in H.26X's latest release, and vice versa.

We now provide a broad overview of the development of some of the most popular video compression standards. For simplicity, our discussion focuses on the MPEG family of video compression standards. Nonetheless, the evolution of standards outlined for MPEG is very close to the trends in the development of the H.26X compression schemes.

The earliest efforts at video compression were based on methods developed for image compression. Specifically, direct application of the Joint Photographic Experts Group (JPEG) image compression standard, used for compression of continuous-tone still images, to video sequences is known as Motion JPEG (MJPEG). This approach is used when random access to each picture is essential in applications such as video editing and enhanced VCR functionality. It is also used in high-quality video applications in the motion picture industry. The use of MJPEG exploits the spatial redundancy of the image. However, this approach fails to benefit from the high temporal redundancy of consecutive image frames in video sequences.

A video compression standard that exploits both spatial and temporal redundancies was proposed by MPEG-1. Its goal was to produce VCR NTSC (352×240) quality video compression to be stored on CD-ROM (CD-I and CD-video format) using a data rate of 1.5 Mbps. This was accomplished by introducing a method of motion compensation to extract temporal redundancy and increase the compression ratio.[1] A detailed presentation of the methods used for motion detection and estimation is provided in Chapter 3.

The next goal of the MPEG community was to develop a broadcast-quality video compression standard. A new method, based on the fundamental concepts present in MPEG-1, had emerged. This method is the well-known MPEG-2 video compression standard. Its popularity and efficiency in high-quality video compression resulted in the expansion of the standard to support higher resolution video formats including HDTV.[2] The HDTV Grand Alliance standard has adopted the MPEG-2 video compression and transport stream (TS) standards in 1996.[3] The MPEG-2 compressed video data rates are in the range of 3–100 Mbps.[4] Additional information on the specific details of the MPEG-1 and MPEG-2 video compression standards is available in Chapter 9.[5]

In its next mission, focused on the MPEG-4 video compression standard, the MPEG community attempted to address low-bandwidth video compression at data rate of 64 kbps that can be transmitted over a single N-Integrated Services Digital Networks (ISDN) B channel. This goal of MPEG-4 had become largely superfluous due to the release of the H.263 suite of video compression standards.

The goal of MPEG-4 has subsequently evolved to the development of flexible scalable extendable interactive compression streams that can be used with any communication network for universal accessibility (e.g., Internet and wireless networks). A dramatic change in approach, emphasizing content-based hierarchical audio–visual object (AVO) representation and composition, was used in the development of MPEG-4. The resulting standard is a genuine multimedia compression standard that supports audio and video as well as synthetic and animated images, text, graphics, texture, and speech synthesis.

Despite the novel approach and initial excitement surrounding the release of the MPEG-4 standard, its use in practical applications has been marginal. The limitations of MPEG-4 stem from the difficulty in efficient extraction of AVOs from the video bit stream. Detection of AVOs requires the use of efficient segmentation and tracking techniques as discussed in Chapters 6 and 7.

---

[1] MPEG-1 relies on the basic principles of the H.261 video compression standard, which was designed for video transmission over ISDN at data rates corresponding to multiples of 64 kbps.

[2] The MPEG-3 video compression standard, which was originally intended for HDTV, was later cancelled.

[3] The HDTV Grand Alliance standard, however, has selected the Dolby Audio Coding 3 (AC-3) audio compression standard.

[4] The HDTV Grand Alliance standard video data rate is approximately 18.4 Mbps.

[5] MPEG-2 has also been adopted as the H.262 video compression standard.

Moreover, the subsequent release of the H.264 video compression standard has been demonstrated to provide superior video compression compared with MPEG-4. Interestingly, the success of H.264 has been reached without the use of AVOs. Instead, the H.264 video compression standard relies primarily on a nested array of adaptive rectangular blocks to achieve high compression.

What ultimately emerged from the MPEG-4 compression standard is an array of various video compression schemes. The MPEG-4 profiles that focus on the use of AVOs for content-based video compression have had very little impact on the commercial world.

The greatest impact of MPEG-4 on practical video communication systems has focused on a couple of specific profiles: One popular implementation of the MPEG-4 standard relies on a version of the standard known as MPEG-4 Part 2—Simple Profile/Advanced Simple Profile (SP/ASP). MPEG-4 SP/ASP has been adopted by various commercial video codecs.[6]

Another profile of the MPEG-4 compression standard that has emerged as a powerful technique is called MPEG-4 Part 10—Advanced Video Coding (AVC). However, this profile has been developed by simply incorporating the H.264 video compression standard as a new part in MPEG-4. The MPEG-4 AVC standard has emerged as the leading method for video communications over wireless networks as well as high-definition DVD.[7] For more information on the details of the MPEG-4 and H.264 video compression standards, refer to the discussion in Chapter 10.

The MPEG-4 SP/ASP and MPEG-4 AVC schemes have made impressive gains in recent years through their adoption in various commercial products. Nonetheless, the classical MPEG-2 scheme remains the most popular compression method for video communications. This popularity is likely to grow further as the transition to widespread use of HDTV broadcasting takes a more prominent foothold on commercial television.[8] It is for this reason that we choose to focus in this presentation on MPEG-2 video communications.

16.2.3 MPEG-2 Video Compression Standard

MPEG-2 video compression relies on block coding based on the Discrete Cosine Transform (DCT). Specifically, each frame is divided into 8×8 blocks, which are transformed by using DCT. Quantization of the transformed blocks is obtained by dividing the

[6]MPEG-4 SP/ASP does not rely on content-based AVOs and is closely related to MPEG-2 and virtually identical to the H.263 video compression standard.

[7]MPEG-4 AVC has been adopted as the video compression standard for video communication to mobile devices in MediaFlo and Digital Multimedia Broadcasting (DMB) and supported by Digital Video Broadcasting-Handheld (DVB-H) as well as high-definition digital disk formats such as HD DVD and Blu-ray Disc.

[8]MPEG-2 is used as the video compression standard for communication between content providers and the MediaFlo network, which is subsequently transcoded to H.264 for one-directional video communications to mobile devices, and is also supported by Digital Video Broadcasting-Handheld (DVB-H) for communication to mobile devices.

transformed pixel values by corresponding elements of a quantization matrix and rounding the ratio to the nearest integer. The transformed and quantized block values are scanned using a zigzag pattern to yield a one-dimensional sequence of the entire frame. A hybrid variable-length coding scheme that combines Huffman coding and run-length coding is used for symbol encoding.

The procedure outlined is used for both intraframe and interframe codings. Intraframe coding is used to represent an individual frame independently. The scheme used for intraframe coding is essentially identical to JPEG image compression. An intraframe-coded picture in the video sequence is referred to as an intra-Picture (I-Picture).

Interframe coding is used to increase compression by exploiting temporal redundancy. Motion compensation is used to predict the content of the frame. Coding of its residual error represents the predicted frame. The frame is divided into 16×16 macroblocks (2×2 blocks). An optimal match of each macroblock in the neighboring frame is determined. A motion vector is used to represent the offset between the macroblock and its "best-match" in the neighboring frame. A residual error is computed by subtracting the macroblock from its "best match." Coding of the residual error image proceeds in the same manner as intraframe coding. Coding of the motion vector field is performed separately using difference coding and variable-length coding. An interframe-coded picture in the video sequence that restricts the search to the previous frame is referred to as a predicted-Picture (P-Picture), whereas those pictures that allow for either the previous or subsequent frames is referred to as a bidirectional-Picture (B-Picture).

Rows of macroblocks in the picture are called slices. The collection of slices forms a picture. Group of pictures (GOP) refer to sequences of pictures. A GOP is used to specify a group of consecutive frames and their picture types. For example, a typical GOP may consist of 15 frames with the following picture types: IBBPBBPBBPBBPBB. This scheme would allow for random access and error propagation that does not exceed intervals of $1/2$ s assuming the video is streamed at a rate of 30 fps.

16.2.4 MPEG-2 Systems Standard

The compressed image and video data are stored and transmitted in a standard format known as a compression stream. The discussion in this section will be restricted exclusively to the presentation of the video compression stream standards associated with the MPEG-2 systems layer: elementary stream (ES), packetized elementary stream (PES), program stream (PS), and TS.

The MPEG-2 systems layer is responsible for the integration and synchronization of the ESs: audio and video streams, as well as an unlimited number of data and control streams that can be used for various applications such as subtitles in multiple languages. This is accomplished by first packetizing the ESs thus forming the PESs. These PESs contain time stamps from a system clock for synchronization.

The PESs are subsequently multiplexed to form a single output stream for transmission in one of two modes: PS and TS. The PS is provided for error-free environments such as storage in CD-ROM. It is used for multiplexing PESs that share a common time

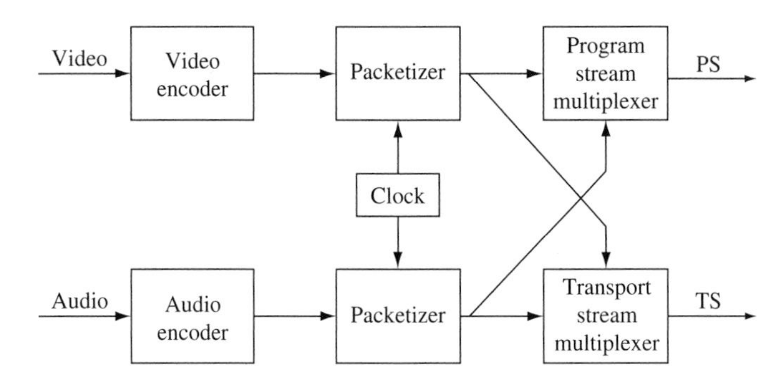

FIGURE 16.2

MPEG-2 audio and video systems layer.

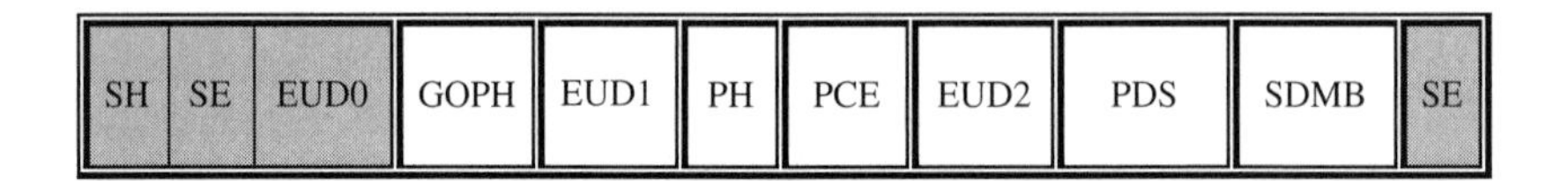

FIGURE 16.3

Video ES format.

base, using long variable-length packets.[9] The TS is designed for noisy environments such as communication over Asynchronous Transfer Mode (ATM) networks. This mode permits multiplexing streams (PESs and PSs) that do not necessarily share a common time base, using fixed-length (188 bytes) packets. An example of the MPEG-2 systems layer illustrating the multiplexing of the packetized audio and video ESs is depicted in Fig. 16.2.

16.2.4.1 *MPEG-2 Elementary Stream*

As indicated earlier, MPEG-2 systems layer supports an unlimited number of ESs. Our focus is centered on the presentation of the ES format associated with the video stream. The structure of the video ES format is dictated by the nested MPEG-2 compression standard: video sequence, GOP, pictures, slices, and macroblocks. The video ES is defined as a collection of access units (pictures) from one source. An illustration of the video ES format is depicted in Fig. 16.3. A corresponding glossary of the video ES format is provided in Table 16.1. Note that the unshaded segment of the video ES format presented in Fig. 16.3 is used to denote that any permutation of the fields within this segment can be repeated as specified by the video compression standard.

[9]The MPEG-2 program stream (PS) is similar to the MPEG-1 systems stream.

TABLE 16.1 Video ES format glossary.

Abbreviation	Function
SH	Sequence header
SE	Sequence extension
EUD0	Extension and user data 0
GOPH	Group of picture header
EUD1	Extension and user data 1
PH	Picture header
PCE	Picture coding extension
EUD2	Extension and user data 2
PDS	Picture data containing slices
SDMB	Slices data containing macroblocks
SE	Sequence end

16.2.4.2 MPEG-2 PES

The MPEG-2 systems layer packetizes all ESs—audio, video, data, and control streams—thus forming the PESs. Each PES is a variable-length packet with a variable format that corresponds to a single ES.

The format of the PES header is defined by the stream ID (SID) used to identify the type of ES. The PES packet length (PESPL) indicates the number of bytes in the PES packet. The scrambling mode is represented by the scrambling control (SC). The PES header data length (PESHDL) indicates the number of bytes in the optional PES header (OPESH) fields, as well as stuffing bytes (SB) used to satisfy the communication network requirements.

The PES header contains time stamps to allow for synchronization by the decoder. Two different time stamps are used: presentation time stamp (PTS) and decoding time stamp (DTS). The PTS specifies the time at which the access unit should be removed from the decoder buffer and presented. The DTS represents the time at which the access unit must be decoded. The DTS is optional, and it is only used if the decoding time differs from the presentation time.[10]

The elementary stream clock reference (ESCR) indicates the intended time of arrival of the packet at the system target decoder (STD). The rate at which the STD receives the PES is indicated by the elementary stream rate (ESR). Error checking is provided by the PES cyclic redundancy check (PESCRC).

The pack header field (PHF) is a PS pack header. The program packet sequence counter (PPSC) indicates the number of system streams. The STD buffer size is specified by the P-STD buffer (PSTDB) field.

A nested representation of the PES header is depicted in Fig. 16.4. The corresponding glossary of the PES header is provided in Table 16.2. Note that the unshaded boxes presented in Fig. 16.4 are used to represent optional fields in the PES header.

[10]This is the situation for MPEG-2 video elementary stream profiles that contain B-Pictures.

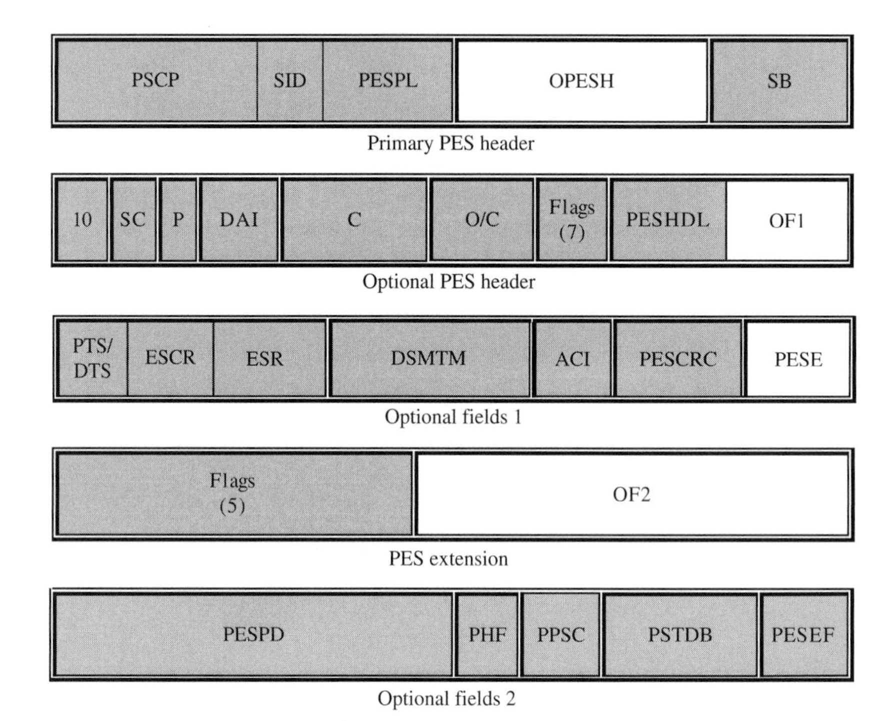

FIGURE 16.4

PES header.

16.2.4.3 *MPEG-2 Program Stream*

A PS multiplexes several PESs, which share a common time base, to form a single stream for transmission in error-free environments. The PS is intended for the storage and retrieval of programs from digital storage media such as CD-ROM. The PS uses relatively long variable-length packet. For a more detailed presentation of the MPEG-2 PS refer to X.

16.2.4.4 *MPEG-2 Transport Stream*

A TS permits multiplexing streams (PESs and PSs) that do not necessarily share a common time base for transmission in noisy environments. The TS is designed for broadcasting over communication networks such as ATM networks. The TS uses small fixed-length packets (188 bytes) that make them more resilient to packet loss or damage during transmission. The TS provides the input to the transport layer in the Open Systems Interconnection (OSI) reference model.[11]

[11]The transport stream (TS), however, is not considered as part of the transport layer.

TABLE 16.2 PES header glossary.

Abbreviation	Function
PSCP	Packet start code prefix
SID	Stream identification
PESPL	Packetized elementary stream packet length
OPESH	Optional packetized elementary stream header
SB	Stuffing bytes
SC	Scrambling control
P	Priority
DAI	Data alignment indicator
C	Copyright
O/C	Original/copy
Flags (7)	7 flags
PESHDL	Packetized elementary stream header data length
OF1	Optional fields 1
PTS/DTS	Presentation time stamps/decoding time stamps
ESCR	Elementary stream clock reference
ESR	Elementary stream rate
DSMTM	DSM trick mode
ACI	Additional copy information
PESCRC	Packetized elementary stream cyclic redundancy check
PESE	Packetized elementary stream extension
Flags (5)	5 flags
OF2	Optional fields 2
PESPD	Packetized elementary stream private data
PHF	Pack header field
PPSC	Program packet sequence counter
PSTDB	P-STD buffer
PESEF	Packetized elementary stream extension field

FIGURE 16.5

TS header.

The TS packet is composed of a four-byte header followed by 184 bytes shared between the variable-length adaptation field (AF) and the TS packet payload. An illustration of the TS header is depicted in Fig. 16.5. The corresponding glossary of the TS header is provided in Table 16.3. Note that the unshaded box appearing in Fig. 16.5 is used to represent the optional AF.

TABLE 16.3 TS header glossary.

Abbreviation	Function
SB	Synchronization byte
TEI	Transport error indicator
PUSI	Payload unit start indicator
TSC	Transport scrambling control
TP	Transport priority
PID	Packet identifier
AFC	Adaptation field control
CC	Continuity counter
AF	Adaptation field (optional)

The TS header includes a synchronization byte (SB) designed for detection of the beginning of each TS packet. The transport error indicator (TEI) points to the detection of an uncorrectable bit error in this TS packet. The payload unit start indicator (PUSI) is used to ascertain if the TS payload contains PES packets or program-specific information (PSI). The packet ID (PID) identifies the type and source of payload in the TS packet. The presence or absence of the AF and payload is indicated by the adaptation field control (AFC). The continuity counter (CC) provides the number of TS packets with the same PID, which is used to determine packet loss.

The optional AF contains additional information that need not be included in every TS packet. One of the most important fields in the AF is the program clock reference (PCR). The PCR is a 42-bit field composed of a 9-bit segment incremented at 27 MHz as well as a 33-bit segment incremented at 90 kHz.[12] The PCR is used along with a voltage controlled oscillator as a time reference for synchronization of the encoder and decoder clocks.

A PES header must always follow the TS header and possible AF. The TS payload may consist of the PES packets or PSI. The PSI provides control and management information used to associate particular ESs with distinct programs. A program is once again defined as a collection of ESs that share a common time base. This is accomplished by means of a program description provided by a set of PSI associated signaling tables (AST): program association tables (PAT), program map tables (PMT), network information tables (NIT), and conditional access tables (CAT). The PSI tables are sent periodically and carried in sections along with cyclic redundancy check (CRC) protection in the TS payload.

An example illustrating the formation of the TS packets is depicted in Fig. 16.6. The choice of the size of the fixed-length TS packets—188 bytes—is motivated by the fact that the payload of the ATM Adaptation Layer-1 (AAL-1) cell is 47 bytes. Therefore, four AAL-1 cells can accommodate a single TS packet. A detailed discussion of the mapping of the TS packets to ATM networks is presented in the next section.

[12]The 33-bit segment incremented at 90 kHz is compatible with the MPEG-1 system clock.

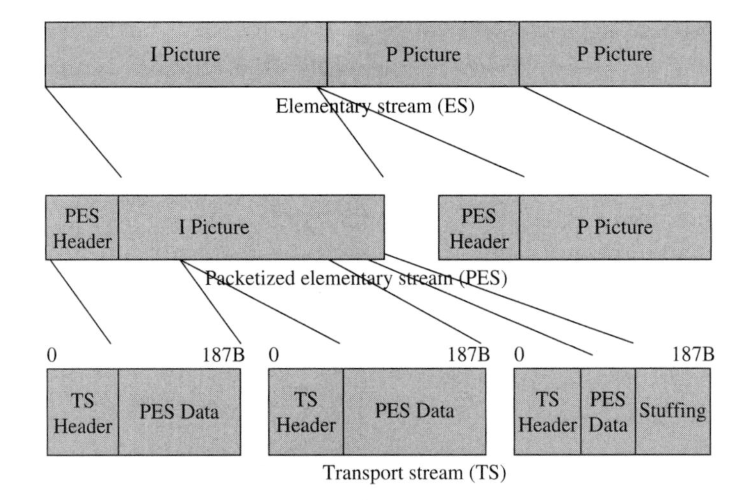

FIGURE 16.6

TS packets.

16.3 VIDEO COMMUNICATION NETWORKS

16.3.1 Introduction

A wide array of communication networks has proliferated over the past few decades. The goal of many communication networks is to provide as much communication bandwidth as possible while controlling the infrastructure costs. Efforts to provide inexpensive communication mediums have focused on exploiting the infrastructure of existing communication systems. For example, the hybrid of fiber optics and coaxial cable used in the cable television system has been adapted for data communications. Similarly, the web of copper wiring used in the telephone system has also been used for digital transmission. Moreover, the traditional use of air as a conduit in wireless systems—radio and television, cellular telephony, etc.—has recently been extended to accommodate high-bandwidth data communications.

Some applications require high-bandwidth communication networks that do not compromise transmission quality to reduce infrastructure costs. Examples of applications that impose severe bandwidth demands are communication backbones for wide and metropolitan area networks. On occasion, high volume traffic in local area networks (LAN) will also require a high-bandwidth communication infrastructure. Communication networks that need extremely high-bandwidth generally rely on fiber optics as the communication medium. A common deployment of high-bandwidth communication networks is based on ATM networks. ATM networks are extremely fast networks that are usually, although not necessarily, implemented using fiber optics. In some networks, ATM provides a protocol stack that characterizes the entire communication network.

In most instances, however, ATM serves to represent the lower level layers in a communication network such as the Internet.

A general design philosophy adopted is to provide very high-bandwidth communication for the networks' backbone and exploit existing infrastructure to connect individual users. This methodology is rooted in economics: the investment costs in installation of a powerful backbone will serve all customers and can be easily recovered. Deployment costs of high-bandwidth communication lines to each individual user, on the other hand, are excessive and therefore avoided. Indeed, tremendously powerful communication backbones have been implemented in the past few decades and are continuously evolving. Practically, one of the main difficulties presented today is the local distribution problem: how to efficiently connect individual customers to the communication networks' backbone? This problem is also colloquially referred to as the "the last mile problem." Various solutions have been proposed by the cable television, telephone, and wireless industries. Cable television and wireless communication systems are inherently broadcast systems, which may pose some limitations for many communication network applications. Telephone systems are based on point-to-point communications, which may be exploited for linking the backbone to customers' homes.

In this section, an overview of some of the main communication networks and their utility for multimedia communication applications is presented. The cable television network—known as the Hybrid Fiber-Coax (HFC) network—is discussed in Section 16.3.2. Adaptation of wireline telephone networks to computer networking through the Digital Subscriber Loop (DSL) protocol is presented in Section 16.3.3. The evolution of various wireless networks to high-bandwidth communication applications is sketched in Section 16.3.4. The widest bandwidth communication conduit is provided by fiber optics, which are discussed in Section 16.3.5. A brief presentation of digital communications based on ISDN is provided in Section 16.3.6. Finally, the use of ATM networks for multimedia communications is discussed in Section 16.3.7. For brevity, this presentation will be restricted exclusively to video communications based on the MPEG-2 compression standard.

16.3.2 Hybrid Fiber-Coax Networks

Cable television providers have installed an extensive communication network for delivery of television channels to the home. The main communication conduit used by the cable television industry is the coaxial cable. Coaxial cables are usually deployed between homes and a central point known as an optical node. Several optical nodes are connected via optical fibers to a head end. The cable television network is thus a mixture of both fiber optics and coaxial cable known as a HFC network.

Bandwidth limitations in HFC networks are primarily due to the coaxial cable. The bandwidth of coaxial cable is either 300–450 MHz or 750 MHz. The number of analog channels carrying a 6-MHz NTSC signal accommodated on coaxial cable is 50–75 or 125 channels, respectively.

Communication networks deployed over existing cable television systems must accommodate both data communications and television broadcasting. Cable television systems rely on the unused frequency in the 5–42 MHz band for upstream channels. Normal cable television channels in the 54–550 MHz region are maintained. Downstream channels are allocated in the frequency range available above 550 MHz.[13]

Downstream channels represent the data using Quadrature Amplitude Modulation (QAM) for signal modulation. Generally, QAM-64 is used to provide data rate of 27 Mbps. At times, the cable quality is sufficiently good to use QAM-256, which allows for about 39 Mbps. Upstream channels, on the other hand, rely on Quadrature Phase Shift Keying (QPSK) for signal modulation. Consequently, only 2 bits per baud are used for upstream communication, whereas either 6 or 8 bits per baud are provided for downstream channels.

The cable television system is a broadcasting system and bandwidth resources must therefore be shared among all customers. Let us assume that 50 channels can be used for data communications and must accommodate no more than 500 customers. In this scenario, a dedicated 4 Mbps data communication channel can be allocated to each home. This communication rate would be sufficient to handle MPEG-2 video streams.

In reality, however, existing cable television systems cannot afford to devote 50 channels exclusively to a limited number of customers not exceeding 500 homes. Most current cable television providers do not guarantee data communication rates above 700 kbps. At these rates, video communications using the MPEG-2 compression standard could not be conducted.

16.3.3 Digital Subscriber Loop

A long tradition has evolved in an effort to use the public switched telephone network (PSTN) for data communications.[14] The main advantage of PSTN is that it is widely accessible to virtually all homes. Modem technology for dial-up service over PSTN has improved and can reach rates of up to 56 kbps. A communication standard—H.324—has been developed for multimedia communications over PSTN. Video communications, however, requires much wider bandwidth using most compression standards.

The telephone industry had invested a tremendous amount of money to build a complex infrastructure that provides copper twisted pair wiring into virtually every home. In an effort to leverage this investment, the telephone industry has proposed a communication standard known as the DSL. The basic idea behind DSL is to present an efficient modulation scheme that will exploit the copper wires for data communications.

Traditionally, the telephone systems impose a filter in the end office that limits voice communications to 4 kHz. DSL circumvents this restriction by switching data signals to avoid the filters in the end office. Fundamental bandwidth limitations are consequently due to the physical properties of the copper twisted pair in the local loop.

The approach taken to the design of DSL uses a concept known as discrete multitone (DMT). The spectrum available on the local loop is about 1.1 MHz. DMT divides the

---

[13] In Europe, the low end of the standard television signal is 65 MHz and channels are 6–8 MHz to accommodate the PAL and SECAM television signals.

[14] The telephone network is also referred to as the Plain Old Telephone System (POTS).

bandwidth among 256 channels. Each channel has a bandwidth of 4.3125 kHz. Channel 0 is reserved for Plain Old Telephone Service (POTS). Channels 1–5 are reserved as guard bands to avoid interference between the voice and data channels. Additionally, one channel is used for upstream control and another for downstream control. The remaining 248 channels are available for data communications. A common split of the data channels is to allocate 32 channels for upstream data and the remaining 216 channels for downstream data. This implementation, which provides higher bandwidth for downstream than upstream data communications, is known as asymmetric DSL (ADSL).

High-bandwidth communication over the channels is achieved by the use of an efficient signal modulation scheme. Each channel provides a sampling rate of 4000 baud. QAM-16 with up to 15 bits per baud is used for signal modulation. For relatively short distances, DSL can provide communication at rates that exceed 8 Mbps. For instance, the ADSL standards ANSI T1.413 and ITU G.992.1 allow for data rates of 8 Mbps downstream and 1 Mbps upstream. Typically, premium service data rates are offered at 1 Mbps downstream and 256 kbps upstream. Standard service is further restricted to 512 kbps downstream and 64 kbps upstream.

Although extremely high rates can be provided over DSL for very short distances, most practical scenarios demand longer transmission lengths. The bandwidth provided by DSL decreases rapidly as the transmission distance increases. Therefore, most telephone companies will only guarantee all users data communications over DSL at rates of 128 kbps. These rates are insufficient for video communications based on most compression standards.

16.3.4 Wireless Networks

Historically, wireless networks date to ancient civilization. Fire signals were used for messaging between hilltops. Modern wireless communications dates back to the Italian physicist Gugliemo Marconi who, in 1901, used a wireless telegraph with Morse code to establish a communication to a ship.

Wireless networks have been developed for many different applications. Traditionally, wireless networks were used to refer to cellular networks for speech communications. Evolution of wireless networks has been designed to accommodate data communications. More recently, wireless networks have been used as LAN and wireless local loops.

Wireless networks were until recently primarily devoted to paging as well as real-time speech communications. First-generation wireless communication networks were analog systems. The most widely used analog wireless communication network is known as the Advanced Mobile Phone Service (AMPS).[15] The AMPS system is based on frequency-division multiple access (FDMA) and uses 832 30 kHz transmission channels in the range of 824–849 MHz and 832 30 kHz reception channels in the range of 869–894 MHz.

Second-generation wireless communication networks are digital systems based on two approaches: time-division multiple access (TDMA) and code-division multiple access (CDMA). Among the most common TDMA wireless communication networks

[15]The AMPS system is also known as TACS and MCS-L1 in England and Japan, respectively.

are the IS-54 and IS-136 as well the Global Systems for Mobile communications (GSM). The IS-54 and IS-136 are dual mode (analog and digital) systems that are backward compatible with the AMPS system.[16] In IS-54 and IS-136, the same 30 kHz channels are used to accommodate three simultaneous users (six time slots) for transmission at data rates of approximately 8 kbps. GSM originated in Europe and is a pure digital system based on both FDMA and TDMA. It consists of 50 200 kHz bands in the range of 900 MHz used to support eight separate connections (eight time slots) for transmission at data rates of 13 kbps.[17]

The second approach to digital wireless communication networks is based on CDMA. The origins of CDMA are based on spread-spectrum methods that date back to secure military communication applications during the Second World War.[18] The CDMA approach uses direct-sequence spread-spectrum (DSSS), which provides for the representation of individual bits by pseudorandom chip sequences. Each station is assigned a unique orthogonal pseudorandom chip sequence. The original bits are recovered by determining the correlation (inner product) of the received signal and the pseudorandom chip sequence corresponding to the desired station. The current CDMA wireless communication network is specified in IS-95.[19] In IS-95, a channel bandwidth of 1.25 MHz is used for transmission at data rates of 8 kbps or 13 kbps.

Efforts at integration of cellular networks to packet-based data communication over wireless networks have been made. This allows for data communication between wireless devices and fixed terminals connected to the Internet. For example, a mobile user could use his laptop to browse the web. Specifically, the General Packet Radio Service (GPRS) wireless access network is an overlay packet network that is used in D-AMPS and GSM systems. GPRS provides data communications at data rates in the range of 9–21.4 kbps using a single time slot. An improved wireless access technology, known as Enhanced Data rates for GSM Evolution (EDGE), can be used to provide data rates in the range of 8.8–59.2 kbps using a single time slot. Use of multiple time slots can increase the data rates as high as 170 kbps.

Plans have been proposed for the implementation of the third-generation wireless communication networks in the International Mobile Communications-2000 (IMT-2000). The motivation of IMT-2000 is to expand mobile communications to multimedia applications as well as to provide access to existing networks (e.g., ATM and Internet). This is accomplished by providing circuit and packet switched channel data connection as well as larger bandwidth used to support much higher data rates. The focus of IMT-2000 is on the integration of several technologies: CDMA-2000, Wideband CDMA (W-CDMA), Universal Wireless Communications-136 (UWC-136), and Wireless Multimedia and Messaging Services (WIMS).

---

[16]The Japanese JDC system is also a dual mode (analog and digital) system that is backward compatible with the MCS-L1 analog system.
[17]The implementation of the GSM system in the range of 1.8 GHz is known as DCS-1800.
[18]In 1940, the actress Hedy Lamarr, at the age of 26, invented a form of spread-spectrum, known as frequency-hopping spread-spectrum (FHSS).
[19]The IS-95 standard has recently been referred to as CDMA-One.

CDMA-2000 is designed to be a wideband synchronous intercell CDMA based network using frequency-division duplex (FDD) mode and is backward compatible with the existing CDMA-One (IS-95). The CDMA-2000 channel bandwidth planned for the first phase of implementation will be restricted to 1.25 MHz and 3.75 MHz for transmission at data rates of up to 1 Mbps. The CDMA-2000 channel bandwidth will be expanded during the second phase of implementation to include 7.5, 11.25, and 15 MHz for transmission that will support data rates that could possibly exceed 2.4 Mbps. CDMA-2000 was proposed by Qualcomm.

W-CDMA is a wideband asynchronous intercell CDMA (with some TDMA options) based network that provides for both FDD and time-division duplex (TDD) operations. W-CDMA is designed to interwork with the existing GSM and provides possible harmonization with WIMS. The W-CDMA channel bandwidth planned for the initial phase of implementation is 5 MHz for transmission at data rates of up to 480 kbps. The W-CDMA channel bandwidth planned for a later phase of implementation will reach 10 MHz and 20 MHz for transmission that will support data rates of up to 2 Mbps. W-CDMA has been proposed by Ericsson and advocated by the European Union, which called it Universal Mobile Telecommunications System (UMTS).

UWC-136 is envisioned to be an asynchronous intercell TDMA based system that permits both FDD and TDD modes. UWC-136 is backward compatible with the current IS-136 and provides possible harmonization with GSM. UWC-136 is a unified representation of IS-136+ and IS-136 High Speed (IS-136 HS). IS-136+ will rely on the currently available channel bandwidth of 30 kHz, for transmission at data rates of up to 64 kbps. The IS-136 HS outdoor (mobile) channel bandwidth will be 200 kHz, for transmission at data rates of up to 384 kbps, whereas the IS-136 HS indoor (immobile) channel bandwidth will be expanded to 1.6 MHz for transmission that will support data rates of up to 2 Mbps. UWC-136 no longer appear to be a serious contender for adoption by industry.

WIMS is planned to be a wideband asynchronous intercell CDMA based system using the FDD operation and is compatible with ISDN. The WIMS channel bandwidth scheduled for the first phase of implementation is 5 MHz, for transmission at data rates of 16 kbps. The WIMS channel bandwidth proposed for the second phase of implementation will expand to 10 and 20 MHz for transmission that will approach 2.4 Mbps. Currently, it does not seem likely that WIMS will be adopted by industry.

The larger bandwidth and significant increase in data rates supported by the various standards in IMT-2000 will facilitate video communication over wireless networks. Moreover, the packet switched channel data connection option provided by the various standards in IMT-2000 will allow for the implementation of many of the methods and protocols used for real-time IP networks for video communication over wireless networks (e.g., real-time transport protocol [RTP]/real-time transport control protocol [RTCP], real-time streaming protocol [RTSP], session initiation protocol [SIP]).

Wireless networks also serve a very different role as LAN. Extensions of the well-known IP and ATM protocols have been adopted for wireless communications and are known as mobile IP and wireless ATM, respectively. Among the most popular LAN currently used is the IEEE 802.11 standard. IEEE 802.11 provides communications at data rates of up to 11 Mbps. It is an enormously popular standard, which has accounted for a high density of hubs scattered in urban areas. For this reason, some have speculated that in the future the IEEE 802.11 standard may serve a role as a wireless network that is not necessarily restricted to its local area.

On a much smaller scale, wireless networks are used to provide interconnection among various computer devices within close physical proximity. This approach allows for ease of operation and avoids the wire connections required by traditional methods. A wireless network called Bluetooth has been adopted as a standard by the computer and communication industry to achieve this goal.[20] Bluetooth technology also served as the basis for the IEEE 802.15 standard for wireless personal area networks.

Wireless local loops are another example of the use of wireless networks for data communications. A common scenario is a wireless transmission between a home and a base station. In this case, wireless networks are used to address the "last mile problem." Data rates are dependent on the distance between the client and the base station. The shorter the distance the higher the data rate that can be provided. Classifications of wireless local loops depend on their radius of service: Multichannel Multipoint Distribution Service (MMDS) and Local Multipoint Distribution Service (LMDS). MMDS provides service across distances in the range of 30 miles. MMDS data rates may not exceed 1 Mbps. It uses 198 MHz in the 2.1 and 2.5 GHz range. LMDS provides service over much shorter distances in the range of 3 miles. LMDS data rates may be as high as 100–600 Mbps. It was allotted 1.3 GHz—the single largest bandwidth allocation by the FCC—in the 28–31 GHz range.[21] The IEEE 802.16 standard has been developed to provide broadband fixed wireless networking capability for LMDS applications.

Another form of wireless networks is provided by satellite communications. Video broadcasting over satellites has been conducted for many years. Both analog and digital video broadcasting have been used over satellite networks. More recent efforts have attempted to use satellites for real-time video communications. Limited success of this endeavor is due to the large number of satellites that are required to be launched into low orbit to reduce the communication delay.

16.3.5 Fiber Optics

There are two main methods provided by the telephone industry for local distribution using fiber optics: fiber to the curb (FTTC) and fiber to the home (FTTH). FTTC requires the installation of optical fibers from the end office to central locations such as residential neighborhoods. These central locations are equipped with a device known as an Optical Network Unit (ONU). The ONU is the termination point of multiple copper local loops connected within the immediate vicinity. The local loops between users and the ONU are sufficiently short that it is now possible to provide much higher communication bandwidth. For example, full-duplex T1 or T2 communication networks can be run over the copper wires for transmission of MPEG-2 video channels.

---

[20] Bluetooth technology was named after Harald Blaatand (Bluetooth) II (940–981) who was a Viking king unified Denmark and Norway.

[21] A similar approach to LMDS is used in Europe in the 40 MHz range.

An even more ambitious design is provided by FTTH. In this scheme, an optical fiber line is deployed directly into each customer's home. Consequently, an OC-1 or OC-3 or higher carrier rates can be accommodated. These rates are extremely high and can be used for virtually any multimedia communication application desired. The prohibitive factor in FTTH is cost. Installation of fiber optics into every home is very expensive. Nonetheless, some new residences and businesses have already been wired and fitted with fiber optic communication lines. Although large-scale deployment of FTTH may not happen for several years, it is clearly our future direction.

16.3.6 Integrated Services Digital Network

The ISDN is the first public digital network. It was designed to support a large variety of date types including data, voice, and video. It is based on circuit-switched synchronous communication. ISDN uses B channels for basic traffic and D channels for return signaling. Each B channel provides data rate of 64 kbps and each D channel is 16 kbps. Multiples of B channels are used to accommodate $p \times 64$ kbps. The basic rate interface (BRI) provides 2B+D channel, which can be used for signal delivery at the rate of 128 kbps. At this level, high-quality video communication cannot be supported. Wider bandwidth communications based on ISDN is available by using the primary rate interface (PRI). PRI communications can rely on up to 24 B channels for transmission at rates of 1.5 Mbps.[22] H.320 provides a communication system standard for audiovisual conferencing over ISDN. The vast majority of videoconferencing and video telephony systems currently used rely on H.320.

ISDN has become known as Narrowband ISDN (N-ISDN). It can be offered over the existing twisted-pair copper wiring used by the telephone industry. A second generation of ISDN has emerged and is known as Broadband ISDN (B-ISDN). It provides transmission channels that are capable of supporting much higher rates.[23] Like N-ISDN, the bandwidth of B-ISDN is specified in terms of multiples of 64 kbps. Whereas N-ISDN is limited to 1–24 multiples of 64 kbps channels, the multiplying factor for B-ISDN ranges from 1 to 65 535. The physical conduit required for B-ISDN is coaxial cable of optical fibers. Efficient implementation of B-ISDN is achieved by adopting ATM packet switching technology.

16.3.7 ATM Networks

ATM, also known as cell relay, is a method for information transmission in small fixed-size packets called cells based on asynchronous time-division multiplexing. ATM technology was proposed as the underlying foundation for the B-ISDN. B-ISDN is an ambitious very high data rate network that will replace the existing telephone system and all specialized networks with a single integrated network for information transfer applications such as VoD, broadcast television, and multimedia communication. These

[22] PRI systems in Europe can use 30 B channels for signal delivery at rates of 1.9 Mbps.

[23] Examples of higher bandwidth channels used in B-ISDN are the H0 channel with a rate of 384 kbps, the H11 channel with a rate of 1.536 Mbps, and the H12 channel with a rate of 1.92 Mbps.

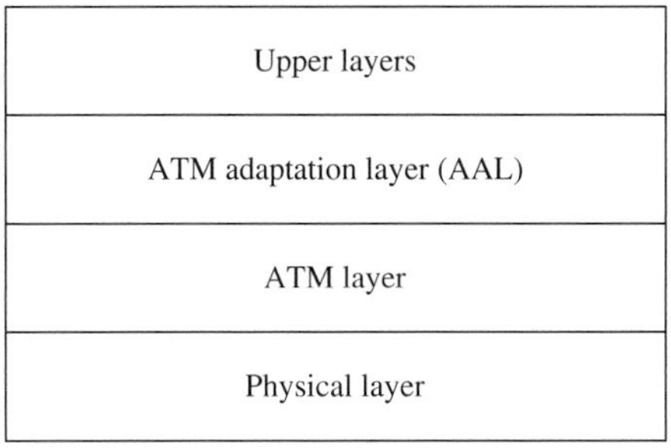

Upper layers

ATM adaptation layer (AAL)

ATM layer

Physical layer

FIGURE 16.7

B-ISDN ATM reference model.

lofty goals not withstanding, ATM technology has found an important niche in providing the bandwidth required for the interconnection of existing LAN, for example, Ethernet.

The ATM cells are 53 bytes long of which 5 bytes are devoted to the ATM header and the remaining 48 bytes are used for the payload. These small fixed-sized cells are ideally suited for the hardware implementation of the switching mechanism at very high data rates. The data rates envisioned for ATM are 155.5 Mbps (OC-3), 622 Mbps (OC-12), and 2.5 Gbps (OC-48).[24]

The B-ISDN ATM reference model is shown in Fig. 16.7. It consists of several layers: physical layer, ATM layer, AAL, and upper layers.[25] This layer can be further divided into the physical medium dependent (PMD) sublayer and the transmission convergence (TC) sublayer. The PMD sublayer provides an interface with the physical medium and is responsible for transmission and synchronization on the physical medium (e.g., SONET or SDH). The TC sublayer converts between the ATM cells and the frames—strings of bits—used by the PMD sublayer. ATM has been designed to be independent of the transmission medium. The data rates specified at the physical layer, however, require category 5 twisted pair or optical fibers.[26]

The ATM layer provides the specification of the cell format and cell transport. The header protocol defined in this layer provides generic flow control, virtual path, channel identification, payload type (PT), cell loss priority, and header error checking. The ATM layer is a connection-oriented protocol that is based on the creation of end-to-end virtual circuits (channels). The ATM layer protocol is unreliable—acknowledgements are not provided—because it was designed for use of real-time traffic such as audio and video over fiber optic networks that are highly reliable. The ATM layer nonetheless provides

[24]The data rate of 155.5 Mbps was chosen to accommodate the transmission of high-definition television (HDTV) and for compatibility with the Synchronous Optical Network (SONET). The higher data rates of 622 Mbps and 2.5 Gbps were chosen to accommodate 4 and 16 channels, respectively.

[25]Note that the B-ISDN ATM reference model layers do not map well into the OSI reference model layers.

[26]Existing twisted-pair wiring cannot be used for B-ISDN ATM transmission for any substantial distances.

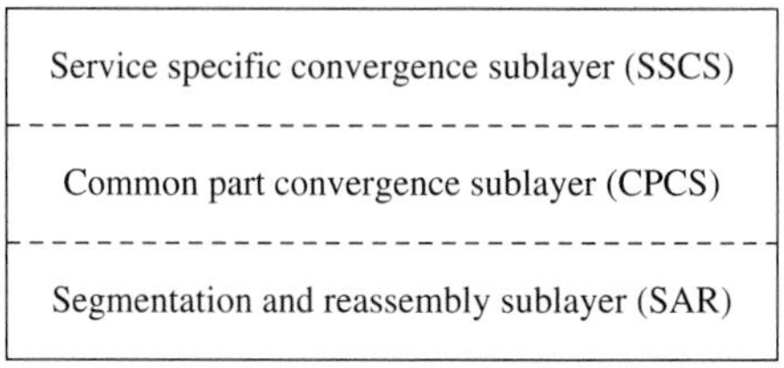

Service specific convergence sublayer (SSCS)
Common part convergence sublayer (CPCS)
Segmentation and reassembly sublayer (SAR)

FIGURE 16.8

ATM adaptation layer (AAL).

quality-of-service (QoS) guarantees in the form of cell loss ratio (CLR), bounds on maximum cell transfer delay (MCTD), cell delay variation (CDV)—known also as delay jitter. This layer also guarantees the preservation of cell order along virtual circuits.

The structure of the AAL is illustrated in Fig. 16.8. This layer can be decomposed into the segmentation and reassembly sublayer (SAR) and the convergence sublayer (CS). The SAR sublayer converts between packets from the CS sublayer and the cells used by the ATM layer. The CS sublayer provides standard interface and service options to the various applications in the upper layers. This sublayer is also responsible for converting between the message or data streams from the applications and the packets used by the SAR sublayer. The CS sublayer is further divided into the common part convergence sublayer (CPCS) and the service specific convergence sublayer (SSCS).

Initially, four service classes were defined for the AAL (Class A–D). This classification has been subsequently modified by the characterization of four protocols: Class A is used to represent real-time constant bit-rate connection-oriented services handled by AAL-1. This class includes applications such as circuit emulation for uncompressed audio and video transmission. Class B is used to define real-time variable bit-rate connection-oriented services given by AAL-2. Among the applications considered by this class are compressed audio and video transmission. Although the aim of the AAL-2 protocol is consistent with the focus of this presentation, we will not discuss it in detail because the AAL-2 standard has not yet been defined. Classes C and D support non-real-time variable bit-rate services corresponding to AAL-3/4.[27] Class C is further restricted to non-real-time variable bit-rate connection-oriented services provided by AAL-5.[28] It is expected that this protocol will be used to transport IP packets and interface to ATM networks. A summary of the ATM adaptation layer service classes and protocols is presented in Table 16.4.

[27]Classes C and D were originally used for the representation of non-real-time (NRT) variable bit-rate (VBR) connection-oriented (CO) and connectionless services handled by AAL-3 and AAL-4, respectively. These protocols, however, were so similar—differing only in the presence or absence of a multiplexing header field—that they eventually decided to merge them into a single protocol provided by AAL-3/4.

[28]A new protocol AAL-5—originally named simple efficient adaptation layer (SEAL)—was proposed by the computer industry as an alternative to the previously existing protocol AAL-3/4, which was presented by the telecommunications industry.

TABLE 16.4 ATM adaptation layer service classes.

Parameters	Service Classes			
	Class A	Class B	Class C	Class D
Timing compensation	Required		Not required	
Bit rate	Constant		Variable	
Connection mode	Connection-oriented		Connectionless	
Applications	Voice/video circuit emulation	VBR video/audio	Frame relay	SMDS data transfer
AAL Type	AAL-1	AAL-2	AAL-3/4 AAL-5	AAL-3/4

As is apparent from the preceding discussion, the main methods available for video communications over ATM are based on AAL-1 and AAL-5. The remainder of this section therefore focus on the mapping of the MPEG-2 TS to the ATM Application Layer (AAL) – AAL-1 and AAL-5.

16.3.7.1 ATM Application Layer-1

The AAL-1 protocol is used for transmission of real-time constant bit-rate connection-oriented traffic. This application requires transmission at constant rate, minimal delay, insignificant jitter, and low overhead.

Transmission using the AAL-1 protocol is in one of two modes: unstructured data transfer (UDT) and structured data transfer (SDT). The UDT mode is provided for data streams where boundaries need not be preserved. The SDT mode is designed for messages where message boundaries must be preserved.

The CS sublayer detects lost and misinserted cells that occur due to undetected errors in the virtual path or channel identification. It also controls incoming traffic to ensure transmission at a constant rate. This sublayer also converts the input messages or streams into 46–47 bytes segments to be used by the SAR sublayer.

The SAR sublayer has a 1-byte protocol header. The convergence sublayer indicator (CSI) of the odd-numbered cells forms a data stream that provides a 4-bit synchronous residual time stamp (RTSP) used for clock synchronization in SDT mode.[29] The timing information is essential for the synchronization of multiple media stream as well as for the prevention of buffer overflow and underflow in the decoder. The sequence count (SC) is a modulo-8 counter used to detect missing or misinserted cells. The CSI and SC fields are protected by the CRC field. An even parity (P) bit covering the protocol header affords additional protection of the CSI and SC fields. The AAL-1 SAR sublayer protocol header

[29]The synchronous residual time stamp (RTSP) method encodes the frequency difference between the encoder clock and the network clock for synchronization of the encoder and receiver clock in asynchronous service clock operation mode despite the presence of delay jitter.

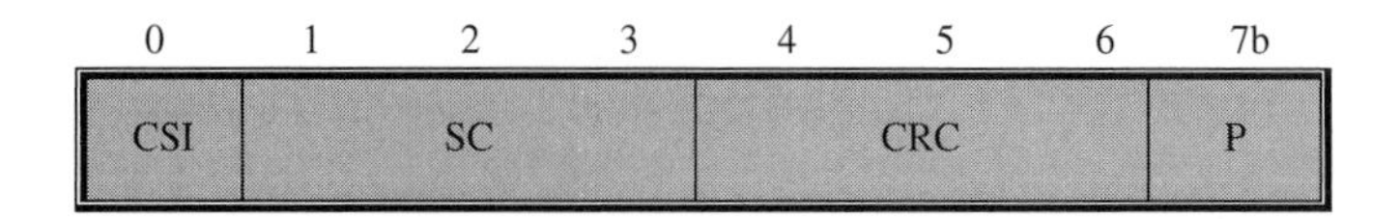

FIGURE 16.9

AAL1 SAR-PDU header.

TABLE 16.5 AAL1 SAR-PDU header glossary.

Abbreviation	Function
CSI	Convergence sublayer indicator
SC	Sequence count
CRC	Cyclic redundancy check
P	Parity (even)

is depicted in Fig. 16.9. A corresponding glossary of the AAL-1 SAR sublayer protocol header is provided in Table 16.5.

An additional 1-byte pointer field is used on every even-numbered cell in STD mode.[30] The pointer field is a number in the range of 0–92 used to indicate the offset of the start of the next message either in its own cell or the one following it to preserve message boundaries. This approach allows messages to be arbitrarily long and need not align on cell boundaries. In this presentation, however, we restrict ourselves to operation in the UDT mode for data streams where boundaries need not be preserved and the pointer field will be omitted.

As we have indicated earlier, the MPEG-2 systems layer consists of 188-bytes fixed-length TS packets. The CS sublayer directly segments each of the MPEG-2 TS packets into four 47-bytes fixed-length AAL-1 SAR payloads. This approach is used when the CLR that is provided by the ATM layer is satisfactory.

An alternative optional approach is used in noisy environments to improve reliability by the use of interleaved Reed–Solomon (128,124) forward error correction (FEC). The CS sublayer groups a sequence of 31 distinct 188-bytes fixed-length MPEG-2 TS packets. This group is used to form a matrix written in standard format (row-by-row) of 47 rows and 124 bytes in each row. Four bytes of the Reed–Solomon (128,124) FEC are appended to each row. The resulting matrix is composed of 47 rows and 128 bytes in each row. This matrix is forwarded to an interleaver that reads the matrix in transposed format (column-by-column) for transmission to the SAR sublayer. The interleaver assures that a cell loss would be limited to the loss of a single byte in each row, which can be recovered by the FEC. A mild delay equivalent to the processing of 128 cells is introduced by the

[30]The high-order bit of the pointer field is currently unspecified and reserved for future use.

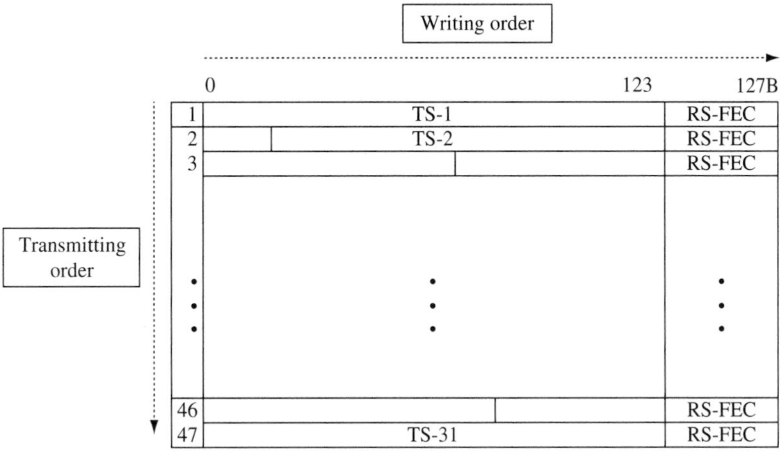

FIGURE 16.10

Interleaved TS (Reed–Solomon FEC).

matrix formation at the transmitter and the receiver. An illustration of the formation of the interleaved Reed–Solomon (128,124) FEC TS packets is depicted in Fig. 16.10.

Whether the interleaved FEC of the TS packets is implemented or direct transmission of the TS packets is used, the AAL-1 SAR sublayer receives 47-bytes fixed-length payloads that are appended by the 1-byte AAL-1 SAR protocol header to form 48-bytes fixed-length packets. These packets serve as payloads of the ATM cells and are attached to the 5-bytes ATM headers to comprise the 53-bytes fixed-length ATM cells. An illustration of the mapping of MPEG-2 systems layer TS packets into ATM cells using the AAL-1 protocol is depicted in Fig. 16.11.

16.3.7.2 ATM Application Layer-5

The AAL-5 protocol is used for non-real-time variable bit-rate connection-oriented traffic. This protocol also offers the option of reliable and unreliable services.

The CS sublayer protocol is composed of a variable-length payload of length not to exceed 65 535 bytes and a variable-length trailer of length 8–55 bytes. The trailer consists of a padding (P) field of length 0–47 bytes chosen to make the entire message – payload and trailer—be a multiple of 48 bytes. The user-to-user (UU) direct information transfer field is available for higher layer applications (e.g., multiplexing). The common part indicator (CPI) field designed for interpretation of the remaining fields in the CS protocol is currently not in use. The length field provides the length of the payload (not including the padding field). The standard 32-bit CRC field is used for error checking over the entire message—payload and trailer. This error checking capability allows for the detection of missing or misinserted cells without using sequence numbers (SNs). An illustration of the AAL-5 CPCS protocol trailer is depicted

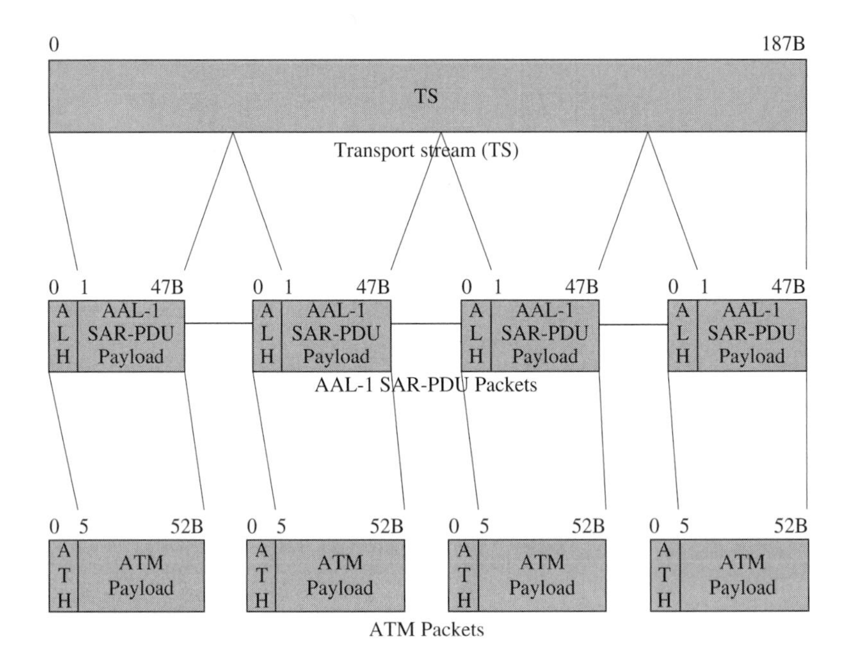

FIGURE 16.11

MPEG-2 TS AAL-1 PDU mapping.

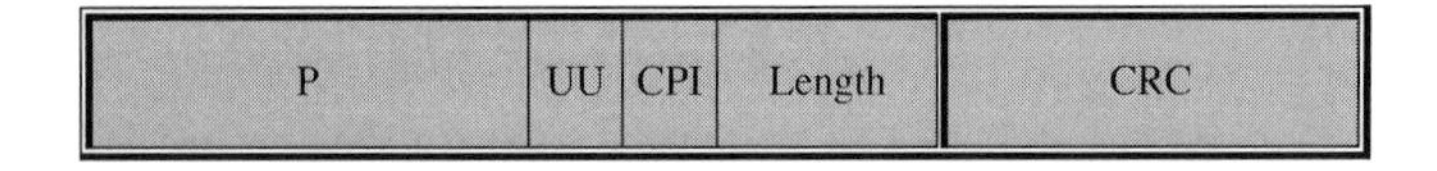

FIGURE 16.12

AAL5 CPCS-PDU trailer.

TABLE 16.6 AAL-5 CPCS-PDU trailer glossary.

Abbreviation	Function
P	Padding
UU	User-to-user direct information transfer
CPI	Common part indicator field
Length	Length of payload
CRC	Cyclic redundancy check

in Fig. 16.12. A corresponding glossary of the AAL-5 CPCS protocol trailer is provided by Table 16.6.

The SAR sublayer simply segments the message into 48-byte units and passes them to the ATM layer for transmission. It also informs the ATM layer that the ATM user-to-user

(AAU) bit in the payload type indicator (PTI) field of the ATM cell header must be set on the last cell to preserve message boundaries.[31]

Encapsulation of a single MPEG-2 systems layer 188-bytes fixed-length TS packet in one AAL-5 CPCS packet would introduce a significant amount of overhead due to the size of the AAL-5 CPCS trailer protocol. The transmission of a single TS packet using this approach to the implementation of the AAL-5 protocol would require five ATM cells in comparison to the four ATM cells needed using the AAL-1 protocol. More than one TS packet must be encapsulated in a single AAL-5 CPCS packet to reduce the overhead.

The encapsulation of more than one TS packet in a single AAL-5 CPCS packet is associated with an inherent packing jitter. This will manifest itself as delay variation in the decoder and may affect the quality of the systems clock recovered when one of the TS packets contains a PCR. To alleviate this problem, the number of TS packets encapsulated in a single AAL-5 CPCS packet should be minimized.[32]

The preferred method adopted by the ATM Forum is based on the encapsulation of two MPEG-2 systems layer 188-bytes TS packets in a single AAL-5 CPCS packet. The AAL-5 CPCS packet payload consequently occupies 376 bytes. The payload is appended to the 8-bytes AAL-5 CPCS protocol trailer (no padding is required) to form a 384-bytes AAL-5 CPCS packet. The AAL-5 CPCS packet is segmented into exactly eight 48-bytes AAL-5 SAR packets, which serve as payloads of the ATM cells and are attached to the 5-bytes ATM headers to comprise the 53-bytes fixed-length ATM cells. An illustration of the mapping of two MPEG-2 systems layer TS packets into ATM cells using the AAL-5 protocol is depicted in Fig. 16.13.

The overhead requirements for the encapsulation of two TS packets in a single AAL-5 CPCS packet are identical to the overhead needed using the AAL-1 protocol—both approaches map two TS packets into eight ATM cells. This approach to the implementation of the AAL-5 protocol is currently the most popular method for mapping MPEG-2 systems layer TS packets into ATM cells.

16.4 INTERNET PROTOCOL NETWORKS

16.4.1 Introduction

An important communication network that has not been discussed previously is the Internet. Many of the networks we have discussed thus far have been characterized by physical layer medium and protocols. The Internet, on the other hand, allows for communication across various networks having different physical medium and lower layer

[31] Note that this approach is in violation of the principles of the open architecture protocol standards—the AAL layer should not invoke decisions regarding the bit pattern in the header of the ATM layer.

[32] An alternative solution to the packing jitter problem, known as PCR-aware packing, requires that TS packets containing a program clock reference (PCR) appear in the last packet in the AAL-5 CPCS packet. This approach is rarely used due to the added hardware complexity in detecting TS packets with a PCR.

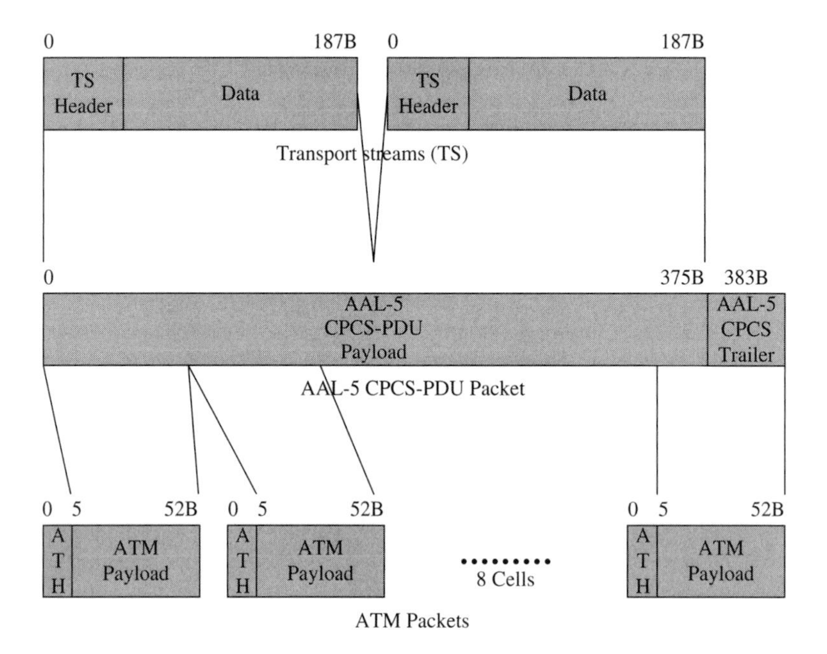

FIGURE 16.13

MPEG-2 TS AAL-5 PDU mapping.

protocols. Communication among these separate networks is facilitated by abstracting the lower layer protocols using a common network protocol known as the IP.

The most commonly used network protocol today is the IPv4 protocol. The IPv4 header consists of a 20-byte fixed header followed by an optional variable-length header. Among the fixed header fields are the version, header length, type of service, packet length, identification and fragmentation information, time to live, transport protocol, header checksum, source and destination addresses.

Popularity of the Internet and forecasts of future applications of the Internet have increased rapidly. Particularly, the convergence of communications, computing, and entertainment has begun. It is likely that in the not so distant future separate applications such as telephony, televisions, and the web will merge into flexible computing systems. We envision that future stationary and mobile telephone and television devices throughout the world will become Internet nodes used for audio and video communications. To accommodate the large number of new nodes on the Internet, a new version of the IP protocol had to emerge. Currently, we are in the midst of a migration to a new network protocol—IPv6.[33]

In 1990, Internet Engineering Task Force (IETF) had begun work on the new IP protocol. Its aim was to provide sufficient addresses to accommodate future growth

[33]The IPv5 denotes an experimental real-time stream protocol that was not widely used.

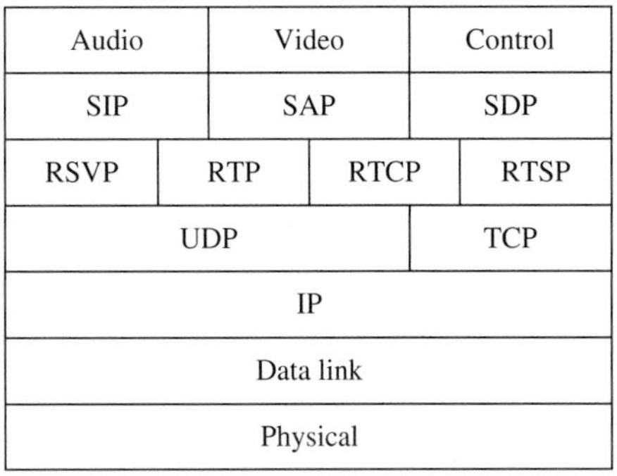

Audio	Video	Control
SIP	SAP	SDP

RSVP	RTP	RTCP	RTSP
UDP			TCP
IP			
Data link			
Physical			

FIGURE 16.14

Video IP stack.

of the Internet and increase the protocol's efficiency and flexibility. The main difference introduced in IPv6 is longer addresses. The new IPv6 addresses are 16-bytes long, whereas the old IPv4 addresses consist of merely 4-bytes. Additionally, the header of the IPv6 protocol is much simpler and more flexible than its predecessor. Reduced number of fields and improved optional field support has resulted in faster router processing time. Moreover, network security has been improved by incorporation of authentication and privacy features.[34] Finally, the QoS attributes were enhanced in the new protocol. The fields of the IPv6 header consist of the version, traffic class, flow label, payload length, next header, hop limit, source and destination addresses.

An illustration of the protocol stack used for video communication over the Internet is depicted in Fig. 16.14. The primary transport layer protocols used over the Internet are the User Datagram Protocol (UDP) and TCP. The UDP is an unreliable connectionless protocol. It is well suited for real-time applications such as audio and video communications that require prompt delivery rather than accurate delivery and flow control. The UDP is restricted to an 8-byte header that contains minimal overhead including the source and destination ports, the length of the packet, and an optional checksum over the entire packet.

The TCP is a far better-known protocol for communication over the Internet. It is a reliable connection-oriented protocol and is used for most current applications used over the Internet. Among the most popular applications are remote login, electronic mail, file transfer, hypertext transfer, etc. These applications require precise delivery rather than timely delivery of the contents. TCP uses sequencing for packet reordering and an acknowledgement and retransmission process—known as Automatic Retransmission Request (ARQ)—for packet recovery. It also relies on a complex flow control scheme to avoid packet congestion. TCP packets have at a minimum a 20-byte header and contain numerous fields including source and destination ports, sequence and acknowledgement

[34] The security features of IPv6 have been incorporated into the IPv4 protocol.

numbers, header length, various flags, window size, checksum, urgent pointer, and optional fields.

In this section, an overview of the protocols used for video communications over the Internet is presented. A discussion of the MBONE is provided in Section 16.4.2. The standard protocol for the transport of real-time data—RTP—is presented in Section 16.4.3. Augmented to the RTP is the standard protocol for data delivery monitoring, as well as minimal control and identification capability, provided by the RTCP, which is presented in Section 16.4.4. An application level protocol for the on-demand control over the delivery of real-time data is provided by the RTSP discussed in Section 16.4.5. A protocol stack architecture designed as an Internet telephony standard known as H.323 is described in Section 16.4.6. A more flexible approach to real-time communications over the Internet has been proposed by the SIP, which is presented in Section 16.4.7. Integrated services architecture and the basic methods used for resource reservation and QoS control are provided by the resource reservation protocol (RSVP), which is discussed in Section 16.4.8. A simpler approach to QoS control provided by the differentiated services architecture and the operation of the DiffServ protocol is presented in Section 16.4.9. For brevity, the protocols discussed in this section will be illustrated by concentrating primarily on MPEG-2 video communications over the Internet.

16.4.2 Multicast Backbone

A critical factor in our ability to provide worldwide multimedia communication is the expansion of the existing bandwidth of the Internet. The NSF has recently restructured its data networking architecture by providing the vBNS. The vBNS currently uses ATM switches and OC-12c SONET fiber optic communications at data rates of 622 Mbps.

The vBNS MBONE—a worldwide digital radio and television service on the Internet—was developed in 1992. MBONE is used to provide global digital multicast real-time audio and video broadcast via the Internet. The multicast process is intended to reduce the bandwidth consumption of the Internet.

Let us consider the broadcast of a television station over the Internet. It is apparent that standard Internet technology would require a separate transmission of the television signal between the station and each television receiver. The amount of bandwidth required by such an approach to global television broadcast would be impossible to accommodate. Tremendous bandwidth losses are incurred by the transmission of multiple copies of the television signal over large segments of the communication network that are shared by multiple users. The aim of MBONE is to propagate a single packet stream over routing path segments that are shared by multiple users. Specifically, packet streams are addressed to user groups and multiple copies generated only in nodes where the routing path is no longer common to all members of the user group.

Implementation of MBONE is through a virtual overlay network on top of the Internet. It consists of islands that support multicast traffic and tunnels that are used to propagate MBONE packets between these islands. The islands are interconnected using multicast routers (mrouters), which are logically connected by tunnels.

The multicast Internet Protocol (multicast IP) was adopted as the standard protocol for multicast applications on the Internet. MBONE packets are transmitted as multicast IP packets between mrouters in different islands. Multicast IP packets are encapsulated within ordinary IP packets and regarded as standard unicast data by ordinary routers along a tunnel.

Operation of MBONE is facilitated by using multicast addresses.[35] This address serves the role of a station frequency in traditional radio broadcasts or channel number in television transmission. Identification of the multicast addresses that will be broadcasted is established by using the Internet Group Management Protocol (IGMP). Periodically, IGMP broadcast packets are sent by each mrouter to determine which hosts would like to receive which multicast addresses. IGMP responses are sent back by the hosts indicating the multicast addresses they wish to receive.

The routing process used by MBONE is the reverse path forwarding routing algorithm which helps prevent flooding. To limit the scope of multicasting, a weight is used in the Time to Live field of the IP header. The weight assigned to a packet is determined by the source, and its value is decremented by the weight of each tunnel it passes. Once the weight of a packet is no longer sufficient to pass through a tunnel, the packet is discarded. Consequently, the region of broadcasting may be limited by adjusting the weight of the transmitted packets.

MBONE applications such as multimedia data broadcasting do not require reliable communication or flow control. These applications do require, however, real-time transmission over the Internet. The loss of an audio or video packet will not necessarily degrade the broadcast quality. Significant jitter delay, on the other hand, cannot be tolerated. The UDP—not the Transmission Control Protocol (TCP)—is consequently used for transmission of multimedia traffic.

16.4.3 Real-Time Transport Protocol

The RTP provides end-to-end network transport functions for the transmission of real-time data such as audio or video over unicast or multicast services independent of the underlying network or transport protocols. Its functionality, however, is enhanced when run on top of the UDP. It is also assumed that resource reservation and QoS have been provided by lower layer services (e.g., RSVP). The RTP protocol, however, does not assume or provide guaranteed delivery or packet order preservation.

RTP services include time stamp packet labeling for media stream synchronization, sequence numbering for packet loss detection, and packet source identification and tracing.

RTP is designed to be a flexible protocol that can be used to accommodate the detailed information required by particular applications. The RTP protocol is, therefore, deliberately incomplete, and its full specification requires one or more companion documents: profile specification and payload format specification. The profile specification

[35] A multicast address is a class D address provided to the source.

document defines a set of PTs and their mapping to payload formats. The payload format specification document defines the method by which particular payloads are carried.

The RTP protocol supports the use of intermediate system relays known as translators and mixers. Translators convert each incoming data stream from different sources separately. An example of a translator is used to provide access to an incoming audio or video packet stream beyond an application-level firewall. Mixers combine the incoming data streams from different sources to form a single stream. An example of a mixer is used to resynchronize an incoming audio or video packet stream from high-speed networks to a lower bandwidth packet stream for communication across low-speed networks.

An illustration of the RTP packet header is depicted in Fig. 16.15. A corresponding glossary of the RTP packet header is provided in Table 16.7. The version number of the

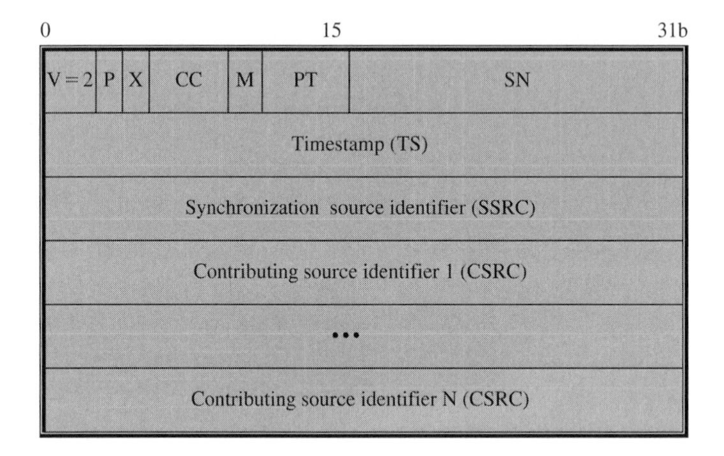

FIGURE 16.15

RTP packet header.

TABLE 16.7 RTP packet header glossary.

Abbreviation	Function
V	Version
P	Padding
X	Extension
CC	Contributing source count
M	Marker
PT	Payload type
SN	Sequence number
TS	Time stamp
SSRC	Synchronization source identifier
CSRC	Contributing source identifier

RTP protocol is defined in the version (V) field. The version number of the current RTP protocol is number two.[36] A padding (P) bit is used to indicate if additional padding bytes, which are not part of the payload, have been appended at the end of the packet. The last byte of the padding field provides the length of the padding field. An extension (X) bit is used to indicate if the fixed header is followed by a header extension. The contributing source count (CC) provides the number (up to 15) of contributing source (CSRC) identifiers that follow the fixed header. A marker (M) bit is defined by a profile for various applications such as the marking of frame boundaries in the packet stream. The PT field provides the format and interpretation of the payload. The mapping of the PT code to payload formats is specified by a profile. An incremental SN is used by the receiver to detect packet loss and restore packet sequence. The initial value of the SN is random to combat possible attacks on encryption. The time stamp (TS) provides the sampling instant of the first byte in the packet derived from a monotonically and linearly incrementing clock for synchronization and jitter delay estimation. The clock frequency is indicated by the profile or payload format specification. The initial value of the time stamp is once again random. The synchronization source (SSRC) field is used to identify the source of a stream of packets from a synchronization source. A translator forwards the stream of packets while preserving the SSRC identifier. A mixer, on the other hand, becomes the new synchronization source and must therefore include its own SSRC identifier. The SSRC field is chosen randomly to prevent two synchronization sources from having the same SSRC identifier in the same session. A detection and collision resolution algorithm prevents the possibility that multiple sources will select the same identifier. The CSRC field designates the source of a stream of packets that has contributed to the combined stream, produced by a mixer, in the payload of this packet. The CSRC identifiers are inserted by the mixer and correspond to the SSRC identifiers of the contributing sources. As indicated earlier, the CC field provides the number (up to 15) of contributing sources.

The quality of real-time multimedia transmission in noisy environments is very poor due to high packet loss rates. This problem can be alleviated by the use of generic FEC to compensate for packet loss of arbitrary real-time media streams supported by a wide variety of error-correction codes (e.g., parity, Reed–Solomon, and Hamming codes). The payload of an FEC packet provides parity blocks obtained by exclusive-or based operations on the payloads and some header fields of several RTP media packets. The FEC packets and media packets are encapsulated and sent as separate RTP streams. This feature implies that FEC is backward compatible with hosts that do not support and simply ignore RTP FEC streams.

The RTP packet header used for the RTP encapsulation of FEC determines the PT field through dynamic out of band means. The SN field is set from an independent SN space—consecutive FEC packets are assigned incremental SNs. The time stamp (TS) field is monotonically increasing and corresponds to the value of the media RTP time stamp at the time that the FEC packet is sent. The source synchronization (SSRC) field is generally

[36] Version numbers 0 and 1 have been used in previous versions of the RTP protocol.

identical to the corresponding field in the RTP media stream. The CSRC field, on the other hand, is omitted. The remaining fields are computed via the protection operation of the FEC.

The RTP encapsulation of FEC requires the use of the RTP FEC header extension following the RTP packet header. The RTP FEC header extension contains the sequence number base (SNB) field, which should be set to the minimum SN of the packets protected by FEC. The FEC may extend over any string that does not exceed 24 packets. The length recovery (LR) field is used to determine the length of any recovered packets. This is accomplished by applying the protection operation to the lengths of the media payloads associated with the FEC packet. An extension (E) bit must currently be set to zero and is reserved for future use of a header extension. The payload type recovery (PTR) field is used to ascertain the PT of any recovered packets. This is obtained by applying the protection operation to the PT fields of the media packet headers associated with the FEC. The mask (M) is a 24-bit field that provides the string of packets that are associated with the FEC. The activation of the nth bit in the mask (M) field is used to indicate that the media packet whose SN corresponds to (SNB + $n-1$) is associated with the FEC. The time stamp recovery (TS) field is used to resolve the time stamps of any recovered packets. This field is computed by applying the protection operation to the time stamp fields of the media packet headers associated with the FEC. An illustration of the RTP generic FEC header is depicted in Fig. 16.16. A corresponding glossary of the generic FEC header is provided in Table 16.8.

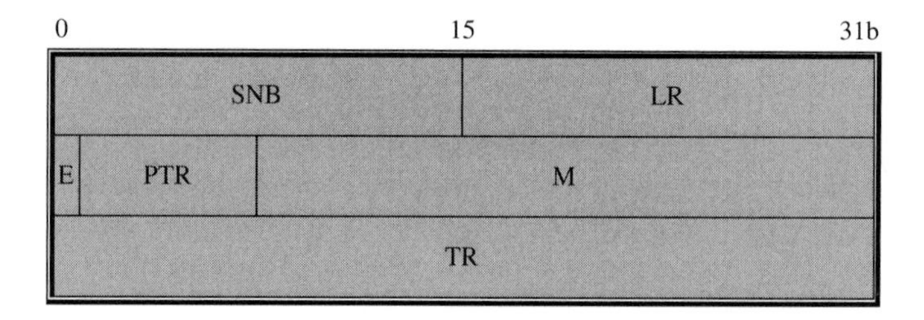

FIGURE 16.16

RTP FEC header extension.

TABLE 16.8 RTP FEC header extension glossary.

Abbreviation	Function
SNB	Sequence number base
LR	Length recovery
E	Extension
PTR	Payload type recovery
M	Mask
TR	Time stamp recovery

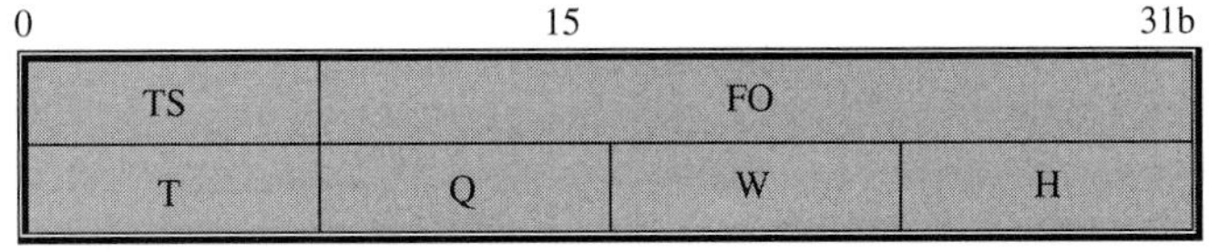

0 15 31b

TS FO

T Q W H

FIGURE 16.17

RTP MJPEG header.

The JPEG standard is used for the compression of continuous-tone still images. A direct extension of the JPEG standard to video compression known as MJPEG is obtained by the JPEG encoding of each individual picture in a video sequence. The RTP payload encapsulation of MJPEG data streams is restricted to the single-scan interleaved sequential DCT operating mode represented in abbreviated format. The RTP header for encapsulation of MJPEG is set using a 90-kHz time stamp. The RTP marker (M) bit must be activated in the last packet of each frame.

The RTP payload encapsulation of MJPEG format requires that an RTP MJPEG header follow each RTP packet header. The RTP MJPEG header represents all of the information that is associated with the JPEG frame headers and scan headers. The RTP MJPEG header contains a type specific (TS) field whose interpretation depends on the value of the type (T) field. The fragment offset (FO) field provides the number of bytes that the current packet is offset in the JPEG frame data. The total length of the JPEG frame data—corresponding to the FO field and the length of the payload data in the current packet—must not exceed 2^24 bytes. The type (T) field specifies information included in the JPEG abbreviated table specification as well as other parameters not defined by JPEG. Types 0–63 are reserved for fixed well-known mappings. Types 64–127 are also reserved for fixed well-known mappings that contain restart markers in the JPEG data. For these types, a restart marker header must appear immediately following the RTP MJPEG header. Types 128–255 are reserved for dynamic mappings defined by a session setup protocol. The quantization tables (Q) field defines the quantization tables for the frame. Q values 0–127 indicate that the type (T) field determines the quantization tables. Q values 128–255 indicate that a quantization table header following the RTP MJPEG header and possible restart marker header is used to explicitly specify the quantization tables. The width (W) and height (H) fields provide the width and height of the image in 8-pixel multiples, respectively. The maximal width and height of the image is restricted to 2040 pixels.[37] Depending on the values of the T and Q fields, the restart marker header and quantization table header may follow the RTP MJPEG header. An illustration and a corresponding glossary of the RTP MJPEG header are provided in Fig. 16.17 and Table 16.9, respectively.

The RTP payload encapsulation of MJPEG format fragments the data stream into packets such that each packet contains an entropy-coded segment of a single-scan frame.

[37]The height (H) field represents the height of a video field in the interlaced mode of motion JPEG.

TABLE 16.9 RTP MPEG header glossary.

Abbreviation	Function
TS	Type specific
FO	Fragment offset
T	Type
Q	Quantization tables
W	Width
H	Height

The payload is started immediately with the entropy-coded scan—the scan header is not present—and terminated explicitly or implicitly with an EOI marker.

The most popular current video compression standards are based on MPEG. RTP payload encapsulation of MPEG data streams can be accomplished in one of two formats: systems stream (SS)—TS and PS—as well as ES. The format used for encapsulation of MPEG SS is designed for maximum interoperability with video communication network environments. The format used for the encapsulation of MPEG SS, however, provides greater compatibility with the Internet architecture including other RTP-encapsulated media streams and current efforts in conference control.[38]

The RTP header for encapsulation of MPEG SS is set as follows: the PT field should be assigned to correspond to the SS format in accordance with the RTP profile for audio and video conferences with minimal control. The marker (M) bit is activated whenever the time stamp is discontinuous. The time stamp (TS) field provides the target transmission time of the first byte in the packet derived from a 90 kHz clock reference, which is synchronized to the system stream PCR or system clock reference (SCR). This time stamp is used to minimize network jitter delay and synchronize relative time drift between the sender and receiver. The RTP payload must contain an integral number of MPEG-2 TS packets—there are no restrictions imposed on MPEG-1 SS or MPEG-2 PS packets.

The RTP header for encapsulation of MPEG ES is set as follows: The PT field should once again be assigned to correspond to the ES format in accordance with the RTP profile for audio and video conferences with minimal control. The marker (M) bit is activated whenever the RTP packet contains an MPEG frame end code. The time stamp (TS) field provides the presentation time of the subsequent MPEG picture derived from a 90-kHz clock reference, which is synchronized to the system stream PCR or SCR.

The RTP payload encapsulation of MPEG ES format requires that an MPEG ES video-specific header follow each RTP packet header. The MPEG ES video-specific header contains a must be zero (MBZ) field that is currently unused and must be set to zero. An indicator (T) bit is used to announce the presence of an MPEG-2 ES video-specific header extension following the MPEG ES video-specific header. The temporal reference

[38]RTP payload encapsulation of MPEG elementary stream (ES) format defers some of the issues addressed by the MPEG systems stream (SS) to other protocols proposed by the Internet community.

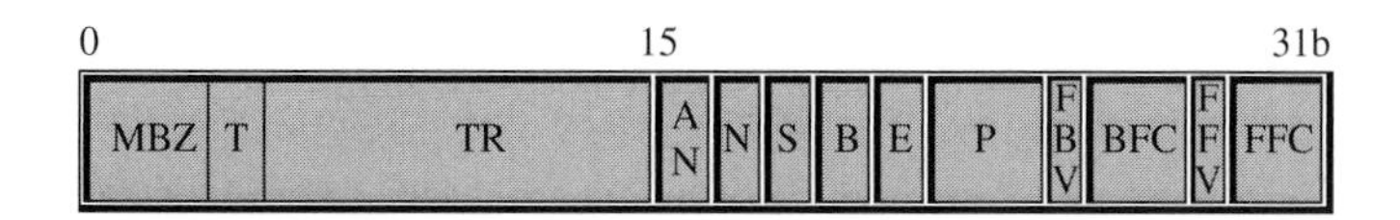

FIGURE 16.18

RTP MPEG ES video-specific header.

TABLE 16.10 RTP MPEG ES video-specific header glossary.

Abbreviation	Function
MBZ	Must be zero
T	Video-specific header extension
TR	Temporal reference
AN	Active N
N	New picture header
S	Sequence header present
B	Beginning of slice
E	End of slice
P	Picture type
FBV	Full pel backward vector
BFC	Backward F code
FFV	Full pel forward vector
FFC	Forward F code

(TR) field provides the temporal position of the current picture within the current GOP. The active N (AN) bit is used for error resilience and is activated when the following indicator (N) bit is active. The new picture header (N) bit is used to indicate parameter changes in the picture header information for MPEG-2 payloads.[39] A sequence header present (S) bit indicates the occurrence of an MPEG sequence header. A beginning of slice (B) bit indicates the presence of a slice start code at the beginning of the packet payload, possibly preceded by any combination of a video sequence header, GOP header, and picture header. An end of slice (E) bit indicates that the last byte of the packet payload is the end of a slice. The picture type (PT) field specifies the picture type – I-Picture, P-Picture, B-Picture, or D-Picture. The full pel backward vector (FBV), backward F code (BFC), full pel forward vector (FFV), and forward F code (FFC) fields are used to provide necessary information for determination of the motion vectors.[40] Figure 16.18 and Table 16.10 provide an illustration and corresponding glossary of the RTP MPEG ES video-specific header, respectively.

[39]The active N (AN) and new picture header (N) indicator bits must be set to 0 for MPEG-1 payloads.

[40]Only the FFV and FFC fields are used for P-Pictures; whereas, none of these fields are used for I-Pictures and D-Pictures.

An illustration of the RTP MPEG-2 ES video-specific header extension is depicted in Fig. 16.19. A corresponding glossary used to summarize the function of the RTP MPEG-2 ES video-specific header extension is provided in Table 16.11. Particular attention should be paid to the composite display flag (D) bit, which indicates the presence of a composite display extension—a 32-bit extension that consists of 12 zeroes followed by 20 bits of composite display information—following the MPEG-2 ES video-specific header extension. The extension (E) bit is used to indicate the presence of one or more optional extensions—quantization matrix extension, picture display extension, picture temporal scalable extension, picture spatial scalable extension, and copyright extension—following the MPEG-2 ES video-specific header extension as well as the composite display extension. The first byte of each of these extensions is a length (L) field that provides the number of 32-bit words used for the extension. The extensions are self-identifying

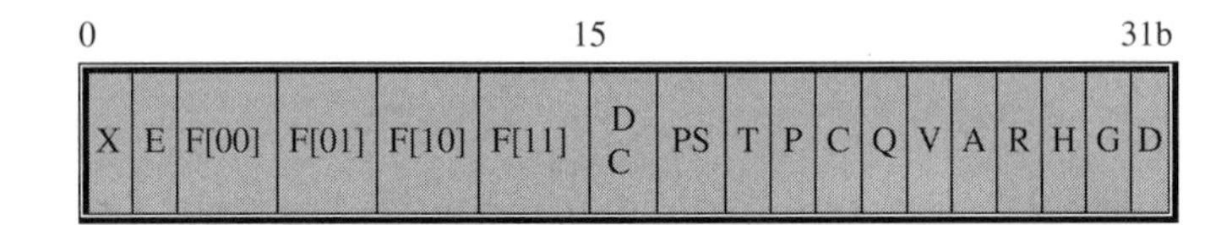

FIGURE 16.19

RTP MPEG-2 ES video-specific header extension.

TABLE 16.11 RTP MPEG-2 ES video-specific header extension glossary.

Abbreviation	Function
X	Unused (zero)
E	Extension
F[00]	Forward horizontal F code
F[01]	Forward vertical F code
F[10]	Backward horizontal F code
F[11]	Backward vertical F code
DC	Intra DC precision (intra macroblock DC difference value)
PS	Picture structure (field/frame)
T	Top field first (odd/even lines first)
P	Frame predicted frame DCT
C	Concealment motion vectors (I-Picture exit)
Q	Q-scale type (quantization table)
V	Intra VLC format (Huffman code)
A	Alternate scan (section/interlaced field breakup)
R	Repeat first field
H	Chroma 420 type (options also include 422 and 444)
G	Progressive frame
D	Composite display flag

because they must also include the extension start code (ESC) and the extension start code ID (ESCID).

The RTP payload encapsulation of MPEG ES format fragments the stream into packets such that the following headers must appear hierarchically at the beginning of a single payload of an RTP packet: MPEG video sequence header, MPEG GOP header, and MPEG picture header. The beginning of a slice—the fundamental unit of recovery—must be the first data (not including any MPEG ES headers) or must follow an integral number of slices in the payload of an RTP packet.

16.4.4 Real-Time Transport Control Protocol

The RTCP augments the RTP protocol to monitor the QoS and data delivery monitoring as well as to provide minimal control and identification capability over unicast or multicast services independent of the underlying network or transport protocols. The primary function of the RTCP protocol is to provide feedback on the quality of data distribution that can be used for flow and congestion control. The RTCP protocol is also used for the transmission of a persistent source identifier to monitor the participants and associate related multiple data streams from a particular participant. The RTCP packets are sent to all participants to estimate the rate at which control packets are sent. An optional function of the RTCP protocol can be used to convey minimal session control information.

The implementation of the RTCP protocol is based on the periodic transmission to all participants in the session of control information in several packet types summarized in Table 16.12. The sender report (SR) and receiver report (RR) provide reception quality feedback and are identical except for the additional sender information that is included for use by active senders. The SR or RR packets are issued depending on whether a site has sent any data packets during the interval since the last two reports were issued. The source description item (SDES) includes items such as canonical end point identifier (CNAME), user name (NAME), electronic mail address (EMAIL), phone number (PHONE), geographic user location (LOC), application or tool name (TOOL), notice/status (NOTE), and private extensions (PRIV). The end of participation (BYE) packet indicates that a source is no longer active. The application specific functions (APP) packet is intended for experimental use as new applications and features are developed.

RTCP packets are composed of an integral number of 32-bit structures and are, therefore, stackable—multiple RTCP packets may be concatenated to form compound

TABLE 16.12 RTCP packet types.

Abbreviation	Function
SR	Sender report
RR	Receiver report
SDES	Source description item (e.g., CNAME)
BYE	End of participation indication
APP	Application specific functions

RTCP packets. RTCP packets must be sent in compound packets containing at least two individual packets of which the first packet must always be a report packet. Should the number of sources for which reports are generated exceed 31—the maximal number of sources that can be accommodated in a single report packet—additional RR packets must follow the original report packet. An SDES packet containing a CNAME item must also be included in each compound packet. Other RTCP packets may be included subject to bandwidth constraints and application requirements in any order, except that BYE packet should be the last packet sent in a given session. These compound RTCP packets are forwarded to the payload of a single packet of a lower layer protocol (e.g., UDP).

An illustration of the RTCP SR and RR packets is depicted in Figs. 16.20 and 16.21, respectively. A corresponding glossary of the RTCP SR and RR packets is provided in Table 16.13. The RTCP SR and RR packets are composed of a header section, zero or more reception report blocks, and a possible profile-specific extension section. The SR packets also contain an additional sender information section.

The header section defines the version number of the RTCP protocol in the version (V) field. The version number of the current RTCP protocol is number two—the same as the version number of the RTP protocol. A padding (P) bit is used to indicate if additional padding bytes, which are not part of the control information, have been appended at the end of the packet. The last byte of the padding field provides the length of the padding field. In a compound RTCP packet, padding should only be required on the last individual packet. The reception report count (RC) field provides the number of reception report blocks contained in the packet. The packet type (PT) field contains the constant 200 and 201 to identify the packet as a SR and RR RTCP packet, respectively. The length (L) field provides the number of 32-bit words of the entire RTCP packet—including the header and possible padding – minus one. The synchronization source (SSRC) field is used to identify the sender of the report packet.

The sender information section appears in the SR packet exclusively and provides a summary of the data transmission from the sender. The network time protocol time stamp (NTPT) indicates the wall clock time at that instant the report was sent.[41] This time stamp along with the time stamps generated by other reports is used to measure the round-trip propagation to the other receivers. The real-time protocol time stamp (RTPT) corresponds to the NTPT provided using the units and random offset used in the RTP data packets. This correspondence can be used for synchronization among sources whose NTPTs are synchronized. The packet count (PC) field indicates the total number of RTP data packets transmitted by the sender since the beginning of the session up until the generation of the SR packet. The octet count (OC) field represents the total number of bytes in the payload of the RTP data packets—excluding header and padding—transmitted by the sender since the beginning of the session up until the generation of the SR packet. This information can be used to estimate the average payload data rate.

---

[41] Wall clock time (absolute time) represented using the Network Time Protocol (NTP) time stamp format is a 64-bit unsigned fixed-point number provided in seconds relative to 0 h UTC on January 1, 1900.

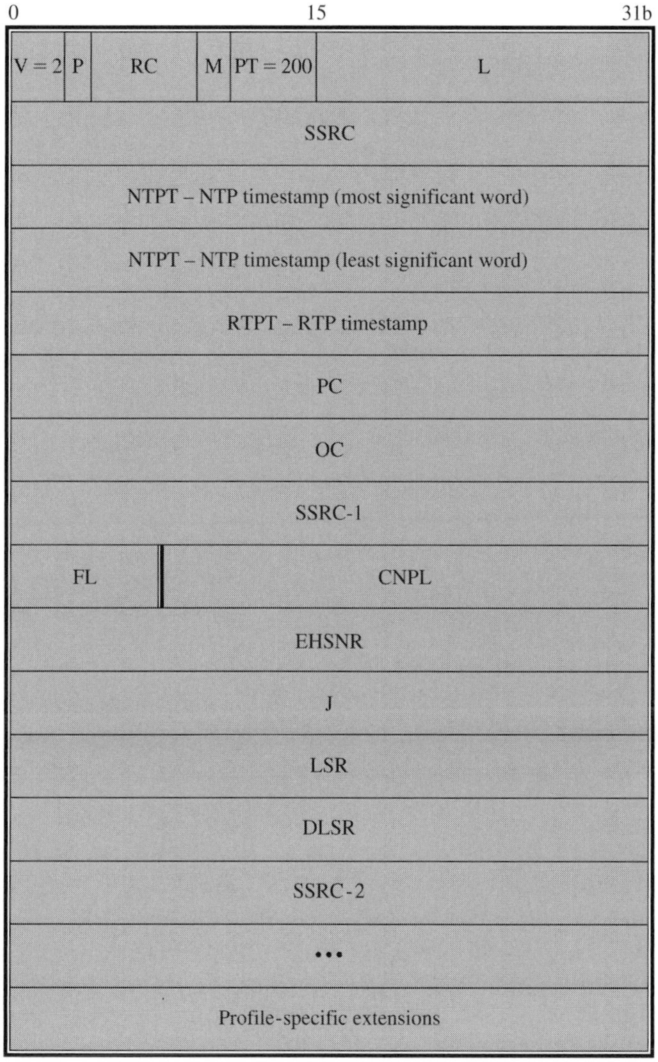

0					15	31b
V = 2	P	RC	M	PT = 200	L	
SSRC						
NTPT – NTP timestamp (most significant word)						
NTPT – NTP timestamp (least significant word)						
RTPT – RTP timestamp						
PC						
OC						
SSRC-1						
FL	CNPL					
EHSNR						
J						
LSR						
DLSR						
SSRC-2						
•••						
Profile-specific extensions						

FIGURE 16.20

RTCP sender report (SR) packet.

All RTCP report packets must contain zero or more reception report blocks

$$J = J + \frac{1}{16}(|D(i, i-1)| - J)$$

corresponding to the number of synchronization sources from which the receiver has received RTP data packets since the last report. These reception report blocks convey

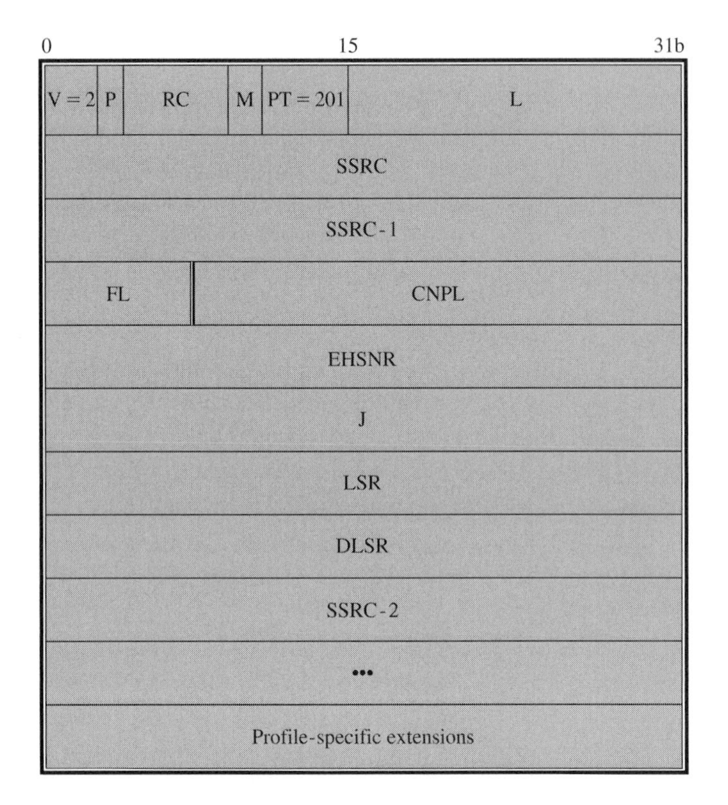

FIGURE 16.21

RTCP receiver report (RR) packet.

statistical data pertaining to the RTP data packets received from a particular synchronization source. The synchronization source (SSRC-N) field is used to identify the *N*th synchronization source to which the statistical data in the *N*th reception report block is attributed. The fraction lost (FL) field indicates the fraction of RTP data packets from the *N*th synchronization source lost since the previous report was sent. This fraction is defined as the number of packets lost divided by the number of packets expected (NPE). The cumulative number of packets lost (CNPL) field provides the total number of RTP data packets from the *N*th synchronization source lost since the beginning of the session. The CNPL is defined as the NPE less the number of packets received. The extended highest sequence number received (EHSNR) field contains the highest SN of the RTP data packets received from the *N*th synchronization source stored in the 16 least significant bits of the EHSNR field. Whereas, the extension of the SN provided by the corresponding count of SN cycles is maintained and stored in the 16 most significant bits of the EHSNR field. The EHSNR is also used to estimate the NPE, which is defined as the last EHSNR less the initial SN received. The interarrival jitter (*J*) field provides an estimate of the statistical variance of the interarrival time of the RTP data packets

TABLE 16.13 RTCP sender report (SR) and receiver report (RR) packet glossary.

Abbreviation	Function
V	Version
P	Padding
RC	Reception report count
PT	Packet type
L	Length
SSRC	Synchronization source identifier (sender)
NTPT	Network time protocol time stamp
RTPT	Real-time transport protocol time stamp
PC	Packet count (sender)
OC	Octet count (sender)
SSRC-N	Synchronization source identifier-N
FL	Fraction lost
CNPL	Cumulative number of packets lost
EHSNR	Extended highest sequence number received
J	Interarrival jitter
LSR	Last sender report time stamp
DLSR	Delay since last sender report time stamp

from the Nth synchronization source. The interarrival jitter (J) is defined as the mean deviation of the interarrival time (D) between the packet spacing at the receiver compared to the sender for a pair of packets, that is, $D(i,j) = [R(j) - R(i)] - [S(j) - S(i)]$, where $S(i)$ and $R(i)$ are used to denote the RTP time stamp from the RTP data packet i and the time of arrival in RTP time stamp units of RTP data packet i, respectively. The interarrival time D is equivalent to the difference in relative transit time for the two packets, that is, $D(i,j) = [R(j) - S(j)] - [R(i) - S(i)]$. An estimate of the interarrival jitter (J) is obtained by the first-order approximation of the mean deviation given by as follows:

$$J = J + \frac{1}{16}(|D(i, i-1)| - J)$$

The estimate of the interarrival jitter (J) is computed continuously as each RTP data packet is received from the Nth synchronization source and sampled whenever a report is issued. The last sender report time stamp (LSR) field provides the NTPT received in the most recent RTCP SR packet that arrived from the Nth synchronization source. The LSR field is confined to the middle 32 bits out of the 64-bit NTPT. The delay since last sender report (DLSR) expresses the delay between the time of the reception of the most recent RTCP SR packet that arrived from the Nth synchronization source and sending the current reception report block. These measures can be used by the Nth synchronization source to estimate the round-trip propagation delay (RTPD) between the sender and the Nth synchronization source. The estimate of the RTPD obtained provided the time of arrival T of the reception report block from the sender is recorded at the Nth synchronization source is given by RTPD $= T - \text{LSR} - \text{DLSR}$.

Figure 16.22 and Table 16.14 provide an illustration and a corresponding glossary of the RTCP SDES packet. The SDES packet is composed of a header followed by zero or more chunks. The SDES packet header version (V), padding (P), and length (L) are used in the same manner as described for the previous RTCP packets. The packet type (PT) contains the constant 202 to identify the packet as an RTCP SDES packet. The source count (SC) provides the number of chunks contained in this SDES packet. Each chunk consists of a SSRC/CSRC identifier followed by zero or more items.

Figure 16.23 and Table 16.15 provide an illustration and a corresponding glossary of the RTCP SDES CNAME item. The CNAME field contains the constant 1 to identify this as an RTCP SDES CNAME packet. The length (L) describes the length of the text field in the user and domain name (UDN) field. The UDN text field is restricted to be no longer than 255 bytes. The format used for the UDN field—"user@host" or "host" if a user is not available—should be derived algorithmically, when possible.

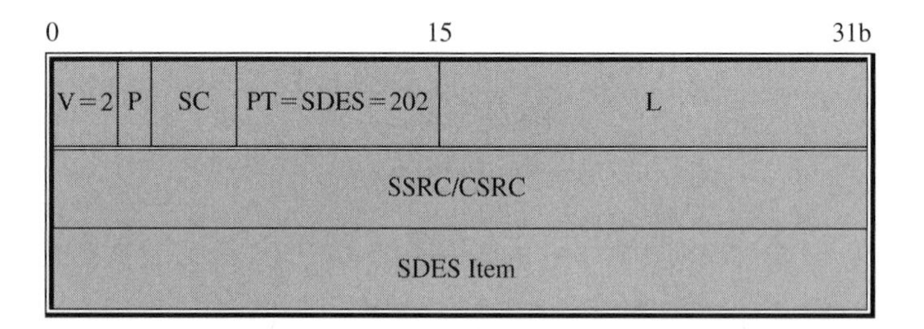

FIGURE 16.22

RTCP source description (SDES) packet.

TABLE 16.14 RTCP source description (SDES) packet glossary.

Abbreviation	Function
V=2	Version
P	Padding
SC	Source count
PT	Packet type
L	Length
SSRC/CSRC	Synchronization source/contributing source identifier
SDES item	Source description item

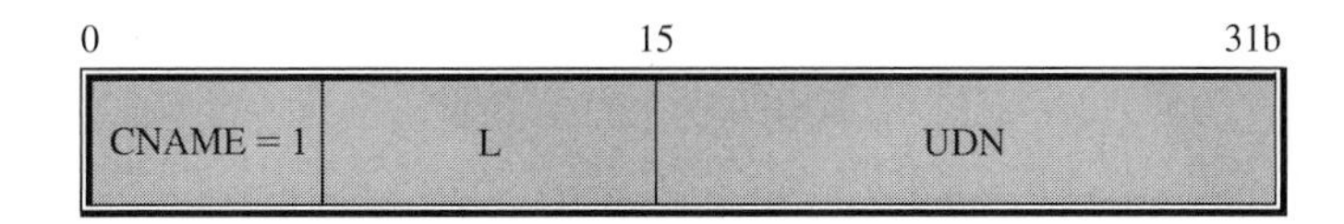

FIGURE 16.23

RTCP SDES canonical end point identifier (CNAME) item.

TABLE 16.15 RTCP SDES canonical end point identifier (CNAME) item glossary.

Abbreviation	Function
CNAME	SDES item type
L	Length
UDN	User and domain name

16.4.5 Real-Time Transport Streaming Protocol

The RTSP is an application level protocol that provides for the on-demand control over the delivery of real-time data. The RTSP protocol is intended for the control of channels and mechanisms used for multiple synchronized data delivery sessions from stored and live sources such as audio and video streams between media servers and clients. Functionally, the RTSP protocol serves the role of a "remote control" of multimedia communication systems – networks and servers.

The RTSP protocol relies on a presentation description to define the set of streams that it controls. These controls support for the following basic operations: (a) retrieval of media from a media server, (b) invitation of a media server to a conference, and (c) addition of media to an existing presentation. Table 16.16 provides a summary of the methods used by the RTSP protocol to perform various operations on the presentation or media streams.[42]

Presentation description of sessions in RTSP uses the session description protocol (SDP). The SDP is a generic textual method for describing the presentation details. It includes the session's name and purpose, streams' transport and media types, and presentations' start and end times.

Control requests and responses using RTSP may be sent over TCP or UDP. The order of arrival of the requests is critical. Therefore, the request header has a *Cseq* field that contains the SNs of the clients' requests. A retransmission mechanism is required in case any requests are lost. The use of UDP is thus limited and may cause severe problems.

Another problem in the use of RTSP is the absence of a mechanism for system recovery. For example, once the client has lost state information about a session, there is

[42]The RTSP protocol is intentionally similar in syntax and operation to HTTP 1.1—RTSP aims to provide the same services to audio and video streams as HTTP does for text and graphics.

TABLE 16.16 RTSP methods.

Abbreviation	Function
Options	Client and server inform each other of nonstandard options
Describe	Retrieves the description of a presentation or media object
Announce	(a) Client to server: posts the description of a presentation or media object
	(b) Server to client: updates the session description in real time
Setup	Specifies the transport mechanism to be used for streamed media
Play	Instructs server to start sending data via the transport mechanism
Pause	Temporarily interrupts stream delivery
Teardown	Stops the stream delivery
Get_Parameter	Retrieves the value of a parameter of a presentation or stream
Set_Parameter	Sets the value of a parameter of a presentation or stream
Redirect	Informs the client that it must connect to another server location
Record	Initiates recording a range of media data

no method to send control requests to the server. A presentation stream may continue to be transmitted to the client unless the session identifier can be recovered. Thus, RTSP implementation requires some other fail–safe method or session control option.

16.4.6 H.323

In 1996, the ITU presented the H.323 protocol stack in an effort to adopt a standard communication protocol stack for visual telephony over communication networks. The H.323 standard provides an architecture for the design of Internet telephony. It is based on the integration of various protocols to support functionality such as speech and video compression, call signaling and control, real-time transport. The H.323 protocol stack, however, does not provide a QoS capability. An illustration of the H.323 protocol stack is depicted in Fig. 16.24.

The H.323 protocol stack relies on the H.26X standards for video compression;[43] similarly, it uses the G.71X and G.72X standards for speech compression.[44] The H.245 call control protocol is adopted to allow terminals to negotiate the compression standards and bandwidth they desire. The Q.931 call signaling protocol is used to perform standard

[43]The H.323 protocol stack must support the H.261 video compression standard with a spatial resolution of QCIF. H.261 provides video compression based on the Discrete Cosine Transform (DCT) representation. Raw video representation of QCIF and CIF formats require uncompressed data rates of 9.1 and 37 Mbps. H.261 video representation of QCIF and CIF streams are compressed to 64 and 384 kbps, respectively. Modern systems that use the H.323 standard use H.263 for video communications.

[44]The H.323 protocol stack is required to be compatible with the G.711 speech compression standard. G.711 encodes a voice stream represented as an 8-bit Pulse Code Modulation (PCM) signal sampled at 8000 samples per second yielding uncompressed speech at 64 kbps. This is the standard currently used for digital transmission of telephone signals over the Public Switched Telephone Network (PSTN). Other speech compression standards such as G.722 and G.728 may also be used by the H.323 communication system.

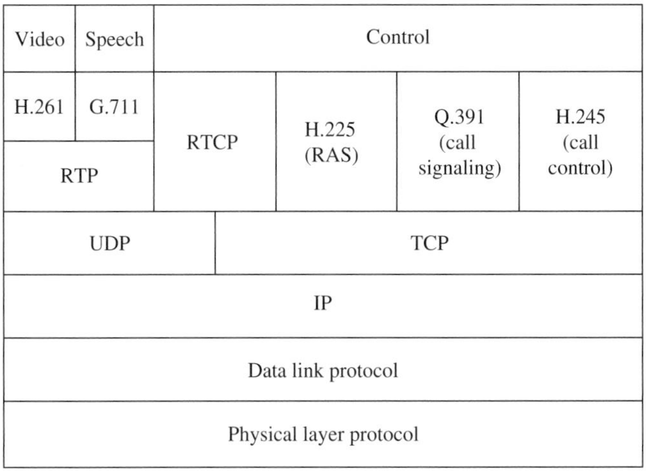

Video | Speech | Control
H.261 | G.711
RTP | RTCP | H.225 (RAS) | Q.391 (call signaling) | H.245 (call control)
UDP | TCP
IP
Data link protocol
Physical layer protocol

FIGURE 16.24

H.323 Protocol stack.

TABLE 16.17 Q.931 call setup messages.

Abbreviation	Function
Setup	PC request sent to the telephone
Call proceeding	Gatekeeper response sent to the PC
Alert	End office response sent to the PC
Connect	End office response sent to the PC

telephony functions such as establish and terminate connections, control dial tones and ringing. The registration/admission/status (RAS) protocol allows the terminals to communicate to a gatekeeper in a LAN. The RTP protocol is used for data transport and is managed by the RTCP protocol. RTCP is also used for audio/video synchronization.

To illustrate the operation of the H.323 protocol, we describe a sequence of messages needed to establish communication when a PC is used to call a telephone. Call setup in the H.323 protocol stack relies on existing telephone network protocols. These protocols are connection-oriented; hence, a TCP connection is required for call setup. Table 16.17 summarizes some of the messages used by the Q.931 call signaling protocol. The PC sends a SETUP message to the gatekeeper over the TCP connection.[45] The gatekeeper responds

[45] Communication between the PC and the gatekeeper within the LAN has already been established prior to the call setup procedure. A gatekeeper discovery packet is broadcast over UDP to determine the gatekeeper's IP address. The RAS protocol is used to send a sequence of messages over UDP to register the PC with the gatekeeper and request bandwidth.

with a CALL PROCEEDING message to acknowledge the receipt of the request. The SETUP message is forwarded by the gatekeeper to the gateway. The gateway contacts the end office associated with the terminal destination. The end office rings the telephone at the terminal destination and sends an ALERT message to the calling PC to inform it that ringing has begun. Once the telephone at the terminal destination has been connected, the end office send a CONNECT message to the calling PC to signal that connection has been established. Once connection has been established, the PC and gateway communicate directly, bypassing the gatekeeper in the calling PC's LAN. When wither the PC or telephone hangs up, the Q.391 call signaling protocol is used to tear down the connection.

16.4.7 Session Initiation Protocol

The IETF proposed the SIP. Its aim was to design a simpler and more flexible method for real-time communication networks than H.323. Instead of an entire protocol stack required by H.323, SIP is a single protocol that is capable of interfacing with existing IPs used for real-time communications. SIP can accommodate two-party, multiparty, and multicast sessions. The sessions may be used for audio, video, or data communications. The functionality of SIP is required to handle the setup and termination of sessions. It is also responsible for providing the services necessary for management of real-time communication sessions such as determining the callee's location and capabilities.

The SIP protocol is modeled after HTTP. A text message containing a method name and various parameters is sent. Table 16.18 lists the six methods defined by the core specification of SIP. A typical payload of SIP messages would be an SDP session description requested by the caller. Connection is established by a three-way handshake: an INVITE message is sent from the caller to the callee. The callee responds with an HTTP reply code. For example, if the callee accepts the call, it responds with a 200 reply code, indicating that the request succeeded. The session is connected once the caller responds to the callee with an ACK message to confirm receipt of the HTTP reply code. Termination may be initiated by a request from either the caller or callee by sending of a BYE message. The session has been terminated when receipt of the BYE message has been acknowledged.

Both UDP and TCP may be used for transport of SIP messages. Reliable transport of SIP messages is inherent when using TCP. In case UDP is used, SIP must provide its

TABLE 16.18 SIP methods.

Abbreviation	Function
INVITE	Request initiation of a session
ACK	Confirm initiation of a session
BYE	Request termination of a session
OPTIONS	Query the capabilities of a host
CANCEL	Cancel a pending request
REGISTER	Inform a redirection server of the current location of a user

own reliability and retransmission mechanism. Nonetheless, SIP transmission over UDP allows for timing and reliability control that results in a superior signaling protocol.

16.4.8 Integrated Services—Resource Reservation Protocol

Efforts at multimedia streaming over communication networks resulted in a QoS architecture known as integrated services. This architecture consists of a collection of flow-based algorithms aimed at both unicast and multicast applications. The main protocol proposed for integrated services architecture is the RSVP.

The RSVP provides an integrated service resource reservation and QoS control. It supports dynamically changing unicast and multicast routing protocols in connectionless heterogeneous networks. An important example of the use of the RSVP protocol is the reservation of bandwidth in routers along the reserved path required to guarantee low packet transfer times and minimal network jitter in multimedia applications such as audio and video communications over MBONE.

The RSVP protocol is based on receiver-initiated reservation requests that are used to establish soft states in the routers along the reserved paths. Any receiver can send a reservation request up the spanning tree provided by a unicast or multicast routing algorithm to the sender. The routing algorithm used to generate the spanning tree is not part of the RSVP protocol. The reservation requests are propagated using the reversed path forwarding algorithm along the spanning tree from the receiver toward the sender. Each router along the propagation path reserves the necessary bandwidth provided sufficient bandwidth is available. The reservation request will propagate all the way back along the spanning tree until it reaches the source or a node that already satisfies the reservation request. At this point, the required bandwidth has been reserved along a path from the sender to the receiver. The senders and receivers must refresh the soft state in the routers along the reserved paths periodically to prevent the timing out of the reservation.

The reservation requests are propagated within the nodes—hosts and routers—to the local decision modules: admission control and policy control. The admission control module determines whether the node has the available resources to accommodate the reservation requested. The policy control module decides whether the receiver has the administrative permission to establish the reservation requested. The resource reservation requested is implemented once approval from the local decision modules has been granted, by a collection of mechanisms known as traffic control: packet classifier and packet scheduler. The packet classifier determines the QoS class for each packet. The packet scheduler decides when each packet must be forwarded to achieve the guaranteed QoS. An illustration of the internal control mechanism of RSVP within the nodes is depicted in Fig. 16.25.

There are two basic message types: RESV and PATH. The RESV messages are reservation requests that are sent by the hosts up the spanning tree provided by the routing protocol toward the sender. These messages create and maintain soft states in each node along the reserved paths. The reservation requests consist of a flow descriptor—a flowspec and a filter spec. The flowspec specifies the desired QoS (Rspec) and data flow (Tspec)

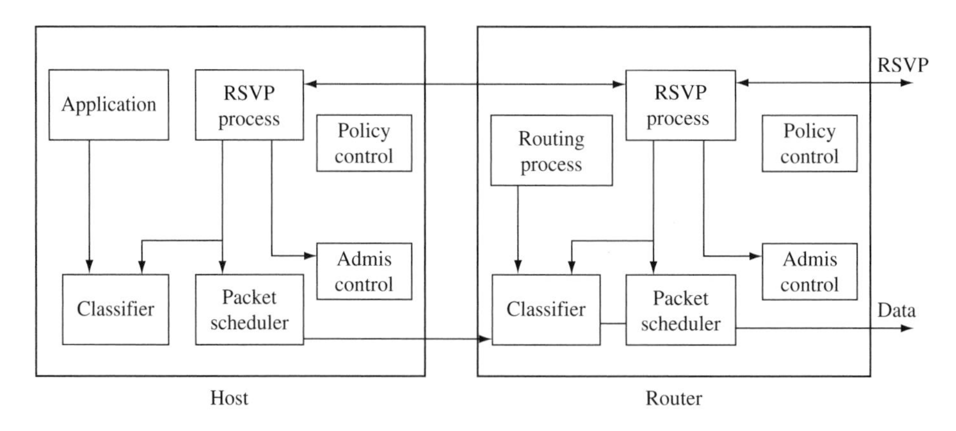

FIGURE 16.25

RSVP in hosts and routers.

parameters, for the packet scheduler. The filter spec defines the set of data packets (flow) that must receive the QoS specified by the flowspec, for the packet classifier.

The PATH messages are transmitted by the hosts down the spanning tree provided by the routing protocol—following the paths of the data flow—toward the receivers. These messages include the information necessary to route the RESV messages in the reverse direction (e.g., the IP address of the previous hop node). A PATH message must also provide a sender template – used to indicate the format of the data packets that will be transmitted by the sender—in the form of a filter spec. An additional parameter that must appear in these messages provides the traffic characteristics of the data flow that will be transmitted by the sender (sender Tspec). An optional package that carries advertising of the predicted end-to-end QoS (Adspec) is provided to the local traffic control, where it is updated and forwarded to other nodes down the spanning tree toward the receiver.[46]

Figure 16.26 illustrates an example of the RSVP message flow. The sender initially generates PATH messages down the spanning tree provided by the routing protocol toward all possible receivers. Each receiver generates RESV messages propagated down the spanning tree, back along the reverse route than that followed by the PATH messages, toward the sender. The RESV messages will propagate all the way back along the spanning tree, provided sufficient resources are available, until it reaches the sender or a node that already satisfies the reservation request. In case sufficient resources are not available at a node, an error message is sent to the receiver and the procedure is aborted. The data packets are transmitted from the sender to the receivers along the same routes followed by the PATH messages. The senders and receivers generate RESV and PATH messages along the reserved paths periodically. Once these periodic messages have not been generated

[46]RSVP supports an enhancement to the basic protocol, known as One Pass With Advertising (OPWA), for the prediction and distribution of end-to-end QoS.

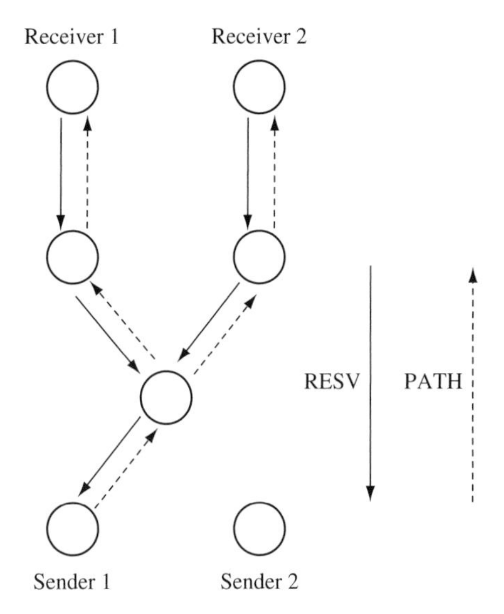

FIGURE 16.26
RSVP message flow.

for a sufficiently long time duration or an explicit teardown instruction has been issued, the reserved path is cancelled.

16.4.9 Differentiated Services—DiffServ

Integrated services present the potential to provide very high QoS for video communication network applications. However, the effort required in setting flow-based resource reservations along the route is enormous. Further, the control signaling required and state maintenance at routers limit the scalability of this approach. Consequently, at present, integrated services are rarely deployed for high QoS video communication networks.

Differentiated services were proposed as a simpler approach to high QoS communication networks. The basic principle behind differentiated services is a class-based approach that relies on local routers. A set of service classes with corresponding forwarding rules is defined. A marker located in a field within the packet header is used to ascertain the level of service. Different levels of service may differ in terms of various QoS parameters such as jitter, delay, packet loss rate, throughput.

Two service classes management schemes are currently used: expedited forwarding and assured forwarding. In expedited forwarding, two classes of service are typically available: best-effort and expedited.[47] The resources devoted to transmission of expedited packets are much better than best-effort packets. A typical implementation of the

[47] Expedited service is also known as premium service.

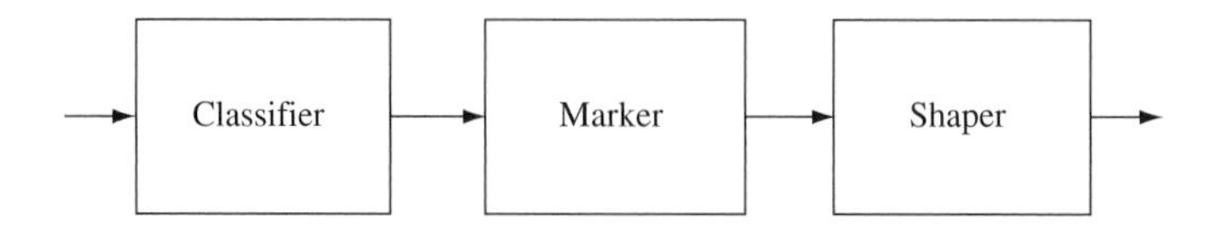

FIGURE 16.27

Differentiated services traffic conditioning functionality.

expedited forwarding scheme would rely on a two-queue structure in the router. Upon arrival of a packet, its class of service is ascertained, and it is queued in the best-effort or expedited queue accordingly. Packets are scheduled for transmission from the two-queue structure according to a policy determined by a weighted fair queue.

Assured forwarding provides a more complex service class management scheme. Four priority classes with separate resources are specified. Additionally, three packet drop rates are defined for each priority class: low, medium, and high. A matrix of four priority classes and three packet drop rates results. Consequently, 12 classes of service are available in assured forwarding.

Typically, interior router processing required for differentiated services is minimal. It consists of a forwarding treatment that is referred to as per-hop behavior (PHB). The PHB includes a queuing discipline and packet dropping rules that provide preferential treatment and buffer control of packets. In addition to PHB mechanisms, boundary routers require traffic conditioning functionality to provide the desired service. Thus, most of the complexity needed for differentiated services resides in boundary routers. The boundary routers functionality can also be provided by the sending host or first-hop router.

A typical procedure used for traffic conditioning functionality in differentiated services is depicted in Fig. 16.27. The *classifier* is used to sort the packets into different priority classes. Separate queues are used to identify the distinct priority classes. For example, in assured forwarding, the classifier would be used to divide the packets among the four priority classes. The *marker* determines the class of service which is marked in a header field. For this purpose, it is suggested to rely on the 8-bit Type of Service field in the IPv4 header.[48] A 6-bit differentiated service (DS) subfield is used for marking class services within the Type of Service field, thus leaving two unused bits. The marker can also be used to remark packets. For instance, packets whose QoS profile has been exceeded or not been met or packets that transmit across the boundary of DS domains may be remarked. The *shaper* is a filter that delays or drops packets to shape the priority streams into desired forms. For example, a leaky bucket or token bucket may be used as the shaper.

Scalability of differentiated services is achieved by implementation on local routers and processing individual packets. Moreover, aggregate flows within the same class of service are treated equally. Further, use of an existing field in the IP header implies that no change in the network protocol is required. For these reasons, differentiated services have become the most widely acceptable QoS mechanism in communication networks.

[48]Differentiated services use the 8-bit Traffic Class field for marking the class of service in the IPv6 header.

16.5 SUMMARY

In this presentation, we have provided a broad overview of video communication networks. The fundamental video compression standards were briefly discussed. The system standard associated with the most widespread video compression standard—MPEG-2—was presented. Future implementation of video communications over various networks – HFC, DSL, wireless, fiber optics, ISDN, and ATM—were presented. A broader topic addressing the issue of video communication over the Internet was also discussed. Multicast video communications over MBONE backbone was introduced. Several protocols—RTP, RTCP, and RTSP—that are essential for efficient and reliable video communication over the Internet were illustrated. Other important efforts to facilitate video communications over the Internet provided by various session layer protocols—H.323 and SIP—were also discussed. QoS architectures based on integrated and differentiated services and their corresponding protocols—RSVP and DiffServ—were finally presented. The entirety of this presentation points to the imminent incorporation of a variety of multimedia applications into a seamless nested array of wireline and wireless communication networks. It is not long before the anticipated integration of the computing, communication, and entertainment industries emerges. Automobiles passengers will be able to view cable television over their laptops while traveling. Airline passengers will be able to use their handheld telephones or PDAs to browse the web and exchange e-mails. Progress toward realization of this vision is currently under way by improvement in network infrastructure and advancements in real-time protocol design.

REFERENCES

[1] I. Brosky. *Wireless: The Revolution in Personal Telecommunications.* Artech House, Boston, MA, 1995.

[2] J. F. K. Buford (ed.), *Multimedia Systems.* Addison-Wesley, Boston, MA, 1994.

[3] D. E. Comer. *The Internet Book.* Prentice-Hall, Upper Saddle River, NJ, 1995.

[4] J. Crowcroft, M. Handley, and I. Wakeman. *Intenetworking Multimedia.* Morgan Kaufmann Publishers, Burlington, MA, 1999.

[5] M. De Prycker. *Asynchronous Transfer Mode.* Ellis Horwood, Chichester, UK, 1993.

[6] F. Fluckiger. *Understanding Networked Multimedia.* Prentice-Hall, Upper Saddle River, NJ, 1995.

[7] V. Garg, K. Smolik, and J. E. Wilkes. *Applications of CDMA in Wireless and Personal Communications.* Prentice-Hall, Upper Saddle River, NJ, 1997.

[8] V. Garg and J. E. Wilkes. *Wireless and Personal Communication Systems.* Prentice-Hall, Upper Saddle River, NJ, 1996.

[9] W. J. Goralski. *Introduction to ATM Networking.* McGraw-Hill, New York, 1995.

[10] R. Handel, M. N. Huber, and S. Schroder. *ATM Concepts, Protocols, and Applications.* Addison-Wesley, Boston, MA, 1994.

[11] M. Handley and J. Crowcroft. *The World Wide Web—Beneath the Surface.* UCL Press, London, 1994.

[12] W. W. Hodge. *Interactive Television.* McGraw-Hill, New York, 1995.

[13] V. Kumar. *MBONE: Interactive Multimedia on the Internet.* New Riders, Indianapolis, IN, 1996.

[14] F. Kuo, W. Effelsberg, and J. J. Garcia-Luna-Aceves. *Multimedia Communications: Protocols and Applications.* Prentice-Hall, Upper Saddle River, NJ, 1998.

[15] O. Kyas. *ATM Networks.* International Thompson Publishing, Andover, UK, 1995.

[16] W. C. Y. Lee. *Mobile Cellular Telecommunications.* McGraw-Hill, New York, 1995.

[17] D. E. McDysan and D. L. Spohn. *ATM—Theory and Application.* McGraw-Hill, New York, 1995.

[18] D. Minoli. *Video Dialtone Technology.* McGraw-Hill, New York, 1995.

[19] D. Minoli and M. Vitella. *ATM & Cell Relay for Corporate Environments.* McGraw-Hill, New York, 1994.

[20] M. Orzessek and P. Sommer. *ATM & MPEG-2: Integrating Digital Video into Broadband Networks.* Hewlett-Packard Professional Books, New York, 1998.

[21] K. R. Rao, Z. S. Bojkovic, and D. A. Milovanovic. *Multimedia Communication Systems: Techniques, Standards, and Networks.* Prentice-Hall, Upper Saddle River, NJ, 2002.

[22] K. R. Rao and J. J. Hwang. *Techniques and Standards for Image, Video, and Audio Coding.* Prentice-Hall, Upper Saddle River, NJ, 1996.

[23] T. S. Rappaport. *Wireless Communications: Principles & Practice.* Prentice-Hall, Upper Saddle River, NJ, and IEEE Press, New York, 1996.

[24] Redl, Weber, and Oliphant. *An Introduction to GSM.* Artech House, Boston, MA, 1995.

[25] M. Y. Rhee. *CDMA Cellular Mobile Communications and Network Security.* Prentice-Hall, Upper Saddle River, NJ, 1998.

[26] B. Sklar. *Digital Communications: Fundamentals and Applications.* Prentice-Hall, Upper Saddle River, NJ, 1998.

[27] P. Smith. *Frame Relay.* Addison-Wesley, Boston, MA, 1993.

[28] J. D. Solomon. *Mobile IP: The Internet Unplugged.* Prentice-Hall, Upper Saddle River, NJ, 1998.

[29] J. D. Spragins, J. L. Hammond, and K. Pawlikowski. *Telecommunications Protocols and Design.* Addison-Wesley, Boston, MA, 1991.

[30] W. Stallings. *ISDN and Broadband ISDN with Frame Relay and ATM.* Prentice-Hall, Upper Saddle River, NJ, 1995.

[31] W. Stallings. *Data and Computer Communications.* Prentice-Hall, Upper Saddle River, NJ, 1997.

[32] R. Steele. *Mobile Radio Communications.* Pentech Press, London, 1992.

[33] R. Steinmetz and K. Nahrstedt. *Multimedia: Computing, Communications, and Applications.* Prentice-Hall, Upper Saddle River, NJ, 1995.

[34] A. S. Tanenbaum. *Computer Networks.* Prentice-Hall, Upper Saddle River, NJ, 1996.

[35] G. Varrall and R. Belcher. *Data Over Radio.* Quantum Publishing, London, 1992.

[36] A. J. Viterbi. *CDMA Principles of Spread-Spectrum Communication.* Addison-Wesley, Boston, MA, 1995.

[37] Y. Wang, J. Ostermann, and Y.-Q. Zhang. *Video Processing and Communications.* Prentice-Hall, Upper Saddle River, NJ, 2002.

[38] E. K. Wesel. *Wireless Multimedia Communications.* Addison-Wesley, Boston, MA, 1997.

[39] C.-H. Wu and J. D. Irwin. *Emerging Multimedia Computer Communication Technologies.* Prentice-Hall, Upper Saddle River, NJ, 1998.

[40] W. Zhou and C.-C. J. Kuo. *Intelligent Systems for Video Analysis and Access Over the Internet.* Prentice-Hall, Upper Saddle River, NJ, 2003.

CHAPTER

Video Security and Protection

17

Min Wu[1] and Qibin Sun[2]

[1] *ECE Department, University of Maryland, College Park, Maryland, USA*
[2] *Hewlett-Packard Inc., Shanghai, China*

17.1 INTRODUCTION

The past decade has witnessed significant advancement in coding and communication technologies for digital multimedia, paving ways to many opportunities for people around the world to acquire, utilize, and share multimedia content [1]. To allow for wider availability of multimedia information and successful commercialization of many multimedia-related services, assuring that multimedia information is used only by authorized users for authorized purposes has become essential. The security and protection of media content can be boiled down to several main issues, including: (1) encryption for confidentiality and access control; (2) authentication for content integrity; (3) embedding fingerprinting for tracing misuse and illicit distribution of content. We will discuss these three issues in this chapter.

17.2 VIDEO ENCRYPTION

We start the discussion by examining a typical use of multimedia illustrated in Fig. 17.1, whereby the owner of the multimedia content wants to securely distribute the content through networks or archive it for future use. With the sophistication of heterogeneous networks and the growing amount of information being generated, it is becoming less efficient for content owners to manage the distribution or archiving process all by themselves. As a result, third-party service providers, equipped with specialized servers, huge disk space, and abundant bandwidth resources, will serve as delegates for content owners to perform content distribution, archiving, search, and retrieval. On one hand, the delegate service providers often need to process the received media data, such as adapting the rate of the media data according to the available bandwidth. On the other hand, the owner does not want to reveal the media content to these delegates because of security

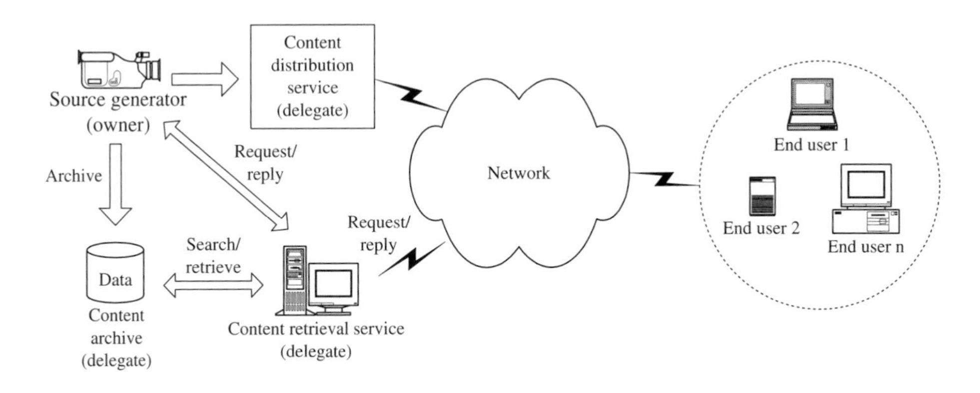

FIGURE 17.1

Typical usage of multimedia content.

and privacy concerns. One such example is privacy-preserving data retrieval using an untrusted server [2].

A common way to achieve content confidentiality is to encrypt the entire multimedia sequence using a cipher, such as DES, AES, or RSA [3, 4]. However, many types of processing, such as rate adaptation for multimedia transmission in heterogeneous networks [5] and DC-image extraction for multimedia content searching [6], cannot be applied directly in the bit stream encrypted by these generic encryption tools or their simple variations. This implies that the processing would still require the delegates to hold the decryption keys to decrypt the content, process the data, and then re-encrypt the content. Because revealing decryption keys to potentially untrustworthy delegates is often not in line with the security requirements of many applications, generic encryption alone is inadequate in the delegate service scenario.

Another unique issue with multimedia encryption is the relation between value, quality, and timeliness of the data. Unlike the "all-or-nothing" protection for generic data, the value of multimedia and in turn its security protection are closely tied with the perceptual quality and the timeliness of the content. A popular sports game, for example, requires paramount protection during the show to ensure revenue from the viewership, but the demands of security reduce quickly or even vanish over the following days. Additionally, high-quality content is often priced at premium rates and has restricted access, while its low-resolution, low-quality versions are moderately priced or given out for free to reach a wide audience [7].

The first step towards addressing the aforementioned issues is to design flexible multimedia encryption schemes that can handle delegate processing and achieve access control by content and quality [8, 9]. We first examine candidate domains in which encryption can be applied to multimedia data and discuss the advantages and limitations. We then present three representative encryption techniques that integrate signal processing and cryptographic techniques, followed by discussions on security evaluation. Finally, we put these building blocks together into a video encryption system and examine the performances under various configurations.

17.2.1 Candidate Domains for Encrypting Multimedia

We examine in this subsection the possible domains in which encryption can be applied to multimedia, along with a review of related work. Using a widely adopted multimedia coding framework, we illustrate the candidate domains for applying encryption to multimedia in Fig. 17.2.

17.2.1.1 Encryption Before and After Coding

According to Fig. 17.2, there are two straightforward places to apply generic encryption to multimedia. The first possibility is to encrypt multimedia samples before any compression (i.e., Stage 1 in Fig. 17.2). The main problem with this approach is that the encryption often significantly changes the statistical characteristics of the original multimedia source, resulting in much reduced compressibility. It is worth noting an interesting theoretical exploration on efficiently compress encrypted data that relate the problem with distributed source coding when decryption and decompression can be jointly carried out [10]. In this case, the key shared between the encryption system and the joint decryption–decompression system can be considered as side information, and the correlation between the key and the encrypted data can be exploited by proper encoding. It can be shown that the maximum compression gain on encrypted data can be the same as compressing the unencrypted data in the case of an ideal Gaussian source. The compression gain, however, would be reduced for more general source that is common in practice, and it cannot easily support many other forms of delegate processing.

The second possibility is to apply generic encryption to the encoded bit stream after compression (i.e., Stages 5 and 6 in Fig. 17.2) [11]. This approach introduces little overhead, but may destroy the structures and syntax readily available in the unencrypted bit stream. Such structures, often indicated by special header/marker patterns, would enable many kinds of processing in delegate service providers and intermediate network links, such as bandwidth adaptation, unequal error protection, and random access [12–15].

As headers and markers are special bit patterns in a compressed bit stream for a variety of purposes [14], a simple way to realize syntax-aware encryption is only to encrypt the content-carrying fields of the compressed multimedia bit stream, such as the fields of motion vectors (MVs) and discrete cosine transform (DCT) coefficients in MPEG video, and keep the structure and headers/markers of the bit stream unchanged [16, 17].

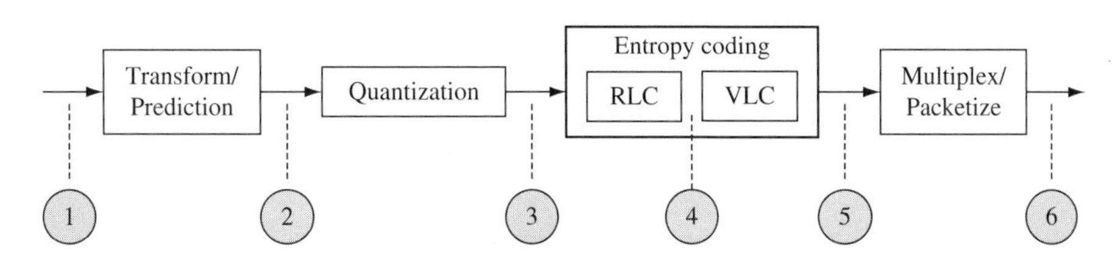

FIGURE 17.2

Candidate domains to apply encryption to multimedia.

A good example of this framework can be found through the emerging JPEG-2000 security (JPSEC) standard [18]. A general limitation of marker-based approach is that transmission of encrypted data streams has to be handled in a special way different from unencrypted data, which requires the support of all major providers of delegate processing units, except for communications within a closed, proprietary system; in addition, only a limited amount of syntax from unencrypted media can be preserved to facilitate a preidentified set of processings. The long and uncertain process of industry standardization can be a hurdle to provide timely and flexible solutions to secure multimedia communications.

17.2.1.2 Encryption at Intermediate Stages of Coding

There has been an interest in studying how to encrypt multimedia data in such a way that the encrypted data can still be represented in a meaningful, standard-compliant format. This is particularly useful for secure delegate services and multimedia communications that prefer handling media streams compliant to certain multimedia coding standards, such as JPEG or MPEG-1/2/4 standard [15, 19]. The encryption is performed in the intermediate stages illustrated in Fig. 17.2. For example, at Stage 2, the MVs in video can be encrypted by applying DES to their codeword indices [15]. At Stage 3, DC and selected AC coefficients in each block of a JPEG image or an MPEG video frame can be shuffled within the block [20], or across blocks but within the same frequency band [21]. At Stage 4, the entropy codeword can be spatially shuffled within the compressed bit stream [19]; the Huffman codewords of coefficients and/or MVs can be encrypted by alternating between several Huffman codebooks in a cryptographically secure fashion [22]. At Stage 5, only intracoded frames and blocks of an MPEG video are selected and encrypted using a classic DES cipher [23] or its variations [11]. Some of these schemes are also known as selective encryption [19, 22, 23], that is, they encrypt only portions of multimedia data stream that carry rich content, in hope of alleviating the problem of high computational complexity and the potential bit rate overhead.

17.2.2 Building Blocks for Media Encryption

In this subsection, we review several representative approaches that have strong roots in signal processing and can serve as building blocks for encrypting video and multimedia data in general.

17.2.2.1 Randomized Entropy Coding

Huffman coding and arithmetic coding are two main entropy coding approaches commonly employed in multimedia compression to encode transform coefficients and prediction parameters. For given source statistics and coding gain, the Huffman codebooks and arithmetic coding rules are not unique. For example, at each bifurcation point when building a Huffman table as illustrated in Fig. 17.3, we can label 1-0 or 0-1 to the two branches without sacrificing decodability or coding efficiency. This is called Huffman tree mutation process [24]. Similar reordering can be applied at each dyadic approximation during arithmetic coding. The upper and lower part of the interval at each partition

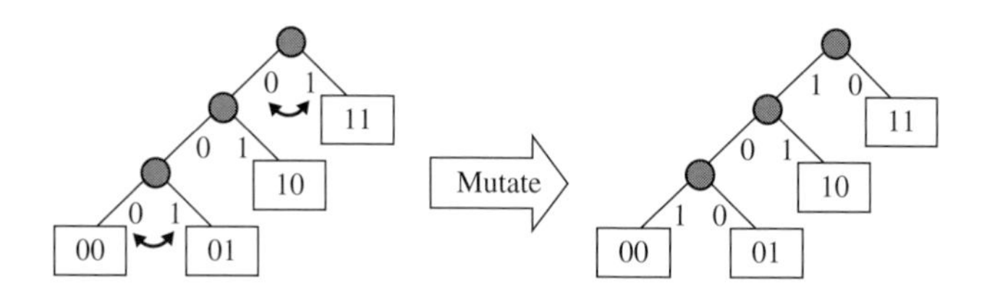

FIGURE 17.3

Illustration of Huffman tree mutation process (from Fig. 17.7 of [24]).

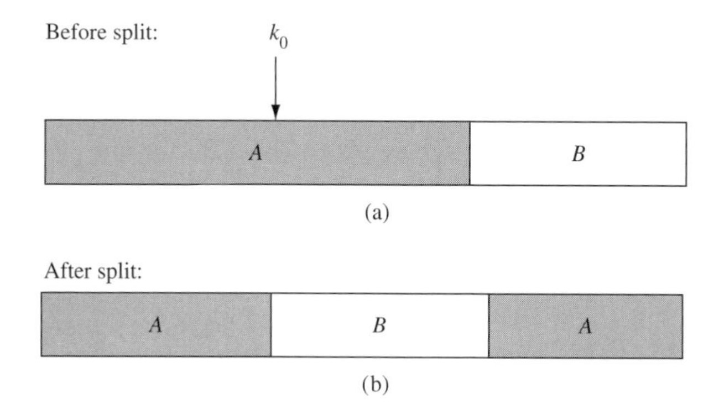

FIGURE 17.4

Illustration of interval splitting in building secure arithmetic coder (from Fig. 17.2 of [26]).

round can be assigned 1-0 or 0-1. These randomizations provide a large pool of candidate codebooks to produce the final codewords of compressed multimedia files. A key can direct the selection of one of these codebooks, or equivalently, the specific route of randomization during entropy coding, in a cryptographically secure fashion [22, 24, 25]. As the mutation does neither change the source's statistics nor the length assignment of codewords, the overall compression performance is approximately the same as the original entropy codebook [24].

The mutation process illustrated earlier is an example of controlled randomization fitting within an established entropy coding framework. For arithmetic coding, the traditional assumption that a contiguous interval is used for each symbol can be lifted to open up new opportunities for randomization [26]. As illustrated in Fig. 17.4, a cryptographic key known both to the encoder and decoder is used to select a splitting point of one of the intervals. The intervals associated with each symbol, which are continuous in a traditional arithmetic coder, are split according to a key known both to the encoder and decoder; the coder uses a generalized constraint that the sum of the lengths of the one or more intervals associated with the same symbol be equal to its probability. Consider a simple example of a binary system with two symbols A and B with $P(A) = 2/3$ and $P(B) = 1/3$. A traditional partitioning would represent A by the range $[0, 2/3)$ and

B by the range $[2/3,1)$. Now instead, if symbol A is represented by the combination of the intervals $[0,1/3)$ and $[2/3,1)$ and symbol B by $[1/3,2/3)$, the overall symbol probabilities are preserved. It can be shown that with such interval splitting, the coding performance is at most one bit larger than traditional arithmetic coding for each of the N-symbol sequence that is encoded together [26]. As sequence length grows, the relative bit rate overhead decreases quickly and becomes negligible. Details on cryptanalysis and implementation complexity of such a secure arithmetic coder by interval splitting can be found in [26].

17.2.2.2 Generalized Index Mapping with Controlled Overhead

Unlike generic data encryption where the encryption output can take values over the entire data space, joint signal processing and cryptographic encryption requires that the encrypted outputs satisfy additional constraints. These constraints are essential to preserve the structure, syntax, and standard compliance that enable delegate processing and leads to communication friendliness.

An example of format-compliant encryption performs encryption on indices of codeword. The basic scheme, as proposed in [15], assigns a fixed-length index to each variable length codeword (VLC) at a chosen stage of media encoding. The concatenated indices are then encrypted, and the encrypted indices are mapped back to codeword domain to form an encrypted bit stream. This approach would work well with such codes as the Huffman codes and the Golomb-Rice codes, which associate each symbol coming from a finite set with a unique codeword of integer length.

To extend the scheme to other VLCs that allow fractional codeword length per symbol (such as the arithmetic codes), we apply the index encryption idea directly to media symbols that take values from a finite set before getting into VLC codeword domain. Examples include working with quantized coefficients and quantized prediction residues (Stage 3 in Fig. 17.2), as well as run-length coding symbols (Stage 4 in Fig. 17.2). The encryption process to produce a ciphertext symbol $X^{(\text{enc})}$ from a clear-text symbol X can be represented as:

$$X^{(\text{enc})} = \text{Encrypt}(X) \triangleq T^{-1}[\mathcal{E}(T(X))], \tag{17.1}$$

where $\mathcal{E}(\cdot)$ is a core encryption primitive such as AES or one-time pad [3], and $T(\cdot)$ represents a codebook that establishes a bijective mapping between all possible symbol values and indices represented by binary strings. The goal of this bijection is to produce fixed-length indices that will be passed to subsequent encryption or decryption. The decryption process has a similar structure:

$$X = \text{Decrypt}\left(X^{(\text{enc})}\right) = T^{-1}\left[\mathcal{D}\left(T\left(X^{(\text{enc})}\right)\right)\right], \tag{17.2}$$

where $\mathcal{D}(\cdot)$ is a core decryption primitive corresponding to $\mathcal{E}(\cdot)$.

As a simple example, we consider encrypting a string of symbols coming from a finite set {A, B, C, D}. The symbol sequence to be encrypted is "ABBDC." We first assign a fixed-length index to each symbol:

$$A \rightarrow [00], \quad B \rightarrow [01],$$
$$C \rightarrow [10], \quad D \rightarrow [11].$$

We then convert the symbol sequence to an index sequence "00 01 01 11 10," and encrypt the index sequence using an appropriate encryption primitive such as a stream cipher (the one-time pad) with a random bit stream [0100 1011 1001 ...]. Finally, we convert the encrypted index sequence "01 01 11 00 00" back to symbol sequence "BBDAA." After encryption, any appropriate VLC coding can be applied to the encrypted symbol sequence. It is worth noting that in such an encryption one input symbol can be mapped to different encrypted cipher-text. For instance, in the previous example the symbol B has appeared in the clear-text sequence twice, the first time it was mapped to B and the second time to D.

In general, when processing a large sequence of symbols, the encryption method by index mapping tends to make the encrypted symbols uniformly distributed, which is good in terms of security [3]. However, the entropy of the encrypted symbols is increased from the unencrypted ones. Because the compressibility of a sequence of symbols using entropy coding depends on the entropy of the source symbols [27], index-mapping encryption would bring a bit-rate overhead in compression. Its impact on the compressibility of the source symbols can be quantified by the changes in average code length before and after encrypting a sequence of symbols. We also want to explore techniques that can keep the bit-rate overhead low.

A common case is to consider compressing the source symbols using a default entropy codebook as provided by many multimedia standards. The default codebook is obtained from a set of representative training samples and is used most of the time for the simplicity of implementation. We denote the probability mass function of the symbols prior to encryption by $\{p_i\}$, that of the symbols after encryption by $\{q_i\}$, and the code length designed for distribution $\{p_i\}$ by $\{l_i\}$. If encryption is performed on an index drawn from the full range of symbol values, the distribution of ciphertext symbols, q, will be uniform over the entire range. Alternatively, if we partition the range of symbol values into mutually exclusive subsets $\{S_j\}$ and restrict the outcome of the encryption of a symbol $x \in S_j$ to be within the subset S_j, that is, $\text{Encrypt}(x) \in S_j$, the distribution q will be a piecewise uniform approximation of p, as illustrated in Fig. 17.5. It can be shown that the changes of the expected code length δL is

$$\delta L = \sum_i (q_i - p_i) l_i = D(p||q) + D(q||r) - D(p||r), \tag{17.3}$$

where $D(\cdot||\cdot)$ represents the Kullback–Leibler divergence and r represents a probability distribution of $\{r_i \triangleq P(R = i) = 2^{-l_i} / \sum_k 2^{-l_k}\}$.

If we partition the symbol range S into more than one subset and restrict the encryption output to be in the same subset as the input symbol, the complexity of a brute-force attack for each symbol is reduced from $2^{|S|}$ to $2^{|S_j|}$, where S_j is the subset to which the symbol belongs. On the other hand, the overhead is also reduced because in the

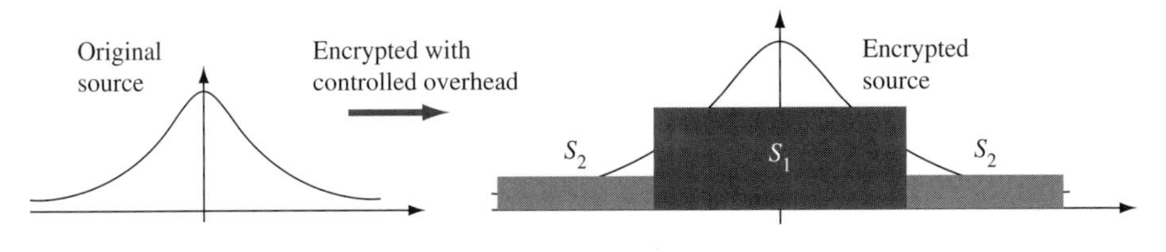

FIGURE 17.5

Index mapping within subsets gives piecewise constant approximation of the distribution.

Kullback–Leibler divergence sense, the distance from the original distribution p to the piecewise uniform distribution q is closer than that to a completely uniform distribution. Thus by controlling the set partitioning, we can adjust the tradeoff between the security and the overhead.

Another scenario of compressing the source symbols uses adaptive or universal entropy coding that adjusts itself to the source's distribution. Arithmetic coding and Lempel-Ziv coding are two such examples. This can be analyzed similarly and the same strategy can be applied to control the bit-rate overhead.

The final result of the relative bit-rate overhead (η) also depends on the ratio of the size of the content to be encrypted (B_1) to the overall size of the stream (B). That is,

$$\eta = \frac{B_1^{(e)} - B_1}{B} \times 100\% = \eta_e \frac{B_1}{B} \times 100\%, \tag{17.4}$$

where $B_1^{(e)}$ denotes the size of the encrypted part and η_e is the relative overhead for the part being encrypted. Even if η_e is large, the overall overhead can be constrained if only a relatively small part of the stream is encrypted.

17.2.2.3 Constrained Shuffling

Random permutation or shuffling is a common cryptographic primitive operation. The temporal characteristic of audio and video data as well as the spatial characteristic of visual data make permutation a natural way to scramble the semantic meaning of multimedia signals [11, 19–21]. Even before the arrival of digital technology, an early work by Cox et al. builds an analog voice privacy system on a subband representation framework and permutes time segments of subband signals across both time and subbands [28]. More sophisticated coding techniques have been employed by modern digital coding systems. Thus to control the bit-rate overhead and allow for delegate processing, random permutation should be performed in a constrained way and in appropriate domains. For example, instead of permuting image samples that destroys intersample correlations and makes an image much less compressible, taking each 8×8 block as a unit in the same way as JPEG compression does and permuting the ordering of these image blocks will keep the bit rate increase low under JPEG compression. Each block may be further rotated by a random multiple of 90° or mirrored [21]. It is also worth mentioning that a major drawback for block shuffling alone lies in the fact that the information within a block is perfectly retained. When block size is not very small, an attacker can exploit the correlation across the blocks (such as the continuity of edges and similarity of colors and textures) and reassemble the shuffled blocks with a moderate number of trials [29]. Therefore, block shuffling alone is often not a secure encryption operation, although it can be applied as a low-complexity, easy-to-implement building block to complement other techniques in the design of an overall encryption system.

Shuffling can be applied not only to samples but also to encoded symbols or codewords. We now use the encryption of scalable video with fine granularity as an example to illustrate a constrained shuffling technique known as the intra bitplane shuffling (IBS). This encryption technique is compatible with fine granularity scalable coding and provides a tool for access control of multimedia content at different quality levels. Fine granularity scalability (FGS) is desirable in multimedia communications to provide a near-continuous tradeoff between bit rate and quality. FGS is commonly achieved by bit plane coding, as used in the embedded zero-tree wavelet (EZW) coder [30] and the MPEG-4 FGS coder [31]. We shall use the MPEG-4 FGS to illustrate the concept and the approach can be extended to other scalable or embedded coders. As surveyed in [5], MPEG-4 FGS is a functionality provided by the MPEG-4 streaming video profile. A video is first encoded into two layers, namely, a base layer that provides a basic quality level at a low bit rate and an enhancement layer that provides successive refinement. The enhancement layer is encoded bit plane by bit plane from the most significant bit plane to the least significant one to achieve FGS. Each bit plane within an image block is represented by (R_i, EOP_i) symbols, where R_i is the run of zeros before the ith "1," and EOP_i is an end-of-plane flag indicating whether the current "1" is the last bit with value 1 in the current bit plane. The run-EOP symbols are encoded using variable-length codes and interleaved with sign bits.

To provide access control to the FGS encoded enhancement layers, the index-based encryption discussed in Section 17.2.2.2 can be applied to each run-EOP symbol, and the overhead can be analyzed using Eq. (17.3). Alternatively, we can shuffle each bit plane according to a set of cryptographically secure shuffle tables. Figure 17.6 illustrates this IBS approach. We perform random shuffling on each bit plane of n bits and the shuffled bit plane will then be encoded using the run-EOP approach. For example, the first unencrypted bit plane [1] in Fig. 17.6 "1 0 1 0 0 0 0 0 0 0" has $n_1 = 2$ bits of value "1" out of a total of $n = 10$ bits, which will lead to $\binom{n}{n_1} = 45$ different permutated patterns. In addition to bit-plane shuffling, the sign bit s_i of each coefficient is randomly flipped according to a pseudorandom bit b_i from a one-time pad, that is, the sign remains the same when $b_i = 0$ and changes when $b_i = 1$.

In general, the bit-rate overhead for each bit plane by the IBS approach depends on the overhead of each run-length symbol and the number of symbols. Proper parameter selections can keep the overhead at a moderate level. Details on the overhead analysis can be found in [9].

[1] We use "the first bit plane" to denote the MSB bit plane throughout this chapter.

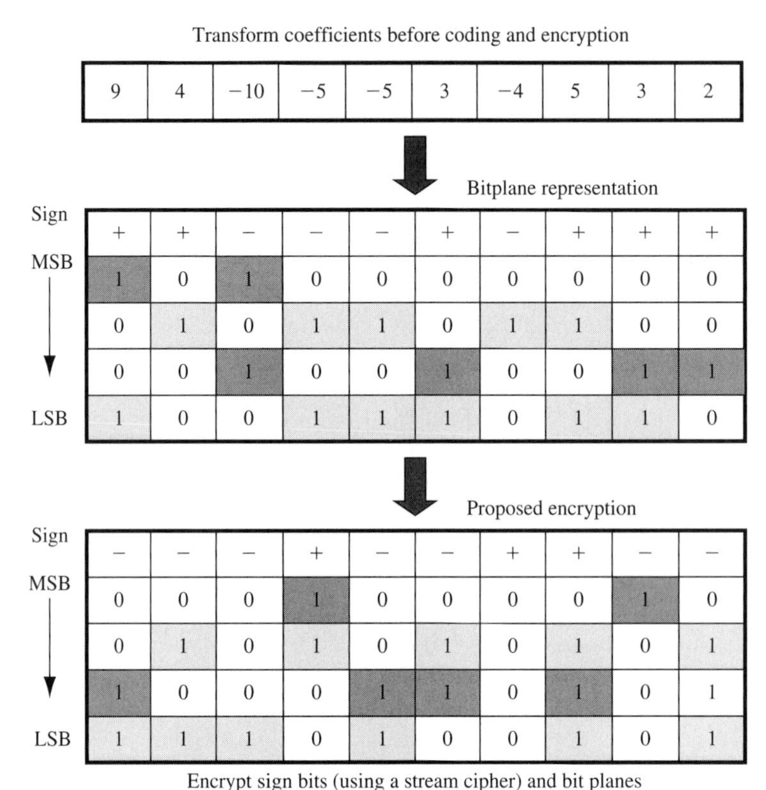

FIGURE 17.6

Illustration of intra bit plane shuffling.

17.2.3 Security Evaluation of Media Encryption

One simple and common way to evaluate the security is to count the number of brute-force trials to break the encryption, which is proportional to min{| clear-text space |, | key space |}, where $|\cdot|$ denotes cardinality. Aside from the brute-force search, there are also notions of security that quantify the security of a system in terms of the amount of resources needed to break it [32, 33]. However, the traditional all-or-nothing situation in generic data security is not always appropriate for measuring the security of multimedia encryption [20, 21, 23]. Beyond the exact recovery from ciphertext, it is also important to ensure partial information that is perceptually intelligible is not leaked out from the ciphertext. Many forms of multimedia data, such as image, video, audio, and speech, contain inherent spatial and/or temporal correlation. The encrypted multimedia content may be approximately recovered based on the syntax, context, and the statistical information known as a priori. This is possible even when the encrypted part is provably secure according to some generic security notions. For example, in MPEG-4 video

(a)

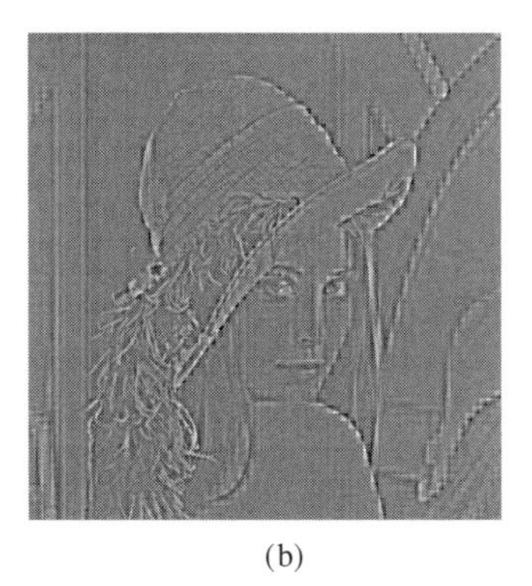

(b)

FIGURE 17.7

(a) Encryption results on the Lena image based on generalized index mapping of DC differential residues; (b) approximated Lena image by setting all DC coefficients to 0.

encryption [19], when MV fields are encrypted and cannot be accurately recovered, a default value 0 can be assigned to all MV fields. This approximation sometimes results in a recovered frame with fairly good quality for frames having a limited amount of motion. Additionally, the statistical information, neighborhood patterns, and/or smoothness criterion can help estimate an unknown area in an image [34] and automatically reorder shuffled image blocks [29]. Although these estimations may not be exact, they can reveal perceptually meaningful information once the estimated signal is rendered or visualized.

As an example, we show in Fig. 17.7(a) the experimental result of encrypting the DC prediction residue of the Lena image. Although the directly rendered version of the encrypted Lena image is highly obscured, an attacker can obtain edge information by setting the DCs to a constant and observing the resulting image shown in Fig. 17.7(b). We can see that the edge and contour of the approximated Lena image is clearly comprehensible, which suggests that it is necessary to encrypt other components in addition to DCs.

Because the value of multimedia content is closely tied with its perceptual quality, such value composition should be reflected in access control and confidentiality protection. This prompts the following framework to evaluate the security of multimedia encryption: after encryption, the encrypted media is first undergone some approximation attacks. We then use perceptual similarity scores to measure the amount of information leakage about the original media data through the approximated media. The results can indicate the security of the encryption scheme against the approximation attacks.

Studies on human visual system have shown that the optical characteristic of eyes can be represented by a low-pass filter [35], and that human eyes can extract coarse visual information in images and videos despite a small amount of noise and geometric distortion. The important information extracted by human visual system includes spatial-luminance information and edge and contour information [36]. Motivated by these studies and the recent work on automated image quality measurement [37–39], we design a luminance similarity score and an edge similarity score (ESS) to reflect the way that human perceives visual information. These scores can quantitatively measure the perception-oriented distance between the clear-text copy of multimedia and the attacker's recovered copy from the encrypted media.

17.2.3.1 *Luminance Similarity Score*

To capture the coarse luminance information, we introduce a block-based luminance similarity score (LSS). We assume that two given images are preprocessed to be aligned and scaled to the same size. These two images are first divided into blocks in the same way, using 8 × 8 or 16 × 16 nonoverlapping blocks. Then the average luminance values of the i-th block from both images, y_{1i} and y_{2i}, are calculated. We define the LSS as

$$\text{LSS} \triangleq \frac{1}{N}\sum_{i=1}^{N} f(y_{1i}, y_{2i}). \tag{17.5}$$

Here, the function $f(x_1, x_2)$ for each pair of average luminance values is defined as

$$f(x_1,x_2) \triangleq \begin{cases} 1 & \text{if } |x_1 - x_2| < \dfrac{\beta}{2}; \\ -\alpha \, \text{round}\left(\dfrac{|x_1 - x_2|}{\beta}\right) & \text{otherwise,} \end{cases}$$

where the parameters α and β control the sensitivity of the score. Because the images under comparison may be corrupted by noise during transmission or be misaligned by a few pixels, such noise and perturbation should be suppressed during similarity estimation. The resistance to minor perturbation and noise can be achieved by appropriately choosing the scaling factor α and the quantization parameter β . In our experiments, α and β are set to 0.1 and 3, respectively. A negative LSS value indicates substantial dissimilarity in the luminance between the two images.

17.2.3.2 *Edge Similarity Score*

The ESS measures the degree of resemblance of the edge and contour information between two images. After the images are partitioned into blocks in the same way as in the LSS evaluation, edge direction classification is performed for each block by extracting the dominant edge direction and quantizing it into one of the eight representative directions that are equally spaced by 22.5°. We use indices 1–8 to represent these eight directions, and use index 0 to represent a block without edge. Denoting e_{1i} and e_{2i} as the edge direction indices for the i-th block in two images, respectively, the ESS for a total of N image blocks is computed as follows:

$$\text{ESS} \triangleq \frac{\sum_{i=1}^{N} w(e_{1i}, e_{2i})}{\sum_{i=1}^{N} c(e_{1i}, e_{2i})}. \tag{17.6}$$

Here, $w(e_1, e_2)$ is a weighting function defined as

$$w(e_1,e_2) \triangleq \begin{cases} 0 & \text{if } e_1 = 0 \text{ or } e_2 = 0, \\ |\cos(\phi(e_1) - \phi(e_2))| & \text{otherwise,} \end{cases}$$

where $\phi(e)$ is the representative edge angle for an index e, and $c(e_1, e_2)$ an indicator function defined as

$$c(e_1, e_2) \triangleq \begin{cases} 0 & \text{if } e_1 = e_2 = 0; \\ 1 & \text{otherwise.} \end{cases}$$

The score ranges from 0 to 1, where 0 indicates that the edge information of the two images is highly dissimilar and 1 indicates a match between the edges in the two images. A special case arises when the denominator in Eq. (17.6) is zero, which happens when both input images are "blank" without any edge. We assign an ESS score of 0.5 to this special case. In our experiments, the input images are partitioned into nonoverlapping 8×8 blocks, and the Sobel operator is used for edge detection [35]. The dominant edge direction of a block is determined by a majority voting inside the block according to the number of pixels associated with each representative direction by the Sobel operator.

17.2.3.3 *Evaluation Framework of Multimedia-Oriented Security*

When evaluating the image similarity, we first calculate the ESS and LSS scores between the attacked/approximated image and the original image, and then compare the scores with two predetermined thresholds, ESS_{th} and LSS_{th}, respectively. An encrypted image/video is said to pass the similarity test against a certain attack if both the ESS and the LSS are lower than the thresholds. In our experiments, we set ESS_{th} to 0.5 and LSS_{th} to 0.

These similarity scores exhibit a tradeoff between capturing coarse semantic information and texture details of images. The sensitivity of the two similarity scores to image details can be controlled by the size of the block partition. When small blocks (e.g., 4×4) are used, a small amount of noise or geometric distortion can result in scores that indicates dissimilarity for two similar images. We refer to this type of misclassification as a miss. From security point of view, such a miss would lead to a security breach. When blocks with larger size (e.g., 32×32) are used, the scores tend to identify some images as similar when their details are different, leading to false alarms. As preventing information leak is the main concern in many access control applications, usually there are relatively stringent requirements on keeping misses as low as possible, while allowing to tolerate a moderate amount of false alarms. Given these considerations, we suggest using 8×8 or 16×16 blocks in block partition.

17.2.4 Video Encryption System Design

In this subsection, we present an example system for video encryption that employs the building blocks discussed earlier. Using this encryption system, several example configurations are presented and the encrypted videos are compared in terms of the security against brute-force and approximation attack, the friendliness to delegate processing, and the compression overhead.

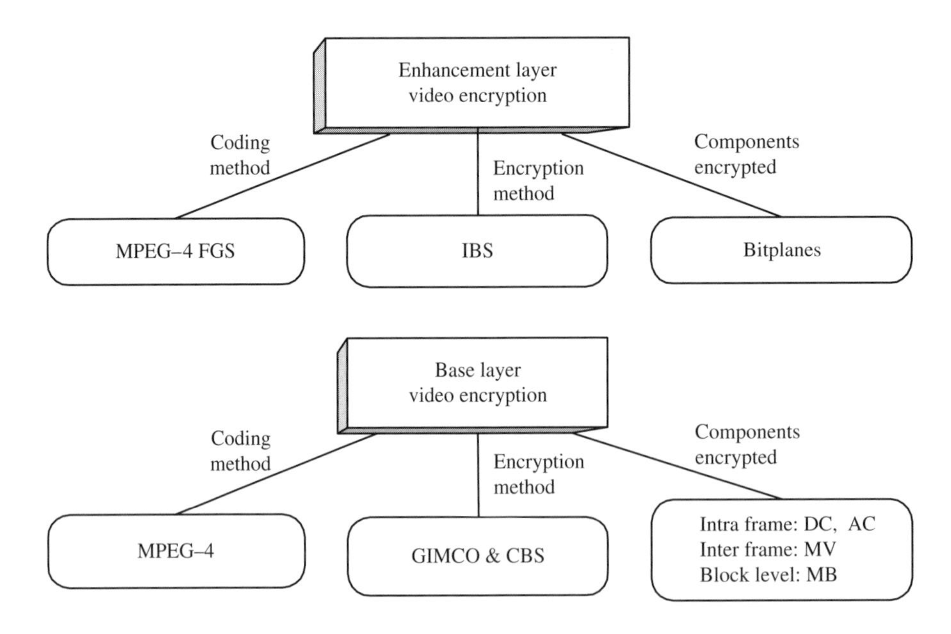

FIGURE 17.8

Video encryption system description. The encryption system is divided into two layers and for each layer candidate encryption components and methods are listed.

17.2.4.1 System Setup

Designed with scalable video in mind, the video encryption system has two layers as shown in Fig. 17.8. The base-layer video is coded with the MPEG-4 standard and the enhancement layer with the MPEG-4 FGS standard. The size of the group of pictures (GOP) is set to 15 and all predicted frames are set to P frames. For each layer, we provide candidate encryption methods and components to be encrypted.

The encryption operations discussed in early sections, namely, the generalized index mapping with controlled overhead (GIMCO), the spatial coded block shuffling (CBS), and the IBS, lend themselves naturally as building blocks for this system. For example, the index mapping encryption can be applied to intrablock DC/AC coefficients and interblock MV residues, the IBS encryption can be applied to FGS bit planes, and the CBS can be applied to macroblock (MB) coding units. We use AES [40] with a 128-bit key to generate the pseudorandom numbers for all encryptions.

17.2.4.2 Base-Layer Video Encryption

Four video clips, from fast motion to slow motion, are used in our experiment. They are the *Football*, the *Coastguard*, the *Foreman*, and the *Grandma*, and each is 40 frames long. Encryption is performed under the settings shown in the left column of Table 17.1, where the encryption of DC, AC, and/or MV is based on the proposed generalized index mapping. The DC and AC encryption ranges are chosen as [−63,64] and [−32,32] with

TABLE 17.1 Encryption and attack settings for security analysis.

	Encryption system settings		Approximation attack settings
(E1)	encrypting intrablock DC residue by index mapping;	(A1)	set all intrablock DC coefficients to 0;
(E2)	encrypting inter block MV residue in the first two P frames of a GOP, and all intra block DC residues;	(A2)	set all intra block DC coefficients to 0 and set the encrypted MV values to 0;
(E3)	encrypting all the components in E2, plus the first two (in the zig-zag scan order) non-zero AC coefficients of intra block;	(A3)	including all the approximations in A2, plus set the encrypted AC coefficients to 0;
(E4)–(E6)	correspond to E1–E3 plus macro-block shuffling in the compressed bit stream, respectively;	(A4)–(A6)	the same as A1–A3, respectively.

set partitioning, respectively; the MV encryption is applied to the first two P frames in each GOP. Additionally, macroblock shuffling in the compressed bit stream is applied to every frame. Each encrypted video complies with the syntax prescribed in the MPEG-4 standard. We also consider approximation attacks to these settings to emulate an adversary's action, and list them in the right column of Table 17.1. The security for these encryption configurations are discussed in the following section.

17.2.4.2.1 Security Against Exact Recovery by Exhaustive Search

To accurately recover an original I frame from the encrypted one, an attacker needs to recover all the DC coefficients. For P frames, recovering the values of MVs is also necessary. In the above configuration, each DC coefficient and MV component has 6 bits encrypted. Each I frame in QCIF format has 2376 equivalent DC bits encrypted and the encrypted MVs in a P frame is equivalent to 1188 bits. Because a 128-bit key is used, the security against exact recovery by exhaustive search is determined by the cryptographic primitive with a 128-bit key.

17.2.4.2.2 Visual Security Against Approximation Recovery

To evaluate the visual security for our encryption system, we first encrypt the test video and then apply approximation attacks to the encrypted video. After that we obtain the ESS and LSS scores of the approximated video and compare them with the thresholds, $ESS_{th} = 0.5$ and $LSS_{th} = 0$. An encryption is considered not secure enough when either score is above the corresponding threshold.

Figure 17.9 shows the video encryption results under different settings for the CoastGuard clip. The results presented are for Y components as they carry most of the information about the video. Visual examination suggests that encrypting DC alone

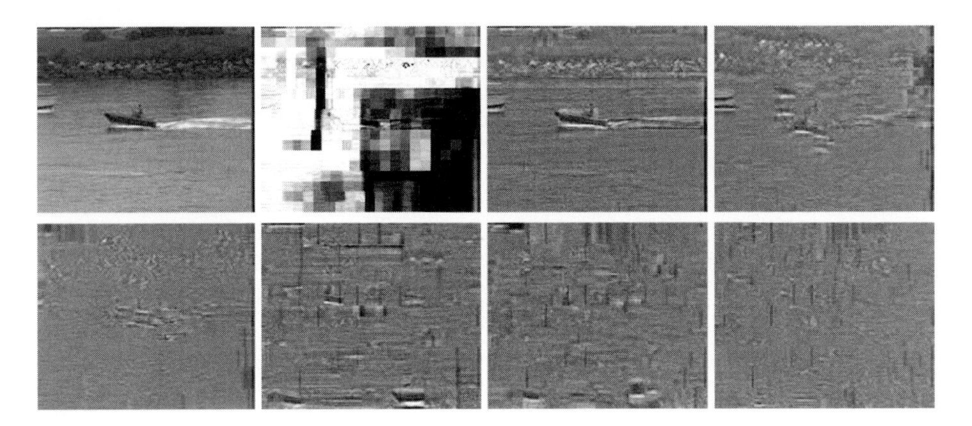

FIGURE 17.9

Encryption results for Coast-guard. The encryption–approximation settings are as follows: (top row, left to right) unencrypted, E1, E1-A1, E2-A2; (bottom row, left to right) E3-A3, E4-A4, E5-A5, E6-A6.

TABLE 17.2 Perception based security measures for video encryption.

Encryption–approximation settings	Football		Grandma		Coastguard		Foreman	
	ESS	LSS	ESS	LSS	ESS	LSS	ESS	LSS
E1-A1	0.70	−0.78	0.64	−2.13	0.79	−1.18	0.71	−1.42
E2-A2	0.53	−0.85	0.46	−2.13	0.43	−1.19	0.43	−1.48
E3-A3	0.53	−0.86	0.30	−2.13	0.40	−1.20	0.40	−1.48
E4-A4	0.12	−0.93	0.05	−2.13	0.07	−1.20	0.07	−1.47
E5-A5	0.13	−0.92	0.05	−2.13	0.06	−1.21	0.06	−1.45
E6-A6	0.12	−0.92	0.04	−2.13	0.04	−1.20	0.05	−1.47

still leaks contour information after approximation attacks, whereas extending encryption to MV and/or some ACs helps diffuse the contour to reduce the information leakage. Furthermore, shuffling self-contained coding unit such as MBs, coupled with the above value encryption, can scramble the content to a completely unintelligible level. Table 17.2 lists the average ESS and LSS (averaged over a total of 40 frames) of the videos after approximation recovery. From the average LSS and ESS scores we can see that, when CBS is not used as an encryption tool, only the LSS score is below its security threshold of 0, and the ESS score is around or above its threshold of 0.5. This confirms that the encryption leaks out shape information and is not secure enough, which we have already observed from Fig. 17.9. Once the CBS is incorporated in the encryption, the ESS and LSS indicate that the encryption is secure against approximations. These results concur with the visual examination.

To examine the detailed ESS scores, we plot the frame-by-frame ESS score of Coastguard under different encryption-attack settings in Fig. 17.10. The top curve is from the

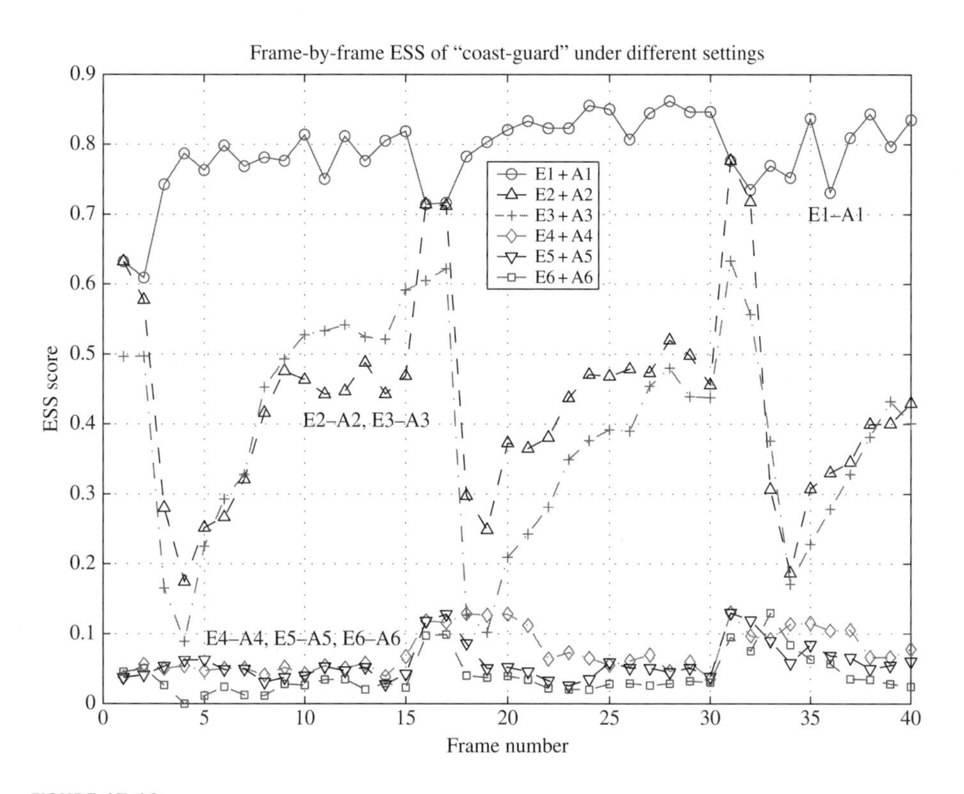

FIGURE 17.10

Frame-by-frame ESS of Coastguard video sequence under different settings. The corresponding settings are listed in Table 17.1.

attacked video with DC encrypted only, which confirms that encrypting DC alone still leaves some contour information unprotected. The two middle curves are the results involving MV encryption for interblocks and AC encryption for intrablocks, where the ESS scores are low at the beginning of a GOP and increase substantially toward the end of the GOP. This is because as it approaches the end of a GOP, motion compensation becomes less effective and the compensation residue provides a significant amount of edge information. Such observation suggests that if we can only afford the bit-rate overhead to encrypt two P frames in a GOP, the two encrypted P frames should be interleaved, such as choosing the 1st and the 8th P frames in a GOP of 15 frames. On the other hand, by incorporating the shuffling of MB coding units, the resulting ESS measurements are consistently around 0.1 or lower.

17.2.4.2.3 Relative Overhead

Table 17.3 lists the compression overhead for four videos under each encryption settings. We can see that the overhead is low for high-complexity, fast-motion video such as the Football and the Foreman, and relatively high for low-complexity, slow motion

TABLE 17.3 Relative compression overhead of the encrypted videos.

	Football (%)	Foreman (%)	Coastguard (%)	Grandma (%)
E1 and E4	1.29	1.75	3.15	6.96
E2 and E5	3.88	6.41	8.74	11.11
E3 and E6	6.47	9.62	11.54	24.61

video such as the Grandma and the Coastguard. As we go from the setting E1 to E3, more components are encrypted and thus the overhead increases. We also see that the CBS approach does not introduce overhead, as shown in setting E4 to E6. Overall, the overhead of 4–11% by the E1, E2, E4, and E5 is comparable to that of a direct adaptation of generic encryption to multimedia. Considering both security and compression overhead, we see that the E5 setting provides a very good tradeoff. This setting is a combination of block shuffling and selective value encryption via generalized index mapping.

17.2.4.3 *Protecting FGS Enhancement Layer Video*

We use 10 frames from the Foreman video sequence to demonstrate the protection of the enhancement data while preserving the FGS characteristics from the source coding. The proposed IBS encryption is applied within each 8×8 block, and the sign bit of each coefficient is encrypted using a stream cipher. To allow for a better visual examination of the protection effects on the enhancement data, we combine the encrypted FGS bit planes with a clear-text base layer.

For most natural images, the coded DCT coefficients have decreasing dynamic range versus the frequency. As such, we emulate an approximation attack, whereby for each significant bit plane, all the "1"s of the bit plane is put to the lowest possible frequency bins. A total of six encryption-attack settings are used, namely:

(a) to shuffle the first bit plane with clear-text base layer,

(b) to approximate the bit plane of (a) with the correct signs,

(c) to approximate the bit plane of (a) with random signs,

(d) to shuffle the first two bit planes,

(e) to approximate the bit planes of (d) with the correct signs,

(f) to approximate the bit planes of (d) with random signs.

Table 17.4 lists the corresponding average PSNR, LSS, and ESS of the videos under the six encryption-attack settings. From the table, we can see the approximation recovery can only reduce a little luminance error in terms of LSS and PSNR in the approximated video compared to the encrypted video, and the edge similarity in terms of ESS remains imperfect and has little improvement after attack. So without knowing the decryption key to the enhancement layer, an attacker cannot obtain a more refined video than

TABLE 17.4 Intra bitplane shuffling and approximation attack.

	(a)	(b)	(c)	(d)	(e)	(f)
PSNR (dB)	28.59	28.76	28.74	27.39	27.87	27.50
LSS	0.28	0.34	0.34	0.28	0.34	0.29
ESS	0.85	0.85	0.85	0.85	0.85	0.85

the base-layer video using approximations from the encrypted enhancement layer. This demonstrates that the proposed method can encrypt the premium quality version of the content in a FGS compatible way, and protect them at specific strength and with separate keys as needed.

17.3 VIDEO AUTHENTICATION

Authentication, by the definition from cryptography [4], is a process determined by the authorized receivers, and perhaps the arbiters, that a particular data is most probably sent by the authorized transmitter and haven't subsequently been altered or substituted for. Authentication is usually associated with data integrity and nonrepudiation (i.e., source identification) because these issues are very often related to each other: data that have been altered effectively should have a new source; and if the source cannot be determined, then the question of alteration cannot be settled (without reference to the original source).

Digital signature techniques are the natural tools for addressing authentication issues. To date, many signature-based authentication technologies particularly public key infrastructure (PKI) have been incorporated into the international standards (e.g., X.509) and state-of-arts network protocols (e.g., secure socket layer – SSL), for the purposes of data integrity and source identification. It would be great if we could extend digital signature schemes from data applications to multimedia applications. First the system security of a digital signature scheme, which is a very important issue in data authentication, has been well studied in cryptography. Second the current security protocols in network, which work for data exchange/streaming, do not need to be redesigned for multimedia exchange/streaming. Therefore, in this section we focus on the discussion of authenticating multimedia content based on digital signature schemes, though digital watermarking techniques might be employed in some specific schemes.

17.3.1 Background

The basic diagram of a digital signature scheme is shown in Fig. 17.11. Given a video with arbitrary size, applying cryptographic hashing on the video to obtain its MAC (message authentication code) that is usually hundreds bits in length (e.g., 128 bits with MD5 algorithm and 160 bits with SHA-1 algorithm [4]), signing on the MAC to generate the crypto signature of the video by using the sender's private key, and sending the video

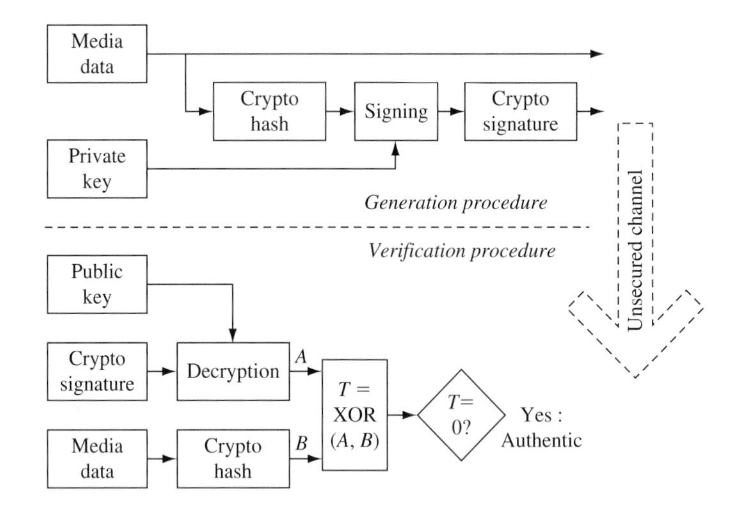

FIGURE 17.11

Block diagram of cryptographic signature schemes.

together with the signature to the recipient. At the receiver site, the authenticity of the video is verified through the following steps: applying the same hash function, as used at the sending site, to obtain a MAC A. Decrypting the received signature using the sender's public key to obtain another MAC B. Comparing A and B bit by bit: the received video will be deemed unauthentic if any discrepancies, even one bit difference, occur.

However, in real applications of video streaming over the networks, the video to be sent is often required to be transcoded to adapt to various channel capacities (e.g., network bandwidth) as well as terminal capacities (e.g., computing and display power) [41]. Note that we essentially regard transcoding as the process of converting a compressed bit stream into lower rates. Such transcoding poses new challenges on authentication due to (1) the distortions introduced during video transcoding and network; (2) flexibilities of various video transcoding methods like dropping video frames or dropping video packets.

In this section, we first present a content-level video authentication system, which is robust to frame resizing, frame dropping, requantization, and their combinations. We then present a stream-level video authentication scheme, which is robust to packet loss due to either bandwidth constraints or network reliability. All the schemes achieve an end-to-end authentication that is independent of specific transcoding design and balances the system performance in an optimal way.

17.3.2 Content Level Authentication

A typical application scenario for video streaming and transcoding is illustrated in Fig. 17.12. Considering the variations in the transmission channels and the terminals, we would argue that a robust and secure authentication scheme for video transcoding should satisfy the following prerequisites.

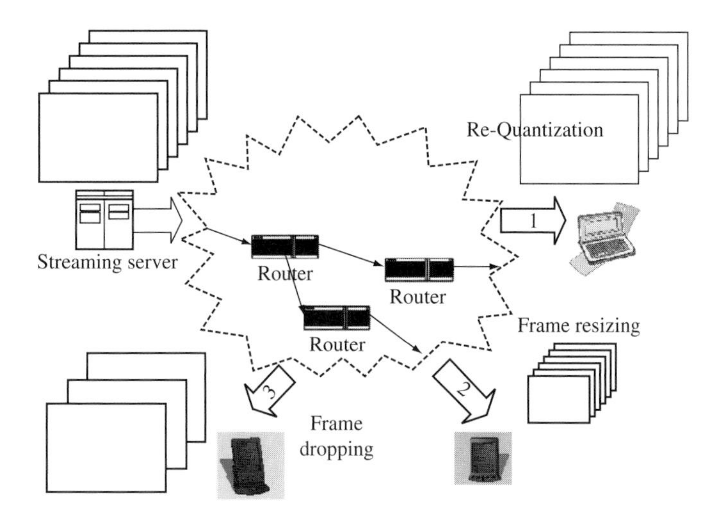

FIGURE 17.12

Typical transcoding methods for scalable video streaming.

- Robustness: The authentication scheme must be robust to the video transcoding approaches, namely video requantization, frame resizing, frame dropping, and their combinations.
- Security: The authentication scheme must be secure enough to prevent the possible malicious attacks such as frame insertion/removal/replacement, or some in-frame modifications (e.g., content copy-paste) which intend to change the meaning of the video content.
- Efficiency: The authentication scheme should be very efficient. This is especially important for video applications because of its computation complexity and transmission cost.
- Independency: The authentication scheme should be independent of the specific network infrastructure and protocol used for video streaming. For example, considering that the coded video could be transcoded either at the server site before streaming or at some intermediate network nodes (e.g., routers) during streaming, this would make end-to-end authentication a requirement for the scheme to achieve.

System diagram of the proposed system (signing part) is shown in Fig. 17.13. The signing operations are performed in the DCT domain to reduce system computation complexity. With reference to Fig. 17.13, three inputs for video signing are as follows: the video sender's private key, the authentication strength, and possible transcoding approaches such as frame dropping, resizing, requantization, or a combination of them. Here the authentication strength means protecting the video content to a certain degree (i.e., the video will not be deemed as authentic if it is transcoded beyond this degree).

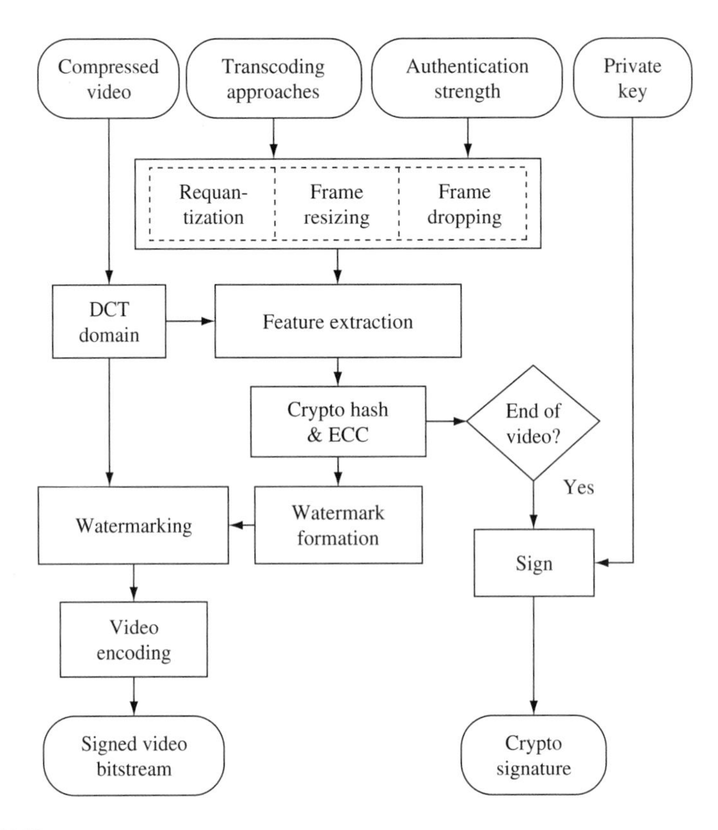

FIGURE 17.13

System diagram of the authentication solution robust to video transcoding.

Here we mainly use the quantization step size to control the authentication strength [42]. Based on the given transcoding approaches and the authentication strength, we extract the invariant features from DCT coefficients. Such frame-based features are cryptographically hashed, concatenated, and then coded by a forward error correction (FEC) scheme and embedded back into the DCT domain as a watermark. Note that the selected watermarking scheme is also required to be robust to the predefined transcoding approaches as well as the authentication strength. The watermarked video content is entropy coded again to form the signed video bit stream. In addition, the cryptographic hashing is recursively operated frame by frame, till the end of the video. The video sender's private key is used to sign on the final hash value to generate the cryptographic signature of this video. The signature, together with the watermarked video, is sent to the recipient to prove the authenticity of the video. At the receiver site, the verification of video authenticity is actually an inverse process of video signing using the video sender's public key.

The video signing and verification are both performed in the compressed domain to reduce system computation, as shown in Fig. 17.14. The input MPEG bit stream is

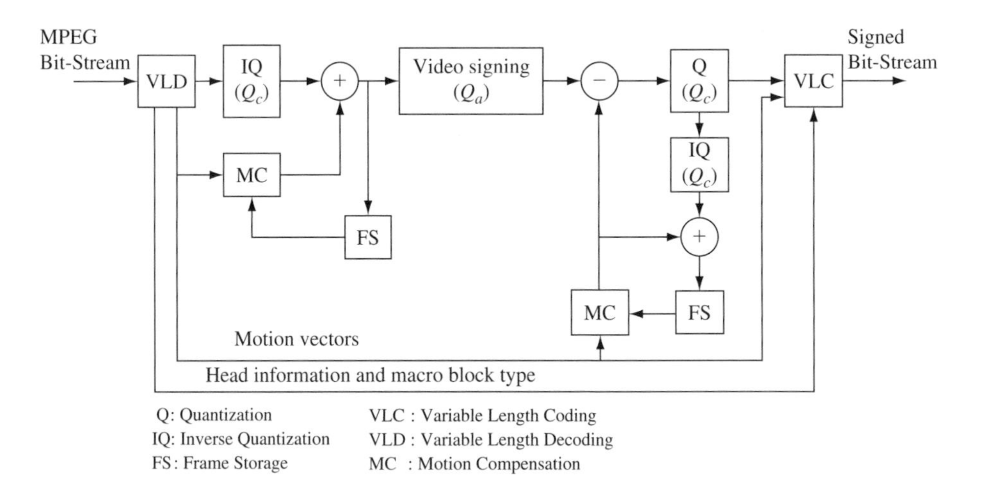

FIGURE 17.14

Illustration of the video signing process.

partially decoded to obtain DCT coefficients. The video signing operation (i.e., feature extraction, watermarking and cryptographic hashing, etc.) is then performed on the DCT coefficients, frame by frame. Finally the signed video is re-encoded using the MVs from the original MPEG bit stream. Note that during the whole process we keep the MVs unchanged to avoid motion reestimation, although such a process may slightly degrade the quality of the newly coded video, because the watermarked DCT coefficients are different from the original DCT coefficients used for estimating the MVs. However, motion estimation is very time-consuming. Hence, skipping it (Fig. 17.14) will greatly reduce system computation. Note that the scheme shown in Fig. 17.14 also has the function on drift compensation. It means that the distortion caused by watermarking will be "compensated" by re-encoding the B and P frames.

Considering the fact that different transcoding approaches affect the robustness of the extracted feature and watermarking in different ways, we shall address them separately. If the video transcoding is a combination of the above-mentioned three approaches, the selected features should be an intersection of each extracted feature set. To describe the proposed authentication system in a clearer way, we now give an example on how to address the robustness to frame-dropping-based transcoding. More detailed solutions can be found in [43].

One possible transcoding for bit-rate reduction is frame dropping. For instance, the original video sequence "Salesman" encoded at 64 kb/s with 30 frames per second (fps) can be transcoded to a new version of 32 kb/s by dropping to 10 fps. To have low computation, low-memory usage, and high visual quality, the state-of-the-art frame-dropping-based video transcoding is usually performed in compressed domain and the frames are dropped in a flexible way. For instance, an original video with the frames like $I_1 B_2 B_3 P_4 B_5 B_6 P_7 B_8 B_9 \;\; P_{10} B_{11} B_{12} I_{13}$ could be transcoded to a new one whose frames are $I_1B_2P_4B_5P_7B_8P_{10}B_{11}I_{13}$ (i.e., linear dropping) or $I_1B_2B_3P_4B_5B_6P_7B_8P_{10}I_{13}$ (i.e., nonlinear dropping). Therefore, the proposed robust video authentication solution should meet these transcoding requirements. It means, if the frames of a video are received incomplete only because of transcoding, we would still like to be able to authenticate this video as if it is the same as the original one whose frames were not lost. This defines resistance to frame loss in a strong sense: a video frame is either dropped or authenticable.

We solved this problem based on the concept of FEC. The idea is to pay extra transmission cost (i.e., multiple times of transmitted crypto hashes) for authentication robustness. We further resolve this extra payload problem by embedding those hash values generated from other frames into current frame using watermarking. If some frames are dropped during transcoding, we can obtain their corresponding crypto hashes from other frames and the whole authentication process for other frames can still be continuously executed. The process flow is illustrated in Fig. 17.15. The basic idea is to append several MACs from other video frames to the current frame: if the current frame is lost, its MAC can be correctly obtained from other frames so that signature verification on the whole video can still be carried out. Obviously, these solutions will result in an extra transmission cost that depends on the rate of packet loss. Such extra payload could be resolved by adopting digital watermarking [43].

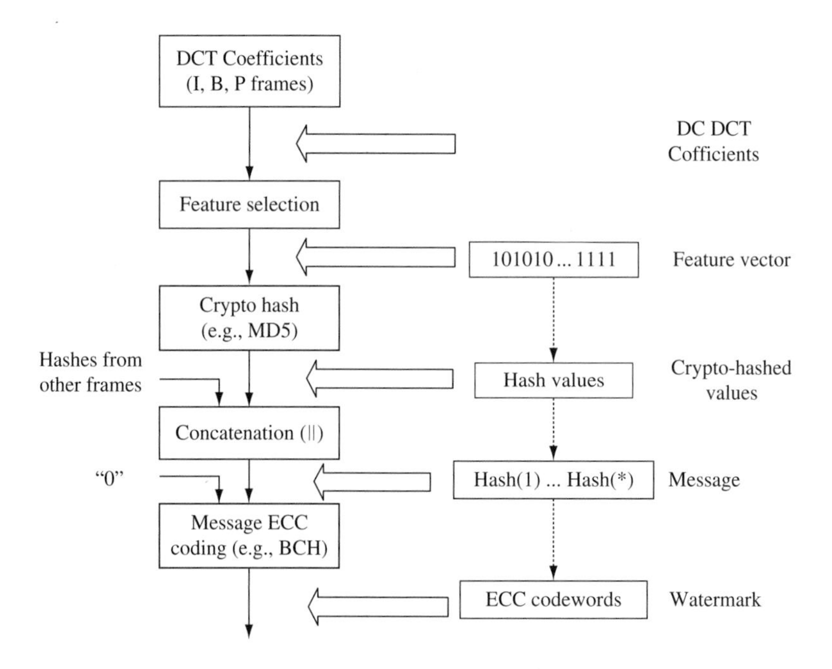

FIGURE 17.15

Authentication robust to frame-dropping.

17.3.3 Stream Level Authentication

The second class of possible solutions to achieve end-to-end media stream authentication is to directly authenticate at the stream or packet level. The system security can be mathematically proven as it is based on conventional data security approaches, though its system robustness is not as strong as content-based authentication. For example, it may only be robust to packet loss, and not to other manipulations.

Many solutions for data stream authentication have been proposed [41]. Those previous approaches assume and treat all packets as if they are of equal importance, which generally is not true for media packets. For example, packets containing P-frame coded video data are typically more important than those containing B-frame coded video data. In other words, a small number of packets are much more important than the rest of the packets. Note that this characteristic is often exploited via unequal error protection to transport media data over lossy networks. Similarly, stream authentication can also utilize this characteristic by trading off authentication redundancy based on packet importance: increasing the redundancy degree for more important packets, so as to increase their verification probability, and reducing it for the less important packets that have a smaller effect on reconstructed visual quality. We believe that this approach can be more practically useful for media applications than conventional authentication approaches which do not account for the varying importance of each media packet. Further, in contrast to generic data stream authentication where verification probability is deemed as the primary performance measure to be optimized, for media stream authentication the media quality of the authenticated media often is a more important metric. Therefore, we believe that media quality is a more important metric for optimization than verification probability.

In this section, we describe authentication-aware R-D optimized streaming for authenticated video. Based on each packet's importance in terms of both video quality and authentication dependencies, this technique computes a packet transmission schedule that minimizes the expected end-to-end distortion of the authenticated video at the receiver subject to a constraint on the average transmission rate. This work was motivated by recent advances in Rate-Distortion Optimized (RaDiO) [44] streaming techniques, which compute a packet transmission policy that minimizes the expected end-to-end distortion at the receiver subject to a constraint on the average transmission rate.

In the following, we assume the case of pre-encoded video to be streamed, for example for video on demand services. Given a compressed video with associated authentication information, the first step is to compute the important quantities associated with each packet. The distortion increment, packet size, and display time are the same as in conventional RaDiO techniques [44]. The overhead size can be computed from the topology of the authentication graph. Second, at every transmission opportunity, the R-D optimization process selects the best packet(s) for transmission based on their parameters. For example, packets with higher importance (distortion increment + authentication importance) and smaller size (packet size + overhead size) are assigned more transmission opportunities. In summary, we formulate a rate-distortion-authentication optimization problem to minimize the expected distortion of the authenticated video at the receiver, subject to the constraint on average transmission rate. Please recall that unlike conventional RaDiO where all packets received before their associated playout deadline contribute to improve the media quality, in our case only the received and authenticated

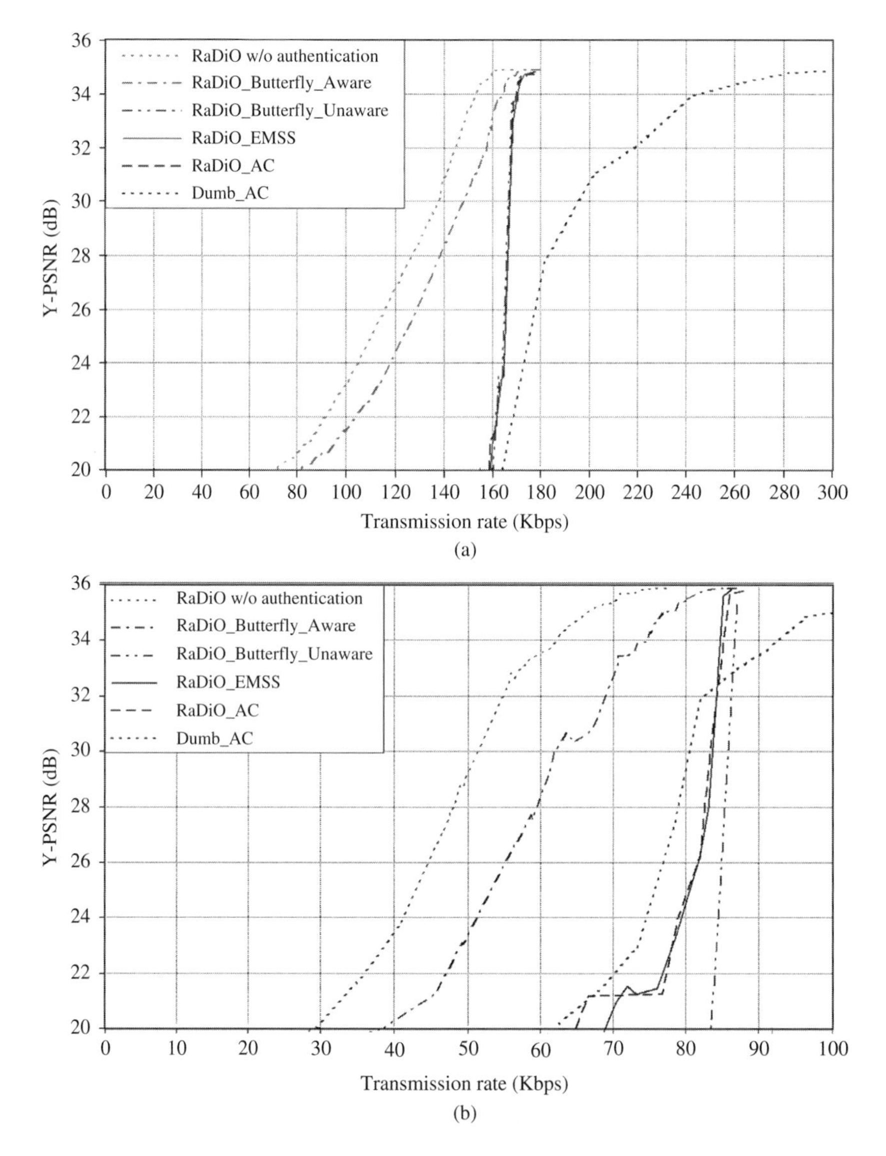

FIGURE 17.16

R-D curves for the following systems when streaming over a network with 3% packet loss and time-varying delay.

packets contribute, that is, a packet which is received but not authenticated is equivalent to being lost.

Further information about the algorithm is given in [45], here we highlight the algorithms performance via simulation results using the latest video compression standard H.264 [46]. In Fig. 17.16 we plot the R-D performance with 3% packet loss and time-varying delay. RaDiO implements the original RaDiO without authentication, whose performance is used as the upper bound for all other systems. Dumb_AC implements a straightforward transmission of video packets protected with Augmented Chain that is claimed optimal for generic data streaming [47]. Authentication-aware RaDiO streaming, incorporating joint optimization of RaDiO and authentication, and using Butterfly dependency graph for authentication is examined in RaDiO_Butterfly_Aware [48]. It is used to illustrate the performance achievable by an authentication-aware RaDiO technique. RaDiO_Butterfly_Unaware (i.e., no joint optimization between RaDiO and authentication) implements authentication-unaware RaDiO with Butterfly Authentication. It is the same as RaDiO_Butterfly_Aware except that it uses authentication-unaware RaDiO, and therefore the gap in performance between these two can be used to estimate the gain of "authentication awareness." RaDiO_EMSS and RaDiO_AC implement authentication-unaware RaDiO with EMSS [49] and Augmented Chain, respectively.

RaDiO_Butterfly_Aware outperforms all schemes, because it computes the transmission policy based on both packets' distortion increments and authentication importance. At low bandwidths, the authentication-unaware RaDiO fails as its R-D curve drops quickly to unacceptable levels. Nevertheless, at the same low bandwidth the proposed authentication-aware RaDiO provides an R-D curve that drops gracefully in parallel with the upper bound, given by RaDiO for unauthenticated video. However, we still notice that there is a performance gap between RaDiO and RaDiO_Butterfly_Aware (which is larger than the 8 kb/s rate for authentication overhead) and how to reduce the gap is an issue for future investigation.

17.4 VIDEO FINGERPRINTING FOR TRAITOR TRACING

With the advances in broadband communication and compression technologies, an increasing amount of video is being shared among groups of users through the Internet and other broadband channels. In the meantime, piracy becomes increasingly rampant as users can easily duplicate and redistribute the received video to a large audience. Digital fingerprinting is an emerging technology to protect multimedia content from such unauthorized dissemination, whereby a unique ID representing each user, called digital fingerprint, is embedded in his/her copy [50, 51]. When a copy is leaked, the embedded fingerprint can help trace back to the source of the leak. Adversaries may apply various attacks to remove the fingerprints before redistribution. *Collusion* is a powerful multiuser attack, where a group of users combine their copies of the same content to generate a new version with fingerprint attenuated or removed. In addition to resistance against attacks, three aspects of system efficiency need to be considered when designing an anticollusion fingerprinting system, namely, the efficiency in constructing, detecting and distributing fingerprinted signals. Construction efficiency concerns the computational complexity involved during the generation of fingerprinted content; detection efficiency is related to the detection computational complexity; and the distribution efficiency refers to the amount of bandwidth consumed during the transmission of all the fingerprinted signals through cable or wireless network. Given the high data volume associated with video data, these efficiency issues are particularly important to protect video distribution to a large number of potential users.

17.4.1 The Background

A multimedia fingerprinting system for trace and track purposes builds on top of robust data embedding methods that are capable of withstanding adversaries' attacks to remove the embedded fingerprint. In this aspect, embedded fingerprinting is closely related to digital watermarking, and the fingerprints are sometimes referred to as "forensic watermarks." In contrast to having a single marked copy available to adversaries in most watermarking applications, the presence of multiple distinctly marked copies in most fingerprinting applications prompts the additional challenges of collusion resilient designs to combat collusion attacks and identify colluders. Many techniques have been proposed for embedding information in multimedia signals [52]. Discussions on basic data embedding can be found in the watermarking chapter of the *Essential Guide to Image Processing* [39]. Here we briefly review the spread spectrum additive embedding technique and its role in robustly embedding fingerprint signals into multimedia.

17.4.1.1 Spread Spectrum Embedding and Fingerprinting

Spread spectrum embedding has proven robust against a number of signal processing operations (such as lossy compression, format conversions, and filtering) and attacks [53, 54]. With appropriately chosen features and additional alignment procedures, the spread spectrum watermark can survive moderate geometric distortions, such as rotation, scale, shift, and cropping [55, 56]. Information theoretic studies also suggest that it is nearly capacity optimal when the original host signal is available in detection [57, 58]. The combination of robustness and capacity makes spread spectrum embedding a promising technique for protecting multimedia.

Spread spectrum embedding borrows ideas from spread spectrum modulation [59]. The basic process of spread spectrum embedding consists of the following steps. The first step is to identify and compute proper features that will carry watermark signals. Depending on the application and design requirements, the features can be signal samples, transform coefficients (such as DCT and wavelet coefficients), or other functions of the media content. Next, we generate a watermark signal and apply perceptual models to tune its strength to ensure imperceptibility. Typically, we construct the watermark to resemble weak noise and cover a broad spectrum as well as a large region of the content. Finally, we add the watermark to the feature signal, replace the original feature signal with the watermarked version, and convert it back to the signal domain to obtain a watermarked signal. The detection process for spread spectrum watermarks begins with extracting features from a media signal in question. Then the similarity between the features and a watermark is examined to determine the existence or absence of the watermark in the media signal. A correlation similarity measure is commonly used, often in conjunction with preprocessing (such as whitening) and normalization to achieve reliable detection [52].

A straightforward way of applying spread spectrum watermarking to fingerprinting is to use mutually orthogonal watermarks as fingerprints to identify each user [60, 61]. The orthogonality allows for distinguishing the fingerprints to the maximum extent. The simplicity of encoding and embedding orthogonal fingerprints makes them attractive to applications that involve a small group of users. The orthogonality may be approximated using random number generators to produce independent watermark signals for different users. A second option of using spread spectrum watermarking is to employ code modulation. Code modulation allows fingerprint designers to design more fingerprints for a given fingerprint dimensionality by constructing each user's fingerprint signal as a linear combination of orthogonal noise-like basis signals. For a large number of users, the detection complexity of coded fingerprinting can be much lower than the orthogonal construction that is proportional to the number of users. We will discuss fingerprint construction issues in more detail in the next subsection.

17.4.1.2 Collusion Attacks

Understanding the threats and strategies from adversaries is an important step to build effective fingerprinting systems for content protection. As mentioned earlier, collusion attack is a powerful attacks mounted by multiple users. During a collusion attack, a group of colluders holding differently fingerprinted versions of the same content examine their different copies and attempt to create a new signal that will no longer be tied to any of the colluders. There are several types of collusion attacks [51, 62]. One method is simply to synchronize the fingerprinted copies and take a linear combination of them, which is referred to as linear collusion attack. Another collusion attack, referred to as the copy-and-paste attack or interleaving attack, involves users taking portions of each of their media signals and assemble them together into a new version. Other attacks may employ nonlinear operations, for example, following order statistics such as taking the maximum or median of the values of corresponding components of individual copies. Considering the fairness among colluders each of whom would not want to assume higher risk of being caught than others, each colluder would contribute a similar amount of share, for example, by averaging their signals, leading to the averaging collusion attack.

Research has shown that for orthogonal fingerprinting, interleaving collusion, and many variants of order, statistics-based nonlinear collusion have a similar effect on the detection of fingerprints to collusion by averaging and possibly followed by additive noise [63]; therefore, if the overall distortion level introduced by collusion attacks is the same, similar detection performance can be observed despite the differences in the specific form of collusion. If the fingerprint is constructed in modules such as through coding, the simple collusion model of averaging plus additive noise can also well approximate many collusion attacks for systems where all logical components are spread over the hosting signal in such a way that adversaries can only distort them as a whole and cannot alter them individually [64]. On the other hand, for many code-based systems whereby various code segments of the fingerprints are embedded in different parts of the hosting signal, different collusion attacks may have different levels of effectiveness [65]. We shall take a closer look in later subsections at how to construct collusion resilient fingerprints in this situation.

Additionally, the designer should consider whether or not the original content is available during the detection phase of the fingerprinting application. We will refer to nonblind detection as the process of detecting the embedded watermarks with the assistance of the original content, and blind detection as the process of detecting the embedded watermarks without the knowledge of the original content. Nonblind fingerprint detection requires considerable storage resources and a method for recognizing the content from a database, but the available original unmarked content can serve as a reference to help overcome attacks caused by misalignment and provide higher strength ratio between the fingerprint and the equivalent noise caused by attacks and distortions to lead to a reliable detection decision. Blind detection, on the other hand, allows for distributed detection scenarios or the use of web crawling programs, although the robustness against severe attacks is limited. We shall focus on nonblind detection scenarios in this chapter.

It is worth mentioning that another class of collusion attack, which is sometimes referred to as intracontent collusion, may be mounted against fingerprints by a single user by replacing each segment of the content signal with another, seemingly similar segment from different spatial or temporal regions of the content. This is particularly of concern in the protection of video signals, whereby adjacent frames within a scene appear very similar to each other. We will address this collusion in Section 17.4.4.

17.4.2 Coded Fingerprinting

As mentioned earlier, the potential efficiency in fingerprint construction and detection makes coded fingerprints attractive than the simple orthogonal construction for applications that involve a large number of users and high-bandwidth data such as video. A typical framework for code-based multimedia fingerprinting includes a code layer and a spread spectrum based embedding layer [65], as illustrated in Fig. 17.17. At code layer, each codeword is assigned to one user as his/her fingerprint. For anticollusion purposes, fingerprint code can be constructed based on error correcting code (ECC) to have large minimum distance so that the fingerprint codewords for different users are well separated [66]. To embed a codeword, we first partition the host signal into L nonoverlapped segments, which can be one frame or a group of frames of video, with one segment corresponding to one symbol. For each segment, we generate q mutually orthogonal spread spectrum sequences $\{\mathbf{u}_1, \ldots, \mathbf{u}_q\}$ with equal energy to represent the q possible symbol values in the alphabet. Each user's fingerprint sequence is constructed by concatenating the spreading sequences corresponding to the symbols in his/her codeword. The fingerprint sequence is then perceptually shaped and added to the host signal through spread spectrum embedding [53] to form the final fingerprinted signal.

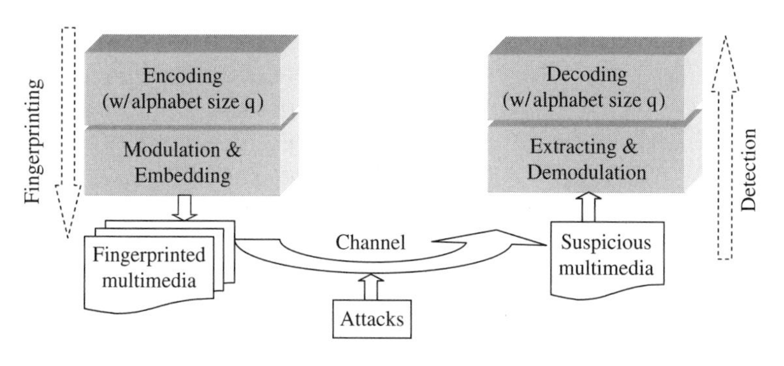

FIGURE 17.17

Framework of coded fingerprinting.

After receiving the fingerprinted copies, users may collaborate and mount collusion attacks. A widely considered collusion model in coded fingerprinting is interleaving collusion [66], whereby each colluder contributes a nonoverlapped set of segments (corresponding to symbols) and these segments are assembled to form a colluded copy. Another major type of collusion as discussed earlier is done in the signal domain; a typical example is the averaging collusion [67], whereby colluders average the corresponding components in their copies to generate a colluded version. The averaging collusion can be modeled as:

$$\mathbf{z} = \frac{1}{K} \sum_{j \in S_c} \mathbf{s_j} + \mathbf{x} + \mathbf{d}, \tag{17.7}$$

where $\mathbf{z}$ is the colluded signal, $\mathbf{x}$ is the host signal, $\mathbf{d}$ represents additional noise, $\mathbf{s}_j$ represents the fingerprint sequence for user j, S_c is the colluder set, and K is the number of colluders. For simplicity in analysis, we assume that the additional noise under both types of collusions follows *i.i.d.* Gaussian distribution. Studies in [63] have shown that a number of other collusions based on order statistics, such as minimum and min-max collusion attacks, can be well approximated by averaging collusion plus additive white Gaussian noise.

At the detector side, our goal is to catch at least one of the colluders with a high probability. For every segment of the test sequence, we correlate the signal with each of the q possible sequences representing the alphabet and determine the symbol as the one with the highest correlation. Under nonblind detection as the original hosting signal is often available to detectors in fingerprinting applications, the detection statistic for the kth segment is

$$T_s(k,i) = \frac{(\mathbf{z}_k - \mathbf{x}_k)^T \mathbf{u}_i}{\sqrt{\|\mathbf{u}_i\|^2}}, \quad i = 1,2,\ldots,q, \tag{17.8}$$

where $\mathbf{z}_k$ and $\mathbf{x}_k$ represent the kth segment of the colluded signal and that of original signal, respectively; and the extracted symbol from kth segment is $\hat{i} = \arg\max_{i=1,2,\ldots,q} T_s(k,i)$. The symbols detected from each segment will form the extracted codeword. We proceed to the code layer and apply a decoding algorithm to identify the colluder whose codeword has the most matched symbols with the extracted symbol sequence.

Alternatively, we can employ a soft detector to correlate the entire test signal directly with every user's fingerprint signal $\mathbf{s}_j$. The user whose fingerprint has the highest correlation with the test signal is identified as the colluder, that is, $\hat{j} = \arg\max_{j=1,2,\ldots,N_u} T_N(j)$, where N_u is the total number of users. Here, the detection statistic $T_N(j)$ is defined as:

$$T_N(j) = \frac{(\mathbf{z}-\mathbf{x})^T \mathbf{s}_j}{\sqrt{\|\mathbf{s}_j\|^2}} \quad j = 1, 2, \ldots, N_u, \tag{17.9}$$

where $\mathbf{z}$ and $\mathbf{x}$ represent the colluded signal and the original signal as a whole, respectively. This matched-filter detector takes advantage of the soft information from the embedding layer and provides better detection performance than the hard detection.

17.4.2.1 Considerations in Code Parameters

A common practice in fingerprint code design treats the symbols contributed from other colluders as errors, and makes the minimum distance between codewords large enough to tolerate the errors. The minimum distance requirement ensures that the best match with a colluded codeword (referred to as the descendant) comes from one of the true colluders. The traceability code for resisting c colluders, or c-TA code in short, is such an example [68]. Under the attack model by interleaving collusion, a c-TA code can be constructed using an ECC if its minimum distance D satisfies [68]

$$D > \left(1 - \frac{1}{c^2}\right) L, \tag{17.10}$$

where L is the code length and c is the colluder number. As the distortions and attacks mounted by adversaries on the fingerprinted multimedia can lead to errors in detecting fingerprint code symbols, these errors and erasures should be accounted for and the above code parameter requirements can be extended accordingly [69].

Overall, the ECC-based fingerprint code calls for an ECC with larger minimum distance to tolerate more colluders. Among ECC constructions, Reed–Solomon codes have the minimum distance that achieves the Singleton bound [70] and is widely used in the existing coded fingerprinting works [66, 68]. We employ a q-ary Reed–Solomon code with code length L to construct a c-TA code. The parameters of the L-tuple Reed–Solomon code for N_u users should satisfy [69]

$$N_u = q^t, \quad \text{and} \quad t = \lceil \frac{L}{c^2} - \frac{c+1}{c^2} L_{FA} \rceil, \tag{17.11}$$

where L_{FA} is an auxiliary parameter indicating the number of symbol errors the code is designed to tolerate.

Owing to a relatively small alphabet size q compared to the number of users N_u as well as one symbol being put in one nonoverlapping media segment, the ECC fingerprinting has the potential to generate and distribute fingerprinted media in an efficient way. For example, for each frame of a video, a total of q copies carrying q different symbol values can be generated beforehand; a fingerprinted copy for any user can then be quickly obtained by assembling appropriate copies of the frames together according to the fingerprint codeword assigned to him/her. The small alphabet size also keeps the computational complexity of fingerprint detection lower than the orthogonal fingerprinting approach. The analysis in [65] has shown that the ECC fingerprinting has computational complexity of $O(qN + N_u L)$ compared with $O(N_u N)$ of orthogonal fingerprinting, where N is the number of host signal samples. In many practical fingerprinting applications for small user groups, we generally have $N_u << N$ to ensure fingerprints be reliably embedded in multimedia data. This suggests that the first term qN of the demodulation process to extract the embedded symbols is dominant, and the overall computational complexity becomes $O(qN)$. Proper coding also allows for accommodating a large number of users, in which case N_u can become comparable to N and we need to consider both terms in the complexity expression.

17.4.2.2 Analysis on Collusion Resistance

Consider an ECC-based fingerprinting system employing a L-tuple code with minimum distance D over q-ary alphabet to represent N_u users. Under the (symbol wise) interleaving collusion, the colluders exploit the fingerprint pattern and contribute segment by segment with each segment carrying one symbol. In contrast, averaging collusion does not rely on the fingerprint pattern and simply takes the average value of each signal component, thus allowing each of the L symbols from every user to leave some traces on the final colluded signal. These two collusion attacks have different effects on colluder detection, and can be analyzed based on signal detection theory [65].

To illustrate the overall collusion resistance, we consider an example system with the parameters chosen as follows. For a system holding N_u users, analysis has shown that a larger L and a smaller t are preferred to obtain better collusion resistance under interleaving and averaging collusion [65]. Because t can only take integer values, we take $t = 2$ to obtain a nontrivial Reed–Solomon code construction. This also determines q since $q^t = N_u$. On the other hand, larger L results in a smaller segment size for a given host signal, which may lead to a higher error probability in symbol detection. Typically a segment size of 1000 can provide reliable symbol detection. With an additional condition that $L \leq q$, we choose L to be a number smaller than but close to q. In our example considering a total of $N_u = 1024$ users and a host signal with $N = 3 \times 10^4$ embeddable components, we choose $L = 30$ and use a Reed–Solomon code with parameters of $q = 32$ and $D = 29$. According to Eq. (17.10), the code level alone can only assure resisting up to five users' interleaving collusion; on the other hand, the correlation between fingerprint sequences is only 0.03, suggesting that it should have similar performance to orthogonal fingerprinting under averaging collusion.

We show the analytical approximation of P_d for the ECC-based fingerprinting under interleaving and averaging collusion with the above settings in Figs. 17.18(a) and (b), respectively. The Watermark-to-Noise-Ratio (WNR) ranges from 0 dB to -20 dB, which includes the scenarios from severe distortion to mild distortion. The performance of

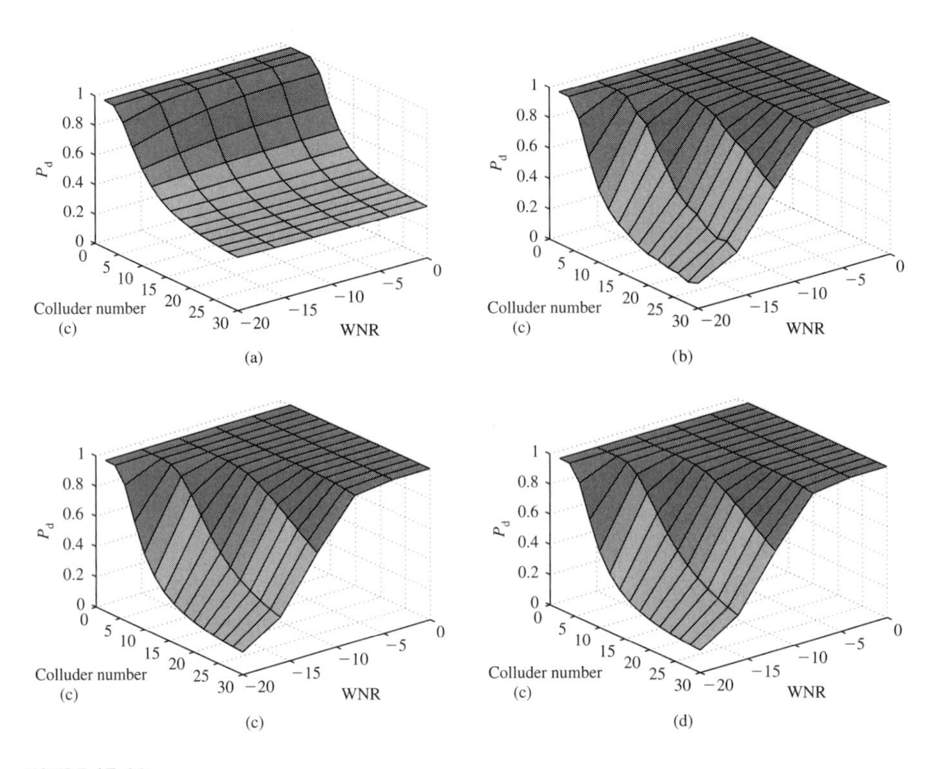

FIGURE 17.18

Performance of ECC-based fingerprinting under (a) interleaving collusion; (b) averaging collusion; and of orthogonal fingerprinting under (c) interleaving collusion and (d) averaging collusion.

orthogonal fingerprinting from [67] are shown in Figs. 17.18(c) and (d) for interleaving collusion and averaging collusion, respectively. Comparing Figs. 17.18(b) and (d), we see that under averaging collusion, the orthogonal fingerprinting and the ECC-based fingerprinting constructed earlier have similar colluder identification performance. They both can resist at least a few dozens colluders' averaging attack under high WNR and about half dozen's under very low WNR. Thus from colluders' point of view, averaging collusion for an ECC-based fingerprinting system is not a very effective strategy. However, under interleaving collusion, we observe from Figs. 17.18(a) and (c) a huge gap on the collusion resistance between the two systems. For orthogonal fingerprinting, the probability of colluder detection under interleaving collusion is the same as that under averaging collusion owing to the orthogonal spreading; at WNR = 0 dB, the P_d remains close to 1 when c is around a few dozens. On the other hand, the detection probability of the ECC-based fingerprinting drops sharply when more than seven colluders come to create an interleaved copy, even when WNR is high. Thus from colluders' point of view, interleaving collusion is an effective strategy to circumvent the protection.

17.4.2.3 Boosting Collusion Resistance by Permuted Subsegment Embedding

The drastic difference in the collusion resistance against averaging and interleaving collusions of ECC-based fingerprinting inspires us to look for an improved fingerprinting method, for which the interleaving collusion would have a similar effect to averaging collusion. Careful examination on the two types of collusion shows that the difference in the resistance against them comes from the amount of role given to the embedding layer to play. The segment-wise interleaving collusion is equivalent to the symbol-wise interleaving collusion on the code level, because each colluded segment comes from just one user. The collusion resilience primarily relies on what is provided by the code layer and almost bypasses the embedding layer. Because of the limited alphabet size, the chance for the colluders to interleave their symbols and to create a colluded fingerprint close to the fingerprint of an innocent user is so high that it would require a large minimum distance in the code design, if to handle this on the code level alone. This means that either codes representing a given number of users can resist only a small number of colluders, or codes can represent only a small total number of users. On the other hand, for averaging collusion, every colluder contributes his/her share in every segment. Through a correlation detector, the collection of such contribution over the entire test signal leads to high expected correlation values when correlating with the fingerprints from the true colluders, and to low expected correlation values when correlating with the fingerprints from innocent users. In other words, the embedding layer contributes to defending against the collusion. This suggests that more closely considering the relationship among fingerprint encoding, embedding, and detection is helpful to improve the collusion resistance against interleaving collusion.

The basic idea of the improved algorithm is to prevent the colluders from using the whole segment that carries one symbol as an interleaving unit and to exploit the code-level limitation. We accomplish this by making each colluded segment contains multiple colluders' contribution. The solution builds upon the existing code construction and performs two important additional steps that we collectively refer to as Permuted Subsegment Embedding [65]. As shown in Fig. 17.19, consider as before a fingerprint signal generated by concatenating the appropriate sequences corresponding to the symbols in a user's codeword. We first partition each segment of the fingerprint signal into β subsegments, giving a total of βL subsegments. We then randomly permute these subsegments according to a secret key to obtain the final fingerprint signal to represent the user. In detection, the extracted fingerprint sequence is first inversely permuted and then the correlator Eq. (17.9) is applied to identify the colluder.

With subsegment partitioning and permutation, each colluded segment after interleaving collusion most likely contains subsegments from multiple users. To correlation-based detectors (including both hard and soft detection on the symbol level), this would have a similar effect to what averaging collusion brings. Because averaging collusion is far less effective from the colluders' point of view, the permuted subsegment embedding can greatly improve the collusion resistance of ECC-based fingerprinting under interleaving collusion. Even if the colluders know the actual size of a segment or a subsegment, the

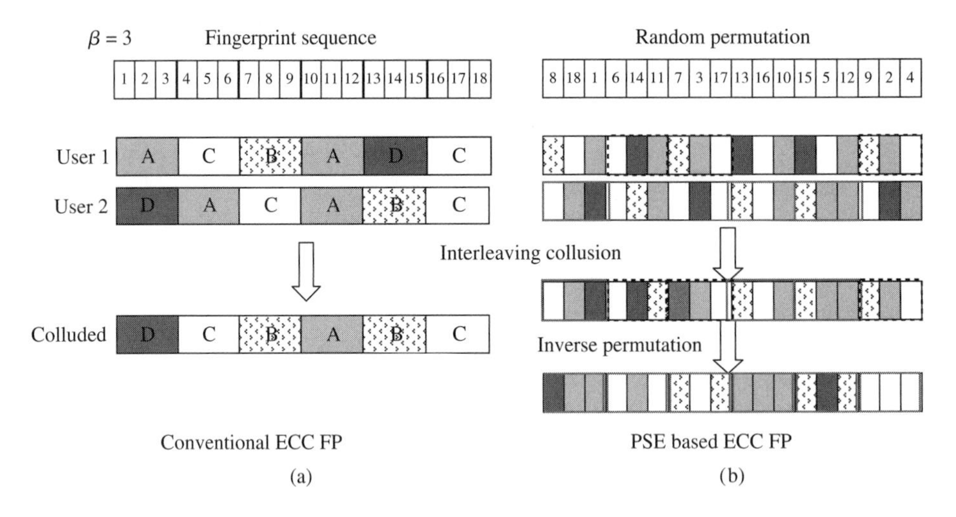

FIGURE 17.19

Illustration of the permutated subsegment embedding for ECC-based fingerprinting: (a) the conventional ECC-based fingerprinting; (b) After permutated subsegment embedding.

permutation unknown to them prevents them from creating a colluded signal with the equivalent effect of symbol interleaving in the code domain.

17.4.2.4 *Efficient Detection via Trimming for Large Scale Fingerprinting System*

A video program contains a large amount of data and the potential users can be on the order of 10 million. Thus, fingerprinting video requires high collusion resistance as well as efficient generation, detection, and distribution for many applications. Among the existing fingerprinting schemes, most of them consider an experimental settings with user number on the order of a few thousand and a small collusion group around 10 colluders. However, in real applications, such as cable TV, the user number can be as high as 10–100 million. The potential collusion group may involve hundreds of colluders. Most of the existing schemes cannot reach such large user size and high collusion resistance requirements. For example, to hold 10 million users, one of the widely cited schemes that were designed for generic data [71] gives a code on the order of 10^7 bits that needs 22-h video for embedding, and it can only resist 10 users' collusion. On the other hand, the orthogonal fingerprinting can be scaled up to hold 10 million users with collusion resistance of 100, but the detection computational complexity increases linearly with the number of users and it becomes prohibitively high for large scale system. In contrast, analysis has shown that the improved ECC fingerprinting discussed in the previous subsection can potentially protect a video signal as short as 10 min and resist more than 100 users' collusion out of 100 million users [72]. This inspires us to explore the application of ECC fingerprinting onto video signals with large user group.

The high data volume in a video stream provides seemingly abundant spaces for data embedding and offers high degrees of freedom in choosing how to fingerprint. We need to determine how to apply ECC fingerprinting onto video signals to achieve the large user scale. Recall that the computational complexity of the detection is the sum of two terms: qN for demodulation and $N_u L$ for colluder identification. When the number of users scales up to millions, the second term $N_u L$ becomes a nontrivial part of the total computational complexity. Proper code structure can significantly reduce the computational complexity, for example, through a trimming process based on the detection results on predefined symbol positions, which we refer to as trimming positions. We first calculate the correlation statistics T^s_{ij} for the segments Ψ corresponding to trimming positions with every possible spreading sequence. That is:

$$T^s_{ij} = \frac{(\mathbf{z}_i - \mathbf{x}_i)^T \mathbf{u}_j}{\|\mathbf{u}_j\|}, \quad i \in \Psi; \quad j = 1,2,\ldots,q. \tag{17.12}$$

where $\mathbf{z}_i$ and $\mathbf{x}_i$ are the ith segment of the test signal and original signal, respectively, and $\mathbf{u}_j$ is the spreading sequence for symbol j. Then, for each trimming position $i \in \Psi$, we pick the symbols that have higher statistic than a threshold h as candidate symbols:

$$S_i = \left\{ j | T^s_{ij} > h \right\}, \quad i \in \Psi. \tag{17.13}$$

The codewords that match candidate symbols in S_i for all the positions in Ψ are put into a suspicious codeword set $W = \{w | w_i \in S_i, i \in \Psi\}$. Finally, we apply matched filter detection of Eq. (17.9) within the suspicious set W to identify the colluder.

The computational complexity of this scheme is determined by the number of trimming positions. If k' symbol positions are used for trimming, the resulting computational complexity for colluder identification can be reduced from $O(q^k L)$ to $O(q^{k-k'} L)$, and the reduction is $q^{k'}$ folds. For example, in a system holding around 1000 users with $q = 32$ and $k = 2$, when we use all the information symbols for trimming, we can obtain more than 1000 times reduction on the computational complexity. The detection computational complexity can be more significantly reduced for a large scale system with large q and k'.

17.4.3 Experimental Results of Video Fingerprinting

Putting together the building blocks that we have seen in previous sections, we examine the overall performance of an example video fingerprinting system. The test video signal is obtained from [73] and has VGA size of 640×480. The total number of users that we target at is on the order of 10 million and colluder number is around 100. We choose a Reed–Solomon code with $q = 64$, $k = 4$, and $L = 63$, which leads to the number of users $N_u = 1.6 \times 10^7$. The expansion parameter γ in the efficient detection is set at 3, and the first four symbol positions are selected for trimming. Thus the equivalent codeword length is 71.

During the fingerprint embedding, each frame is transformed into DCT domain. Fingerprint sequences are embedded into these DCT coefficients through additive

(a) (b) (c)

FIGURE 17.20

Experimental results: (a) original frame; (b) fingerprinted frame before attack, with PSNR of 32 dB; (c) fingerprinted frame after 3 Mbps MPEG compression.

embedding with perceptual scaling. The host video signal is chosen to have 852 frames with about 12 frames for each codeword symbol. For simplicity, we repeatedly embed the same fingerprint sequence into every group of frames consisting of six consecutive frames. The issue of intravideo collusion attack will be discussed in Section 17.4.4. Subsegment partition factor β is set as 24. Figures 17.20(a)–(c) show the 500th frame in the original, fingerprinted, and compressed video sequences. Figure 17.21 shows the simulation results on the probability of catching one colluder, P_d, versus colluder number, K, under averaging and interleaving collusion attacks followed by MPEG compression. The curves are obtained by averaging the results of 50 iterations.

The results shown in Fig. 17.21 is encouraging in that we are able to hold more than 10 million users and resist more than 100 users' averaging collusion and 60 users' interleaving collusion within less than 30 s video. The resistance can be further improved by increasing the video sequence length and employing a larger β value for the permuted subsegment embedding, and trading off the reduced efficiency in distributing fingerprinted signals. On the other hand, without the joint coding and embedding approach, the system can only resist about two users' collusion as indicated by the dash line in Fig. 17.21(a). We can see that the joint coding and embedding strategy can help to overcome the code-level limitation and substantially improve the collusion resistance at an affordable computational complexity. We further decrease the compression bit rate down to 1.5 Mbps that is of VCD quality and examine the collusion resistance. Even under this quality, the collusion resistance only reduces a little under interleaving collusion down to 50 colluders. The resistance under averaging collusion is till higher than 100 colluders.

17.4.3.1 ***Results under Nonlinear Collusion Attacks***

The experimental and analytical results that we have presented so far are based on averaging collusion and interleaving collusion. Our next experiments examine the collusion resistance under Min-Max attack, which can be used as a representative of nonlinear collusion [63]. In the Min-Max attack, colluders choose the average of the minimum and maximum values of their copies in each DCT coefficient position to generate the colluded version. MPEG-2 compression is further applied to the colluded signal. Figure 17.21(a)

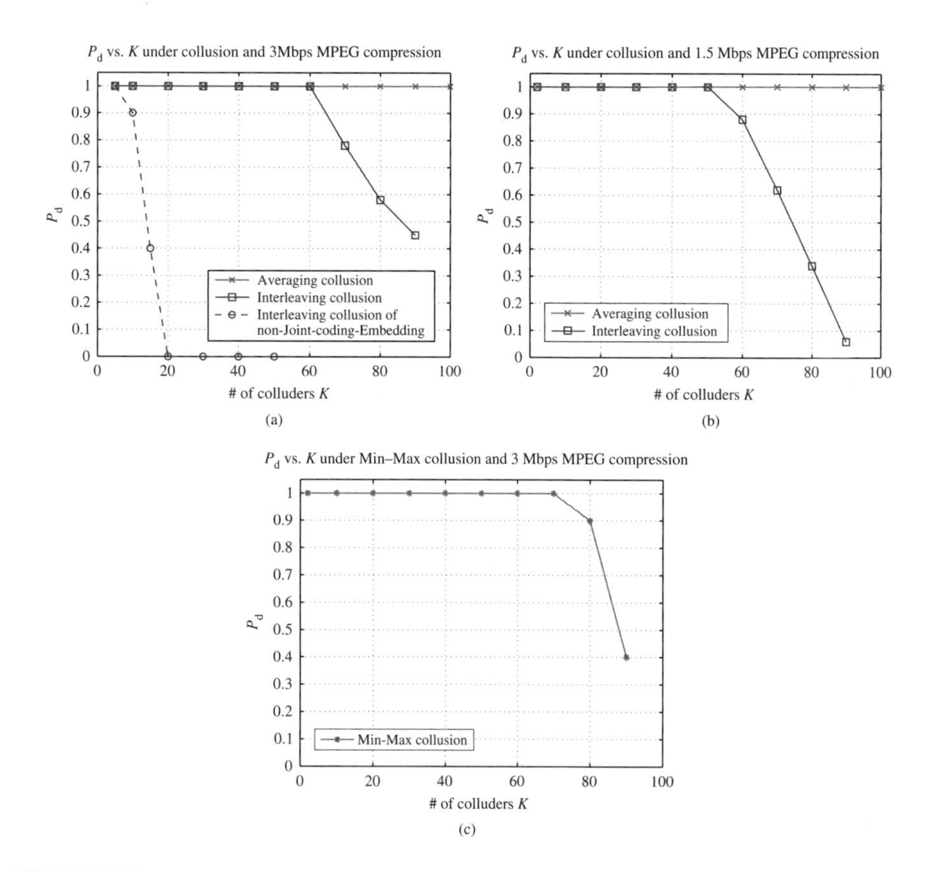

FIGURE 17.21

Probability of catching one colluder P_d versus colluder number c under averaging and interleaving collusion followed by (a) 3 Mbps, (b) 1.5 Mbps MPEG-2 compression, and (c) under Min-Max collusion followed by 3 Mbps MPEG-2 compression.

shows the 500th frame of the colluded video after compression, and Fig. 17.21(b) shows the detection probability P_d versus the colluder number under the Min-Max attack followed by 3 Mbps MPEG-2 compression. The results show that under this nonlinear collusion attack, the collusion resistance is around 80 colluders, which further demonstrates the effectiveness of the proposed large scale fingerprinting. Its performance gap compared with that under averaging collusion is mainly because the Min-Max collusion introduces higher distortion on the colluded signal than the averaging collusion [63]. The increased distortion results in a lower collusion resistance of the system to Min-Max collusion attack.

It is worth mentioning that the computational complexity of order-statistics-based nonlinear collusion significantly increases because of the sorting involved. In the examined experimental settings, the nonlinear collusion attack from 80 colluders requires 12 h while the averaging collusion only needs 70 min. This high computational complexity can also help to deter the colluders from employing the nonlinear collusion on video especially for a large colluder group.

17.4.4 Intravideo Collusion

The large amount of data in video is a double-edged sword as it also benefits the attackers. Given one copy of the fingerprinted video, an attacker may apply multiple-frame collusion [74], whereby several frames are used to estimate and eventually remove the fingerprint. One possible implementation of such intravideo attacks is that the attacker may average several frames that have "visually similar" content but are embedded with independent fingerprint sequences. By collecting enough frames, this averaging operation can successfully remove the embedded fingerprint at a possible expense of reduced visual quality. Furthermore, this attack can be extended to object-based collusion, where similar objects are identified and averaged or swapped to circumvent the detection. Another possible attack is that the attacker may identify several "visually dissimilar" frames or regions embedded with the same fingerprint sequences and average these frames to estimate the embedded fingerprint. This estimated fingerprint sequence will then be subtracted from the fingerprinted signal to obtain an approximation of the original frame. In each of the above mentioned cases, the attacker can succeed by attacking just one fingerprinted copy without help from other colluders. Therefore, the design and embedding of the fingerprint sequence should be robust to these intravideo attacks.

A basic principle to resist these attacks is to embed fingerprint sequences based on the content of the video, that is similar fingerprint sequences are embedded in frames/objects with similar content and different fingerprint sequences are embedded in frames with different content [75]. As consecutive frames in one scene are visually similar, we can repeatedly embed the same fingerprint sequence into those consecutive frames, for example, using a group-of-picture (GoP) of half dozen frames that belong to the same scene as a unit [76]. For visually dissimilar frames that are assigned similar fingerprint sequences based on ECC construction, we need to modify the fingerprint sequences to make them more distinct. To achieve this goal, we can use a visual hash of each frame to adjust the fingerprint sequences. Details can be found in [72] on how to construct fingerprint sequences that have correlation changing linearly with the Hamming distance between two frames' hashes.

REFERENCES

[1] A. Puri and T. Chen. *Multimedia Systems, Standards, and Networks.* Marcel Dekker Inc., New York, 2000.

[2] H. Hacigümüs, B. R. Iyer, and S. Mehrotra. Efficient execution of aggregation queries over encrypted relational databases. *Lecture Notes in Computer Science 2973*, 125–136, Springer-Verlag, New York, 2004.

[3] W. Trappe and L. C. Washington. *Introduction to Cryptography with Coding Theory*. Prentice Hall, Upper Saddle River, NJ, 2001.

[4] A. J. Menezes, P. C. van Oorschot, and S. A. Vanstone. *Handbook of Applied Cryptography*. CRC Press, Boca Raton, FL, 1996.

[5] W. Li. Overview of fine granularity scalability in MPEG-4 video standard. *IEEE Trans. Circuits Syst. Video Technol.*, 11(3):301–317, 2001.

[6] B.-L. Yeo and B. Liu. Rapid scene analysis on compressed video. *IEEE Trans. Circuits Syst. Video Technol.*, 5(6):533–544, 1995.

[7] F. C. Mintzer, L. E. Boyle, A. N. Cazes, B. S. Christian, S. C. Cox, F. P. Giordano, H. M. Gladney, J. C. Lee, M. L. Kelmanson, A. C. Lirani, K. A. Magerlein, A. M. B. Pavani, and F. Schiattarella. Toward on-line, worldwide access to Vatican library materials. *IBM J. Res. Dev.*, 40(2):139–162, 1996.

[8] T. Lookabaugh, and D. C. Sicker. Selective encryption for consumer applications. *IEEE Comm. Mag.*, 42(5):124–129, 2004.

[9] Y. Mao and M. Wu. A joint signal processing and cryptographic approach to multimedia encryption. *IEEE Trans. Image Process.*, 15(7):2061–2075, 2006.

[10] M. Johnson, P. Ishwar, V. Prabhakaran, D. Schonberg, and K. Ramchandran. On compressing encrypted data. *IEEE Trans. Signal Process.*, 52(10, part 2):2992–3006, 2004. Special Issue on Secure Media.

[11] L. Qiao and K. Nahrstedt. Comparison of MPEG encryption algorithms. *Inter. J. Comput. Graphics*, 22(3), 1998.

[12] N. Yeadon, F. Garcia, D. Hutchison, and D. Shepherd. Continuous media filters for heterogeneous internetworking. *Proc. of SPIE*, Multimedia Computing and Networking (MMCN'96), San Jose, January 1996.

[13] M. Wu, R. Joyce, H.-S. Wong, L. Guan, and S.-Y. Kung. Dynamic resource allocation via video content and short-term traffic statistics. *IEEE Trans. Multimed*, 3(2):186–199, 2001.

[14] Y. Wang, S. Wenger, J. Wen, and A. Katasggelos. Error resilient video coding techniques. *IEEE Signal Process. Mag.*, 14(4):61–82, 2000.

[15] J. Wen, M. Muttrell, and M. Severa. Access control of standard video bitstreams. *Proc. Inter. Conf. Media Future*, Florence, Italy, May 2001.

[16] S. Wee and J. Apostolopoulos. Secure scalable streaming enabling transcoding without decryption. *Proc. IEEE Inter. Conf. Image Process.*, Thessaloniki, Greece, October 2001.

[17] C. Yuan, B. Zhu, Y. Wang, S. Li, and Y. Zhong. Efficient and fully scalable encryption for MPEG-4 FGS. *Proc. IEEE Inter. Symp. Circuits Syst.*, Bangkok, Thailand, May 2003.

[18] Susie J. Wee and John G. Apostolopoulos. Secure transcoding with JPSEC confidentiality and authentication. *Proc. IEEE Inter. Conf. Image Process.*, Singapore, 577–580, 2004.

[19] J. Wen, M. Severa, W. Zeng, M. H. Luttrell, and W. Jin. A format-compliant configurable encryption framework for access control of video. *IEEE Trans. Circuits Syst. Video Technol.*, 12(6):545–557, 2002.

[20] L. Tang. Methods for encrypting and decrypting MPEG video data efficiently. *Proc. 4th ACM Inter. Conf. Multimed.*, 219–229, Boston, MA, November 1996.

[21] W. Zeng and S. Lei. Efficient frequency domain video scrambling for content access control. *Proc. ACM Multimed.*, Orlando, FL, November 1999.

[22] C.-P. Wu and C.-C. Kuo. Efficient multimedia encryption via entropy codec design. In *SPIE Inter. Symposium on Electronic Imaging, Proc. SPIE*, Vol. 4314, San Jose, CA, January 2001.

[23] T.-L. Wu and S. F. Wu. Selective encryption and watermarking of MPEG video. *Inter. Conf. on Image Science, Systems, and Technology (CISST'97)*, Las Vegas, NV, 1997.

[24] C.-P. Wu and C.-C. J. Kuo. Design of integrated multimedia compression and encryption systems. *IEEE Trans. Multimed.*, 7(5):828–839, 2005.

[25] M. Grangetto, E. Magli, and G. Olmo. Multimedia selective encryption by means of randomized arithmetic coding. *IEEE Trans. Multimed.*, 8(5): 905–917, 2006.

[26] H. Kim, J. Wen, and J. D. Villasenor. Secure arithmetic coding. *IEEE Trans. Signal Process.*, 55(5):2263–2272, 2007.

[27] T. M. Cover and J. A. Thomas. *Elements of Information Theory*, 2nd ed. John Wiley & Sons, Inc., New York, 1991.

[28] R. V. Cox, D. E. Bock, K. B. Bauer, J. D. Johnston, and J. H. Snyder. The analog voice privacy system. *AT&T Tech. J.*, 66:119–131, 1987.

[29] A. Pal, K. Shanmugasundaram, and N. Memon. Automated reassembly of fragmented images. *Proc. Int. Conf. Multimed. Expo*, Baltimore, MD, July 2003.

[30] J. M. Shapiro. Embedded image coding using zerotrees of wavelet coefficients *IEEE Trans. Signal Process.*, 41(12):3445–3462, 1993.

[31] H. M. Radha, M. van der Schaar, and Y. Chen. The MPEG-4 fine-grained scalable video coding method for multimedia streaming over IP. *IEEE Trans. Multimed.*, 3(1):53–68, 2001.

[32] M. Bellare. Practice-oriented provable security. *Proc. First Int. Workshop Inf. Secur. (ISW 97)*. Springer-Verlag, New York, 1998. Lecture Notes in Computer Science No. 1396.

[33] M. Bellare, A. Desai, E. Jokipii, and P. Rogaway. A concrete security treatment of symmetric encryption. *Proc. 38th Symp. Found. Comput. Sci.*, IEEE, 1997.

[34] Y. Wang, Q.-F. Zhu, and L. Shaw. Maximally smooth image recovery in transform coding. *IEEE Trans. Commun.*, 41(10):1544–1551, 1993.

[35] A. K. Jain. *Fundamentals of Digital Image Processing*. Prentice Hall, Upper Saddle River, NJ, 1988.

[36] D. J. Field, A. Hayes, and R. F. Hess. Contour integration by the human visual system: evidence for a local 'association field'. *Vision Res.*, 33(2):173–193, 1993.

[37] Z. Wang and A. C. Bovik. A Universal image quality index. *IEEE Signal Process. Lett.*, 9(3):81-84, 2002.

[38] Z. Wang, L. Lu, and A. C. Bovik. Video quality assessment based on structural distortion measurement. *Signal Process.: Image Commun.*, 19(1), January, 2004.

[39] A. C. Bovik, editor. *The Essential Guide to Image Processing*. Elsevier, Burlington, MA, 2008.

[40] J. Daemen and V. Rijmen. AES Proposal: Rijndael. Available at *http://csrc.nist.gov/CryptoToolkit/aes/rijndael/Rijndael-ammended.pdf*

[41] Q. Sun, J. Apostolopoulos, C. W. Chen, and S.-F. Chang, Quality-optimized and secure end-to-end authentication for media delivery. In Special issue on recent advances in distributed multimedia communications. *Proc. IEEE*, 96(1):97–111, 2008.

[42] C.-Y. Lin and S.-F. Chang. A robust image authentication method distinguishing JPEG compression from malicious manipulation. *IEEE Trans. Circuits Syst. Video Technol.*, 11(2):153–168, 2001.

[43] Q. Sun, D. He, and Q. Tian. A secure and robust authentication scheme for video transcoding. *IEEE Trans. Circuits Syst. Video Technol.*, 16(10):1232–1244, 2006.

[44] P. A. Chou and Z. Miao. Rate-distortion optimized streaming of packetized media. *IEEE Trans. Multimed.*, 8(2):390–404, 2006.

[45] Z. Zhang, Q. Sun, W.-C. Wong, J. Apostolopoulos, and S. Wee. Rate-distortion optimized streaming of authenticated video. *IEEE Trans. Circuits Syst. Video Technol. (CSVT)*, 17(5):544–557, 2007.

[46] G. Sullivan and T. Wiegand. Video compression–from concepts to the H.264/AVC Standard. *Proc. IEEE*, 93(1):18–31, 2005.

[47] P. Golle and N. Modadugu. Authenticating streamed data in the presence of random packet loss. *ISOC Network Distrib. Syst. Secur. Symp.*, 13–22, 2001.

[48] Z. Zhang, Q. Sun, and W.-C. Wong. A proposal of butterfly-graph based stream authentication over lossy networks. In *Proc. IEEE Int. Conf. Multimed. Expo (ICME)*, 2005.

[49] A. Perrig, R. Canetti, J. Tygar, and D. Song. Efficient authentication and signing of multicast streams over lossy channels. In *Proc. IEEE Symp. Secur. Privacy*, 56–73, 2000.

[50] K. J. R. Liu, W. Trappe, Z. J. Wang, M. Wu, and H. Zhao. *Multimedia Fingerprinting Forensics for Traitor Tracing.* EURASIP Book Series on Signal Processing and Communications, Hindawi Publishing Co., New York, 2005.

[51] M. Wu, W. Trappe, Z. Wang, and K. J. R. Liu, Collusion resistant fingerprinting for multimedia. *IEEE Signal Process. Mag.*, 15–27, 2004.

[52] I. Cox, J. Bloom, and M. Miller. *Digital Watermarking: Principles & Practice.* Morgan Kaufmann Publishers, Burlington, MA, 2001.

[53] I. Cox, J. Kilian, F. Leighton, and T. Shamoon. Secure spread spectrum watermarking for multimedia. *IEEE Tran. Image process.*, 6(12):1673–1687, 1997.

[54] C. Podilchuk and W. Zeng. Image adaptive watermarking using visual models. *IEEE J. Sel. Areas Commun. (JSAC)*, 16(4):525–538, 1998.

[55] C.-Y. Lin, M. Wu, Y.-M. Lui, J. A. Bloom, M. L. Miller, and I. J. Cox. Rotation, scale, and translation resilient public watermarking for images. *IEEE Trans. Image Process.*, 10(5):767–782, 2001.

[56] J. Lubin, J. Bloom, and H. Cheng. Robust, content-dependent, high-fidelity watermark for tracking in digital cinema. *Secur. Watermark. Multimed. Contents V, Proc. of SPIE*, Vol. 5020, January 2003.

[57] P. Moulin and J. A. O'Sullivan. Information-Theoretic Analysis of Information Hiding. Preprint, September 1999, Revised December 2001. Available at http://www.ifp.uiuc.edu/~moulin/paper.html.

[58] B. Chen and G. W. Wornell. Quantization index modulation: A class of provably good methods for digital watermarking and information embedding. *IEEE Trans. Info. Theory*, 47:1423–1443, 2001.

[59] J. G. Proakis. *Digital Communications.* McGraw-Hill, 4th ed., 2000.

[60] F. Ergun, J. Kilian, and R. Kumar. A note on the limits of collusion-resistant watermarks. In *Eurocrypt '99.* 140–149, 1999.

[61] J. Kilian, T. Leighton, L. Matheson, T. Shamoon, R. Tarjan, and F. Zane. Resistance of digital watermarks to collusive attacks. *Proc. IEEE Int. Symp. Inf. Theory*, pp. 271, August 1998. Full version is available as Princeton CS TR-585–98.

[62] H. S. Stone. Analysis of attacks on image watermarks with randomized coefficients. Tech. Rep., 96–045, NEC Research Institute, Princeton, NJ, 1996.

[63] H. Zhao, M. Wu, Z. J. Wang, and K. J. R. Liu. Forensic analysis of nonlinear collusion attacks for multimedia fingerprinting. *IEEE Trans. Image Process.*, 14(5):646–661, 2005.

[64] W. Trappe, M. Wu, Z. J. Wang, and K. J. R. Liu. Anti-collusion fingerprinting for multimedia. *IEEE Trans. on Signal Process.*, 51(4):1069–1087, 2003. Special issue on Signal Processing for Data Hiding in Digital Media & Secure Content Delivery.

[65] S. He and M. Wu. Joint coding and embedding techniques for multimedia fingerprinting, *IEEE Trans. Info. Forensics Secur.*, 1(2):231–247, 2006.

[66] R. Safavi-Naini and Y. Wang. Collusion secure q-ary fingerprinting for perceptual content. *Secur Privacy Digit. Rights Manage. (SPDRM'01)*, 57–75, 2002.

[67] Z. J. Wang, M. Wu, H. Zhao, W. Trappe, and K. J. R. Liu. Anti-collusion forensics of multimedia fingerprinting using orthogonal modulation. *IEEE Trans. Image Process.*, 14(6):804–821, 2005.

[68] J. N. Staddon, D. R. Stinson, and R. Wei. Combinatorial properties of frameproof and traceability codes. *IEEE Trans. Inf Theory*, 47(3):1042–1049, 2001.

[69] S. He and M. Wu. Performance study of ECC-based collusion-resistant multimedia fingerprinting. In *Proc. 38th CISS*, 827–832, March 2004.

[70] S. B. Wicker. *Error Control Systems for Digital Communication and Storage.* Prentice Hall, Upper Saddle River, NJ, 1995.

[71] D. Boneh and J. Shaw. Collusion-secure fingerprinting for digital data. *IEEE Trans. Inf. Theory*, 44(5):1897–1905, 1998.

[72] S. He and M. Wu. Collusion-resistant video fingerprinting for large user group. *IEEE Trans. Info. Forensics Secur.*, 2(4):697–709, 2007.

[73] Sample video sequences from Technische Universitat Munchen, Germany. Available at http://www.ldv.ei.tum.de/page70?LANG=EN.

[74] K. Su, D. Kundur, and D. Hatzinakos. Statistical invisibility for collusion-resistant digital video watermarking, *IEEE Trans. Multimed.*, 7(1):43–51, 2005.

[75] M. D. Swanson, B. Zhu, and A. T. Tewfik. Multiresolution scene-based video watermarking using perceptual models, *IEEE J. Selected Areas Commun.*, 16(4):540–550, 1998.

[76] M. Wu, H. Yu, and B. Liu. Data hiding in image and video: Part-II – designs and applications. *IEEE Trans. Image Process.*, 12(6):696–705, 2003.

[77] A. V. Oppenheim and J. S. Lim. The importance of phase in signals. *Proc. IEEE*, 69(5):529–541, 1981.

CHAPTER

Wireless Video Streaming

18

Fan Zhai[1], Peshala Pahalawatta[2], and Aggelos K. Katsaggelos[3]

[1] *Video Technology, DSP Systems Texas Instruments, Dallas, Texas, USA*
[2] *Image Technology Research, Dolby Laboratories Inc, Burbank, California, USA*
[3] *Department of Electrical Engineering and Computer Science, Northwestern University, Evanston, Illinois, USA*

18.1 INTRODUCTION

The compression or coding of a signal (e.g., speech, text, image, video) has been a topic of great interest for a number of years. Source compression is the enabling technology behind the multimedia revolution we are experiencing. The two primary applications for data compression are storage and transmission.

Video transmission is the topic of this chapter. Video transmission applications have various forms such as multimedia telephony, multimedia messaging service (MMS), video on demand, and streaming. Different applications have different delay constraints and different quality of service (QoS) requirements, and thus call for different coding schemes and transmission techniques. In this chapter, we mainly discuss the study of video streaming over wireless channels, where error-resilient source coding is generally required since streaming services usually use unreliable transmission protocols [1]. In addition, power efficiency is the key for successful deployment of wireless video streaming applications. Particularly, we focus on how video signals should be efficiently coded and transmitted over wireless channels to achieve the best possible quality.

In an increasing number of applications, video is transmitted to and from portable wireless devices, such as cellular phones, laptop computers connected to wireless local area networks (WLANs), and cameras in surveillance and environmental tracking systems. For example, the dramatic increase in bandwidth brought by new technologies such as the present third-generation (3G) and the emerging fourth-generation (4G) wireless systems, and the IEEE 802.11 WLAN standards is beginning to enable video streaming in personal communications. Although wireless video communications are highly desirable in many applications, there are two major limitations in any wireless system: the hostile radio environment, including noise, time-varying channels, and abundant electromagnetic interference, and dependence of mobile devices typically on a battery with a limited energy supply. Such limitations are especially of a concern for video transmission because of the high bit rate and high energy consumption rate in encoding and transmitting video bitstreams. Thus, the efficient use of bandwidth and energy becomes highly important in the deployment of wireless video applications.

To design an energy-efficient communication system, the first issue is to understand how energy is consumed in mobile devices. Energy in mobile devices is mainly used for computation, transmission, display, and driving speakers. Among those, computation and transmission are the two largest energy consumers. During computation, energy is used to run the operating system software and to encode and decode the audio and video signals. During transmission, energy is used to transmit and receive the RF audio and video signals. It should be acknowledged that computation has always been a critical concern in wireless communications. For example, energy-aware operating systems have been studied to efficiently manage energy consumption by adapting the system behavior and workload based on the available energy, job priority, and constraints. Computational energy consumption is especially a concern for video transmission, because motion estimation and compensation, forward and inverse DCT (IDCT) transformations, quantization, and other components in a video encoder, all require a significant number of calculations. Energy consumption in computation has been addressed in [2], where a power-rate-distortion model has been proposed to study the optimal trade-off between computation power, transmission rate, and video distortion. Nonetheless, advances in VLSI design and integrated circuit (IC) manufacturing technologies have led to ICs with increasingly high integration densities while using less power. According to Moore's Law, the number of transistors on an IC doubles every 1.5 years. As a consequence, the energy consumed in computation is expected to become a less significant fraction of the total energy consumption. Therefore, we concentrate primarily on transmission energy.

In this chapter, we focus on the last hop of a wireless network. Specifically, we focus on the situation where video is captured and transmitted from a mobile wireless device to the base station, which is likely to be the bottleneck of the whole video transmission system. Such mobile wireless devices typically rely on a battery with a limited energy supply. In this case, the efficient utilization of transmission energy is a critical design consideration [3]. Thus, the overall problem is how to encode a video source and transmit it to the base station in an energy efficient and error resilient way. To study error resiliency, we need to understand how video is encoded and transmitted over wireless channels.

In a video communication system, the video is first compressed and then segmented into fixed or variable length packets and multiplexed with other types of data, such as audio. Unless a dedicated link that can provide a guaranteed QoS is available between the source and the destination, data bits or packets may be lost or corrupted due to either traffic congestion or bit errors due to impairments of the physical channels. Such is the case, for example, with the current Internet and wireless networks. Due to its best effort design, the current Internet makes it difficult to provide the QoS, such as bandwidth, packet loss probability, and delay needed by video communication applications. Fading, multipath, and shadowing effects, in wireless channels result in a much higher bit error

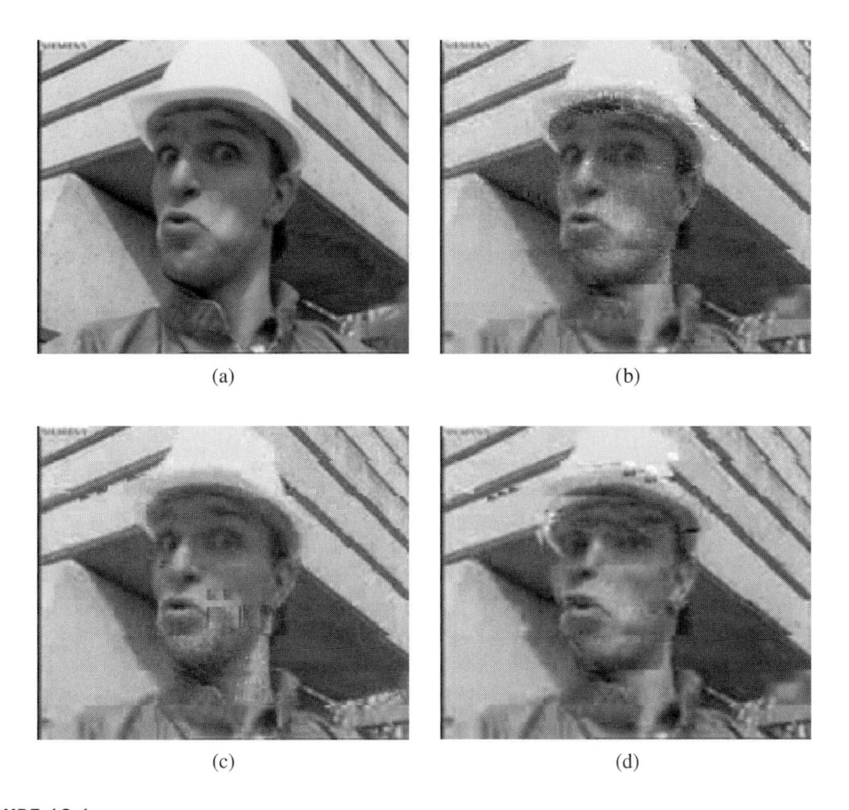

FIGURE 18.1

Illustration of the effect of channel errors on a video stream compressed using the H.263 standard: (a) Original frame; Reconstructed frame at (b) 3% packet loss (c) 5% packet loss (d) 10% packet loss (QCIF Foreman sequence, frame 90, coded at 96 kbps and frame rate 15 fps).

rate (BER) and consequently an even lower throughput than wired links. Figure 18.1 illustrates the effect of channel errors on a typical compressed video sequence in the presence of packet loss.

Due to the "unfriendliness" of the channel to the incoming video packets, the packets have to be protected so that the best possible quality of the received video is achieved at the receiver. A number of techniques, which are collectively called error resilient techniques, have been devised to combat transmission errors. They can be grouped into [4] (i) those introduced at the source and channel coder to make the bit stream more resilient to potential errors; (ii) those invoked at the decoder upon detection of errors to conceal the effects of errors, and (iii) those which require interactions between the source encoder and decoder so that the encoder can adapt its operations based on the loss conditions detected at the decoder.

A number of reasons make the error resiliency problem a challenging one. First, compressed video streams are sensitive to transmission errors because of the use of predictive coding and variable-length coding (VLC) by the source encoder. Due to the use of spatiotemporal prediction, a single bit error can propagate in space and time. Similarly, because of the use of VLCs, a single bit error can cause the decoder to loose synchronization so that even successfully received subsequent bits become unusable. Second, both the video source and the channel conditions are time varying, and therefore, it is not possible to derive an optimal solution for a specific transmission of a given video signal. Finally, severe computational constraints are imposed for real-time video communication applications.

The development of error resilient approaches or approaches for increasing the robustness of the multimedia data to transmission errors is a topic of utmost importance and interest. To make the compressed bit stream resilient to transmission errors, redundancy must be added into the stream. Such redundancy can be added either by the source or by the channel coder. Shannon said it fifty years ago [5] that source coding and channel coding can be separated for optimal performance communication systems. The source coder should compress a source to a rate below the channel capacity while achieving the smallest possible distortion, and the channel coder should add redundancy through forward error correction (FEC) to the compressed bit stream to enable the correction of transmission errors. The application of Shannon's separation theory resulted in major advances on source coding (e.g., rate-distortion optimal coders and advanced entropy coding algorithms) and channel coding (e.g., Reed–Solomon codes, Turbo codes, and Tornado codes). The separation theory not only promises that the separate design of source and channel coding does not introduce any performance sacrifice but it also greatly reduces the complexities of a practical system design. However, the assumptions on which separation theory is based (infinite length codes, delay, and complexity) may not hold in a practical system. This leads to the joint consideration of source coding and channel coding, referred to as joint source-channel coding (JSCC), to deal with the time-varying channel and time-varying source. JSCC can greatly improve the system performance when there are, for example, stringent end-to-end delay constraints or implementation complexity concerns.

Recently, several adaptation techniques have been proposed specifically for energy efficient wireless video communications [6–8]. A trend in this field of research is the joint adaptation of source coding and transmission parameters based on the time-varying source content and channel conditions, which generally requires a "cross-layer" optimization perspective [9]. Specifically, the lower layers in a protocol stack, which directly control transmitter power, need to obtain knowledge of the importance level of each video packet from the video encoder located at the application layer. In addition, it can also be beneficial if the source encoder is aware of the estimated channel state information (CSI) that is available at the lower layers, as well as of which channel parameters at the lower layers can be controlled so that it can make smart decisions in selecting the source coding parameters to achieve the best video delivery quality. For this reason, joint consideration of video encoding and power control is a natural way to achieve the highest efficiency in transmission energy consumption.

Our purpose in this chapter is to review the basic elements of some of the more recent approaches towards JSCC for wireless video transmission systems. As mentioned above, we limit our discussion to the wireless video streaming application, which generally require relatively strict end-to-end delay constraints. The rest of the chapter is organized as follows. We first provide basic rate-distortion (RD) definitions in addressing the need for JSCC in Section 18.2. We then describe the basic components in a video compression and transmission systems in Section 18.3, followed by a discussion on channel coding techniques that are widely used for video communications in Section 18.4. In Section 18.5, the JSCC problem formulation is presented, followed by examples of several practical implementations. In Section 18.6, we cover distributed multimedia communications over different wireless network architectures, including point-to-multipoint networks, peer-to-peer (P2P) networks, and multihop networks. Finally, Section 18.7 contains concluding remarks.

18.2 ON JOINT SOURCE-CHANNEL CODING

Due to the high bandwidth of the raw video data, lossy compression is generally required to reduce the source redundancy for video transmission applications. It is well known that the theoretical bound for lossless compression is the entropy of the source. In the same way entropy determines the lowest possible rate for lossless compression, RD theory [5, 10] addresses the same question for lossy compression.

18.2.1 Rate-Distortion Theory

A high-level block diagram of a communication system is shown in Fig. 18.2. In this diagram, X and $\hat{X}$ represent, respectively, the source and reconstructed signals, X_s and $\hat{X}_s$ the source encoder output and the source decoder input, and X_c and $\hat{X}_c$ the channel encoder output and the channel decoder input.

Rate is defined as the average number of bits used to represent each sample value. The general form of the source distortion D can be written as

$$D = \sum_{x_i, \hat{x}_j} d(x_i, \hat{x}_j) p(x_i) p(\hat{x}_j | x_i), \tag{18.1}$$

where $d(x_i, \hat{x}_j)$ is the distortion metric, $p(x_i)$ the source density, and $p(\hat{x}_j | x_i)$ the conditional probability of the reconstructed value $\hat{x}_j \in \hat{X}$, given knowledge of the input to the encoder $x_i \in X$. Since the source probabilities are solely determined by the source and the distortion metric is decided ahead of time depending on the application, the distortion

FIGURE 18.2

Block diagram of a communication system.

is a function only of the conditional probabilities $p(\hat{x}_j|x_i)$, which represent a description of the compression scheme. We can therefore define the set Π_D of compression schemes, which will result in distortion D less than some value D^* as

$$\Pi_D = \left\{\{p(\hat{x}_j|x_i)\} : D(\{p(\hat{x}_j|x_i)\}) \leq D^*\right\}. \tag{18.2}$$

The central entity of RD theory is the RD function $R(D)$, which provides the theoretical information bound on the rate necessary to represent a certain source with a given fidelity or average distortion. It is given by [5]:

$$R(D) = \min_{\{p(\hat{x}_j|x_i)\}\in\Pi_D} I(X;\hat{X}), \tag{18.3}$$

where $I(X;\hat{X})$ is the average mutual information between X and $\hat{X}$.

The RD function $R(D)$ has some nice properties, for example, it is a nonincreasing convex function of D. It can be used to find the minimum bit rate for a certain source with a given distortion constraint, as expressed by (18.3). Conversely, it can also be used to determine the information theoretical bounds for the average distortion subject to a source rate constraint, R^*, via the distortion-rate function, defined as

$$D(R) = \min_{\{p(\hat{x}_j|x_i)\}\in\Pi_R} D(X;\hat{X}), \tag{18.4}$$

where R is the average source rate and

$$\Pi_R = \left\{\{p(\hat{x}_j|x_i)\} : R(\{p(\hat{x}_j|x_i)\}) \leq R^*\right\}. \tag{18.5}$$

Note that the $D(R)$ function may be more widely applicable in practical image/video communication systems since, as a practical matter, the aim is usually to deliver the best quality image/video subject to a certain channel bandwidth. RD theory is of fundamental importance in that it conceptually provides the information bounds for lossy data compression. However, RD theory only tells us that $R(D)$ exists but not how to achieve it in general (except for a few special cases such as the case of a Gaussian source). In addition, as already mentioned, it only addresses source coding. In a communication system where the channel may introduce errors, we need ideal channel coding to ensure that $X_s = \hat{X}_s$ in order for (18.4) to be valid, where $D(R)$ is interpreted as the end-to-end distortion and R^* as the channel capacity. Thus, there are several practical issues that prevent us from obtaining the RD function or designing a system to achieve it. In the following two sections, we discuss these issues.

18.2.2 Operational Rate-Distortion Theory

From (18.3), we see that $R(D)$ is evaluated based on information which may not always be available in practice. First, the source probability density function required to evaluate the distortion, which in turn is needed in (18.3) or (18.4), is not typically available for any given video segment. If a specific model is to be used, in general, it does not capture well the spatiotemporal dynamics of the given video. Second, the minimization in (18.3) or (18.4) is over all possible coders or conditional distributions $p(\hat{x}_j|x_i)$, which belong to Π_D or Π_R, respectively. In a practical coding environment, however, we want to use RD theory to allocate resources to a specific video segment using a specific coder. In this case, we usually only have a finite admissible set for $p(\hat{x}_j|x_i)$, as defined by the finite set of encoding modes (e.g., intra or inter encoding, quantization step size, etc.). In addition, it is usually difficult or simply impossible to find closed-form expressions for the $R(D)$ or $D(R)$ functions for general sources. For these reasons, one has to resort to numerical algorithms to specify the operational RD (ORD) function.

Let $\mathcal{S}$ be the set of admissible source coding parameters or modes, q, of a given coder. Then, since the structure of the coder has been determined, each of the parameter choices will lead to a pair of rate and distortion values. The lower bound of all these RD pairs is referred to as the ORD function. The set of source coding parameters that result in the ORD function can be formally defined as

$$\mathcal{U}_{\text{ORDF}} = \{q : q \in \mathcal{S}, R(q) \geq R(p) \Rightarrow D(q) < D(p), \forall p \in \mathcal{S}\}. \tag{18.6}$$

Clearly, the ORD function should be lower bounded by the RD function because the former is obtained using a specific coder and the finite set $\mathcal{S}$, which is a subset of the set of source codes that achieve all conditional distributions $p(\hat{x}_j|x_i)$.

According to (18.6), the ORD function is a strictly decreasing function although it may not be convex. However, if the size of the admissible set $\mathcal{S}$ is sufficiently large, the ORD function will be closely approximated by the convex hull of all the operating points, which is a convex function. The resulting convex hull approximation solution is very close to the optimal solution if the set $\mathcal{S}$ is sufficiently large, which is usually the case in video compression.

18.2.3 Practical Constraints in Video Communications

We can see from (18.3) that the process of finding the optimal compression scheme requires searching over the entire set of conditional probabilities that satisfy the distortion constraint shown in (18.2). Under the assumption that the source encoder output is identical to the source decoder input, that is, $X_s = \hat{X}_s$, the problem becomes a pure source coding problem since the conditional probabilities $p(\hat{x}_j|x_i)$ have nothing to do with the channel. However, such an assumption requires an ideal channel coding scheme such that error free transmission of the source output over a noisy channel with source bit rate $R(D)$ less than the channel capacity C, that is, $R(D) < C$, can be guaranteed.

Such ideal channel coding generally requires infinite length code words, which can only be realized without complexity and delay constraints, both of which are important factors in practical real-time systems. Due to these constraints, most practical channel coding schemes cannot achieve channel capacity and do not provide an idealized error free communication path between the source and the destination. For this reason, the overall distortion between X and $\hat{X}_c$ consists of both source distortion and channel distortion. Minimizing the total distortion usually requires joint design of the source and channel encoders, which is referred to as JSCC.

At the receiver side, gains may be realized by jointly designing the channel and source decoders, which is referred to as *joint source-channel decoding.* In using joint source-channel decoding, the channel decoder does not make *hard* decisions on the output $\hat{X}_s$. Instead, the decoder makes *soft* decisions to allow the source decoder to make use of information such as the signal-to-noise ratio (SNR) of the corrupted code. Alternatively, such soft decisions can be regarded as hard decisions plus a confidence measure. Soft-decision processing used in joint source channel decoding can usually help improve the coding gain by about 2 dB compared to hard-decision processing [11]. In this chapter, we focus only on JSCC.

18.2.4 Illustration

In Fig. 18.3, we plot several ORD functions for a practical system at different channel error rates to illustrate the basic idea of JSCC. When the channel is error free, increased bit rate leads to decreased distortion, as in standard RD theory. This is illustrated by the lowest curve in the figure, in which the lowest distortion is obtained by using the largest available source bit rate, represented by the point (R1, D1). However, when channel errors are present, this trend may not hold since the overall distortion now (the vertical axis in Fig. 18.3) consists of both source and channel distortion. Assuming that the transmission rate is fixed, as more bits are allocated to source coding (the horizontal axis in Fig. 18.3), fewer bits will be left for channel coding, which leads to less protection and higher channel distortion. As shown in Fig. 18.3, an optimal point exists for which the allocation of the available fixed transmission rate bits are optimally allocated between source and channel.

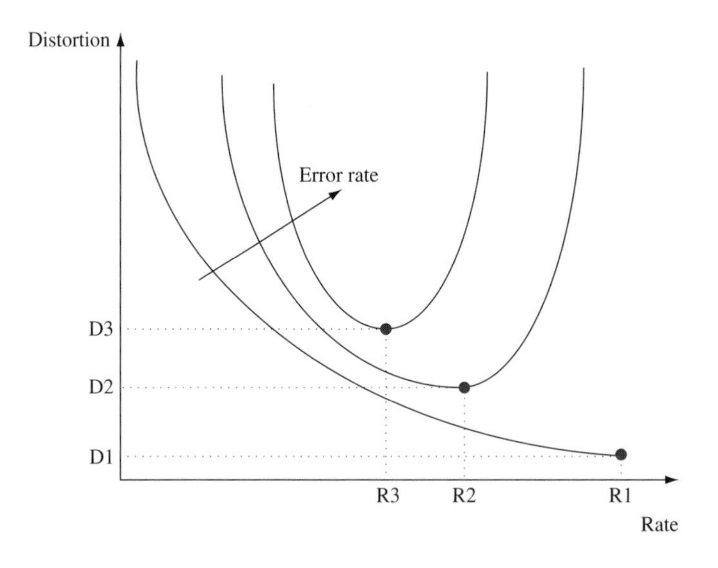

FIGURE 18.3

Illustration of joint source-channel coding.

Note that different channel error rates result in different optimal allocations. This is indicated by the points (R2, D2) and (R3, D3) on the two curves with different channel error rates.

The trade-off between source and channel coding has been studied from a theoretical standpoint based on the use of vector quantizers [12, 13]. In general, JSCC is accomplished by designing the quantizer and entropy coder jointly for given channel characteristics, as in [12, 14]. There is a substantial number of research results in this area. Interested readers can refer to [15] for a comprehensive review on this topic. In this chapter, we focus on the specific application of JSCC in image and video communications, where JSCC usually faces three tasks: finding an optimal bit allocation between source coding and channel coding for given channel loss characteristics; designing the source coding to achieve the target source rate; and designing the channel coding to achieve the required robustness [16, 17]. These tasks, although stated separately, are inter-related, forming the backbone of the integrated nature of JSCC.

We have discussed the basic concept underlying JSCC and its significance in image/video communications. Next, we will first provide an overview of the video compression and transmission systems. Then, we will highlight the key components of JSCC for video applications, such as the different forms of error resilient source coding, channel codes used to deal with different types of channel errors, the general problem formulation, and the general solution approach. In addition to the commonly used video compression standards, such as MPEGX and H.26X, we also briefly discuss wavelet and subband-based video compression schemes, since they are also widely used.

18.3 VIDEO COMPRESSION AND TRANSMISSION

18.3.1 Video Transmission System

We begin by providing a brief high-level overview of a packet-based video transmission system. Some of the major conceptual components found in such a system are shown in Fig. 18.4.

Most practical communication networks have limited bandwidth and are lossy by nature. Facing these challenges, the *video encoder* has two main objectives: to compress the original video sequence and to make the encoded sequence resilient to errors. Compression reduces the number of bits used to represent the video sequence by exploiting both temporal and spatial redundancy. On the other hand, to minimize the effects of losses on the decoded video quality, the sequence must be encoded in an error resilient way. A recent review of error resilient video coding techniques can be found in [4]. The source bit rate is shaped or constrained by a rate controller that is responsible for allocating bits to each video frame or packet. This bit rate constraint is set based on the estimated CSI reported by the lower layers, such as the application and transport layers. It is mentioned here that the system in Fig. 18.4 is a simplified version of a seven or a five layer Open Systems Interconnection (OSI) reference model. For example, in both OSI models, the network, data link, and physical layers are below the transport layer. In the

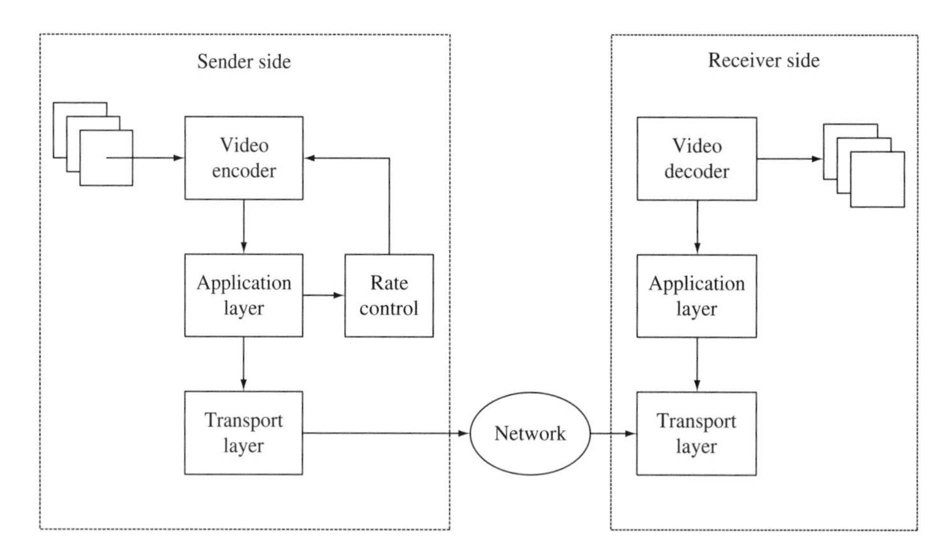

FIGURE 18.4

Video transmission system architecture.

subsequent sections, we will be referring to the various layers, since allowing the various layers to exchange information leads to cross-layer design of a video communication system, which is a central theme in this chapter.

In Fig. 18.4, the *network* block represents the communication path between the sender and the receiver. This path may include routers, subnets, wireless links, etc. The network may have multiple channels (e.g., a wireless network) or paths (e.g., a network with path diversity), or support QoS (e.g., integrated services or differentiated services networks). Packets may be dropped in the network due to congestion or at the receiver due to excessive delay or unrecoverable bit errors in a wireless network. To combat packet losses, parity check packets used for FEC may be generated at the application/transport layer. In addition, lost packets may be retransmitted if the application allows.

For many source-channel coding applications, the exact details of the network infrastructure may not be available to the sender and they may not always be necessary. Instead, what is important in JSCC is that the sender has access to or can estimate certain network characteristics, such as the probability of packet loss, the transmission rate, and the round-trip-time (RTT). In most communication systems, some form of CSI is available at the sender, such as an estimate of the fading level in a wireless channel or the congestion over a route in the Internet. Such information may be fed back from the receiver and can be used to aid in the efficient allocation of resources.

On the receiver side, the transport and application layers are responsible for depacketizing the received transport packets, channel decoding (if FEC is used), and forwarding the intact and recovered video packets to the video decoder. The video decoder then decompresses the video packets and displays the resulting video frames in real time (i.e., the video is displayed continuously without interruption at the decoder). The video decoder typically employs error detection and concealment techniques to mitigate the effects of packet loss. The commonality among all error concealment strategies is that they exploit correlations in the received video sequence to conceal lost information.

18.3.2 Video Compression Basics

In this section, we focus on one of the most widely used video coding techniques, that of hybrid block-based motion-compensated (HBMC) video coding. There are two main families of video compression standards: the H.26X family and the Moving Picture Experts Group (MPEG) family. These standards are application-oriented and address a wide range of issues such as bit rate, complexity, picture quality, and error resilience.

The newest standard is H.264/AVC, aiming to provide the state-of-the-art compression technologies. It is the result of the merger between the MPEG-4 group and the ITU H.26L committee in 2001, known as JVT (Joint Video Team), and is a logical extension of the previous standards adopted by the two groups. Thus, it is also called H.264, AVC, or MPEG-4 part 10 [18]. For an overview and comparison of the video standards, see [19]. It is important to note that all standards specify the decoder only, that is, they standardize the syntax for the representation of the encoded bit stream and define the decoding process but leave substantial flexibility in the design of the encoder. This approach to standardization allows for maximizing the latitude in optimizing the encoder for specific applications [18].

All the above-mentioned video compression standards are based on the HBMC approach and share the same block diagram, as shown in Fig. 18.5. Each video frame is presented by block-shaped units of associated luma and chroma samples (16×16 region) called MBs (macroblocks).

As shown in Fig. 18.5(a), the core of the encoder is motion-compensated prediction (MCP). The first step in MCP is motion estimation (ME), aiming to find the region in the previously reconstructed frame that best matches each MB in the current frame.[1] The offset between the MB and the prediction region is known as the motion vector. The motion vectors form the motion field, which is differentially entropy encoded. The second step in MCP is motion compensation (MC), where the reference frame is predicted by applying the motion field to the previously reconstructed frame. The prediction error, known as the displaced frame difference (DFD), is obtained by subtracting the reference frame from the current frame.

Following MCP, the DFD is processed by three major blocks, namely, transform, quantization, and entropy coding. The key reason for using a transform is to decorrelate the data so that the associated energy in the transform domain is more compactly represented and thus the resulting transform coefficients are easier to encode. The discrete cosine transform (DCT) is one of the most widely used transforms in image and video coding due to its high transform coding gain and low computational complexity.

[1]The H.264/AVC standard supports more flexibility in the selection of motion-compensation block sizes and shapes than any previous standards, with a minimum luma motion-compensation block size as small as 4×4.

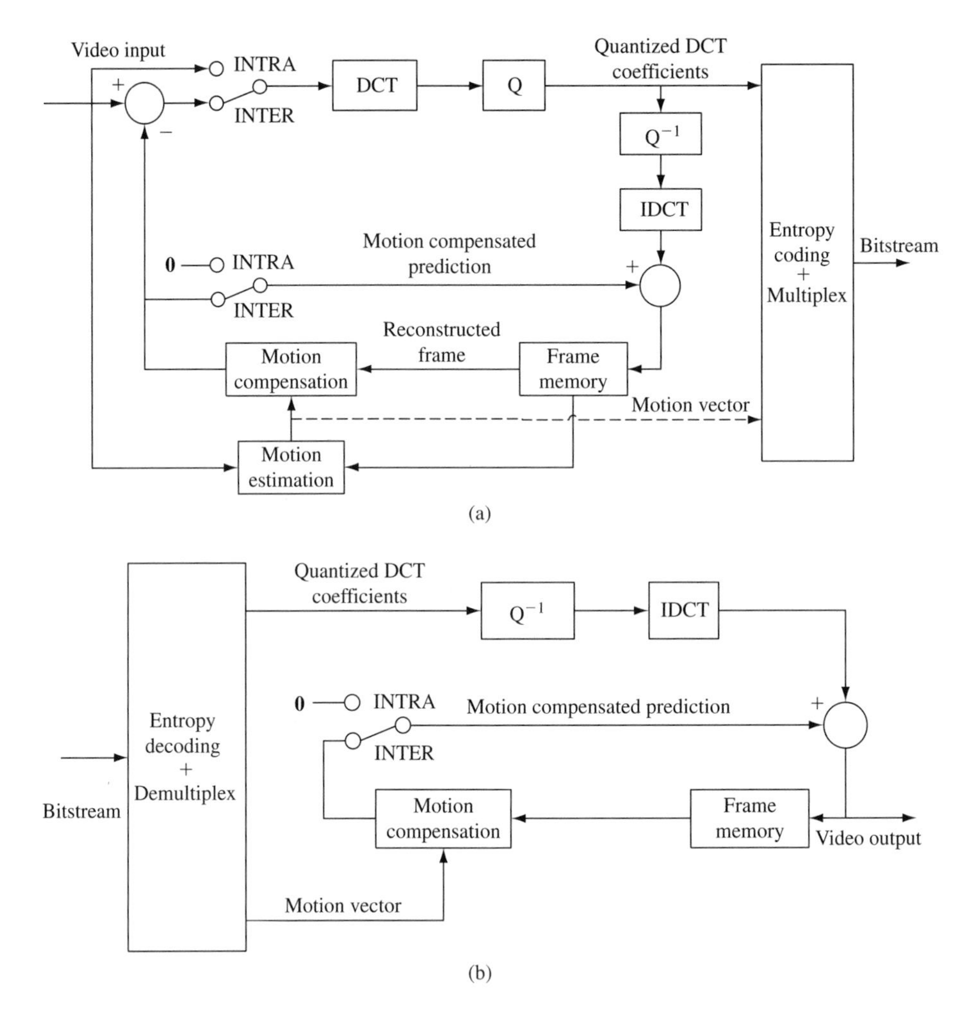

FIGURE 18.5

Hybrid block-based motion-compensated video (a) encoder and (b) decoder.

Quantization introduces loss of information and is the primary source of the compression gain. Quantized coefficients are entropy encoded, for example, using Huffman or arithmetic coding. The DFD is first divided into 8×8 blocks, and the DCT is then applied to each block, with the resulting coefficients quantized.[2] In most block-based motion-compensated (BMC) standards, a given MB can be intraframe coded, interframe coded

[2] While all major prior video coding standards use a transform block size of 8×8, the H.264/AVC standard is based primarily on a 4×4 transform. In addition, instead of DCT, a separable integer transform with similar properties to a 4×4 DCT is used.

using MCP or simply replicated from the previously decoded frame. These prediction modes are denoted as Intra, Inter, and Skip modes. Quantization and coding are performed differently for each MB according to its mode. Thus, the coding parameters for each MB are typically represented by its prediction mode and the quantization parameter.

At the decoder, as shown in Fig. 18.5(b), the IDCT is applied to the quantized DCT coefficients to obtain a reconstructed version of the DFD; the reconstructed version of the current frame is obtained by adding the reconstructed DFD to the MCP of the current frame based on the previously reconstructed frame.

As H.264/AVC is the latest video compression standard and represents the state-of-the-art technology in this area, we briefly highlight its major advancements with respect to its predecessors. Compared to its predecessors, H.264/AVC improves in two aspects: compression efficiency and error resiliency. We discuss the error resiliency aspects in Section 18.3.5. H.264 improves on compression efficiency through the use of variable block sizes for MC. This enables more localized motion prediction and better exploitation of the temporal redundancies in the sequence during compression. This can also be an important feature when combined with temporal error concealment techniques that use the motion vectors of neighboring blocks to estimate the motion vector of a lost block. Since, each macroblock can contain up to 16 motion vectors, more flexibility is available in choosing a candidate concealment motion vector for a lost macroblock. Another improvement in H.264 is that the accuracy of ME is improved to quarter-pixel resolution for the luma component of the video signal. H.264 also provides for the use of an in-loop adaptive deblocking filter. In-loop refers to the fact that the deblocking filter is applied to each picture prior to using the picture as a reference for prediction of subsequent pictures in the sequence. In addition to improving perceptual quality, the deblocking filter also improves compression efficiency since it generates a more accurate reference frame that can be used for MC of subsequent frames. Another significant compression feature of H.264/AVC is that it uses spatial prediction from neighboring macroblocks for the encoding of intra macroblocks. Then, only the residual signal is transform coded. While this form of prediction achieves greater compression efficiency, it can lead to error propagation if the spatially neighboring macroblocks are intercoded and use unreliable reference blocks for MC. Therefore, in packet lossy channels, a constrained intraprediction mode, which stipulates that only pixels from other intracoded macroblocks can be used for intraspatial prediction, can be employed at the encoder. Unlike in previous video compression standards, H.264 allows for MC of bipredictive (B) coded pictures from previously coded B pictures in the sequence. In addition to greater compression efficiency, this also enables temporal scalability through hierarchical biprediction, which is a useful feature for scalable video coding (SVC).[3]

An important advancement in video compression related to H.264/AVC is the recently standardized SVC extension [20]. SVC results in a bit stream that offers progressive refinement of video quality by dynamically adapting the source rates to changing network and channel conditions. SVC uses hierarchical prediction for progressive refinement

[3] H.264 also enables hierarchical prediction structures with inter (P) coded pictures.

of temporal resolution. Quality (SNR) scalability of the bit stream is obtained using a coarse-grained scalability (CGS) approach in which each additional enhancement layer corresponds to a specified increase in quality. Another approach termed medium-grained scalability (MGS) is also available in SVC by which greater flexibility in quality refinement is offered. Spatial scalability is also offered in SVC through the use of a multilayered structure. SVC uses a number of interlayer prediction schemes to improve the compression efficiency of the spatially scalable bit stream. Overall, the SVC extension provides a number of significant advancements over previous SVC efforts and provides a promising platform for the development of network/channel adaptive wireless video transmission schemes.

During the last decade, the discrete wavelet transform (DWT) and subband decomposition have gained increased popularity in image coding due to the substantial contributions in [21, 22], JPEG2000 [23], and others. Recently, there has also been active research applying the DWT to video coding [24–27]. Among the above studies, 3D wavelet or subband video codecs have received special attention due to their inherent feature of full scalability [26, 27]. Until recently, the disadvantage of these approaches has been their poor coding efficiency caused by inefficient temporal filtering. A major breakthrough which has greatly improved the coding efficiency and led to renewed efforts toward the standardization of wavelet-based scalable video coders has come from the contributions of combining lifting techniques with 3D wavelet or subband coding [28, 29].

18.3.3 Channel Models

The development of mathematical models that accurately capture the properties of a transmission channel is a very challenging but extremely important problem. For video applications, two fundamental properties of the communication channel are the probability of packet loss and the delay needed for each packet to reach the destination. In wired networks, channel errors usually appear in the form of packet loss and packet truncation. In wireless networks, besides packet loss and packet truncation, bit error is another common source of error. Packet loss and truncation are usually due to network traffic and clock drift, whereas bit corruption is due to the noisy air channel [30].

18.3.3.1 Internet

In the Internet, queuing delays experienced in the network can be a significant delay component. The Internet, therefore, can be modeled as an independent time-invariant packet erasure channel with random delays, as in [31]. In real-time video applications, a packet is typically considered lost and discarded if it does not arrive at the decoder before its intended playback time. Thus, the packet loss probability is made up of two components: the packet loss probability in the network and the probability that the packet experiences excessive delay. Combining these two factors, the overall probability of loss for packet k is given by

$$\rho_k = \epsilon_k + (1 - \epsilon_k) P\{\Delta T_n(k) > \tau\},$$

where ϵ_k is the probability of packet loss in the network, $\Delta T_n(k)$ is the network delay for packet k, and τ is the maximum allowable network delay for this packet.

18.3.3.2 Wireless Channel

Compared to their wireline counterparts, wireless channels exhibit higher BERs, typically have a smaller bandwidth, and experience multipath fading and shadowing effects. At the IP level, the wireless channel can also be treated as a packet erasure channel, as it is seen by the application. In this setting, the probability of packet loss can be modeled by a function of transmission power used in sending each packet and the CSI. Specifically, for a fixed transmission rate, increasing the transmission power will increase the received SNR and result in a smaller probability of packet loss. This relationship could be determined empirically or modeled analytically. For example, in [32], an analytical model based on the notion of outage capacity is used. In this model, a packet is lost whenever the fading realization results in the channel having a capacity less than the transmission rate. Assuming a Rayleigh fading channel, the resulting probability of packet loss is given by

$$\rho_k = 1 - \exp\left(\frac{1}{P_k S(\theta_k)}(2^{R/W} - 1)\right),$$

where R is the transmission rate (in source bps), W the bandwidth, P_k the transmission power allocated to the kth packet, and $S(\theta_k)$ the normalized expected SNR given the fading level, θ_k. Another way to characterize channel state is to use bounds for the BER with regard to a given modulation and coding scheme; for example, in [33, 34], a model based on the error probability of binary phase-shift keying (BPSK) in a Rayleigh fading channel is used.

18.3.4 End-to-End Distortion

In an error prone channel, the reconstructed images at the decoder usually differ from those at the encoder due to random packet losses, as shown in Fig. 18.6. Even with the same channel characteristics, the reconstruction quality at the decoder may vary greatly based on the specific channel realization, as indicated in Figs 18.6(c) and 18.6(d). In this case, the most common metric used to evaluate video quality in communication systems is the expected end-to-end distortion, where the expectation is with respect to the probability of packet loss. The expected distortion for the kth packet can be written as

$$E[D_k] = (1 - \rho_k)E[D_{R,k}] + \rho_k E[D_{L,k}], \tag{18.7}$$

where $E[D_{R,k}]$ and $E[D_{L,k}]$ are the expected distortion when the kth source packet is either received correctly or lost, respectively, and ρ_k is its loss probability. $E[D_{R,k}]$ accounts for the distortion due to source coding as well as error propagation caused by interframe coding, while $E[D_{L,k}]$ accounts for the distortion due to concealment. Predictive coding and error concealment both introduce dependencies between packets.

FIGURE 18.6

(a) Illustration of effect of random channel errors to a video stream compressed using the H.263 standard. (a) Original frame; (b) Reconstructed frame at the encoder; and reconstructed frame at the decoder (c) simulation 1 (d) simulation 2 (QCIF Foreman sequence, frame 92, coded with 96 kbps, frame rate 15 fps, and packet loss probability 10%).

Because of these dependencies, the distortion for a given packet is a function of how other packets are encoded as well as their probability of loss. Accounting for these complex dependencies is what makes the calculation of the expected distortion a challenging problem.

Methods for accurately calculating the expected distortion have recently been proposed [16, 35]. With such approaches, it is possible, under certain conditions, to accurately compute the expected distortion with finite storage and computational complexity by using per-pixel accurate recursive calculations. For example, in [35], a powerful algorithm called ROPE is developed, which efficiently calculates the expected mean squared error by recursively computing only the first and second moments of each pixel in a frame. Model-based distortion estimation methods have also been proposed (e.g., [36–38]), which are useful when the computational complexity and storage capacity are limited.

18.3.5 Error Resilient Source Coding

If source coding removes all the redundancy in the source symbols and achieves entropy, a single error occurring at the source will introduce a great amount of distortion. In other words, ideal source coding is not robust to channel errors. In addition, designing an ideal or near-ideal source coder is complicated, especially for video signals, as they are usually not stationary, they have memory, and their stochastic distribution may not be available during encoding (especially for live video applications). Thus, redundancy certainly remains after source coding. JSCC should not aim to remove the source redundancy completely but should make use of it and regard it as an implicit form of channel coding [15].

The redundancy added should prevent error propagation, limit the distortion caused by packet losses, and facilitate error detection, recovery, and concealment at the receiver. To "optimally" add redundancy during source coding so as to maximize error resilience efficiency, error resilient source coding should adapt to the application requirements such as computational capacity, delay requirements, and channel characteristics.

As discussed above, MC introduces temporal dependencies between frames, which leads to errors in one frame propagating to future frames. In addition, the use of spatial prediction for the coding of the DC coefficients and motion vectors introduces spatial dependencies within a picture. Because of the use of MC, an error in one part of a picture will not only affect its neighbors in the same picture but also the subsequent frames. The solution to error propagation is to terminate the dependency chain. Techniques such as reference picture selection (RPS), intra-MB insertion, independent segment decoding, video redundancy coding (VRC), and multiple description coding (MDC) are designed for this purpose. A second approach towards error resilience is to add redundancy at the entropy coding level. Examples include reversible VLCs (RVLCs), resynchronization and data partitioning techniques, which can help limit the error propagation effect to a smaller region of the bit stream once the error is detected. The third type of error resilient source coding tools help with error recovery or concealment of the error effects, such as flexible macroblock ordering (FMO).

Finally, scalable coding or layered video coding, although designed primarily for the purpose of transmission, along with computation and display scalability in heterogeneous environments, produces a hierarchy of bit streams, where the different parts of an encoded stream have unequal contributions to the overall quality. Scalable coding has inherent error resilience benefits, especially if the layered property can be exploited in transmission, where, for example, available bandwidth is partitioned to provide unequal error protection (UEP) for different layers with different importance. This approach is commonly referred to as *layered coding with transport prioritization* [39]. Temporal, spatial, and SNR scalability are supported by many standards, that is, in H.263+ Annex O [40], the Fine Granular Scalability (FGS) in MPEG-4 [41], H.264/AVC [42], and the SVC extension of H.264/AVC [20].

Next, we briefly highlight the error resilience features offered by H.263+, H.263++, and H.264/AVC. The general techniques described above will not be repeated even if they are covered by the two standards.

18.3.5.1 Slice Structure

This mode is defined in H.263+ Annex K replacing the GOB concept in baseline H.263. Each slice in a picture consists of a group of MBs, and these MBs can be arranged either in scanning order or in a rectangle shape. The reason why this mode provides error resilience can be justified in several ways. First, slices are independently decodable without using information from other slices (except for the information in the picture header), which helps limit the region affected by errors and reduce error propagation. Second, the slice header itself serves as a resynchronization marker, thus, further reducing the loss probability for each MB. Third, slice sizes are highly flexible and can be sent and received in any order relative to each other, which can help minimize latency in lossy environment. For these reasons, this mode is also defined in the H.264/AVC standard.

18.3.5.2 Independent Segment Decoding

The independent segment decoding mode is defined in H.263+ Annex R. In this mode, picture segment (defined as a slice, a GOB, or a number of consecutive GOBs) boundaries are enforced by not allowing dependencies across the segment. This mode limits error propagation between well-defined spatial parts of a picture, thus enhancing the error resiliency capabilities.

18.3.5.3 Reference Picture Selection

The RPS mode is defined in H.263+ Annex N, which allows the encoder to select an earlier picture rather than the immediate previous picture as the reference in encoding the current picture. The RPS mode can also be applied to individual segments rather than full pictures. The information as to which picture is selected as the reference is conveyed in the picture/segment header. The VRC technique discussed above can be achieved through this mode.

When a feedback channel is available, the error resilience capability can be greatly enhanced by using this mode. For example, if the sender is informed by the receiver through a NACK that one frame is lost or corrupted during transmission, the encoder may choose not to use this picture for future prediction and instead choose an unaffected picture as the reference. If a feedback channel is not available, this mode can still be used to provide error resiliency capability, for example, through the VRC scheme discussed above.

18.3.5.4 Flexible Macroblock Ordering

In the H.264/AVC standard, each *slice group* is a set of MBs defined by a macroblock to slice group map, which specifies the slice group to which each macroblock belongs. The MBs in one slice group can be in any scanning pattern, for example, interleaved mapping, the group can contain one or more foreground and background slices and the mapping can even be a checker-board type mapping. In addition, each slice group itself can be partitioned into one or more slices, following the raster scan order. Thus, FMO provides a very flexible tool to group MBs from different locations into one slice group, which can help deal with error-prone channels.

Note that besides the above description, the FEC mode (Annex H)[4] is also designed for supporting error resilience [40]. As FEC refers to channel coding, we will discuss it in detail in the next section (as we can see here, the interface between source coding and channel coding becomes vague). As mentioned above, H.263+ Annex O defines temporal, spatial, and SNR scalability, and H.263++ defines data partitioning and RVLC modes in providing error resilience.

As for the H.264/AVC standard, it defines two new frame types called SP- and SI-frames, which can provide functionalities such as bit-stream switching, splicing, random access, error recovery, and and error resiliency [43]. This new feature can ensure drift-free switching between different representations of a video content that use different data rates. H.264/AVC also defines a new mode called *redundant pictures*, which increases error robustness by sending an additional representation of regions of pictures for which the primary representation has been lost. In addition to the above error, resiliency tools defined in video coding layer (VCL), there are some other advanced features to improve error resilience defined in the network abstraction layer (NAL), such as parameter set and NAL unit syntax structures. In using the parameter set structure, the key information such as sequence header or picture header can be separately handled in a more flexible and specialized manner so as to be robust to losses. A detailed discussion can be found in [18].

18.4 CHANNEL CODING

In this section, we discuss the channel coding techniques that are widely used for the transmission of images and video. Two basic techniques used for video transmission are FEC and Automatic Repeat reQuest (ARQ). Each has its own benefits with regard to error robustness and network traffic load [44, 45].

As the name indicates, FEC refers to techniques in which the sender adds extra information known as check or parity information to the source information to make the transmission more robust to channel errors; the receiver analyzes the parity information to locate and correct errors. FEC techniques have become an important channel coding tool used in modern communication systems. One advantage of FEC techniques is that they do not require a feedback channel. In addition, these techniques improve system performance at significantly lower cost than other techniques aiming to improve channel SNR, such as increased transmitter power or antenna gain [15].

Of the two error correction techniques, FEC is usually preferred in real-time video applications because of the delay requirements of these applications. Also, ARQ may not be appropriate for multicast scenarios due to their inherent scalability problems [4, 17]. This is because retransmission typically benefits only a small portion of receivers while all others wait unproductively, resulting in poor throughput. For these reasons, FEC-based

[4]The FEC mode is designed for Integrated Service Digital Network (ISDN) by using the BCH (511, 492) FEC code.

techniques are currently under consideration by the Internet Engineering Task Force (IETF) as a proposed standard in supporting error resilience [46].

The error detection/correction capability of FEC is limited due to the restrictions on the block-size dictated by the application's delay constraints. In addition, the appropriate level of FEC usually depends heavily on the accurate estimation of the channel's behavior. ARQ, on the other hand, can automatically adapt to the channel loss characteristics by transmitting only as many redundant packets as are lost. Compared to FEC, ARQ can usually achieve a level closer to channel capacity. Of course, the trade-off is that larger delays are introduced by ARQ. Thus, if the application has a relatively loose end-to-end delay constraint (e.g., on-demand video streaming), ARQ may be better suited. Even for real-time applications, delay constrained application-layer ARQ has been shown to be useful in some situations [31, 44, 47].

18.4.1 Forward Error Correction

The choice of the FEC method depends on the requirements of the system and the nature of the channel. For video communications, FEC can usually be applied across packets (at the application or transport layer) and within packets (at the link layer) [48]. In interpacket FEC, parity packets are usually generated in addition to source packets to perform cross-packet FEC, which is usually achieved by erasure codes. At the link layer, redundant bits are added within a packet to perform intrapacket prediction from bit errors.

The Internet can usually be modeled as a packet erasure channel [17, 30, 31]. For Internet applications, many researchers have considered using erasure codes to recover packet losses [49]. With such approaches, a video stream is first partitioned into segments and each segment is packetized into a group of m packets. A block code is then applied to the m packets to generate additional l redundant packets (also called parity packets) resulting in a n-packet block, where $n = m + l$. With such a code, the receiver can recover the original m packets if a sufficient number of packets in the block are received. The most commonly studied erasure codes are Reed–Solomon (RS) codes [50]. They have good erasure correcting properties and are widely used in practice, as for example in storage devices (VCD, DVD), mobile communications, satellite communications, digital television, and high-speed modems (ADSL) [49]. Another class of erasure codes that have recently been considered for network applications are Tornado codes, which have slightly worse erasure protecting properties but can be encoded and decoded much more efficiently than RS codes [39].

RS codes are a subset of BCH codes and are linear block codes. An RS code is represented as RS(n, m) with s-bit symbols, where m is the number of source symbols and $l = n - m$ is the number of parity symbols. Figure 18.7 shows a typical RS code word. RS codes are based on Galois fields (GF) or finite fields. RS codes with code words from GF(q) have length equal to $q - 1$. Given a symbol size s, the maximum code word length for an RS code is $n = 2^s - 1$. A popular RS code is chosen from the field GF($2^8 - 1$), since each symbol can be represented as a byte. For the detailed encoding and decoding operation rules and implementations in hardware or software, refer to [51, 52] for a comprehensive tutorial.

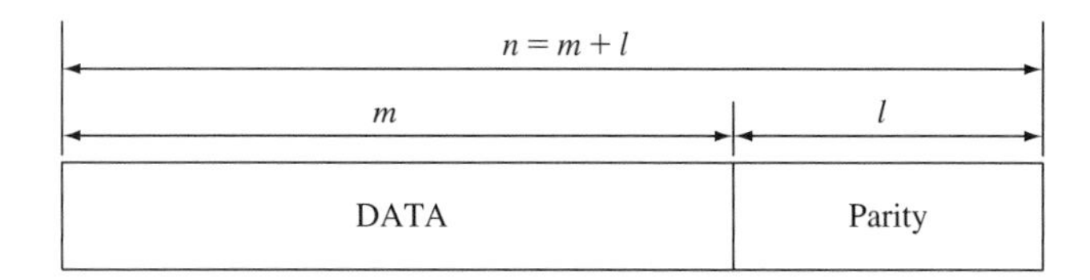

FIGURE 18.7

Illustrate of RS(n,m) code word.

An RS code can be used to correct both errors and erasures (an erasure occurs when the position of an error symbol is known). An RS(n, m) decoder can correct up to $(n-m)/2$ errors or up to $(n-m)$ erasures, regardless of which symbols are lost. The code rate of an RS(n, m) code is defined as m/n. The protection capability of an RS code depends on the block size n and the code rate m/n. These are limited by the extra delay introduced by FEC. The block size can be determined based on the end-to-end system delay constraints.

The channel errors in wired links are typically in the form of packet erasures, so an RS(n, m) code applied across packets can recover up to $(n-m)$ lost packets. Thus, the block failure probability (i.e., the probability that at least one of the original m packets is in error) is $P_b(n,m) = 1 - \sum_{j=0}^{n-m} P(n,j)$, where $P(n,j)$ represents the probability of j errors out of n transmissions. As for wireless channels, channel coding is applied within each packet to provide protection. Source bits in a packet are first partitioned into m symbols, and then $(n-m)$ parity symbols are generated and added to the source bits to form a block. In this case, the noisy wireless channel causes symbol errors within packets (but not erasures). As a result, the block error probability for an RS(n, m) code can be expressed as $P_b(n,m) = 1 - \sum_{j=0}^{(n-m)/2} P(n,j)$.

Another popular type of codes used to perform intrapacket FEC is rate-compatible punctured convolutional (RCPC) codes [48], first introduced in [53]. These codes are easy to implement and have the property of being rate compatible, that is, a lower rate channel code is a prefix of a higher rate channel code. A family of RCPC codes is described by the mother code of rate $1/N$ and memory M with generator tap matrix of dimension $N \times (M+1)$. Together with N, the puncturing period G determines the range of code rates as $R = G/(G+l)$, where l can vary between 1 and $(N-1)G$. RCPC codes are punctured codes of the mother code with puncturing matrices $\boldsymbol{a}(l) = (a_{ij}(l))$ (of dimension $N \times G$), with $a_{ij}(l) \in (0,1)$ and 0 denoting puncturing.

The decoding of convolutional codes is most commonly achieved through the Viterbi algorithm, which is a maximum-likelihood sequence estimation algorithm. The Viterbi upper bound for the bit error probability is given by

$$p_b \leq \frac{1}{G} \sum_{d=d_{\text{free}}}^{\infty} c_d p_d,$$

where d_{free} is the free distance of the convolutional code, which is defined as the minimum Hamming distance between two distinct code words, p_d the probability that the wrong path at distance d is selected, and c_d the number of paths at Hamming distance d from the all-zero path. d_{free} and c_d are parameters of the convolutional code, whereas p_d depends on the type of decoding (soft or hard) and the channel. Both the theoretical bounds of BER and the simulation methods to calculate BER for RCPC codes can be found in [52, 53].

18.4.2 Retransmission

Due to the end-to-end delay constraint of real-time applications, retransmissions used for error control should be delay-constrained. Various delay-constrained retransmission schemes for unicast and multicast video are discussed in [17]. In this chapter, we focus on the unicast case, where the delay-constrained retransmissions can be classified into sender-based, receiver-based, and hybrid control, according to [17].

We illustrate the basic idea of receiver-based retransmission control in Fig. 18.8(a), where T_c is the current time, D_s is a slack term, and $T_d(n)$ is the scheduled playback time for packet n. The slack term D_s is introduced to take into account the error in estimating the RTT and other processing time, such as error correction and decoding. For a detected loss of packet n, if $T_c + RTT + D_s < T_d(n)$, which means if the retransmitted packet n can arrive at the receiver before its playback time, the receiver sends a retransmission request of packet n to the sender. This is the case depicted in Fig. 18.8(a) for packet 2.

Different from the receiver-based control, in sender-based control, decisions are made at the sender end. The basic idea is illustrated in Fig. 18.8(b), where T_0 is the estimated forward-trip-time, and $T'_d(n)$ is an estimate of $T_d(n)$. If $T_c + T_0 + D_s < T'_d(n)$ holds, it can be expected that the retransmitted packet n will arrive at the receiver in time for playback. The hybrid control is a direct combination of the receiver-based and sender-based control so that better performance can be achieved at the cost of higher complexity. After laying out the basic concept of delay-constrained retransmission, we next discuss how retransmission techniques are implemented in a network.

Delay-constrained retransmission can be implemented in multiple network layers. First, it is well known that the Transport Control Protocol (TCP) is a reliable end-to-end transport protocol that provides reliability by means of a window-based positive acknowledgement (ACK) with a go-back-N retransmission scheme in the IP suite [54].

In an IP-based wireless network for the emerging 3G and 4G systems, such as CDMA2000, transport packets are transferred to the radio link control (RLC) frames and further to the Medium Access Control (MAC) frames in the link layer. 3G and 4G systems allow both RLC frame retransmissions and MAC frame retransmissions [55]. The current wireless local area network (WLAN) standard IEEE 802.11 also allows MAC frame retransmission [9]. Compared to transport-layer retransmission TCP, link-layer and MAC-layer retransmission techniques introduce smaller delays because the lower layers react to the network faster than the upper layers [56]. Due to the strict end-to-end delay constraint, TCP is usually not preferred for real-time video communications.

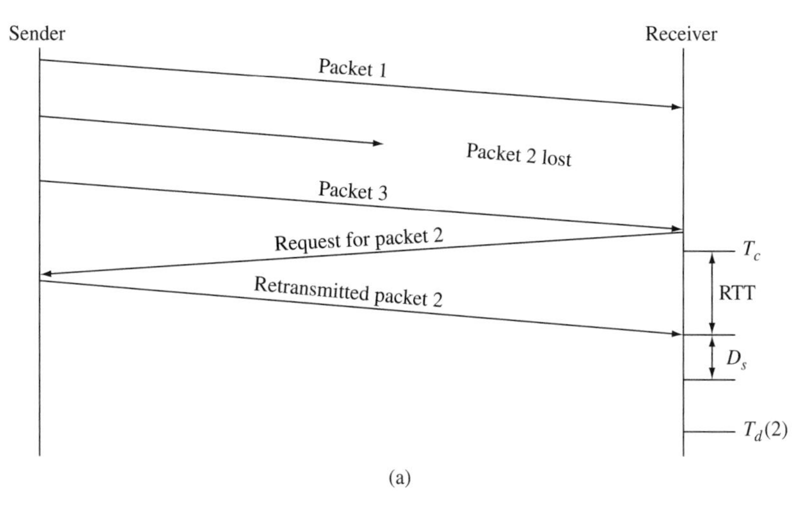

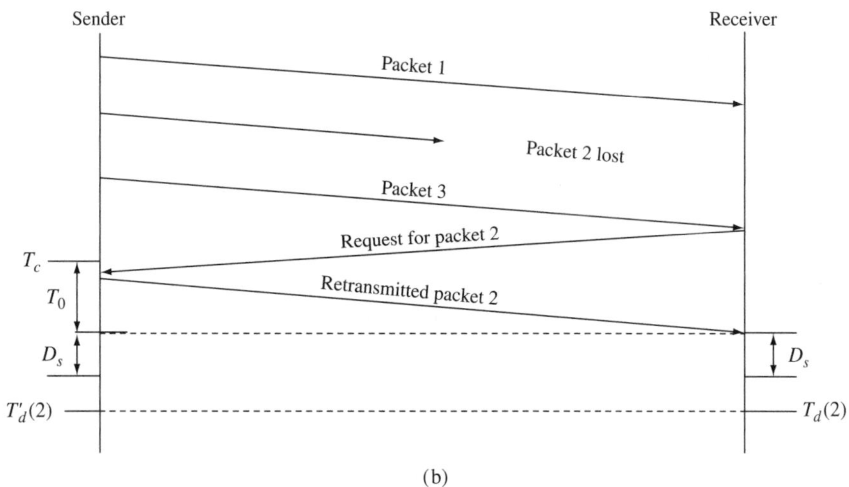

FIGURE 18.8

Timing diagram for delay-constrained retransmission: (a) receiver-based control, (b) sender-based control (adapted from [17]).

However, because delay-constrained retransmission at the link and MAC layers introduce much shorter delays, they are widely used in real-time video communications [9]. For example, researchers have been studying how many retransmissions in the MAC layer are appropriate for multimedia transmission applications to achieve the best trade-off between error correction and delay [9, 56, 57].

18.5 JOINT SOURCE-CHANNEL CODING

The trade-off between source and channel coding has been studied from a theoretical standpoint based on the use of vector quantizers in [12, 13]. In general, JSCC is accomplished by designing the quantizer and entropy coder jointly for given channel characteristics, as in [12, 14]. There is a substantial number of research results in this area. Interested readers can refer to [15] for a comprehensive review on this topic. In this section, we focus on the specific application of JSCC in image and video communications, where JSCC usually faces three tasks: finding an optimal bit allocation between source coding and channel coding for given channel loss characteristics; designing the source coding to achieve the target source rate; and designing the channel coding to achieve the required robustness [16, 17]. These tasks, although stated separately, are inter-related, forming the backbone of the integrated nature of JSCC.

JSCC for image/video transmission applications has been traditionally studied as a typical optimal bit allocation problem. As discussed above, error resilient source coding and channel coding are both error control mechanisms available at the sender. JSCC for image/video transmission applications can also be regarded as techniques that aim to allocate the available resources to these two components to provide the best end-to-end video quality. Next, we first present the general problem formulation for optimal bit allocation and then we focus on these optimal resource allocation techniques.

As a preliminary matter of a formal approach to problem solving, several factors need to be clarified. An appropriate system performance evaluation metric should first be selected. Second, the constraints need to be specified. Third, a model of the relationship between the system performance metric and the set of adaptation parameters needs to be established. The final step is to find the best combination of adaptation parameters that maximizes the system performance while meeting the required constraints. Keeping those four steps in mind, we next present a formal approach to formulate and provide solutions to the JSCC problem.

18.5.1 Problem Formulation

A commonly used criterion for the evaluation of the system performance is the expected distortion. The expectation is required due to the stochastic nature of the channel. As shown in (18.7), in calculating the expected distortion for each source packet, the two distortion terms, $E[D_{R,k}]$ and $E[D_{L,k}]$, and the loss probability for the source packet ρ_k need to be determined. The two distortion terms depend on the source coding parameters such as quantization stepsize and prediction mode, as well as the error concealment schemes used at the decoder. The relationship between the source packet loss probability and channel characteristics depends on the specific packetization scheme, the channel model, and the adaptation parameters chosen.

Let $\mathcal{S}$ be the set of source coding parameters and $\mathcal{C}$ the channel coding parameter. Let $\boldsymbol{s} = \{s_1, \ldots, s_M\} \in \mathcal{S}^M$ and $\boldsymbol{c} = \{c_1, \ldots, c_M\} \in \mathcal{C}^M$ denote, respectively, the vector of source coding parameters and channel coding parameters for the M packets in one video frame or a group of frames. The general problem formulation then is to minimize the total expected distortion for the frame(s), given the corresponding bit-rate constraint [6], that is,

$$\begin{aligned} &\min_{\boldsymbol{s}\in\mathcal{S}^M,\boldsymbol{c}\in\mathcal{C}^M} E[D(\boldsymbol{s},\boldsymbol{c})] \\ &\text{s.t.} \quad R(\boldsymbol{s},\boldsymbol{c}) \leq R_0, \end{aligned} \tag{18.8}$$

where $R(\boldsymbol{s},\boldsymbol{c})$ represents the total number of bits used for both source and channel coding, and R_0 the bit-rate constraint for the frame(s). The bit-rate constraint is usually obtained based on the estimated channel throughput. Note that since video packets are usually of different importance, the optimal solution will result in an UEP cross video packets.

As shown in Fig. 18.6, with the same channel characteristics, different simulations may diverge to a large extent with regard to reconstruction quality. A novel approach called variance-aware per-pixel optimal resource allocation (VAPOR) is proposed in [58] to deal with this. Besides the widely used expected distortion metric, the VAPOR approach aims to limit error propagation from random channel errors by accounting for both the expected value and the variance of the end-to-end distortion when allocating source and channel resources. By accounting for the variance of the distortion, this approach increases the reliability of the system by making it more likely that what the end user sees, closely resembles the mean end-to-end distortion calculated at the transmitter.

This type of constrained problem can be solved in general using the Lagrangian relaxation method; that is, instead of the original problem, the following problem is solved

$$\min_{\boldsymbol{s}\in\mathcal{S}^M,\boldsymbol{c}\in\mathcal{C}^M} J(\boldsymbol{s},\boldsymbol{c},\lambda) = \min_{\boldsymbol{s}\in\mathcal{S}^M,\boldsymbol{c}\in\mathcal{C}^M} \{E[D(\boldsymbol{s},\boldsymbol{c})] + \lambda R(\boldsymbol{s},\boldsymbol{c})\}. \tag{18.9}$$

The solution of (18.8) can be obtained, within a convex hull approximation, by solving (18.9) with the appropriate choice of the Lagrange multiplier, $\lambda \geq 0$ so that the bit-rate constraint is satisfied. The difficulty in solving the resulting relaxed problem depends on the complexity of the interpacket dependencies. Depending on the nature of such dependencies, an iterative descent algorithm based on the method of alternating variables for multivariate minimization [59] or a deterministic dynamic programming algorithm [60] can be used to efficiently solve the minimization problem.

The JSCC problem formulation (18.8) is general for the fact that both the source coding and channel coding can take a variety of forms, depending on the specific application. For example, when FEC is used, the packet loss probability becomes a function of the FEC choice. The details of this model will depend on how transport packets are formed from the available video packets [34]. In addition to FEC, retransmission-based error control may be used in the form of ARQ protocols. In this case, the decision whether to retransmit a packet or to send a new one forms another channel coding parameter, which also affects the probability of loss as well as the transmission delay. When considering the transmission of video over a network, a more general JSCC scheme may cover modulation and demodulation [61], power adaptation [32], packet scheduling [62], and data rate adaptation [62]. These adaptation components can all be regarded as channel coding parameters. Source coding parameters, on the other hand, can be in the form of mode selection [16, 32, 34], packetization [9], intra-MB refreshment rate [37], and entropy coding mechanism [12]. By solving problem (18.8) and selecting the source coding and channel coding parameters within their sets, we can obtain the optimal trade-off among all those adaptation components. We next provide examples of the applications of JSCC to video transmission in different network infrastructures.

18.5.2 Internet Video Transmission

For video transmission over the Internet, channel coding usually takes the form of FEC and/or ARQ at the transport layer. FEC is usually preferred for applications that impose a relatively short end-to-end delay constraint. Joint source coding and FEC has been extensively studied in the literature [33, 49, 63–65]. Such studies focus on the determination of the optimal bit allocation between source coding and FEC. In [66], the authors introduced the integrated joint source-channel coding (IJSCC) framework, where error resilient source coding, channel coding, and error concealment are jointly considered in a tractable optimization setting. In using the IJSCC framework, an RD optimized hybrid error control scheme has been presented in [66], which results in the optimal allocation of bits among source, FEC, and ARQ.

18.5.2.1 Joint Source Coding and FEC

As mentioned above, the appropriate way of calculating the loss probability per packet depends on the chosen FEC, as well as the way transport packets are formed from the available video packets. Next, we show one example where the source packet is a video slice (a group of blocks).

Figure 18.9 illustrates a packetization scheme for a frame, where one row corresponds to one packet. In this packetization scheme, one video slice is directly packetized into one transport packet by the attachment of a transport packet header. Since the source packet sizes (shown by the shaded area in Fig. 18.9) are usually different, the maximum packet size of a block (a group of packets protected by one RS code) is determined first, and then all packets are padded with stuffing bits in the tail part to make their sizes equal. The stuffing bits are removed after the parity codes are generated and thus are not transmitted. The resulting parity packets are all of the same size (maximum packet size mentioned above). Each source packet in Fig. 18.9 is protected by an RS(N, M) code, where M is the number of video packets and N is the number of total packets including parity packets.

In this case, the channel coding parameter $\boldsymbol{c}$ would be the choice of the RS code rate. If we take the source coding parameter $\boldsymbol{s}$ as the prediction mode and quantizer for each video packet, by solving (18.8), we can obtain the optimal JSCC solution, that is, the optimal bit allocation as well as the optimal error resilient source coding and FEC.

To illustrate the advantage of the IJSCC approach, we compare two systems: (i) system 1, which uses the proposed framework to jointly consider error resilient source coding

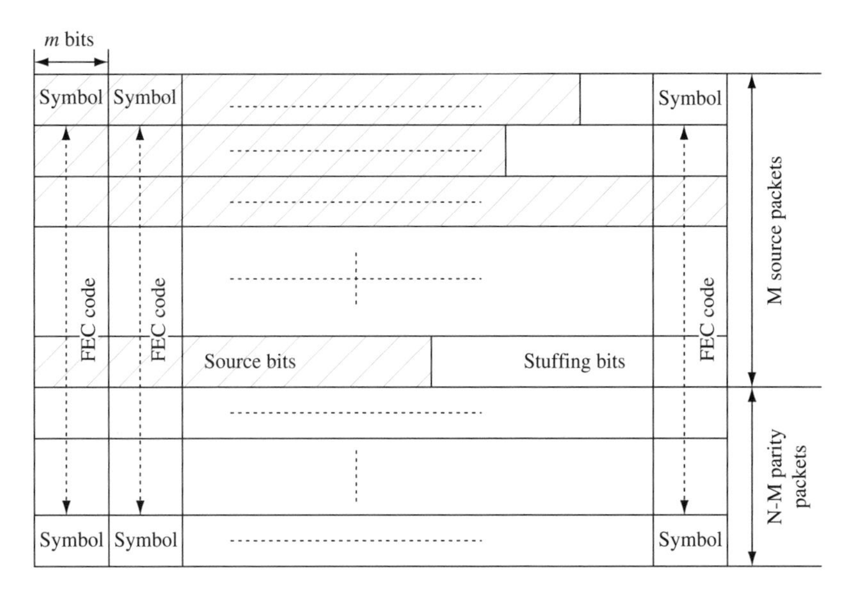

FIGURE 18.9

Illustration of a packetization scheme for interpacket FEC.

and channel coding; (ii) system 2, which performs error resilient source coding but with fixed rate channel coding. Note that system 2 is also optimized, that is, it performs optimal error resilient source coding to adapt to the channel errors (with fixed rate channel coding).

In Fig. 18.10, the performance of the two systems is compared, using the QCIF Foreman test sequence coded by an H.263+ codec at transmission rate 480 kbps and frame rate 15 fps. Here, we plot the average PSNR in dB versus different packet loss rates. It can be seen in Fig. 18.10 that system 1 outperforms system 2 at different preselected channel coding rates. In addition, system 1 is above the envelope of the four performance curves of system 2 by 0.1–0.4 dB. This is due to the flexibility of system 1, which is capable of adjusting the channel coding rate in response to the CSI as well as the varying video content.

18.5.2.2 Joint Source Coding and Hybrid FEC/ARQ

When considering the use of both FEC and ARQ, the channel coding parameter $\boldsymbol{c}$ includes the FEC rate chosen to protect each packet and the retransmission policy for each lost packet. Hybrid FEC/retransmission has been considered in [31], where a general cost-distortion framework was proposed to study several scenarios such as DiffServ (Differentiated Services), sender-driven retransmission, and receiver-driven retransmission. In [66], optimal error control is performed by jointly considering source coding with hybrid FEC and sender-driven application-layer selective retransmission. This study is carried out with the use of (18.8), with a sliding window scheme in which lost packets are

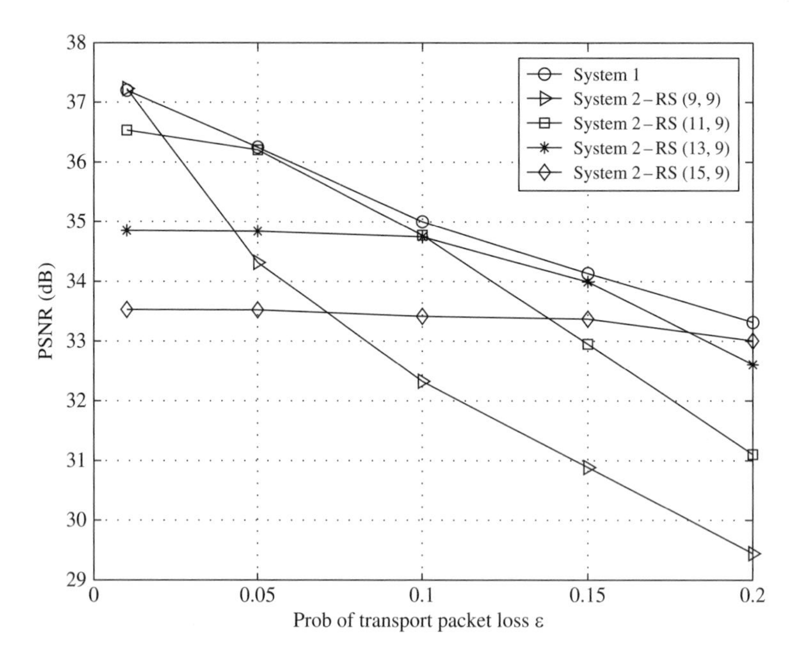

FIGURE 18.10

Average PSNR versus transport packet loss probability (QCIF Foreman sequence, transmission rate 480 kbps, coded at 15 fps).

selectively retransmitted according to a RD optimized policy. Simulations in [66] show that the performance advantage in using either FEC or selective retransmission depends on the packet loss rate and the RTT. In that work, the proposed hybrid FEC and selective retransmission approach is able to derive the benefits of both approaches by adapting the type of error control based on the channel characteristics.

A receiver-driven hybrid FEC/pseudo-ARQ mechanism is proposed for Internet multimedia multicast in [45]. In that work, the sender multicasts all the source layers and all the channel layers (parity packets obtained by using RS coding similar to what we have discussed in the previous section) to separate multicast groups. Each user computes the optimal allocation of the available bit rate between source and channel layers based on its estimated channel bandwidth and packet loss probability, and joins the corresponding multicast group. This is achieved through a pseudo-ARQ system, in which the sender continuously transmits delayed parity packets to additional multicast group, and the receivers can join or leave a multicast group to retrieve the lost information up to a given delay bound. Such a system looks like ordinary ARQ to the receiver and an ordinary multicast to the sender. This can be characterized as JSCC with receiver feedback. More specifically, the optimal JSCC is obtained by solving (18.8) at the receiver side, where the source coding parameter is the number of source layers and the channel coding parameter is the number of channel layers.

18.5.3 Wireless Video Transmission

Wireless video communications is a broad, active, and well-studied field of research [7, 67]. Recently, several adaptation techniques have been proposed specifically for energy efficient wireless video communications. A trend in this field of research is the joint adaptation of source coding and transmission parameters based on the time-varying source content and channel conditions. The general JSCC framework (18.8) therefore encompasses these techniques with an additional constraint on the total energy consumed in delivering the video sequence to the end user. Correspondingly, the channel coding parameters would cover more general channel adaptation parameters such as the transmission rate, physical-layer modulation modes, and the transmitter power.

18.5.3.1 Joint Source Coding and FEC

As with Internet video transmission, the problem of joint source coding and FEC for wireless video communications focuses on the optimal bit allocation between source and channel coding by solving (18.8). The difference is that due to the different types of channel errors (bit errors) in a wireless channel, FEC is achieved by adding redundant bits within packets to provide intrapacket protection. RCPC and RS codes are widely used for this purpose. Recent studies have considered using turbo codes [68, 69] due to their ability to achieve capacity close to Shannon's bound [70]. As mentioned above, this topic has been extensively studied. Some examples are provided in the following.

Optimal bit allocation has been studied in [64] based on a subband video codec. A binary symmetric channel (BSC) with an AWGN model has been considered for simulations. The source coding parameters are the bit rate of the source subband and the channel coding parameters are the FEC parameter for each subband. A similar problem has been studied for video transmission over a Rayleigh fading wireless channel in [33] based on an H.263+ SNR scalable video codec. In that work, Universal RD Characteristics (URDC) of the source scheme are used to make the optimization tractable. Both studies use RCPC codes to achieve the intrapacket FEC. RS codes are used to perform channel coding in [37] for video transmission over a random BSC. Based on their proposed RD model, the source coding parameter is the intra-MB refreshment rate and the channel coding parameter is the channel rate.

An adaptive cross-layer protection scheme is presented in [9] for robust scalable video transmission over 802.11 wireless LANs. In this study, the video data is pre-encoded using an MPEG-4 FGS coder and thus source coding adaptation is not considered. The channel parameters considered in this study include adaptation components in various layers in the protocol stack, including the application layer FEC (RS codes are used to achieve inter-packet protection), the MAC retransmission limit, and the packet sizes. These channel parameters are adaptively and jointly optimized for efficient scalable video transmission.

As for image transmission, joint source channel coding algorithms have been studied in [71] for efficient progressive image transmission over a BSC. In this study, two UEP schemes have been studied: a progressive rate-optimal scheme and a progressive distortion-optimal scheme. With a given transmission bit budget, in the progressive rate-optimal scheme, the goal is to maximize the average of the expected number of correctly received source bits over a set of intermediate rates. In the progressive optimal-distortion problem, the goal is to minimize the average of the expected distortion over a set of intermediate rates. Embedded coder such as the SPIHT coder [22] and JPEG2000 is used for source coding. As for channel coding, an RCPC code and a rate-compatible punctured turbo code are used for the rate-optimal scheme and distortion-optimal scheme, respectively. Thus, the source coding parameter is the number of embedded bits allocated to each packet, and the channel coding parameter is the selected channel rates for the protection of each packet. Although the proposed schemes have slightly worse performance at a few rates close to the target rate compared with the schemes that are maximizing the correctly received source bits or minimizing the expected distortion at the target transmission rate, the proposed schemes have better performance at most of the intermediate rates.

18.5.3.2 Joint Source Coding and Power Adaptation

Joint source coding and power adaptation techniques deal with the varying error sensitivity of video packets by adapting the transmission power per packet based on the source content and CSI. In other words, these techniques use transmission power as part of an UEP mechanism. In this case, the channel coding parameter is the power level for each video packet. Video transmission over CDMA networks using a scalable source coder (3-D SPIHT) along with error control and power allocation is considered in [72]. A joint source coding and power control approach is presented in [73] for optimally allocating source coding rate and bit energy normalized with respect to the multiple-access interference noise density in the context of 3G CDMA networks. In [32], optimal mode and quantizer selection are considered jointly with transmission power allocation.

To illustrate some advantages of joint adaptation of the source coding and transmission parameters in wireless video transmission systems, we present experimental results that are discussed in detail in [32]. We compare a joint source coding and transmission power allocation (JSCPA) approach, that is, the approach described by (18.8), with an independent source coding and power allocation (ISCPA) approach in which $\boldsymbol{s}$ and $\boldsymbol{c}$ are independently adapted. In Fig. 18.11, we plot the expected PSNR per frame of both approaches for the Foreman test sequence coded at 15 fps. It is important to note that both approaches use the same transmission energy and delay per frame.

As shown in Fig. 18.11, the JSCPA approach achieves significantly higher quality (expected PSNR) per frame than the ISCPA approach. Because the video encoder and the transmitter operate independently in the ISCPA approach, the relative importance of each packet, that is, their contribution to the total distortion, is unknown to the transmitter. Therefore, the transmitter treats each packet equally and adapts the power to maintain a constant probability of packet loss. The JSCPA approach, on the other hand, is able to adapt the power per packet, and thus the probability of loss, based on the relative importance of each packet. For example, more power can be allocated to packets that are difficult to conceal. As shown in Fig. 18.11, the PSNR improvement is the greatest during periods of high activity. For example, around frame 100, there is a scene change in which the camera pans from the foreman to the construction site. During this time, the JSCPA

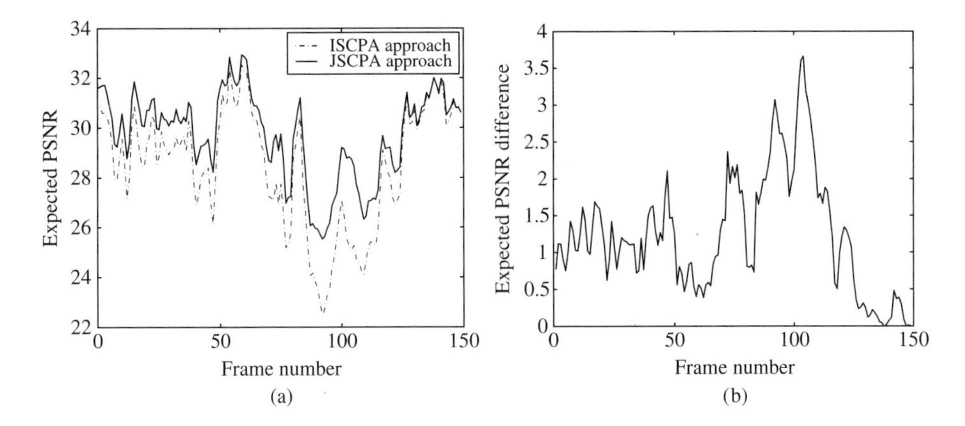

FIGURE 18.11

(a) Expected PSNR per frame for the ISCPA and JSCPA approaches. (b) Difference in expected PSNR between the two approaches (adapted from [32]).

approach achieves PSNR improvements of up to 3.5 dB. This gain comes from the ability of the JSCPA approach to increase the power while decreasing the number of bits sent to improve the reliability of the transmission. The ISCPA scheme is unable to adapt the protection level and thus incurs large distortion during periods of high source activity.

We show the visual quality comparison of the two approaches in Fig. 18.12. An expected reconstructed frame is shown from the Foreman sequence when the same amount of energy is consumed in the two approaches. It can be clearly seen that the JSCPA approach achieves a much better video reconstruction quality than the ISCPA approach.

As mentioned in Section 18.5.1, the VAPOR approach is used to limit error propagation by accounting for not only the mean but also the variance of the end-to-end distortion [58]. In Fig. 18.13, we compare a series of reconstructed frames at the decoder for the minimum expected distortion (MED) approach (18.8) and VAPOR approach using the same amount of transmission energy for the Silent sequence. These images are for a single-channel loss realization when the same MBs are lost in both schemes. We can clearly see the advantage of using the VAPOR approach. For example, the error occurring at frame 109 persists until frame 123 in the MED approach while it has been quickly removed by the VAPOR approach.

18.5.3.3 Joint Source-Channel Coding and Power Adaptation

In an energy-efficient wireless video transmission system, transmission power needs to be balanced against delay to achieve the best video quality. Specifically, for a given transmission rate, increasing the transmission power will decrease BER, resulting in a smaller probability of packet loss. On the other hand, for a fixed transmission power, increasing the transmission rate will increase the BER but decrease the transmission

(a)

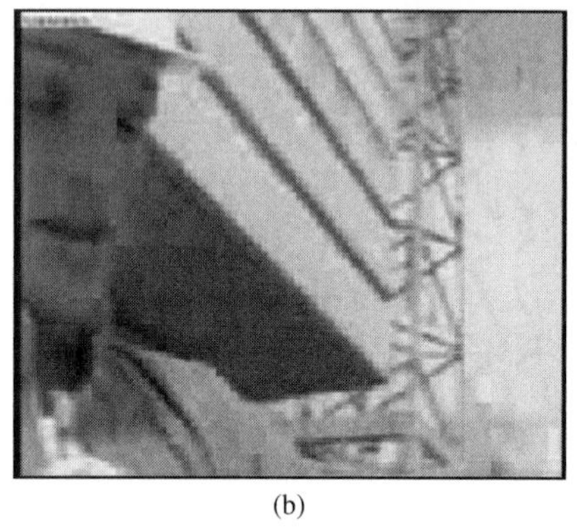

(b)

(c)

FIGURE 18.12

Frame 184 in the Foreman sequence. (a) Original frame; (b) expected frame at the decoder using the JSCPA approach; and (c) expected frame at the decoder using the ISCPA approach (adapted from [32]).

delay needed for a given amount of data (or allow more data to be sent within a given time period). Therefore, to efficiently use resources such as energy and bandwidth, those two adaptation components should be designed jointly.

In [74], the study is based on progressive image and video transmission, and error resilient source coding is achieved through optimized transport prioritization for layered video. Joint source-channel coding and processing power control for transmitting layered video over a 3G wireless network is studied in [75]. A joint FEC and transmission power allocation scheme for layered video transmission over a multiple user CDMA network is proposed in [72] based on the 3D-SPIHT codec. Source coding and error concealment are not considered in that work.

Source-channel coding and power adaptation can also be used in a hybrid wireless/wireline network, which consists of both wireless and wired links. An initial investigation of this topic is described in [34], where lower layer adaptation includes interpacket FEC at the transport layer and intrapacket FEC at the link layer, which are used

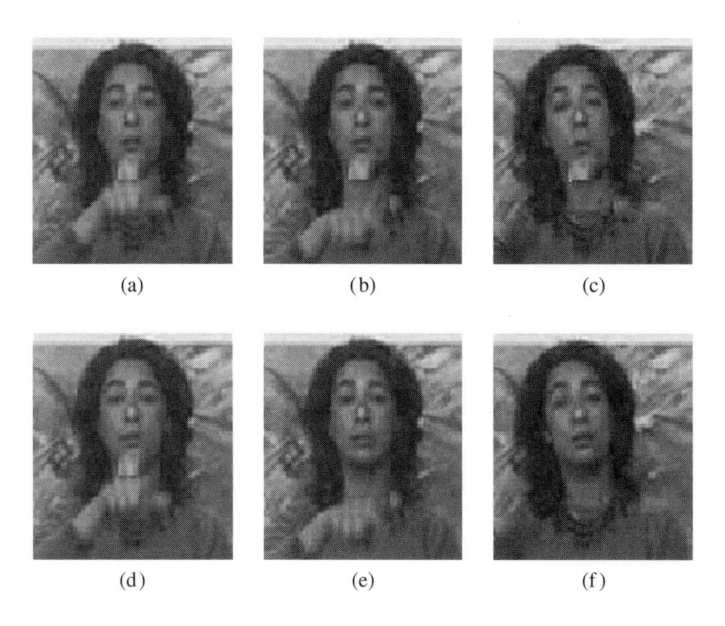

FIGURE 18.13

Illustration of error propagation effect using the QCIF Silent test sequence. MED approach: frame number (a) 109, (b) 110, (c) 123. VAPOR approach: frame number (d) 109, (e) 110, (f) 123 (adapted from [58]).

to combat packet losses in the wired line and bit errors in the wireless link, respectively. In addition to channel coding, power is assumed to be adjustable in a discrete set at the physical layer. The selection of channel codes and power adaptation is jointly considered with source coding parameter selection to achieve energy-efficient communication.

18.5.3.4 Joint Source Coding and Data Rate Adaptation

Joint source coding and transmission rate adaptation has also been studied as a means of providing energy efficient video communications. To maintain a certain probability of packet loss, the energy consumption increases as the transmission rate increases [76]. Therefore, to reduce energy consumption, it is advantageous to transmit at the lowest rate possible [77]. In addition to affecting energy consumption, the transmission rate determines the number of bits that can be transmitted within a given period of time. Thus, as the transmission rate decreases, the distortion due to source coding increases. Joint source coding and transmission rate adaptation techniques adapt the source coding parameters and the transmission rate to balance energy consumption against end-to-end video quality.

In [62], the authors consider optimal source coding and transmission rate adaptation. In this work, each video packet can consist of different number of MBs and can be transmitted at different rates. In addition, each packet has the option to stay idle at the transmitter instead of being transmitted immediately after it leaves the encoder buffer, when the current channel condition is very poor. The channel coding parameters $\boldsymbol{c}$ then includes the selection of the number of MBs in each video packet, the transmission rate, and transmission schedule (waiting time at the transmitter) for each packet. The source coding parameters are the quantizer and prediction mode for each MB in each video packet. The goal is then to jointly select the source coding and channel coding parameters for each packet with the objective of minimizing the total expected energy required to transmit the video frame subject to both an expected distortion constraint and a delay per packet constraint, through solving the minimum cost optimization formulation. In this work, stochastic Dynamic Programming is used to find an optimal source coding and transmission policy based on a Markov state channel model. A key idea in this work is that the performance can be improved by allowing the transmitter to suspend or slowdown transmissions during periods of poor channel conditions, as long as the delay constraints are not violated.

18.6 DISTRIBUTED MULTIMEDIA COMMUNICATIONS

In the previous sections of this chapter, we mainly focused on point-to-point video transmission, where two single end points are communicating with each other. In this section, we cover distributed multimedia communications over different wireless network architectures, including point-to-multipoint networks, P2P networks, and multihop networks. We first study video streaming over multiuser networks, in which we discuss in detail how optimal resource allocation can be performed in this paradigm. We then briefly review the prevalent mobile TV broadcasting standards, with focus on the error resiliency and power efficiency features. We also briefly review video streaming applications over some emerging network architectures including P2P networks and multihop networks.

18.6.1 Video Streaming over Multiuser Networks

In a point-to-multipoint network, video can be transmitted from a single transmitter to multiple receivers. The new generation of cellular standards, such as HSDPA (High-Speed Downlink Packet Access) [78] and IEEE 802.16 (*WiMAX*) [79], are geared to achieve the required data rates to support multiuser video transmission. For example, Multimedia Broadcast/Multicast Service (MBMS) is one of the four types of visual content delivery services and technologies defined in Third-Generation Partnership Project (3GPP) standards [80, 81]. In addition, as the combination of Wi-Fi and 3G cellular networks is providing a realistic and comfortable solution beyond 3G, where 54 Mbps in hot spots and several hundred kilobits per second with wide coverage are available, the popularity of video broadcast/multicast applications has gained an increasing momentum (e.g., the iPhone released in July 2008 supports both Wi-Fi and 3G).

In multiuser video transmission, the inherent heterogeneity of video data and the diversity of channel and network conditions experienced by the different clients makes it difficult to achieve bandwidth efficiency and service flexibility. Overcoming the challenges requires the efficient scheduling and allocation of system resources, such as transmission power and bandwidth, and also, the consideration of source content.

Recent developments in scheduling and resource allocation exploit the time-varying nature of the wireless channel to maximize network throughput. These methods rely on the multiuser diversity gain achieved by allocating a majority of the available resources to users with good channel quality who can support higher data rates [82, 83]. In addition to high overall throughput, the resource allocation must also ensure fairness across users [84–87]. Most of the existing schemes can be generalized as gradient-based scheduling policies [88, 89]. Gradient-based policies define a user utility as a function of a QoS measure, such as throughput, and then maximize a weighted sum of the users' data rates, where the weights are determined by the gradient of a utility function. For example, choosing the weights to be the reciprocals of the long-term average throughputs of each user leads to a proportionally fair scheduling scheme [86]. Choosing the weights based on the delay of the head-of-line (HOL) packet of each user's transmission queue leads to a delay-sensitive scheduling scheme [87]. Gradient-based policies rely only on the instantaneous channel states of each user to determine the resource allocations and do not assume any knowledge of the underlying channel state distributions.

Packet scheduling schemes for video streaming that also take into account the video content can be found in [90–92]. In [90], a heuristic approach is used to determine the importance of frames across users based on the frame type (I, P, or B) or their position in a group of pictures. In [91], a concept of incrementally additive distortion among video packets, introduced in [31], is used to determine the importance of video packets for each user. Scheduling across users, however, is performed using conventional, content-independent techniques. In [92], the priority across users is determined as a combination of a content-aware importance measure similar to that in [91], and the delay of the HOL packet for each user. At each time slot, all the resources are dedicated to the user with the highest priority.

The scheduling problem for video broadcasting has also been studied in terms of the tradeoff of distortion versus delay in [93, 94] according to their proposed Multiple Distortion Measures. In this work, the video source is pre-encoded using H.264/MPEG-4 SVC, and channel decisions are whether to serve the HOL packet by transmitting/retransmitting it or to service the following packet by dropping it at each time slot. The optimal trade-off has been casted as a stochastic shortest path problem and solved using dynamic programming methods.

In [95], a content-based utility function is integrated with the utility-based framework of [88] to form a content-aware packet scheduling technique. The key to deriving the utility function is to appropriately prioritize the video packets according to their "importance." The importance is measured in terms of the distortion of the received video signal taking into account the concealment of packet losses. The scheduling algorithm maximizes the sum of the instantaneous data rates assigned to each user at each scheduling opportunity, which are a function of the channel parameters (c_i) assigned to each user i, weighted by the gradients of the distortion-based utility function. This can be written as,

$$\max_{c \in C} \sum_{i=1}^{K} w_i u_i r_i(c_i, e_i), \tag{18.10}$$

where $\boldsymbol{c} = \{c_1, c_2, \ldots, c_K\}$ contains the channel parameter set for all K users, u_i is the reduction in distortion after error concealment, in the video sequence of user i due to the transmission of an additional video packet, w_i denotes an additional fairness parameter, and $r_i(c_i, e_i)$ denotes the achievable information rate for user i given the channel parameters, c_i, and the channel fading state, e_i. $\boldsymbol{C}$ represents the set of all possible channel parameter allocations, which is constrained by the requirements of the system. The optimization is performed at each transmission opportunity to determine the transmission rate provided to each user. Video packets in the transmission buffer for each user are transmitted according to their priority, and any packets whose decoding deadlines have expired are removed from the transmission buffer and considered lost at the decoder. In a TDM/CDMA-based system such as HSDPA, the controllable channel parameters can consist of the number of spreading codes assigned to each user, n_i and the allocated transmission power per user, p_i, at each scheduling opportunity, and then, (18.10) is of the form

$$\max_{(\boldsymbol{n},\boldsymbol{p})\in \boldsymbol{C}} \sum_{i=1}^{K} w_i u_i \log\left(1 + \frac{p_i e_i}{n_i}\right), \tag{18.11}$$

where the channel parameters, $\boldsymbol{n} = \{n_1, n_2, \ldots, n_K\}$, $\boldsymbol{p} = \{p_1, p_2, \ldots, p_K\}$ are constrained by the maximum number of spreading codes available in the system (15 for HSDPA) and the maximum transmission power. A more detailed description of such a scheme is given in [95].

In [96], the above approach is extended to the case when only an imperfect estimate of the channel state is available. In that case, in addition to video packets being dropped from the transmission buffer due to the expiry of packet deadlines, random losses can also occur in the wireless channel. Such random losses combined with error concealment at the decoder make it impossible for the scheduler to determine the actual distortion of the sequence at the receiver. Instead, the scheduler computes an expected distortion using a per-pixel estimation technique [35]. The expected distortion is a function of both the transmission rate (number of packets transmitted), as well as the packet loss rate. An outage capacity model [97] is used to determine the probability of channel loss based on the estimated channel state. In that case, the probability of loss is dependent on the allocated resources and the transmission rate. Therefore, assuming a system as in (18.11), the problem of minimizing the expected distortion across all users can be written as

$$\min_{(\boldsymbol{n},\boldsymbol{p})\in \boldsymbol{C},\boldsymbol{r}} \sum_{i=1}^{K} E\{D_i[r_i, \varepsilon_i(n_i, p_i, r_i, e_i)]\}, \tag{18.12}$$

where $E\{D_i\}$ represents the expected distortion at user i's decoder, $\boldsymbol{r} = \{r_1, r_2, \ldots, r_K\}$ is the vector of transmission rates to each user, and ε_i represents the packet loss rate, which is in turn dependent on the channel parameters and allocated transmission rate.

In [96], the above optimization is solved using a two-step approach that consists of first fixing the probability of loss and calculating the optimal resource parameters for the fixed probability of loss and then optimizing the transmission rate given the resource parameters.

Figure 18.14 shows the average and variance of received PSNR obtained in a multiuser video transmission system using the above scheduling and resource allocation scheme. The system uses 7 QCIF sequences encoded such that the average PSNR without losses is 35 dB. The resource constraints are set similar to an HSDPA system with a maximum transmission power of 25 W. Four different scheduling methods are compared. ED gradient denotes the scheme described above. ED gradient with fixed ε denotes the scheme described above but using a fixed value of ε of 0.1 instead of optimizing the transmission rate allocation over ε. The queue-length-based scheme also fixes ε but uses the queue length of each user's transmission queue instead of the expected distortion gradient to determine the utility function. The final scheme is a maximum throughput based scheme (max C/I) that only considers each user's channel state for scheduling decisions. The figures show that the content-dependent schemes significantly outperform the content-independent schemes such as queue length and max C/I. The max C/I scheme shows a significantly larger variance across users than the others. The content-dependent schemes show the smallest variance across users.

Figure 18.15 provides an example comparison of the output from the content-dependent expected distortion gradient based scheme and the content-independent queue-length-based scheme. Similar schemes to those above can also be used with scalable video coding techniques [98, 99]. The schemes described in this section attempt to tackle the problem of cross-layer design for wireless multiuser systems while taking into account the actual perceived video quality to the end user. They simplify the multiuser

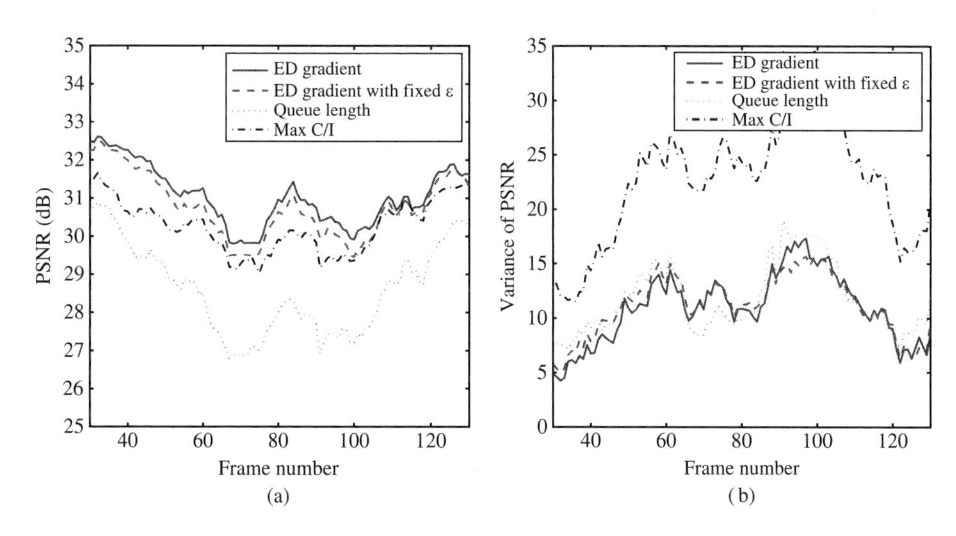

FIGURE 18.14

(a) Average PSNR over all users. (b) Variance of PSNR across users and channel realizations.

FIGURE 18.15

Three frames for comparison of content-aware scheme (a), (b), (c), and content-independent queue-length-based scheme (d), (e), (f) under the same channel realization.

resource allocation problem by using a gradient-based scheduling policy. Results such as those illustrated above indicate that content-dependent scheduling and resource allocation provide great potential for improving the QoS in multiuser video transmission systems. Interesting new challenges also occur when similar schemes are extended to uplink multiuser video transmission, video transmission in multihop networks, etc.

18.6.2 Mobile TV Standards

Mobile television services have been gaining popularity and are expected to grow considerably in the coming years. Currently, there are four prevalent mobile TV standards known as Digital Multimedia Broadcasting (DMB), DVB-H Digital Video Broadcasting - Handheld (DVB-H), OneSeg, and MediaFLO. DMB was developed and is used in South Korea, while OneSeg was developed and is used in Japan. MediaFLO is Qualcomm's proprietery technology and is leading the North American market as it has been deployed by Verizon and AT&T. DVB-H, mainly developed by Nokia and officially endorsed by the European Union, has started to get momentum in North America in addition to Europe and is likely about to become a globally dominant standard.

In this subsection, our goal is not to cover in detail all the standards but to highlight some of the key technologies of error resiliency and power efficiency defined in these mobile TV broadcasting systems. The reader can then have a better idea of how to evaluate the trade-offs to achieve the best video quality from the JSCC or optimal resource allocation perspective. We next briefly review the DVB-H and MediaFLO standards.

18.6.2.1 DVB-H

The DVB-H system is defined based on the existing DVB-T (DVB-Terrestrial) standard for fixed and in-car reception of digital TV. It is a superset of the DVB-T system, with additional features to meet the specific requirements of the error resiliency and power efficiency for handheld receivers. For example, the main additional elements at the link layer are time slicing and additional FEC coding, also known as multiprotocol encapsulation (MPE) FEC, which provides equal error protection (EEP) to the transmitted media streams.

Time slicing, which is based on time-multiplexed transmission of different services, is used to reduce power consumption for small handheld terminals. Data are transmitted in bursts at high bit rates in small time slots, allowing the receiver to be switched off in inactive periods. With MPE-FEC, the IP datagrams of each time-sliced burst are protected by Reed–Solomon parity codes, calculated from the IP datagrams of the burst. The front-end receiver only switches on for the time interval when the burst data is on air. Time slicing can significantly reduce the average power at the front-end receiver (up to about 90–95%) and also enable smooth and seamless frequency handover when a user changes service cell area [100]. In addition, because there is no requirement for fixed burst sizes or fixed time between bursts, a variable bit-rate-coded video stream can therefore use a variable burst size and/or a variable time between bursts. Note that in DVB-H, the use of time slicing is mandatory while in MPE-FEC, it is optional. Also note that the MPE-FEC module offers, in addition to error correction in the physical layer transmission, a complementary FEC function that allows the receiver to cope with particularly difficult reception situations.

In addition to the link layer extensions, DVB-H defines four additional features at the physical layer with respect to DVB-T to further improve the receiver performance and design flexibility for handheld applications. For example, the optional new orthogonal frequency division multiplexing (OFDM), 4K mode provides a good compromise between high-speed, small-area 2K single-frequency networks (SFNs) and the slower, larger-area 8K mode. It also gives additional flexibility to network design. Readers can refer to [101] for details or to [100] for a review.

18.6.2.2 MediaFLO

MediaFLO is Qualcomm's proprietery technology to broadcast data and media streams, as well as IP datacast application data such as stock market quotes, sports scores, and weather reports in real time and non-real-time to mobile devices. FLO in MediaFLO stands for forward link only, meaning that the data transmission path is one way from the transmitter to the FLO device. The MediaFLO system transmits data on a frequency separate from the frequencies used by current cellular networks. More specifically, the FLO physical layer targets transmission in the VHF/UHF/L-band frequency bands, over channel bandwidths of 5, 6, 7, and 8 MHz [102]. It can achieve an average bit rate of 200–250 kbps, which can provide a reasonably good video reception quality, especially on handheld devices that have relatively small screens, thanks to the use of advanced video coding codecs such as H.264/AVC.

OFDM is used at the physical layer with each subcarrier modulated by a quadrature amplitude modulated (QAM) symbol. The use of OFDM enables higher capacity compared with single-carrier modulation methods. Moreover, all the transmitters send the same signal and use the same frequency. This allows the mobile devices to decode the signal from more than one transmitter in the same way as if it were a delayed version from the same transmitter. Similar to DVB-H, this SFN scheme avoids explicit handoff operation.

Forward error correction is used in MediaFLO for the benefits of OFDM over time/frequency-selective channels. The FEC design is based on a concatenated coding scheme, consisting of an outer Reed–Solomon erasure correcting code and an inner parallel concatenated convolutional code (PCCC), also known as turbo code [102]. The utilization of RS codes is to exploit the time diversity of the packets within a superframe, while the inner turbo codes used for the MAC information and RS-coded parity packets exploit the frequency-diversity inherent in the channel. The use of turbo coding is significantly beneficial for an OFDM system when the channel has spectral nulls, which are likely to occur in an SFN environment. Both the RS codes and the Turbo codes provide a few coding rate options to choose from depending on the network conditions and specific application requirements. In addition, MediaFLO supports regular and layered subcarrier modulation based on QPSK and 16-QAM. All these choices for FEC codes and constellation/modulation modes are certainly optimization knobs that a system designer can use to optimize the system performance so as to achieve the best video quality and minimize the transmission power consumption.

With regard to power efficiency, the interlace structure defined in MediaFLO can save power consumption, as the receiver only needs to demodulate the required interlace subsets [103]. In addition, the implementation of statistical multiplexing enables the receiver to demodulate and decode only the Multicast Logical Channels (MLCs) of interest, which provides significant power savings. The layered modulation design can certainly achieve significant power savings in a wide coverage area in which the users usually have wide range of reception signal strength. In this case, it is desirable to use base layer modulation for the users with relatively poor reception while using enhancement layer modulation for the remaining users.

18.6.3 Peer-to-Peer Internet Video Broadcasting

Unlike standard TV broadcasting, video broadcasting over the Internet can be very costly if it is based on centralized media server architectures. In addition, it can only support a limited number of users due to the bandwidth limitation of the network and the media server itself. The P2P network architecture provides a solution to this problem [104, 105]. It differs from the conventional server-client model in that it does not have the notion of clients or servers but only equal peer nodes that simultaneously function as both *clients and* servers to the other nodes on the network; that is, each user of P2P video streaming uploads the content it downloaded for reviewing at the same time. As a result, P2P networks make use of the cumulative network bandwidth, which eliminates the need to use costly centralized media servers to distribute the broadcast and results in extremely efficient utilization of the network bandwidth. In addition, such technology does not require support from Internet routers and network infrastructure, and is consequently extremely cost-effective and easy to deploy [104]. P2P technologies provide a number of unique advantages such as ubiquity, resilience, and scalability, making them ideal for a wide variety of distributed multimedia applications.

P2P live streaming systems have been widely studied from the protocol design point of view and often consider media streams as generic data sources. They have emerged as important technology for a wide range of applications such as file download and voice-over-IP. However, video broadcast applications pose very different challenges than those applications. Specifically, video broadcast imposes stringent real-time performance requirements in terms of bandwidth and latency [104].

Recently, the topic of how to efficiently encode and transmit video by taking advantage of the particular features of distribution topology of P2P networks has gained considerable attention [105]. For example in [105], the authors claimed that application-layer multicast and video transport should be considered jointly to achieve better performance in P2P video streaming systems. In addition, video coding and streaming need to be both content-adaptive and network-adaptive. For example, at the source coding side, MDC together with path diversity can naturally fit within the P2P paradigm. At the channel coding side, UEP through FEC and/or retransmission considering the unequal importance of the video stream can be certainly beneficial. We agree with the authors in [105] that these topics will be of central importance for the second generation of P2P video broadcast. We even believe that the joint consideration of the source coding and channel coding is able to explore the full potential of P2P video broadcast.

18.6.4 Video Streaming over Multihop Wireless Networks

Multihop wireless networks such as mobile ad hoc networks, mesh networks, and wireless sensor networks are another key area of interest in video streaming. Some of the challenges in this area are a result of the architecture of such wireless networks, which can consist of a mixture of fixed and mobile wireless nodes. Mobility implies that the routing and resource allocation in the network will need to dynamically adapt to changing conditions as links between nodes are established and removed depending on their changing locations and time-varying channel conditions. Other challenges are a result of the video content. For example, real-time video traffic imposes stringent delay constraints that must be met for individual data packets. In addition to low delay, video traffic requires higher overall data throughput than other types of data, which is difficult to achieve given the low and time-varying data rates achievable in wireless networks. Also, the data rate requirements are highly dependent on the particular video content being transmitted and can vary with each video flow. Therefore, an important challenge is to allocate resources across the flows such that fairness and high QoS is maintained across multiple flows while efficiently using the available resources such as power and bandwidth in the network.

Due to the challenges mentioned above, the efficient transmission of video streams over wireless multihop networks requires the joint control of parameters across different layers of the protocol stack within each network node as well as across multiple nodes.

A good overview of the cross-layer optimization techniques used in multihop wireless networks is provided in [106]. Recent work in the area of cross-layer optimized video transmission for ad hoc wireless networks can be found in [107], where a scheme is described that determines the optimal routing paths as well as retransmission limits for delay constrained video packets transmitted over a multihop wireless network. Some initial work on multiuser video streaming over multihop wireless networks is presented in [108].

Methods that use path diversity are also well suited for these applications [109–111]. MDC techniques, however, can lead to higher source encoding rates as they are less capable of exploiting correlations that exist in the video data. In [109, 110], a multipath streaming method is proposed that relates the end-to-end video quality to the congestion caused by transmitting the video. This enables the design of a flow-based "congestion-distortion optimized" scheme that takes into account the effects of source encoding as well as packet losses due to congestion at the network node.

There still remain many open challenges in this area, especially when dealing with multiple video streams. Important unresolved challenges include that of reducing the complexity of the cross-layer optimization schemes, improving energy efficiency, and ensuring security of the content.

18.7 DISCUSSION

While application of Shannon's separation theorem leads to the introduction of redundancy only during channel coding for achieving error-free transmission, this is not the case under real-time constraints. Redundancy needs to be introduced during both source and channel encoding in a judicious way. Furthermore, a well-designed decoder can recover some of the lost information using error-concealment techniques. When a feedback channel is available, a retransmission protocol can be implemented, offering a different means for improving the error resiliency of the video communication system.

In this chapter, JSCC for video communications has been discussed. We have used the term "channel encoding" in a general way to include modulation and demodulation, power adaptation, packet scheduling, and data rate adaptation. We provided an overview of the state-of-the-art implementations of JSCC in various network infrastructures. Although the most recent video coding standards H.263, MPEG4, and H.264 provide a number of error resilient tools, there are a number of resource allocation problems that need to be resolved to efficiently use such tools. In addition, there is a plethora of issues that need to be addressed by considering new system structures.

As mentioned earlier, cross-layer design is a general term, which encompasses JSCC, and represents the current state-of-the-art. To efficiently use limited network resources, the video transmission system needs to be adaptive to the changing network conditions. In the traditional layered protocol stack, each layer is optimized or adapted to the changing network conditions independently. The adaptation, however, is very limited due to the limited conversations between layers. Cross-layer design aims to improve the system's overall performance by jointly considering multiple protocol layers. The studies on this topic so far not only show the necessity to use the joint design of multiple layers but also point out the future direction of network protocol suite development to better support video communications over the current best effort networks.

Cross-layer design is a powerful approach to account for different types of channel errors in a hybrid wireless/wireline network that consists of both wired and wireless links. An initial investigation of this topic is described in [34], where lower layer adaptation includes interpacket FEC at the transport layer and intrapacket FEC at the link layer, which are, respectively, used to combat packet losses in the wired line and bit errors in the wireless link. Such channel adaptations are jointly designed with mode selection in source coding to achieve optimal UEP by solving (18.8).

Overall, the topic addressed in this chapter does not represent mature technology yet. Although technologies providing higher bit rates and lower error rates are continuously being deployed, higher QoS will inevitably lead to higher user demands of service, which for video applications translates to higher resolution images of higher visual quality.

REFERENCES

[1] F. Zhai and A. K. Katsaggelos. Joint source-channel video transmission. In: Al Bovik, editor, *Synthesis Lectures on Image, Video, and Multimedia Processing Series*, Morgan & Claypool Publishers, San Rafael, CA, 2007.

[2] Z. He, Y. Liang, and I. Ahmad. Power-rate-distortion analysis for wireless video communication under energy constraint. In *Proc. SPIE Visual Commun. Image Process.*, San Jose, CA, 2004.

[3] N. Bambos. Toward power-sensitive network architectures in wireless communications: Concepts, issues, and design aspects. *IEEE J. Select. Areas Commun.*, 18:966–976, 2000.

[4] Y. Wang, G. Wen, S. Wenger, and A. K. Katsaggelos. Review of error resilience techniques for video communications. *IEEE Signal Process. Mag.*, 17:61–82, 2000.

[5] C. E. Shannon. Coding theorems for a discrete source with a fidelity criterion. *IRE Int. Conv. Rec.*, 142–163, 1959, Part 4.

[6] A. K. Katsaggelos, Y. Eisenberg, F. Zhai, R. Berry, and T. N. Pappas. Advances in efficient resource allocation for packet-based real-time video transmission. *Proc. IEEE*, 93:135–147, 2005.

[7] B. Girod and N. Farber. Wireless video. In A. Reibman and M.-T. Sun, editors, *Compressed Video Over Networks*, 124–133. Marcel Dekker, New York, 2000.

[8] A. K. Katsaggelos, F. Zhai, Y. Eisenberg, and R. Berry. Energy efficient video coding and delivery. *IEEE Wireless Commun. Mag.*, 12:24–30, 2005.

[9] M. var der Schaar, S. Krishnamachari, S. Choi, and X. Xu. Adaptive cross-layer protection strategies for robust scalable video transmission over 802.11 WLANs. *IEEE J. Select. Areas Commun.*, 21:1752–1763, 2003.

[10] C. E. Shannon. A mathematical theory of communication. *Bell Syst. Tech. J.* 27:379–423, 623–656, 1948.

[11] B. Sklar. *Digital Communications: Fundamentals and Applications*, 2nd ed. Prentice-Hall Inc., Englewood Cliffs, NJ, 2001.

[12] N. Farvardin and V. Vaishampayan. Optimal quantizer design for noisy channels: An approach to combined source-channel coding. *IEEE Trans. Inform. Theory*, IT-38:827–838, 1987.

[13] F. Farvardin. A study of vector quantization for noisy channels. *IEEE Trans. Inform. Theory.*, 36:799–809, 1990.

[14] A. Kurtenbach and P. Wintz. Quantizing for noisy channels. *IEEE Trans. Commun. Technol.*, COM-17:291–302, 1969.

[15] R. E. Van Dyck and D. J. Miller. Transport of wireless video using separate, concatenated, and joint source-channel coding. *Proc. IEEE*, 87:1734–1750, 1999.

[16] D. Wu, Y. T. Hou, B. Li, W. Zhu, Y.-Q. Zhang, and H. J. Chao. An end-to-end approach for optimal mode selection in Internet video communication: Theory and application. *IEEE J. Select. Areas Commun.*, 18(6):977–995, 2000.

[17] D. Wu, Y. T. Hou, and Y.-Q. Zhang. Transporting real-time video over the Internet: Challenges and approaches. *Proc. IEEE* 88:1855–1877, 2000.

[18] T. Wiegand, G. J. Sullivan, G. Bjntegaard, and A. Luthra. Overview of the H.264/AVC video coding standard. *IEEE Trans. Circ. Syst. Video Technol.*, 13:560–576, 2003, Special issue on the H.264/AVC video coding standard.

[19] T. Wiegand, H. Schwarz, A. Joch, F. Kossentini, and G. J. Sullivan. Rate-constrained coder control and comparison of video coding standards. *IEEE Trans. Circ. Syst. Video Technol.*, 13:688–703, 2003, Special issue on the H.264/AVC video coding standard.

[20] H. Schwarz, D. Marpe, and T. Wiegand. Overview of the scalable video coding extension of the h.264/avc standard. *IEEE Trans. Circ. Syst. Video Technol.*, 17:1103–1120, 2007, Invited paper.

[21] J. M. Shapiro. Embedded image coding using zerotrees of wavelet coefficients. *IEEE Trans. Signal Process.*, 41:3445–3463, 1993.

[22] A. Said and W. Pearlman. A new, fast, and efficient image codec based on set partitioning in hierarchical trees. *IEEE Trans. Circ. Syst. Video Technol.*, 6:243–250, 1996.

[23] *JPEG-2000 VM3.1 A Software*, ISO/IECJTC1/SC29/WG1 N1142, 1999.

[24] K. Shen and E. J. Delp. Wavelet based rate scalable video compression. *IEEE Trans. Circ. Syst. Video Technol.*, 9:109–122, 1999.

[25] Y.-Q. Zhang and S. Zafar. Motion-compensated wavelet transform coding for color video compression. *IEEE Trans. Circ. Syst. Video Technol.*, 2:285–296, 1992.

[26] J. R. Ohm. Three-dimensional subband coding with motion compensation. *IEEE Trans. Image Proc.*, 3:559–571, 1994.

[27] S. Choi and J. W. Woods. Motion-compensated 3-D subband coding of video. *IEEE Trans. Image Proc.*, 8:155–167, 1999.

[28] A. Secker and D. Taubman. Motion-compensated highly scalable video compression using an adaptive 3D wavelet transform based on lifting. In *Proc. IEEE Int. Conf. Image Process. (ICIP)*, Thessaloniki, Greece, 2001.

[29] J.-R. Ohm. Motion-compensated wavelet lifting filters with flexible adaptation. In *Proc. Tyrrhenian Int. Workshop Digital Commun.*, Capri, Italy, 2002.

[30] D. A. Eckhardt. *An Internet-style approach to managing wireless link errors*, Ph.D Thesis, Carnegie Mellon University, Pittsburgh, PA, 2002.

[31] P. A. Chou and Z. Miao. Rate-distortion optimized streaming of packetized media. *IEEE Trans. Multimed.*, 8(2):390–404, 2006.

[32] Y. Eisenberg, C. E. Luna, T. N. Pappas, R. Berry, and A. K. Katsaggelos. Joint source coding and transmission power management for energy efficient wireless video communications. *IEEE Trans. Circuits Syst. Video Technol.*, 12(6):411–424, 2002.

[33] L. P. Kondi, F. Ishtiaq, and A. K. Katsaggelos. Joint source-channel coding for motion-compensated DCT-based SNR scalable video. *IEEE Trans. Image Process.*, 11:1043–1052, 2002.

[34] F. Zhai, Y. Eisenberg, T. N. Pappas, R. Berry, and A. K. Katsaggelos. Joint source-channel coding and power adaptation for energy-efficient wireless video communications. *Signal Process. Image Commun.*, 20/4:371–387, 2005.

[35] R. Zhang, S. L. Regunathan, and K. Rose. Video coding with optimal inter/intra-mode switching for packet loss resilience. *IEEE J. Select. Areas Commun.*, 18:966–976, 2000.

[36] G. Côté, S. Shirani, and F. Kossentini. Optimal mode selection and synchronization for robust video communications over error-prone networks. *IEEE J. Select. Areas Commun.*, 18:952–965, 2000.

[37] Z. He, J. Cai, and C. W. Chen. Joint source channel rate-distortion analysis for adaptive mode selection and rate control in wireless video coding. *IEEE Trans. Circ. Syst. Video Technol.*, 12:511–523, 2002.

[38] T. Wiegand, N. Farber, and B. Girod. Error-resilient video transmission using long-term memory motion-compensated prediction. *IEEE J. Select. Areas Commun.*, 18(6):1050–1062, 2000.

[39] A. Albanese, J. Blomer, J. Edmonds, M. Luby, and M. Sudan. Priority encoding transmission. *IEEE Trans. Inform. Theory*, 42:1737–1744, 1996.

[40] ITU-T. *Video Coding for Low Bitrate Communication.* ITU-T Recommendation H.263, 1998, Version 2.

[41] *Coding of audio-visual objects, Part 2-visual: Amendament 4: streaming video profile*, ISO/IEC 14496-2/FPDAM4, 2000.

[42] ITU-T. *Draft ITU-T Recommendation and Final Draft International Standard of Joint Video Specification (ITU-T Rec. H.264/ISO/IEC 14 496-10 AVC*, JVT of ISO/IEC MPEG and ITU-T VCEG, JVT G050, 2003.

[43] M. Karczewicz and R. Kurceren. The SP- and SI- frames design for H.264/AVC. *IEEE Trans. Circ. Syst. Video Technol.*, 13:637–644, 2003, Special issue on the H.264/AVC video coding standard.

[44] F. Hartanto and H. R. Sirisena. Hybrid error control mechanism for video transmission in the wireless IP networks. In *Proc. IEEE Tenth Workshop Local Metropolitan Area Networks (LANMAN'99)*, 126–132. Sydney, Australia, 1999.

[45] P. A. Chou, A. E. Mohr, A. Wang, and S. Mehrotra. Error control for receiver-driven layered multicast of audio and video. *IEEE Trans. Multimed.*, 108–122, 2001.

[46] J. Rosenberg and H. Schulzrinne. An RTP payload format for generic forward error correction. Tech. Rep., Internet Engineering Task Force, Request for Comments (Proposed Standard) 2733, 1999.

[47] B. J. Dempsey, J. Liebeherr, and A. C. Weaver. On retransmission-based error control for continuous media traffic in packet-switched networks. *Comput. Netw. ISDN Syst.*, 28:719–736, 1996.

[48] N. Celandroni and F. Pototì. Maximizing single connection TCP goodput by trading bandwidth for BER. *Int. J. Commun. Syst.*, 16:63–79, 2003.

[49] M. Gallant and F. Kossentini. Rate-distortion optimized layered coding with unequal error protection for robust Internet video. *IEEE Trans. Circ. Syst. Video Technol.*, 11(3):357–372, 2001.

[50] I. S. Reed and G. Solomon. Polynomial codes over certain finite fields. *SIAM J. Appl. Math.*, 8:300–304, 1960.

[51] S. B. Wicker and V. K. Bhargava. *Reed-Solomon Codes and Their Applications.* John Wiley & Sons Inc., 1999.

[52] J. G. Proakis. *Digital Communications.* McGraw-Hill, New York, 2000.

[53] J. Hagenauer. Rate-compatible punctured convolutional codes (RCPC codes) and their applications. *IEEE Trans. Commun.*, 36:389–400, 1988.

[54] D. E. Comer. *Internetworking with TCP/IP*, vol. 1, Prentice-Hall, Upper Saddle River, NJ, 1995.

[55] P. Luukkanen, Z. Rong, and L. Ma. Performance of 1XTREME system for mixed voice and data communications. In *Proc. IEEE Inf. Conf. Commun.*, 1411–1415, Helsinki, Finland, 2001.

[56] A. Chockalingam and G. Bao. Performance of TCP/RLP protocol stack on correlated fading DS-CDMA wireless links. *IEEE Trans. Veh. Tech.*, 49:28–33, 2000.

[57] H. Zheng. Optimizing wireless multimedia transmissions through cross layer design. In *Proc. IEEE*, vol. 1, 185–188, Baltimore, MD, 2003.

[58] Y. Eisenberg, F. Zhai, T. N. Pappas, R. Berry, and A. K. Katsaggelos. VAPOR: Variance-aware per-pixel optimal resource allocation. *IEEE Trans. Image Process.*, 15:289–299, 2006.

[59] R. Fletcher. *Practical Methods of Optimization*, 2nd ed. Wiley, New York, 1987.

[60] D. P. Bertsekas. *Dynamic Programming: Deterministic and Stochastic methods.* Prentice-Hall, 1987.

[61] Y. Pei and J. W. Modestino. Multi-layered video transmission over wireless channels using an adaptive modulation and coding scheme. In *Proc.2001 IEEE Int. Conf. Image Process.*, Thesaloniki, Greece, 2001.

[62] C. E. Luna, Y. Eisenberg, R. Berry, T. N. Pappas, and A. K. Katsaggelos. Joint source coding and data rate adaption for energy efficient wireless video streaming. *IEEE J. Select. Areas Commun.*, 21:1710–1720, 2003.

[63] G. Davis and J. Danskin. Joint source and channel coding for Internet image transmission. In *Proc. SPIE Conf. Wavelet Appl. Digital Image Process. XIX*, Denver, CO, 1996.

[64] G. Cheung and A. Zakhor. Bit allocation for joint source/channel coding of scalable video. *IEEE Trans. Image Process.*, 9:340–356, 2000.

[65] J. Kim, R. M. Mersereau, and Y. Altunbasak. Error-resilient image and video transmission over the Internet using unequal error protection. *IEEE Trans. Image Process.*, 12:121–131, 2003.

[66] F. Zhai, Y. Eisenberg, T. N. Pappas, R. Berry, and A. K. Katsaggelos. Rate-distortion optimized hybrid error control for real-time packetized video transmission. *IEEE Trans. Image Process.*, 15:40–53, 2006.

[67] D. Wu, Y. Hou, and Y.-Q. Zhang. Scalable video coding and transport over broad-band wireless networks. *Proc. IEEE*, 89:6–20, 2001.

[68] C. Lee and J. Kim. Robust wireless video transmission employing byte-aligned variable-len Turbo code. In *Proc. SPIE Conf. Visual Commun. Image Process.*, 2002.

[69] B. Sklar and F. J. Harris. The ABCs of linear block codes. *IEEE Signal Process. Mag.*, 14–35, 2004.

[70] C. Berrou, A. Glavieux, and P. Thitimajshima. Near Shannon limit error-correcting coding and decoding: Turbo codes. In *Proc. IEEE Int. Conf. Commun.*, 1064–1070. Geneva, Switzerland, 1993.

[71] V. Stanković, R. Hamzaoui, Y. Charfi, and Z. Xiong. Real-time unequal error protection algorithms for progressive image transmission. *IEEE J. Select. Areas Commun.*, 31:1526–1535, 2003.

[72] S. Zhao, Z. Xiong, and X. Wang. Joint error control and power allocation for video transmission over CDMA networks with multiuser detection. *IEEE Trans. Circ. Syst. Video Technol.*, 12:425–437, 2002.

[73] Y. S. Chan and J. W. Modestino. A joint source coding-power control approach for video transmission over CDMA networks. *IEEE J. Select. Areas Commun.*, 21:1516–1525, 2003.

[74] S. Appadwedula, D. L. Jones, K. Ramchandran, and L. Qian. Joint source channel matching for wireless image transmission. In *Proc. IEEE Int. Conf. Image Process. (ICIP)*, vol. 2, 137–141, Chicago, IL, 1998.

[75] Q. Zhang, W. Zhu, and Y.-Q. Zhang. Network-adaptive scalable video streaming over 3G wireless network. In *Proc. IEEE Int. Conf. Image Process. (ICIP)*, vol. 3, 579–582, Thessaloniki, Greece, 2001.

[76] R. G. Gallager. Energy limited channels: Coding, multi-access and spread spectrum. Tech. Rep., M.I.T. LIDS-P-1714, 1987.

[77] A. El Gamal, C. Nair, B. Prabhakar, E. Uysal-Biyikoglu, and S. Zahedi. Energy-efficient scheduling of packet transmissions over wireless networks. In *Proc. INFOCOM'02*, 2002.

[78] *High Speed Downlink Packet Access; Overall Description*, 2006, TS 25.308 v7.0.0.

[79] *IEEE Standard for Local and Metropolitan Area Networks; Part 16: Air Interface for Fixed and Mobile Broadband Wireless Access Systems*, 2005, 802.16e.

[80] 3GPP. Multimedia Broadcast/Multicast Service (MBMS). Tech. Rep., Tech. spec., Group Services and System Aspects, 2004, TS 26.346.

[81] M. Etoh and T. Yoshimura. Wireless video applications in 3G and beyond. *IEEE Wireless Commun. Mag.*, 12:66–72, 2005.

[82] R. Knopp and P. A. Humblet. Information capacity and power control in single-cell multiuser communications. In *Proc. IEEE Int. Conf. Commun.*, vol. 1, 331–335, 1995.

[83] S. Shakkottai, T. S. Rappaport, and P. C. Karlsson. Cross-layer design for wireless networks. *IEEE Commun. Mag.*, 41(10):74–80, 2003.

[84] S. Shakkottai, R. Srikant, and A. Stolyar. Pathwise optimality and state space collapse for the exponential rule. In *Proc. IEEE Int. Symp. Inf. Theory*, 379, 2002.

[85] P. Liu, R. Berry, and M. L. Honig. Delay-sensitive packet scheduling in wireless networks. In *Proc. IEEE Wireless Commun. Networking*, vol. 3, 1627–1632, 2003.

[86] A. Jalali, R. Padovani, and R. Pankaj. Data throughput of CDMA-HDR a high efficiency - high data rate personal communication wireless system. In *Proc. VTC*, Spring 2000.

[87] M. Andrews, K. Kumaran, K. Ramanan, A. Stolyar, P. Whiting, and R. Vijayakumar. Providing quality of service over a shared wireless link. *IEEE Commun. Mag.*, 39(2):150–154, 2001.

[88] R. Agrawal, V. Subramanian, and R. Berry. Joint scheduling and resource allocation in CDMA systems. In *Proc. 2nd Workshop on Modeling and Optimization in Mobile, Ad Hoc, and Wireless Networks (WiOpt '04)*, 2004.

[89] K. Kumaran and H. Viswanathan. Joint power and bandwidth allocation in downlink transmission. *IEEE Trans. Wireless Commun.*, 4(3):1008–1016, 2005.

[90] R. S. Tupelly, J. Zhang, and E. K. P. Chong. Opportunistic scheduling for streaming video in wireless networks. In *Proc. Conf. Inf. Sci. Syst.*, 2003.

[91] G. Liebl, T. Stockhammer, C. Buchner, and A. Klein. Radio link buffer management and scheduling for video streaming over wireless shared channels. In *Proc. Packet Video Workshop*, 2004.

[92] G. Liebl, M. Kalman, and B. Girod. Deadline-aware scheduling for wireless video streaming. In *Proc. IEEE Int. Conf. Multimed. Expo*, 2005.

[93] C. W. Chan, N. Bambos, S. Wee, and J. Apostolopoulos. Optimal scheduling of media packets with multiple distortion measures. In *Proc. IEEE ICME*, Beijing, China, 2007.

[94] C. W. Chan, N. Bambos, S. Wee, and J. Apostolopoulos. Wireless video broadcasting to diverse users. In *Proc. IEEE ICC*, Beijing, China, 2008.

[95] P. V. Pahalawatta, R. Berry, T. Pappas, and A. Katsaggelos. Content-aware resource allocation and packet scheduling for video transmission over wireless networks. *IEEE J. Selected Areas Commun. Special Issue on Cross-Layer Optim. Wireless Multimed. Commun.*, 25(4):749–759, 2007.

[96] E. Maani, P. V. Pahalawatta, R. Berry, T. Pappas, and A. Katsaggelos. Resource allocation for multiuser downlink video transmission over wireless lossy networks. *To appear IEEE Trans. Image Process.*

[97] L. Ozarow, S. Shamai, and A. D. Wyner. Information theoretic considerations for cellular mobile radio. *IEEE Trans. Vehicular Technol.*, 43(2):359–378, 1994.

[98] P. V. Pahalawatta, T. Pappas, R. Berry, and A. K. Katsaggelos. Content-aware resource allocation for scalable video transmission to multiple users over a wireless network. In *Proc. IEEE Int. Conf. Acoustics, Speech Signal Process.*, vol. 1, I 853–I 856, 2007.

[99] X. Ji, J. Huang, M. Chiang, and F. Catthoor. Downlink OFDM scheduling and resource allocation for delay constraint SVC streaming. In *IEEE Int. Conf. Commun.*, 2512–2518, 2008.

[100] G. Faria, J. A. Henriksson, E. Stare, and P. Talmola. Dvb-h: Digital broadcast services to handheld devices. *Proc. IEEE*, 94:194–209, 2006.

[101] European Telecommunications Standards Institute (ETSI). Digital video broadcasting (DVB); transmission system for handheld terminals (DVB-H). Tech. Rep., European Standard EN 302 304, 2004, version 1.1.1.

[102] M. R. Chari, F. Ling, A. Mantravadi, R. Krishnamoorthi, R. Vijayan, G. K. Walker, and R. Chandhok. Flo physical layer: An overview. *IEEE Trans. Broadcast.*, 53:145–160, 2007.

[103] Y. Li. Pilot symbol aided channel estimation for OFDM in wireless systems. *IEEE Trans. Vehicle Technol.*, 49:1207–1215, 2000.

[104] J. Liu, S. Rao, B. Li, and H. Zhang. Opportunities and challenges of peer-to-peer Internet video broadcast. *Proc. IEEE*, 96:11–24, 2008.

[105] E. Setton, P. Baccichet, and B. Girod. Peer-to-peer live multicast: A video perspective. *Proc. IEEE*, 96:25–38, 2008.

[106] Q. Zhang and Y-Q. Zhang. Cross-layer design for QOS support in multihop wireless networks. *Proc. IEEE*, 96(1):64–76, 2008.

[107] Y. Andreopoulos, N. Mastronarde, and M. Van der Schaar. Cross-layer optimized video streaming over wireless multihop mesh networks. *IEEE J. Selected Areas Commun.*, 24(11):2104–2115, 2006.

[108] H.-P. Shiang and M. van der Schaar. Multi-user video streaming over multi-hop wireless networks: A distributed, cross-layer approach based on priority queueing. *IEEE J. Selected Areas Commun. Special Issue on Cross-Layer Opt. Wireless Multimed. Commun.*, 25(4):770–785, 2007.

[109] E. Setton, T. Yoo, X. Zhu, A. Goldsmith, and B. Girod. Cross-layer design of ad hoc networks for real-time video streaming. *IEEE Wireless Commun.*, 59–65, 2005.

[110] E. Setton, X. Zhu, and B. Girod. Congestion-optimized multi-path streaming of video over ad hoc wireless networks. In *Proc. IEEE Int. Conf. Multimed. Expo*, 2004.

[111] W. Wei and A. Zakhor. Multiple tree video multicast over wireless ad hoc networks. *IEEE Trans. Circuits Syst. Video Technol.*, 17(1):2–15, 2007.

CHAPTER

Video Surveillance

19

Joonsoo Lee and Al Bovik

The University of Texas at Austin, USA

19.1 INTRODUCTION

Video surveillance has emerged as one of the most active application-oriented research topics in image analysis and processing. The growing level of uncertain security in our world, along with cheaper and more "intelligent" digital camera technology is spurring great interest in developing video-based systems capable of observing and interpreting specific scenes, with the aim of detecting anomalies or events that may affect the safety, security, economics, or other vital aspect of human activity. Naturally, the possibilities involved are extremely broad, and we cannot hope to encompass either the breadth or depths of video surveillance within a single chapter. Moreover, the technology of the field is dynamically changing at a rapid pace, both at the hardware levels (smart camera technology), the network level (visual sensor networks are emerging rapidly), and at the software level. It is in the end that we will focus our attention in this chapter.

Video surveillance is founded on video analysis, which, in this context, is more specifically termed *video content analysis* or *video analytics*. It may be regarded as a specific application of machine vision, yet the wide range of algorithm sophistication and intelligence, and the (typically) passive observational setting loosely defines this application field. The array of video analytics is quite diverse and often domain-dependent, with practical applications ranging from seemingly simple (yet deceptively difficult) applications, such as parking lot security, to intermediate difficulty, such as counting people in a queue or passing through a door, or cars passing through a toll gate, to challenging applications including urban-traffic monitoring and airport surveillance, which might include algorithms for detecting suspicious persons or recognizing people from their faces or other biometrics. Biometrics, which involves the recognition of persons according to some physical attribute such as their facial appearance, their fingerprint or iris, or their retinal pattern, is not covered in this chapter. Rather, these topics are covered in detail in several chapters of the previous volume, *The Essential Guide to Image Processing*, and in Chapter 20 of the current volume.

Naturally, the sophistication of the video analytics algorithm required for a given surveillance application depends on the complexity of the information that is required to be extracted from the scene, and by the complexity of the spatiotemporal information contained in the target scene. It also depends on whether the scene is dynamically changing or whether it is static and of known appearance. For example, a system for conducting video surveillance of an ATM security area will deal with a mostly static scene; an outdoor parking lot surveillance system will contend with a deceptively complex scene, owing to changing lighting and weather conditions; an airport video security system could be expected to handle quite complex scenes containing numerous moving objects and people of constantly changing and novel attributes.

As a segue into developing some general principles that might be of use to the designer of a video surveillance system, we will first define some important and common target application areas and generally quantify the complexity of the visual content contained in those scenes. Then, we will define generic procedural steps used in the design of video surveillance systems, and we will seek to broadly generalize and categorize the types of video analytic tools used in each step of the algorithms used in these systems. Typically, the overall software application will consist of a suite of sequenced video analytics algorithms applied to the video data.

19.2 CATEGORIZING APPLICATIONS, TARGET SCENES, AND VIDEO ANALYTICS

To give the reader a general sense of the scope of the video surveillance question, and also to assist in directing the designer towards a solution of their own problem. In the following, we categorize the video surveillance problem by application, by the type of scene environment that is typically encountered, and by the type and complexity of algorithm(s) that might be required.

19.2.1 Video Surveillance Applications

Video surveillance is a very broad concept, and the applications that might fall into this topic are limited only by the imagination of video system designers. However, at the scale of significantly funded research and development, we can attempt a fairly coarse classification of applications by type into a few broad categories, which are not ordered in any particular way, other than by an approximate relevance to daily, general application.

19.2.1.1 Outdoor Perimeter Security

This involves cameras being strategically placed to image parking lots, building perimeters, and other open spaces. Factors affecting the complexity of this problem include weather, lighting, the number of expected and unexpected objects or people that might be encountered, the types of object motion that might occur, and the degree of sophistication of the information that is required of the algorithm. Figure 19.1 of a parking lot area from ref. [1] depicts an image that exemplifies this scenario.

FIGURE 19.1

A parking lot; a typical scenario for outdoor video surveillance. This would in category (S2 [see Section 19.2.2.2]), where the scene may be expected to change moderately over time.

19.2.1.2 Indoor Perimeter Security

Here the surveillance cameras might be deployed in airports, hallways, train stations or other public areas, or inside private-sector buildings or private residences. These problems are not affected as much by weather conditions, yet the lighting may still vary and the number of objects or people that might be encountered may vary from just a few, to many. Figure 19.2 from ref. [2] depicts an indoor surveillance scene of this type.

19.2.1.3 Retail

This is a fairly specific category, wherein a retail outlet may be interested in automatically counting the number of people in a store or queued for check-out, for detecting suspicious activities associated with shoplifting or for inventory control. Depending on the specificity and required accuracy, the video analytics may range from fairly simple (e.g., people counting) to quite complex (activity analysis). Figure 19.3 (courtesy of Chesapeake Systems Service, Inc. [3]) depicts a possible retail surveillance situation. There would be little activity in this scene at most times.

19.2.1.4 Automotive

Video surveillance for automotive applications can involve cameras attached to the exterior or the interior of vehicles, and might include assistance with parallel parking; detection of sleepy drivers, automatic vehicle spacing detection; or intervehicle calculations or communication.

FIGURE 19.2

An indoor scene; a typical scenario for indoor video surveillance. This could fall into either category (S1 [see Section 19.2.2.1]) or category (S2 [see Section 19.2.2.2]), depending on the frequency and diversity of objects and people moving through the scene, and whether the scene can be expected to change much over time. The moving sunlight might be a consideration.

19.2.1.5 ***Transportation Infrastructure***

This is a very broad and collectively complex category that involves the deployment of cameras as part of the roadway infrastructure. The applications involved can include vehicle counting or traffic density calculation; license plate detection and recognition; red light or speed violation detection; traffic flow feedback control; accident reporting, or stolen car detection, and so on. Moreover, the traffic conditions to be handled can be extremely diverse, ranging from quiet suburban or rural settings, to high-volume but regular highway traffic, to highly chaotic and dynamic urban-traffic scenarios. Figure 19.4 depicts a typical highway traffic surveillance image, taken from a California Department of Transportation Caltrans live camera [4].

FIGURE 19.3

Possible retail surveillance application. A camera is located in the back room of a store. This scene would probably fall into category (S1 [see Section 19.2.2.1]).

19.2.1.6 ***Financial***

This is a familiar application whereby digital security cameras are deployed in banks or ATM lobbies for the purpose of monitoring the safety of financial customers.

19.2.1.7 ***Military***

This is probably the oldest category of video surveillance application, with a rich literature that has developed over the past 40+ years. Generally, the problem involved is the detection of enemy vehicles, weapons, or personnel using airborne, spaceborne, or land-based cameras, often at high speeds, with the intention of acting on the information gleaned in a rapid manner to achieve strategic or tactical goals. This area of video surveillance is, generally, extremely diverse since it can involve imaging at any number of wavelengths or combination thereof, since the environmental conditions may have an infinite variability, and since the objects being sought may also vary considerably, might either be in complex motion or be camouflaged. Yet, the area is also specific because of the limited scope of military aims. This area is sometimes (incompletely) called *automatic target recognition*. Naturally, it is not possible to cover much of this area within the limited confines of this chapter, although many of the video analytics techniques discussed are naturally applicable.

FIGURE 19.4

A typical highway surveillance scene. This scene may contain unpredictable background changes owing to weather, time of day, seasonal changes. The movement of objects, however, is predictable, except in unusual circumstances, such as an accident.

19.2.1.8 Medical

This is an exciting and relatively new area of video surveillance, wherein digital video analytics are used to assist with surgery, either on-site or remotely; for video telemedicine, including video telemonitoring of patients. Naturally, multiple imaging modalities involving different types of radiation might be involved. The match of video surveillance with the rich and broad area of medical imaging provides enormous opportunities, which largely remain to be developed.

Video surveillance can be applied to a wide variety of areas, and the above list is hardly exhaustive. Among the abovementioned sample application domains, some have already been aggressively researched, with developed products, and a strong customer base. Many others remain in a theoretical stage of development.

19.2.2 Video Surveillance Target Scenes

Earlier, we have pointed out that each category of video surveillance applications can operate in widely diverse environmental conditions. The complexity of a video analytics algorithm for video surveillance, regardless of the task required, depends greatly on the complexity of the target scenes being imaged. In an effort to increase understanding of the diverse range of target scenes that might be encountered, we offer a broad categorization of common target scenes encountered in video surveillance applications. The following three categories represent a general guide to the different levels of dynamic activity that might be encountered in a scene being imaged in a video surveillance application. Of course, the actual division of target scenes by complexity could be arbitrarily finer; however, our goal is to give a general sense of the types of scenes that are encountered and how they affect the design of video analytics algorithms for video surveillance applications. Generally, the reader may observe that increased levels of activity within the scene lead to the need for increased levels of analysis to understand what is happening in the scene.

19.2.2.1 (S1) Static Scene Surveillance

In this type of application, the scene being imaged is essentially unchanging beyond some small level of activity. The scene is assumed to contain few moving macroscopic objects, although it is likely that some movement exists, for example, dust motes, small insects, small disturbances related to air movement, and small variations in lighting (an important general consideration, since lighting changes can significantly affect a scene's appearance). A likely application for which this assumption is made would be indoor security surveillance, where any significant scene change might give rise to an alarm.

19.2.2.2 (S2) Moderately Changing Scene Surveillance

In this type of application, the scene being imaged contains some activity, which is moderate or sporadic, and which might be predictable. An example would be outside security surveillance of a parking lot with neighboring trees which might sway, or an indoor security surveillance application where objects might be regularly displaced, where some human activity occurs, or where the lighting changes, for example, as the sunlight changes entering a window on the scene. These types of changes must generally be *accommodated* in some way, either by modeling them, by ignoring them, or by labeling them based on low-level image processing (e.g., identifying the size or type of movement of a moving object as being uninteresting).

19.2.2.3 (S3) Crowded/Dynamic Scene Surveillance

In this type of application, the scene being imaged contains a significant amount of activity which might, or might not, be separate from any objects of interest. The objects may be variable in their numbers, motion, velocities, and trajectories. A good example would be security surveillance in a crowded subway or airport; or on a crowded street filled with vehicles, people, and other objects; a people counting system in a crowded retail store, or a scene that contains considerable and unpredictable light variation from reflections, shadows, flashes, and so on. Often the different types of motions might occur in the same scene. Any analysis of scenes of this type will require a considerable step up in complexity to accommodate the dynamism of the scene, and which might require recognizing, or at least classifying the different types of objects and or motions that are under surveillance.

Again, the above categories are for general convenience in discussion, and do not represent any formal categorization that is agreed-upon by the video surveillance community. However, we will find them useful to refer to in our discussion of video analytics algorithms for video surveillance.

19.2.3 Video Analytics for Video Surveillance

Video surveillance systems generally include many different steps of processing, which often operate in sequence. There are no algorithms that are common to all, or even to most video surveillance applications. Nevertheless, there are video analytics algorithms that are used by many applications. In the following discussion, we will make a distinction between a *video surveillance algorithm* and a *video analytics algorithm.* Specifically, a video surveillance algorithm broadly solves a video surveillance problem at the software level, and likely makes use of multiple video analytics algorithms. For example, a video surveillance algorithm may detect unexpected vehicles, people, or other objects in a parking lot. By contrast, we will associate the term video analytics algorithm with a module that is used as part of a video surveillance algorithm. Such an algorithm is often associated with a low-level process such as extracting edges or computing motion primitives. It may also be an intermediate level algorithm that operates on extracted low-level primitives, such as a shape classification algorithm, or it might be a high-level algorithm that accepts a variety of extracted low-level features, and sues them to classify an object in some way, either by type, or for the purpose of recognition. Again, as with most of the image analysis area, it is difficult to make specific categorizations of video analytics algorithms as monolithic, meaning that the algorithm does not contain subalgorithms that are also video analytic algorithms. Rather than seeking a philosophical inquiry into this question, we take the pragmatic approach of identifying common video analytic algorithms types that are used in practical and emerging video surveillance systems.

Video analytics algorithms are sometimes associated with an application category being addressed or with a scene environment being operated in. An example might be skin tone calculations used in face recognition. Generally, however, it is difficult to make blanket statements regarding the use of any particular video analytics algorithm. Therefore, in the course categorization of different types of video analytics algorithms for video surveillance we make in the following, we have not attempted to associate the algorithms with any specific application domains (except by example). Nor have we attempted to be exhaustive, since one can envision innumerable applications tasks, subtasks, and philosophical and creative approaches to a given problem. However, the video analytics categories that follow are ordered by a very coarse approximation to their intrinsic sophistication or difficulty.

19.2.3.1 Change and Motion Detection

Perhaps the simplest types of video analytics algorithms are those that seek to detect changes in intensity (or color) along the temporal direction. This usually involves a local between-frame differencing operation to enhance the amount of change, often preceded by a smoothing operation to control noise or scale. Typically, a thresholding operation might be applied to fix the level of motion sensitivity. Such a tool might be used in any video surveillance situation, and for example, might be used in a simple way in S1 (see Section 19.2.2.1) mentioned earlier. Chapter 3 of this *Guide* covers this topic in detail.

19.2.3.2 Motion Estimation

Estimation of the trajectories of image intensities, which is called motion estimation, represents a level of complexity higher than motion detection. As described in Chapter 3, the goal is to calculate video motion vectors, which can then be used in a variety of ways, for example, to segment the video into objects, as described in Chapter 6 of this *Guide.*

19.2.3.3 Object Detection/Analysis

This is related to motion detection and motion estimation but with an addition level of sophistication. The temporal changes in the scene would be analyzed to produce general size attributes, or to compute bounding boxes, approximate object boundaries, or other simple primitives. Simple spatial processing steps might also be applied, such as edge detection or texture analysis, as described in the companion volume, *The Essential Guide to Image Processing.*

19.2.3.4 Simple Motion Analysis

The general motion of groups of pixels that have been segmented into likely objects can be computed from the motion vectors. Object velocity, size, direction, and other attributes might be used to determine the relative importance of the moving objects that have been determined to be within a scene. It may also be of interest to track the motion of objects along many frames (see Chapter 7).

19.2.3.5 Simple Shape Analysis

Detected objects may be further analyzed to compute both spatial/temporal *features* of interest. Spatial features might include the aspect ratio of the object; the direction of its principal axis; textural or color features, curvature descriptors of its boundary, the direction of acceleration of its path, and so on. Both static and motion information can be used to compute simple shape information.

In all of the above, the surveillance application is likely a simple one, where the goal of the application is to simply determine the presence of an unexpected (or expected) object or objects, or perhaps to use simple shape and motion attributes of the detected objects to produce a decision regarding the nature of the object. These kinds of algorithms would likely be used in application scenarios S1 (see Section 19.2.2.1) and S2 (see Section 19.2.2.2), and as precursors to further processing in S3 (see Section 19.2.2.3).

19.2.3.6 Shape Recognition and Categorization

At this level of processing, basic shapes, or shape categories of increased complexity are recognized. For example, spatiotemporal analysis might be used to detect moving objects, then segment them according to their motion vectors, then identify them as having a specific general category of shape, using processes in the above categories. Subsequent processing in this category could range from recognizing that an object obeys certain size and motion behaviors, such as a vehicle on a known street, to more complex processing, such recognizing whether an object is a person with several moving parts, which might be in any pose or viewed from many different angles. This level of processing can be either model-based, wherein the objects being analyzed are compared to stored templates, or feature-based, where objects features are detected, functionally estimated or fit with curves, or some combination.

The complexity of shape recognition is greatly affected by the scene. For example, recognition of a "walking person" in a static, known scene (S1 [see Section 19.2.2.1]) is not very difficult, but recognition of the same in a scene containing multiple objects that may occlude one another can be extremely difficult. We do not cover algorithms involving this level of processing in this *Guide.*

19.2.3.7 Computation of Three-Dimensional Shape Attributes

Three-dimensional (3D) information about the objects in a scene can be computed, with some effort, using motion information and or by using multiple cameras. If more than one camera is used, with known relative positions, and/or if analysis of the motion of object structures over multiple frames is accomplished, then object shapes attributes, including whole 3D surfaces and their locations and poses in 3D space, can be computed from the videos. The difficulty of these processes is greatly affected by the scene complexity, and the cost of this type of processing can be considerable. However, the added degree of information is very rich and can be used to more accurately detect, localize, and recognize objects.

19.2.3.8 Recognition of Specific Objects

This very high level of processing goes beyond shape recognition, since the goal is to recognize specific instantiations of objects, such as individual human faces or even facial expressions, vehicle types, specific actions such as fence-climbing, and so on. The complexity and difficulty of this problem varies tremendously, and is highly dependent on the scene characteristics and the objects being recognized. For example, recognition of faces that are framed (the subject is told to stand in a certain spot and hold their head a certain way) remains a difficult problem of which much progress has been made. Finding and recognizing faces in a crowd is a problem whose solution remains elusive (see Chapter 20).

19.3 REVIEW OF VIDEO ANALYTIC ALGORITHMS

Naturally, there are a great number of different types of algorithms that are used in applications [5], and we cannot hope to cover them all. Our goal in this chapter is twofold: First, to survey and describe some of the most general and commonly used video analytics algorithms that are used as parts of larger integrated video surveillance systems, and second, to survey and describe video analytics solutions that video surveillance system designers have proposed to solve specific problems, but that may have general applicability to the broader field.

19.3.1 Motion and Change Detection

A large number of techniques for temporal change detection or for motion detection have been developed in the field of video surveillance, as well as in other fields such as remote sensing, medical diagnosis, and so on. Unlike remote sensing, where the changes to be detected occur over large periods of time, and only a couple of images in temporal sequence might be available, in video surveillance applications there is generally a live stream of video delivering at a rate of 30 frames per second. Of course, handling such a large amount of data requires ample computational resources, and in any particular application, the capability of accomplishing effective surveillance will depend greatly on the resources available. Ordinarily, the types of dynamic events of interest that might occur in a scene will be rapid and transient, suggesting that it is expedient to handle as many frames as possible.

Many techniques of change detection have been proposed in the literature. The question arises as to whether change detection is the same as motion detection; of course, intensity changes in a video stream are not necessarily due to the motion of objects within the field of view. There may also be lighting changes, moving shadows, and changes due to noise. However, gradual lighting changes can often be accounted for, and most rapid, significant changes arise from moving objects. In any case, there is not a clear demarcation between methods for change detection against motion detection, except where higher-level processes (such as segmentation) may be involved. As such, we will consider these collectively, while noting that the goal is usually to detect moving objects of interest.

The goal of motion detection is to extract changing regions, which might be associated with moving objects, from a sequence of images. The simplest method of achieving this goal is to calculate the difference between adjacent images based on the assumption that the intensity or color changes will be greatest on the pixels where there are moving objects. However, such a simplifying assumption does not generally work well in practical environments. First, moving objects may give rise to both gradual and sudden illumination changes; sudden changes being concentrated near the boundaries of the moving objects. Second, there may be significant intensity changes in the "background" (nonobject) regions, owing to the motion of objects of little interest, such as the swaying branches on a tree.

19.3.1.1 Temporal Differencing

This is a fundamental conceptual technique for change or motion detection. Intensity changes between image frames are enhanced by taking the simple difference between them on a pixel-wise basis, then comparing the differences with predefined (or possibly adaptive) threshold values. If $I_k(i,j)$ is the video sequence intensity value at frame k and at pixel coordinates (i,j), then compute the absolute difference and compare to threshold value τ.

$$|I_k(i,j) - I_{k-1}(i,j)| > \tau$$

This technique is very simple to implement, but it generally does a poor job of extracting relevant changing pixels of interest. Moreover, it is highly sensitive to noise, minor motions of uninteresting objects, and minor motions of the camera owing to vibrations. A variety of methods can be used to ameliorate these problems, such as attempting to locate groupings of pixels that are spatially connected and that fall above a (perhaps lowered) threshold, or by discounting the movement of objects known to be uninteresting (the swaying branch problem) or other such (largely *ad hoc*) methods. The effects of noise can be reduced by applying spatial smoothing or by computing the differences over more than three frames, which can produce more robust results [6].

19.3.1.2 Adaptive Background Modeling

Statistical approaches are often found to be valuable in distinguishing between moving or changing objects of interest and an uninteresting background. This technique generates a statistical model for the background of scenes, and classifies the pixels into background and foreground based on the model. The background model may also be dynamically modified to handle dynamic scenes. This approach generally provides improved results when the background is relatively stationary. Of course, the performance of this approach depends on how well the model fits the real background, and by how much the background statistics differ from the foreground objects of interest. The following are a variety of proposed techniques using different assumptions on the background statistics.

19.3.1.2.1 Single Gaussian Distribution on Background

In this technique, background pixels are modeled by a Gaussian distribution [7] that adapts to slow changes by recursively updating the model. The model captures over time in the intensity (or color) value of a pixel. The mean vector and covariance of the Gaussian distribution is estimated during a training stage (on test videos) and is continuously updated during the application stage to adapt to slow changes in the background.

Letting $I_k(i,j)$ be the pixel intensity (or vector color) values at coordinate (i,j) in a video frame at time k, $\mu_k(i,j)$ be the mean estimated at that location and time (mean vector in the case of color), and $\mathbf{K}_k(i,j)$ be the covariance matrix at that location and time. For an intensity-only sequence, $\mathbf{K}_k(i,j)$ is the scalar variance $\sigma_k(i,j)$; else it is the covariance between the three-color components (RGB, YUV, etc). The Gaussian model density is

$$f_{I_k(i,j)}(q) = \frac{\exp\left\{-\frac{1}{2}\left[q-\mu_k(i,j)\right]^T \mathbf{K}_k^{-1}(i,j)\left[q-\mu_k(i,j)\right]\right\}}{(2\pi)^{m/2}|\mathbf{K}_k(i,j)|^{1/2}},$$

where $m = 1$ in the case of an intensity-only video sequence, and $m = 3$ for color video.

Each pixel is determined *not* to be background if the likelihood value associated with its neighborhood is small. This may be interpreted as occurring when the value of the density function evaluated at the pixel's gray level falls below a predefined threshold value; equivalently, if the distance between the pixel value and the local mean is greater than some (predefined) multiple of the variance, then pixel is classified as "foreground."

Since the statistical moments are computed from and updated from prior instants at each spatial coordinate, then assignment of a pixel as "foreground" presumes the possibility that the pixel is part of a moving object. Pixels that are collectively connected can be clustered into changing or moving objects by analyzing their connectivity ("connected component analysis," or "region labeling," see Chapter 4 of *The Essential Guide to Image Processing*).

The means and variances of the background model are updated by a simple adaptive filter to accommodate gradual changes in lighting or similar phenomena:

$$\mu_{k+1}(i,j) = \alpha I_k(i,j) + (1-\alpha)\mu_k(i,j)$$

$$\sigma_{k+1}^2(i,j) = \alpha\left[I_k(i,j) - \mu_k(i,j)\right]^2 + (1-\alpha)\sigma_k^2(i,j)$$

This technique is able to handle slow changes in the background but may present difficulties in handling the small, irrelevant, or uninteresting background motions, such as swaying branches or drifting clouds.

19.3.1.2.2 Multiple Gaussian Distributions on Background and Foreground Objects

This technique uses a single Gaussian distribution on background pixels, as well as multiple Gaussian distributions on foreground objects. Pixels are classified as background or foreground objects by finding the model with the lowest log likelihood:

$$d_k^c = -\frac{1}{2}\left[I_k(i,j) - \mu_k^c(i,j)\right]^T\left[\mathbf{K}_k^c(i,j)\right]^{-1}\left[I_k(i,j) - \mu_k^c(i,j)\right] - \frac{1}{2}\ln\left|\mathbf{K}_k^c(i,j)\right| - \frac{m}{2}\ln(2\pi)$$

where as in Section 19.3.1.2.1, $I_k(i,j)$ is the pixel value at frame time k, but here $\mu_k^c(i,j)$ is the mean for class c at the same location and time, $\mathbf{K}_k^c(i,j)$ is the covariance matrix of class k at the same location and time, and m is the dimension of the pixel value, which can be intensity or color, such as YUV, as in ref. [8].

In ref. [8], the mean and variance are updated in a slightly different way than in Section 19.3.1.2.1. The new model mean and covariance for updating a pixel at frame $k+1$ that was determined to be of class c in frame k is

$$\mu_{k+1}^c(i,j) = E^c\left[I_{k+1}(i,j)\right]$$

$$\mathbf{K}_{k+1}^c = E^c\left[I_{k+1}(i,j)I_{k+1}(i,j)^T\right] - \mu_k^c\left(\mu_k^c\right)^T,$$

where E^c denotes expectation using the distribution of class c. This technique does a better job of handling and identifies distinct changing or moving objects but still may have difficulty in handling uncertain background movements.

19.3.1.2.3 A Mixture of Several Gaussian Distributions on Background

This technique augments the single Gaussian model for dynamic background scenes, where complex, variable surfaces are present and where there may be frequent lighting changes. This technique assumes a mixture of several Gaussian distributions on the background [9] only. Given class weights w^c, the distribution is

$$f_{I_k(i,j)}(q|c) = \sum_{c=1}^{C} w^c \frac{\exp\left\{-\frac{1}{2}\left[q-\mu_k^c(i,j)\right]^T\left[\mathbf{K}_k^c\right]^{-1}\left[q-\mu_k^c(i,j)\right]\right\}}{(2\pi)^{m/2}\left|\mathbf{K}_k^c\right|^{1/2}}.$$

In ref. [9], a K-means approximation is used. Each new pixel value in frame k is checked against the set of C existing Gaussian distributions until the best match is found, provided that the pixel value falls within one standard deviation of one of the C distributions. If none of the distributions match the current pixel value in this sense, then the least probable distribution is replaced by a new distribution generated by the current pixel value. The weights for each distribution are updated only when there is a match. This technique appears to improve the performance of background modeling but still is not guaranteed to completely handle small background motions.

19.3.1.2.4 Mixture of Gaussian Distributions on Background and Foreground

All of the pixels in each video frame may be modeled as a mixture of some Gaussian distribution regardless of an identification as background or foreground [10]. For simplicity, the authors in ref. [10] assume only three classes: road, shadow, and vehicle, since this technique was designed for a vehicle tracking problem. However, the method is extensible to other, similar problems. Given class c, the model for a mixture of three Gaussian distributions is defined as

$$f_{I_k(i,j)}(q|c) = \sum_{c=\{\text{road,shadow,vehicle}\}}^{3} w^c \cdot \frac{\exp\left\{-\frac{1}{2}\left[q-\mu_k^c(i,j)\right]^T \left[\mathbf{K}_k^c\right]^{-1}\left[q-\mu_k^c(i,j)\right]\right\}}{(2\pi)^{m/2}\left|\mathbf{K}_k^c\right|^{1/2}},$$

where w^c is the weight for class c. In ref. [10], the parameters are estimated by a variation of the standard expectation maximization (EM) algorithm. Since the standard EM algorithm may have unrealistic computation requirements for real-time processing, an incremental (heuristic) EM algorithm is used in ref. [10].

19.3.1.2.5 A One-Step Wiener Prediction Filter

In the approach, a prediction $\hat{I}_k(i,j)$ is made of each pixel value $I_k(i,j)$ in frame k, calculated as the output of a Wiener filter [12] on P previous pixel values at the same spatial coordinates:

$$\hat{I}_k(i,j) = -\sum_{p=1}^{P} a_p I_{k-p}(i,j),$$

where a_p are the filter coefficients. The coefficients are computed from sample covariance values. If an actual pixel value differs from its predicted value by more than four times the expected error, it is determined to be part of a changing object. Further classification methods, such as spatial proximity/connectivity may be applied to augment the pixel-level classification in ref. [12].

19.3.1.2.6 A Nonparametric Kernel Density

This technique uses a nonparametric kernel density [13] estimate instead of using a mixture of Gaussian distributions. A sum-of-Gaussians kernel function is defined as an instantaneous density function at each spatial coordinate at frame time k:

$$f_{I_k(i,j)}\left[q|I_p(i,j);p=k-n,\ldots,k-1\right]=\frac{1}{N}\sum_{p=k-N}^{k-1}\frac{\exp\left\{-\frac{1}{2}\left[q-I_p(i,j)\right]^T\Sigma_k^{-1}\left[q-I_p(i,j)\right]\right\}}{(2\pi)^{m/2}\left|\Sigma_k\right|^{1/2}},$$

where N is the number of previous samples at each spatial coordinate that is used to estimate the current pixel value, and Σ_k is the kernel function bandwidth at frame time k. This model generalizes the Gaussian mixture model, and allows more accurate estimation of the background, while requiring only recent information from the sequence. A pixel intensity is unlikely based on the instantaneous density function is classified as a possible foreground changing or moving object.

19.3.1.2.7 A Bimodal Distribution Constructed from Order Statistics

Although the above techniques attempt to handle dynamic background scenes, all have difficulty handling small, frequent, or repetitive background motions since they all use a unimodal distribution model. A variation that seeks to modify this represents each pixel by three values that are order statistics of the local training data: the local minimum and maximum intensities, and the maximum intensity frame-difference. If an observed pixel $I_k(i,j)$ value differs by more than a fixed number of maximum difference levels from either the minimum or the maximum, then it is considered as a moving foreground object. Since this approach implicitly assumes a bimodal distribution model [14], it can deliver better performance than the techniques that assume a unimodal Gaussian distribution.

19.3.1.2.8 Dynamic Texture

The dynamic texture approach [15] represents an image sequence as an autoregressive, moving-average process with unknown inputs. In ref. [15], a technique is proposed for estimating the state parameters of the textured background, which is presumed to be dynamic, in the presence of possibly mobile foreground objects. A robust Kalman filter is used to iteratively update the state of the dynamic texture ARMA model and to determine a mask image for the foreground objects.

Figure 19.5 summarizes the various categories of scene background models that are encountered and gives a rough sense of the required modeling complexity that is required to handle scenes as a function of background complexity.

19.3.1.3 Optical Flow

Optical flow, or motion estimation, is a fundamental method of calculating the motion of image intensities, which may be ascribed to the motion of objects in the scene. Optical flow is an extremely fundamental concept that is utilized in one form or another in most video-processing algorithms. Chapter 3 of this *Guide* is devoted to this topic and so we will not explain it here. Optical-flow methods are based on computing estimates of the motion of the image intensities over time in a video. The flow fields can then be

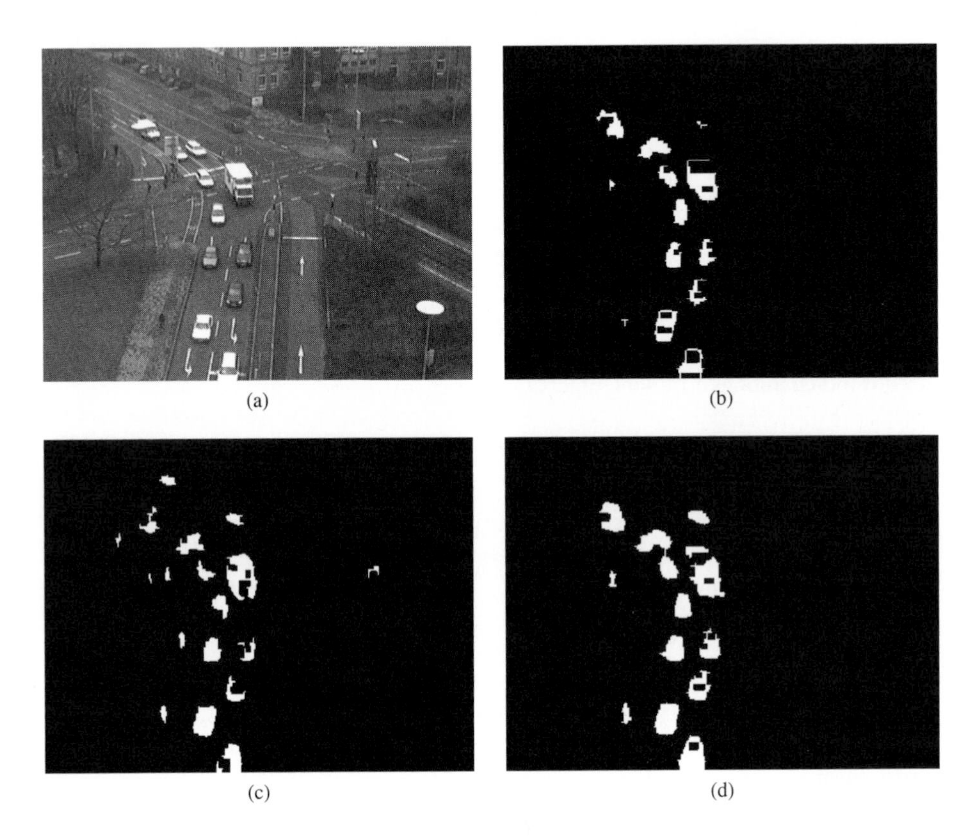

FIGURE 19.5

Background modeling algorithms. (a) Original image from a video sequence [11], (b) temporal differencing, (c) single Gaussian distribution on background, (d) a mixture of several Gaussian distributions on background.

analyzed to produce segmentations into regions, which might be associated with moving objects. Methods for motion-based video segmentation are detailed in Chapter 6 of this *Guide.*

Optical-flow or motion-estimation algorithms can be used to detect and delineate independently moving objects, even in the presence of camera motion. Of course, optical-flow-based techniques are computationally complex, and hence require fast hardware and software solutions to implement. Since optical flow is fundamentally a differential quantity, estimation of it is highly susceptible to noise; ameliorating the noise sensitivity can imply increases in complexity. Therefore, smart camera-based video surveillance systems that use optical-flow calculations of some type must be equipped with substantial computational resources, for example, a dedicated DSP chip, FPGA, or other special-purpose acceleration device. Indeed, it is because such technologies are being introduced to smart camera technology (along with wireless-networking capabilities) that video surveillance systems having significant intelligent capability are being envisioned and realized.

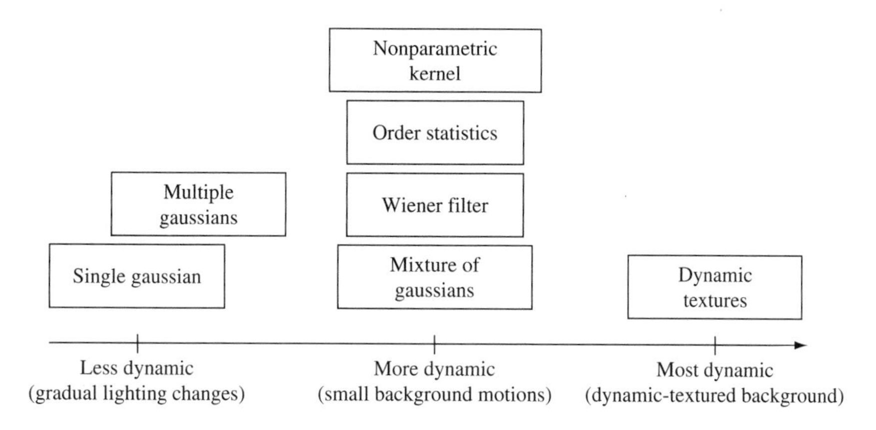

FIGURE 19.6

Summary of appropriate background models as a function of scene complexity.

TABLE 19.1 Summary of motion/change detection tools.

	Background Modeling	Temporal Differencing	Optical Flow
Advantage	Works for relatively static backgrounds	Adaptable to dynamic backgrounds; easy to implement	Effective on dynamic backgrounds
Disadvantage	Must select/decide among various statistical modeling assumptions	Does a poor job of entire moving objects; very noise sensitive	Computationally complex; highly noise-sensitive
Complexity	Moderate	Simple	Complex
Usage	Common	Moderate	Infrequent
Augments/ augmented by	Statistical modeling and higher level postprocessing	Connected component analysis	Special hardware

Table 19.1 tabulates the tradeoffs of the various types of motion and change detection methods, broadly categorized as background modeling, temporal differencing, and optical flow as we have done in the preceding.

19.3.2 Object Detection

Most change-detection and motion-detection algorithms operate below the object level, meaning that there is no attempt made to first segment the video frames into objects. However, some methods attempt higher-level processing (region-level, object-level, and frame-level). For example, pixels classified as possible moving foreground objects may be clustered to form blobs, boxes, or simple point groupings. These are considered as possible moving objects, then further analyzed and classified into a categories based on the analysis and the application. Naturally, correct classification of moving objects is essential to be able to accomplish further processing, such as object tracking, behavioral analysis, or motion-based recognition.

There are two main approaches for analyzing moving objects for classification: shape-based and motion-based approaches.

19.3.2.1 Shape-Based Approaches

Descriptions of shape information can be extracted in a great variety of ways from clusters of points, boxes, and blobs. Most of these are application-specific; for example, if the expected object is a tank, a nose, or a suitcase, then different approaches to object detection will be taken. These could include methods for edge detection and contour analysis (see *The Essential Guide to Image Processing*), or regional segmentation followed by analysis of regional descriptors. Assuming that a region has been segmented into a putative moving blob–by associating connected pixels with similar motion vectors, for example, these blobs can then be analyzed in a great variety of ways.

Perhaps, the simplest blob shape descriptors are *blob dispersedness, blob area,* and *blob aspect ratio*. These simple features are easy to calculate, are relatively robust against noise and imperfect blob classification results, and are often surprisingly effective as features for classification, despite their simplicity [6].

The area of a blob within an image (area in image space) can be a useful piece of information, although it does not necessarily convey direct information about the 3D dimensions of the real-world object. The area in the image space is easily calculated as a pixel count, which can, if needed, be converted into true image space dimensions by taking into account the size of the pixels relative to the sensor dimensions at the optical plane. The 2D image area can also be related to a projection of the silhouette of the 3D object, if desired. If an image blob is defined by $I_{\mathbf{B}} = \{I(i,j) : (i,j) \in \mathbf{B}\}$, where $\mathbf{B}$ is the set of pixel coordinates comprising the blob area, then

$$\text{Area of } I_{\mathbf{B}} = c(\mathbf{B}) = \sum_{(i,j)\in\mathbf{B}} 1_{\mathbf{B}}(i,j),$$

where $c(\mathbf{B})$ is the cardinality (set counting) operator, $1_{\mathbf{B}}(i,j)$ is the indicator function of $\mathbf{B}$, namely, $1_{\mathbf{B}}(i,j) = 1$, when $(i,j) \in \mathbf{B}$, and $1_{\mathbf{B}}(i,j) = 0$ otherwise.

The perimeter of the object may also be found once a decision on what constitutes a boundary pixel is made. A boundary pixel might be defined as an object pixel laying next to a nonobject pixel, but this requires imposing a definition of connectivity as well. Usually, pixels are assumed to be *8-connected* as described in *The Essential Guide to Image Processing*, meaning the four up/down, left/right as well as four diagonal neighbors are regarded as connected. In any case, if the perimeter pixels of the object associated with blob $\mathbf{B}$ have the coordinates in the set $\mathbf{P}(I_{\mathbf{B}}) = \{I(i,j) : (i,j) \in \mathbf{P} \subset \mathbf{B}\}$, then

$$\text{Perimeter of } I_{\mathbf{B}} = c(\mathbf{P}) \leq c(\mathbf{B}).$$

It is, of course, plausible that the area and perimeter turn out to be the same if the object is curvilinear, for example.

More valuable shape measures are *relative*, meaning they compare various dimensions of the object. A simple one is

$$\text{Dispersedness of } I_{\mathbf{B}} = \frac{(\text{Perimeter of } I_{\mathbf{B}})^2}{\text{Area of } I_{\mathbf{B}}} = \frac{c^2(\mathbf{P})}{c(\mathbf{B})}.$$

The *aspect ratio* of a moving object is also a useful measure. The aspect ratio is the ratio between two orthogonal dimensions of the object. Which two dimensions to be used may vary with the application; these could be the horizontal and vertical dimensions, or they could be the ratios of the principal and minor axes of the object.

Knowledge of the specifics of an application can also be used to good effect. For example, if it is known that the objects that might be encountered have certain shapes or classes of shapes, then simple shape parameters such as those above, or features extracted from curves, such as curvature, can be used as features. For example, in ref. [16], extracted moving objects are compared with a small number of simple predefined shapes.

19.3.2.2 Motion-Based Approaches

It has been mentioned that it is also possible to detect moving objects (as opposed to moving pixels) by the analysis of motion. The simplest way is to associate collections of neighboring pixels having similar motion vectors. However, this approach is usually inadequate in isolation; by injecting knowledge of the *types* of behavior that might be expected, significant improvement in performance might be expected. For example, the articulated nature of human body movement exhibits behavioral regularities that can be used to deduce the presence of humans, or to analyze human behavior. Moreover, by tracking a moving object over time, the self-similarity of its shape and motion can be used to analyze the object's shape and to predict further motions. Periodic or regular motions can be analyzed through time-frequency analysis [17] and to perform classification based on this behavior.

An important aspect of moving objects is their rigidity (or lack thereof) as time passes. Rigidity can simplify the analysis of motion but being able to analyze the objects that are nonrigid is an important goal of many applications. Indeed, determining rigidity (versus malleability) can be an important part of the analysis. The difference between the motion of individual pixels in a blob, and the average motion (optical flow) of the blob, which may be termed *residual flow*, can be used to analyze rigidity. Rigid objects will have little residual flow. This rigidity measure can be used as a key feature in deciding whether pixels are associated as moving objects [18].

19.3.3 Object Tracking

Object tracking represents a higher-level of processing than does either object detection or motion detection, since it implies that the object to be tracked has been detected, and that motion information has been extracted which can be used to predict and track the future locations of the object. Object tracking, generally, consists of three subtasks: (i) prediction of object locations in future frames using a model of objects and their motions in the current and past frames; (ii) object matching in future frames using the prediction, and (iii) updating of the object model as the future frames are received and processed.

Object-tracking techniques are closely related to object-detection methods, and indeed, are often specifically designed to operate with specific-object-detection algorithms. Our goal in this chapter is to describe generically applicable approaches, and so we discuss object-tracking techniques that are generally applicable.

Since there are a large number of object-tracking techniques available, with different input assumptions, processing methods, and output characteristics and format, it is useful to roughly group them by category. For example, some algorithms operate on pixel groupings or blobs, while others operate on contours. We divide object-tracking algorithms into four categories according to the way in which objects are represented: as connected regions, as active contours, by features extracted from the objects, and by using specific-object models.

19.3.3.1 Region-Based Tracking

Region-based-tracking techniques consider objects as connected image regions that have been marked as likely moving objects. In this general approach, any method for detecting objects by finding regions of neighboring pixels with similar motions techniques may be used. However, dynamic background subtraction is quite commonly used. Since the image regions detected as moving objects are usually blobs, region-based tracking can be thought of as blob tracking. The position and velocity of blobs are tracked frame by frame. An extremely effective algorithm for such a tracking task is the Kalman filter, which is a recursive filter used for estimating the state of a dynamic system from a series of possibly incomplete and noisy measurements. In the object-tracking task, a frame is the state, and the position and velocity are measurements. The Kalman filter is an object-tracking task widely used in a multitude of applications.

19.3.3.1.1 Kalman-Based Tracking

Since the Kalman filter is used in a large number of object-tracking techniques other than region-based techniques, it is worthwhile to detail its basic idea [19]. For simplicity, we describe the lowest-order Kalman filter, which requires only the most recent filter output to be used. The response or current true state of a Kalman filter at time k evolves from the state at time $(k-1)$ according to

$$x_k = F_k x_{k-1} + B_k u_k + w_k,$$

where F_k is the state transition model, B_k is the input control model, u_k is the control, and w_k is the process noise, modeled as zero-mean: $w_k \sim \aleph(0, Q_k)$. A measurement at time k of the true state x_k is made according to

$$z_k = H_k x_k + v_k,$$

where H_k is the observation model and v_k is the observation noise: $v_k \sim \aleph(0, R_k)$. The Kalman filter proceeds in two distinct phases: *predict* and *update.*

In the prediction phase, estimates from previous time are used to estimate the current state. In the update phase, measurement information from the current time is used to refine the prediction, yielding a more accurate estimate. This property enables it to be used in a great variety of object-tracking techniques.

The state of the Kalman filter is represented by two variables: $\hat{x}_{k|k}$, the estimate of the state at time k, and $P_{k|k}$, the error covariance matrix. The prediction and update stages are described by the following equations.

Prediction:

$$\hat{x}_{k|k-1} = F_k \hat{x}_{k-1|k-1} + B_k u_k$$

$$P_{k|k-1} = F_k \hat{P}_{k-1|k-1} F_k^T + Q_k$$

Update:

$$\tilde{y}_k = z_k - H_k \hat{x}_{k|k-1}$$

$$S_k = H_k P_{k|k-1} H_k^T + R_k$$

$$K_k = P_{k|k-1} H_k^T S_k^{-1}$$

$$\hat{x}_{k|k} = \hat{x}_{k|k-1} + K_k \tilde{y}_k$$

$$P_{k|k} = (I - K_k H_k) P_{k|k-1}$$

When object tracking from a video is accomplished using the Kalman filter principle, specific spatial coordinates of the object are assigned state variables, x_k. For example, the spatial coordinates of the corners of a bounding box of the object or the spatial coordinates of the centroid of the object might be used. The velocity and position of an object at the current time are predicted by the previous updated model and are refined by the current measurements. The current model is then updated to compute future predictions.

19.3.3.1.2 Tracking Based on Kalman Filter Extensions

Although many methods for object tracking are based on Kalman filtering, direct application of the Kalman principle may be limiting as it does not support multiple alternative motion hypotheses, such as merge, where two distinct moving objects are merged into a single unit, and split, where a moving object is split into two or more objects [6]. In ref. [6], the predict and update notions of the Kalman filter are extended to handle cases in which there is matching ambiguity between multiple moving objects. The basic algorithm is as follows.

(i) *Prediction*: Predict object position and velocity and generate a bounding box around each. Any detected moving regions in future frames whose centroid falls within the bounding box are considered as candidates for matching.

(ii) *Matching*: Calculate the correlations between an object template associated with each object, and the candidates within the associated bounding box; find the best match. The positions and velocities are refined according to the new match.

(iii) *Update*: Based on a hypothesis test, analyzes the tracking pattern and decides among match, merge, split, and no match. Then, update the object-track model for the next prediction.

Region-based tracking techniques are relatively simple and require no processing after motion/object detection. However, these techniques, generally, perform poorly in the presence of cluttered background and interaction between multiple moving objects. As in ref. [6], some variations may be able to improve the performance of the region-based techniques in complex scenes.

19.3.3.2 Active Contour-Based Tracking

Active contour-based tracking techniques represent objects by their outline contours. These boundary contours are dynamically updated over successive frames. These techniques require that object contours be found from the video frames, which might be accomplished directly, or which might use the result of a motion-based object detection system, although the former is more efficient. Two popular approaches for active contour-based tracking are the *velocity snake model* [20] and the *geodesic active contour model*, which uses level sets [21].

19.3.3.2.1 Tracking using the Velocity Snake Model

The velocity snake model [20, 22] utilizes the well-known active contour extraction or *snake* algorithm proposed in ref. [23]. The snake is an energy-minimizing spline guided by external constraint forces and influenced by image forces that pull it toward features such as lines and edges. Thus, the snake is an active contour that locks onto nearby edges and localizes them. The basic energy functional of the snake model is

$$E^*_{\text{snake}} = \int_0^1 \{E_{\text{int}}[v(s)] + E_{\text{image}}[v(s)] + E_{\text{con}}[v(s)]\}\, ds,$$

where $v(s) = [x(s), y(s)], s \in [0, 1]$ is the parameterized active contour, E_{int} is the internal energy of the spline due to bending, E_{image} introduces the image forces, and E_{con} are the external constraint forces.

For object tracking, as boundaries or edges are detected, an edge strength functional is used as the image force: $E_{\text{image}} = -|\nabla I(x, y)|^2$. Since the snake model can track object boundaries in successive frames, it may be applied to the object-tracking problem. In ref. [20], this is accomplished by associating the time-axis with updates to the snake model, and by using the velocity of the objects as the external constraint forces. The energy functional of the resulting *velocity snake model* is

$$E^*_{\text{vs}} = \frac{1}{2}\int_0^1 \left[w_1 \parallel v_s \parallel^2 + w_2 \parallel v_{ss} \parallel^2\right] ds - \int_0^1 \parallel \nabla I_t(x, y) \parallel ds + \frac{1}{2}\int_0^1 \mu \parallel v_t \parallel^2 ds,$$

where $v(s, t) = [x(s, t), y(s, t)]$ is the parameterized active contour that evolves over time t, $v_s = \partial v/\partial s$, $v_{ss} = \partial^2 v/\partial s^2$, $v_t = \partial v/\partial t$, and w_1, w_2, and μ are weights. The model is made more robust against biases in the contour position by introducing the Kalman filter principle to the problem, which yields a stochastic version of the velocity snake model. This approach exhibits good performance for object-tracking tasks, although it is sensitive to the accuracy of the initial contour, and it encounters problems in the presence of interacting moving objects.

19.3.3.2.2 Tracking using Geodesic Active Contours

The geodesic active contour model [24] was introduced as a geometric alternative to the snake model. The goal of the model is to find the curve that minimizes the energy functional

$$E[C(p)] = \int_0^1 g\left\{|\nabla I\left[C(p)\right]|\right\} |\dot{C}(p)| dp,$$

where $C(p) = [x(p), y(p)], p \in [0,1]$ is the parameterized contour being tracked, $\dot{C}(p) = \partial C/\partial p$, and g is a monotonically decreasing function. In this formulation, the detection of object contours is equivalent to finding the geodesic curve that best takes into account the desired image characteristics, which is edge strength.

The advantage that this model has over the standard snake model is that it is not dependent on the curve parameterization. Also, by introducing a level set formulation, it is able to handle topological changes such as merge and split, which is a significant advantage over the velocity snake model. The level set formulation converts the family of moving curves $C(p,t)$ into a family of one-parameter evolving surfaces $\phi(x,y,t)$. The curve $C(p)$ may change its topology as the function ϕ evolves.

The energy functional for the object-tracking problem is

$$E[C(p)] = \int_0^1 \left(\gamma g\left\{I_D[C(p)], \sigma_D\right\} + (1-\gamma) g\left\{|\nabla I[C(p);t]|, \sigma_T\right\}\right) |\dot{C}(p)| dp,$$

where I_D is the contour image, γ is a weight function, and σ_D, σ_T are the variances of Gaussian functions g that express the motion and the tracking information. The tracking performance is improved in the presence of topological changes (objects being merged or split), yet it may still fail to recover object location in the presence of background clutter that may interact with the moving objects.

Active contour-based techniques describe objects more simply than region-based methods. The approach can be used to track objects even in the presence of occlusions and other object interactions. However, while region-based methods require that moving regions be detected, active contour methods require that moving contours be found, or at least an initial contour for each object. Active contour methods are sensitive to the initial contour estimate and to the initial track parameters, which makes it difficult to fully automate the process.

19.3.3.3 Feature-Based Tracking

Feature-based methods track objects by matching the features between images, instead of matching the whole regions or entire contours. Elements are first extracted from moving object regions, these elements are then clustered into features that are used for tracking. We classify feature-based tracking techniques into three categories according to the types of features that are used are as follows: global, local, and combined features.

19.3.3.3.1 Tracking using Global Features

Global features that can be used for object tracking are such gross object attributes as the object centroid, the value of the perimeter, the object area, areas, color, and so on. The effectiveness of the features chosen greatly depends on the type of objects being tracked and the operating environment. However, global features have the advantage of being, generally, less susceptible to noise but also have the disadvantage of supplying detailed information such as shape or shape change.

- Centroid—The centroid has been used as a feature for object tracking in ref. [25]. Instead of tracking individual pixels of an object blob, the velocity and position of the centroid of the detected blob is tracked. An approach such as this does not require that shape computations be made because the centroid may be taken as simply that of an associated group of connected and moving pixels.
- Color—Object color was used as a feature in ref. [26]. Color is an attractive attribute since it is a robust area attribute. In ref. [26], object/motion detection using the methods outlined above is not done; instead, compact color regions are extracted by clustering pixels based on their RGB values and by their spatial proximity. The Viterbi algorithm was utilized to track these compact color regions. The algorithm considers every possible sequence of color clusters, then, it computes the Euclidean distance of color means between temporally adjacent clusters in any given sequence. Those having the minimum color distance from previous clusters are chosen as correct clusters in the current frame.

19.3.3.3.2 Tracking using Local Features

Good examples of local object features that are useful for tracking are line segments, curve segments, corner vertices, and so on. Corner features were used in ref. [27]. First, moving object regions are detected. A corner detector is applied that discovers regions where there is high gradient energy in orthogonal directions. Once the corners are detected, a Kalman filter is used to track the corners over time. The tracked corner features can be regrouped into objects by identifying corners that are associated with the same object, namely, are spatially related and have similar motion. This approach was shown to be effective for vehicle tracking in ref. [27].

19.3.3.3.3 Tracking using Combined Features

Combinations of global and local features offer the opportunity of exploiting the best properties of each. In ref. [28], each moving object is expressed as a labeled graph. Fiducial patches are selected as nodes of the graph, and labels are derived from a set of features, including color, edge strength, texture, and geometric structure. The color features used by these authors are H and S from HIS color space. The texture features are Gabor wavelet coefficients [29–31]. The edge features are local mean intensity differences, while shape features are computed that reflect the geometric distribution of neighboring nodes. To elaborate on the shape measure, assume a node A with neighboring nodes $B_1, \ldots, B_n$. A reference vector associated with node A is defined as the average of the vectors $\overline{AB_k}(k = 1, \ldots, n)$. The average absolute differences between each vector $\overline{AB_k}$ and the reference vector is used as the shape measure of the node A. For example, the shape measure is 0 when neighboring nodes are evenly spaced on a straight line, and increases as neighboring nodes are scattered asymmetrically. This shape measure is robust against translation and rotation. An energy functional is formulated that is used in minimization procedure to dynamically find the best match of the graph to the image as the video evolves.

19.3.3.4 ***Model-Based Tracking***

Model-based tracking techniques track objects by matching object models to the image data. The object models are developed using knowledge of expected object shape, environment conditions, expected motions, or any other relevant physical data regarding the application at hand. This may include descriptions of the types of shape changes that might be expected, which is an attribute that is not easily handled by the prior methods that we have discussed in the preceding. Of course, shape change is very difficult concept to get a handle on unless there is some foreknowledge of the types of changes that are expected, as in objects that are semirigid and have reticulated motions, such as humans.

The most common objects that are tracked are humans and vehicles. In the former case, the objects are not rigid, but they are semirigid with predictable reticulated motions. In the latter case, the objects are rigid and move with considerably regularity. Indeed, many models have been developed for humans and vehicles, and used for tracking them, in some cases accounting for expected shape changes, such as the motions of the legs and arms during walking. These techniques are far from generic. However, since they are popular in application, we briefly review different types of models used for humans and vehicles.

19.3.3.4.1 **Model-Based Human Tracking**

A great variety of assumptions and simplifications have been used in constructing models of the human body for video analytics. These usually seek to use primitive information that does not require maintaining the many details presented by the appearance of humans as they move. For example, stick figure representations are often used, for example, in ref. [32], to build a hierarchical model of human dynamics encoded using hidden Markov models (HMMs). A novel moving-contour representation was used in ref. [33], using a 2D cardboard human body model, whereby the human limbs were represented by a set of jointed planar ribbons. More elaborate modeling can be done in 3D space; for example, a set of elliptical cylinders was used to model a human body as connected and separately moving 3D volumes in ref. [34].

19.3.3.4.2 Model-Based Vehicle Tracking

Since a vehicle is largely a rigid body, models for vehicles for tracking are relatively simple compared to those used to model human body tracking. The 3D wire-frame model is very common and popular for vehicle tracking [35]. This model represents a vehicle as a set of line segments (or wires). Constraints on motion, for example, that vehicles are restricted to move on the ground plane along mostly smooth trajectories, can be used to greatly reduce the degree of freedom of the vehicle model.

The wire-frame model is an excellent example of effective use of prior knowledge of the 3D structure of the objects being tracked. Such approaches are intrinsically robust. These techniques also perform relatively well in the presence of occlusion and in handling interactions between multiple objects.

19.3.4 Behavioral Analysis Tools

The analysis of the motion behavior of objects can be a valuable tool for understanding the video content of a scene, and the actions and behaviors of the objects and/or people in it. The analysis of behavior can be viewed as a form of classification of time-varying feature data. It may be of interest to match an unknown test sequence to a group of known reference sequences that representing typical behaviors. Thus, it may be seen that video object behavior may also be viewed as a type of motion analysis that is based on high-level models. Many approaches have been devised for training on and test sequences containing motion behavior. Naturally, most of these were developed within the context of specific applications.

19.3.4.1 Dynamic Time Warping

Dynamic time warping is a template-based dynamic programming matching technique that is widely used for speech recognition but has been deployed for recognizing human gestures [36]. It is conceptually simple, and its performance is generally robust, provided that the extracted matching features are accurate, and the behavioral model is sufficiently sophisticated.

19.3.4.2 Finite-State Machine

The states of a *finite-state machine* are used to determine which reference sequence matches a given test sequence. In ref. [37], the structure of human gestures is analyzed and implemented by an equivalent of a finite state machine. In this approach, there is learning phase required, but the gesture models must be very carefully constructed.

19.3.4.3 Hidden Markov Models

A HMM is a stochastic version of the finite-state machine. It allows for more complex analysis of video sequences containing spatiotemporal complexity. HMMs generally outperform dynamic time warping on undivided time-series data, and because of this, are widely used for behavioral analysis. HMMs have been used for a variety of human behavior analysis scenarios, for example, interactions such as one human following another or meetings between humans [38].

19.3.4.4 Neural Network

Neural networks are also very popular tools for video content analysis. The *time delay neural network* is a common technique that is used to analyze time-varying data and has been used for human behavior analysis from video. In ref. [39], a time delay neural network is used to accomplish hand-gesture recognition. Of course, the success of approaches of this depends very heavily on the selection of good features with strong discriminatory power.

The previous methods for video behavior analysis that we have briefly outlined all require some type of supervised learning. This is acceptable when the types of objects, scenes, and motions are already known or can be accurately modeled. *Self-organizing neural networks* are useful when the object motions are unrestricted. For example, in ref. [40], the movements of objects are analyzed using extracted sequences of flow vectors consisting of the positions and velocities of the objects, using a self-organizing neural network. A model of the probability distribution functions of instantaneous movements and partial trajectories is automatically generated by tracking objects, and the probability distribution functions are estimated by neural networks using vector quantization. This method thereby recognizes atypical motions of interest.

19.3.5 Gait Recognition

An important specific type of video-based behavior analysis is human gait recognition. Such methods are potentially important for any kind of video surveillance application where specific humans are being looked for (since human gait has individual characteristics), or where certain human gait behavior is being looked for, for example, representative of dangerous or hostile behavior. Again, although this is a specific application discipline, we select generic techniques used in human gait analysis for review.

19.3.5.1 Model-Based Approaches

Parameters from predefined models, such as joint trajectories, limb lengths, and their angular speed, can be measured and used for identification in model-based approaches. Tracking the human body in 3D space is still a difficult problem. Although joint angles and their relative movements are powerful descriptors of human gaits, it is difficult to identify joints and their constituent limbs from video sequences.

19.3.5.1.1 Articulated Pendulum Model

This technique models gait as the movement of an articulated pendulum, and use the dynamic *Hough transform* to extract the lines representing the thigh in each frame. A least squares method is used to postprocess the data in order to fill in missing points occurring from self-occlusion of the legs. Finally, the thigh angle is estimated as a function of time. The magnitude spectrum of the function, weighted by phase is used as a feature for gait identification [41].

19.3.5.1.2 Dynamically Coupled Oscillator

This technique utilizes biomechanically motivated *dynamically coupled oscillator model* [42] of walking and running. This anatomical model accounts for the motions of the hips, thighs, and lower legs. Template matching is to extract rotation angles, and then the phase-weighted magnitude spectrum is used for gait recognition.

19.3.5.2 Spatiotemporal Motion-Based Approaches

In this general class of methods, motions are characterized by the overall 3D spatiotemporal data volume spanned by a moving person in a video sequence. These approaches consider motion holistically in order to characterize its spatiotemporal distributions. This approach is able to capture both spatial and temporal information describing human gaits. The computational complexity is generally low, and the implementations relatively simple.

19.3.5.2.1 Translation and Time Pattern-Based Method

One of the earliest spatiotemporal motion-based methods was proposed in ref. [43]. In translation and time (XT) space, motions of the head and legs exhibit significantly different patterns. These patterns are processed to determine the bounding box of a moving body, and are then fitted to a five-stick model. A velocity-normalized fitted model can be used to generate gait features from this model.

In ref. [44], once motion detection is accomplished, XYT patterns (2D space and 1D time) are fitted with a smooth spatiotemporal surface. This surface is represented as a combination of standard parametric surface and a deviation surface. Six parameters are used to represent the standard parametric surface: The initial and final x position, the head and toe y position, and the period and phase of the gait. These parameters are estimated from the motion-detected image sequence. To recover a more accurate spatiotemporal surface, the deviation surface is involved, which is initially set to zero.

19.3.5.2.2 HMM-Based Method

In this approach, a HMM is used to represent and recognize gaits. A set of key frames that occur during a human walk cycle is selected, and the widths of the binary silhouettes of the walking figure in those frames are chosen as input features. Then, a low-dimensional measurement vector is produced using the distance between a given image and the preselected set of key frames. These measurement vectors are the inputs that are used to train the HMMs. The approach is robust to speed changes and generally exhibits good recognition rates. However, changes in clothing and illumination can present difficulties [45].

19.3.5.3 Statistical Approaches

In this category of methods for human gait recognition, motions are represented by statistical descriptions. Statistical approaches can be made relatively robust to noise and to change of time interval. Also, the computational complexity is relatively low.

19.3.5.3.1 Velocity-Moment-Based Method

In ref. [46], a velocity-moment-based method is used to describe object motions in image sequences. The velocity moment features are used to recognize people. Under the assumption that people walk in a frontal-parallel direction towards the camera, the silhouette region is divided into seven sub-regions. A set of moment-based region features is used to recognize people–and to predict gender.

19.3.5.3.2 Parametric Eigenspace Method

In ref. [47], a parametric eigenspace transformation and a canonical space analysis of gaits is used to reduce computational cost, to improve robustness of gait estimation, and to obtain better discrimination. This approach is similar to the eigenspace representation of images used for face recognition [48], since it attempts to reduce the feature space dimensionality while optimizing the discrimination power of the features.

19.3.5.4 Physical Parameter-Based Approaches

The geometric structural properties of the human body are useful features for statistically characterizing the gait pattern of a person. The parameters may include, for example, height, weight, cadence, and stride. This approach is quite natural and intuitive and also independent of viewing angles since such parameters are estimated in the 3D settings. But, since this approach requires 3D settings, it greatly depends on the use of machine vision techniques to recover the parameters. Also, the parameters may adequately discriminate a large population.

19.3.5.4.1 Walking-Activity-Based Method

This method does not directly analyze the dynamics of various gait patterns but instead utilizes walking activities to recover the static body parameters of walking, such as the vertical distance between the head, the feet, and the pelvis and the distance between the left and right feet. An assessment test was done in ref. [49] using a confusion matrix to predict how well these features are able to perform on identification over a large population.

19.3.5.5 Language-Based Approaches

This approach attempts to represent higher-level human activities, such as composite actions and interactions. A human activity is usually composed of multiple subevents, where each subevent might be further decomposed into subevents. To analyze and recognize complex human activities, hierarchical models of human activities must be defined. Language-based representations are a convenient means for defining these hierarchical structures.

19.3.5.5.1 Context-Free-Grammar-Based Method

This is a general framework for representing and recognizing high-level human activities in a hierarchical way. It is a context-free grammar (CFG)–based representation scheme composed of multiple layers. The layers defined are the body-part extraction layer, the pose layer, the gesture layer, and the action/interaction layer. In the first three layers, complex human activities in a scene are translated into sequences of predefined poses of each body part for each person. At the action/interaction layer, actions and interactions are represented semantically by the CFG. This method effectively represents not only human interactions between people [50], but also human interactions with objects [51].

Other biometrics can further improve recognition robustness and reliability when combined with gait information. Integration of face recognition and gait recognition is a good example. This may require multiple cameras, so viewpoint normalization is necessary. Currently, human gait recognition is very much in the early-to-middle stages of development, and many problems remain unsolved. Moreover, the computational complexity of most approaches remains high relative to the needs of practical applications.

19.3.6 Face Recognition

Human faces are the most important biometrics for human recognition. Face recognition has been intensively researched in a variety of directions, and may be viewed as the top level function of the video surveillance problem. Many techniques have been presented for face recognition over the last couple of decades and a large number of research surveys are available. Face recognition from still images is covered in detail in the companion volume, Chapter 24 of *The Essential Guide to Image Processing*. The cogent and more involved topic of face recognition and tracking from video sequences is covered in the next chapter (Chapter 20) of the current volume. Some key survey papers include [52–56].

19.4 CONCLUSION

Hopefully, the reader has discovered that video surveillance is an active application area as well as one of the emerging major research topics in area of video processing and analysis. No doubt, the reader has also gained a sense that the problems faced are somewhat daunting, although progress has been made in many areas and practical systems are viable for many problems. Yet, many problems remain open, and hopefully, readers of this book will be motivated to attack them. While a large amount of theoretical improvement has been made in recent years owing to the popularity of the topic and the hard work of many researchers, there are still significant challenges that must be met before practical automatic digital video surveillance becomes viable for most open and unconstrained applications. The field is important, since video surveillance has great potential to improve public and private safety, and to help improve national security.

REFERENCES

[1] Image from *IEEE Intl. Workshop Performance Evaluation of Tracking and Surveillance, 2001*, http://ftp.pets.rdg.ac.uk/PETS2001/.

[2] Image from *CAVIAR: Context Aware Vision using Image-based Active Recognition*, http://homepages.inf.ed.ac.uk/rbf/CAVIAR/.

[3] Image from Chesapeake Systems Service, Inc., http://www.csscorp.us/.

[4] Image from California Department of Transportation *Caltrans Live Traffic Cameras*, http://video.dot.ca.gov.

[5] W. Hu, T. Tan, L. Wang, and S. Maybank. A survey on visual surveillance of object motion and behaviors. *IEEE Trans. Syst. Man Cybern.*, 34(3), 2004.

[6] R. T. Collins, A. J. Lipton, T. Kanade, H. Fujiyoshi, D. Duggins, Y. Tsin, D. Tolliver, N. Enomoto, O. Hasegawa, P. Burt, and L. Wixson. A system for video surveillance and monitoring. *Tech. Rep.* CMU-RI-TR-00-12, Robotics Institute, Carnegie Mellon University, Pittsburgh, PA, May 2000.

[7] A. Cavallaro and T. Ebrahimi. Video object extraction based on adaptive background and statistical change detection. *Proc. SPIE Vis. Commun. Image Process.*, 465–475, February 2001.

[8] C. Wren, A. Azarbayejani, T. Darrell, and A. Pentland. Pfinder: Real-time tracking of the human body. *IEEE Trans. Pattern Anal. Mach. Intell.*, 19(7), 1997.

[9] C. Stauffer and W. E. L. Grimson. Adaptive background mixture models for real-time tracking. *Proc. IEEE Int. Conf. Comput. Vis. Pattern Recognit.*, 1999.

[10] N. Friedman and S. Russell. Image segmentation in video sequences: A probabilistic approach. *Proc. Conf. Uncertain. Artif. Intell.*, 175–181, 1997.

[11] Institut für Algorithmen und Kognitive Systeme (IAKS) at Universitat of Universitat Karlsruhe, http://i21www.ira.uka.de/image_sequences/.

[12] K. Toyama, J. Krumm, B. Brumitt, and B. Meyers. Wallflower: Principles and practice of background maintenance. *Proc. Int. Conf. Comput. Vis.*, 255–261, 1999.

[13] A. Elgammal, D. Harwood, and L. Davis. Non-parametric model for background subtraction. *Proc. Eur. Conf. Comput. Vis.*, 751–767, 2000.

[14] I. Haritaoglu, D. Harwood, and L. S. Davis. W4: Real-time surveillance of people and their activities. *IEEE Trans. Pattern Anal. Mach. Intell.*, 22(8), 2000.

[15] J. Zhong and S. Sclaroff. Segmenting foreground objects from a dynamic textured background via a Kalman filter. *Proc. Int. Conf. Comput. Vis.*, 44–50, 2003.

[16] Y. Kuno, T. Watanabe, Y. Shimosakoda, and S. Nakagawa. Automated detection of human for visual surveillance system. *Proc. Int. Conf. Pattern Recognit.*, 865–869, 1996.

[17] R. Culter and L. S. Davis. Robust real-time periodic motion detection, analysis, and applications. *IEEE Trans. Pattern Anal. Mach. Intell.*, 22:781–796, 2000.

[18] A. J. Lipton. Local application of optic flow to analyse rigid versus non-rigid motion. *Proc. Int. Conf. Comput. Vis. Workshop Frame Rate Vis.*, 1999.

[19] T. J. Broida and R. Chellappa. Estimation of object motion parameters from noisy images. *IEEE Trans. Pattern Anal. Mach. Intell.*, 8(1):90–99, 1986.

[20] N. Peterfreund. The velocity snake. *IEEE Proc. Workshop Nonrigid Articulated Motion*, 70–79, 1997.

[21] V. Caselles, R. Kimmel, and G. Sapiro. Geodesic active contours. *Proc. IEEE Int. J. Comput. Vis.*, 22(1):61–79, 1997.

[22] N. Peterfreund. Robust tracking of position and velocity with Kalman snakes. *IEEE Trans. Pattern Anal. Mach. Intell.*, 21(6), 1999.

[23] M. Kass, A. Witkin, and D. Terzopoulos. Snakes: Active contour models. *Int. J. Comput. Vis.*, 1:321–331, 1987.

[24] N. Paragois and R. Deriche. Geodesic active contours and level sets for the detection and tracking of moving objects. *IEEE Trans. Pattern Anal. Mach. Intell.*, 22(3):266–280, 2000.

[25] R. Polana and R. Nelson. Low level recognition of human motion. *Proc. IEEE Workshop Motion Non-Rigid Articulated Objects*, 77–82, 1994.

[26] B. Schiele. Model-free tracking of cars and people based on color regions. *Image Vis. Comput.*, 24(11):1172–1178, 2006.

[27] J. Malik and S. Russell. Traffic surveillance and detection technology development: New traffic sensor technology final report. *Tech. Rep.*, UCB-ITS-PRR-97-6, Institute of Transportation Studies, The University of California at Berkeley, January, 1997.

[28] D.-S. Jang and H.-I. Choi. Active models for tracking moving objects. *Intern. J. Pattern Recognit.*, 33:1135–1146, 2000.

[29] M. Clark, A. C. Bovik, and W. S. Geisler. Texture segmentation using Gabor modulation/demodulation. *Pattern Recognit. Lett.*, 6:261–267, 1987.

[30] M. Clark and A. C. Bovik. Experiments in segmenting texton patterns using localized spatial filters. *Pattern Recognit.*, 22(6):707–717, 1989.

[31] A. C. Bovik, M. Clark, and W. S. Geisler. Multichannel texture analysis using localized spatial filters. *IEEE Trans. Pattern Anal. Mach. Intell.*, 12(1):55–73, 1990.

[32] I. A. Karaulova, P. M. Hall, and A. D. Marshall. A hierarchical model of dynamics for tracking people with a single video camera. *Proc. Br. Mach. Vis. Conf.*, 262–352, 2000.

[33] S. X. Ju, M. J. Black, and Y. Yacoob. Cardboard people: A parameterized model of articulated image motion. *Proc. Int. Conf. Automat. Face Gesture Recognit.*, 38–44, 1996.

[34] K. Rohr. Towards model-based recognition of human movements in image sequences. *CVGIP: Image Underst.*, 59(1):94–115, 1994.

[35] T. N. Tan, G. D. Sullivan, and K. D. Baker. Model-based localization and recognition of road vehicles. *Int. J. Comput. Vis.*, 29(1):22–25, 1998.

[36] A. F. Bobick and A. D. Wilson. A state-based approach to the representation and recognition of gesture. *IEEE Trans. Pattern Anal. Mach. Intell.*, 19:1325–1337, 1997.

[37] A. D. Wilson, A. F. Bobick, and J. Cassell. Temporal classification of natural gesture and application to video coding. *Proc. IEEE Conf. Comput. Vis. Pattern Recognit.*, 948–954, 1997.

[38] N. M. Oliver, B. Rosario, and A. P. Pentland. A Bayesian computer vision system for modeling human interactions. *IEEE Trans. Pattern Anal. Mach. Intell.*, 22:831–843, 2000.

[39] M.-H. Yang and N. Ahuja. Extraction and classification of visual motion pattern recognition. *Proc. IEEE Conf. Comput. Vis. Pattern Recognit.*, 892–897, 1998.

[40] N. Johnson and D. Hogg. Learning the distribution of object trajectories for event recognition. *Image Vis. Comput.*, 14(8):609–615, 1996.

[41] D. Cunado, M. S. Nixon, and J. N. Carter. Using gait as a biometric: via phase-weighted magnitude spectra. *Proc. Int. Conf. Audio Video Based Biometric Person Authentication*, 95–102, 1997.

[42] C.-Y. Yam, M. S. Nixon, and J. N. Carter. Extended model-based automatic gait recognition of walking and running. *Proc. Int. Conf. Audio Video Based Biometric Person Authentication*, 277–283, 2001.

[43] S. A. Niyogi and E. H. Adelson. Analyzing and recognizing walking figures in XYT. *Proc. IEEE Conf. Comput. Vis. Pattern Recognit.*, 469–474, 1994.

[44] S. A. Niyogi and E. H. Adelson. Analyzing gait with spatio-temporal surface. *Proc. IEEE Workshop Motion Non-Rigid Articulated Objects*, 64–69, 1994.

[45] A. A. Kale, N. Cuntoor, V. Kruger, and A. N. Rajagopalan. Gait-based recognition of humans using continuous HMMs. *Proc. Int. Conf. Automat Face Gesture Recognit*, 336–341, 2002.

[46] J. D. Shutler, M. S. Nixon, and C. J. Harris. Statistical gait description via temporal moments. *Proc. IEEE Southwest Symp. Image Anal. Interpretation*, 291–295, 2000.

[47] H. Murase and R. Sakai. Moving object recognition in eigenspace representation: Gait analysis and lip reading. *Pattern Recognit. Lett.*, 17(2):155–162, 1996.

[48] M. A. Turk and A. P. Pentland. Face recognition using eigenfaces. *Proc. IEEE Conf. Comput. Vis. Pattern Recognit.*, 586–591, 1991.

[49] A. F. Bobick and A. Y. Johnson. Gait recognition using static, activity-specific parameters. *Proc. IEEE Conf. Comput. Vis. Pattern Recognit.*, 423–430, 2001.

[50] M. S. Ryoo and J. K. Aggarwal. Recognition of composite human activities through context-free grammar based representation. *Proc. IEEE Conf. Comput. Vis. Pattern Recognit.*, 1709–1719, 2006.

[51] M. S. Ryoo and J. K. Aggarwal. Hierarchical recognition of human activities interacting with objects. *Proc. IEEE Conf. Comput. Vis. Pattern Recognit.*, 1–8, 2007.

[52] R. Challappa, C. K. Wilson, and S. Sirohey. Human and machine recognition of faces: A survey. *Proc. IEEE*, 83(5):705–741, 1995.

[53] R. Gross, J. Shi, and J. F. Cohn. Quo vadis face recognition?. *Tech. Rep.* CMU-RI-TR-01-17, Robotics Institute, Carnegie Mellon University, Pittsburgh, PA, June 2001.

[54] W. Y. Zhao and R. Chellappa. Image-based face recognition: Issues and methods. In *Image Recognition and Classification: Algorithms, Systems, and Applications*, 375–402, CRC Press, Boca Raton, FL, 2002.

[55] W. Y. Zhao, R. Chellappa, P. J. Phillips, and A. Rosenfeld. Face recognition: A literature survey. *ACM Comput. Surv.*, 399–458, 2003.

[56] S. J. Gupta, M. K. Markey, and A. C. Bovik. Advances and challenges in 3D and 2D + 3D human face recognition. In E. A. Zoeller, editor, *Pattern Recognition Research Horizons*, 161–200, Nova Science Publishers, Hauppauge, NY, 2008.

CHAPTER

Face Recognition from Video

20

Shaohua Kevin Zhou[1], Rama Chellappa[2], and Gaurav Aggarwal[2]

[1] *Siemens Corporate Research, Princeton, New Jersey, USA*
[2] *Center for Automation Research (CfAR) and Department of Electrical and Computer Engineering, University of Maryland, College Park, Maryland, USA*

20.1 INTRODUCTION

Although face recognition (FR) from a single still image has been studied extensively [1, 2], FR based on a video sequence is an emerging topic, evidenced by the growing increase in the literature. It is predictable that with the ubiquity of video sequences, FR based on video sequences will become more and more popular. In this chapter, we also address FR based on a group of still images (also referred to as multiple still images). Multiple still images are not necessarily from a video sequence; they can come from multiple independent still captures.

It is obvious that multiple still images or a video sequence can be regarded as a single still image in a degenerate manner. More specifically, suppose that we have a group of face images $\{\mathsf{y}_1, \ldots, \mathsf{y}_T\}$ and a single-still-image-based FR algorithm $\mathcal{A}$ (or the base algorithm), we can construct a recognition algorithm based on multiple still images or a video sequence by fusing multiple base algorithms denoted by $\mathcal{A}_i$s. Each $\mathcal{A}_i$ takes a different single image y_i as input. The fusion rule can be additive, multiplicative, and so on.

Even though the fusion algorithm might work well in practice, clearly, the overall recognition performance solely depends on the base algorithm, and hence, designing the base algorithm $\mathcal{A}$ (or the similarity function k) is of ultimate importance. However, the fused algorithms neglect additional properties manifested in multiple still images or video sequences. Generally speaking, algorithms that judiciously exploit these properties will perform better in terms of recognition accuracy, computational efficiency, etc.

There are three additional properties available from multiple still images and/or video sequences:

1. Set of observations: This property is directly exploited by the fused algorithms. One main disadvantage may be the ad hoc nature of the combination rule. However, theoretical analysis based on a set of observations can be performed. For example, a set of observations can be summarized using quantities like matrix, probability density function (PDF), manifold, etc. Hence, corresponding knowledge can be used to match two sets.
2. Temporal continuity/Dynamics: Successive frames in the video sequences are continuous in the temporal dimension. Such continuity, coming from facial expression, geometric continuity related to head and/or camera movement, or photometric continuity related to changes in illumination, provides an additional constraint for modeling face appearance. In particular, temporal continuity can be further characterized using kinematics. For example, facial expression and head movement of an individual during a certain activity result in structured changes in face appearance. Modeling of such a structured change (or dynamics) further regularizes FR.
3. 3D model: The property outlines the fact that different images/frames of a video come from the same 3D object. This means that we are able to reconstruct a 3D model from a group of still images and a video sequence. Recognition can then be based on the 3D model. Using the 3D model provides possible invariance to pose and illumination.

Clearly, the first and third properties are shared by multiple still images and video sequences. The second property is solely possessed by video sequences. We will elaborate these properties in Section 20.2 and review approaches that exploit the properties.

Section 20.3 defines a general framework called probabilistic identity characterization that deals with a group of images, either from multiple image acquisitions or from a video. Section 20.4 then presents two instances of probabilistic identity characterization: subspace identity encoding for FR from a group of images acquired at different illumination conditions and poses and simultaneous tracking and recognition of face from video. Finally, Section 20.5 gives a system identification approach for video-based FR.

FR mainly involves the following three tasks [3]:

1. Verification: The recognition system determines whether the query face image and the claimed identity match.
2. Identification: The recognition system determines the identity of the query face image by matching it with a database of images with known identities, assuming that the query identity is among the known identities belonging to the database.
3. Watch list: The recognition system first determines whether the identity of the query face image is in the stored watch list and, if yes, then identifies the individual.

TABLE 20.1 Recognition configurations based on a single still image, multiple still images, and a video sequence.

Probe/Gallery	A Single Still Image	A Group of Still Images	A Video Sequence
A single still image	*sStill-to-sStill*	*mStill-to-sStill*	*Video-to-sStill*
A group of still images	*sStill-to-mStill*	*mStill-to-mStill*	*Video-to-mStill*
A video sequence	*sStill-to-Video*	*mStill-to-Video*	*Video-to-Video*

Among the three tasks, the watch list task is the most difficult one. This chapter focuses only on the identification task.

In particular, we here follow a FR test protocol FERET [4] widely observed in the FR literature. FERET assumes availability of the following three sets, namely one training set, one gallery set, and one probe set. The training set is used to design the recognition algorithm. Typically, the training set does not overlap with the gallery and probe sets. The gallery and probe sets are used in the testing stage. The gallery set contains images with known identities and the probe set with unknown identities. The algorithm associates descriptive features with the images in the gallery and probe sets and determines the identities of the probe images by comparing their associated features with those features associated with gallery images.

According to the imagery used in the gallery and probe sets, we can define the following nine recognition configurations as in Table 20.1. For instance, the *mStill-to-Video* setting uses multiple still images for each individual in the gallery set and a video sequence for each individual in the probe set. The FERET test investigated the *sStill-to-sStill* recognition setting, and the FR Grand Challenge [5] studies three settings: *sStill-to-sStill*, *mStill-to-sStill*, and *mStill-to-mStill*.

20.2 PROPERTIES AND LITERATURE REVIEW

The multiple still images and video sequences are different from a single still image because they possess additional properties not present in a still image. In particular, three properties manifest themselves, which motivate various approaches recently proposed in the literature. Below, we analyze the three properties one by one.

20.2.1 Set of Observations

This is the most commonly used feature of multiple still images and video sequence. If only this property is considered, a video sequence reduces to a group of still images with the temporal dimension stripped. In other words, every video frame is treated as a still image. Another implicit assumption is that all face images are normalized before subject to subsequent analysis.

As mentioned earlier, the combination rules are rather ad hoc, which leaves room for a systematic exploration of this property. This leads to investigating systematic representations to summarize a set of observations $\{y_1, y_2, \ldots, y_T\}$. Once an appropriate representation is fixed, a recognition algorithm can be accordingly designed.

Various ways of summarizing a set of observations have been proposed; accordingly these approaches can be grouped into four categories.

20.2.1.1 One Image or Several Images

A set of observations $\{y_1, y_2, \ldots, y_T\}$ is summarized into one image $\hat{y}$ or several images $\{\hat{y}_1, \hat{y}_2, \ldots, \hat{y}_m\}$ (with $m < T$). For instance, one can use the mean or the median of $\{y_1, y_2, \ldots, y_T\}$ as the summary image $\hat{y}$. Clustering techniques can be invoked to produce multiple summary images $\{\hat{y}_1, \hat{y}_2, \ldots, \hat{y}_m\}$. In terms of recognition, we can simply apply the still-image-based FR algorithm based on $\hat{y}$ or $\{\hat{y}_1, \hat{y}_2, \ldots, \hat{y}_m\}$. This applies to all nine recognition settings listed in Table 20.1.

20.2.1.2 Matrix

A set of observations $\{y_1, y_2, \ldots, y_T\}$ forms a matrix[1] $Y = [y_1, y_2, \ldots, y_T]$. The main advantage of using the matrix representation is that we can rely on the rich literature of matrix analysis. For example, various matrix decompositions can be invoked to represent the original data more efficiently. Metrics measuring similarity between two matrices can then be used for recognition.

This applies to the *mStill-to-mStill*, *Video-to-mStill*, *Video-to-mStill*, and *Video-to-Video* recognition settings. Suppose that the nth individual in the gallery set has a matrix $X^{[n]}$, we determine the identity of a probe matrix Y as

$$\hat{n} = \arg \min_{n=1,2,\ldots,N} d(Y, X^{[n]}), \tag{20.1}$$

where d is a matrix distance function.

Yamaguchi et al. [6] proposed the so-called *Mutual Subspace Method* (MSM). In this method, the matrix representation is used and the similarity function between two matrices is defined as the angle between two subspaces of the matrices (also referred to as principal angle or canonical correlation coefficient). Suppose that the columns of X and Y represent two subspaces U_X and U_Y, the principle angle θ between the two subspaces is defined as

$$\cos(\theta) = \max_{u \in U_X} \max_{v \in U_Y} \frac{u^T v}{\sqrt{u^T u}\sqrt{v^T v}}. \tag{20.2}$$

It can be shown that the principle angle θ is equal to the largest singular value of the matrix $U_X^T U_Y$ where U_X and U_Y are orthogonal matrices encoding the column bases of the X and Y matrices, respectively.

---

[1]Here we assume that each image y_i is "vectorized."

In general, the leading singular values of the matrices $U_X^T U_Y$ define a series of principal angles $\{\theta_k\}$'s.

$$\cos(\theta_k) = \max_{u \in U_X} \max_{v \in U_Y} \frac{u^T v}{\sqrt{u^T u}\sqrt{v^T v}} \tag{20.3}$$

subject to

$$u^T u_i = 0, v^T v_i = 0, \ i = 1, 2, \ldots, k-1. \tag{20.4}$$

Yamaguchi et al. [6] recorded a database of 101 individuals exhibiting variations in facial expression and pose. They discovered that the MSM is more robust to noisy input image or face normalization error than the still-image-based method that is referred to as conventional subspace method (CSM) in [6]. The similarity function of the MSM is shown to be more stable and consistent than that of the CSM.

Wolf and Shashua [7] extended the computation of the principal angles into a nonlinear feature space $\mathcal{H}$ called reproducing kernel Hilbert space (RKHS) [8] induced by a positive definite kernel function. This kernel function represents an inner product between two vectors in the nonlinear feature space that is mapped from the original data space (say $\mathcal{R}^d$) via a nonlinear mapping function ϕ. Suppose that $x, y \in \mathcal{R}^d$, the kernel function is

$$k(x, y) = \phi(x)^T \phi(y). \tag{20.5}$$

This is known as "kernel trick": once the kernel function k is specified, no explicit form of the function ϕ is required. Therefore, as long as we are able to cast all computations into inner product, we can invoke the "kernel trick" to lift the original data space to an RKHS. Since the mapping function ϕ is nonlinear, the nonlinear characterization of the data structure is captured to some extent. Popular choices for the kernel function k are polynomial kernel, radial basis function (RBF) kernel, neural network kernel, etc. Refer to [8] for their definitions.

Kernel principal angles between two matrices X and Y are then based on their "kernelized" versions $\phi(X)$ and $\phi(Y)$. A "kernelized" matrix $\phi(X)$ of $X = [x_1, x_2, \ldots, x_n]$ is defined as $\phi(X) = [\phi(x_1), \phi(x_2), \ldots, \phi(x_n)]$. The key is to evaluate the matrix $U_{\phi(X)}^T U_{\phi(Y)}$ defined in RKHS. In [7], Wolf and Shashua showed the computation using the "kernel trick."

Another contribution of Wolf and Shashua [7] is that they further proposed a positive kernel function taking the matrix as input. Given such a kernel function, it can be readily plugged into a classification scheme such as a support vector machine (SVM) [8] to take advantage of the SVM's discriminative power. FR using multiple still images, coming from a tracked sequence, was studied: the proposed kernel principal angles slightly outperform other nonkernel versions.

Zhou [9] systematically investigated the kernel functions taking matrix as input (also referred to as matrix kernels). More specifically, the following two functions are kernel functions.

$$k_\bullet(X, Y) = tr(X^T Y), \ k_\star(X, Y) = det(X^T Y), \tag{20.6}$$

where *tr* and *det* are matrix trace and determinant. Using these as building blocks, Zhou [9] constructed more kernels based on the column basis matrix, the "kernelized" matrix, and the column basis matrix of the "kernelized" matrix.

$$k_{U\bullet}(X,Y) = tr\left(U_X^T U_Y\right),\ k_{U\star}(X,Y) = det(U_X^T U_Y), \tag{20.7}$$

$$k_{\phi\bullet}(X,Y) = tr(\phi(X)^T\phi(Y)),\ k_{\phi\star}(X,Y) = det(\phi(X)^T\phi(Y)), \tag{20.8}$$

$$k_{U\phi\bullet}(X,Y) = tr\left(U_{\phi(X)}^T U_{\phi(Y)}\right),\ k_{U\phi\star}(X,Y) = det\left(U_{\phi(X)}^T U_{\phi(Y)}\right). \tag{20.9}$$

20.2.1.3 PDF

In this rule, a set of observations $\{y_1, y_2, \ldots, y_T\}$ is regarded as independent realizations drawn from an underlying distribution. PDF estimation techniques such as parametric, semiparametric, and nonparametric methods [10] can be used to learn the distribution.

In the *mStill-to-mStill*, *Video-to-mStill*, *mStill-to-Video*, and *Video-to-Video* recognition settings, recognition can be performed by comparing distances between PDFs [11], such as Bhattacharyya and Chernoff distances and Kullback–Leibler (KL) divergence. More specifically, suppose that the nth individual in the gallery set has a PDF $p^{[n]}(x)$, we determine the identity of a probe PDF $q(y)$ as

$$\hat{n} = \arg \min_{n=1,2,\ldots,N} d(q(y), p^{[n]}(x)), \tag{20.10}$$

where d is a probability distance function.

In the *mStill-to-sStill*, *Video-to-sStill*, *sStill-to-mStill*, and *sStill-to-Video* settings, recognition becomes a hypothesis testing problem. For example, in the *sStill-to-mStill* setting, if we can summarize the multiple still images in query into a pdf, say $q(y)$, then recognition is to test which gallery image $x^{[n]}$ is mostly likely to be generated by $q(y)$.

$$\hat{n} = \arg \max_{n=1,2,\ldots,N} q(x^{[n]}). \tag{20.11}$$

Notice that this is different from the *mStill-to-sStill* setting, where each gallery object has a density $p^{[n]}(y)$, then given a probe single still image y, recognition checks the following:

$$\hat{n} = \arg \max_{n=1,2,\ldots,N} p^{[n]}(y). \tag{20.12}$$

Equation (20.12) is the same as the probabilistic interpretation of still-image-based recognition, except that the density $p^{[n]}(y)$ for a different n can have a different form. In such case, we in principle no longer need a training set in the *sStill-to-sStill* setting.

Shakhnarovich et al. [12] proposed to use multivariate normal density for summarizing face appearances and the KL divergence or relative entropy for recognition. The KL divergence between two normal densities $p_1 = N(\mu_1, \Sigma_1)$ and $p_2 = N(\mu_2, \Sigma_2)$ can be explicitly computed as [11]

$$\mathrm{KL}(p_1||p_2) = \int_{\mathrm{x}} p_1(\mathrm{x}) \log \frac{p_1(\mathrm{x})}{p_2(\mathrm{x})} d\mathrm{x} \tag{20.13}$$

$$= \frac{1}{2} \log\left(\frac{|\Sigma_2|}{|\Sigma_1|}\right) + \frac{1}{2} tr\left((\mu_1 - \mu_2)^{\mathsf{T}} \Sigma_2^{-1} (\mu_1 - \mu_2) + \Sigma_1 \Sigma_2^{-1}\right) - \frac{d}{2},$$

where d is the dimensionality of the data. One disadvantage of the KL divergence is that it is asymmetric. To make it symmetric, one can use

$$J_D(p_1, p_2) = \int_{\mathrm{x}} (p_1(\mathrm{x}) - p_2(\mathrm{x})) \log \frac{p_1(\mathrm{x})}{p_2(\mathrm{x})} d\mathrm{x} = \mathrm{KL}(p_1||p_2) + \mathrm{KL}(p_2||p_1) \tag{20.14}$$

$$= \frac{1}{2} tr\left((\mu_1 - \mu_2)^{\mathsf{T}} \left(\Sigma_1^{-1} + \Sigma_2^{-1}\right) (\mu_1 - \mu_2) + \Sigma_1 \Sigma_2^{-1} + \Sigma_2 \Sigma_1^{-1}\right) - d.$$

Shakhnoarovich et al. [12] achieved better performance than Yamaguchi et al. [6] who used the MSM approach on a data set including 29 subjects.

Other than KL divergence, probabilistic distance measures such as Chernoff distance and Bhattacharyya distance can also be used. The Chernoff distance is defined and computed in the case of normal density as

$$J_C(p_1, p_2) = -\log\left\{\int_{\mathrm{x}} p_1^{\alpha_2}(\mathrm{x}) p_2^{\alpha_1}(\mathrm{x}) d\mathrm{x}\right\} \tag{20.15}$$

$$= \frac{1}{2} \alpha_1 \alpha_2 (\mu_1 - \mu_2)^{\mathsf{T}} [\alpha_1 \Sigma_1 + \alpha_2 \Sigma_2]^{-1} (\mu_1 - \mu_2) + \frac{1}{2} \log \frac{|\alpha_1 \Sigma_1 + \alpha_2 \Sigma_2|}{|\Sigma_1|^{\alpha_1} |\Sigma_2|^{\alpha_2}},$$

where $\alpha_1 > 0$, $\alpha_2 > 0$, and $\alpha_1 + \alpha_2 = 1$. When $\alpha_1 = \alpha_2 = 1/2$, the Chernoff distance reduces to the Bhattacharyya distance.

In [13], Jebara and Kondon proposed the probability product kernel function

$$k(p_1, p_2) = \int_{\mathrm{x}} p_1^r(\mathrm{x}) p_2^r(\mathrm{x}) d\mathrm{x}, \; r > 0. \tag{20.16}$$

When $r = 1/2$, the kernel function k reduces to the so-called Bhattacharyya kernel since it is related to the Bhattacharyya distance. When $r = 1$, the kernel function k reduces to the so-called expected likelihood kernel. In practice, we can simply use the kernel function k as a similarity function.

However, the Gaussian assumption can be ineffective when modeling the nonlinear face appearance manifold. To absorb the nonlinearity, mixture models or nonparametric densities are used in practice. For such cases, one has to resort to numerical methods for computing the probabilistic distances. Such computation is not robust since two approximations are invoked: one in estimating the density and the other one in evaluating the numerical integral.

In [14], Zhou and Chellappa modeled the nonlinearity through a different approach: kernel methods. As mentioned earlier, the essence of kernel methods is to combine a linear algorithm with a nonlinear embedding, which maps the data from the original vector space to a nonlinear feature space called the RKHS. But no explicit knowledge of the nonlinear mapping function is needed as long as the involved computations can be cast into inner product evaluations. Since a nonlinear function is used, albeit in an implicit fashion, Zhou and Chellappa [14] achieved a new approach to study these distances and investigate their uses in a different space. To be specific, analytic expressions for probabilistic distances that account for nonlinearity or high-order statistical characteristics of the data are derived. On a data set involving subjects presenting appearances with pose and illumination variations, the probabilistic distance measures performed better than their nonkernel counterparts.

Recently, Arandjelović and Cipolla [15] used resistor-average distance (RAD) for video-based recognition.

$$\mathrm{RAD}(p_1,p_2) = (\mathrm{KL}(p_1||p_2)^{-1} + \mathrm{KL}(p_2||p_1)^{-1})^{-1}. \tag{20.17}$$

Further, computation of the RAD was conducted on the RKHS to account for the nonlinearity of face manifold. Some robust techniques such as synthesizing images to account for small localization errors and RANSAC algorithms to reject outliers were introduced to achieve improved performance. They [16] further extended their work to use the symmetric KL divergence distance between two mixture-of-Gaussian densities.

20.2.1.4 *Manifold*

In this rule, a set of face appearances form a highly nonlinear manifold $\mathcal{P}$. Manifold learning has recently attracted a lot of attention. After characterizing the manifold, FR reduces to (i) comparing two manifolds if we are in the *mStill-to-mStill, Video-to-mStill, mStill-to-Video*, and *Video-to-Video* settings and (ii) comparing distances from one data point to different manifolds if we are in the *mStill-to-sStill, Video-to-sStill, sStill-to-mStill*, and *sStill-to-Video* settings.

For instance, in the *Video-to-Video* setting, galley videos are summarized into manifolds $\{\mathcal{P}^{[n]}; n = 1,2,\ldots,N\}$. For the probe video that is summarized into a manifold $\mathcal{Q}$, its identity is determined as

$$\hat{n} = \arg \min_{n=1,2,\ldots,N} d(\mathcal{Q},\mathcal{P}^{[n]}), \tag{20.18}$$

where d calibrates the distance between two manifolds.

In the *Video-to-sStill* setting, for the probe still image y, its identity is determined as

$$\hat{n} = \arg \min_{n=1,2,\ldots,N} d(\mathsf{y},\mathcal{P}^{[n]}), \tag{20.19}$$

where d calibrates the distance between a data point to a manifold.

Fitzgibbon and Zisserman [17] computed a joint manifold distance to cluster appearances. A manifold is captured by subspace analysis, which is fully specified by a mean and a set of basis vectors. For example, a manifold $\mathcal{P}$ can be represented as

$$\mathcal{P} = \{\mathsf{m}_p + \mathsf{B}_p\mathsf{u} | \mathsf{u} \in \mathcal{U}\}, \tag{20.20}$$

where m_p is the mean and B_p encodes the basis vectors. In addition, the authors invoked an affine transformation to overcome geometric deformation. The joint manifold distance between $\mathcal{P}$ and $\mathcal{Q}$ is defined as

$$d(\mathcal{P},\mathcal{Q}) = \min_{\mathsf{u,v,a,b}} \|T(\mathsf{m}_p + \mathsf{B}_p\mathsf{u},\mathsf{a}) - T(\mathsf{m}_q + \mathsf{B}_q\mathsf{v},\mathsf{b})\|^2 + E(\mathsf{a}) + E(\mathsf{b}) + E(\mathsf{u}) + E(\mathsf{v}), \tag{20.21}$$

where $T(\mathsf{x},\mathsf{a})$ transforms image x using the affine parameter a and $E(\mathsf{a})$ is the prior cost incurred by invoking the parameter a.

In experiments, Fitzgibbon and Zisserman [17] performed automatic clustering of faces in feature-length movies. To reduce the lighting effect, the face images are high-pass filtered before a clustering step. The authors reported that sequence-to-sequence matching presents a dramatic computational speedup when compared with pairwise image-to-image matching.

Identity surface is a manifold, proposed by Li et al. in [18], that depicts face appearances presented in multiple poses. The pose is parameterized by yaw α and tilt θ. Face image at (α,θ) is first fitted to a 3D point distribution model and an active appearance model. After pose estimation, the face appearance is warped to a canonical view to provide a pose-free representation from which a nonlinear discriminatory feature vector is extracted. Suppose that the feature vector is denoted by f, the function $\mathsf{f}(\alpha,\theta)$ defines the identity surface that is parameterized by the pose θ. In practice, since only a discrete set of views are available, the identity surface is approximated by piece-wise planes. The distance between two manifolds $\mathcal{P} = \{\mathsf{f}_p(\alpha,\theta)\}$ and $\mathcal{Q} = \{\mathsf{f}_q(\alpha,\theta)\}$ is defined as

$$d(\mathcal{Q},\mathcal{P}) = \int_\alpha \int_\theta w(\alpha,\theta) d(\mathsf{f}_q(\alpha,\theta),\mathsf{f}_p(\alpha,\theta)) d\alpha d\theta, \tag{20.22}$$

where $w(\alpha,\theta)$ is a weight function.

A video sequence corresponds to a trajectory traced out in the identity surface. Suppose that video frames sample the pose space at $\{\alpha_j,\theta_j\}$, the following distance $\sum_j w_j d(\mathsf{f}_q(\alpha_j,\theta_j),\mathsf{f}_p(\alpha_j,\theta_j))$ is used for video-based FR. In the experiments, 12 subjects were involved and a 100% recognition accuracy was achieved.

20.2.2 Temporal Continuity/Dynamics

Property *P*1 strips the temporal dimension available in the video sequence. In property *P*2, we bring back the temporal dimension, and hence, the property *P*2 only holds for video sequence.

Successive frames in a video sequence are continuous in the temporal dimension. The continuity arising from dense temporal sampling is two-fold: the face movement is continuous and the change in appearance is continuous.

Temporal continuity provides an additional constraint for modeling face appearance. For example, smoothness of face movement is used in face tracking. As mentioned earlier, it is implicitly assumed that all face images are normalized before utilization of the property *P*1: set of observations. For the purpose of normalization, face detection is independently applied on each image. When temporal continuity is available, tracking can be applied instead of detection to perform normalization of each video frame.

Temporal continuity also plays an important role for recognition. Recently psychophysical evidence [19] reveals that moving faces are more recognizable. In addition to temporal continuity, face movement and face appearance often follow certain kinematics. In other words, changes in movement and appearance are not random. Understanding kinematics is also important for FR.

Simultaneous tracking and recognition is an approach proposed by Zhou et al. [20] that systematically studied how to incorporate temporal continuity in video-based recognition. Zhou et al. modeled two tasks involved, namely tracking and recognition, in a probabilistic framework using a time series. This will be elaborated in Section 20.4.

Lee et al. [21] performed a video-based FR using probabilistic appearance manifolds. The main motivation is to model appearances under pose variation, that is, a generic appearance manifold consists of several pose manifolds. Since each pose manifold is represented using a linear subspace, the overall appearance manifold is approximated by piecewise linear subspaces. The learning procedure is based on face exemplars extracted from a video sequence. K-means clustering is first applied and then for each cluster principal component analysis is used for a subspace characterization.

In addition, the transition probabilities between pose manifolds are also learned. The temporal continuity is directly captured by the transition probabilities. In general, the transition probabilities between neighboring poses (such as frontal pose to left pose) are higher than those between far-apart poses (such as left pose to right pose). Recognition also reduces to computing a posterior distribution.

Lee et al. compared three methods that use temporal information differently: the proposed method with learned transition matrix, the proposed method with uniform transition matrix (meaning that temporal continuity is lost), and majority voting. The proposed method with learned transition matrix achieved a significantly better performance than the other two methods.

Liu and Chen [22] used adaptive hidden Markov models (HMM) to depict the dynamics. The HMM is a statistical tool to model time series and is represented by $\lambda = (\mathsf{A}, \mathsf{B}, \pi)$, where A is the state transition probability matrix, B is the observation PDF, and π is the initial state distribution. Given a probe video sequence Y, its identity is determined as

$$\hat{n} = \arg \max_{1,2,\ldots,N} = p(\mathsf{Y}|\lambda_n), \tag{20.23}$$

where $p(\mathsf{Y}|\lambda_n)$ is the likelihood of observing the video sequence Y given the model λ_n. In addition, when certain conditions hold, HMM λ_n was adapted to accommodate the appearance changes in the probe video sequence that results in improved modeling over time. Experimental results on various data sets demonstrated the advantages of using adaptive HMMs.

Aggarwal et al. [23] proposed a system identification approach for video-based FR. The face sequence is treated as a first-order autoregressive and moving averaging (ARMA) random process. Once the system is identified or each video sequence is associated with its ARMA parameters, video-to-video recognition uses various distance metrics constructed based on the parameters. Section 20.5 details this approach.

In [24], Turaga et al. revisited the system identification formulation in [23] from an analytic Grassmann manifold perspective. Using the rigorous derivations of the Grassmann manifold (e.g., Procrustes distance and kernel density estimation for the manifold), Turaga et al. improved the FR performance by a large margin.

Facial expression analysis is also related to temporal continuity/dynamics but not directly related to FR. Examples of expression analysis include [25, 26]. A review of face expression analysis is beyond the scope of this chapter.

20.2.3 3D Model

This means that we are able to reconstruct a 3D model from a group of still images and a video sequence. This leads to the literature of light field rendering that takes multiple still images as input and structure from motion (SfM) that takes a video sequence as input. Even though SfM has been studied for more than two decades, current SfM algorithms are not reliable enough for accurate 3D model reconstruction. Researchers therefore incorporate or solely use prior 3D face models (that are acquired beforehand) to derive the reconstruction result. In principle, 3D model provides the possibility of resolving pose and illumination variations.

The 3D model possesses two components: geometric and photometric. Geometric component describes the depth information of the face and photometric component depicts the texture map. The SfM algorithm is more focused on recovering the geometric component, whereas the light field rendering method is more on recovering the photometric component.

Recognition can then be performed directly based on the 3D model. More specifically, for any recognition setting, suppose that galley individuals are summarized into 3D models $\{\mathcal{M}^{[n]}; n = 1,2,\ldots,N\}$. For multiple observations of a probe individual that are summarized into a 3D model $\mathcal{N}$, its identity is determined as

$$\hat{n} = \arg \min_{n=1,2,\ldots,N} d(\mathcal{N},\mathcal{M}^{[n]}), \tag{20.24}$$

where d calibrates the distance between two models.

For one probe still image y, its identity is determined as

$$\hat{n} = \arg \min_{n=1,2,\ldots,N} d(\mathsf{y},\mathcal{M}^{[n]}), \tag{20.25}$$

where d calibrates the cost of generating a data point from a model.

There is a large body of literature on SfM. However, the current SfM algorithms do not reliably reconstruct the 3D face model. There are three difficulties in the SfM algorithm. The first lies in the ill-posed nature of the perspective camera model that results in instability of the SfM solution. The second is that the face model is not a truly rigid model, especially when facial expression and other deformations are present. The final difficulty is related to the input to the SfM algorithm. This is usually a sparse set of feature points provided by a tracking algorithm that itself has its own flaws. Interpolation from a sparse set of feature points to a dense set is very inaccurate.

To alleviate the first difficulty, orthographic and paraperspective models are used to approximate the perspective camera model. Under such approximate models, the ill-posed problem becomes well-posed. In Tomasi and Kanade [27], the orthographic model was used and a matrix factorization principle was discovered. The factorization principle was extended to the paraperspective camera model in Poelman and Kanade [28].

The second difficulty is often resolved by imposing a subspace constraint on the face model. Bregler et al. [29] proposed to regularize the nonrigid face model by using the linear constraint. It was shown that factorization can still be obtained. Brand [30] considered such factorization under uncertainty. Xiao et al. [31] discovered a closed form solution to nonrigid shape and motion recovery.

Interpolation from a sparse set to a dense depth map is always a difficult task. To overcome this, a dense face model is used instead of interpolation. However, the dense face model is only a generic model and hence may not be appropriate for a specific individual. Bundle adjustment [32, 33] is a method that adjust the generic model directly to accommodate the video observation. Roy-Chowdhury and Chellappa [34] took a different approach for combining the 3D face model recovered from the SfM algorithm with the generic face model.

The SfM algorithm mainly recovers the geometric component of the face model, that is, the depth value of every pixel. Its photometric component is naively set to the appearance in one reference video frame. Image-based rendering methods, on the other hand, directly recover the photometric component of the 3D model. Light field rendering [35, 36] in fact bypasses the stage of recovering the photometric component of the 3D model but rather recovers the novel views directly. The light field rendering methods [35, 36] relax the requirement of calibration by a fine quantization of the pose space and recover a novel view by sampling the captured data that form the so-called light field. The "eigen" light field approach developed by Gross et al. [37] assumes a subspace assumption of the light field. In Zhou and Chellappa [38], the light field subspace and the illumination subspace are combined to arrive at a bilinear analysis. Another line of research relates to 3D model recovery using the visual hull methods [39, 40]. But the visual hull method assumes that the shape of the object is convex, which is not always satisfied by the human face, and also requires accurate calibration information. Direct use of visual hull for FR is not found in the literature.

To characterize both the geometric and photometric components of the 3D face model, Blanz and Vetter [41] fitted a 3D morphable model to a single still image. The 3D morphable model uses a linear combination of dense 3D models and texture maps. In principle, the 3D morphable model can be fitted to multiple images. The 3D morphable model can be thought of as an extension of 2D active appearance model [42] to 3D, but the 3D morphable model uses dense 3D models. Xiao et al. [31] proposed to combine a linear combination of 3D sparse model and a 2D appearance model.

Although there is significant interest in recovering the 3D model, directly performing FR using the 3D model is a recent trend [43–46]. Blanz and Vetter [41] implicitly did so by using the combining coefficients for recognition. Beumier and Acheroy [43] conducted matching based on 2D sections of the facial surface. Mavridis et al. [44] used 3D+color camera to perform FR. Bronstein et al. [45] used a 3D face model for compensating the effect of facial expression in FR. However, the above approaches use the 3D range data as input. Because in this chapter we are mainly interested in FR from multiple still images or video sequence, a thorough review of FR based on 3D range data is beyond its scope.

20.3 A GENERAL FRAMEWORK OF PROBABILISTIC IDENTITY CHARACTERIZATION

We consider FR based on a group of images. In terms of the transformations embedded in the group or the temporal continuity between the transformations, the group can be either independent or not. Examples of the independent group (I-group) are face databases that store multiple appearances for one object. Examples of the dependent group are video sequences. If temporal information is stripped, the video sequences reduce to I-groups. Whenever we refer to video sequences, we imply dependent groups of images.

We attempt to propose a unified framework [47] that possesses the following features:

- It processes either a single image or a group of images (including the I-group and video sequences) in a universal manner.
- It handles the localization problem, pose, illumination, and expression (PIE) variations.
- The identity description could be either discrete or continuous. The continuous identity encoding typically arises from a subspace modeling.
- It is probabilistic and integrates all the available evidences.

We elaborate the proposed framework and point out its properties and connections with various approaches in Section 20.3. In Section 20.4, we substantiate the framework with two instances: FR from a group of still images and from a video sequences.

Suppose α represents the identity in an abstract manner. It can be either discrete- or continuous-valued. If we have an N-class problem, α is discrete taking value in $\{1,2,\ldots,N\}$. If we associate the identity with image intensity or feature vectors derived from say subspace projections, α is continuous-valued. Given a group of images $\mathbf{y}_{1:T} \doteq \{\mathbf{y}_1,\mathbf{y}_2,\ldots,\mathbf{y}_T\}$ containing the appearances of the same but unknown identity, *probabilistic identity characterization* is equivalent to finding the *posterior probability* $p(\alpha|\mathbf{y}_{1:T})$.

As the image only contains a transformed version of the object, we also need to associate with it a transformation parameter θ, which lies in a transformation space Θ. Here, the term "transformation" is a loose word to model the variations involved, be it warping, pose, illumination, or expressions. The transformation space Θ is usually application-dependent. Affine transformation is often used to compensate for the localization problem. To handle illumination variation, the estimates of lighting direction are used. If pose variations are involved, a 3D transformation is needed or a discrete set is used if we quantize the continuous view space. Suppose that the dimension of the transformation space Θ is r.

We assume that the prior probability of α is $\pi(\alpha)$, which is assumed to be, in practice, a *noninformative* prior. A noninformative prior is uniform in the discrete case and treated as a constant, say 1, in the continuous case.

The key to our probabilistic identity characterization is as follows:

$$\begin{aligned} \mathrm{p}(\alpha|\mathbf{y}_{1:T}) \propto \pi(\alpha)\mathrm{p}(\mathbf{y}_{1:T}|\alpha) &= \pi(\alpha)\int_{\theta_{1:T}} \mathrm{p}(\mathbf{y}_{1:T}|\theta_{1:T},\alpha)\mathrm{p}(\theta_{1:T})\mathrm{d}\theta_{1:T} \\ &= \pi(\alpha)\int_{\theta_{1:T}} \prod_{t=1}^{T} \mathrm{p}(\mathbf{y}_t|\theta_t,\alpha)\mathrm{p}(\theta_t|\theta_{1:t-1})\mathrm{d}\theta_{1:T}, \end{aligned} \tag{20.26}$$

where the following rules, namely (a) *observational conditional independence* and (b) *chain rule*, are applied:

$$(a)\ \mathrm{p}(\mathbf{y}_{1:T}|\theta_{1:T},\alpha) = \prod_{t=1}^{T} \mathrm{p}(\mathbf{y}_t|\theta_t,\alpha); \tag{20.27}$$

$$(b)\ \mathrm{p}(\theta_{1:T}) = \prod_{t=1}^{T} \mathrm{p}(\theta_t|\theta_{1:t-1});\ \mathrm{p}(\theta_1|\theta_0) \doteq \mathrm{p}(\theta_1). \tag{20.28}$$

Equation (20.26) involves two key quantities: the *observation likelihood* $\mathrm{p}(\mathbf{y}_t|\theta_t,\alpha)$ and the *state transition probability* $\mathrm{p}(\theta_t|\theta_{1:t-1})$. The former is essential to the recognition task, the ideal case being that it possesses a discriminative power in the sense that it always favors the correct identity and rejects others; the latter is also very helpful especially when processing video sequences, which constrains the search space.

We now study two special cases of $\mathrm{p}(\theta_t|\theta_{1:t-1})$.

I-Group In this case, the transformations $\{\theta_t; t = 1,2,\ldots,T\}$ are independent of each other, that is,

$$\mathrm{p}(\theta_t|\theta_{1:t-1}) = \mathrm{p}(\theta_t). \tag{20.29}$$

Equation (20.26) then becomes

$$\mathrm{p}(\alpha|\mathbf{y}_{1:T}) \propto \pi(\alpha)\prod_{t=1}^{T}\int_{\theta_t} \mathrm{p}(\mathbf{y}_t|\theta_t,\alpha)\mathrm{p}(\theta_t)\mathrm{d}\theta_t. \tag{20.30}$$

In this context, the probability $\mathrm{p}(\theta_t)$ can be regarded as a prior for θ_t, which is often assumed to be Gaussian with mean $\hat{\theta}_t$ or noninformative.

The most widely studied case in the literature is when $T = 1$, that is, there is only a single image in the group. Due to its importance, sometime we will distinguish it from the I-group (with $T > 1$) depending on the context. We will present in Section 20.3.1 the shortcomings of many contemporary approaches.

It all boils down to how to compute the integral in (20.30). However, in real applications, it is difficult to directly compute it and numerical techniques are often used.

Video Sequence In the case of a video sequence, temporal continuity between successive video frames implies that the transformations $\{\theta_t; t = 1, 2, \ldots, T\}$ follow a Markov chain. Without loss of generality, we assume a first-order Markov chain, that is,

$$\mathrm{p}(\theta_t|\theta_{1:t-1}) = \mathrm{p}(\theta_t|\theta_{t-1}). \tag{20.31}$$

Equation (20.26) becomes

$$\mathrm{p}(\alpha|\mathbf{y}_{1:T}) \propto \pi(\alpha) \int_{\theta_{1:T}} \prod_{t=1}^{T} \mathrm{p}(\mathbf{y}_t|\theta_t, \alpha)\mathrm{p}(\theta_t|\theta_{t-1})\mathrm{d}\theta_{1:T}. \tag{20.32}$$

The difference between (20.30) and (20.32) is whether the product lies inside or outside the integral. In (20.30), the product lies outside the integral, which divides the quantity of interest into "small" integrals that can be computed efficiently, while (20.32) does not have such a decomposition, causing computational difficulty.

Difference from Bayesian Estimation Our framework is very different from the traditional Bayesian parameter estimation setting, where a certain parameter β is estimated from the i.i.d. observations $\{\mathbf{x}_1, \mathbf{x}_2, \ldots, \mathbf{x}_T\}$ generated from a parametric density $\mathrm{p}(\mathbf{x}|\beta)$. If we assume that β has a prior probability $\pi(\beta)$, then the posterior probability $\mathrm{p}(\beta|\mathbf{x}_{1:T})$ is computed as

$$\mathrm{p}(\beta|\mathbf{x}_{1:T}) \propto \pi(\beta)\mathrm{p}(\mathbf{x}_{1:T}|\beta) = \pi(\beta) \prod_{t=1}^{T} \mathrm{p}(\mathbf{x}_t|\beta) \tag{20.33}$$

and used to derive the parameter estimate $\hat{\beta}$. One should not confuse our transformation parameter θ with the parameter β. Notice that β is fixed in $\mathrm{p}(\mathbf{x}_t|\beta)$ for different ts. However, each $\mathbf{y}_t$ is associated with a θ_t. Also, α is different from β in the sense that α describes the identity and β helps describe the parametric density.

To make our framework more general, we can also incorporate the β parameter by letting the observation likelihood be $\mathrm{p}(\mathbf{y}|\theta, \alpha, \beta)$. Equation 20.26 then becomes

$$\begin{aligned} \mathrm{p}(\alpha|\mathbf{y}_{1:T}) &\propto \pi(\alpha)\mathrm{p}(\mathbf{y}_{1:T}|\alpha) \\ &= \pi(\alpha) \int_{\beta, \theta_{1:T}} \mathrm{p}(\mathbf{y}_{1:T}|\theta_{1:T}, \alpha, \beta)\mathrm{p}(\theta_{1:T})\pi(\beta)\mathrm{d}\theta_{1:T}\mathrm{d}\beta \\ &= \pi(\alpha) \int \prod_{t=1}^{T} \mathrm{p}(\mathbf{y}_t|\theta_t, \alpha, \beta)\mathrm{p}(\theta_t|\theta_{1:t-1})\pi(\beta)\mathrm{d}\theta_{1:T}\mathrm{d}\beta, \end{aligned} \tag{20.34}$$

where $\theta_{1:T}$ and β are assumed to be statistically independent. In this chapter, we will focus only on (20.26) as if we already know the true parameter β in (20.34). This greatly simplifies our computation.

20.3.1 Recognition Setting and Issues

Equation (20.26) lays a theoretical foundation, which is universal for all recognition settings: (i) recognition is based on a single image (an I-group with $T = 1$), an I-group with $T \geq 2$, or a video sequence; (ii) the identity signature is either discrete- or continuous-valued; and (iii) the transformation space takes into account all available variations, such as localization and variations in illumination and pose.

20.3.1.1 Discrete Identity Signature

In a typical pattern recognition scenario, say an N-class problem, the identity signature for $\mathbf{y}_{1:T}$, $\hat{\alpha}$ is determined by the Bayesian decision rule:

$$\hat{\alpha} = \arg \min_{\{1,2,\ldots,N\}} \mathsf{p}(\alpha|\mathbf{y}_{1:T}). \tag{20.35}$$

Usually $\mathsf{p}(\mathbf{y}|\theta,\alpha)$ is a class-dependent density either prespecified or learned.

20.3.1.2 Continuous Identity Signature

If the identity signature is continuous-valued, two recognition schemes are possible. The first is to derive a point estimate $\hat{\alpha}$ (e.g., conditional mean, mode) from $\mathsf{p}(\alpha|\mathbf{y}_{1:T})$ to represent the identity of the image group $\mathbf{y}_{1:T}$. Recognition is performed by matching $\hat{\alpha}$s belonging to different groups of images using a metric $\mathsf{k}(.,.)$. Say $\hat{\alpha}_1$ is for group 1 and $\hat{\alpha}_2$ for group 2, the point distance

$$\hat{\mathsf{k}}_{1,2} \doteq \mathsf{k}(\hat{\alpha}_1,\hat{\alpha}_2) \tag{20.36}$$

is computed to characterize the difference between groups 1 and 2.

Instead of comparing the point estimates, the second scheme directly compares different distributions that characterize the identities for different groups of images. Therefore, for two groups 1 and 2 with the corresponding posterior probabilities $\mathsf{p}(\alpha_1)$ and $\mathsf{p}(\alpha_2)$, we use the following expected distance

$$\bar{\mathsf{k}}_{1,2} \doteq \int_{\alpha_1}\int_{\alpha_2} \mathsf{k}(\alpha_1,\alpha_2)\mathsf{p}(\alpha_1)\mathsf{p}(\alpha_2)d\alpha_1 d\alpha_2. \tag{20.37}$$

Ideally, we wish to compare the two probability distributions using quantities such as the KL distance. However, computing such quantities is numerically prohibitive when α is of high dimensionality.

The second scheme is preferred as it uses complete statistical information, while the first one based on point estimates uses partial information. For example, if only the conditional mean is used, the covariance structure or higher-order statistics are not used.

However, there are circumstances when the first scheme is appropriate, the posterior distribution $\mathrm{p}(\alpha|\mathbf{y}_{1:T})$ is highly peaked or even degenerate at $\hat{\alpha}$. This might occur when (i) the variance parameters are taken to be very small or (ii) we let T go to ∞, that is, keep observing the same subject for a long time.

To evaluate the expected distance $\bar{\mathrm{k}}$, we resort to importance sampling [48]. Other sampling techniques such as Monte Carlo Markov chain [48] can also be applied. Suppose that say for group 1, the importance function is $\mathrm{q}_1(\alpha_1)$, and weighted sample set is $\left\{\alpha_1^{(i)}, w_1^{(i)}\right\}_{i=1}^{I}$, the expected distance is approximated as

$$\bar{\mathrm{k}}_{1,2} \simeq \frac{\sum_{i=1}^{I}\sum_{j=1}^{J} w_1^{(i)} w_2^{(j)} \mathrm{k}\left(\alpha_1^{(i)}, \alpha_2^{(j)}\right)}{\sum_{i=1}^{I} w_1^{(i)} \sum_{j=1}^{J} w_2^{(j)}}. \tag{20.38}$$

The point distance is approximated as

$$\hat{\mathrm{k}}_{1,2} \simeq \mathrm{k}\left(\frac{\sum_{i=1}^{I} w_1^{(i)}\alpha_1^{(i)}}{\sum_{i=1}^{I} w_1^{(i)}}, \frac{\sum_{j=1}^{J} w_2^{(j)}\alpha_2^{(j)}}{\sum_{j=1}^{J} w_2^{(j)}}\right). \tag{20.39}$$

20.3.1.3 The Effects of the Transformation

Even though recognition based on a single image has been attempted for a long time, most efforts assume only one alignment parameter $\hat{\theta}$ and compute the probability $\mathrm{p}(\mathbf{y}|\hat{\theta},\alpha)$. Any recognition algorithm computing some distance measure can be thought of as using a properly defined Gibbs distribution. The underlying assumption is that

$$\mathrm{p}(\theta) = \delta(\theta - \hat{\theta}), \tag{20.40}$$

where $\delta(.)$ is an impulse function. Using (20.40), (20.30) becomes

$$\mathrm{p}(\alpha|\mathbf{y}) \propto \pi(\alpha) \int_{\theta} \mathrm{p}(\mathbf{y}|\theta,\alpha)\delta(\theta - \hat{\theta})d\theta = \pi(\alpha)\mathrm{p}(\mathbf{y}|\hat{\theta},\alpha). \tag{20.41}$$

Incidentally, if the Laplace's method [48] is used to approximate the integral $\int_{\theta} \mathrm{p}(\mathbf{y}|\theta,\alpha)\mathrm{p}(\theta)d\theta$ and the maximizer $\hat{\theta}_{\alpha} = \arg\min_{\theta} \mathrm{p}(\mathbf{y}|\theta,\alpha)\mathrm{p}(\theta)$ does not depend on α, say $\hat{\theta}_{\alpha} = \hat{\theta}$, then

$$\begin{aligned}\mathrm{p}(\alpha|\mathbf{y}) &\propto \pi(\alpha) \int_{\theta} \mathrm{p}(\mathbf{y}|\theta,\alpha)\mathrm{p}(\theta)d\theta \\ &\simeq \pi(\alpha)\mathrm{p}(\mathbf{y}|\hat{\theta},\alpha)\mathrm{p}(\hat{\theta})\sqrt{(2\pi)^r/|\mathrm{I}(\hat{\theta})|} \\ &\propto \pi(\alpha)\mathrm{p}(\mathbf{y}|\hat{\theta},\alpha),\end{aligned} \tag{20.42}$$

where $\mathrm{I}(\theta)$ is an $r \times r$ matrix (r is the dimension of θ) whose ijth element is

$$\mathrm{I}_{ij}(\theta) = -\frac{\partial^2 \log \mathrm{p}(\theta)}{\partial\theta_i \partial\theta_j}. \tag{20.43}$$

This gives rise to the same decision rule as implied by (20.41) and also partly explains why the simple assumption (20.40) can work in practice.

The alignment parameter is therefore very crucial for obtaining good recognition performance. Even a slightly erroneous $\hat{\theta}$ may affect the recognition system significantly. It is very beneficial to have a continuous density $\mathrm{p}(\theta)$ such as a Gaussian or even be noninformative since marginalization of $\mathrm{p}(\theta,\alpha|\mathbf{y})$ over θ yields a robust estimate of $\mathrm{p}(\alpha|\mathbf{y})$.

In addition, our Bayesian framework also provides a way to estimate the best alignment parameter using the posterior probability:

$$\mathrm{p}(\theta|\mathbf{y}) \propto \int_{\alpha} \mathrm{p}(\mathbf{y}|\theta,\alpha)\pi(\alpha)d\alpha. \tag{20.44}$$

20.3.1.4 Asymptotic Behaviors

When we have an I-group or a video sequence, we are often interested in discovering the asymptotic (or large-sample) behaviors of the posterior distribution $\mathrm{p}(\alpha|\mathbf{y}_{1:T})$ when T is large. In [20], the discrete case of α in a video sequence is studied. However, it is very challenging to extend this study to a continuous case. Experimentally (refer to Section 20.4.1.2), we find that $\mathrm{p}(\alpha|\mathbf{y}_{1:T})$ becomes more and more peaked as T increase, which seems to suggest a degenerancy in the true value α_{true}.

20.4 INSTANCES OF PROBABILISTIC IDENTITY CHARACTERIZATION

20.4.1 FR from a Group of Still Images

The main challenge is to specify the likelihood $\mathrm{p}(\mathbf{y}|\theta,\alpha)$. Practical considerations require that (i) the identity encoding coefficient α should be compact so that our target space where α resides is low dimensional and (ii) α should be invariant to transformations and tightly clustered so that we can safely focus on a small portion of the space.

Inspired by the popularity of subspace analysis, we assume that the observation $\mathbf{y}$ can be well explained by a subspace, whose basis vectors are encoded in a matrix denoted by B, that is, there exist linear coefficients α such that $\mathbf{y} \approx \mathrm{B}\alpha$. Clearly, α encodes the identity. However, the observation under the transformation condition (parameterized by θ) deviates from the canonical condition (parameterized by say $\bar{\theta}$) under which the B matrix is defined. To achieve an identity encoding that is invariant to the transformation, there are two possible ways. One way is to inverse-warp the observation $\mathbf{y}$ from the transformation condition θ to the canonical condition $\bar{\theta}$ and the other way is to warp the basis matrix B from the canonical condition $\bar{\theta}$ to the transformation condition θ. In practice, inverse-warping is typically difficult. For example, we cannot easily warp an off-frontal view to a frontal view without explicit 3D depth information that is unavailable. Hence, we follow the second approach, which is also known as *analysis-by-synthesis* approach. We denote the basis matrix under the transformation condition θ by B_{θ}.

20.4.1.1 Subspace Identity Encoding—Invariant to Localization, Illumination, and Pose

The localization parameter, denoted by ε, includes the face location, scale, and in-plane rotation. Typically, an affine transformation is used. We absorb the localization parameter ε in the observation using $\mathcal{T}\{\mathbf{y};\varepsilon\}$, where the $\mathcal{T}\{.;\varepsilon\}$ is a localization operator, cropping the region of interest and normalizing it to match with the size of the basis.

The illumination parameter, denoted by λ, is a vector specifying the illuminant direction (and intensity if required). The pose parameter, denoted by v, is a continuous-valued random variable. However, practical systems [37, 49] often discretize this due to the difficulty in handling 3D to 2D projection. Suppose the quantized pose set is $\{1,2,\ldots,V\}$. To achieve pose invariance, we concatenate all the images [37, 50] $\{\mathbf{y}^1,\mathbf{y}^2,\ldots,\mathbf{y}^V\}$ under all the views and a fixed illumination λ to form a very long vector $\mathsf{Y}^\lambda = [\mathbf{y}^{1,\lambda},\mathbf{y}^{2,\lambda},\ldots,\mathbf{y}^{V,\lambda}]^\mathsf{T}$. To further achieve invariance to illumination, we invoke the Lambertian reflectance model, ignoring the shadow pixels. Now, λ is actually a 3D vector describing the illuminant. We now follow [50] to derive a bilinear algorithm that is summarized as follows.

Since all $\mathbf{y}^v$s are illuminated by the same λ, the Lambertian model gives

$$\mathsf{Y}^\lambda = \mathsf{W}\lambda. \tag{20.45}$$

Following [51, 52], we assume that

$$\mathsf{W} = \sum_{i=1}^{m} \alpha^i \mathsf{W}_i, \tag{20.46}$$

and we have

$$\mathsf{Y}^\lambda = \sum_{i=1}^{m} \alpha^i \mathsf{W}_i \lambda, \tag{20.47}$$

where W_is are illumination-invariant bilinear basis and $\alpha = [\alpha^1,\alpha^2,\ldots,\alpha^m]^\mathsf{T}$ provides an illuminant-invariant identity signature. The bilinear basis can be easily learned as shown in [51–53]. Thus α is also pose-invariant because, for a given view v, we take the part in Y corresponding to this view and still have

$$\mathbf{y}^{\lambda,v} = \sum_{i=1}^{m} \alpha^i \mathsf{W}_i^v \lambda, \tag{20.48}$$

where W_i^v take the part in W_i corresponding to view v.

In summary, the basis matrix B_θ for $\theta = (\varepsilon,\lambda,v)$ with ε absorbed in $\mathbf{y}$ is expressed as $\mathsf{B}_{\lambda,v} = [\mathsf{W}_1^v\lambda,\mathsf{W}_2^v\lambda,\ldots,\mathsf{W}_m^v\lambda]$.

We focus on the following likelihood:

$$\mathsf{p}(\mathbf{y}|\theta) = \mathsf{p}(\mathbf{y}|\varepsilon,\lambda,v,\alpha) = \mathsf{Z}_{\lambda,v,\alpha}^{-1} \exp\{-\mathsf{D}(\mathcal{T}\{\mathbf{y};\varepsilon\},\mathsf{B}_{\lambda,v}\alpha)\}, \tag{20.49}$$

where $D(\mathbf{y}, B_\theta\alpha)$ is some distance measure and $Z_{\lambda,\upsilon,\alpha}$ is the so-called partition function which acts as a normalization role. In particular, if we take D as

$$D(\mathcal{T}\{\mathbf{y};\varepsilon\}, B_{\lambda,\upsilon}\alpha) = (\mathcal{T}\{\mathbf{y};\varepsilon\} - B_{\lambda,\upsilon}\alpha)^T\Sigma^{-1}(\mathcal{T}\{\mathbf{y};\varepsilon\} - B_{\lambda,\upsilon}\alpha)/2, \tag{20.50}$$

with a given Σ (say $\Sigma = \sigma^2 I$ where I is an identity matrix), then (20.49) becomes a multivariate Gaussian and the partition function $Z_{\lambda,\upsilon,\alpha}$ does not depend on the parameters any more. However, even though (20.49) is a multivariate Gaussian, the posterior distribution $p(\alpha|\mathbf{y}_{1:N})$ is no longer Gaussian.

20.4.1.2 Experimental Results

We use a portion of the "illum" subset of the PIE database [54]. This part includes 68 subjects under 12 lighting sources and at 9 poses. In total, we have $68 \times 12 \times 9 = 7344$ images. Figure 20.1 shows one PIE subject under illumination and pose variations.

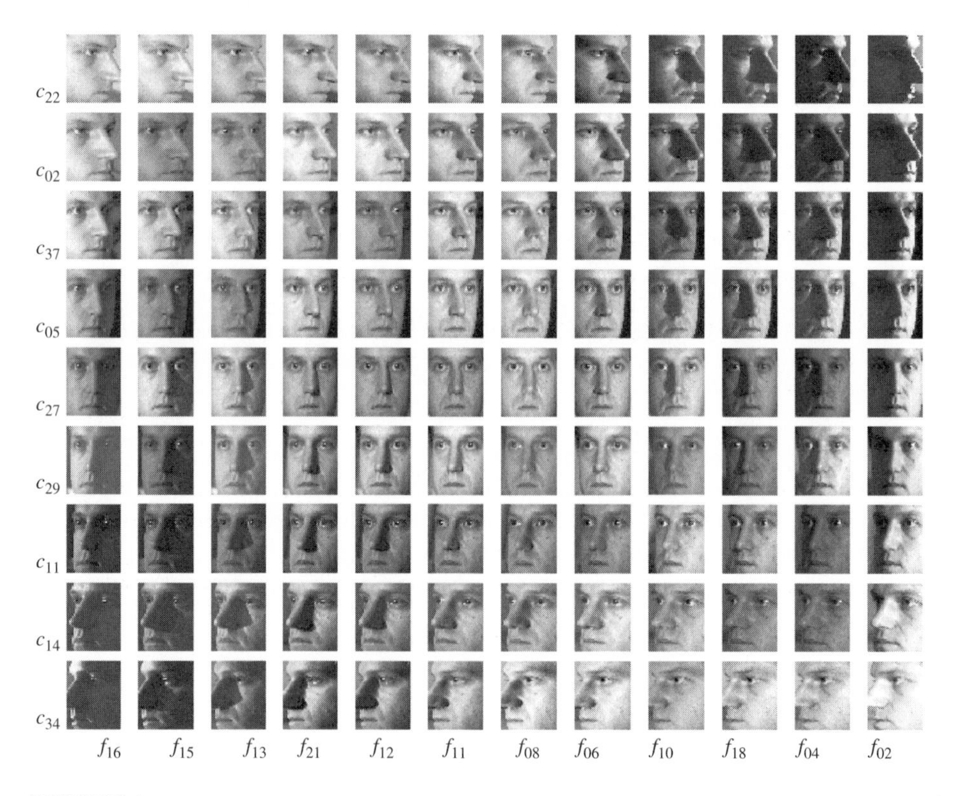

FIGURE 20.1

Examples of the face images of one PIE [54] subject under the different illumination and poses. The images are of size 48×40.

We randomly divide the 68 subjects into two parts. The first 34 subjects are used in the training set and the remaining 34 subjects are used in the gallery and probe sets. It is guaranteed that there is no identity overlap among the training, gallery, and probe sets.

During training, the images are preprocessed by aligning the eyes and mouth to desired positions. No flow computation is carried on for further alignment. After the preprocessing step, the face image is of size 48 by 40, that is, $d = 48 \times 40 = 1920$. Also, we only study grayscale images by taking the average of the red, green, and blue channels of their color versions.

The training set is used to learn the basis matrix B_θ or the bilinear basis W_is. As mentioned before, θ includes the illumination direction λ and the view pose υ, where λ is a continuous-valued random vector and υ is a discrete random variable taking values in $\{1, 2, \ldots, V\}$ with $p = 9$.

The images belonging to the remaining 34 subjects are used as gallery and probe sets. To form a gallery set of the 34 subjects, for each subject, we use an I-group of 12 images under all the illumination conditions under one pose υ_p (e.g., one row of Fig. 20.1); to form a probe set, we use I-groups under the other pose υ_g. We mainly concentrate on the case with $\upsilon_p \neq \upsilon_g$. Thus, we have $9 \times 8 = 72$ tests, with each test giving rise to a recognition score. The 1-NN (nearest neighbor) rule is applied to find the identity for a probe I-group.

During testing, we no longer use the preprocessed images and therefore the unknown transformation parameter includes the affine localization parameter, the light direction, and the discrete pose. The prior distribution $\mathsf{p}(\varepsilon_t)$ is assumed to be Gaussian, whose mean is found by a background subtraction algorithm and whose covariance matrix is manually specified. Numerical computation is conducted as in [47]. The metric $\mathsf{k}(.,.)$ actually used in our experiments is the correlation coefficient:

$$\mathsf{k}(\mathbf{x}, \mathbf{y}) = \{(\mathbf{x}^\mathsf{T}\mathbf{y})^2\}/\{(\mathbf{x}^\mathsf{T}\mathbf{x})(\mathbf{y}^\mathsf{T}\mathbf{y})\}. \tag{20.51}$$

Figure 20.2 shows the marginal posterior distribution of the first element α^1 of the identity variable α, that is, $\mathsf{p}(\alpha^1|\mathbf{y}_{1:T})$, with different Ts. From Fig. 20.2, we notice that (i) the posterior probability $\mathsf{p}(\alpha^1|\mathbf{y}_{1:T})$ has two modes, which might fail those algorithms that use a point estimate, and (ii) it becomes more peaked and tightly-supported as N increases, which empirically supports the asymptotic behavior mentioned in Section 20.3.1.

Figure 20.3 shows the recognition rates for all the 72 tests. In general, when the poses of the gallery and probe sets are far apart, the recognition rates decrease. The best gallery sets for recognition are those in frontal poses and the worst gallery sets are those in profile views.

For comparison, Table 20.2 shows the average recognition rates for four different methods: our two probabilistic approaches using $\bar{\mathsf{k}}$ and $\hat{\mathsf{k}}$, respectively, the principal component approach (PCA) approach [55], and the statistical approach [12] using the KL distance. When implementing the PCA approach, we learned a generic face subspace from all the training images, ignoring their illumination and pose conditions, whereas implementing the KL approach, we fit a Gaussian density on every I-group

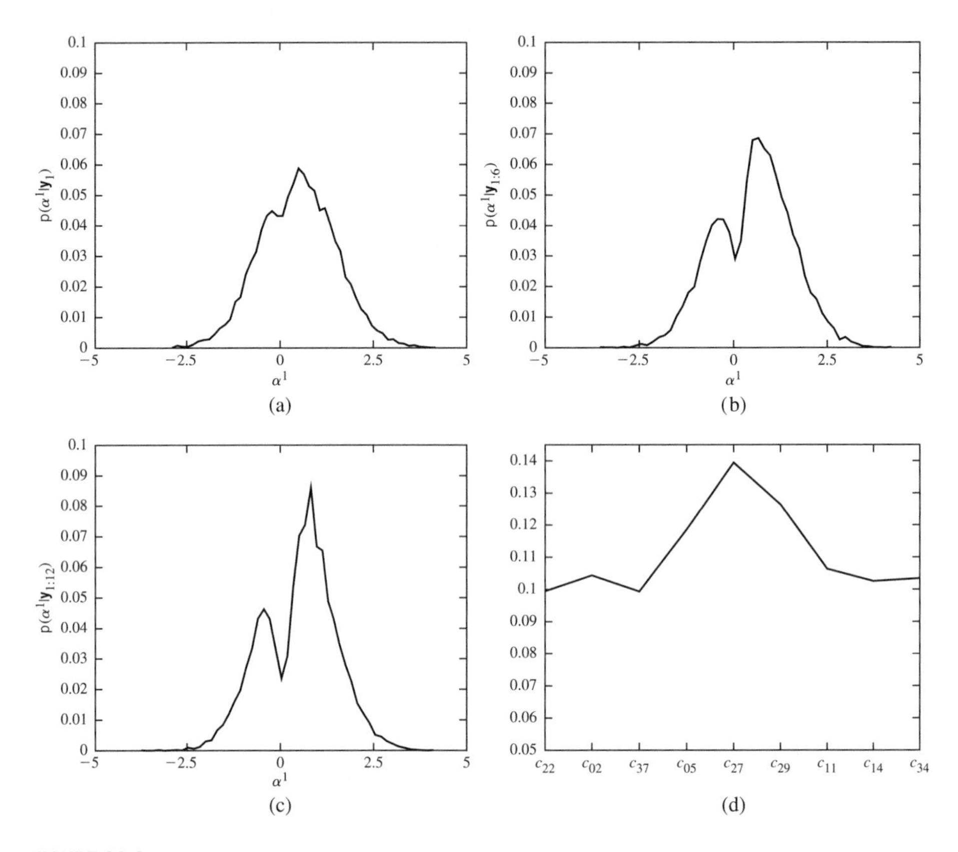

FIGURE 20.2

The posterior distributions $p(\alpha^1|\mathbf{y}_{1:T})$ with different Ts: (a) $p(\alpha^1|\mathbf{y}_1)$, (b) $p(\alpha^1|\mathbf{y}_{1:6})$, (c) $p(\alpha^1|\mathbf{y}_{1:12})$, and (d) the posterior distribution $p(v|\mathbf{y}_{1:12})$. Notice that $p(\alpha^1|\mathbf{y}_{1:T})$ has two modes and becomes more peaked as T increases.

and the learning set is not used. Our approaches significantly outperform the other two approaches due to transformation-invariant subspace modeling. The KL approach [12] performs even worse than the PCA approach simply because no illumination and pose learning is used in the KL approach, whereas the PCA approach has a generic learning algorithm based on image ensembles taken under different illumination conditions and poses (though this specific information is ignored).

As mentioned earlier in Section 20.3.1.3, we can infer the transformation parameters using the posterior probability $p(\theta|\mathbf{y}_{1:T})$. Figure 20.2 also shows the obtained $p(v|\mathbf{y}_{1:12})$ for one probe I-group. In this case, the actual pose is $v = 5$ (i.e., camera c_{27}), which has the maximum probability in Fig. 20.2(d). Similarly, we can find an estimate for ε, which is quite accurate as the background subtraction algorithm already provides a clean position.

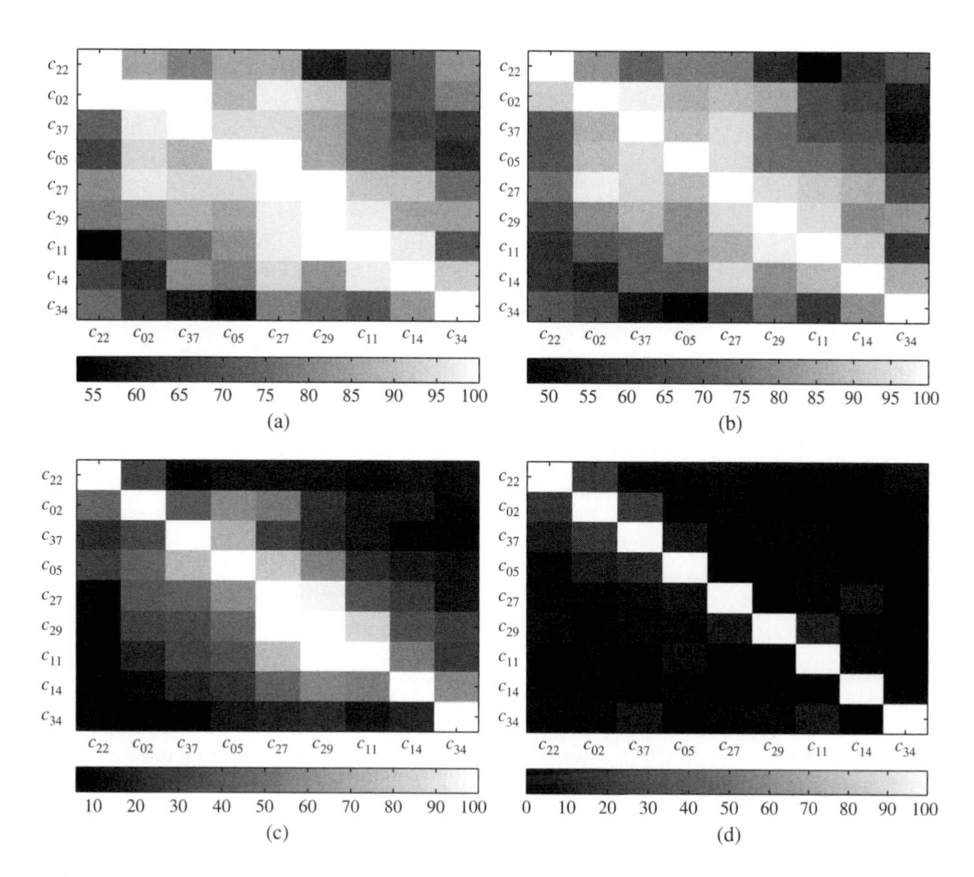

FIGURE 20.3

The recognition rates of all tests. (a) Our method based on $\bar{k}$. (b) Our method based on $\hat{k}$. (c) The PCA approach [55]. (d) The KL approach. Notice the different ranges of values for different methods and the diagonal entries should be ignored.

TABLE 20.2 Recognition rates of different methods.

Method	$\bar{k}$	$\hat{k}$	PCA	KL [12]
Recognition Rate (top 1)	82%	76%	36%	6%
Recognition Rate (top 3)	94%	91%	56%	15%

20.4.2 FR from a Video Sequence

FR from a video inevitably requires solving both tracking and recognition tasks. Visual tracking models the interframe appearance differences, and visual recognition models the appearance differences between the video frames and gallery images. Simultaneous tracking and recognition [20, 56] provides a mechanism of jointly modeling interframe appearance differences, and the appearance differences between video frames and gallery images. Here, we focus on the case when the identity variable is discrete. In addition, we assume that the gallery set consists of images $\{\mathbf{h}_1, \mathbf{h}_2, \ldots, \mathbf{h}_N\}$, with each $\mathbf{h}_\alpha$ for the αth individual.

20.4.2.1 Simultaneous Tracking and Recognition

It is easy to show that the derivation of probabilistic identity characterization is equivalent to a time series state space model consisting of the following three components: the motion transition equation, the identity equation, and the observation likelihood. This defines the recognition task as a statistical inference problem, which can be solved using particle filters. We briefly review the state space model.

Motion Transition Equation In its most general form, the motion model can be written as

$$\theta_t = g(\theta_{t-1}, u_t); \quad t \geq 1, \tag{20.52}$$

where u_t is *noise* in the motion model, whose distribution determines the motion state transition probability $p(\theta_t|\theta_{t-1})$. The function $g(.,.)$ characterizes the evolving motion and it could be a function learned offline or given a priori. One of the simplest choice is an additive function, that is, $\theta_t = \theta_{t-1} + u_t$.

Choice of θ_t is application dependent. The affine motion parameter is often used when there is no significant pose variation in the video sequence. However, if a 3D face model is used, 3D motion parameters should be used accordingly. In the experiments presented below, we set θ_t as the affine motion parameter and $g(.,.)$ as the additive function.

Identity Equation Assuming that the identity does not change as time proceeds, we have

$$\alpha_t = \alpha_{t-1}, \quad t \geq 1. \tag{20.53}$$

In practice, one may assume a small transition probability between identity variables to increase the robustness.

Observation Likelihood In the simplest form, we assume that the transformed observation is a noise-corrupted version of some still template in the gallery, that is,

$$z_t = \mathcal{T}\{\mathbf{y}_t; \theta_t\} = \mathbf{h}_{\alpha_t} + \mathbf{v}_t, \quad t \geq 1, \tag{20.54}$$

where $\mathbf{v}_t$ is the *observation noise* at time t, whose distribution determines the likelihood $p(\mathbf{y}_t|\alpha_t, \theta_t)$.

Particle Filter for Solving the Model We assume statistical independence among all noise variables and prior knowledge on the distributions $p(\theta_0)$ and $p(\alpha_0)$ (uniform prior in fact). Given this model, our goal is to compute the posterior probability

$p(\alpha_t|\mathbf{y}_{1:t})$. It is in fact a probability mass function (PMF) since α_t only takes values from $\mathcal{N} = \{1,2,\ldots,N\}$, as well as a marginal probability of $p(\alpha_t,\theta_t|\mathbf{y}_{1:t})$, which is a mixed-type distribution. Therefore, the problem is reduced to computing the posterior probability.

Since the model is nonlinear and non-Gaussian in nature, there is no analytic solution. We invoke a particle filter [57–59] to provide numerical approximations to the posterior distribution $p(\alpha_t,\theta_t|\mathbf{y}_{1:t})$. Also, for this mixed-type distribution, we can greatly improve the computational load by judiciously using the discrete nature of the identity variable as in [20]. We [20] also theoretically justified the evolving behavior of the recognition density $p(\alpha_t|\mathbf{y}_{1:t})$ under a weak assumption.

20.4.2.2 Experimental Results

In the experiments presented below, we use video sequences with subjects walking in a slant path toward the camera. There are 30 subjects, each having one face template. There is one face gallery and one probe set. The face gallery is shown in Fig. 20.4. The probe contains 30 video sequences, one for each subject. Figure 20.4 gives some example frames extracted from one probe video. As far as the imaging conditions are concerned, the gallery is very different from the probe, especially in lighting. This is similar to the "fc" test protocol of the FERET test [4]. These images/videos were collected, as part of the HumanID project, by researchers in National Institute of Standards and Technology and University of South Florida.

Case 1: Tracking and Recognition using Laplacian Density We first investigate the performance under the following setting: we use an affine motion parameter, a time-invariant first-order Markov Gaussian motion transition model, and a "truncated" Laplacian observation likelihood as follows.

$$p_1(\mathbf{y}_t|\alpha_t,\theta_t) = \mathrm{LAP}(\|\mathcal{T}\{\mathbf{y}_t;\theta_t\} - \mathbf{h}_{\alpha_t}\|;\sigma_1,\tau_1) \tag{20.55}$$

where $\|.\|$ is sum of absolute distance, σ_1 and λ_1 are manually specified, and

$$\mathrm{LAP}(x;\sigma,\tau) = \begin{cases} \sigma^{-1}\exp(-x/\sigma) & \text{if } x \leq \tau\sigma \\ \sigma^{-1}\exp(-\tau) & \text{otherwise} \end{cases}. \tag{20.56}$$

The recognition decision is based on (20.35). Table 20.3 shows that the recognition rate is very poor, only 13% of the time the top match is the correct match. The main reason is that the "truncated" Laplacian density is not able to capture the appearance difference between the probe and the gallery, thereby indicating a need for improved appearance modeling. Nevertheless, the tracking accuracy[2] is reasonable with 83% successfully tracked because we are using multiple face templates in the gallery to track the

[2] We inspect the tracking results by imposing the minimum mean square error (MMSE) motion estimate on the final frame as shown in Fig. 20.4 and determine if tracking is successful or not for this sequence. This is done for all sequences, and tracking accuracy is defined as the ratio of the number of sequences successfully tracked to the total number of all sequences.

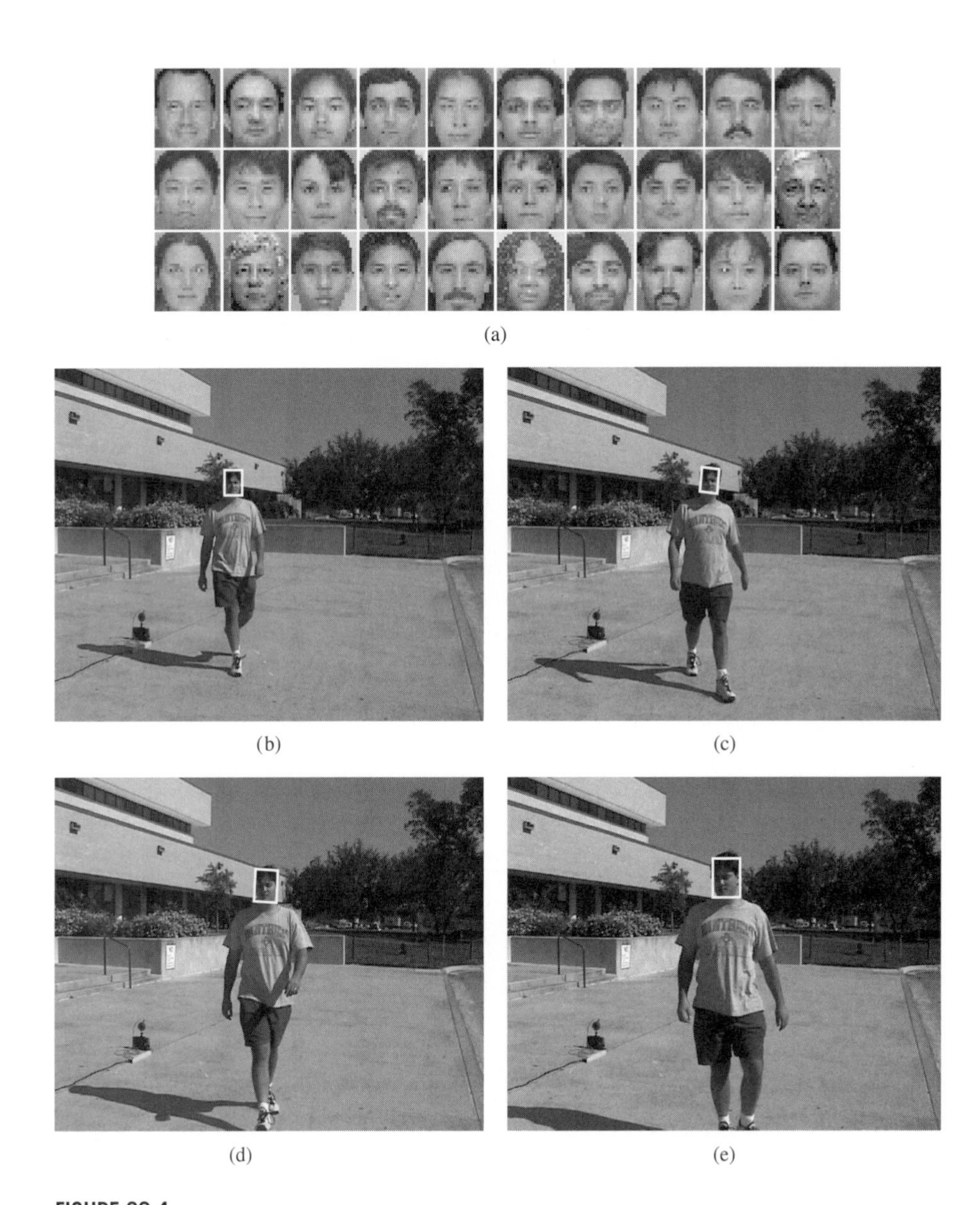

FIGURE 20.4

(a) The face gallery with image size being 30×26. (b–e) Four example frames in one probe video with image size being 720×480 while the actual face size ranges approximately from 20×20 in the first frame to 60×60 in the last frame. Notice the significant illumination variations between the probe and the gallery.

specific face in the probe video. After all, faces in both the gallery and the probe sets belong to the same class of human face and it seems that the appearance change is within the class range.

Case 2: Pure Tracking using Laplacian Density In Case 2, we measure the appearance change within the probe video as well as the noise in the background. To this end, we introduce a dummy template $\mathbf{T}_0$, a cut version in the first frame of the video. Define the observation likelihood for tracking as

$$\mathsf{p}_2(\mathbf{y}_t|\theta_t) = \mathsf{LAP}(\|\mathcal{T}\{\mathbf{y}_t;\theta_t\} - \mathbf{T}_0\|;\sigma_2,\tau_2), \tag{20.57}$$

where σ_2 and τ_2 are set manually. The other settings, such as motion parameter and model, are the same as in Case 1. We still can run the particle filter algorithm to perform pure tracking.

Table 20.3 shows that 87% are successfully tracked by this simple tracking model, which implies that the appearance within the video remains similar. Figure 20.5(a) shows the posterior probability $\mathsf{p}(\alpha_t|\mathbf{y}_{1:t})$ for the video sequence in Fig. 20.4. Starting from uniform $\mathsf{p}(\alpha_0) = N^{-1}$, the posterior probability for the correct identity approaches one as time proceeds and all others decrease to zero. This evolving behavior is characterized by the notion of entropy as shown in Fig. 20.5(b). Also, tracking results, inferred from $p(\theta_t|\mathbf{y}_{1:t})$, are illustrated in Fig. 20.4.

Case 3: Tracking and Recognition using Probabilistic Subspace Density As mentioned in Case 1, we need a new appearance model to improve the recognition accuracy. We use the approach suggested by Moghaddam [60] due to its computational efficiency and high recognition accuracy. However, in our implementation, we model only intra-personal variations instead of both intra/extra-personal variations for simplicity.

We need at least two facial images for one identity to construct the intra-personal space (IPS). Apart from the available gallery, we crop out the second image from the video ensuring no overlap with the frames actually used in probe videos. Figure 20.6(a) shows a list of such images. Compare with Fig. 20.4 to see how the illumination varies between the gallery and the probe.

We then fit a probabilistic subspace density [60] on top of the IPS. It proceeds as follows: a regular PCA is performed for the IPS. Suppose the eigensystem for the IPS

TABLE 20.3 Algorithm performances for five cases.

Case	Case 1	Case 2	Case 3	Case 4	Case 5
Tracking accuracy	83%	87%	93%	100%	NA
Recognition within top 1 match	13%	NA	83%	93%	57%
Recognition within top 3 matches	43%	NA	97%	100%	83%

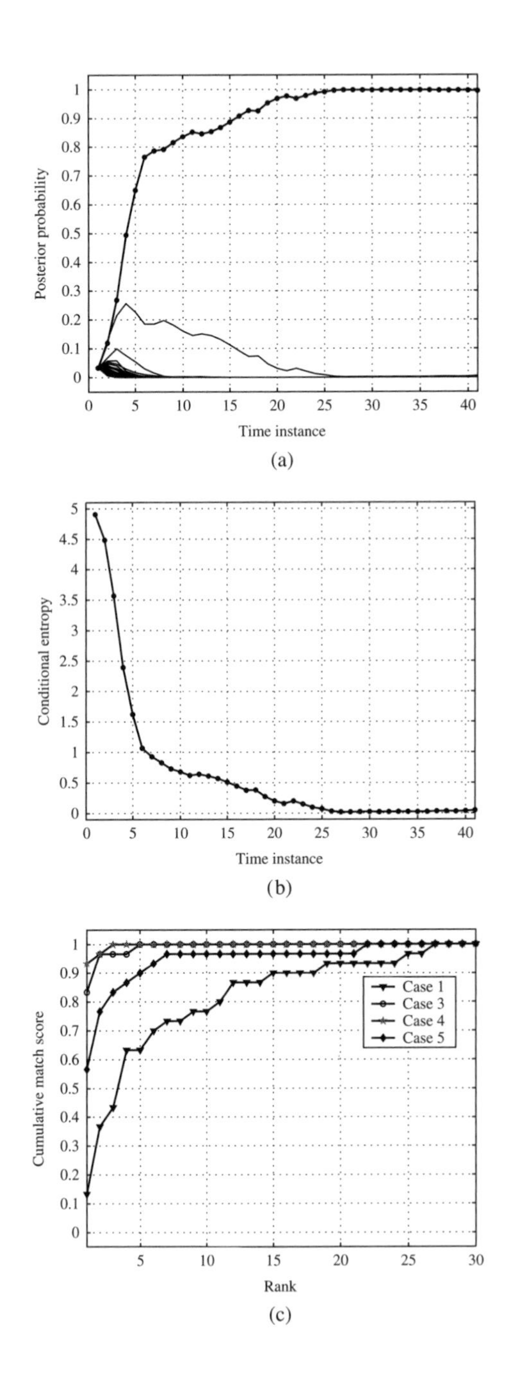

FIGURE 20.5

(a) Posterior probability $p(\alpha_t|\mathbf{y}_{1:t})$. (b) Entropy of posterior probability $p(\alpha_t|\mathbf{y}_{1:t})$. (c) Cumulative match curves for the data set.

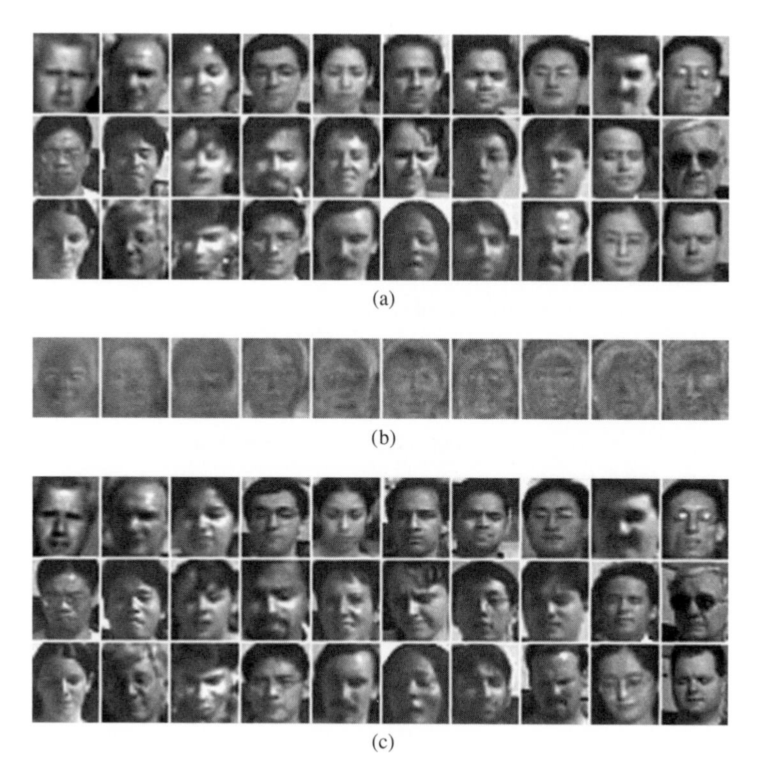

FIGURE 20.6

(a) The second facial images for training probabilistic density. (b) The top 10 eigenvectors for the IPS. (c) The facial images cropped out from the largest frontal view.

is $\{(\lambda_i, \mathbf{e}_i)\}_{i=1}^{d}$, where d is the number of pixels and $\lambda_1 \geq \cdots \geq \lambda_d$. Only top s principal components corresponding to top s eigenvalues are then kept while the residual components are considered as isotropic. We refer the reader to the original paper [60] for full details. Figure 20.6(b) shows the eigenvectors for the IPS. The density is written as follows:

$$\mathrm{Q}_{\mathrm{IPS}}(\mathbf{x}) = \left\{ \frac{\exp(-\frac{1}{2}\sum_{i=1}^{s}\frac{y_i^2}{\lambda_i})}{(2\pi)^{s/2}\prod_{i=1}^{s}\lambda_i^{1/2}} \right\} \left\{ \frac{\exp(-\frac{\epsilon^2}{2\rho})}{(2\pi\rho)^{(d-s)/2}} \right\}, \tag{20.58}$$

where $y_i = \mathbf{e}_i^T \mathbf{x}$ for $i = 1,2,\ldots,s$ is the ith principal component of x, $\epsilon^2 = \|\mathbf{x}\|^2 - \sum_{i=1}^{s} y_i^2$ is the reconstruction error, and $\rho = \left(\sum_{i=s+1}^{d}\lambda_i\right) / (d-q)$. It is easy to write the likelihood as follows:

$$\mathrm{p}_3(\mathbf{y}_t|\alpha_t, \theta_t) = \mathrm{Q}_{\mathrm{IPS}}(\mathcal{T}\{\mathbf{y}_t;\theta_t\} - \mathbf{h}_{\alpha_t}). \tag{20.59}$$

Table 20.3 lists the performance by using this new likelihood measurement. It turns out that the performance is significantly better than Case 1, with 93% tracked successfully and 83% recognized as the top 1 match. If we consider the top 3 matches, 97% are correctly identified.

Case 4: Tracking and Recognition using Combined Density In Case 2, we have studied appearance changes within a video sequence. In Case 3, we have studied the appearance change between the gallery and the probe. In Case 4, we attempt to take advantage of both cases by introducing a combined likelihood defined as follows:

$$p_4(\mathbf{y}_t|\alpha_t,\theta_t) = p_3(\mathbf{y}_t|\alpha_t,\theta_t)p_2(\mathbf{y}_t|\theta_t) \tag{20.60}$$

Again, all other settings are as in Case 1. We now obtain the best performance so far: no tracking error, 93% are correctly recognized as the first match, and no error in recognition is seen when the top 3 matches are considered.

Case 5: Still-to-still FR To make a comparison, we also performed an experiment on still-to-still FR. We selected the probe video frames with the best frontal face view (i.e., biggest frontal view) and cropped out the facial region by normalizing with respect to the eye coordinates manually specified. This collection of images is shown in Fig. 20.6(c), and it is fed as probes into a still-to-still FR system with the learned probabilistic subspace as in Case 3. It turns out that the recognition result is 57% correct for the top 1 match and 83% for the top 3 matches. The cumulative match curves for Case 1 and Cases 3–5 are presented in Fig. 20.5(c). Clearly, Case 4 is the best among all. We also implemented the original algorithm by Moghaddam [60], that is, both intra/extra-personal variations are considered, the recognition rate is similar to that obtained in Case 5.

20.5 A SYSTEM IDENTIFICATION APPROACH

20.5.1 The ARMA Model

In this section, we use a stochastic process to characterize a moving face sequence. In particular, we use a first-order ARMA model as

$$x_{t+1} = A\,x_t + v_t,\; y_t = C\,x_t + w_t, \tag{20.61}$$

where $v_t \sim \mathcal{N}(0,Q)$ and $w_t \sim \mathcal{N}(0,I)$. System identification is equivalent to estimating the parameters A, C, Q from the observations $\{y_1, y_2, \ldots, y_\tau\}$. This formulation has similarities with the pioneering work by Ali [61], where he addresses the problem of estimation and prediction for stationary spatial-temporal processes. Ali also uses a simultaneous linear model to represent spatial-temporal processes.

The closed-form solution can be derived [62]. Let $x_t \in \mathcal{R}^n, Y^\tau = [y_1, \ldots, y_\tau] \in \mathcal{R}^{m\times\tau}$ with $\tau > n$, then for $\{t = 1 \ldots \tau\}$, Eq. (20.61) can be written as

$$Y^\tau = C\,X^\tau + W^\tau; \quad C \in \mathcal{R}^{m\times n}, \tag{20.62}$$

where X and W are defined in a manner similar to Y. If singular value decomposition (SVD) of Y^τ is $Y^\tau = U\Sigma V^T$, where Σ is a diagonal matrix, $U \in \mathcal{R}^{m \times n}$, $U^T U = I$, $V \in \mathcal{R}^{\tau \times n}$, and $V^T V = I$, then

$$\hat{C}(\tau) = U; \quad \hat{X}(\tau) = \Sigma V^T, \tag{20.63}$$

$$\hat{A}(\tau) = \Sigma V^T D_1 V (V^T D_2 V)^{-1} \Sigma^{-1}, \tag{20.64}$$

where $D_1 = \begin{pmatrix} 0 & 0 \\ I_{\tau-1} & 0 \end{pmatrix}$ and $D_2 = \begin{pmatrix} I_{\tau-1} & 0 \\ 0 & 0 \end{pmatrix}$, and

$$\hat{Q}(\tau) = \frac{1}{\tau - 1} \sum_{i=1}^{\tau-1} \hat{v}_i \hat{v}_i^T, \tag{20.65}$$

where $\hat{v}_t = \hat{x}_{t+1} - \hat{A}(\tau)\hat{x}_t$, give a closed-form solution (suboptimal in the sense of Frobenius).

20.5.2 Framework for Recognition

Given gallery and probe face videos, the model parameters for each one of them are estimated. The gallery model, which is *closest* to the probe model, is assigned as the identity of the probe. We here discuss the metrics used to measure this degree of similarity.

Computing the L_2-norm of the difference between corresponding model matrices as a measure of distance will not suffice as it implicitly ignores the underlying geometry of the subspaces which is non-Euclidean. We make use of subspace angles between ARMA models for this case. We follow the mathematical formulation given in [63] to compute these angles. The subspace angles are defined as the principal angles between the column spaces generated by the observability matrices of the two matrices extended with the observability matrices of the corresponding inverse models. Principal angles between two subspaces are the angles between their principal directions.

To estimate the distance between two models, we need certain distance measures based on the computed subspace angles. There are several distance metrics based on subspace angles between ARMA models. The first one is due to Martin [64] and can be written as

$$d_M(M1, M2)^2 = ln \prod_{i=1}^{n} \frac{1}{\cos^2 \theta_i}, \tag{20.66}$$

where M_1 and M_2 are two ARMA models and θ_is are the subspace angles between them. Other distance measures include gap and Frobenius norm-based distances defined as

$$d_g(M1, M2) = \sin \theta_{\max}, \tag{20.67}$$

$$d_f(M_1, M_2)^2 = 2 \sum_{i=1}^{n} \sin^2 \theta_i. \tag{20.68}$$

There is another distance described in [65], which is the largest principal angle between the two models. In our experiments, all these metrics give similar recognition performance.

20.5.3 Experiments, Results, and Discussion

We conducted FR experiments using the proposed framework on two data sets. The first one is same as the one used by Li and Chellappa in [66]. It has face videos for 16 subjects with 2 sequences per subject. In these sequences, the subjects arbitrarily move their heads and change their expressions. The illumination conditions for the two sequences of each subject were quite different. For each subject, one sequence was put in the gallery while the other formed a probe. The second data set (obtained from UCSD/Honda) is the one used by Lee et al. in [21]. With this data set, we have a gallery of size 15 and probe containing 30 video sequences. In each video, the subject moves his or her face in an arbitrary sequence of 2D and 3D rotations while changing facial expression and speed. There is even partial occlusion in a few frames of several video sequences. The illumination conditions vary significantly among the various sequences. Although both the data sets used are small, we consider them good tests for our algorithm because of the extreme pose and expression variations and varying illumination as is evident from Fig. 20.7.

Our experiment broadly consists of three steps: preprocessing, model estimation, and recognition. The preprocessing step involves cropping out the face from each frame of the video sequence. We use a variant of KLT tracker to track the nose tip location and an edge-based rough pose estimator. The nose tip location gives an idea about the location of the face, while the pose information helps in getting the expanse of the face image relative to the nose. Figure 20.7 shows few of the images cropped using this automatic method.

We got recognition performance of more than 90% (15/16 for the first data set and 27/30 for the second). These numbers are very promising given the extent of pose and expression variations in the video sequences. The results reported in [21] are on perframe basis and are not directly comparable even though one of the data sets used is the same.

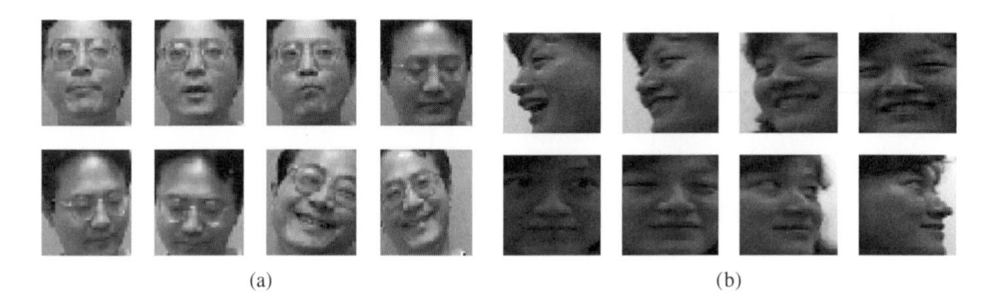

FIGURE 20.7

Few cropped faces from a video sequence in (a) the first data set and (b) the UCSD/Honda data set.

20.6 CONCLUSIONS

This chapter presented a review of FR from video or multiple still images. We then proposed a unified framework in which many approaches can be cast. Two instances were discussed as examples of the framework: one on recognition from groups of still images under variations and the other on recognition from video sequences within which tracking and recognition interact. A system identification approach to video-based FR was also discussed.

REFERENCES

[1] R. Chellappa, C. L. Wilson, and S. Sirohey. Human and machine recognition of faces: A survey. *Proc. IEEE*, 83:705–740, 1995.

[2] W. Zhao, R. Chellappa, A. Rosenfeld, and P. J. Phillips. Face recognition: A literature survey. *ACM Comput. Surveys*, 12, 2003.

[3] P. J. Phillips, P. Grother, R. J. Micheals, D. M. Blackburn, E. Tabbssi, and M. Bone. Face recognition vendor test 2002: evaluation report. *NISTIR 6965*, http://www.frvt.org, 2003.

[4] P. J. Phillips, H. Moon, S. Rizvi, and P. J. Rauss. The FERET evaluation methodology for face-recognition algorithms. *IEEE Trans. Patt. Anal. Mach. Intell.*, 22:1090–1104, 2000.

[5] Face Recognition Grand Challenge. http://bbs.bee-biometrics.org.

[6] O. Yamaguchi, K. Fukui and K. Maeda. Face recognition using temporal image sequence. *Proc. Int. Conf. Automat. Face and Gesture Recognit.*, Nara, Japan, 1998.

[7] L. Wolf and A. Shashua. Kernel principal angles for classification machines with applications to image sequence interpretation. *IEEE Comp. Soc. Conf. Comput. Vis. Patt. Recognit.*, Madison, WI, 2003.

[8] B. Schölkopf and A. Smola. *Learning with Learning.* MIT Press, Cambridge, MA, 2002.

[9] S. Zhou. Matrix and kernel method. http://www.cfar.umd.edu/~shaohua/papers/mtxker.pdf.

[10] R. O. Duda, P. E. Hart, and D. G. Stork. *Patt. Classification.* Wiley-Interscience, New York, 2001.

[11] P. Devijver and J. Kittler. *Patt. Recognition: A Statistical Approach.* Prentice Hall International, Upper Saddle River, NJ, 1982.

[12] G. Shakhnarovich, J. Fisher, and T. Darrell. Face recognition from long-term observations. *Eur. Conf. Comput. Vis.*, Copenhagen, Denmark, 2002.

[13] T. Jebara and R. Kondor. Bhattacharyya and expected likelihood kernels. *Conf. Learn. Theory, COLT*, 2003.

[14] S. Zhou and R. Chellappa. From sample similarity to ensemble similiarity: Probabilistic distance measures in reproducing kernel Hilbert space. *IEEE Trans. Patt. Anal. Mach. Intell.*, 28:917–929, 2007.

[15] O. Arandjelović and R. Cipolla. Face recognition from face motion manifolds using robust kernel resistor-average distance. *IEEE Workshop on Face Process. Video*, Washington D.C., USA, 2004.

[16] O. Arandjelović, G. Shakhnarovich, J. Fisher, R. Cipolla, and T. Darrell. Face recognition with image sets using manifold density divergence. *Proc. IEEE Conf. Comput. Vis. Patt. Recognit.*, vol. 1, 581–588, San Diego, USA, June 2005.

[17] A. Fitzgibbon and A. Zisserman. Joint manifold distance: a new approach to appearance based clustering. *IEEE Comput. Soc. Conf. Comput. Vis. Patt. Recognit.*, Madison, WI, 2003.

[18] Y. Li, S. Gong, and H. Liddell. Constructing face identity surface for recognition. *Int. J. Comput. Vis.*, 53(1):71–92, 2003.

[19] B. Knight and A. Johnston. The role of movement in face recognition. *Vis. Cogn.*, 4:265–274, 1997.

[20] S. Zhou, V. Krueger, and R. Chellappa. Probabilistic recognition of human faces from video. *Comput. Vis. Image Underst.*, 91:214–245, 2003.

[21] K. Lee, M. Yang, and D. Kriegman. Video-based face recognition using probabilistic appearance manifolds. *IEEE Comput. Soc. Conf. Comput. Vis. Patt. Recognit.*, Madison, WI, 2003.

[22] X. Liu and T. Chen. Video-based face recognition using adaptive hidden markov models. *Proc. IEEE Comput. Soc. Conf. Comput. Vis. Patt. Recognit.*, Madison, WI, 2003.

[23] G. Aggarwal, A. Roy-Chowdhury, and R. Chellappa. A system identification approach for video-based face recognition. *Proc. Int. Conf. Patt. Recognit.*, Cambridge, UK, 2004.

[24] P. Turaga, A. Veeraraghavan, and R. Chellappa. Statistical analysis on Stiefel and Grassmann manifolds with applications in computer vision. *IEEE Comput. Soc. Conf. Comp. Vis. Patt. Recognit.*, Anchorage, Alaska, 2008.

[25] M. J. Black and Y. Yacoob. Recognizing facial expressions in image sequences using local paramterized models of image motion. *Int. J. Comput. Vis.*, 25:23–48, 1997.

[26] Y. Tian, T. Kanade, and J. Cohn. Recognizing action units of facial expression analysis. *IEEE Trans. Patt. Anal. Mach. Intell.* 23:1–19, 2001.

[27] C. Tomasi and T. Kanade. Shape and motion from image streams under orthography: a factorization method. *Int. J. Comput. Vis.*, 9(2):137–154, 1992.

[28] C. Poelman and T. Kanade. A paraperpective factorization method for shape and motion recovery. *IEEE Trans. Patt. Anal. Mach. Int.*, 19(3):206–218, 1997.

[29] C. Bregler, A. Hertzmann, and H. Biermann. Recovering nonrigid 3D shape from image streams. *IEEE Comput. Society Conf. Comput. Vis. Patt. Recognit.*, 2000.

[30] M. E. Brand. Morphable 3D models from video. *Proc. IEEE Conf. Comput. Vis. Patt. Recognit.*, Hawaii, 2001.

[31] J. Xiao, J. Chai, and T. Kanade. A closed-form solution to non-rigid shape and motion recovery *Eur. Conf. Comput. Vis.*, 2004.

[32] P. Fua. Regularized bundle adjustment to model heads from image sequences without calibrated data. *Int. J. Comput. Vis.*, 38:153–157, 2000.

[33] Y. Shan, Z. Liu, and Z. Zhang. Model-based bundle adjustment with applicaiton to face modeling. *Proc. Int. Conf. Comp. Vis.*, 645–651, 2001.

[34] A. Roy Chowdhury and R. Chellappa. Face reconstruction from video using uncertainty analysis and a generic model. *Comput. Vis. Image Underst.*, 91:188–213, 2003.

[35] S. J. Gortler, R. Grzeszczuk, R. Szeliski, and M. Cohen., The lumigraph., *Proc. SIGGRAPH*, 43–54, New Orleans, LA, USA, 1996.

[36] M. Levoy and P. Hanrahan. Light field rendering. *Proc. SIGGRAPH*, 31–42, New Orleans, LA, USA, 1996.

[37] R. Gross, I. Matthews, and S. Baker. Eigen light-fields and face recognition across pose. *Proc. Int. Conf. Automat. Face and Gesture Recognit*, Washington, D.C., 2002.

[38] S. Zhou and R. Chellappa. Image-based face recognition under illumination and pose variations. *J. Opt. Soc. Am. (JOSA), A*, 22:217–229, 2005.

[39] A. Laurentini. The visual hull concept for silhouette-based image understanding. *IEEE Trans. Patt. Anal. Mach. Intell.*, 16(2):150–162, 1994.

[40] W. Matusik, C. Buehler, R. Raskar, S. Gortler, and L. McMillan. Image-based visual hulls., *Proc. SIGGRAPH*, 369–374, New Orleans, LA, USA, 2000.

[41] V. Blanz and T. Vetter. Face recognition based on fitting a 3D morphable model. *IEEE Trans. Patt. Anal. Mach. Intell.*, 25:1063–1074, 2003.

[42] T. F. Cootes, G. J. Edwards, and C. J. Taylor. Active appearance models. *IEEE Trans. Patt. Anal. Mach. Intell.*, 23(6):681–685, 2001.

[43] C. Beumie and M. P. Acheroy. Automatic face authentication from 3D surface. *Proc. of British Mach. Vis. Conf.*, 449–458, 1998.

[44] M. Mavridis, F. Tsalakanidou, D. Pantazis, S. Malassiotis, M. G. Strintzis. The HISCORE face recognition applicaiton: Affordable desktop face recognition based on a novel 3D camera. *Proc. Intl. Conf. Augmented Virtual Environ. 3D Imag.*, 2001.

[45] A. M. Bronstein, M. M. Bronstein and R. Kimmel. Expression-invariant 3D face recognition. *Proc. Audio and Video-based Biometric Personal Authentication*, 62–69, 2003.

[46] K. Boywer, K. Chang, and P. Flynn. A survey of approaches and challenges in 3D and multi-modal 3D+2D face recognition. *Comput. Vis. Image Understanding*, 101(1):1–15, 2006.

[47] S. Zhou and R. Chellappa. Probabilistic identity characterization for face recognition. *Proc. IEEE Comput. Soc. Conf. Comput. Vis. Patt. Recognit.*, Washington D.C., USA, June 2004.

[48] C. Robert. *Monte Carlo Statistical Methods.* Springer, New York, 2004.

[49] T. Cootes, K. Walker, and C. Taylor. View-based Active appearance models. *Proc. Int. Conf. Automatic Face and Gesture Recognit.*, Grenoble, France, 2000.

[50] S. Zhou and R. Chellappa. Illuminating light field: Image-based face recognition across illuminations and poses. *Proc. Int. Conf. Automat. Face and Gesture Recognit.*, Seoul, Korea, May 2004.

[51] S. Zhou, R. Chellappa, and D. Jacobs. Characterization of human faces under illumination variations using rank, integrability, and symmetry constraints. *Eur. Conf. Comput. Vis.*, Prague, Czech, May 2004.

[52] S. Zhou, G. Aggarwal, R. Chellappa, and D. Jacobs. Appearance characterization of linear Lambertian objects, generalized photometric stereo and illumination-invariant face recognition. *IEEE Trans. Patt. Anal. Mach. Intell.*, 29:230–245, 2007.

[53] W. T. Freeman and J. B. Tenenbaum. Learning bilinear models for two-factor problems in vision. *Proc. IEEE Comput. Soc. Conf. Comput. Vis. Patt. Recognit.*, Puerto Rico, 1997.

[54] T. Sim, S. Baker, and M. Bast. The CMU pose, illuminatin, and expression (PIE) database. *Proc. Automat. Face and Gesture Recognit.*, 53–58, Washington, D.C., 2002.

[55] M. Turk and A. Pentland. Eigenfaces for recognition. *J. Cogn. Neurosci.*, 3:72–86, 1991.

[56] S. Zhou, R. Chellappa, and B. Moghaddam. Visual tracking and recognition using appearance-adaptive models in particle filters. *IEEE Trans. Image Process.*, 11:1434–1456, 2004.

[57] A. Doucet, N. d. Freitas, and N. Gordon. *Sequential Monte Carlo Methods in Practice.* Springer-Verlag, New York, 2001.

[58] G. Kitagawa. Monte carlo filter and smoother for non-gaussian nonlinear state space models. *J. Comput. Graph. Stat.*, 5:1–25, 1996.

[59] J. S. Liu and R. Chen. Sequential monte carlo for dynamic systems. *J. Am. Stat. Assoc.*, 93:1031–1041, 1998.

[60] B. Moghaddam. Principal manifolds and probabilistic subspaces for visual recognition. *IEEE Trans. Patt. Anal. Mach. Intell.*, 24:780–788, 2002.

[61] M. M. Ali. Analysis of stationary spatial-temporal processes: Estimation and prediction. *Biometrika*, 66:513–518, 1979.

[62] S. Soatto, G. Doretto, and Y. Wu. Dynamic textures. *Proc. Intl. Conf. Comp. Vis.*, 2001.

[63] K. D. Cock and D. B. Moor. Subspace angles and distances between ARMA models. *Proc. Intl. Symp. Math. Theory of Networks and Syst.*, 2000.

[64] R. J. Martin. A metric for ARMA processes. *IEEE Trans. Signal Process.*, 48:1164–1170, 2000.

[65] A. Weinstein. Almost invariant submanifolds for compact group actions. *Berkeley CPAM Preprint Series*, 1999.

[66] B. Li and R. Chellappa. A generic approach to simultaneous tracking and verification in video. *IEEE Trans. Image Process.*, 11(5):530–554, 2002.

CHAPTER

Audiovisual Speech Processing

21

Petar S. Aleksic[1], Gerasminos Potamianos[2], and Aggelos K. Katsaggelos[3]

[1] *Google Inc.,*
[2] *IBM T.J. Watson Research Center,*
[3] *Northwestern University*

21.1 INTRODUCTION

With the increasing use of computers in everyday life, the challenging goal of achieving natural, pervasive, and ubiquitous human–computer interaction (HCI) has become very important, affecting, for example, productivity, customer satisfaction, and accessibility, among others. In contrast to the current prevailing HCI paradigm that mostly relies on locally tied, single-modality and computer-centric input/output, future HCI scenarios are envisioned where the computer fades into the background, accepting and responding to user requests in a human-like behavior, and at the user's location. Not surprisingly, speech is viewed as an integral part of such HCI, conveying not only user linguistic information, but also emotion, identity, location, and computer feedback [1].

However, although great progress has been achieved over the past decades, computer processing of speech still lags significantly compared to human performance levels. For example, automatic speech recognition (ASR) lacks robustness to channel mismatch and environment noise [1, 2], underperforming human speech perception by up to an order of magnitude even in clean conditions [3]. Similarly, text-to-speech (TTS) systems continue to lag in naturalness, expressiveness, and, somewhat less, in intelligibility [4]. Furthermore, typical real-life interaction scenarios, where humans address other humans in addition to the computer, may be located in a variable far-field position compared to the computer sensors, or utilize emotion and nonacoustic cues to convey a message, prove insurmountably challenging to traditional systems that rely on the audio signal alone. In contrast, humans easily master complex communication tasks by utilizing additional channels of information whenever required, most notably the visual sensory channel. It is therefore only natural that significant interest and effort has recently been focused on exploiting the visual modality to improve HCI [5–9]. In this chapter, we review such efforts with emphasis on the main techniques employed in the extraction and integration of the visual signal information into speech processing HCI systems.

Of central importance to human communication is the visual information present in the face. In particular, the lower face plays an integral role in the production of human speech and of its perception, both being audiovisual in nature [9, 10]. Indeed, the visual modality benefit to speech intelligibility has been quantified as back as in 1954 [11]. Furthermore, bimodal integration of audio and visual stimuli in perceiving speech has been demonstrated by the McGurk effect [12]: when for example a person is presented with the audio stimulus /baba/ superimposed on a video of moving lips uttering the sound /gaga/, the person perceives the sound /dada/. Visual speech information is especially critical to the hearing impaired: mouth movement plays an important role in both sign language and simultaneous communication between the deaf [13].

Face visibility benefits speech perception due to the fact that the visual signal is both correlated to the produced audio signal and also contains complementary information to it [14–16]. The former allows the partial recovery of the acoustic signal from visual speech [17], a process akin to speech enhancement when the audio is corrupted by noise [18, 19]. The latter is due to at least the partial visibility of the place of articulation, through the tongue, teeth, and lips, and can help disambiguate speech sounds that are highly confusable from acoustics alone; for example, the unvoiced consonants /p/ (a bilabial) and /k/ (a velar), among others [16]. Not surprisingly, these observations have motivated significant research over the past 20 years on the automatic recognition of visual speech, also known as automatic speechreading, and its integration with traditional audio-only systems, giving rise to audiovisual ASR [20–47].

In addition to improving speech perception, face visibility provides direct and natural communication between humans. Computers, however, typically utilize audio-only TTS synthesis to communicate information back to the user in a manner that lags in naturalness, expressiveness, and intelligibility compared to human speech. To address these shortcomings, much research work has recently focused on augmenting TTS systems by synthesized visual speech [7, 48]. Such systems generate synthetic talking faces that can be directly driven by the acoustic signal or the required text, providing animated or photo-realistic output [5, 47–62]. The resulting systems can have widely varying HCI applications, ranging from assistance to hearing impaired persons, to interactive computer-based learning and entertainment.

It is worth noting that face visibility plays additional important roles in human-to-human communication by providing speech segmental and source localization information, as well as by conveying speaker identity and emotion. All are very important to HCI, with obvious implications to ASR or TTS, among others. A number of recently proposed techniques utilize visual-only or joint audiovisual signal processing for speech activity detection and source localization [63–67], identity recognition from face appearance or visual speech [8, 68–76], and visual recognition and synthesis of human facial emotional expressions [77, 78]. In all cases, the visual modality can significantly improve audio-only systems.

To automatically process and incorporate the visual information into the aforementioned speech-based HCI technologies, a number of steps are required that are surprisingly similar across them. Central to all technologies is the feature representation of visual speech and its robust extraction. In addition, appropriate integration of the

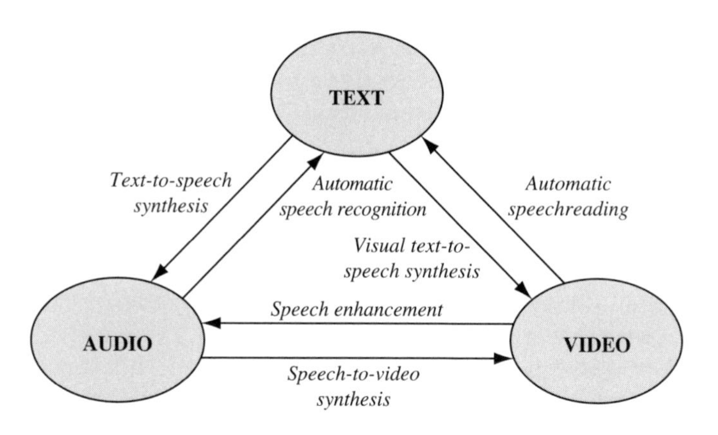

FIGURE 21.1

Conversions and interactions between the acoustic, visual, and text representations of speech, that are the focus of this work (adapted from [5]).

audio and visual representations is required for audiovisual ASR, speaker recognition, speech activity detection, and emotion recognition, to ensure improved performance of the bimodal systems over audio-only baselines. In a number of technologies, this integration occurs by exploiting audiovisual signal correlation: for example, audio enhancement using visual information, speech-to-video synthesis, and detection of synchronous audiovisual sources (localization). Finally, unique to audiovisual TTS and speech-to-video synthesis is the generation of the final video from the synthesized visual speech representation (facial animation). The similarities between the required processing components is reinforced in Fig. 21.1, where conversions and interactions between the acoustic, visual, and textual representation of speech are graphically depicted.

In this chapter, we review these main processing components, and we discuss their application to speech-based HCI, with main emphasis on ASR, TTS, and speaker recognition. In particular, in Section 21.2, we focus on visual feature extraction. Section 21.3 is devoted to the main audiovisual integration strategies, with Section 21.4 concentrating on their application to audiovisual ASR. Section 21.5 addresses audiovisual speech synthesis, whereas Section 21.6 discusses audiovisual speaker recognition. Finally, Section 21.7 touches upon additional applications such as speaker localization, speech activity detection, and emotion recognition, and provides a summary and a short discussion.

21.2 ANALYSIS OF VISUAL SIGNALS

The first critical issue in the design and implementation of audiovisual speech systems for HCI is the choice of visual features and their robust extraction from video. Visual speech information is mostly contained in the speaker's mouth region, therefore, typically, the visual features consist of appropriate representations of mouth appearance and/or shape. Indeed, the various sets of visual features proposed in the literature over the last 20 years for visual speech processing applications are generally grouped into three categories [21]: (a) low-level or appearance-based features, such as transformed vectors of the mouth region pixel intensities using, for example, image compression techniques [22–31, 54, 70, 72]; (b) high-level or shape-based features, such as geometric- or model-based representations of the lip contours [30–41, 52, 54, 68, 73]; and (c) features that are a combination of both appearance and shape [29–31].

The choice of visual features clearly mandates the face, lip, or mouth-tracking algorithms required for their extraction, but is also a function of video data quality and resource constraints in the audiovisual speech application. For example, only a crude detection of the mouth region is sufficient to obtain appearance visual features, requiring as little as tracking the face and the two mouth corners. Such steps become even unnecessary if a properly head-mounted video camera is used for data capture, as in [46]. In contrast, a more computationally expensive lip-tracking algorithm is additionally required for shape-based features, being infeasible in videos that contain low-resolution faces. Needless to say, robust tracking of the face, lips, or the mouth region is of paramount importance for utilizing the benefit of visual speech in HCI. In the following, we review such tracking algorithms, before proceeding with a brief description of some commonly used visual features.

21.2.1 Face Detection, Mouth, and Lip Tracking

Face detection has attracted significant interest in the literature [79–83]. In general, it constitutes a difficult problem, especially in cases where the background, head pose, and lighting are varying. Some reported systems use traditional image processing techniques for face detection, such as color segmentation, edge detection, image thresholding, template matching, or motion information in image sequences [83], taking advantage of the fact that many local facial subfeatures contain strong edges and are approximately rigid.

However, the most widely used techniques follow a statistical modeling approach of face appearance to obtain a binary classification of image regions into the face and nonface classes. Such regions are typically represented as vectors of grayscale or color image pixel intensities over normalized rectangles of a predetermined size, often projected onto lower dimensional spaces, and are defined over a "pyramid" of possible locations, scales, and orientations in the image [79]. These regions can be classified using one or more techniques, such as neural networks, clustering algorithms along with distance metrics from the face or nonface spaces, simple linear discriminants, support vector machines (SVMs), and Gaussian mixture models (GMMs), for example [79–81]. An alternative popular approach uses a cascade of weak classifiers instead, that are trained using the AdaBoost technique and operate on local appearance features within these regions [82]. Notice that if color information is available, certain image regions that do not contain sufficient number of skin-tone-like pixels can be eliminated from the search [79].

Once face detection is successful, similar techniques can be used in a hierarchical manner to detect a number of interesting facial features such as the mouth corners, eyes, nostrils, chin, and so forth. The prior knowledge of their relative position on the face can simplify the search task. Such features are needed to determine the mouth region-of-interest (ROI) and help to normalize it by providing head-pose information. Additional lighting normalization is often applied to the ROI before appearance-based feature extraction (see also Fig. 21.2(a)).

Once the ROI is located, a number of algorithms can be used to obtain lip contour estimates. Some popular methods for this task are snakes [84], templates [85], and active shape and appearance models [86]. A snake is an elastic curve represented by a set of control points, and it is used to detect important visual features, such as lines, edges, or contours. The snake control point coordinates are iteratively updated, converging towards a minimum of the energy function, defined on basis of curve smoothness constraints and a matching criterion to desired features of the image [84]. Templates are parametric curves that are fitted to the desired shape by minimizing an energy function, defined similarly to snakes. Examples of lip contour estimation using a gradient vector field (GVF) snake and two parabolic templates are depicted in Fig. 21.2(b) [40].

In contrast, active shape models (ASMs) are statistical models obtained by performing a principal component analysis (PCA) on vectors containing the coordinates of a training set of points that lie on the shapes of interest, such as the lip inner and outer contours (see also Fig. 21.2(c)). Such vectors are projected onto a lower dimensional space defined by the eigenvectors that correspond to the largest PCA eigenvalues, representing the axes of genuine shape variation. Active appearance models (AAMs) are an extension to ASMs that, in addition to the shape-based model, use two more PCAs: the first captures the appearance variation of the region around the desired shape (for example, of vectors of image pixel intensities within the face contours, as shown in Fig. 21.2(d)), whereas the final PCA is built on concatenated weighted vectors of the shape and appearance representations. AAMs thus remove the redundancy due to shape and appearance correlation,

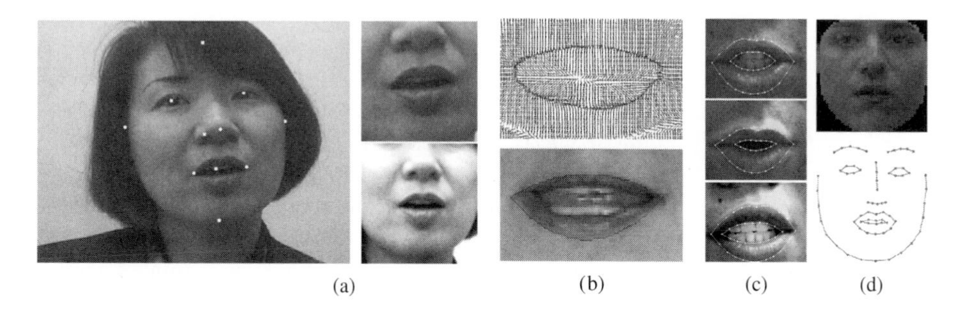

FIGURE 21.2

Mouth appearance and shape tracking for visual feature extraction: (a) Eleven detected facial features using the appearance-based approach of [79]. Two corresponding mouth region-of-interests of different sizes and normalization are also depicted [44]. (b) Lip contour estimation using a gradient vector field snake (upper: the snake's external force field is depicted) and two parabolas (lower) [40]. (c) Three examples of lip contour extraction using an active shape model [31]. (d) Detection of face appearance (upper) and shape (lower) using active appearance models [29].

and they create a single model that compactly describes shape and the corresponding appearance deformation. ASMs and AAMs can be used for tracking lips or other shapes by means of the algorithm proposed in [86]. The technique assumes that, given small perturbations from the actual fit of the model to a target image, a linear relationship exists between the difference in the model projection and image and the required updates to the model parameters. Fitting the models to the image data can be done iteratively, as in [29], or by the downhill simplex method, as in [31]. Examples of lip and face contour estimation by means of ASMs and AAMs are depicted in Figs. 21.2(c) and (d), respectively.

21.2.2 Visual Features

In the appearance-based approach to visual feature extraction, the pixel-value based, low-level representation of the mouth ROI is considered as informative for speechreading. Such ROI is extracted by the algorithms discussed earlier and is typically a rectangle containing the mouth, possibly including larger parts of the lower face, such as the jaw and cheeks [44], or could even be the entire face [29] (see also Figs. 21.2(a) and (d)). Sometimes, it is extended into a three-dimensional rectangle, containing adjacent frame ROIs, in an effort to capture dynamic speech information [24]. Alternatively, the ROI can correspond to a number of image profiles vertical to the estimated lip contour as in [31], or be just a disc around the mouth center [23]. By concatenating the ROI pixel values, a feature vector $\mathbf{x}_t$ is obtained that is expected to contain most visual speech information (see Fig. 21.3).

Typically, however, the dimensionality d of the ROI vector $\mathbf{x}_t$ becomes prohibitively large for successful statistical modeling of the classes of interest, such as subphonetic classes via hidden Markov models (HMMs) for audiovisual ASR [87]. For example, in the case of a 64×64 pixel grayscale ROI, $d = 4096$. Therefore, appropriate, lower dimensional transformations of $\mathbf{x}_t$ are used as features instead. In general, a $D \times d$-dimensional linear transform matrix $\mathbf{P}$ is sought, such that the transformed data vector $\mathbf{y}_t = \mathbf{x}_t\mathbf{P}$ contains most speechreading information in its $D \ll d$ elements (see also Fig. 21.3). Matrix $\mathbf{P}$ is typically borrowed from the image compression and pattern classification literatures and is often obtained based on a number of training ROI vectors. Examples of such transforms are the PCA, also known as "eigenlips," used in the literature for speechreading [22–24, 31], visual TTS [54], speaker recognition [69], the discrete cosine transform (DCT) [23–26], the discrete wavelet transform (DWT) [24], linear discriminant analysis (LDA) [44, 70], and the maximum likelihood linear transform (MLLT) [44, 72]. Often, such transforms are applied in a cascade [44, 70]. Notice that some are amenable to fast algorithmic implementations. Coupled with the fact that a crude ROI extraction can be achieved by utilizing computationally inexpensive face detection algorithms, appearance-based features allow visual speech representation in real time [46]. Their performance, however, degrades under intense head-pose and lighting variations [46].

In contrast to appearance-based features, high-level shape-based feature extraction assumes that most speechreading information is contained in the shape (inner and outer contours) of the speaker lips, or more generally, in the face contours [29]. As a result, such

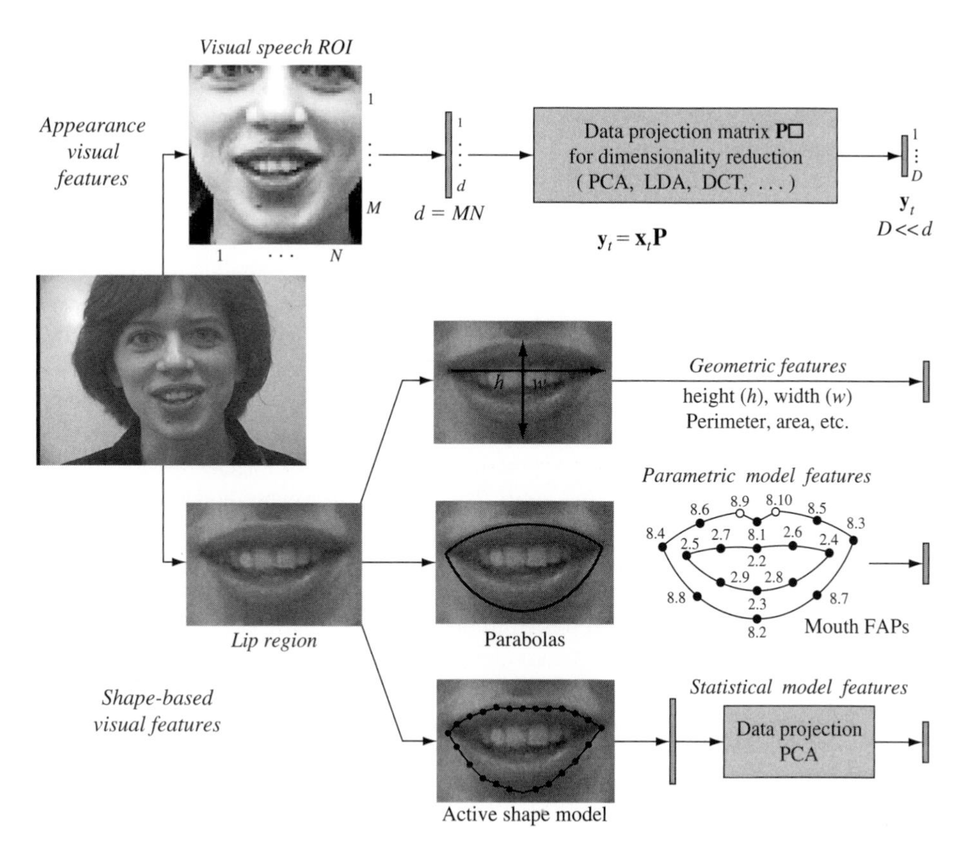

FIGURE 21.3

Various visual speech feature representation approaches discussed in this section: appearance-based (upper) and shape-based features (lower) that may utilize lip geometry, parametric, or statistical lip models.

features achieve a compact representation of visual speech using low-dimensional vectors and are invariant to head pose and lighting. However, to ensure good performance, their extraction requires robust lip-tracking, which often proves difficult and computationally intensive in realistic scenarios.

In general, high-level visual features are divided into geometric and model-based (see also Fig. 21.3). The former represent features that are meaningful to humans and can be readily extracted from the lip inner and outer contours, such as the height, width, perimeter, and area within the contour. Such features contain significant visual speech information, and have been successfully used in speechreading [32–38], visual speech synthesis [5, 54], and speaker recognition [38]. Additional visual features can be derived from the lip contours, such as lip image moments and lip contour Fourier descriptors, that are invariant to affine image transformations [24, 36]. Alternatively, high-level visual features can be model-based, typically obtained in conjunction with one of the parametric or statistical lip-tracking algorithms discussed earlier in Section 21.2.1. In the parametric approach, the template parameters that track the lips, or in the same manner, the tracking snake's control points or radial vectors, can be directly employed as visual speech features [30, 41]. Similarly, ASMs can be used as visual features by applying the model PCA on the vector of point coordinates of the tracked lip contour [31, 68].

In a related, recently introduced approach [40], a standard parametrization of the outer lip contour by means of a subset of facial animation parameters (FAPs) [88] is used to provide visual speech features. FAPs describe facial movement and are used in the MPEG-4 audiovisual object-based video representation standard to control facial animation, together with the so-called facial definition parameters that describe the face shape. There exist 68 FAPs, divided into 10 groups, depending on the particular region of the face that they are located (see also Fig. 21.4). Of particular interest to visual speech applications are the "group 8" parameters, which describe outer lip contour movement [40]. Additional speech information is contained in "group 2" parameters that correspond to inner lip and jaw motion, "group 6" ones that describe the tongue, and less so, in cheek movement captured by "group 5" FAPs.

Clearly, appearance- and shape-based visual features are quite different in nature, coding low- and high-level information about the speaker's face and lip movements. Not surprisingly, combinations of features from both categories have been suggested in the literature. In most cases, features of each type are just concatenated, as in [30, 31], where PCA appearance features are combined with snake-based features or ASMs, respectively. A different approach to combining the two classes of features is to create a single model of face shape and appearance using the AAM [86], discussed earlier. The final model PCA can be applied on the vector of the tracked shape and its corresponding appearance representations, to provide a set of visual features [29]. Finally, it is interesting to note that features from both categories can be used in a hierarchical manner. For example, in the visual text-to-speech (VTTS) synthesis reported in [54], visual unit selection occurs on the basis of the appearance representation of candidate mouth shapes within a set determined by their geometric shape features (see also Section 21.5.3).

In typical speech-based HCI, visual features are used in conjunction with audio features obtained from the acoustic waveform. Such features, for example, could be mel-frequency cepstral coefficients (MFCCs) or linear prediction coefficients (LPCs) and are mostly extracted at a 100 Hz rate [1, 87]. In contrast, visual features are generated at the much lower video frame or field rate. They can however be easily postprocessed (up-sampled) by linear interpolation to achieve audiovisual feature synchrony at the audio rate, and thus simplify audiovisual integration as discussed in Section 21.3 [44]. In addition to interpolation, a number of visual feature postprocessing methods play a critical role in enhancing the performance of visual speech processing systems. The most important such techniques concern capturing the visual speech dynamics. Similarly to audio-only systems, this can be achieved by augmenting the "static" (frame-based) visual feature vector by its first- and second-order derivatives, which are computed over a short temporal window centered at the current video frame [87]. Alternatively, a "dynamic" feature vector can be obtained by training an LDA matrix to project the concatenation of neighboring visual feature vectors onto a lower dimensional space [44]. LDA can also be

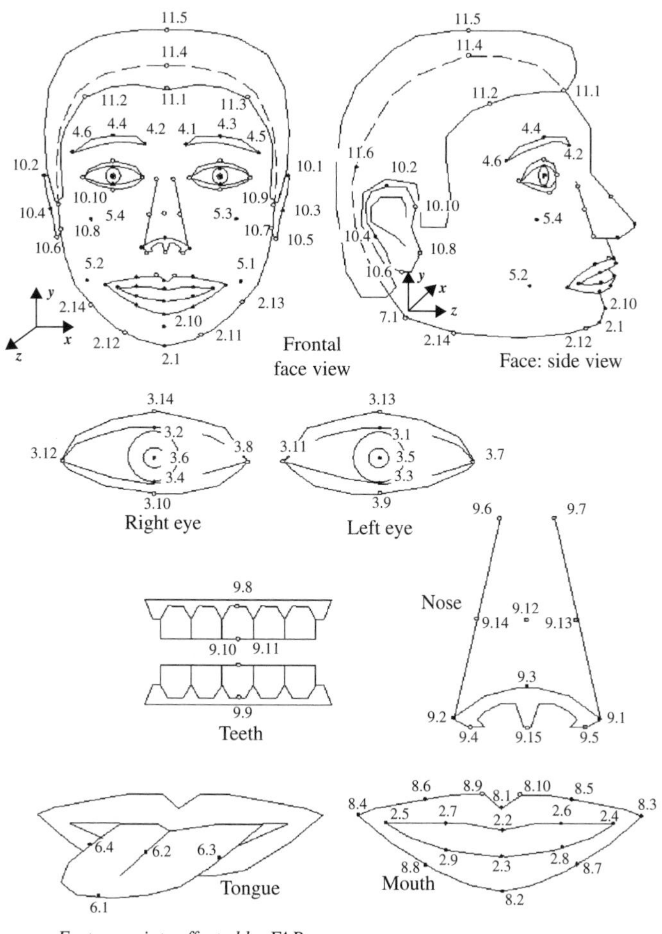

FIGURE 21.4

The facial animation control points supported by the MPEG-4 video representation standard [88]. Facial animation parameters (FAPs) describe the movement of 68 of these control points. There exist 10 FAP groups, with groups 8, 2, 6, and 5 being of interest in shape-based visual speech feature extraction [40, 52, 73].

followed by a feature space rotation matrix (MLLT) to improve statistical modeling of the extracted features [44]. Mean normalization of the visual feature vector can also contribute to improved performance, by reducing variability due to illumination, for example. Finally, feature selection within a larger pool of candidate features can also be considered as a form of postprocessing. A case of such selection for automatic speechreading appears in [37].

In summary, a number of approaches are viable for extracting and representing visual speech information. Unfortunately however, limited only work exists in the literature in comparing their relative performance. Most such comparisons are in the context of automatic speechreading and audiovisual ASR, where features within the same category (appearance- or shape-based) are usually investigated [23, 24, 27, 29, 37]. Occasionally, features across categories are compared, but in most cases with inconclusive results [24, 29, 30, 45]. Thus, the question of what are the most appropriate visual speech features that are sufficiently speaker-independent and robust to visual environment and head-pose variation remains to a large extent unresolved. Nevertheless, as the results in subsequent sections demonstrate, the specific implementations of both appearance- and shape-based systems, which are considered in this chapter and reviewed next, suffice to benefit a number of speech-related HCI technologies under somewhat constrained visual conditions. In practice, factors such as computational requirements, video quality, and the visual environment could determine the most suitable approach in a particular application.

21.2.3 Two Visual Feature Extraction Systems

In this chapter, we will be further considering two particular implementations of visual feature extraction, when reporting audiovisual speech processing results. The first is the appearance-based system developed at IBM Research. The system is depicted in Fig. 21.5(a), in parallel with its complementary audio processing module, as used for providing time-synchronous bimodal feature vectors for audiovisual ASR [44]. With minor modifications, it is also used for audiovisual speaker recognition [72] and noisy audio feature enhancement assisted by the visual observations [19]. Given the video of the speaker's face, the system first detects the face and 26 facial landmark points using the statistical tracking algorithm of [79], thus allowing the extraction of a normalized 64×64-pixel grayscale ROI (see also Fig. 21.2(a)). A two-dimensional, separable DCT is subsequently applied on the ROI vector, and the top 100 coefficients (in terms of energy) are retained. The feature vector dimensionality is further reduced to 30 by means of an intraframe LDA/MLLT. Following some of the postprocessing steps discussed earlier, a 41-dimensional dynamic visual speech vector $\mathbf{o}_{v,t}$ is extracted at each time instant t at a 100 Hz rate, synchronized with 60-dimensional MFCC-based audio features $\mathbf{o}_{a,t}$.

The second system, developed at Northwestern University (NWU), is shape-based and uses a set of FAPs [88] as visual features (see Fig. 21.5(b)). The system first employs a template to track the speaker's nostrils, thus determining the approximate mouth location. Subsequently, the outer lip contour is tracked using a combination of a GVF and a parabolic template (see also Fig. 21.2(b)). Following the outer lip contour detection and tracking, 10 FAPs describing the outer lip shape ("group 8" FAPs [88]) are extracted from the resulting lip contour (see also Figs. 21.3 and 21.4). These are placed into a feature vector that is subsequently projected by means of PCA onto a two-dimensional space [40]. The resulting visual features are augmented by their first and second derivatives providing an six-dimensional dynamic visual speech vector $\mathbf{o}_{v,t}$. These features are interpolated to the 90 Hz frame rate of 39-dimensional, MFCC-based audio features [87].

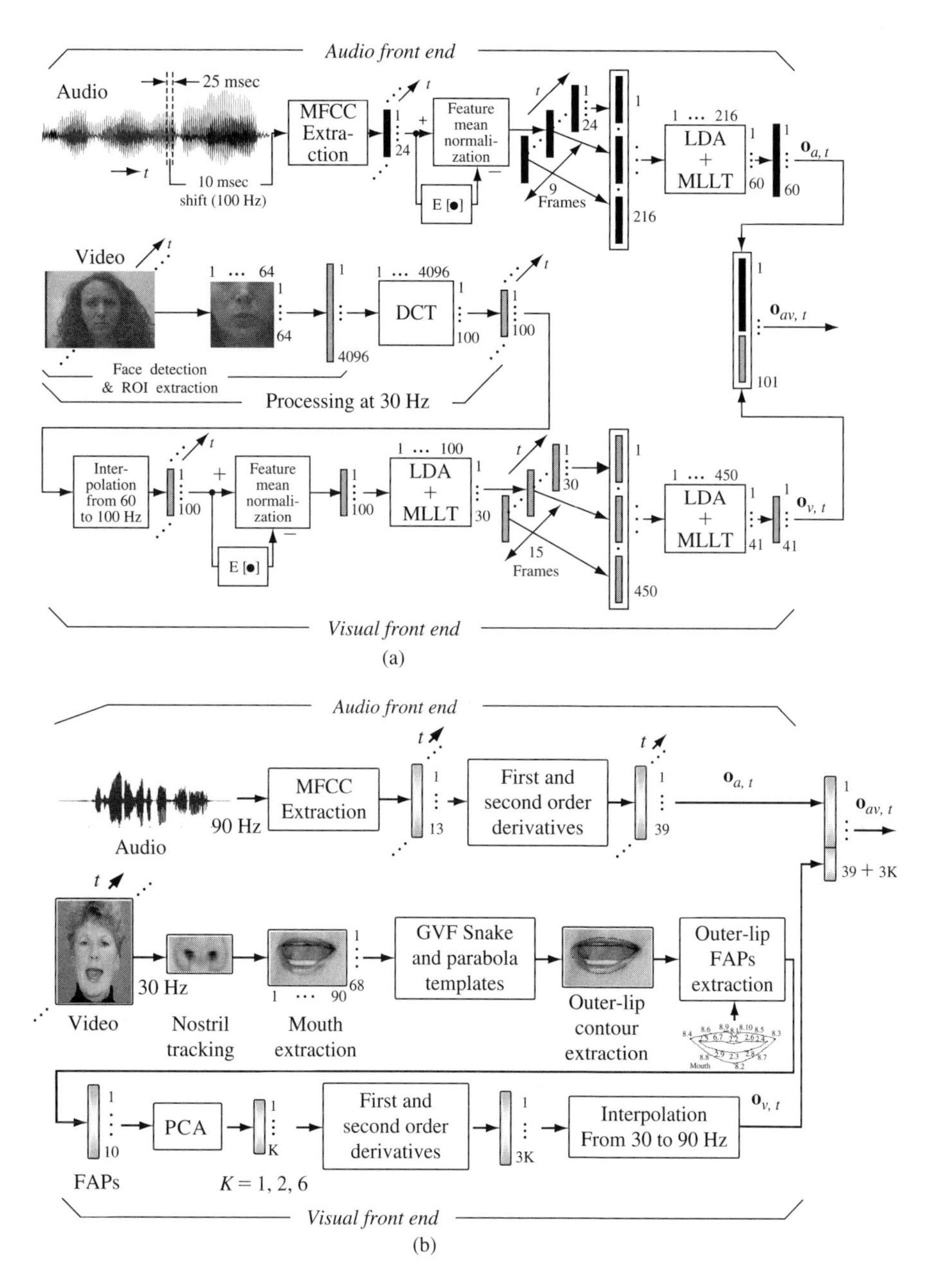

FIGURE 21.5

Two implementations of visual feature extraction, depicted schematically in parallel with the audio front end, as used for audiovisual ASR experiments in this chapter: (a) the appearance-based visual front end system of IBM Research, also employed for bimodal speaker recognition [72] and audio enhancement [19]; (b) the shape-based system of Northwestern University [40], also used for speech-to-video synthesis [52] and audiovisual speaker recognition [73].

The combined features are used for a number of audiovisual speech applications such as ASR, speech-to-video synthesis, and speaker-recognition [40, 52, 73].

21.3 AUDIOVISUAL INFORMATION FUSION

The second critical issue in the design of audiovisual speech processing systems is the integration of the available modality representations. To justify the complexity and cost of incorporating the visual modality into HCI, integration strategies should ensure that the performance of the multimodal system exceeds that of its single-modality counterpart, hopefully by a significant amount. For example, one would expect that the transcription accuracy of an audiovisual ASR system greatly surpasses that of the audio-only system, especially in noisy environments, or that audiovisual TTS is perceived as more friendly, intelligible, and natural than a synthetic voice-only system in subjective evaluation tests. In this section, we review the main concepts and techniques that are essential to successful audiovisual integration in speech-based HCI.

In this chapter, we are interested in a number of diverse bimodal technologies, with main emphasis on ASR, TTS, and speaker recognition. Clearly, the characteristics and requirements of each technology differ significantly, therefore it is natural that, among them, so do the modality integration methods. Nevertheless, a number of themes are similar in at least some of the technologies, thus allowing a common framework in their review. For example, central to ASR, TTS, and text-dependent speaker recognition algorithms is the notion of speech classes underlying the acoustic and visual representations. Of course, different types of classes are required for a number of other audiovisual applications, such as the general speaker recognition problem, emotion detection, audiovisual localization, and so forth. The second common theme across the technologies of interest is the issue of combining the acoustic and visual feature streams using classifiers designed to outperform their single-modality counterparts. The choice of classifiers and algorithms for feature and classifier fusion are clearly central to the design of audiovisual ASR and speaker recognition systems, among others. Finally, techniques for exploiting the correlation between the two signals are also of interest, and in this chapter are considered in the context of speech-to-video synthesis, discussed in Section 21.5.4.

21.3.1 Speech Classes in Audiovisual Integration

The basic unit that describes how speech conveys linguistic information is the phoneme. For American English, there exist approximately 42 such units [89], generated by specific positions or movements of the vocal tract articulators. However, because only a small part of the vocal tract is visible, not every phoneme pair can be disambiguated by the video information alone. The number of visually distinguishable units is therefore much smaller. Such units are referred to as visemes in the audiovisual speech processing and human perception literature [9, 10, 16].

Importantly, visemes capture "place" of articulation information [14, 16], that is, they describe where the constriction occurs in the mouth, and how mouth parts, such as the lips, teeth, tongue, and palate, move during speech articulation. As a result, many consonant phonemes with identical "manner" of articulation, which are therefore difficult to distinguish based on acoustic information alone, may differ in the place of articulation, and thus be visually identifiable; for example, the two nasals /m/ (a bilabial) and /n/ (an alveolar). In contrast, phonemes /m/ and /p/ are easier to perceive acoustically than visually, because they are both bilabial, but differ in the manner of articulation, instead.

Various mappings between phonemes and visemes can be found in the literature. In general, they are derived by human speechreading studies, but they can also be generated using statistical clustering techniques [37]. There is no universal agreement about the exact grouping of phonemes into visemes, although some clusters are well-defined; for example, the bilabial group {/p/, /b/, /m/}. All its three members are articulated at the same place (lips), thus appearing visually the same. A particular phoneme-to-viseme grouping is depicted in Table 21.1 [44].

In audio-only speech applications, the set of classes of interest in technologies such as ASR, text-dependent speaker recognition, and TTS most often consist of subphonetic units. Such classes are designed by clustering the possible phonetic contexts (triphones, for example) by means of a decision tree, to allow coarticulation modeling [87, 89]. Occasionally, subword units are employed in specific, small-vocabulary tasks. Naturally therefore, in visual speech applications one could consider visemic subphonetic classes, obtained for example by decision tree clustering based on visemic context. Indeed, visemic classes have been occasionally used in ASR [32, 43], and of course play a central role in visual synthesis systems (see Section 21.5). However, the use of different classes for the audio and visual components complicates audiovisual integration, especially in ASR and

TABLE 21.1 A 42 phoneme to 12 viseme mapping of the HTK phone set [87].

Viseme Class	Phonemes in Cluster
	/ao/, /ah/, /aa/, /er/, /oy/, /aw/, /hh/
Lip-rounding	/uw/, /uh/, /ow/
based vowels	/ae/, /eh/, /ey/, /ay/
	/ih/, /iy/, /ax/
Alveolar-semivowels	/l/, /el/, /r/, /y/
Alveolar-fricatives	/s/, /z/
Alveolar	/t/, /d/, /n/, /en/
Palato-alveolar	/sh/, /zh/, /ch/, /jh/
Bilabial	/p/, /b/, /m/
Dental	/th/, /dh/
Labio-dental	/f/, /v/
Velar	/ng/, /k/, /g/, /w/

text-dependent speaker-recognition. For such applications, identical classes are used for both speech modalities, most often subphonetic classes.

21.3.2 Classifiers in Speech Applications

In typical speech technologies discussed in this chapter, the classes of interest are hidden. We denote such unknown classes by $c \in \mathcal{C}$. For example, in the case of ASR, $\mathcal{C}$ represents a set of subphonetic or subword units, as discussed earlier. Classification of a sequence of such units gives rise to recognized words, based on a phonetic dictionary for the ASR task vocabulary. In the synthesis systems discussed in Section 21.5, set $\mathcal{C}$ can contain all candidate concatenative units, or describe a set of quantized representations of the signal to be synthesized. For speaker identification, $\mathcal{C}$ corresponds to the enrolled subject population, possibly augmented by a class denoting the unknown subject, whereas for authentication, $\mathcal{C}$ reduces to a two-member set. In the particular case of text-dependent speaker recognition, $\mathcal{C}$ can be considered as the product space between the set of speakers and the set of phonetic based units.

The hidden classes are observed only through the signal representation, namely a series of extracted feature vectors. We denote such vectors by $\mathbf{o}_{s,t}$, and their sequence over an interval $\mathcal{T}$ by $\mathbf{O}_s = \{\mathbf{o}_{s,t}, t \in \mathcal{T}\}$, where $s \in \mathcal{S}$ denotes the available modality; for example $\mathcal{S} = \{a, v, f\}$, in the speaker-recognition system of [74], that is based on audio, visual-labial, and face-appearance input.

A number of methods can then be used to model the association between the unknown classes and the observed feature vectors. Most such approaches are statistical in nature and provide a conditional probability measure for $\Pr(\mathbf{o}_{s,t}|c)$ or $\Pr(c|\mathbf{o}_{s,t})$; for example, artificial neural networks (ANNs), used for automatic speechreading in [22, 23, 33] and visual speech synthesis in [51, 61], or SVMs, as in [43]. Alternatively, the space of possible observation vectors is discretized through the process of vector quantization (VQ), as in [39, 62]. Then, the statistical model provides conditional probabilities of the form $\Pr(q(\mathbf{o}_{s,t})|c)$, where $q(\bullet)$ belongs to a discrete set of codebooks.

In most practical cases though, a Gaussian mixture density is assumed, namely

$$\Pr(\mathbf{o}_{s,t}|c) = \sum_{k=1}^{K_{s,c}} w_{s,c,k} \mathcal{N}(\mathbf{o}_{s,t}; \mathbf{m}_{s,c,k}, \mathbf{s}_{s,c,k}), \tag{21.1}$$

resulting in the GMM classifier. In (21.1), $K_{s,c}$ denotes the number of mixture weights $w_{s,c,k}$, which are positive and add to one, and $\mathcal{N}(\mathbf{o}; \mathbf{m}, \mathbf{s})$ represents a multivariate normal distribution with mean $\mathbf{m}$ and a covariance matrix $\mathbf{s}$, typically considered as diagonal. Emission probability model (21.1) is therefore described by parameter vector

$$\mathbf{b}_s = \left[\{[w_{s,c,k}, \mathbf{m}_{s,c,k}, \mathbf{s}_{s,c,k}], k = 1, \ldots, K_{s,c}, c \in \mathcal{C}\}\right], \tag{21.2}$$

for a particular modality s.

This model is sufficient to address problems where a single underlying class c is assumed to generate the entire observation sequence $\mathbf{O}_s$, and conditional independence of the observations holds. This is the case in most text-independent speaker recognition systems, for example. The model then allows maximum-a-posteriori estimation of the unknown class, as

$$\hat{c} = \arg\max_{c \in \mathcal{C}} \Pr(c) \prod_{t \in \mathcal{T}} \Pr(\mathbf{o}_{s,t}|c), \tag{21.3}$$

where $\Pr(c)$ denotes the class prior.

Model (21.3) is however inappropriate for applications where a temporal sequence of interacting states is assumed to generate the series of observations, as is the case in ASR, speech synthesis, and text-dependent speaker recognition. There, HMMs are widely used. In generating the observed sequence in modality s, the HMM assumes a sequence of hidden states sampled according to the transition probability parameter vector $\mathbf{a}_s = [\{\Pr(c'|c''), c', c'' \in \mathcal{C}\}]$. The states subsequently "emit" the observed features with class-conditional probability given by (21.1). The HMM parameter vector $\mathbf{p}_s = [\mathbf{a}_s, \mathbf{b}_s]$ is typically estimated iteratively, using the expectation–maximization (EM) algorithm [87, 89], as

$$\mathbf{p}_s^{(j+1)} = \arg\max_{\mathbf{p}} Q(\mathbf{p}_s^{(j)}, \mathbf{p}|\mathbf{O}_s), \quad j = 0, 1, \ldots. \tag{21.4}$$

In (21.4), $\mathbf{O}_s$ consists of all feature vectors in the training set, and $Q(\bullet, \bullet|\bullet)$ represents the EM algorithm auxiliary function, defined as in [89]. Alternatively, discriminative training methods can be used [89]. Once the model parameters are estimated, HMMs can be used to obtain the hidden classes of interest, also known as the "optimal state sequence" $\mathbf{c} = \{c_t, t \in \mathcal{T}\}$, given an observation sequence $\mathbf{O}_s$ over interval $\mathcal{T}$; namely,

$$\hat{\mathbf{c}} = \arg\max_{\mathbf{c} \in \mathcal{C}^{|\mathcal{T}|}} \Pr(\mathbf{O}_s, \mathbf{c}), \tag{21.5}$$

where

$$\Pr(\mathbf{O}_s, \mathbf{c}) = \prod_{t \in \mathcal{T}} \Pr(c_t|c_{t-1}) \Pr(\mathbf{o}_{s,t}|c_t). \tag{21.6}$$

In practice, the Viterbi algorithm is used for solving (21.5), based on dynamic programming [87, 89].

Recently, interest in using dynamic Bayesian networks (DBNs) as classifiers in speech applications has emerged. A DBN is a way to extend Bayes nets to model probability distributions over semi-infinite collections of random variables [90]. For example, an HMM is a simple DBN in which at each time instance, the Bayes net has a hidden discrete variable and an observed discrete or continuous variable. By having access to a general Bayesian network at each time step, complex and intricate dependencies can be used. For example, the GMM used at time t may depend not only on nearby contextual factors but external contexts or contextual factors on varying time scales that may affect the pronunciation (i.e., an external measure of the rate of speech or the current word's semantic importance) [91]. The training and evaluation of a DBN mimics that of an HMM in that a generalized EM algorithm is used during training and the Viterbi algorithm is used during evaluation [92]. DBNs offer a promising and flexible speech modeling framework, but their development is not as mature as HMMs. Even so, recent works have shown the promise of DBNs in relation to HMMs [93–95].

21.3.3 Feature and Classifier Fusion

The aforementioned presentation assumes that only one observation stream, $\mathbf{o}_{s,t}$, is provided. In practical audiovisual speech applications though, multiple streams are available, which result in multimodal observations $\mathbf{o}_t = \{\mathbf{o}_{s,t}, s \in \mathcal{S}\}$, assuming time-synchronous stream feature representations; for example, in the case of audiovisual ASR, $\mathbf{o}_{av,t} = [\mathbf{o}_{a,t}, \mathbf{o}_{v,t}]$. As mentioned earlier, integrating such multimodal information into systems that outperform their single-modality counterparts constitutes a major focus of audiovisual speech research.

Indeed, various information fusion algorithms have been considered in the literature, differing both in their basic design, as well as in the terminology used [8, 21, 35, 44, 47]. In this chapter, we adopt a broad grouping of such techniques into feature fusion and decision fusion methods [44]. The first are based on training a single classifier on the multimodal feature vector $\mathbf{o}_t$, or on any appropriate transformation of it [34, 35, 44]. In contrast, decision fusion algorithms utilize each single-modality classifier output to jointly estimate the hidden classes of interest. Typically, this is achieved by linearly combining the class-conditional observation log-likelihoods of the individual classifiers into a joint audiovisual classification score, using appropriate weights that capture the reliability of each single-modality classifier, or data stream [21, 28, 31–33]. The two approaches are schematically depicted in Fig. 21.6, in the case of one observation stream available for each of the audio and visual modalities.

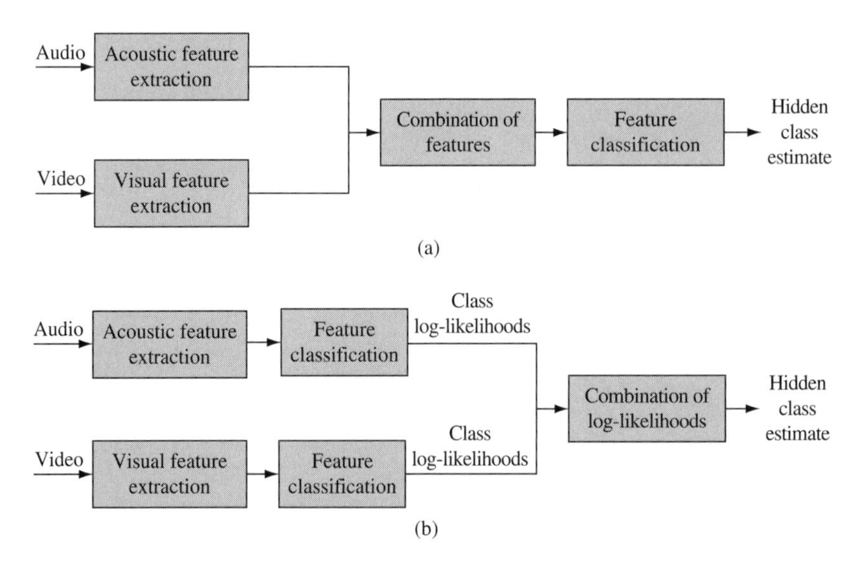

FIGURE 21.6

Block diagrams of the (a) feature fusion and (b) decision fusion approaches to audiovisual integration.

Audiovisual feature fusion techniques include plain feature concatenation [34], feature weighting [35, 47], both also known as direct identification fusion [35], as well as the "dominant" and "motor" recording fusion [35]. The latter seek a data-to-data mapping of either the visual features into the audio space, or of both modality features to a new common space, followed by linear combination of the resulting features. Audio feature enhancement on the basis of audiovisual features (for example, using regression, as in [18]) also falls within this category of fusion. Another interesting feature fusion technique, proposed for audiovisual ASR in [44], seeks a discriminant projection of the concatenated bimodal vector $\mathbf{o}_{av,t}$ onto a lower dimensional space for improved statistical modeling. The projected vector $\mathbf{o}_{d,t} = \mathbf{o}_{av,t}\mathbf{P}_{av}$ is modeled using the single-stream HMM of (21.1) and (21.6), where $\mathbf{P}_{av}$ is a cascade of an LDA projection and MLLT rotation (see also Section 21.2.3).

Although many feature fusion techniques result in improved system performance [44], they cannot explicitly model the reliability of each modality. Such modeling is extremely important, because of the varying speech information content of the audio and visual streams. The decision fusion framework, on the other hand, provides a mechanism for capturing these reliabilities, by borrowing from classifier combination theory, an active area of research with many applications [96].

Various classifier combination techniques have been considered for audiovisual speech applications, including for example a cascade of fusion modules, some of which possibly using only rank-order classifier information about the hidden classes of interest [20, 32]. However, by far the most commonly used decision fusion techniques belong to the paradigm of classifier combination using a parallel architecture, adaptive combination weights, and class score level information. These methods derive the most likely hidden class by linearly combining the log-likelihoods of the single-modality classifier decisions, using appropriate weights [28, 31, 34–36]. This corresponds to the adaptive product rule in the likelihood domain [96], and it is also known as the separate identification model for audiovisual fusion [32, 35].

In the most common application of this approach to audiovisual speech systems, the combination occurs at the observation frame level, resulting in the multimodal class-conditional

$$\Pr(\mathbf{o}_t | c) = \prod_{s \in \mathcal{S}} \Pr(\mathbf{o}_{s,t} | c)^{\lambda_{s,c,t}}, \tag{21.7}$$

for all hidden classes $c \in \mathcal{C}$. Notice that (21.7) does not represent a probability distribution in general, and should be viewed as a "score," when used in conjunction with (21.3) and (21.5). In (21.7), $\lambda_{s,c,t}$ denote the stream exponents (weights), that are nonnegative, and model stream reliability as a function of modality s, state c, and utterance frame (time) t. These are typically constrained to sum to one or $|\mathcal{S}|$, and are often set to global, modality-only dependent values, $\lambda_s \leftarrow \lambda_{s,c,t}$, for all c and t.

Joint model (21.7) can be used, for example, in audiovisual speaker recognition in conjunction with the GMM of (21.1) and (21.3), as in [72, 73], as well as for audiovisual ASR [25, 28, 31], resulting in the so-called multistream HMM (see also (21.1) and (21.6)). Notice that (21.7) also provides a framework to incorporate feature fusion; for example, in [44], the discriminant feature vector $\mathbf{o}_{d,t}$ is used as one of two or three streams for audiovisual ASR together with audio and possibly visual features. The approach is referred to as "hybrid" fusion.

Training the parameters of (21.7) requires additional steps, compared to (21.4). For example, in the particular case of two observation streams (audio and visual), each modeled by a single-stream HMM classifier with identical set of classes, the multistream HMM parameter vector becomes (see also (21.1), (21.2), and (21.7))

$$\overline{\mathbf{p}}_{av} = [\mathbf{p}_{av}, \lambda_a, \lambda_v], \quad \text{where } \mathbf{p}_{av} = [\mathbf{a}_{av}, \mathbf{b}_a, \mathbf{b}_v].$$

This consists of the HMM transition probabilities $\mathbf{a}_{av}$ and the emission probability parameters $\mathbf{b}_a$ and $\mathbf{b}_v$ of its single-stream components. The parameters of $\mathbf{p}_{av}$ can be estimated separately for each stream component using the EM algorithm, namely (21.4) for $s = a, v$, and subsequently, by possibly setting the joint HMM transition probability vector equal to the audio-one, that is, $\mathbf{a}_{av} = \mathbf{a}_a$. The alternative is to jointly estimate parameters $\mathbf{p}_{av}$, to enforce state synchrony in training. In the latter scheme, the EM based parameter reestimation becomes [87]

$$\mathbf{p}_{av}^{(j+1)} = \arg\max_{\mathbf{p}} Q(\overline{\mathbf{p}}_{av}^{(j)}, \mathbf{p} | \mathbf{O}_{av})$$

(see also (21.4)). The two approaches thus differ in the E-step of the EM algorithm. In both separate and joint HMM training, in addition to $\mathbf{p}_{av}$, the stream exponents λ_a and λ_v need to be obtained. This can be performed using discriminative training methods, simple parameter search on a grid, or mappings of signal quality measures to exponent values [25, 28, 33–35, 44].

Finally, of particular interest to audiovisual ASR is the level at which the stream log-likelihoods are combined. The use of HMMs allows likelihood recombination at a coarser level than the HMM state, for example at the phone or word boundary. Product or coupled HMMs are typically employed for the task, as in [26, 31, 44]. Such models allow state-level asynchrony between the acoustic and visual observations within the phone or word, forcing their synchrony at the unit boundaries instead. Product HMMs consist of composite audiovisual states, as depicted in Fig. 21.7, thus resulting in a much larger state space compared to multistream models fused at the state level, as in (21.7). To avoid undertraining, the single-stream emission probability components of the observation class-conditionals are tied along identical visual and audio states (see also Fig. 21.7).

Audio and visual recognition log-likelihoods can also be combined at the utterance level. This approach can easily be applied on small-vocabulary tasks, where likelihoods can be calculated for each word, based on the acoustic and visual observations. However, the number of possible hypotheses on large-vocabulary and continuous speech recognition tasks becomes prohibitively large. In such cases, recombination is usually limited to N-best hypotheses, generated either by the audio-only system or obtained as the union of audio- and visual-only N-best hypotheses. These are then rescored by combining the log-likelihoods generated using audio and visual HMMs [34, 36].

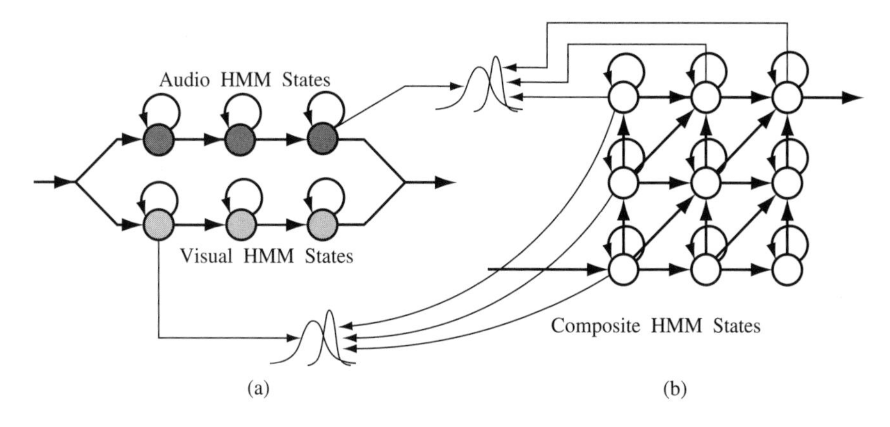

FIGURE 21.7

(a) Phone-synchronous two-stream HMM with three states per phone and modality. (b) Its equivalent product HMM; the single-stream emission probabilities are tied for states along the same row (column) to the corresponding audio (visual) state probabilities of form (21.1).

21.4 AUDIOVISUAL AUTOMATIC SPEECH RECOGNITION

As discussed in the Introduction, visual speech plays an important role in human speech perception, improving speech intelligibility especially in noise [9–11]. A number of reasons were cited there, the most important being the fact that visual speech information contains both correlated and complementary information to the acoustic signal. Not surprisingly, such information can be beneficial to ASR as well. Incorporating visual speech information into ASR is generally viewed as a very promising approach for improving speech recognition robustness to noise [1, 2], and bridging the gap between human and automatic performance [3]. Naturally therefore, significant research has recently focused in this area.

In 1984, Petajan [20] developed the first audiovisual ASR system. He used image thresholding to obtain binary mouth images from the input video, which were subsequently analyzed to derive mouth height, width, perimeter, and area, to be used as visual features in speech recognition. He first reported visual-only speech recognition results of isolated words within a 100-word vocabulary, using dynamic time warping [89]. In addition, he combined the acoustic and visual speech recognizers in a serial fashion to improve ASR performance: The visual speech system was used to rescore several N-best word hypotheses, as obtained by audio-only ASR, in order to generate the final bimodal recognition result.

A number of researchers have developed audiovisual ASR systems since [20–47]. Their systems vary in a number of areas, which have been discussed in detail in Sections 21.2 and 21.3. In summary, variations can be found in: the visual front end design, with some works adopting appearance-based features [22–31], whereas other researchers considering shape-based techniques [30–39], or even combinations of the two approaches

[29–31]; the choice of classes used in the recognition process, for example subphonetic [29, 40, 44], subword [28, 31, 34], or viseme-based [32, 43]; the employed recognition method, such as ANNs [22, 23, 33], SVMs [43], simple weighted distances used with VQ [20], and HMMs with various emission probability models [29, 31, 36, 39, 44]; and finally, the approach of integrating the audio and visual observation streams, generally grouped into feature fusion [34, 35, 44, 47] and decision fusion methods [25, 26, 28, 31–36]. Overall, the reported bimodal systems show improved performance compared to audio-only ASR for the recognition tasks considered: such are typically small-vocabulary tasks, for example isolated words [31], connected digits [28], or closed-set sentences [37], with large-vocabulary tasks recently reported [40, 44].

In the remainder of the section, we briefly review corpora commonly used for audio-visual ASR research, and we present experimental results on some of them using the IBM and NWU systems, previously discussed in Section 21.2.3. These results clearly demonstrate the benefit of incorporating the visual modality into ASR.

21.4.1 Bimodal Corpora for ASR

In contrast to the abundance of audio-only corpora, there exist only a few databases suitable for audiovisual ASR research. This is because the field is relatively young, but also due to the fact that audiovisual corpora pose additional challenges concerning database collection, storage, distribution, and privacy. Most commonly used databases in the literature are the product of efforts by few university groups or individual researchers with limited resources, and as a result, they contain small number of subjects, have relatively short duration, and mostly address simple recognition tasks, such as small-vocabulary ASR of isolated or connected words [8, 21]. Examples of such popular datasets in audio-visual ASR research are the CUAVE corpus containing connected digit strings [36], the AMP/CMU database of 78 isolated words [47], the Tulips1 set of four isolated digits [27], and the digit portion of the (X)M2VTS corpora, more often used in speaker recognition experiments (see Section 21.6). Additional datasets exist that are suitable for recognition of isolated nonsense words consisting of vowel-consonant combinations [34], connected letter strings in English [28] and German [22, 23], as well as continuous large-vocabulary speech [37] (see also [40, 97]).

A number of proprietary corpora have also been recorded by many groups, including recent work at IBM Research [44, 46]. There, a number of databases have been collected containing large subjects populations (50–290 subjects), uttering both large-vocabulary speech and connected-digit strings [46]. The corpora have been recorded in four different audiovisual conditions, to benchmark the performance of the IBM appearance-based visual front end. Three of the sets contain frontal full-face videos, and correspond to increasingly more challenging visual domains: the first was collected in a quiet studio-like environment, using a high-quality camera, uniform lighting and background, and relatively stable frontal subject head pose. The second corpus was recorded using a portable collection system on a laptop, with quarter-frame resolution video captured via an inexpensive webcam and audio by the built-in PC microphone. The database subjects were typically recorded in their own offices with varying lighting, background, and head pose. The third set was recorded in an automobile, both stationary and moving at approximately 30 or 60 mph, that was equipped with a wideband microphone and a lipstick-style camera. Compared to the previous two databases, the lighting, background, and head pose vary significantly, therefore this database represents the most challenging set. Finally, to study the benefits of direct visual ROI capture, a fourth set was recorded by means of a specially designed audiovisual wearable headset with an infrared camera housed inside its boom. This device provides high-quality visual data of the mouth ROI, being relatively insensitive to head- pose and lighting variations [46]. The video frame rate of all corpora is 30 Hz, and—with the exception of the office data – the resolution is 704 × 480 pixels. Typical frames of all four sets are depicted in Fig. 21.9.

21.4.2 Experimental Results

We now proceed to experimentally demonstrate the benefit of visual speech to ASR. In the first set of experiments, the IBM appearance-based audiovisual ASR system (see also Fig. 21.5(a)) with two-stream HMM-based decision fusion is applied to the four corpora depicted in Fig. 21.9. A number of connected-digits recognition results are reported in Table 21.2 in terms of word error rate (WER), %, using a multispeaker training-testing scenario. In addition to visual-only recognition, audio-only and AV ASR results are depicted for two acoustic conditions: the original recorded audio, as well as artificially corrupted audio by nonstationary babble speech noise. The noise level varies per database, with the experiment designed to result in audio-only WER of about 25% for all four corpora. In addition to the WER results, the approximate relative % reduction in WER, achieved by incorporating the visual modality into ASR, is shown for both acoustic conditions.

Table 21.2 demonstrates two major points [46]: first, that the ASR gains due to visual speech are large, even for the relatively clean acoustic conditions of the original data. Such benefits become dramatic at high noise levels, reaching for example a relative 69% WER

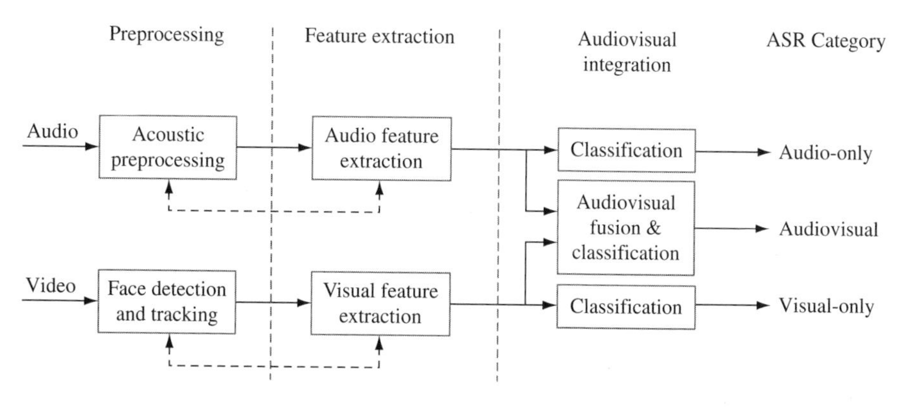

FIGURE 21.8

Block diagram of an audiovisual ASR system.

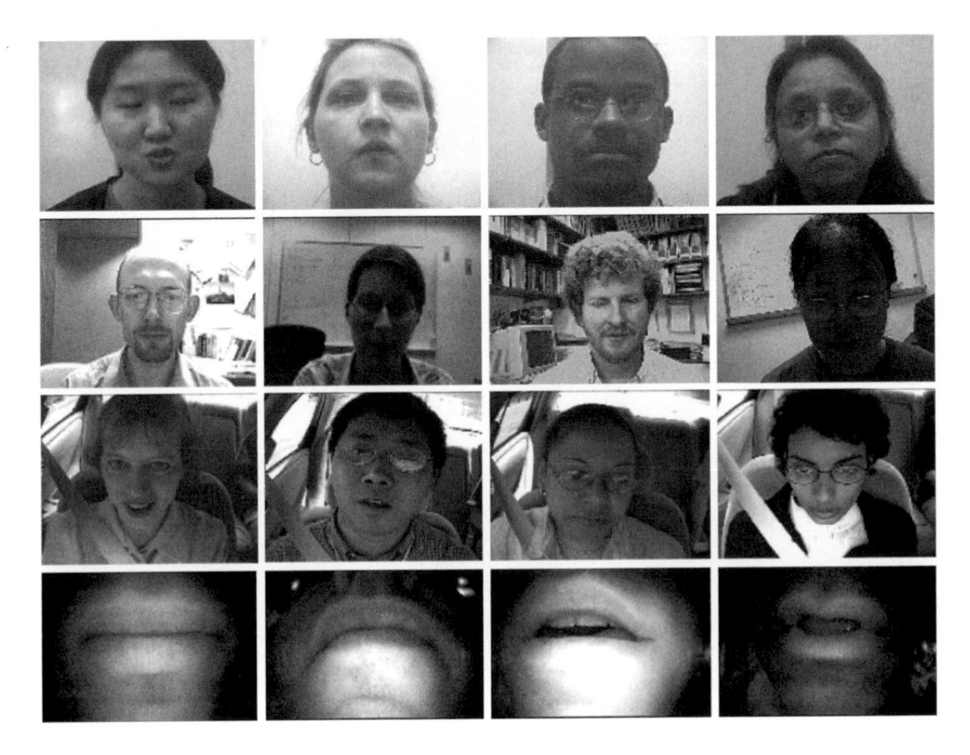

FIGURE 21.9

Example frames from the four IBM audiovisual ASR corpora discussed in Sections 21.4.1 and 21.4.2. Top-to-bottom: Full-face data collected in the studio-like, office, and car environments; Bottom line: ROI-only data captured by a specially designed headset [46].

reduction for the headset data. It is interesting to note that these gains hold even though the visual-only performance is significantly worse than audio-only ASR (e.g., 15–25 times worse in WER for the particular tasks). Second, as the visual environment becomes more challenging, due to head-pose and lighting variation, both visual-only performance and ASR gains degrade. For example, the visual-only WER is only 21.3% for the headset corpus, but 68.7% in the automobile data. Clearly, under challenging visual conditions, the performance of appearance-level visual features suffers.

The aforementioned observations carry through to large-vocabulary ASR as well. This is partially demonstrated in Fig. 21.10(a), where speaker-independent, large vocabulary (>10 k words) continuous speech recognition results are depicted for the studio-quality database using three fusion techniques for audiovisual ASR over a wide range of acoustic signal-to-noise-ratio (SNR) conditions. The best results are obtained by a hybrid fusion approach that uses the two-stream HMM (AV-MS) framework to combine audio features with fused audiovisual discriminant features (AV-Discr.), achieving for example an 8 dB "effective SNR" performance gain at 10 dB, as depicted in Fig. 21.10(a) [44].

TABLE 21.2 Connected-digit recognition on the four IBM databases of Fig. 21.9 [46].

Database		Clean			Noisy		
	VI	AU	AV	%	AU	AV	%
Studio	27.44	0.84	0.66	21	24.56	10.66	58
Office	43.33	2.51	1.96	22	24.91	14.73	41
Car	68.75	2.83	2.38	16	25.89	16.22	37
Headset	21.35	1.33	0.94	29	25.23	7.92	69

Audiovisual (AV) versus audio-only (AU) word error rate (WER), %, is depicted for clean and artificially corrupted data using HMMs trained on clean data. The approximate % relative improvement due to the visual modality is also shown for each condition (%), as well as the visual-only (VI) WER.

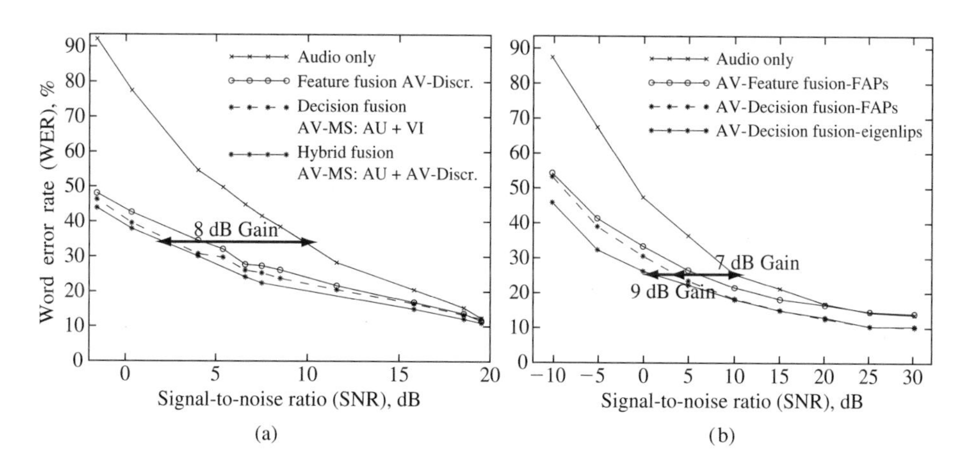

FIGURE 21.10

Large-vocabulary, audiovisual ASR results using the IBM (left) and NWU (right) systems. In both cases, audio-only and audiovisual WER, %, are depicted versus audio channel SNR for HMMs trained in matched noise conditions. The effective SNR gains are also shown with reference to the audio-only WER at 10 dB. Notice that the axes ranges in the two plots differ. In more detail: (a) In the IBM system, appearance-based visual features are combined with audio using three different techniques. Reported resuts are on the IBM, studio-quality database [44]. (b) In the NWU system, shape-based (FAPs) and appearance-based (eigenlips) visual features are combined with audio features by means of feature or decision fusion. Reported resuts are on the Bernstein lip-reading corpus [97].

Similar conclusions are reached when using the NWU audiovisual ASR system that employs shape-based visual features, obtained by PCA on FAPs of the outer and inner lip contours, or appearance-based visual features (eigenlips) [40,45] (see also Section 21.2.2). A summary of single-speaker, large-vocabulary ($\approx$1k words) recognition experiments using the Bernstein lipreading corpus [97] is depicted in Fig. 21.10(b). There, audio-only WER, %, is compared to audiovisual ASR performance over a wide range of acoustic SNR conditions (−10 to 30 dB), obtained by corrupting the original signal with white Gaussian noise. It can be clearly seen in Fig. 21.10(b) that considerable ASR improvement is achieved, compared to the audio-only performance, for all noise levels tested when visual speech information is utilized. The best results are obtained by a decision fusion approach that uses the two-stream HMM (AV-MS) framework to combine audio features with appearance-based features (eigenlips), achieving a 9 dB "effective SNR" performance gain at 10 dB, as depicted in Fig. 21.10(b). The "effective SNR" performance gain at 10 dB when shape-based features were used was 7 dB. Of particular interest is to compare performance of shape- and appearance-based visual features. Figure 21.10(b) demonstrates that the appearance-based system outperformed shape-based systems for SNR levels below 10 dB; however, the performance of both systems was similar for SNR levels larger than 10 dB. In addition, it is shown in [40, 45] that inner-lip FAPs do not provide as much speechreading information as the outer-lip FAPs. However, when both inner- and outer-lip FAPs are used as visual features, the performance of the audiovisual ASR system improves as compared to when only the outer-lip FAPs are used [40]. Note that these results are consistent with investigations of inner versus outer lip geometric visual features for automatic speechreading [24].

21.5 AUDIOVISUAL SPEECH SYNTHESIS

Audiovisual speech synthesis is a topic at the intersection of a number of areas including computer graphics, computer vision, image and video processing, speech processing, physiology, and psychology. Audiovisual speech synthesis systems automatically generate either voice and facial animation from arbitrary text (audiovisual TTS or visual TTS (VTTS)), or facial animation from arbitrary speech (speech-to-video synthesis). A view of an animated face, be it text- or speech-driven, can significantly improve intelligibility of both natural and synthetic speech, especially under nonideal acoustic conditions. Moreover, facial expressions and prosodic information can signal emotions, add emphasis to speech, and support dialog interaction.

Audiovisual speech synthesis systems have numerous applications related to human communication and perception, including tools for the hearing impaired, multimodal virtual agent-based user interfaces (desktop assistants, email messengers, newscasters, online shopping agents, etc.), computer-based learning, net-gaming, advertizing, and entertainment. For example, facial animation generated from telephone speech by a speech-to-video synthesis system could greatly benefit the hearing impaired, whereas an email service that transforms text and emoticons (facial expressions coded into a certain series of keystrokes) from text into an animated talking face could personalize and improve the email experience. Audiovisual speech synthesis is also suitable for wireless communication applications. Indeed, some face animation technologies have very low-bandwidth transmission requirements, utilizing a small number of animation control parameters. New mobile technology standards allow large-bandwidth multimedia applications, thus enabling the transmission of full synthetic video, if desired.

Two critical topics in the design and performance of audiovisual speech synthesis systems are modeling the speech coarticulation and the animation of the face. Various approaches exist in the literature for addressing these issues and are presented in detail in the next two sections. Following their review, VTTS and speech-to-video synthesis are discussed, and evaluation results of the visual speech synthesis system developed at NWU are presented.

21.5.1 Coarticulation Modeling

Coarticulation refers to changes in speech articulation (acoustic or visual) of the current speech segment (phoneme or viseme) due to neighboring speech. In the visual domain, this phenomenon arises because the visual articulator movements are affected by the neighboring visemes. Addressing this issue is crucial to visual speech synthesis, since, to achieve realistic facial animation, the dynamic properties and timing of the articulatory movements need to be proper. A number of methods have been suggested in the literature to model coarticulation. In general, they can be classified into rule-based and data-based approaches and are reviewed next.

Techniques in the first category define rules to control the visual articulators for each speech segment of interest, which could be phonemes, bi-, or triphones. For example, Löfquist proposed an "articulatory gesture" model [98]. He suggested utilizing dominance functions, defined for each phoneme, which increase and decrease over time during articulation, to model the influence of the phoneme on the movement of articulators. Dominance functions corresponding to the neighboring phonemes will overlap, therefore, articulation at the current phoneme will depend not only on the dominance function corresponding to the current phoneme, but also on the ones of the previous and following phonemes. In addition, it is proposed that each phoneme has a set of dominance functions, one for each articulator (lips, jaw, velum, larynx, tongue, etc.), because the effect of different articulators on neighboring phonemes is not the same. Dominance functions corresponding to various articulators may differ in offset, duration, and magnitude. In [49], Cohen and Massaro implemented Löfqvist's gestural theory of speech production, using negative exponential functions as a general form for dominance functions. In their system, the movement of articulators that correspond to a particular phoneme is obtained by spatially and temporally blending (using dominance functions) the effect of all neighboring phonemes under consideration. In other rule-based coarticulation modeling approaches, Pelachaud et al. [56] clustered phonemes into visemes with different deformability ranks, while Breen et al. [57] directly used context in the units employed for synthesis, by utilizing static context-dependent visemes. Overall, rule-based methods allow for incremental improvements by refining the articulation models of particular phonemes, which can be advantageous in certain scenarios.

In contrast to rule-based techniques, data-based coarticulation models are derived after training (optimizing) a number of model parameters on an available audiovisual database. Various such models have been considered for this purpose in the literature, for example ANNs and HMMs [51, 52]. Data-based coarticulation models can also be obtained using a concatenative approach [53, 54], where a database of video segments corresponding to context-dependent visemes is created using the phoneme-level transcription of a training audiovisual database. The main advantage of data-driven methods is that they can capture subtle details and patterns in the data, which are generally difficult to model by rules. In addition, retraining for a different speaker or language can be automated. Several approaches for generating visual speech parameters from the acoustic speech representation using data-driven coarticulation models have been investigated in the literature [51–54].

21.5.2 Facial Animation

The face is a complex structure consisting of bones, muscles, blood vessels, skin, cartilage, and so forth. Developing a facial animation system is therefore an involved task, requiring a framework for describing the geometric surfaces of the face, its skin color, texture, and animation capabilities. Several computer facial animation systems have been reported in the literature, that can be classified as model-based (also known as knowledge-based) or image-based.

In the model-based facial animation approach, a face is modeled as a 3D object, and its structure is controlled by a set of parameters. The approach has become popular due to the MPEG-4 facial animation standard [88], and it consists of the following three steps: designing the 3D facial model; digitizing a 3D mesh; and animating the 3D mesh to simulate facial movements. In the first step, a 3D model that captures the facial geometry is created. Most models describe the facial surface using a polygonal mesh (see also Fig. 21.11(a)). This method is frequently used because of its simplicity and availability of graphics hardware for efficient rendering of polygon surfaces. The facial surface should not be oversampled, because that would lead to computationally expensive facial animation. The polygons must also be laid out in a way that permits the face to flex and change shape naturally. In the second step, a digitized 3D facial mesh is constructed. This is typically achieved by obtaining the subject's facial geometry using 3D photogrammetry or a 3D scanner. Finally, in the third step, the 3D mesh is animated to simulate facial movements. During animation, the face surface is deformed by moving the vertices of the polygonal mesh, keeping the network topology unchanged.

The motion of the vertices is driven by a set of control parameters. These are mapped to vertex displacements based on interpolation, direct parameterization, pseudomuscular deformation, or physiological simulation. In the interpolation approach, a number of key frames, usually corresponding to visemes and facial expressions, are defined, and their vertex positions are stored. The frames in-between key frames are generated by interpolation, because all possible linear combinations of key frames are represented by the control parameter space. The main advantages of this approach are simplicity and its support by commercial animation packages. However, the disadvantages lie in the fact that facial feature motion is typically nonlinear, and that the number of achievable facial

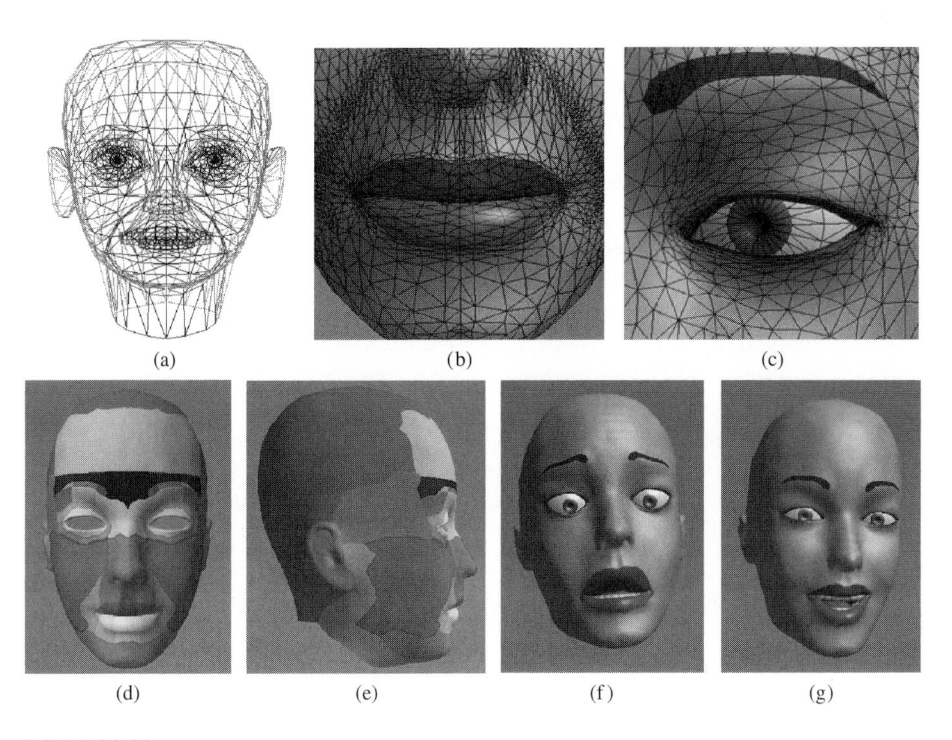

FIGURE 21.11

MPEG-4 compliant facial animation: (a) depicts a polygonal mesh [99]; (b,c) depict detailed structure of the most expressive face regions; (d,e) show how the 3D surface is divided into areas corresponding to feature points affected by FAPs; and (f,g) depict synthesized expressions of fear and joy ((b–g) correspond to model "Greta," reproduced with permission from [59]).

expressions is limited by the number of employed key frames. In the direct parameterization approach, basic geometric transformations, such as translation, rotation, and scaling, are used to describe vertex displacements. Pseudomuscular models, on the other hand, use facial muscle structure to model deformations. The space of allowable deformations is reduced by knowledge of the human face anatomic limitations. Muscles are modeled with one end affixed to the bone structure of the skull and the other end attached to the skin. Finally, modeling the skin with three spring-mass layers has also been used to develop more detailed physiological models. The main advantage of this approach is the improved realism over purely geometric facial modeling techniques.

The majority of model-based facial animation systems used today are extensions to Parke's work [58]. His model utilizes a parametrically controlled polygon topology, where the face is constructed from a network of approximately 900 surfaces, arranged and sized to match the facial contours. Large polygons are employed in flattered regions of the face, while small ones are used in high curvature areas. Face animation is controlled by a set of about 50 parameters, 10 of which drive the articulatory movements involved in speech production. In related work [59], Pasquariello and Pelachaud developed a 3D facial model, named "Greta," consisting of 15000 polygons (see Figs. 21.11(b) and (c)). Greta is compliant with the MPEG-4 standard [88], and able to generate, animate, and render in real-time the structure of a proprietary 3D model. The model uses the pseudomuscular approach to describe face behavior, and it includes features such as wrinkles, bulges, and furrows to enhance its realism. In particular, a great level of detail is devoted to the facial regions that contain most speechreading and expression information, such as the mouth, eyes, forehead, and the nasolabial furrow (see Figs. 21.11(b) and (c)). Furthermore, to achieve more control on the polygonal lattice, the 3D model surface is divided into areas that correspond to feature points affected by FAPs (see also Fig. 21.4 and Figs. 21.11(d) and (e)). Examples of facial animation employing the Greta model to display fear and joy expressions are shown in Figs. 21.11(f) and (g).

In contrast to the model-based techniques discussed earlier, the image-based facial animation approach relies mostly on image processing algorithms [53–55]. There, most of the work is performed during a training process, through which a database of video segments is created. Thus, unlike model-based approaches that use a static facial image, image-based techniques use multiple facial images, being able to capture subtle face deformations that occur during speech. Image-based facial animation consists of the following steps: recording of the video of the subject; video segmentation into animation groups; and animation of the model by concatenating various animation groups. In the first step, video of the subject uttering nonsense syllables or sentences in a controlled environment is recorded. In the second step, the recorded video is analyzed, and video segments consisting of phone, triphone, or word boundaries are identified. Finally, in the third step, the video segments are concatenated to realize the animation. Interpolation and morphing are usually employed to smooth transitions between boundary frames of the video segments.

Several examples of image-based facial animation can be found in the literature. For example, in [55], Ezzat and Poggio report a visual TTS system, named "MikeTalk," which uses 52 viseme images, representing 24 consonants, 12 monophthongs, and 16 diphthongs. To generate smooth transitions between the viseme images, morphing is employed. However, the system processes the mouth area only, and does not synthesize head movements or facial expressions. In [53], Bregler et al. report a speech-to-video synthesis system that also employs image-based facial animation. Their system utilizes existing footage to create video of a subject uttering words that were not spoken in the original footage. In the analysis stage, time-alignment of the speech is performed (using HMMs trained on the TIMIT database) to obtain phonetic labels, which are consequently used to segment the video into triphones. Only the mouth area is processed and then reimposed with new articulation into the original video sequence. Triphone videos and the phoneme labels are stored in the video model. In the synthesis stage, morphing and stitching are used to perform time-alignment of triphone videos, time-alignment of the lips to the utterance, illumination matching, and combination of lips and the background. The main disadvantages of this approach lie in the size of the triphone video database, and in the fact that only the mouth area is processed. Other facial parts such as eyes and eyebrows that carry important conversational information are not considered. To overcome these shortcomings, Cosatto and Graf [54] decompose their facial model into separate parts. The decomposed head model contains a "base face," which covers the area of the whole face, and serves as a substrate onto which the facial parts are integrated. The facial parts are the mouth with cheeks and jaw, the eyes, and the forehead with eyebrows. Each part is modeled separately, therefore the number of stored image samples in the database is kept at a manageable level. This allows for independent animation of various areas of the face, therefore increasing the number of free parameters in the animation system and the amount of conversational information contained in the facial animation.

21.5.3 Visual Text-to-Speech

A general block diagram of a text-driven facial animation system is shown in Fig. 21.12 [100]. The input text is first processed by a natural language processor (NLP), which analyzes it at various linguistic levels to produce phonetic and prosodic information. The latter refers to speech properties, such as stress and accent (at the syllable or word level), and intonation and rhythm, which describe changes in pitch and timing across words and utterances. NLP can also produce visual prosody, which conveys information about facial expressions (e.g., anger, disgust, happiness, sadness, etc.). The generated phonetic and prosodic information can then be used by the speech synthesis module to produce an acoustic speech signal. Similarly, the face synthesizer uses viseme information, obtained through phoneme-to-viseme mapping, and visual prosody to produce a visual speech signal.

A number of researchers have used this approach for VTTS, in conjunction with their techniques for coarticulation modeling and facial animation. For example, in [60], Cohen and Massaro used Parke's facial model as the basis for their text-driven speech synthesizer. Their main improvements over Parke's model were the inclusion of tongue and their extensive study of coarticulation, which they integrated into face animation [49]. For the "MikeTalk" system [55], discussed earlier, Ezzat and Poggio manually extracted a set of

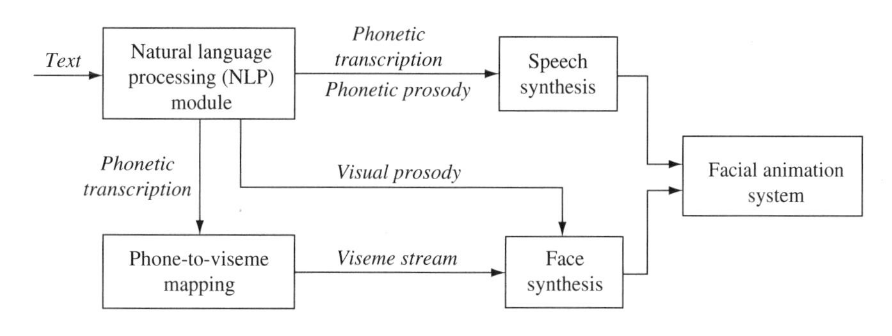

FIGURE 21.12

Components of typical VTTS synthesis systems.

viseme images from the recordings of a subject enunciating 40–50 words. They assumed a one-to-one mapping between phonemes and visemes, and modeled a viseme with a static lip shape image, instead of using a sequence of images. Subsequently, they reduced the number of visemes to 16, and constructed a database of 256 optical flow vectors that specified the transitions between all possible viseme images. Finally, they employed a TTS system to translate text into a phoneme stream with duration information, and used it to generate a sequence of viseme images for face animation, synchronized with TTS-produced speech.

In other work [54], Cossato and Graf developed a TTS system using their decomposed facial model, discussed in the previous section. They first recorded a video database consisting of common triphones and quadri-phones uttered by a subject. Then, they extracted and processed mouth images from the video, obtaining both geometric and PCA visual features, and subsequently parametrized and stored them in bins located on a multidimensional grid within the geometric feature space. To reduce storage requirements, within each bin, they discarded mouth images "close" to others in the PCA space, using a VQ scheme based on Euclidean distance. For synthesis, they employed a coarticulation model, similar to the one in [49], to obtain a smooth trajectory in the geometric feature space, based on the target phonetic sequence and a mapping between visemes and their "average" geometric feature representation. To generate the mouth region animation, they sampled the resulting trajectory at the video rate, and, at each time instant, they chose the closest grid point, providing in this manner a set of candidate mouth bitmaps located within the corresponding bin. Next, they utilized the Viterbi algorithm to compute the lowest-cost path through a graph, having as nodes the candidate images at each time instant. For the transition cost between nodes (mouth images) at consecutive times, they used their Euclidean distance in the PCA space, setting this cost to zero in case the images correspond to neighboring frames of the original video. The resulting path provided the final mouth sequence animation.

21.5.4 Speech-to-Video Synthesis

Speech-to-video synthesis systems exploit the correlation between acoustic and visual speech, to synthesize a visual signal from the available acoustic signal (see Fig. 21.13). Several approaches for speech-to-video synthesis have been reported in the literature, using methods, such as VQ, ANNs, or HMMs (see also Section 21.3.2). In general, these techniques can be classified into regression- and symbol-based approaches, and are briefly reviewed in this section.

Regression-based methods establish a direct continuous association between acoustic and visual features. VQ and ANNs are commonly used for this task, with the former constituting a simpler approach. In the training phase of VQ-based speech-to-video synthesis, an acoustic codebook is first constructed using clustering techniques. The codebook allows classifying audio features into a small number of classes, with the visual features associated with each class averaged to produce a centroid to be used in synthesis. At the synthesis stage, the acoustic parameters at a given instant are compared against all possible acoustic classes. The class located closest to the given parameters is selected,

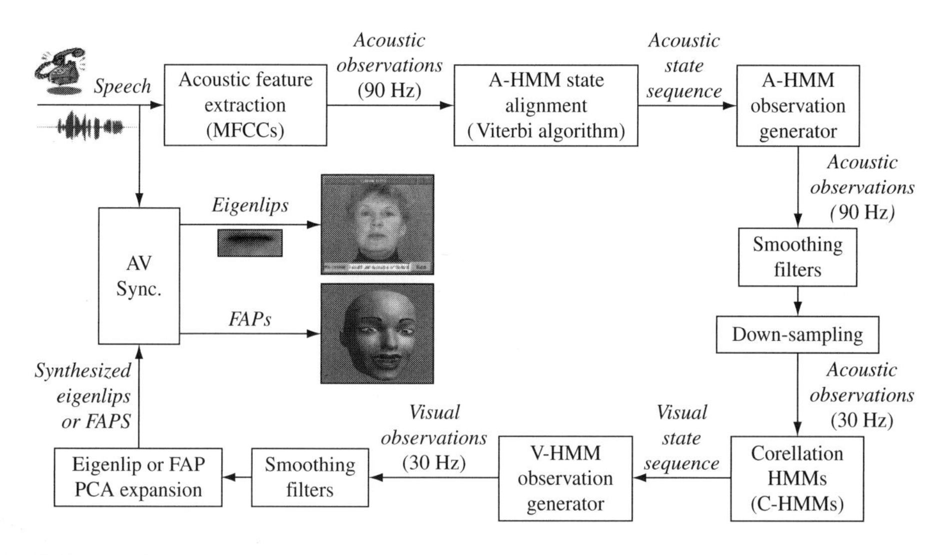

FIGURE 21.13

The speech-to-video synthesis systems developed in [50, 52] utilize narrowband speech to generate two possible visual representations: eigenlips that can be superimposed on frontal face videos for animation, or FAPs that can be used to drive an MPEG-4 compliant facial animation model.

and the corresponding visual centroid is employed to drive the facial animation. In ANN-based speech-to-video synthesis, the acoustic and visual speech features correspond to the input and output network nodes, respectively. In the training phase, the network weights are adjusted using the back-propagation algorithm. For synthesis, at each time instant, the speech features are presented to the network input, with the visual speech parameters generated at the output nodes of the ANN.

The work reported in [51] constitutes a typical example of the regression-based approach. There, Morishima and Harashima investigated the use of VQ and ANNs for predicting facial features from audio. They considered 16-dimensional LPC vectors and eight facial feature points (located on the lips, jaw, and ears) as the representations of the acoustic and visual signals, respectively. In their VQ scheme, they created a five-bit codebook to allow mapping of the acoustic to the visual parameters, while in their ANN-based algorithm, they employed a three-layer ANN architecture. In related work, Lavagetto [61] proposed using six independent time-delay neural networks with four layers, each accepting identical acoustic feature input (12-dimensional LPC vectors), but generating as output individual parameters of a geometric visual speech representation.

In contrast to regression-based techniques, in the symbol-based approach the acoustic signal is first transformed into an intermediate discrete representation consisting of a sequence of subphonetic or subword units. HMMs are typically used for this purpose, because they provide explicit phonetic information, which can help in the analysis of coarticulation effects. Reported HMM-based systems vary in two basic aspects: the units used for recognition (i.e., what do the HMM states represent; see also Sections 21.3.1 and 21.3.2), and the method for synthesizing the visual parameter trajectories from the recognized HMM state sequence. Examples of such systems are provided next.

Simons and Cox [62] developed an HMM-based speech-driven synthetic head. They analyzed a small number of phonetically rich sentences to obtain several acoustic and visual training vectors. They used VQ to produce audio and visual codebooks of sizes 64 and 16, respectively. Then, they created a fully-connected 16-state discrete HMM, each state representing a particular vector quantized mouthshape, and producing the 64 possible audio codewords. The HMM transition and observation probabilities were trained on the basis of the joint audiovisual vector-quantized data. Subsequently, the trained HMM was employed in synthesis by means of the Viterbi algorithm, generating the most likely visual state sequence (hence, visual representation), given the input audio observations.

Chen and Rao [5] trained continuous whole-word HMMs using audiovisual observations (henceforth such HMMs are referred to as AV-HMMs). They used the width and height of the outer lip contour as visual features, and 13 MFCCs as acoustic features. Subsequently, they built for each word an acoustic HMM (A-HMM), which had the same transition matrix and initial state distribution as the corresponding AV-HMM (see Eqs. (21.1), (21.2), and (21.4)). The state acoustic observation pdf for each particular A-HMM state was derived by integrating the AV-HMM observation pdf over the visual parameters. In the synthesis phase, the A-HMMs and the acoustic observations were first used, employing the Viterbi algorithm, to obtain the optimal acoustic state sequence. Next, assuming that the AV-HMM state sequence is the same as the A-HMM state sequence, they estimated for each state the corresponding visual feature vector, using AV pdfs and the acoustic observations.

Bregler et al. [53] created an HMM-based speech-driven facial animation system called "Video Rewrite." They first trained an A-HMM system on the TIMIT database, and used it to segment the audio portion of a joint audiovisual database into triphones. The visual segments, time-synchronous to the resulting triphones, were then stored into a video database, indexed by the corresponding tri-visemes. At the synthesis stage, given the input acoustic signal, they first obtained its phonetic level transcription using the A-HMM system. Subsequently, they used the concatenative approach to synthesis with visual segments selected from the created video database. Various cost metrics were considered for the segment selection, and the Viterbi algorithm was used to obtain the optimal sequence of video segments. Finally, they used warping techniques to smooth the selected video segments and synchronize them with the speech signal.

Finally, two systems were developed at NWU [50, 52], using two different visual speech representations, eigenlips and FAPs. PCA was performed on both FAPs and mouth images to obtain visual features of lower dimensionality. MFCCs were used as acoustic features in both systems. The block diagram of the developed systems is shown in Fig. 21.13. The two systems utilized continuous A-HMMs, visual HMMs (V-HMMs) and correlation HMMs (C-HMMs). In this approach, the A-HMMs and the Viterbi algorithm were used to realize the audio state sequence that best described the acoustic observations extracted from the input narrowband speech signal. The A-HMM observation generator then used the means corresponding to each resulting A-HMM state to produce speaker-independent observations (see also (21.1), (21.2), and (21.6) in Section 21.3.2). Smoothing and down-sampling were subsequently used to obtain acoustic observations at the video rate (30 Hz), whereas the C-HMM system mapped the generated acoustic observations, using the Viterbi algorithm, into a visual state sequence. Finally, the visual state sequence and the V-HMM observation generator were employed to produce visual observations.

Two key elements of the NWU systems were the C-HMM training procedure and model architecture. To ensure that the C-HMMs were capable of approximating the optimal visual state sequence given the acoustic observations, they were built with the same topology and identical state transition and initial probabilities as the V-HMMs. As a result of the aforementioned constraints, only the C-HMM observation pdfs had to be estimated during training (see also Section 21.3.2). In more detail, the C-HMMs were trained using the following procedure [50, 52]: In the first step, A-HMMs and V-HMMs were independently trained using the TIMIT corpus and the visual part of the Bernstein database, respectively. The two HMMs had different topologies to account for the unequal audio and video observation rates. Next, in the second step, the trained A-HMMs and V-HMMs in conjunction with the Viterbi algorithm were used to force-align acoustic and visual training data, respectively, and generate corresponding acoustic and visual state sequences. In the third step, down-sampled acoustic observation sequences were generated using the acoustic state sequences obtained in the second step. In the fourth step, the visual state sequence generated in the second step was utilized as a constraint to distribute the down-sampled acoustic observations among the C-HMM states. Finally, reestimation of the C-HMM observation pdfs was carried out. This training procedure generated C-HMMs capable of producing, in conjunction with V-HMMs, visual state sequences and estimates of the visual articulatory movements from down-sampled acoustic observations.

21.5.5 Visual Speech Synthesis Evaluation

Evaluating visual synthesis systems is extremely important to benchmark algorithmic improvements, assess the suitability of specific databases for training data-driven techniques, and quantify the benefit of incorporating the visual modality over traditional audio-only synthesis, for example. There exist both objective and subjective methods to evaluate visual speech synthesis. The former typically compare the difference between a set of synthesized and recorded test sequences, in terms of mean squared error or other distance metrics in the visual speech representation space, or, alternatively, report ASR performance on the synthesized test set [52]. Although relatively easy to perform, objective evaluation does not necessarily indicate how two systems will be relatively received by human users in practice. Subjective testing is instead required for such assessments [7, 48].

In general, subjective evaluation of visual speech synthesis performance should be application-dependent. Such tests should be developed with the goal of evaluating a number of issues, for example the degree of realism in the animation, user satisfaction, and the effectiveness in communicating the intended message. In particular, effectiveness in communicating and especially intelligibility of a talking head should be of primary importance. Intelligibility evaluation approaches aim at measuring either phoneme identification performance (recognition of vowels and consonants) or speechreading performance (recognition of isolated words or sentences), by human subjects.

In this section, we provide an example of an intelligibility subjective evaluation in the case of the eigenlips-based, speech-to-video synthesis system developed at NWU and discussed earlier. In these tests [48, 50], several subjects have been presented with three types of stimuli: (a) audio-only signal; (b) audio, supplemented by the synthesized video signal; and (c) audio, together with the original video footage. In all cases, the audio (and video, where applicable) utterances were from the Bernstein lipreading corpus [97], and the intelligibility experiments were conducted with the audio corrupted by additive, white Gaussian noise, resulting in speech signals of −5 dB, −10 dB, and −15 dB SNRs (see Fig. 21.14). The audio-only word recognition accuracy achieved by the subjects was 92.2%, 66.8%, and 11.7%, at the three SNR levels, respectively. The word recognition accuracies improved significantly, when synthesized video was also presented to the subjects, reaching 97.9%, 87.5%, and 46.1%. Subjective tests under scenarios (a) and (b)

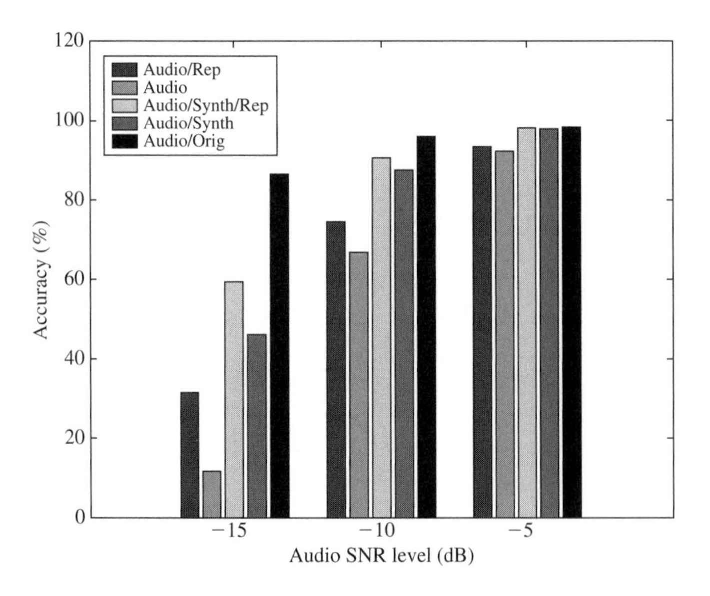

FIGURE 21.14

Intelligibility-based subjective evaluation of the speech-to-video synthesis system developed at Northwestern University [48, 50]. Human speech perception is compared using audio-only versus audio with synthesized video and versus audio with natural video of the lip region. For the first two conditions, results for repetitive presentation of the stimuli to the subjects are also given ("Rep"). Experiments are reported over three acoustic noise conditions.

were also performed using sets that contained certain number of repeated utterances used throughout all the tests. The word recognition accuracies improved when repeated utterances were used, especially for the SNR of −15 dB, indicating that the subjects used prior knowledge to assist the transcription of the repeated utterances. However, these results were still inferior to human speech perception word recognition accuracies obtained using natural instead of synthesized video, namely 98.3%, 95.9%, and 86.5%, respectively. Clearly, these subjective tests suggest that animated faces obtained using visual speech synthesizers can improve speech intelligibility, especially under noisy conditions (see also [7]).

21.6 AUDIOVISUAL SPEAKER RECOGNITION

Audiovisual speaker recognition (also known as audiovisual biometrics) systems utilize acoustic and visual information to perform automatic person recognition. A person recognition system should be capable of rejecting claims from impostors, persons not registered with the system, and accepting claims from the clients, persons registered with the system. Person recognition can be classified into two problems: person identification and person verification (authentication) [68]. Person identification is the problem of determining the identity of a person (who the person is) from a closed set of candidates, whereas person verification refers to the problem of determining whether a person is who s/he claims to be. There are a number of systems that require person recognition to reliably determine the identity of persons requesting their services. Applications that can employ person recognition systems include automatic banking, computer network security, information retrieval, secure building access, and so forth. Personal property, such as cell phones, PDAs, laptops, cars, and so forth, could also have built-in person recognition systems that would prevent impostors from using them.

Biometrics, or biometric recognition, refers to utilizing physiological and behavioral characteristics for automatic person recognition. Traditional person identification methods, including knowledge-based (e.g., passwords, PINs) and token-based (e.g., ATM or credit cards, and keys) do not provide reliable performance. Passwords can be compromised, while keys and cards can be stolen or duplicated. Identity theft is one of the fastest growing crimes in the United States. Unlike knowledge- and token-based information, biometric characteristics cannot be forgotten or easily stolen. There are many different biometric characteristics, that can be used in person recognition systems, including fingerprints, palm prints, hand and finger geometry, hand veins, iris and retinal scans, infrared thermograms, DNA, ears, faces, gait, voice, signature, and so forth (see Fig. 21.15) [71, 72, 76, 101].

Each biometric characteristic has its own advantages and disadvantages and there is no single modality that performs the best for all applications. The choice of biometric characteristics depends on many factors including the best achievable performance, uniqueness, robustness to noise, cost of biometric sensors, invariance of characteristics with time, robustness to attacks, population coverage, scalability, and so forth. All of these

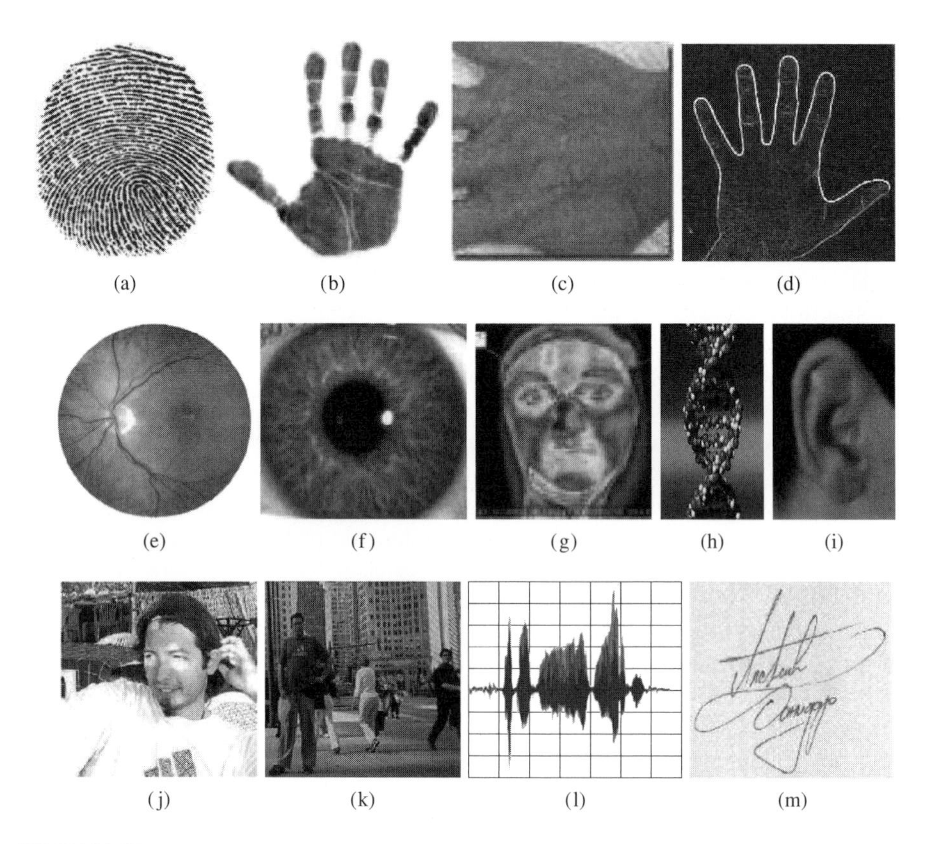

FIGURE 21.15

Biometric characteristics: (a) fingerprints; (b) palm print; (c) hand veins; (d) hand and finger geometry; (e) retinal scan; (f) iris; (g) infrared thermogram; (h) DNA; (i) ears; (j) face; (k) gait; (l) speech; (m) signature.

factors are usually considered when choosing the most appropriate biometric characteristics for a certain application. In addition, there are a number of biometric applications for which it is desirable to use nonintrusive and user-friendly methods for extraction of biometric features. Developing such biometric systems makes biometric technology more socially acceptable and accelerates its integration into every day life.

A person's voice and face are biometric characteristics that are easily collected and natural to the user. These characteristics can be utilized for nonintrusive person recognition. The recent technology advances decreased the cost of audio and video biometric sensors and opened a door to audiovisual biometrics. Acoustic and visual biometric characteristics can contain static and dynamic information. LPCs, MFCCs, and their derivatives, are commonly used as acoustic features in speaker recognition systems. Visual features can describe only the mouth region (visual–labial features) or the whole face (visual–facial features). Both mouth and face can be represented using shape-based features or appearance-based features (see also Section 21.2). Shape-based labial features include lip-contour shape and geometric features, while shape-based facial features include ASMs, facial feature geometry, elastic graphs, and so forth. Labial and facial appearance-based features are obtained using image projections such as LDA, PCA, DCT, and so forth, on mouth or face images. Facial features can also be classified as global or local if the face is represented by only one feature vector or by multiple vectors each representing local information. Face images used for extraction of visual features can be visible or infrared, 2D or 3D, and so forth.

Although single-modality biometric systems can achieve high performance in some cases, they are usually not robust to noise and do not meet the needs of many potential person recognition applications. Speaker recognition systems that rely only on audio data are sensitive to microphones (headset, desktop, telephone, etc.), acoustic environment (car, plane, factory, etc.), and channel noise (telephone lines, VoIP, etc.). On the other hand, systems that rely only on visual data can be sensitive to visual noise (lightning changes, poor video quality, occlusion, segmentation errors, etc.). To improve the robustness of biometric systems, multisamples (multiple samples of the same biometric characteristic), multi-algorithms (multiple algorithms with the same biometric sample), and multimodal (different biometric characteristics) biometric systems have been developed. The advantage of multimodal biometric systems lies in their robustness, because different modalities can provide independent (complementary) information. Different modalities are combined to eliminate problems characteristic of single modalities. It has been shown that using multiple biometric modalities improves the performance of a biometric system [71–73, 76].

In audiovisual speaker recognition systems [102], speech is utilized together with either static (visual features obtained from a single face image) or dynamic (video sequences of the face or the mouth area) visual information to improve speaker recognition performance (see Fig. 21.16). Audiovisual speaker recognition systems can also utilize all three modalities [74]. Utilizing dynamic visual information also significantly

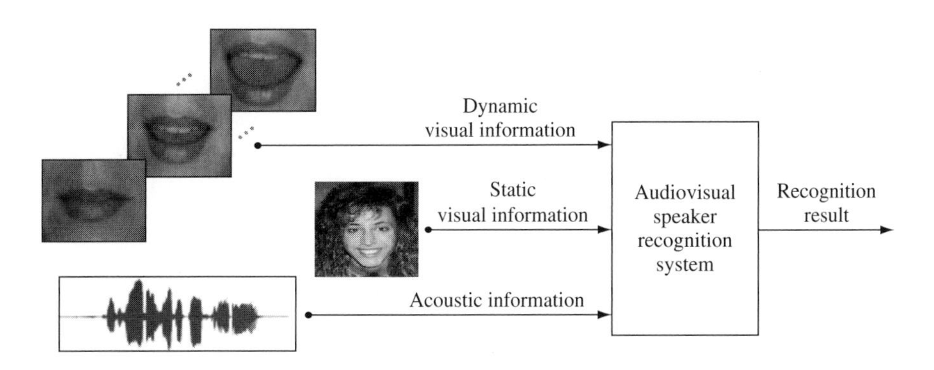

FIGURE 21.16

Block diagram of an audiovisual speaker recognition system that utilizes static (face image) and dynamic (visual speech) visual information together with acoustic information.

reduces chances of impostor attacks (spoofing). Audio-only and static-image-based (face recognition) speaker recognition systems are susceptible to impostor attacks if the impostor possesses a photograph and/or speech recordings of the client. However, it is considerably more difficult for an impostor to impersonate both acoustic and dynamical visual information simultaneously. Audiovisual speaker recognition holds promise for wider adoption because of the low-cost of audio and video sensors and the ease of acquiring audio and video signals (even without assistance from the client) [103].

Audiovisual speaker recognition systems can be either text-dependent, where speech used for training and testing is constrained to be the same, or text-independent, where speech used for testing is unconstrained. The methods for modeling speakers based on their audiovisual biometric data are usually statistical in nature. Such approaches include, ANNs, SVMs, GMMs, HMMs, and so forth (see also Section 21.3.2). HMMs represent the most commonly used approach for speaker recognition.

In speaker identification systems, the objective is to determine the class $\hat{c}$, corresponding to the enrolled person or the impostor, that best matches the unknown person's audiovisual biometric data $\mathbf{o}_{av,t}$, that is

$$\hat{c} = \arg\max_{c \in C} \Pr(c|\mathbf{o}_{av,t}),$$

where C denotes the set of classes corresponding to all speakers in the database and the impostor, and $\Pr(c|\mathbf{o}_{av,t})$ the conditional probability that biometric observations $\mathbf{o}_{av,t}$ were generated by the statistical model for the class c.

In speaker verification systems there are only two classes, and it is necessary to determine whether the class corresponding to the general population (w), or the class corresponding to the true claimant (c), best matches the claimant's biometric observations. The similarity measure (D) can be defined as the likelihood-ratio between the speaker set and the world set, that is

$$D = \log \Pr(c|\mathbf{o}_{av,t}) - \log \Pr(w|\mathbf{o}_{av,t}).$$

If D is larger than an a priori defined verification threshold the claim is accepted; otherwise it is rejected.

In text-dependent speaker recognition systems, subphonetic units are usually modeled. In that case, the set of classes can be considered as the product space between the set of speakers and the set of phonetic-based units. HMMs are commonly used for text-dependent speaker recognition, through modeling the phonetic units by Gaussian mixture densities (see (21.1) in Section 21.3.2). In text-independent systems single-state HMMs (GMMs) can be used to model speakers. In this case a single GMM is assumed to generate the entire audiovisual observation sequence.

The performance of identification systems is usually reported in terms of identification error or rank-N correct identification rate. The latter is defined as the probability that the correct match of the unknown person's biometric data is in the top N similarity scores (this scenario corresponds to the identification system that is not fully automated, needing human intervention or additional identification systems applied in cascade). Two commonly used error measures for verification performance are false acceptance rate (FAR), where – an impostor is accepted, – and false rejection rate (FRR), where – a client is rejected. They are defined by

$$\text{FAR} = I_{\text{A}}/I \times 100\% \quad \text{FRR} = C_{\text{R}}/C \times 100\%,$$

where I_{A} denotes the number of accepted impostors, I the number of impostor claims, C_{R} the number of rejected clients, and C the number of client claims. There is a trade-off between FAR and FRR, which is controlled by the choice of the verification threshold. It is usually chosen according to certain FAR and FRR requirements, based on results obtained through experiments on the evaluation set. The choice of the verification threshold clearly depends on the application and costs assigned to each of the error measures. For example, systems that control access to a highly secure area, or manage banking transactions, would require very low FAR at the expense of increased FRR. On the other hand, systems that control tolls or gym access would avoid putting their legitimate customers in inconvenient situations by requiring low FRR, at the expense of increased FAR. The receiver operator curve (ROC) or the detection error trade-off (DET) curve can be used to graphically represent the trade-off between FAR and FRR [104]. DET and ROC depict FRR as a function of FAR in a log, and linear scale, respectively. The detection cost function (DCF) is a measure derived from FAR and FRR according to

$$\text{DCF} = \text{Cost(FR)}P(\text{client})\text{FRR} + \text{Cost(FA)}P(\text{impostor})\text{FAR}$$

where P(client) and P(impostor) are the prior probabilities that a client or an impostor will use the system, respectively, whereas Cost(FA) and Cost(FR) represent respectively the costs of false acceptance and false rejection. Half total error rate (HTER) [104, 105] is a special case of DCF when the prior probabilities are equal to 0.5 and the costs equal to 1. Verification system performance is also often reported using a single measure either by choosing the threshold for which FAR and FRR are equal, resulting in the equal error rate (EER), or by choosing the threshold that minimizes DCF (or HTER). The appropriate threshold can be found either using the test set (providing biased results) or a separate validation set [105].

Performance of audiovisual speaker recognition systems strongly depends on the choice and accurate extraction of the visual features, and the information fusion approach. Acoustic and visual observations can either be combined to form joint audiovisual observations or utilized as separate observation streams. Information fusion approaches commonly used for fusion of audio and visual biometric information are discussed in more detail in Section 21.3.3 and in [71, 72, 102, 106]. A number of audiovisual speaker recognition systems that utilize various types of visual features and audiovisual information fusion strategies have been reported in the literature [70, 72–76, 102]. With respect to the type of acoustic and visual information they use, person recognition systems can be classified into audio-only [107], visual-only-static (only visual features obtained from a single face image are used) [108], visual-only-dynamic (visual features containing temporal information obtained from video sequences are used) [109], audiovisual-static [76, 106], and audiovisual-dynamic [70, 72–75].

Brunelli and Falavigna [76] developed a text-independent speaker identification system that combines audio-only speaker identification and face recognition systems. The two systems provide five classifiers, two acoustic and three visual. Two acoustic classifiers correspond to two sets of acoustic features (static and dynamic) derived from the short time spectral analysis of the speech signal. Their audio-only speaker identification system is based on VQ. Three visual classifiers correspond to the visual classifying features extracted from three regions of the face: eyes, nose, and mouth. The individually obtained classification scores are combined using a weighted geometric average. The identification rate of the integrated system is 98%, compared to 88% and 91% rates obtained by the audio-only speaker recognition and face recognition systems, respectively.

Aleksic and Katsaggelos [73] developed an audiovisual speaker recognition system that utilized 13 MFCC coefficients and their first-and second-order derivatives as acoustic features. A visual feature vector consisting of 10 FAPs that describe the movement of the outer-lip contour [88] was projected by means of the PCA onto a three-dimensional space. The resulting visual features were augmented with first-and second-order derivatives providing nine-dimensional dynamic visual feature vectors. They used a feature fusion integration approach and single-stream HMMs to integrate acoustic and visual information. Speaker verification and identification experiments were performed using audio-only and audiovisual information, under both clean and noisy audio conditions at SNRs ranging from 0 dB to 20 dB. Speaker identification and verification results obtained, expressed in terms of the identification error and EER, are shown in Table 21.3. Significant improvement in performance over audio-only speaker recognition system was achieved, especially under noisy acoustic conditions. For instance, the identification error was reduced from 53.1%, when audio-only information was utilized, to 12.82%, when audiovisual information was employed at 0 dB SNR.

Jourlin et al. [75] developed an audiovisual speaker verification system that utilizes both acoustic and visual dynamic information. Their 39-dimensional acoustic features consist of LPC coefficients and their first-and second-order derivatives. They use 14 lip-shape parameters, 10 intensity parameters and the scale as visual features, resulting in a 25-dimensional visual feature vector. They utilize HMMs and the decision fusion

TABLE 21.3 Speaker identification and verification errors obtained when audio-only (AU) or audiovisual (AV) biometric data was utilized.

SNR	Identification error [%]		Verification error (EER) [%]	
	AU	AV	AU	AV
Clean	5.13	5.13	2.56	1.71
20	19.51	7.69	3.99	2.28
10	38.03	10.26	4.99	2.71
0	53.10	12.82	8.26	3.13

integration approach to perform audio-only, visual-only, and audiovisual experiments. The audiovisual score is computed as a weighted sum of the audio and visual scores. Their results demonstrate a reduction of FA from 2.3% when the audio-only system is used to 0.5% when the multimodal system is used.

Chaudhari et al. [72] developed an audiovisual speaker identification and verification system, which modeled reliability of the audio and video information streams with parameters that were time-varying and context dependent. The acoustic features consisted of 23 MFCC coefficients, while visual features consisted of 24 DCT coefficients obtained by applying DCT on the ROI extracted by means of a face tracking algorithm. They utilized GMMs to model speakers, and parameters that depended on time, modality, and speaker to model stream reliability. The system that utilized time dependent stream weights achieved an EER of 1.04%, compared to 1.71% , 1.51%, and 1.22%, of the audio-only, video-only, and audiovisual (feature fusion) systems, respectively.

Wark et al. [70] employed multistream HMMs to develop text-independent AV speaker verification and identification systems tested on the M2VTS database. They utilized MFCCs as acoustic features and lip contour information obtained after applying PCA and LDA, as visual features. They trained the system in clean conditions and tested it in degraded acoustic conditions. At low SNRs, the audiovisual system achieved significant performance improvement over the audio-only system and also outperformed the visual-only system, while at high SNRs the performance was similar to the performance of the audio-only system.

Dieckmann et al. [74] developed a system that used visual features obtained from all three modalities, face, voice, and lip movement. The identification error decreased to 7% when all three modalities were used, compared to 10.4%, 11%, and 18.7%, when voice, lip movements, and face visual features were used individually.

In summary, there is a need for resources for advancing and accessing audiovisual speaker recognition systems. Publicly available multimodal corpora that better reflects realistic conditions, such as acoustic noise and lighting changes, would help in investigating robustness of audiovisual systems. In addition, standard experiments and evaluation procedures should be defined to enable fair comparison of different systems. Baseline algorithms and systems could also be chosen and made available to facilitate separate investigation of effects that factors, such as, the choice of acoustic and visual features, the information fusion approach, and classification algorithms, have on system performance.

21.7 SUMMARY AND DISCUSSION

In this chapter, we have focused on how the joint processing of visual and audio signals, both generated by a talking person, can provide valuable speech information to benefit a number of audiovisual speech processing applications, crucial to HCI. We first concentrated on the analysis of visual signals, and described various possible ways of representing and extracting the speech information available in them. We then discussed how the obtained visual features can complement features extracted (by well-studied methods) from the acoustic signal, and how the two modality representations can be fused together to allow joint audiovisual speech processing. The general bimodal integration framework was subsequently applied to three problems, namely ASR, talking face synthesis, as well as speaker identification and authentication. In all three cases, we discussed issues specific to the particular application, reviewed several relevant systems that have been reported in the literature, and presented results using the implementations developed at IBM Research and/or Northwestern University. The experimental results demonstrated the importance of utilizing visual speech information, especially in the presence of acoustic noise.

As we mentioned in the Introduction, there exist a number of additional applications that can benefit from the joint processing of audio and visual signals. Examples of such are emotion recognition, speaker detection and localization, speech activity detection, and enhancement of the acoustic signal or of its corresponding audio features. Because of lack of space, we only briefly address some of them in the following.

Automatic emotion recognition has many potential applications in HCI, for example by indirectly providing valuable user input to dialogue management. Both facial expression and voice reflect the emotional state of a person, thus bimodal processing is a sensible approach. In the visual domain, there exist six basic facial expressions: happiness, anger, sadness, fear, surprise, and disgust. Their study is enabled by the facial action coding system (FACS), that provides standardized coding of changes in facial motion through 46 action units, which describe basic facial movements [110]. The FACS is based on muscle activity, and captures in detail the effect of each action unit on the visual face features. Commonly used visual features for automatic emotion recognition include lip and eyebrow movements, whole face images, optical flow and so forth [77, 78]. For the classification process, both spatial and spatiotemporal approaches can be used. In the former, visual features obtained from single face images are employed, whereas spatiotemporal approaches utilize features extracted from each frame of the video sequence of interest. Typically, in facial expression recognition systems, ANNs are used to perform spatial classification, whereas HMMs are frequently employed in the spatiotemporal approach [77]. Visual systems can of course be combined with audio-only emotion recognizers, using the audiovisual integration framework of Section 21.3. In this case, typically used audio features include the acoustic signal energy, pitch contour statistics, and so forth.

Among additional joint audiovisual processing applications, speaker detection and tracking is especially useful in environments such as conference rooms, where multiple persons are present, and signals from both video cameras and microphone arrays are available. In such occasions, speaker detection and tracking can be performed using acoustically-guided cameras, visually-guided microphone arrays, or through joint audiovisual tracking [63–65]. Of particular importance to speech applications is also the detection of synchronous audiovisual sources in the presence of multiple speakers in the scene, as is often the case in broadcast videos. Joint audiovisual speech activity localization can benefit from the fact that the two modalities are correlated, and, for example, can be quantified using mutual information of the two signals [66]. Furthermore, visual information, such as user pose and proximity to a computer or kiosk, as well as mouth movement, can be used to flag speech intent [67], or augment acoustic cues for speech activity detection. The resulting systems will be robust to environmental noise and are expected to eventually make the "push-to-talk" button in present automatic speech recognizers obsolete. Another application that exploits the correlation between the audio and visual speech signals is the bimodal enhancement of audio. There, acoustic information is restored using the video of the speaker's mouth region in conjunction with the corrupted audio signal. The enhancement can occur either in the signal space or the audio feature space, utilizing linear or nonlinear techniques [18, 19]. Such an approach is beneficial, for example, when the amount of visual data available for training is insufficient to obtain visual-only speech models, thus not allowing audiovisual ASR by means of the fusion techniques discussed in Section 21.3.

Clearly, the field of joint audiovisual signal processing is a very new, active, and exciting topic of research and development. Indeed, there are a number of major accomplishments, some of which have been described in this chapter in the context of speech applications for HCI. Concerning the practical deployment of these technologies, several obstacles have been slowly lifting, with audiovisual speech processing systems starting to exhibit real-time performance and improved robustness [46]. Nevertheless, various research issues remain open to further investigation. Such are, for example, the design of a truly speaker-independent, high-performing visual feature representation with improved robustness to the visual environment and user behavior, possibly employing three-dimensional face information, as well as the development of improved audiovisual integration algorithms that will allow unconstrained audiovisual asynchrony modeling and robust, localized reliability estimation of the signal information content, to name a few. Clearly, further research is required to advance the field, and for audiovisual signal processing to become widespread in practice. The ground is fertile for additional major accomplishments and revolutionary future multimodal technologies and applications, promising to improve HCI and, with that, life quality.

REFERENCES

[1] D. O'Shaughnessy. Interacting with computers by voice: Automatic speech recognition and synthesis. *Proc. IEEE*, 91(9):1272–1305, 2003.

[2] R. Stern, A. Acero, F.-H. Liu, and Y. Ohshima. Signal processing for robust speech recognition In C.-H. Lee, F. K. Soong, and Y. Ohshima, editors. *Automatic Speech and Speaker Recognition. Advanced Topics*, ch. 15, 357–384. Norwell, MA: Kluwer Academic Pub., 1997.

[3] R. P. Lippmann. Speech recognition by machines and humans. *Speech Commun.*, 22(1):1–15, 1997.

[4] R. van Bezooijen and V. J. Heuven. Assessment of synthesis systems In D. Gibbon, R. Moore, and R. Winski, editors. *Handbook of Standards and Resources for Spoken Language Systems*, ch. 12, 481–563. New York, NY: Mouton de Gruyter, 1997.

[5] T. Chen and R. R. Rao. Audio-visual integration in multimodal communication. *Proc. IEEE*, 86(5): 837–852, 1998.

[6] S. Oviatt, P. Cohen, L. Wu, J. Vergo, L. Duncan, B. Suhm, J. Bers, T. Holzman, T. Winograd, J. Landay, J. Larson, and D. Ferro. Designing the user interface for multimodal speech and pen-based

gesture applications: State-of-the-art systems and research directions. *Hum. Comput. Interact.* 15(4): 263–322, 2000.

[7] J. Schroeter, J. Ostermann, H. P. Graf, M. Beutnagel, E. Cosatto, A. Syrdal, A. Conkie, and Y. Stylianou. Multimodal speech synthesis. In *Proc. Int. Conf. Multimedia Expo*, 571–574. New York, NY, July 30–August 2, 2000.

[8] C. C. Chibelushi, F. Deravi, and J. S. D. Mason. A review of speech-based bimodal recognition. *IEEE Trans. Multimedia*, 4(1):23–37, 2002.

[9] D. G. Stork and M. E. Hennecke, editors. *Speechreading by Humans and Machines.* Berlin, Germany: Springer, 1996.

[10] R. Campbell, B. Dodd, and D. Burnham, editors. *Hearing by Eye II: Advances in the Psychology of Speechreading and Auditory-Visual Speech.* Hove, United Kingdom: Psychology Press Ltd., 1998.

[11] W. H. Sumby and I. Pollack. Visual contribution to speech intelligibility in noise. *J. Acoustical Soc. Am.*, 26(2):212–215, 1954.

[12] H. McGurk and J. MacDonald. Hearing lips and seeing voices. *Nature*, 264:746–748, 1976.

[13] M. Marschark, D. LePoutre, and L. Bement. Mouth movement and signed communication. In R. Campbell, B. Dodd, and D. Burnham, editors. *Hearing by Eye II: Advances in the Psychology of Speechreading and Auditory-Visual Speech*, ch. 13, 245–266. Hove, United Kingdom: Psychology Press Ltd., 1998.

[14] A. Q. Summerfield. Some preliminaries to a comprehensive account of audio-visual speech perception. In R. Campbell and B. Dodd, editors. *Hearing by Eye: The Psychology of Lip-Reading*, 3–51. London, United Kingdom: Lawrence Erlbaum Associates, 1987.

[15] H. Yehia, P. Rubin, and E. Vatikiotis-Bateson. Quantitative association of vocal-tract and facial behavior. *Speech Commun.*, 26(1–2):23–43, 1998.

[16] D. W. Massaro and D. G. Stork. Speech recognition and sensory integration. *Am. Sci.* 86(3):236–244, 1998.

[17] J. P. Barker and F. Berthommier. Estimation of speech acoustics from visual speech features: A comparison of linear and non-linear models. In *Proc. Conf. Audio-Visual Speech Processing*, 112–117. Santa Cruz, CA, August 7–9, 1999.

[18] L. Girin, J.-L. Schwartz, and G. Feng. Audio-visual enhancement of speech in noise. *J. Acoust. Soc. Am.*, 109(6):3007–3020, 2001.

[19] S. Deligne, G. Potamianos, and C. Neti. Audio-visual speech enhancement with AVCDCN (audio-visual codebook dependent cepstral normalization). In *Proc. Int. Conf. Spoken Lang. Process.*, 1449–1452. Denver, CO, September 16–20, 2002.

[20] E. Petajan. Automatic lipreading to enhance speech recognition. Ph.D. dissertation, University of Illinois at Urbana-Champaign, Urbana, IL, 1984.

[21] M. E. Hennecke, D. G. Stork, and K. V. Prasad. Visionary speech: Looking ahead to practical speechreading systems. In D. G. Stork and M. E. Hennecke, editors. *Speechreading by Humans and Machines*, 331–349. Berlin, Germany: Springer, 1996.

[22] C. Bregler and Y. Konig. 'Eigenlips' for robust speech recognition. In *Proc. Int. Conf. Acoust., Speech, Signal Process.*, Adelaide, Australia, 669–672. April 19–22, 1994.

[23] P. Duchnowski, U. Meier, and A. Waibel. See me, hear me: Integrating automatic speech recognition and lip-reading. In *Proc. Int. Conf. Spoken Lang. Process.*, Yokohama, Japan, 547–550. September 18–22, 1994.

[24] G. Potamianos, H. P. Graf, and E. Cosatto. An image transform approach for HMM based automatic lipreading. In *Proc. Int. Conf. Image Process.*, vol. 1, Chicago, IL, 173–177. October 4–7, 1998.

[25] S. Nakamura, H. Ito, and K. Shikano. Stream weight optimization of speech and lip image sequence for audio-visual speech recognition. In *Proc. Int. Conf. Spoken Lang. Process.*, vol. 3, Beijing, China, 20–23. October 16–20, 2000.

[26] A. V. Nefian, L. Liang, X. Pi, X. Liu, and K. Murphy. Dynamic Bayesian networks for audio-visual speech recognition. *EURASIP J. Appl. Signal Process.*, 2002 (11):1274–1288, 2002.

[27] M. S. Gray, J. R. Movellan, and T. J. Sejnowski. Dynamic features for visual speech-reading: A systematic comparison. In M. C. Mozer, M. I. Jordan, and T. Petsche, editors. *Advances in Neural Information Processing Systems*, vol. 9, 751–757. Cambridge, MA: MIT Press, 1997.

[28] G. Potamianos and H. P. Graf. Discriminative training of HMM stream exponents for audio-visual speech recognition. In *Proc. Int. Conf. Acoust., Speech, Signal Process.*, 3733–3736. Seattle, WA, May 12–15, 1998.

[29] I. Matthews, G. Potamianos, C. Neti, and J. Luettin. A comparison of model and transform-based visual features for audio-visual LVCSR. In *Proc. Int. Conf. Multimedia Expo*, Tokyo, Japan, August 22–25, 2001.

[30] G. Chiou and J.-N. Hwang. Lipreading from color video. *IEEE Trans. Image Process.*, 6(8): 1192–1195, 1997.

[31] S. Dupont and J. Luettin. Audio-visual speech modeling for continuous speech recognition. *IEEE Trans. Multimedia*, 2(3):141–151, 2000.

[32] A. Rogozan and P. Deléglise. Adaptive fusion of acoustic and visual sources for automatic speech recognition. *Speech Commun.*, 26(1–2):149–161, 1998.

[33] M. Heckmann, F. Berthommier, and K. Kroschel. Noise adaptive stream weighting in audio-visual speech recognition. *EURASIP J. Appl. Signal Process.*, 2002(11):1260–1273, 2002.

[34] A. Adjoudani and C. Benoît. On the integration of auditory and visual parameters in an HMM-based ASR. In D. G. Stork and M. E. Hennecke, editors. *Speechreading by Humans and Machines*, 461–471. Berlin, Germany: Springer, 1996.

[35] P. Teissier, J. Robert-Ribes, and J. L. Schwartz. Comparing models for audiovisual fusion in a noisy-vowel recognition task. *IEEE Trans. Speech Audio Process.*, 7(6):629–642, 1999.

[36] S. Gurbuz, Z. Tufekci, E. Patterson, and J. N. Gowdy. Application of affine-invariant Fourier descriptors to lipreading for audio-visual speech recognition. In *Proc. Int. Conf. Acoust., Speech, Signal Process.*, 177–180. Salt Lake City, UT, May 7–11, 2001.

[37] A. J. Goldschen, O. N. Garcia, and E. D. Petajan. Rationale for phoneme-viseme mapping and feature selection in visual speech recognition. In D. G. Stork and M. E. Hennecke, editors. *Speechreading by Humans and Machines*, 505–515. Berlin, Germany: Springer, 1996.

[38] X. Zhang, C. C. Broun, R. M. Mersereau, and M. Clements. Automatic speechreading with applications to human-computer interfaces. *EURASIP J. Appl. Signal Process.*, 2002(11):1228–1247, 2002.

[39] P. L. Silsbee and A. C. Bovik. Computer lipreading for improved accuracy in automatic speech recognition. *IEEE Trans. Speech Audio Process.*, 4(5):337–351, 1996.

[40] P. S. Aleksic, J. J. Williams, Z. Wu, and A. K. Katsaggelos. Audio-visual speech recognition using MPEG-4 compliant visual features. *EURASIP J. Appl. Signal Process.*, 2002(11):1213–1227, 2002.

[41] D. Chandramohan and P. L. Silsbee. A multiple deformable template approach for visual speech recognition. In *Proc. Int. Conf. Spoken Lang. Process.*, 50–53. Philadelphia, PA, October 3–6, 1996.

[42] S. M. Chu and T. S. Huang. Audio-visual speech modeling using coupled hidden Markov models. In *Proc. Int. Conf. Acoust., Speech, Signal Process.*, 2009–2012. Orlando, FL, May 13–17, 2002.

[43] M. Gordan, C. Kotropoulos, and I. Pitas. A support vector machine-based dynamic network for visual speech recognition applications. *EURASIP J. Appl. Signal Process.*, 2002(11):1248–1259, 2002.

[44] G. Potamianos, C. Neti, G. Gravier, A. Garg, and A. W. Senior. Recent advances in the automatic recognition of audiovisual speech. *Proc. IEEE*, 91(9):1306–1326, 2003.

[45] P. S. Aleksic and A. K. Katsaggelos. Comparison of low- and high-level visual features for audio-visual continuous automatic speech recognition. In *Proc. Int. Conf. Acoust., Speech, Signal Process.*, vol. 5, 917–920. Montreal, Canada, May 17–21, 2004.

[46] G. Potamianos, C. Neti, J. Huang, J. H. Connell, S. Chu, V. Libal, E. Marcheret, N. Haas, and J. Jiang. Towards practical deployment of audio-visual speech recognition. In *Proc. Int. Conf. Acoust., Speech, Signal Process.*, vol. 3, 777–780. Montreal, Canada, May 17–21, 2004.

[47] T. Chen. Audiovisual speech Processing Lip reading and lip synchronization *IEEE Signal Process. Mag.*, 18(1):9–21, 2001.

[48] J. J. Williams, A. K. Katsaggelos, and D. C. Garstecki. Subjective analysis of an HMM-based visual speech synthesizer. In *Proc. SPIE Conf. Hum. Vis. Electronic Imag.*, vol. 4299, 544–555. San Jose, CA, January 21–25, 2001.

[49] M. M. Cohen and D. W. Massaro. Modeling coarticulation in synthetic visual speech. In M. Magnenat-Thalmann and D. Thalmann, editors. *Models and Techniques in Computer Animation*, 141–155. Tokyo, Japan: Springer-Verlag, 1993.

[50] J. J. Williams and A. K. Katsaggelos. An HMM-based speech-to-video synthesizer. *IEEE Trans. Neural Networks*, 13(4):900–915, 2002.

[51] S. Morishima and H. Harashima. A media conversion from speech to facial image for intelligent man-machine interface. *IEEE J. Select. Areas Commun.*, 9(4):594–600, 1991.

[52] P. S. Aleksic and A. K. Katsaggelos. Speech-to-video synthesis using MPEG-4 compliant visual features. *IEEE Trans. Circuits Syst. Video Technol.*, 14(5):682–692, 2004.

[53] C. Bregler, M. Covell, and M. Slaney. Video rewrite: Driving visual speech with audio. In *Proc. Int. Conf. Computer Graphics Interact. Techniques*, 353–360. Los Angeles, CA, August 3–8, 1997.

[54] E. Cosatto and H. P. Graf. Photo-realistic talking-heads from image samples. *IEEE Trans. Multimedia*, 2(3):152–163, 2000.

[55] T. Ezzat and T. Poggio. MikeTalk: A talking facial display based on morphing visemes. In *Proc. Comput Animation*, Philadelphia, PA, 96–98. June 1998.

[56] C. Pelachaud, N. Badler, and M. Steedman. Linguistic issues in facial animation. In N. Magnenat-Thalmann and D. Thalmann, editors. *Computer Animation*, 15–30. Berlin, Germany: Springer-Verlag, 1991.

[57] A. P. Breen, E. Bowers, and W. Welsh. An investigation into the generation of mouth shapes for a talking head. In *Proc. Int. Conf. Spoken Lang. Process.*, vol. 4, 2159–2162. Philadelphia, PA, October 3–6, 1996.

[58] F. I. Parke. Parameterized models for facial animation. *IEEE Comput. Graph. Appl.*, 2(9):61–68, 1982.

[59] S. Pasquariello and C. Pelachaud. Greta: A simple facial animation engine. In *6th Online World Conference on Soft Comput. Ind. Appl.*, September 2001.

[60] D. W. Massaro and M. M. Cohen. Perception of synthesized audible and visible speech. In *Psychol. Sci.*, 1:55–63. 1990.

[61] F. Lavagetto. Time-delay neural networks for estimating lip movements from speech analysis: A useful tool in audio/video synchronization. *IEEE Trans. Circuits Syst. Video Technol.*, 7(5): 786–800, 1997.

[62] A. D. Simons and S. J. Cox. Generation of mouthshapes for a synthetic talking head. *Proc. Inst. Acoustics*, 12:475–482, 1990.

[63] U. Bub, M. Hunke, and A. Waibel. Knowing who to listen to in speech recognition: Visually guided beamforming. In *Proc. Int. Conf. Acoust., Speech, Signal Process.*, 848–851. Detroit, MI, May 9–12, 1995.

[64] C. Wang and M. S. Brandstein. Multi-source face tracking with audio and visual data. In *Proc. Works. Multimedia Signal Process.*, 475–481. Copenhagen, Denmark, September 13–15, 1999.

[65] D. N. Zotkin, R. Duraiswami, and L. S. Davis. Joint audio-visual tracking using particle filters. *EURASIP J. Appl. Signal Process.*, 2002(11):1154–1164, 2002.

[66] G. Iyengar, H. J. Nock, and C. Neti. Audio-visual synchrony for detection of monologues in video archives. In *Proc. Int. Conf. Acoust., Speech, Signal Process.*, vol. 5, 772–775. Hong Kong, China, 2003.

[67] P. De Cuetos, C. Neti, and A. Senior. Audio-visual intent to speak detection for human computer interaction. In *Proc. Int. Conf. Acoust., Speech, Signal Process.*, 1325–1328. Istanbul, Turkey, June 5–9, 2000.

[68] J. Luettin. Speaker verification experiments on the XM2VTS database. IDIAP Research Institute, Martigny, Switzerland, Research Report 99-02, January 1999.

[69] B. Maison, C. Neti, and A. Senior. Audio-visual speaker recognition for broadcast news: some fusion techniques. In *Proc. Works. Multimedia Signal Process.*, 161–167. Copenhagen, Denmark, September 13–15, 1999.

[70] T. Wark, S. Sridharan, and V. Chandran. Robust speaker verification via fusion of speech and lip modalities. In *Proc. Int. Conf. Acoust., Speech, Signal Process.*, 3061–3064. Phoenix, AZ, March 15–19, 1999.

[71] C. Sanderson and K. K. Paliwal. Information fusion and person verification using speech and face information. IDIAP Research Institute, Martigny, Switzerland, Research Report 02-33, September 2002.

[72] U. V. Chaudhari, G. N. Ramaswamy, G. Potamianos, and C. Neti. Information fusion and decision cascading for audio-visual speaker recognition based on time-varying stream reliability prediction. In *Proc. Int. Conf. Multimedia Expo*, vol. 3, 9–12. Baltimore, MD, July 6–9, 2003.

[73] P. S. Aleksic and A. K. Katsaggelos. An audio-visual person identification and verification system using FAPs as visual features. In *Proc. Works. Multimedia User Authentication*, 80–84. Santa Barbara, CA, December 11–12, 2003.

[74] U. Dieckmann, P. Plankensteiner, and T. Wagner. SESAM: A biometric person identification system using sensor fusion. *Pattern Recognit. Lett.*, 18(9):827–833, 1997.

[75] P. Jourlin, J. Luettin, D. Genoud, and H. Wassner. Acoustic-labial speaker verification. *Pattern Recognit. Lett.*, 18(9):853–858, 1997.

[76] R. Brunelli and D. Falavigna. Person identification using multiple cues. *IEEE Trans. Pattern Anal. Machine Intell.*, 17(10):955–966, 1995.

[77] I. Cohen, N. Sebe, A. Garg, L. S. Chen, and T. S. Huang. Facial expression recognition from video sequences: temporal and static modeling. *Comput. Vis. Image Understand.*, 91(1–2):160–187, 2003.

[78] P. S. Aleksic and A. K. Katsaggelos. Automatic facial expression recognition using facial animation parameters. *IEEE Trans. Inf Forensics Secur*, 1(1):3–11, 2006.

[79] A. W. Senior. Face and feature finding for a face recognition system. In *Proc. Int. Conf. Audio Video-based Biometric Person Authentication*, 154–159. Washington, DC, March 22–23, 1999.

[80] H. A. Rowley, S. Baluja, and T. Kanade. Neural network-based face detection. *IEEE Trans. Pattern Anal. Machine Intell.*, 20(1):23–38, 1998.

[81] K.-K. Sung and T. Poggio. Example-based learning for view-based human face detection. *IEEE Trans. Pattern Anal. Machine Intell.*, 20(1):39–51, 1998.

[82] P. Viola and M. Jones. Rapid object detection using a boosted cascade of simple features. In *Proc. Conf. Computer Vision Pattern Recognit.*, 511–518. Kauai, HI, December 11–13, 2001.

[83] H. P. Graf, E. Cosatto, and G. Potamianos. Robust recognition of faces and facial features with a multi-modal system. In *Proc. Int. Conf. Systems, Man, Cybernetics*, 2034–2039. Orlando, FL, 1997.

[84] M. Kass, A. Witkin, and D. Terzopoulos. Snakes: Active contour models. *Int. J. Comput. Vis.*, 4(4):321–331, 1988.

[85] A. L. Yuille, P. W. Hallinan, and D. S. Cohen. Feature extraction from faces using deformable templates. *Int. J. Comput. Vis.*, 8(2):99–111, 1992.

[86] T. F. Cootes, G. J. Edwards, and C. J. Taylor. Active appearance models. In *Proc. Europ. Conf. Comput. Vis.*, 484–498. Freiburg, Germany, 1998.

[87] S. Young, D. Kershaw, J. Odell, D. Ollason, V. Valtchev, and P. Woodland. *The HTK Book.* United Kingdom: Entropic Ltd., 1999.

[88] I. S. Pandzic and R. Forchheimer, editors. *MPEG-4 Facial Animation: The Standard, Implementation and Applications.* United Kingdom: John Wiley and Sons, 2002.

[89] J. R. Deller, Jr., J. G. Proakis, and J. H. L. Hansen. *Discrete-Time Processing of Speech Signals.* Englewood Cliffs, NJ: Macmillan Publishing Company, 1993.

[90] K. P. Murphy. Dynamic Bayesian networks: Representation, inference, and learning. Ph.D. dissertation, University of California, Berkeley, CA, 2002.

[91] J. A. Bilmes and C. Bartels. Graphical model architectures for speech recognition. *Signal Process. Mag. IEEE*, 22(5):89–100, 2005.

[92] G. G. Zweig. Speech recognition with dynamic bayesian networks. Ph.D. dissertation, University of California, Berkeley, CA, 1998.

[93] K. Saenko and K. Livescu. An asynchronous DBN for audio visual speech recognition. In *Proc. Spoken Lang. Techn. Works.*, December 2006.

[94] G. Lv, D. Jiang, R. Zhao, and Y. Hou. Multi-stream asynchrony modeling for audio-visual speech recognition. In *Proc. Int. Symp. Multimedia*, 37–44, 2007.

[95] L. H. Terry and A. K. Katsaggelos. A phone-viseme dynamic Bayesian network for audio-visual automatic speech recognition. In *Proc. Int. Conf. Pattern Recognit.*, December 2008.

[96] A. K. Jain, R. P. W. Duin, and J. Mao. Statistical pattern recognition: A review *IEEE Trans. Pattern Anal. Mach. Intell.*, 22(1):4–37, 2000.

[97] L. E. Bernstein. Lipreading corpus V-VI: Disc 3. Gallaudet University, Washington, D.C., 1991.

[98] A. Löfqvist. Speech as audible gestures. In W. J. Hardcastle and A. Marchal, editors *Speech Production and Speech Modelling*, 289–322. Dordrecht, The Netherlands: Kluwer, 1990.

[99] G. A. Abrantes. FACE - Facial Animation System, version 3.3.1. Instituto Superior Tecnico, 1997–98.

[100] T. Dutoit. *An Introduction to Text-To-Speech Synthesis.* Dordrecht, The Netherlands: Kluwer Academic Pub., 1997.

[101] A. K. Jain, A. Ross, and S. Prabhakar. An introduction to biometric recognition. *IEEE Trans. Circuits Syst. Video Technol.*, 14(1):4–20, 2004.

[102] P. S. Aleksic and A. K. Katsaggelos. Audio-visual biometrics. *Proc. IEEE*, 94(11):2025–2044, 2006.

[103] J. D. Woodward. Biometrics: privacy's foe or privacy's friend? *Proc. IEEE*, 85(9):1480–1492, 1997.

[104] F. Cardinaux, C. Sanderson, and S. Bengio. User authentication via adapted statistical models of face images. *IEEE Trans. Signal Process.*, 54(1):361–373, 2006.

[105] S. Bengio and J. Mariethoz. The expected performance curve: a new assessment measure for person authentication. In *Speaker Lang. Recognit. Works. (Odyssey)*, 279–284. Toledo, Spain, 2004.

[106] C. Sanderson and K. K. Paliwal. Identity verification using speech and face information. *Digital Signal Process.*, 14(5):449–480, 2004.

[107] J. P. Campbell. Speaker recognition: A tutorial. *Proc. IEEE*, 85(9):1437–1462, 1997.

[108] W.-Y. Zhao, R. Chellappa, P. J. Phillips, and A. Rosenfeld. Face recognition: A literature survey. *ACM Computing Survey*, 399–458, 2003.

[109] J. Luettin, N. A. Thacker, and S. W. Beet. Speaker identification by lipreading. In *Proc. Int. Conf. Spoken Lang. Process.*, vol. 1, 62–65. Philadelphia, PA, October 3–6, 1996.

[110] P. Ekman and W. Friesen. *Facial Action Coding System.* Palo Alto, CA: Consulting Psychologists Press Inc., 2003.

Index

Page numbers followed by "f" indicate figures and "t" indicate tables.

A

AAL. *see* ATM adaptation layer
Absolute temporal information (ATI) filter, 424
Active appearance models (AAMs), 694
Active learning, efficient annotation through, 466–467
Active shape models (ASMs), 694
Activity-dominant vector selection (ADVS), 381
Adaptive 5/3 MCTF, 355, 356f, 356t
Adaptive background modeling, 630
Adaptive filtering, 83
Adaptive intra refresh (AIR), 385
Advanced mobile phone service (AMPS), 488
Advanced video coding (AVC), 8
Aliasing, 17, 102, 241, 243
Analog-to-digital (A/D) conversion, 3
Analysis-by-synthesis, 670
ANNs. *see* Artificial neural networks
API. *see* Application programming interface
Apparent motion, 31, 46
Application programming interface (API)
  high-level APIs, 401–402
  low-level kernels, 402–404
Arbitrary slice order (ASO), 311
Arithmetic coding, 530, 534
ARMA model. *see* Autoregressive and moving averaging model
ARPANET, 474
ARQ. *see* Automatic repeat request
Articulated pendulum model, 645
Artifact reduction
  in frequency domain, 83–84
  in spatial domain, 83
Artificial neural networks (ANNs) in speech-to-video synthesis, 719
ASMs. *see* Active shape models
ASR. *see* Automatic speech recognition
Asymmetric DSL (ADSL), 488
Asymptotic behaviors, 670
Asynchronous transfer mode networks. *see* ATM networks
ATM adaptation layer (AAL)
  AAL-1
    convergence sublayer, 495–496
    segmentation and reassembly sublayer, 495, 496f
  AAL-5
    common part convergence sublayer, 497–499, 498f
    convergence sublayer, 497
    segmentation and reassembly sublayer, 498
  service classes, 494, 495t
  structure of, 494, 494f
ATM networks, 492–495
  AAL-1, 495–497
  AAL-5, 497–499
Audio marker, 440, 460
Audio track, 456–457
Audiovisual biometrics. *see* Audiovisual speaker recognition
Audiovisual information fusion
  classifiers in speech applications, 702–704
  feature and classifier fusion in, 704–707
  speech classes in, 700–702
Audiovisual objects (AVOs), 298–299, 477–478
Audiovisual scene in MPEG-4, 298, 299f
Audiovisual speaker recognition, 723–729
  characteristics of, 724f
Audiovisual speech synthesis, 712–723
  coarticulation modeling in, 713–714
  facial animation in, 714–717
  speech-to-video, 718–721
  visual speech evaluation in, 721–723
  VTTS in, 717–718
Authentication, 545
  content level, 546–550
  stream level, 541–553
Automatic motion segmentation, examples for, 165–169
Automatic repeat request (ARQ), 501, 589–590
Automatic speech recognition (ASR)
  bimodal corpora for, 708–709
  experimental results, 709–712
Autoregressive and moving averaging (ARMA) model, 30–31

B

Background scenes
  Gaussian distribution on, 630–632, 634f
  statistical model for, 630
Bandwidth, 485, 502

Bayes theorem, 156
Bayesian estimation, 667–668
Bayesian object tracking, 197–198
in Kalman filters, 198–202
in particle filters, 202–203
B-frame coded video data, 551
Bilinear interpolation, 381
Bimodal distribution, 633
Binary hypothesis testing, 32–33
Binary phase-shift keying (BPSK), 585
Binary symmetric channel (BSC), 599
Bit allocation model, 384
Bit error rate (BER), 573
Bit rate control
in video encoding, 383
in video transcoding, 383–384
Bit rate reduction, video transcoding for, 369–370, 370f, 380
Black frame, 450
Blind VQA, 418
Block layer in H.261 video encoder, 259, 262
Block transform coding, 248–250
Block-based motion compensated (BMC) standards, 582
Block-based video coding, 391–392
Block-matching algorithms (BMA), 59–61, 60f, 254
estimation criterion, 59–60
search methods, 60–61
Blotch detection system, 85–88
Blotch detector, pixel-based, 86
Blotch removal system, 84–94
Blueray, 1
Bluetooth, 491
BMA. *see* Block-matching algorithms
Body animation in MPEG-4, 306
Body animation parameters (BAPs), 306
Body definition parameters (BDPs), 306
Broadband ISDN (B-ISDN), 492
ATM reference model, 493, 493f

C

C6000™ DSP core, 398
CABAC. *see* Context-adaptive binary arithmetic coding
Cable television systems, 486–487
Camera model, 113–114
Camera motion, 447–448, 449f
effect of, 114–115
Candidate domains for multimedia encryption, 529–530, 529f
CANDIDE face model, 207
Captions in video, 453–455, 453f
CAVLC. *see* Context-adaptive variable-length coding
CCDs. *see* Charge coupled devices
CDMA, 488–489
CDMA-2000, 490
Change detection, 626, 629–634, 635t
Channel coding, 589–593
forward error correction (FEC), 574, 590–592, 591f
retransmissions, 592–593
Channel models, 584–585
internet, 584
wireless channel, 585
Channel state information (CSI), 574
Charge coupled devices (CCDs), 242, 243
Chroma-keying, 164
Chrominance, 422–423
CIF. *see* Common Interchange Format
Cluster detection, 454
Coarse-grained scalability (CGS) approach, 584
Coaxial cables, 486
Coded block pattern (CBP) parameter in H.261 video encoder, 264
Coded block shuffling (CBS), 540
Coded fingerprinting, 556, 557f
parameters, 558–559
Code-division multiple access. *see* CDMA
Coding artifact reduction, 82–84
Coding efficiency, 232
Collusion attack, 555–556, 564
Collusion resistance
analysis on, 559
boosting, 561–562
Color segmentation algorithm, 181–182
Common Interchange Format (CIF), 244
Communication networks, video, 485–486
ATM, 492–499
based on packet switching, 474
digital subscriber loop, 487–488
fiber optics, 491–492
hybrid fiber-coax, 486–487
ISDN, 492
quality of services (QoS) for, 521, 523
wireless networks, 488–491
Communication system, 575f
Compressed-domain features, 450–452
Compression ratio, 232
Compression stream, 479
Computer processing unit (CPU), 398
concurrent processing between DMA and, 407–408, 407f

Conferencing, video, 327
Content level authentication, 546–550
Content-based features
audio and language, 456–457
embedded video features, 458–459
object detection, 452
articulated objects, 456
captions and graphics, 453–456
human subjects, 452–453
potential query applications with, 465t
rule-based features
adjacent similar scenes, 457–458
introduction scenes, 457
short successive scenes, 458
Content-based video retrieval, 464–465
Context-adaptive binary arithmetic coding (CABAC), 320, 338
Context-adaptive variable-length coding (CAVLC), 320
Context-free-grammar (CFG)-based method, 647
Continuous time-varying imagery, 16–20, 16f, 20f
Continuous-valued video, 3
Contrast masking property, 427–428
Conversion
frame-rate, 22
sampling structure, 20, 26–27
Coring
definition of, 79
functions, 80f
C-reference code, 411, 413
Cryptographic signature schemes, 546f
Cumulative number of packets lost (CNPL), 514

D

2D face detector, 206
Data partitioning (DP), 396
DBNs. *see* Dynamic Bayesian networks
DCF. *see* Detection cost function
DCT. *see* Discrete cosine transform
Deblocking filter, 317–318, 317f
Delay jitter, 494
Delayed frame memory
in intraframe encoding mode, 235
motion compensation (MC), 236
motion estimation, 235–236
Dense motion, 63–64
DET curve. *see* Detection error trade-off curve
Detection cost function (DCF), 727
Detection error trade-off (DET) curve, 727
DFD. *see* Displaced frame difference
DFT. *see* Discrete Fourier transform
Differential pulse code modulation (DPCM), 333
coding method, 307
predictive encoder, 247, 248
Differentiated services (DiffServ), 523–525
assured forwarding, 524
expedited forwarding, 523–524
scalability of, 524
traffic conditioning functionality, 524f
Digital fingerprint, 553
Digital multimedia broadcasting (DMB), 608
Digital multimedia products, handheld and portable, 389, 390f
Digital signal processing (DSP), 8
Digital still images, 2
Digital storage media control commands (DSM-CC), 278
Digital subscriber loop (DSL), 487–488
Digital video, 1
applications and technology, 2, 232
processing, 2
signals and formats, 241–245, 242f, 244t, 245t
Digital Video Broadcasting-Handheld (DVB-H), 608, 609
Digital video quality (DVQ) metric, 421
Digital video research field, terminology used in, 440–442
Digital video streams, 6
Digital video transcoding. *see* Video transcoding
Direct memory access (DMA), 393, 407
concurrent processing between CPU and, 407–408, 407f
Direct-sequence spread-spectrum (DSSS), 489
Discrete cosine transform (DCT), 82, 84, 234, 248–250, 376–377, 421, 439, 451, 581
DFT *versus*, 250f
gross quantization of, 251f
requantization of, 371, 371f
domain
inverse motion compensation (IMC), 375–377, 376f
video transcoding, 375, 375f, 378, 378f, 381
Discrete Fourier transform (DFT), 53, 249, 250f
Discrete memoryless source (DMS), 246
Discrete multitone (DMT), 487
Discrete wavelet transform (DWT), 79, 584
Discrete-valued video, 3
Displaced frame differences (DFD), 335, 581
Distortions
spatial, 425–426, 428
temporal, 425–426, 428

DPCM. *see* Differential pulse code modulation
Drift, 372
Drift-free video transcoding, 372, 375
DSL. *see* Digital subscriber loop
DVB-H. *see* Digital Video Broadcasting-Handheld
DVQ metric. *see* Digital video quality metric
Dynamic Bayesian networks (DBNs), 703
Dynamic texture, 633
Dynamic time warping, 644
Dynamically coupled oscillator model, 646

E

ECCs. *see* Error correction codes
Edge profile, example of, 318f
Edge similarity score (ESS), 537–539
Eigenfaces, sample, 194, 194f
EKF. *see* Extended Kalman filters
Embedded video codec
  coding quality and computational complexity of, 394
  design flow
    chip architecture, 397–400
    codec algorithms, 400–401
    concurrent processing, 407–408
    Golden C implementation, 404–405
    kernel optimization and integration, 406–407
    modularity and APIs definition, 401–404, 401f
    overall optimization, 408–410
    platform-specific development and porting, 405–406
    stress and conformance testing, 410–411
  in handheld and portable multimedia products, 390f
  mode-decision process, 395
  motion vector estimation in, 394–395
  rate control, 395
  rate-distortion (R-D) optimization, 396
  requirements and constraints, 393–397
Embedded video features, 458–459
Embedded zero block coder (EZBC), 331, 332f, 348
  coding process, 349–351
  context modeling, 351–352, 352f
  definitions, 349–350
  frequency roll-off, 353–355, 355f
  packetization and, 353, 353f
  pseudo-code, 351
  and scalability, 352–353
Embedded zero-tree wavelet (EZW), 535
Emission probabilities, single-stream, 707f
Emission probability model, 702
Encrypting multimedia
  before and after coding, 529–530
  candidate domains for, 529–530, 529f
  intermediate stages of coding, 530
Encryption, 527
  and attack settings, 541t
  index mapping, 532–534, 534f
  media, 530–539
  system design, 539, 540f
    base-layer, 540
    system setup, 540
Enhanced Data rates for GSM Evolution (EDGE), 489
Entropy coding, 235, 246–248, 530–532
  in H.264/AVC, 320
Entropy encoders, 246, 247
Equal error protection (EEP), 609
Error correction codes (ECCs), 239, 259, 556
  based fingerprinting system, 559–560
Error-resilience
  in MPEG-2, 289
  in MPEG-4, 308
  source coding, 587–589
  techniques, 573
ESS. *see* Edge similarity score
Euler-Lagrange equations, 35, 42, 62
Expectation maximization (EM) algorithm, 632
Extended Kalman filters (EKF), 132, 198–202
EZBC. *see* Embedded zero block coder

F

Face detection, 692–694
Face recognition (FR), 648
  framework for, 683–684
  from group of still images, 670–675
  rates of, 675f, 675t
  still-to-still images, 682
  from video sequences, 653, 675–682
Face tracking technique, template-based, 196
Facial animation
  in audiovisual speech synthesis, 714–717
  control points of, 697f
  in MPEG-4, 305–306, 306f, 715f
Facial animation parameters (FAPs), 305, 696
Facial definition parameters (FDPs), 305
False acceptance rate (FAR), 727
False detection probability, 88
  correct detection *versus,* 89f
False rejection rate (FRR), 727

FAPs. *see* Facial animation parameters
FAR. *see* False acceptance rate
Fast video-transcoding architectures, 372–375
  DCT domain video transcoder, 375, 375f
  pixel domain open-loop transcoder, 372, 373f
  pixel domain video transcoder, 373f
FDMA, 488–489
Feature, image and video, 442
  compressed-domain, 450–452
  content-based. *see* Content-based features
  statistical. *see* Statistical features
Feature based methods
  full reference VQA system, 422–425
  video retrieval, 464
Feature points, 165
FEC. *see* Forward error correction
Feedback control signaling (FCS) methods, 385
FGS. *see* Fine granularity scalability
Fiber optics, 491–492
Fidelity range extensions (FRExt), 296
Fine granularity scalability (FGS), 535, 587
  coding, 307
  enhancement layer video, protecting, 544–545
Finite impulse response (FIR) filter, 421
Finite-state machine, 644
Flexible macroblock ordering (FMO), 587, 588–589
Flicker parameter
  estimation of, 98–99
  smoothing and interpolating, 100
Forward dominant vector selection (FDVS), 381
Forward error correction (FEC), 548
  channel coding, 574, 590–592, 591f
  joint source coding and, 596–597, 599–600
  in MediaFLO, 610
  Reed-Solomon, 496, 497f
  RTP encapsulation of, 505–506
Fourier transform analysis, spatiotemporal sampling structures, 14
FR. *see* Face recognition
Frame/field adaptive coding in H.264/AVC, 320–322
Frame-rate conversion, 22
Frames, 4
Frequency division duplex (FDD), 490
Frequency division multiple access. *see* FDMA
Frequency domain. *see also* Spatial domain
  artifact reduction in, 83–84
FRR. *see* False rejection rate
Full reference VQA algorithms, 418, 422

G

Gabor filter, 421, 427–429, 428f, 430f
Gabor wavelets, 193
Galois fields (GF), 590
Gaussian distribution, 630–632
Gaussian filter, 443
Gaussian smoothing, 96
General packet radio service (GPRS), 385, 489
Generalized index mapping with controlled overhead (GIMCO), 532–534, 540
Geodesic active contour model, 641
Gibbs distribution, 33, 161
GIMCO. *see* Generalized index mapping with controlled overhead
Global flow models, 118–119
Global motion compensation (GMC), 56
Global systems for mobile communications (GSM), 489
GMC. *see* Global motion compensation
Golden C, 404–405
GOP. *see* Group of pictures
GPRS. *see* General packet radio service
Gradient-based techniques for motion estimation, 54
Graphics detection, 455–456, 456f
Group of blocks (GOB) layer in H.261 video encoder, 260–261
Group of pictures (GOP), 391, 479
  in MPEG-1, 271–273, 271f–273f
GSM. *see* Global systems for mobile communications

H

H.245 call control protocol, 518
H.323 protocol stack, 518–520, 519f
H.264 standard, 369, 392, 478
H.320 standard, 492
H.324 standard, 487
H.261 video encoder, 232, 258–265, 263t
  data hierarchy, 259, 261, 261f
  implementation, 259, 260f
H.26X standards, 476, 518
Haar filters, 332, 335–336
Haar MC-EZBC coder, 334, 343f
Haar MCTF, 333, 338, 342–348
Hammersley-Clifford theorem, 34
H.264/AVC, 333, 581, 583
  block diagram of, 311f
  coding performance, 325
  enhanced motion compensation prediction model, 314–317, 315f, 316f

H.264/AVC (*continued*)
in-loop deblocking filter, 317–318, 317f
multireference picture prediction in, 316f
slices and slice groups, 312–313, 313f
spatial directional intra prediction, 313–314
technical overview, 310–312
transform, quantization, and scanning, 319–320
visual profiles of, 322, 323t
HCI. *see* Human-computer interaction
HDTV. *see* High-definition television
Heterogeneous video transcoding, 377–379
macro-block-coding type decision, 383
MV estimation, 379–381
spatial resolution reduction, 381–382
HFC networks. *see* Hybrid fiber-coax networks
Hidden Markov models (HMMs), 643–644, 646, 662, 703, 706
in speech-to-video synthesis, 720–721
Hierarchical variable-size block matching (HVSBM), 334, 335f, 337f
High-bandwidth communication, 485–486, 488
High-definition television (HDTV), 1
and IMAX formats, 21
picture resolution, 238
terrestrial transmission, 231
High-definition television (HDTV) Grand Alliance standard, 477
Highest confidence first (HCF) algorithm, 42, 55, 64
Highlight candidate, 441, 462
Highlight group, 441, 462
HMMs. *see* Hidden Markov models
Homography, 116, 132
Hough transform methods, 153–154
Hue-saturation components, 181, 184
Huffman coding, 235, 247, 479, 530
Huffman tree mutation process, 530, 531f
Human body
cylinder-based volumetric representation of, 214f
hierarchic representation of, 210f
polygonal representation of, 213f
stick model of, 213f
Human gait recognition, 645–648
language-based approaches, 647
model-based approaches, 645–646
physical parameter-based approaches, 647
spatiotemporal motion-based approaches, 646
statistical approaches, 646–647
Human visual system (HVS), 235, 243
modeling based methods, 419–421
IQA systems, 420–421, 420f
temporal mechanisms in, 421
VQA systems, 420–421, 421f
motion processing in, 426
and video compression techniques, 237
Human-computer interaction (HCI), 689
HVS. *see* Human visual system
HVSBM. *see* Hierarchical variable-size block matching
Hybrid block-based motion-compensated (HBMC), 581, 582f
Hybrid coders, 333, 357
Hybrid fiber-coax (HFC) networks, 486–487
Hypothesis testing, 32–33
with adaptive threshold, 38–41
fixed-threshold, 36–38
Hysteresis thresholding, 88

I

IBS. *see* Intra bitplane shuffling
ICMs. *see* Iterated conditional modes
Identity equation, 676
Identity signature
continuous, 668
discrete, 668
IEC. *see* International Electrotechnical Commission
IEEE 802.11 standard, 490–491
IEEE 802.15 standard, 491
IEEE 802.16 standard, 491
IGMP. *see* Internet group management protocol
Image features, 115–116
Image motion parameters, 130
Image quality assessment (IQA), HVS-based, 420–421, 420f
Image sequences, 12
estimation of motion on, 99–101
mean of corrupted and corrected, 101f
variance of corrupted and corrected, 101f
Image stabilization, 122–134
definition of, 122
with IMU, 130–133
with metadata, 128–130
three dimensional, 137
Imagery, time-varying, 11, 13
Imaging geometry of camera model, 114f
IMU. *see* Inertial measurement unit
Independent source coding and power allocation (ISCPA) approach, 600, 601f, 602f
Index mapping encryption, 532–534, 534f

Indexing, 456, 468–469
motion characteristics in, 446
problem in, 464
structural information, 456
Indoor video surveillance, 621, 622f
Inertial measurement unit (IMU), 130
Infinite impulse response (IIR) filters, 421
Insect navigation
control of flight speed, 112
stabilization, 112
Insects
centering behavior of, 111–112
collision avoidance in, 111–112
Integrated circuit (IC), 572
Integrated services. *see* Resource reservation protocol
Integrated Services Digital Network (ISDN), 6, 258, 492
Intensity flicker, 71
correction, 97–101
definition of, 97
parameter estimation, 98–99, 100f
Interarrival jitter, 515
Interframe coding, 479
Interlaced scanning, 4
International Consultative Committee for Radio (CCIR), 244
International Electrotechnical Commission (IEC) standards, 256
International Mobile Communications-2000 (IMT-2000), 489, 490
International Standards Organization (ISO), 233, 256, 475–476
International Telecommunications Union (ITU), 232, 244, 256, 258, 476
Internet, 474, 499–500, 584
transport layer protocols over, 501
video broadcasting, 610
video transmission, 596–598
Internet Engineering Task Force (IETF), 500, 590
Internet group management protocol (IGMP), 503
Internet protocol networks. *see* IP networks
Interpicture coding, 297
Interpicture prediction, 297
Interpolation
motion-compensated, 24–26, 26f, 45f
pure temporal, 23–24, 24f
spatiotemporal, 28
Intra bitplane shuffling (IBS), 535, 536f, 540
Intraframe coding, 479
Intra-personal space (IPS), 679
Intrapicture coding, 297
Intravideo collusion, 566
Inverse discrete cosine transform (IDCT), 249
operation, 374, 375
transformations, 572
Inverse discrete Fourier transform (IDFT), 249
Inverse motion compensation (IMC), 373
DCT domain, 375–377, 376f
Inverse operators in intraframe encoding mode, 235
IP networks, 499–502
DiffServ, 523–525
H.323, 518–520
multicast backbone, 502–503
real-time transport control protocol, 511–517
real-time transport protocol, 503–511
real-time transport streaming protocol, 517–518
resource reservation protocol, 521–523
session initiation protocol, 520–521
IP stack, 501f
IPS. *see* Intra-personal space
IPv4, 500–501
IPv6, 500–501
IQA. *see* Image quality assessment
IS-54, 489
IS-95, 489
IS-136, 489
ISDN. *see* Integrated Services Digital Network
ISO. *see* International Standards Organization
Iterated conditional modes (ICMs), 42, 64
Iterated weighted least squares technique, 121
Iteration algorithm, two-step, 162–164
ITU. *see* International Telecommunications Union

J

Joint Photographic Experts Group standard. *see* JPEG standard
Joint source coding
and data rate adaptation, 603–604
and FEC, 596–597, 599–600
and hybrid FEC/ARQ, 597–598
and power adaptation, 600–601
Joint source coding and transmission power allocation (JSCPA) approach, 600, 601f, 602f
Joint source-channel coding (JSCC), 574, 594
channel models, 584–585
error resilient source coding, 587–589
illustration, 578–579, 578f
internet video transmission, 596–598

Joint source-channel coding (JSCC) (*continued*)
operational rate-distortion theory, 576–577
and power adaptation, 601–603
practical constraints in video communications, 577–578
problem formulation, 594–596
rate-distortion theory, 575
wireless video transmission, 599–604
Joint video team (JVT), 258, 296
JPEG-2000 security (JPSEC) standard, 530
JPEG standard, 235, 476, 507
JSCC. *see* Joint source-channel coding

K

Kalman filters (KF), 179, 638
Bayesian object tracking in, 198–202
extended, 198–202
tracking based on, 638–639
Karhunen-Loeve transform (KLT), 248
Kernel-level optimization, 406–407
Key frame, 440
extraction, 459–460
KF. *see* Kalman filters
Kinescope moiré
phenomenon, 71
removal of, 101–103
K-means method, 151–153
Kullback–Leibler divergence, 533

L

Laplacian density, tracking and recognition using, 677–679
Lattices, 12, 13f, 15f
Layered structure of film, 95f
LCC. *see* Linear correlation coefficient
LeGall and Tabatabai (LGT) 5/3 filters, 355–357
Lempel-Ziv coding, 534
LGT. *see* LeGall and Tabatabai 5/3 filters
Light emitting diodes (LED), 177
Likelihood ratio, 33
Linde-Buzo-Gray (LBG) design, 253
Linear basis function model, 120, 120f
Linear correlation coefficient (LCC), 432
Linear filters, 423
temporally averaging, 72–75
temporally recursive, 75–76
Lip tracking, 692–694
LLMMSE estimator. *see* Local linear minimum mean squared error estimator
Lloyd-Max quantizers, 252
Local area networks (LAN), 485, 488, 490
Local linear minimum mean squared error (LLMMSE) estimator, 76
Local multipoint distribution service (LMDS), 491
Log-likelihood function, 154
Lossy operation, 232
LSS. *see* Luminance similarity score
Lubin model, 421
Luminance, 422–423
Luminance similarity score (LSS), 538

M

Macroblock (MB), 391
in H.261 video encoder, 259, 262
Macroblock-adaptive field/frame coding (MB-AFF), 320
Macroblock-coding type decision, 383
Mahalanobis distance in object tracking, 197
Manifold mosaics, 122
MAP. *see* Maximum a posteriori probability
Markov random field (MRF), 33–34, 41–42, 90
Markov source models, 247, 248
Marr–Hildreth edge detector, 219
Maximum likelihood (ML)
estimation, 34
segmentation, 154–156
color regions based, 158–160
framework, 161–162
performance of, 165–166, 168f, 169f
Maximum a posteriori probability (MAP), 32, 34, 41–42
dense motion, 63–64
segmentation, 156–158
variational formulation, 42–43
MBONE. *see* Multicast backbone
MC-EZBC. *see* Motion-compensated-embedded zero-block coder
MCP. *see* Motion-compensated prediction
MCTF. *see* Motion-compensated temporal filtering
Media encryption
building blocks for, 530–536
constrained shuffling, 534–536
index mapping, 532–534, 534f
randomized entropy coding, 530–532
security evaluation of, 536–539
MediaFLO, 609–610
system architecture, 368–369, 368f
Median filtering, 37
Medium access control (MAC) frames, 592
Medium-grained scalability (MGS), 584
Memory access, 406

Mesh object coding in MPEG-4, 304–305, 305f
Message authentication code (MAC), 545
Millions instruction per second (MIPS), 393, 396
Min-Max attack, 564
MJPEG, 476–477
  RTP header for encapsulation of, 507
    elementary stream (ES), 508–511
    systems stream (SS), 508
ML. *see* Maximum likelihood
MMF. *see* Multistage median filter
Mobile devices, 572
Mobile IP, 490
Mobile TV standards, 608–610
Model consistency check, 122
Model fitting, 122
Model-based coding in MPEG-4, 305–306
Mosaic formation, 123f, 125f
Mosaicing, 127–134
  definition of, 122
  video, 124–127
Motion, notation and preliminaries, 32–35
Motion based video integrity evaluation (MOVIE) index, 427
  Gabor filter, 427–429, 428f
  spatial, 428, 430, 431f
  temporal, 428–430, 431f
Motion blocks, categories of, 342, 343
Motion compensation (MC), 236, 253–256, 581
Motion constraint equation, 49
Motion detection, 31, 32, 35, 44f, 626, 629–634, 635t
  comparison methods of, 43–44
  hypothesis testing. *see* Hypothesis testing
  MAP in, 32, 34, 41–42
  MRF in, 33–34, 41–42
Motion estimation (ME), 25, 31, 44, 235–236, 253–256, 255f, 581, 627
  algorithms, 45, 164, 166
    block matching, 59–61
    dense motion, 63–64
    global, 56–58, 58f
    optical flow via regularization, 62–63
    phase correlation, 61–62
  comparison methods, 64–65
  and compensation for MCTF, 333–335
  connected and unconnected blocks, 335–336, 336f
  criteria for, 51, 52f
    Bayesian, 54
    frequency-domain, 52–53
    pixel-domain, 51–52
    search strategies, 54
  goals of, 45
  gradient-based techniques, 54
  in MC-EZBC, 346–347
  models in, 45–51
    observation models, 49–51
    spatial, 46–47
    temporal, 47
  region of support, 48–49
  regularization, 53
  using chroma for, 337–338
  variational formulations, 34–35
Motion inpainting, 92
Motion JPEG. *see* MJPEG
Motion modeling based methods, 425–431
  Area MT/V5, 426
  for VQA, 427–431
Motion parameter space, clustering in, 151–154
Motion segmentation, 31
  dominant, 147–148
  multiple motions, 150–151
  using two frames, 148–149
Motion super-resolution, 134–137
  definition of, 134
  registration, 135
Motion texture coding in MPEG-4, 301–302
Motion tracking in digital video, 175
Motion transition equation, 676
Motion vector coding
  layered structure of, 347–348
  in MPEG-4, 301
  scalable, 345–348
Motion vector interpolation. *see* Motion vector repair
Motion vector matching, 156
Motion vector repair, 88–91
Motion vectors (MVs), 269, 377, 439, 529
  analysis of, 451
  bit stream for, 345
  estimation
    in embedded video codec, 394–395
    for spatial resolution reduction, 379–380, 379f
    for temporal resolution reduction, 380–381, 380f
  in H.264/AVC, 314
  scan/spatial prediction for, 345–346, 346f
  symbols, alphabet general partition of, 346–347
Motion-compensated interpolation, 24–26, 26f
Motion-compensated prediction (MCP), 581
  bi-directional, 268–269, 268f

Motion-compensated prediction (MCP) (*continued*)
  frame-based and field-based, 283–284, 283f, 284f
  with half-pixel accuracy, 269–270, 270f
  in H.264/AVC, 314–317, 315f, 316f
Motion-compensated subband/wavelet coders, 331
Motion-compensated temporal filtering (MCTF)
  bidirectional, 342, 344f
  improvements to, 338–342
  invertible half-pixel accurate, 339
  lifting implementation, 339–340
  motion estimation and compensation for, 333–342, 334f–337f, 339f
  noninvertible approach to, 338
  with OBMC, 345
  subpixel interpolation, 340–341
Motion-compensated-embedded zero-block coder (MC-EZBC)
  comparison of different interpolation filters, 341–342
  motion estimation in, 346–347
  and overlapped block motion compensation, 344
  rate-distortion curves of, 342, 343f
  structure of, 332f
  visual results with, 357, 359, 359f
Motion-field model, 161–162
Mouth, 692–694
  region-of-interest (ROI), 693
MOVIE index. *see* Motion based video integrity evaluation index
Moving Picture Experts Group. *see* MPEG
Moving pictures quality metric (MPQM), 421
MPEG
  encoding standards, 231
  and ITU-T video encoder/decoder, 392f
  MPEG-1 standard, 231, 235, 257, 477
    background and structure of, 267–268
    bit-stream structures, 276–277, 276f
    compared with H.261, 268–270, 270f, 274–275
    simulation model, 275
    target applications and requirements, 268
  MPEG-2 standard, 231, 247, 257, 333, 473, 476–480
    audio and video, 480f
    background and structure of, 277–278
    compared with MPEG-1, 281–291
    data partitioning, 288–289
    elementary stream (ES), 480, 480f
    frame/field DCT, 284–285, 285f
    input resolutions and formats, 280–281, 281f
    packetized elementary stream (PES), 481, 482f, 483t
    profiles and levels, 278–280, 280t
    program stream (PS), 482
    and system bit stream structures, 290–291
    target applications and requirements, 278–279
    test model developed in, 289–290
    and tools for error-resilience, 289
    transcoded to MPEG-4 bit stream, 377, 378f
    transport stream (TS), 482–484, 483f, 484t, 485f, 498f, 499, 500f
  MPEG-4 standard, 231, 257, 300f, 477–478
    advanced video coding (AVC), 478
    compression performance, 323–326
    FGS coder, 535
  MPEG-2 video transcoded to, 377, 378f
    part 2, 298–310, 323–325, 324f
    part 10: H.264/AVC, 310–323, 311f, 313f, 314f–317f, 323t, 325–326, 325f
    simple profile/advanced simple profile (SP/ASP), 296, 301, 478
    video applications, 326–327
    visual profiles, 308, 309t
  MPEG-7 standard, 231, 257
  MPEG-21 standard, 258
MPQM. *see* Moving pictures quality metric
MRF. *see* Markov random field
MSM. *see* Mutual subspace method
Multicast backbone (MBONE), 474, 502–503
Multicast IP, 503
Multicast logical channels (MLCs), 610
Multichannel multipoint distribution service (MMDS), 491
Multidimensional signals, 2
Multihop wireless networks, 611–612
Multimedia Broadcast/Multicast Service (MBMS), 604
Multimedia content, typical use of, 527, 528f
Multimedia content description interface, 257
Multimedia encryption, 528
  candidate domains for, 529–530
Multimedia messaging service (MMS), 571
Multimodal analysis, 438, 459
Multiple description coding (MDC), 587
Multiple still images
  properties of, 654–665

temporal continuity/dynamics of, 661–663
three dimensional model, 663–665
Multiprocessor architectures, 412
Multiprotocol encapsulation (MPE), 609
Multiresolution filters, 79–82
Multistage median filter (MMF), 77
spatiotemporal windows in, 78f
Mutual subspace method (MSM), 656
MVs. *see* Motion vectors

N

Name-It system, 465
Narrowband ISDN (N-ISDN), 492
National Science Foundation Network (NSFNET), 474
National Science Foundation (NSF), 474–475
National Television Systems Committee (NTSC), 238
Natural language processor (NLP), 717
Network abstraction layer (NAL), 589
Network time protocol time stamp (NTPT), 512
Neural networks
self-organizing, 645
time delay, 645
NEWPRED, 308
NLP. *see* Natural language processor
Noise filter operation, 73
Noise reduction, 72
Noiseless source coding theorem, 247
Nonparametric kernel density, 632
Nyquist sampling theorem, 241, 242f

O

Object detection, 175, 635–637
Object tracking, 637
active contour-based tracking, 640
algorithms, 175, 179, 180
applications for, 175–176
articulated, 210f
2D, 219–220
3D, 209–212
3D example techniques, 216–219
3D modeling, 212–214, 213f, 214f
image cues, 215–216
kinematic and motion constraints, 214–215
Bayesian, 197–198
combined features for, 642
contour-based, 188–190
2D rigid, 180
3D rigid, 205–209, 207f, 208f
feature point-based, 190–193, 192f
feature-based methods, 641–643
global features for, 642
Kalman-based tracking, 638–639
local features for, 642
methods of, 178
model-based, 176, 643–644
occlusion handling in, 203–205
region-based, 181–187, 182f, 183f, 638
template-based, 193–194, 194f
videophone application, 176
Object-based representation in MPEG-4, 298–299
OBMC. *see* Overlapped block motion compensation
Observation likelihood, 676
Off-chip memory, 393, 405
On-chip memory, 393, 405
Open Systems Interconnection (OSI), 482, 579
Operational RD (ORD) function, 577
Optical character recognition (OCR), 455
Optical flow, 161, 163
algorithms, 119–122, 633–635
measuring distance by integrating, 112–113
Optical flow segmentation. *see* Motion segmentation
Optical node, 486
Order-statistic filters, 76–79
Orthogonal frequency division multiplexing (OFDM), 609
OSI. *see* Open Systems Interconnection
Outliers, 57
Overlapped block motion compensation (OBMC), 344
incorporation into MCTF, 345

P

Packet switching, communication network based on, 474
Parallel concatenated convolutional code (PCCC), 610
Parametric eigenspace transformation, 647
Particle filters, 676–677
Bayesian object tracking in, 202–203
PCA. *see* Principal component approach
PDF. *see* Probability density function
Peak signal-to-noise ratio (PSNR), 233, 324f, 336, 374, 432–433
Perceptual distortion metric (PDM), 421
Perceptual evaluation of video quality (PEVQ) model, 422

Perceptual video quality measure (PVQM), 422, 426
PEVQ model. *see* Perceptual evaluation of video quality model
Pfinder, tracking system, 185
P-frame coded video data, 551
Phase correlation method, 61–62
Phoneme mapping, 701, 701t
Photoconductor storage tubes, 243
Pictorial Transcripts system, 465
Picture layer in H.261 video encoder, 259
Pixel domain video transcoding, 372, 373f, 377, 378f, 381
Plain old telephone service (POTS), 6, 488
Play/break segmentation, 460, 462
POCS method. *see* Projection onto convex sets method
Posterior distributions, 674f
Posterior probability, entropy of, 680f
POTS. *see* Plain old telephone service
Predictive coders, 246
Principal component approach (PCA), 673, 696
Probabilistic identity characterization, 665–682
  recognition setting and issues in, 668–670
  video sequences in, 667
Probabilistic subspace density, tracking and recognition using, 679–682
Probability density function (PDF), 658–660
Progressive scanning, 4
Projection onto convex sets (POCS) method, 83
Pseudo-perspective model, 119
PSNR. *see* Peak signal-to-noise ratio
Public key infrastructure (PKI), 545
Public switched telephone networks (PSTN), 258, 487
Pure temporal interpolation, 23–24, 24f
PVQM. *see* Perceptual video quality measure

Q

Q.931 call signaling protocol, 518–520
QCIF. *see* Quarter Common Interchange Format
Q-scale, 395
Quadrature amplitude modulation (QAM), 487
Quadrature phase shift keying (QPSK), 487
Qualitative mosaics, 122
Quantization, 3, 583
Quantization error, 371, 375
Quantizer in intraframe encoding mode, 235
Quarter Common Interchange Format (QCIF), 244
Query-concept learner, 466

R

Radio link control (RLC) frames, 592
Random field source models, 248
Rank order difference (ROD) detector, 86
Rate-compatible punctured convolutional (RCPC) codes, 591
Rate-distortion optimized (RaDiO), 551, 553
Rate-distortion (RD), 239, 383, 575
  embedded video codec, 396
  operational theory, 576–577
  theory, 575
Real-time interface (RTI), 278
Real-time protocol time stamp (RTPT), 512
Real-time transport control protocol (RTCP), 511–517
  packet types, 511, 511t
  receiver report packet, 512, 514f, 515t
  sender report packet, 512, 513f, 515t
  source description (SDES)
    canonical end point identifier (CNAME) item, 516, 517f, 517t
    packet, 516, 516f, 516t
Real-time transport protocol (RTP), 503–511
  contributing source (CSRC) identifier, 505
  FEC header extension, 506f, 506t
  MJPEG header, 507, 507f, 508t
    elementary stream, 508–511
    systems stream, 508
  MPEG ES video-specific header, 508–509, 509f, 509t
  MPEG-2 ES video-specific header, 510, 510f, 510t
  packet header, 504f, 504t
  synchronization source (SSRC) identifier, 505
  time stamp (TS), 505
Real-time transport streaming protocol (RTSP), 517–518
  methods, 518f
Receiver operator curve (ROC), 727
Reduced reference VQA algorithms, 418
Redundant pictures, 589
Reed–Solomon forward error correction (FEC), 496, 497f
Reed–Solomon (RS) codes, 558, 590
Reference picture selection (RPS), 587, 588
Region-of-interest (ROI), mouth, 693, 693f
Relevance vector machine (RVM), 217
Requantization error, 371
Resolutions, high definition (HD), 411–413
Resource reservation protocol (RSVP), 521–523
  in hosts and routers, 522f

message flow, 522, 523f
message types, 521–522
Restoration
application of, 69
in conditions of difficult object motion, 92–94
Resynchronization marks, 396
Retail video surveillance, 621, 623f
Reversible variable-length codings (RVLCs), 396, 587
ROC. *see* Receiver operator curve
ROD detector. *see* Rank order difference detector
Root mean square error (RMSE), 432
Round-trip propagation delay (RTPD), 515–516
Round-trip-time (RTT), 580
RSVP. *see* Resource reservation protocol
RTCP. *see* Real-time transport control protocol
RTP. *see* Real-time transport protocol
RTSP. *see* Real-time transport streaming protocol
Run-length coding, 479

S

Sampling, 3–6
continuous time-varying imagery, 16–20, 16f, 20f
spatiotemporal structures, 12–15, 13f, 26–27
structure conversion, 20, 26–27
Sarnoff JNDMetrix technology, 421
Satellite communications, 491
Scalability
hybrid, 288
in MPEG-4, 307–308
SNR, 286, 287f
spatial, 287–288, 287f
temporal, 288
Scalable video coding (SVC), 286–288, 287f, 296, 322, 353, 357, 367–368, 583
Scalar quantizers, 252
Scene change, 451
categories, 445–446
detection, 142–143
Scene transition graph (STG), 461
Scratch detection, 104
Scratch removal, 103–104
Scripted content, 440, 459
summarization of, 438
video representation for, 441f, 460–461
SDP. *see* Session description protocol
SDT. *see* Structured data transfer
Security and protection, media content, 527
Segmentation, video
definition of, 141
MAP, 156–158
motion, 146–160
performance measures of, 170
scene change detection, 142–143
semantic video object, 164–165
simultaneous motion estimation and, 160–164
Self-organizing neural networks, 645
Semantic video object segmentation
chroma-keying in, 164
semiautomatic, 164–165
Session description protocol (SDP), 517
Session initiation protocol (SIP), 520–521
methods, 520f
SfM algorithm. *see* Structure from motion algorithm
Shape coding in MPEG-4, 302–303
Shape recognition, 627–628
Shot boundary detection, 142–143, 459
Signal-to-noise-ratio (SNR), 576
effective gains of, 711f, 712
Simoncelli pyramid decomposition, 80, 81f
Simple mosaic, 123. *see also* Static mosaic
Single-frequency networks (SFNs), 609
SIP. *see* Session initiation protocol
SNR. *see* Signal-to-noise-ratio
Society of Motion Picture and Television Engineers (SMPTE) digital recording, 244
Software products for video conferencing applications, 327
SOR. *see* Successive over-relaxation
Spatial change detection using two frames, 144–145
Spatial distortions, 425–426, 428
Spatial domain, artifact reduction in, 83
Spatial motion models, 46–47
Spatial operator in intraframe encoding mode, 234–235
Spatial resolution reduction, 381–382
DCT domain algorithms for, 381, 382f
MV estimation for, 379–380, 379f
pixel averaging method, 381
subsampling method, 381
Spatial segmentation, 146
Spatially coherent blotch, 88
Spatiotemporal change detection, 144–146
Spatiotemporal interpolation, 28, 89
Spatiotemporal multiresolution filtering, 82f
Spatiotemporal noise filtering, 72–82
Spatiotemporal restoration in conditions of difficult object motion, 92–94

Spatiotemporal sampling structures, 12–15
  conversion, 26–27
  deinterlacing, 27
  parameters of, 22t
Spatiotemporal windows, 90
  used in multistage median filter, 78f
Speaker identification system, 728, 728t, 729
Speaker verification system, 728, 728t, 729
Spearman rank order correlation coefficient (SROCC), 432
Spectral repeats, 16
Speech recognition, 457
Speech-to-video synthesis
  ANNs in, 719
  HMMs in, 720–721
  vector quantization in, 719
Spike-detector index (SDI). *see* Blotch detector
Spot-It system, 465
Spread spectrum embedding, 554–555
Sprite coding in MPEG-4, 304, 304f
SROCC. *see* Spearman rank order correlation coefficient
Stabilization
  image, 122–134
    definition of, 122
    with IMU, 130–133
    with metadata, 128–130
    three dimensional, 137
  video, in insect navigation, 111–113
Standard input format (SIF), 244, 270
Static mosaic, 123
Static video, spectrum of, 429f
Statistical features, 442
  alternative
    audio features, 449
    shape and position, 449
    texture, 449
  hierarchical video structure, 450
  image difference
    absolute difference, 443
    histogram difference, 444
  motion analysis, 446–447
    camera motion, 447–448, 449f
    object motion, 448, 449f
  video segmentation, 444–445
    alternative technology, 446
    scene change categories, 445–446
Still texture coding in MPEG-4, 307
Stream level authentication, 541–553
Structural similarity (SSIM) index, 426
Structure from motion (SfM) algorithm, 116–117, 137, 663
Structured data transfer (SDT), 495
Subband/wavelet transform (SWT), 331, 361
Subspace identity encoding, invariant to localization, illumination, and pose, 671
Successive over-relaxation (SOR), 100
Summarization, video
  browsing through highlights-based, 463–464
  browsing through ToC based, 463
  of scripted content, 438
  of unscripted content, 438
Support vector machine (SVM), 187, 217
SVC. *see* Scalable video coding
SVM. *see* Support vector machine
Switching filter, 76
SWT. *see* Subband/wavelet transform

T

Target scenes, video surveillance, 624–626
TCP. *see* Transport control protocol
TDMA, 488
Television, HDTV, 231, 238
Template-matching techniques, 193
Temporal continuity, 661–663
Temporal distortions, 425–426, 428
Temporal dynamics, 661–663
Temporal filter coefficients, 72, 75
Temporal filtering, 420–421
Temporal integration, 145–146, 149–150
Temporal motion models, 47
Temporal resolution reduction, MV estimation for, 380–381, 380f
Text detection, process of, 454, 455f
Texture map coordinate system, 207
Texture synthesis, constrained, 93f
Threading operation, 118
Three-dimensional (3D) motion, 31, 46, 59
Three-dimensional shape attributes, 628
Time delay neural network, 645
Time-division duplex (TDD), 490
Time-division multiple access. *see* TDMA
Time-varying imagery, 11, 13
  continuous, 16–20, 16f, 20f
TMS320C6X, 398–400, 398f, 406
TMS320C54X DSP, 398–400, 399f, 406
TMS320DM270, 408–410, 409f
Transport control protocol (TCP), 501, 592
Transportation infrastructure, 622, 624f
Tree search VQ (TSVQ), 236, 253
TSVQ. *see* Tree search VQ
Two-dimensional (2D) motion, 31, 46, 59

U

UDP. *see* User datagram protocol
UDT. *see* Unstructured data transfer
Unconstrained MCTF (UMCTF), 336
Unequal error protection (UEP), 587
Uniform scalar quantizer, 252
Universal Mobile Telecommunications System (UMTS), 490
Universal Wireless Communications-136 (UWC-136), 490
Unscented Kalman filter (UKF), 202
Unscripted content, 441, 460
- highlights extraction from, 462
- summarization of, 438
- video representation for, 442f, 462

Unstructured data transfer (UDT), 495
User datagram protocol (UDP), 501
UWC-136. *see* Universal Wireless Communications-136

V

VAPOR. *see* Variance-aware per-pixel optimal resource allocation
Variable-length codings (VLCs), 246, 299, 384, 392, 532
- in intraframe encoding mode, 234

Variable-length decoding (VLD), 392
Variance-aware per-pixel optimal resource allocation (VAPOR), 595
Vector quantization (VQ), 236, 252
VelociTI™, 398
Velocity snake model, 640–641
Very long instruction word (VLIW), 398
Very-high-speed Backbone Network Service (vBNS), 475, 502
Video
- analysis of, 459–460
- authentication, 545
- broadcasting over internet, 610
- conferencing, 327
- enhancement, 69
- storage and compression, 439–440

Video analytics, 619–620
- algorithms, 626, 628
- face recognition, 648
- gait recognition, 645–648
- human body models for, 643
- neural networks for, 645
- for video surveillance, 626–628

Video artifacts, removal of, 71, 71f
Video browsing, 438–439
- and retrieval, unified framework for, 468–469, 468f
- table of contents (ToC)-based, 438–439, 463
- using highlights-based summary, 463

Video coding, 176
- block-based, 391–392
- interlaced, 282, 282f
- MPEG-1 standard
  - background and structure of, 267–268
  - compared with H.261, 268–270, 270f, 274–275
  - GOP and I-B-P pictures, 271–273
  - slice, macroblock, and block structures, 273–274, 274f
  - for source input format, 270–271
- MPEG-2 standard
  - background and structure of, 277–278
  - compared with MPEG-1, 281–291
  - input resolutions and formats, 280–281, 281f
  - interlaced *versus* progressive video, 281–282
  - profiles and levels, 278–280, 280t
  - target applications and requirements, 278–279
- scalable
  - coding, 286–288
  - and MCTF, 333–342, 334f–337f, 339f
  - multiple adaptations in, 360–361, 360f, 362f
  - objective and visual comparisons, 357–359, 358f, 359f

Video coding layer (VCL), 589
Video communication services, 476f
Video communications, 577–578
Video compression, 8, 46, 59, 232–237, 234f, 579
- application requirements of, 237–241, 240t
- basics, 581–584
- block transform coding, 248–250
- characteristics and performance, 239
- design and selection, 238–240
- DPCM predictive encoder, 248
- entropy coding, 246–248
- importance of, 237
- motion compensation and estimation, 253–256, 255f
- techniques, 245–256

Video compression standards, 389–390, 475–476
- encoding standards, 256–265
- H.264/AVC standard, 581, 583
- international standards, 233

MPEG-2, 473, 476–477, 478–479
MPEG-4, 477–478
overview, 476–478
Video content
accessing, 438–439, 438f
categories, 438–439
Video cut, 445
Video encoder
design and selection, 232
quantization stage of, 246, 250–253
Video encoding, 231
bit rate control in, 383
Video encryption system, 527, 539, 540f
base-layer, 540
relative compression overhead of, 543, 544t
security measures for, 542t
system setup, 540
Video fingerprinting, 553
collusion attack, 555–556, 564
ECC-based, 559–560
efficient detection via trimming for, 562–563
experimental results of, 563
spread spectrum embedding and, 554–555
Video group, 440
Video inpainting, 91–92
Video marker, 441, 460
Video object plane (VOP), 299, 300f
Video object (VO) coding in MPEG-4, 299–301
Video quality assessment (VQA), 8
applications of, 417–418
feature based methods for, 422–425
full reference algorithms, 418, 422
HVS-based, 420–421, 421f
motion modeling based methods for, 427–431
no reference algorithms, 418
objective, 431
performance of, 431–433, 432t
reduced reference algorithms, 418
subjective, 418
Video quality metric (VQM), 422–423, 426
calibration stage of, 422
feature computation in, 423–424
pooling stage in, 424–425
quality parameters estimation, 424
Video redundancy coding (VRC), 587
Video representation, 439
for scripted content, 441f, 460
groups, 461
scenes, 461
sequential key frames, 460–461
video mosaics, 461
for unscripted content, 442f, 462
audiovisual markers, 462
highlight candidates, 462
highlight groups, 462
play/break segmentation, 462
Video retrieval, 438–439, 463
and browsing, unified framework for, 468–469, 468f
content-based, 464–465
efficient annotation through active learning, 466–467
feature-based, 464
multimedia databases, considerations in, 467–468
query-concept learner, 466
relevance feedback, 465–466
Video scanning, 4f
interlaced, 4
progressive, 4
Video scene, 440
Video segmentation, 444–446
definition of, 141
histogram-based, 445f
MAP, 156–158
motion, 146–160
performance measures of, 170
scene change detection, 142–143
semantic video object, 164–165
simultaneous motion estimation and, 160–164
Video sequences
face recognition from, 653
in probabilistic identity characterization, 667
properties of, 654–665
stabilization on, 133
temporal continuity/dynamics of, 661–663
three dimensional model, 663–665
Video shot, 440
Video streaming, 571
over multihop wireless networks, 611
over multiuser networks, 604–608
Video surveillance, 619
algorithms, 626
applications, 620–624
target scenes, 624–626
video analytics for, 626–628
Video transcoding, 368
architectures for, 372–375
DCT domain, 375, 375f, 378, 378f
performance of, 374f
pixel domain, 372, 373f, 377, 378f

bit rate control in, 383–384
for bit rate reduction, 369–370, 370f, 380
error-resilient, 384–385
heterogeneous. *see* Heterogeneous video transcoding
of intercoded frame, 372
of intracoded frame, 370–372
Video transmission, 6, 571, 579–581
channel models, 584–585
internet, 596–598
OSI in, 579
wireless, 599–604
Video verification and identification (VIVID), mosaics of, 129f, 131
Vinegar syndrome
removal of, 94–97
restoration scheme for, 96f
Viseme mapping, 701, 701t
Visual feature extraction systems
appearance-based, 698, 699f
shape-based, 698, 699f
Visual signals
face detection, 692–694
lip tracking, 692–694
mouth, 692–694
visual features in, 694–698
Visual telephony systems, 258
Visual text-to-speech (VTTS) in audiovisual speech synthesis, 717–718
VIVID. *see* Video verification and identification
VLCs. *see* Variable-length codings
VLD. *see* Variable-length decoding
VQ. *see* Vector quantization
VQA. *see* Video quality assessment
VQEG FR-TV Phase I database, 431–433
VQM. *see* Video quality metric
VTTS. *see* Visual text-to-speech

W

Watermark-to-noise-ratio (WNR), 559
Wavelet filterbank, 237
Wavelet transform encodings, 237
WER. *see* Word error rate
Wideband CDMA (W-CDMA), 490
Wiener filter, 632
three dimensional, 74
WIMS. *see* Wireless Multimedia and Messaging Services
Wireless ATM, 490
Wireless channel, 585
Wireless local area networks (WLANs), 571, 592
Wireless local loops, 491
Wireless Multimedia and Messaging Services (WIMS), 490
Wireless networks, 488–491
analog systems, 488
digital systems, 488–489
dual mode systems, 489
Wireless video transmission, 599–604
WNR. *see* Watermark-to-noise-ratio
Word error rate (WER), 709
audiovisual (AV) *versus* audio-only (AU), 711f, 711t
Word recognition accuracy, audio-only, 722, 722f
World Wide Web (WWW), 474

Z

Zerotree coding method, 307
Zigzag scan, 285